International Review of Cytology
A Survey of Cell Biology

CELL LINEAGE SPECIFICATION AND PATTERNING OF THE EMBRYO

VOLUME 203

International Review of **Cytology**

A Survey of **Cell Biology**

Guest Edited by

Laurence D. Etkin
Department of Molecular Genetics
University of Texas M.D. Anderson
Cancer Center
Houston, Texas

Kwang W. Jeon
Department of Biochemistry
University of Tennessee
Knoxsville, Tennessee

CELL LINEAGE SPECIFICATION AND PATTERNING OF THE EMBRYO

VOLUME 203

ACADEMIC PRESS

A Harcourt Science and Technology Company

San Diego San Francisco New York Boston London Sydney Tokyo

Front cover by Ms. Tammy Hong (paperback edition only).

This book is printed on acid-free paper. ♾

Academic Press
A Harcourt Science and Technology Company
525 B Street, Suite 1900, San Diego, California 92101-4495, USA
http://www.academicpress.com

Academic Press
Harcourt Place, 32 Jamestown Road, London NW1 7BY, UK
http://www.academicpress.com

International Standard Book Number: 0-12-364607-3 (casebound)
International Standard Book Number: 0-12-384621-8 (paperback)

PRINTED IN THE UNITED STATES OF AMERICA
00 01 02 03 04 05 EB 9 8 7 6 5 4 3 2 1

CONTENTS

PART I. SPECIFICATION OF THE CELL LINEAGES

Determinants of Cell and Positional Fate in Ascidian Embryos

William R. Jeffery

RNA Localization and Germ Cell Determination in *Xenopus*

M. Kloc, S. Bilinski, A. P-Y. Chan, L. H. Allen, N. Ruth Zearfoss, and L. D. Etkin

Asymmetric Germ Cell Division and Oocyte Determination during *Drosophila* Oogenesis

Wei Deng and Haifan Lin

The Germline in *C. elegans:* Origins, Proliferation, and Silencing

Geraldine Seydoux and Tim Schedl

Assembly of the *Drosophila* Germ Plasm

Anthony P. Mahowald

Early Events in Mammalian Germ Line

Robert Anderson, Janet Heasman, and Christopher Wylie

PART II. PATTERNING OF THE EMBRYO

The Initial Phase of Embryonic Patterning in Mammals

R. L. Gardner

Cadherin and Catenins, Wnts and Soxs: Embryonic Patterning in *Xenopus*

Allison L. St. Amand and Michael W. Klymkowsky

Establishment of Left–Right Asymmetry

H. Joseph Yost

Initiation and Early Patterning of the Endoderm

Debbie Clements, Maria Rex, and Hugh R. Woodland

Neural Patterning in the Vertebrate Embryo

Curtis R. Altmann and Ali Hemmati-Brivanlou

Cell Biology of Limb Patterning

Scott A. Schaller, Shaoguang Li, Valerie Ngo-Muller, Man-Jong Han, Minori Omi, Rosalie Anderson, and Ken Muneoka

PART III. MECHANISMS TO ESTABLISH POLARITY AND INITIATE CELL FATE DETERMINATION

Defining Cis-Acting Elements and Trans-Acting Factors in RNA Localization

Karina Yaniv and Joel K. Yisraeli

RNA Localization and Translational Regulation during Axis Specification in the *Drosophila* Oocyte

Ramona L. Cooperstock and Howard D. Lipshitz

Translational Control in Vertebrate Development

Cornelia H. de Moor and Joel D. Richter

CONTRIBUTORS

Numbers in parentheses indicate the pages on which the authors' contributions begin.

Luetta H. Allen (63), *Department of Molecular Genetics, M.D. Anderson Cancer Center, University of Texas, Houston, Texas 77030-4009*

Curtis R. Altmann (447), *Rockefeller University, New York, New York 10021-6399*

Robert Anderson (215), *University of Minnesota School of Medicine, Minneapolis, Minnesota 55455*

Rosalie Anderson (483), *Department of Cell & Molecular Biology, Tulane University, New Orleans, Louisianna 70118*

S. Bilinski (63), *Institute of Zoology, Jagiellonian University, Krakow, Poland*

A. P-Y. Chan (63), *Department of Molecular Genetics, M.D. Anderson Cancer Center, University of Texas, Houston, Texas 77030-4009*

Debbie Clements (383), *Department of Biological Sciences, University of Warwick, Coventry CV47AL, United Kingdom*

Ramona L. Cooperstock (541), *Program in Developmental Biology, Hospital for Sick Children Research Institute; and University of Toronto, Toronto, Ontario M5G 1X8, Canada*

Cornelia H. de Moor (567), *Department of Molecular Genetics & Microbiology, University of Massachusetts Medical School, Worcester, Massachusetts 01605*

Wei Deng (93), *Department of Cell Biology, Duke University Medical Center, Durham, North Carolina 27710*

Laurence D. Etkin (63), *Department of Molecular Genetics, M.D. Anderson Cancer Center, University of Texas, Houston, Texas 77030-4009*

R. L. Gardner (233), *Department of Zoology, University of Oxford, Oxford OX13PS, United Kingdom*

Man-Jong Han (483), *Department of Cell & Molecular Biology, Tulane University, New Orleans, Louisianna 70118*

Janet Heasman (215), *Division of Developmental Biology, Childrens Hospital Medical Center, Cincinnati, Ohio 45229-3039*

Ali Hemmati-Brivanlou (447), *Rockefeller University, New York, New York 10021-6399*

William R. Jeffery (3), *Department of Biology, University of Maryland, College Park, Maryland 20742-4415*

M. Kloc (63), *Department of Molecular Genetics, M.D. Anderson Cancer Center, University of Texas, Houston, Texas 77030-4009*

Michael W. Klymkowsky (291), *Department of Molecular, Cellular & Developmental Biology, University of Colorado at Boulder, Boulder, Colorado 80309-0347*

Shaoguang Li (483), *Center for Blood Research, Harvard Medical School, Boston, Massachusetts 02115*

Haifan Lin (93), *Department of Cell Biology, Duke University Medical Center, Durham, North Carolina 27710*

Howard D. Lipshitz (541), *Program in Developmental Biology, Hospital for Sick Children Research Institute; and University of Toronto, Toronto, Ontario M5G 1X8, Canada*

Anthony P. Mahowald (187), *Department of Molecular Genetics and Cell Biology, University of Chicago, Chicago, Illinois 60637*

Ken Muneoka (483), *Department of Cell & Molecular Biology, Tulane University, New Orleans, Louisianna 70118*

Valeri Ngo-Muller (483), *INSERM U129, ICGM Port-Royal, 75014 Paris, France*

Minori Omi (483), *Department of Cell & Molecular Biology, Tulane University, New Orleans, Louisianna 70118*

Maria Rex (383), *Department of Biological Sciences, University of Warwick, Coventry CV47AL, United Kingdom*

Joel D. Richter (567), *Department of Molecular Genetics & Microbiology, University of Massachusetts Medical School, Worcester, Massachusetts 01605*

Scott A. Schaller (483), *Department of Cell & Molecular Biology, Tulane University, New Orleans, Louisianna 70118*

Tim Schedl (139), *Department of Genetics, Washington University School of Medicine, St. Louis, Missouri 63110*

Geraldine Seydoux (139), *Department of Molecular Biology & Genetics, Johns Hopkins University School of Medicine, Baltimore, Maryland 21205-2185*

Allison L. St. Amand (291), *Department of Molecular, Cellular & Developmental Biology, University of Colorado at Boulder, Boulder, Colorado 80309-0347*

Hugh Woodland (383), *Department of Biological Sciences, University of Warwick, Coventry CV47AL, United Kingdom*

Christopher Wylie (215), *Division of Developmental Biology, Childrens Hospital Medical Center, Cincinnati, Ohio 45229-3039*

Karina Yaniv (521), *Department of Anatomy & Cell Biology, Hebrew University Hadassah Medical School, Jerusalem 91120, Israel*

Joel K. Yisraeli (521), *Department of Anatomy & Cell Biology, Hebrew University Hadassah Medical School, Jerusalem 91120, Israel*

H. Joseph Yost (357), *Huntsman Cancer Institute, Center for Children, University of Utah, Salt Lake City, Utah 84112-5550*

N. Ruth Zearfoss (63), *Department of Molecular Genetics, M.D. Anderson Cancer Center, University of Texas, Houston, Texas 77030-4009*

PREFACE

One of the most fascinating and puzzling dilemmas in embryonic development is how embryonic pattern and cell lineages arise. The chapters in this volume describe the historical perspective, critical issues, current and future challenges, and approaches used to address these problems in a variety of different systems. This area is much too broad to include all examples; therefore, we focused on several model organisms and systems that have been analyzed in detail. The contributions emphasize a number of important concepts, such as how polarity is established during oogenesis; the elegant sequential hierarchy by which germ cell determinants become organized in establishment of the germ cell lineage; the different strategies various organisms use to establish the germ cell lineage; similarities and differences in the mechanisms used in embryonic patterning; the mechanisms and machinery by which molecules such as RNAs become asymmetrically segregated; the use of similar signaling pathways in patterning of the dorsal–ventral, right–left asymmetries, embryonic germ layers, and limb and nervous system; and the important link between fundamental biological processes such as translation and RNA localization in regulating axis specification.

There is substantial evidence that in many organisms the localization of maternal components in the form of determinants is important in patterning of the embryo and initiation of cell lineages. In Chapter 1, Jeffery discusses the special qualities and characteristics that have made the ascidians an important model system with which to analyze the nature of cytoplasmic determinants. He presents evidence indicating that such determinants may, in fact, be localized maternal mRNAs for transcription factors and signal transduction components.

Kloc *et al.* (Chapter 2) consider this issue further, discussing how and when the oocyte polarity is first established during oogenesis in *Xenopus laevis.* They describe the major pathways through which different groups of RNAs are localized during oogenesis and consider how these components contribute to the establishment of the germ cell lineage in *Xenopus.* They also present evidence showing that the structures containing putative germ cell determinants, germinal granules, form in a sequential process during oogenesis and that several of the localized mRNAs such as Xcat2 are integral components of the granules. The functions of localized RNAs such as Xdazl, which plays an important role in germ cell specification and movement, and VegT, which functions in patterning of the embryo, are also discussed. In Chapter 3, Deng and Lin consider how early oogenesis in *Drosophila* can be used as a model for studying establishment of cell lineages and patterning. The initial stage of oogenesis, during which the germ-line stem cell divides asymmetrically to produce a daughter stem cell and cystoblast, is a fascinating phenomenon. This self-renewing process provides a continuous source of germ cells for oogenesis and utilizes an intricate and elegant signaling cascade that involves a variety of interesting gene products. Additionally, they discuss how the oocyte is determined and the implication of early events in oogenesis for subsequent patterning of the embryo.

The germ cell lineage is unique in that it is set aside very early during embryogenesis. It can be argued that this early segregation is the ancestral mechanism used in flies, worms, fish, and frogs, while a different strategy has evolved in mammals. In Chapter 4, Seydoux and Schedl focus on molecular mechanisms and events that contribute to the precocious establishment of the germ cell lineage in *Caenorhabditis elegans.* They discuss the embryonic origin of the germ cell lineage, the regulation of proliferation, initiation of meiotic development, and the silencing mechanism used by germ cells. They emphasize the characteristics of *C. elegans* that have led to its usefulness in such studies.

Perhaps one of the best models for studying the process of germ plasm organization, assembly, and function is oogenesis in *Drosophila* (see Chapter 5). Its elegant genetics has made this organism the model for studies of all phases of germ cell biology. Mahowald discusses the historical aspects of the discovery of germ plasm function, the molecular basis for its assembly, and the questions that remain unanswered. The elegant sequential hierarchy by which germ plasm components become organized in the establishment of the germ line may serve as a paradigm by which all lineage determinants may function.

Anderson *et al.* (Chapter 6) discuss mechanisms of germ cell determination, migration, proliferation, and survival in mammals. They present evidence indicating that germ line specification in mammals is probably not initiated by maternally encoded determinants as it is in invertebrates and *Xenopus* but instead arises early in development, near the posterior margins of the embryo, as in the mouse. They indicate that a candidate for such a spatial cue in the proximal epiblast is bone morphogenetic protein-4 (Bmp4).

The next section of the book addresses the question of embryonic patterning. Gardner (Chapter 7) discusses the establishment of the antero–posterior axis of the mammalian fetus. The first morphological sign of this polarity is at the primitive streak stage at the onset of gastrulation. This axis is the initial step in laying down the basic body plan, and Gardner discusses how and when orientation and polarity of the fetal antero–posterior axis are specified. It is widely believed that mammals do not possess maternal information that is involved in patterning and other developmental processes. Gardner presents evidence challenging the view that egg organization is not important in embryonic patterning in mammals.

A contemporary issue concerns the identification and analysis of signal transduction pathways that play a role in patterning of the embryo. The *Xenopus laevis* embryo has been used extensively as a model in such studies. St. Amand and Klymkowsky (Chapter 8) focus on the role of β-catenin–wnt signaling in dorsal–ventral patterning and discuss how this pathway may be regulated and the possible cross-talk with other molecular pathways.

In vertebrates the external body plan is bilaterally symmetric in a plane running through the dorsal–ventral axis. However, many internal organs exhibit a well-conserved left–right asymmetry. In Chapter 9, Yost discusses the fascinating topic of how left–right patterning is established during embryogenesis. Among the important issues considered are (1) the nature of the developmental mechanisms that coordinate pathways specifying left–right identity and antero–posterior and proximal–distal identities, (2) how symmetry is first broken during embryogenesis, and (3) the evolution of left–right asymmetry.

In Chapter 10, Clements *et al.* discuss the formation and patterning of the endoderm in a variety of animal models. This germ layer is quite important because in vertebrates it gives rise to the inner layer of the gut as well as the lungs, thyroid, thymus, liver, and pancreas. The specification, patterning, and differentiation of the endoderm are areas currently receiving a great deal of attention because, in addition to aiding in understanding how the basic vertebrate body plan is established, the information will be important for numerous biomedical applications. In many organisms the endoderm is the region where the primordial germ cells originate, and their specification and differentiation are intimately associated with this germ layer. Clements *et al.* argue that the endoderm may be the most ancient part of the embryo, because the fate of this region is involved in reproduction and feeding. The developmental significance of the endoderm is emphasized through its role as an organizing center that controls early development.

Altman and Hemmati-Brivanlou (Chapter 11) describe the events involved in the antero–posterior, dorsal–ventral, and left–right patterning of the central nervous system. It is clear that many of the same pathways involved in patterning of the early embryo such as FGF and wnt also play important roles in patterning the nervous system. A salient issue will be to understand how these diverse signaling pathways interact both temporally and spatially.

From the analysis of the signal transduction pathways used in patterning the vertebrate limb it becomes very clear that there is a reliance on the same signaling pathways in dorsal–ventral patterning of the embryo, patterning of individual germ layer such as endoderm, left–right asymmetry, and patterning of the nervous system (Chapter 12). Schaller *et al.* envision limb development and patterning as a sequential process that involves the spatial and temporal regulation of cellular proliferation, migration, and adhesion. All of these processes are regulated by signal transduction pathways. The interface between the signaling pathways and the cell biology of limb development is one of the major thrusts of this chapter.

The last portion of the book focuses on the mechanistic aspects through which patterning and cell lineage initiation might be regulated. Yaniv and Yisraeli (Chapter 13) discuss recent advances in the mapping of *cis*-acting signals in the RNA and the identification of the *trans*-acting factors that interact with these elements to localize RNAs in different systems. They emphasize the conservation of these signals and the factors that bind to them in a variety of systems such as *Xenopus* and *Drosophila* oocytes, oligodendrocytes, and fibroblasts. Cooperstock and Lipshitz (Chapter 14) discuss the role of localized RNA in establishing polarity and the antero–posterior and dorso–ventral axes of the *Drosophila* oocytes. They use GURKEN as a model which integrates the mechanisms of translational regulation and localization of RNA in regulating the establishment of polarity and axial patterning. The generality of these mechanisms is also discussed. In Chapter 15, de Moor and Richter discuss how translational regulation plays an important role in regulating vertebrate development. This includes important developmental events such as oocyte maturation, early embryogenesis, and growth and differentiation of the embryonic germ layers. Included in this chapter is a detailed discussion of the role of both 5′ and 3′ untranslated regions and specific factors that bind these regions in mediating translation.

We hope that this volume serves to stimulate readers' interest in the cellular, developmental, and molecular biology of polarity, pattern formation, and axial patterning. We thank all of the contributors for their time and effort in making this volume possible. In addition, we thank Ms. Tammy Hong for the cover design (paperback edition only).

Laurence D. Etkin
Kwang W. Jeon

Part I

Specification of the Cell Lineages

Determinants of Cell and Positional Fate in Ascidian Embryos

William R. Jeffery
Department of Biology, University of Maryland, College Park, Maryland 20742

Ascidians have played a major role in studies to understand the function of cytoplasmic determinants in animal development. Special qualities, including eggs with colored cytoplasmic regions, an invariant cleavage pattern and cell lineage, embryos with low cell numbers, larvae with typical chordate features and only six different tissues, rapid development, and a small genome, combine to make these animals a unique system for studying cytoplasmic determinants. There is evidence for determinants that specify the cleavage pattern; the differentiation of epidermal, endodermal, and muscle cells; and cell movements associated with gastrulation. The muscle determinants appear to be modified in concert with tail and muscle regression in species that have evolved an anural, or tailless, larva. Several lines of evidence suggest that determinants may be localized maternal mRNAs, which encode transcription factors or signal transduction components responsible for initiating differential gene activity. Different approaches and strategies are being used to isolate and characterize the function of these localized maternal mRNAs.

KEY WORDS: Ascidians, Cytoplasmic determinants, Ooplasmic segregation, Cleavage pattern, Specification mechanisms, Gastrulation, Localized mRNAs.

I. Introduction

Two processes generate cell and regional diversity during animal development. The first process involves the activity of cytoplasmic determinants: regulatory factors localized in the egg cytoplasm and distributed to different cell lineages during cleavage. The second process, embryonic induction, involves signaling between embryonic cells. Determinants can act to promote gene expression directly as transcription factors or indirectly by initiating signaling activities, which promote gene expression in neighboring cells. Although cytoplasmic determinants have been recognized as critical factors in early development throughout the Metazoa (Jeffery, 1988), little is known about their identity, mode of localization, and function in systems other than *Drosophila* (Nüsslein-Volhard, 1991) and *Caenorhabditis elegans* (Mello *et al.,* 1992; 1996; Draper *et al.,* 1996).

Here I review the role of cytoplasmic determinants in ascidian development. I begin with a brief introduction to the ascidian system, which has facilitated studies of cytoplasmic determinants for more than a century. I then review current progress on the localization, evolution, and molecular identity of cytoplasmic determinants.

II. The Ascidian System

Ascidians are invertebrate chordates classified in the subphylum Tunicata (or Urochordata). They are hermaphroditic marine animals with a life cycle including a swimming larva and a sessile adult (Fig. 1). Aside from the presence of pharyngeal gill slits and an endostyle, the adult lacks superficial resemblance to other chordates. In contrast, primitive chordate features are apparent in the tadpole larva. These features consist of a head containing a dorsal nervous system and a tail containing a notochord and flanking

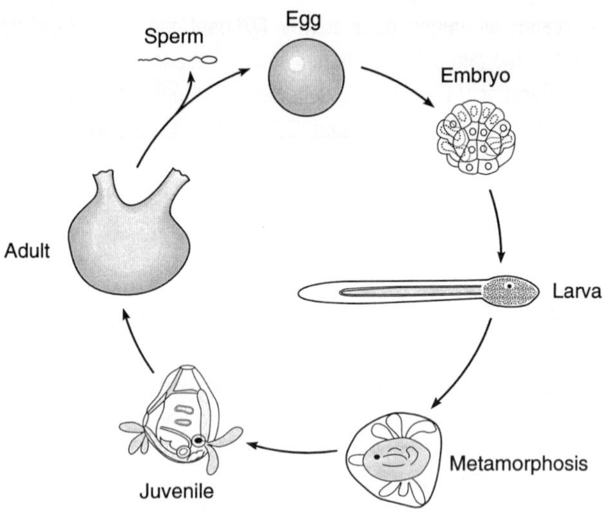

FIG. 1 The ascidian life cycle. From Jeffery (1994).

bands of striated muscle cells (Fig. 2). Indeed, the tadpole larva has been considered to be a paragon of the ancestral chordate (Satoh and Jeffery, 1995; Di Gregorio and Levine, 1998).

A. Attributes of Ascidians

The suitability of ascidians for studying cytoplasmic determinants is based on many favorable attributes. The eggs and embryos of some species have colored cytoplasmic regions, which serve as markers of blastomere identity and developmental fate (Conklin, 1905a). Cleavage is relatively simple, bilateral, rapid, and conserved between different species. Development is exceedingly rapid. In some species a swimming larva is formed only 12 hrs after fertilization. The embryonic cell lineages are well known and invariant both within and between species (Nishida, 1987). The genome is small for a chordate (Lambert and Laird, 1971; see Jeffery, 2000), approximately the same size as the *Drosophila* genome and only about 5% the size of the human genome. The genetic approach is tractable in ascidians because of their short generation times, small genomes, and the potential for simplifying genetic screens due to self-fertilization in hermaphrodites. After languishing for years, ascidians are now poised to make an impact as a genetic system (Moody *et al.,* 1999; Nakatani *et al.,* 1999). Finally, evolutionary studies of development are fostered by the existence of different larval forms in closely related species (Jeffery, 1997).

In contrast to the other model developmental systems, multiple species are being employed to study ascidian development. The molecular biology of ascidian development is being done in *Ciona intestinalis, Halocynthia roretzi, Molgula sp.,* and *Styela clava. Styela* is favored for studies of determinant localization because of the presence of colored egg cytoplasmic regions

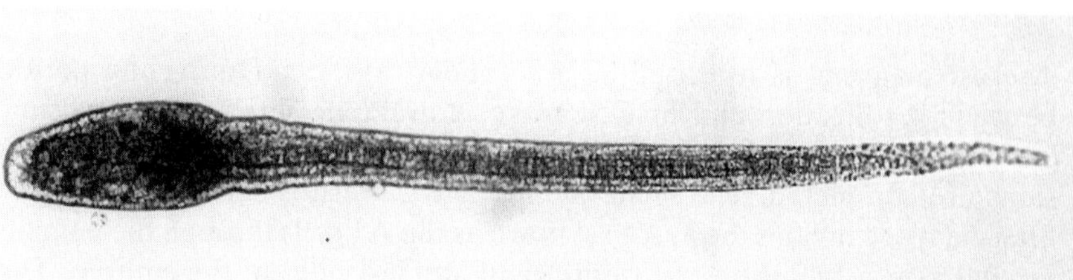

FIG. 2 The tadpole larva of *Styela clava.*

(Conklin, 1905a). *Halocynthia* is preferred for experimental embryology because its eggs and embryos are large and easy to manipulate (Nishida, 1992a). *Phallusia mammillata* is desirable for cell biological studies because of its crystal clear eggs (Speksnijder *et al.,* 1989, 1990). Finally, *Molgula sp.* are used to study the evolution of development because they exhibit radically different larval forms (Jeffery, 1997). The investigation of multiple species makes ascidians one of the few developmental systems in which information about cytoplasmic determinants is available from a comparative perspective.

B. Tadpole Larva

The chordate affinities of ascidians were first recognized by A. O. Kowalevsky (1866), who discovered the tadpole larva. The tadpole larva is bilaterally symmetric with mirror image duplications of cells and tissues (except for the otolith and ocellus; see below) on either side of the body midline (Katz, 1983). The larva contains about 2,500 cells and six different tissues: endoderm, mesenchyme, notochord, epidermis, muscle, and neural tissues. The larval head contains the anterior portion of the nervous system (known as the brain), pockets of mesenchyme cells, and most of the endoderm cells. The brain contains two pigmented sensory organs, the ocellus and the otolith. In some ascidians, the ocellus is reduced (e.g., *Styela;* Fig. 2) or absent (e.g., *Molgula*). At its posterior end, the brain tapers into a spinal cord, which runs down the dorsal side of the tail. The larval tail contains the notochord, consisting of 40 cells arranged in single file. The notochord is flanked dorsally by the spinal cord, ventrally by a strand of endoderm cells, and bilaterally by three rows of striated muscle cells. The larva is covered by a thin layer of epidermal cells.

C. Cleavage Plan and Cell Lineage

The ascidian egg is fertilized as a secondary oocyte. During the period between fertilization and first cleavage, a dramatic series of cytoplasmic movements known as ooplasmic segregation occurs as the maturation divisions are completed. The bilateral cleavage pattern is shown in Figure 3. First cleavage divides the zygote (known as the A1 cell) through the animal-vegetal axis, establishing the future right and left sides of the embryo. The first two cells are the AB2 cell, located on the left side of the embryo, and the $\underline{\text{AB2}}$ cell, located on the right side of the embryo (Fig. 3A). The second cleavage passes through the animal-vegetal axis, perpendicular to the first cleavage, dividing the embryo into the A3, $\underline{\text{A3}}$, B3, and $\underline{\text{B3}}$ blastomeres

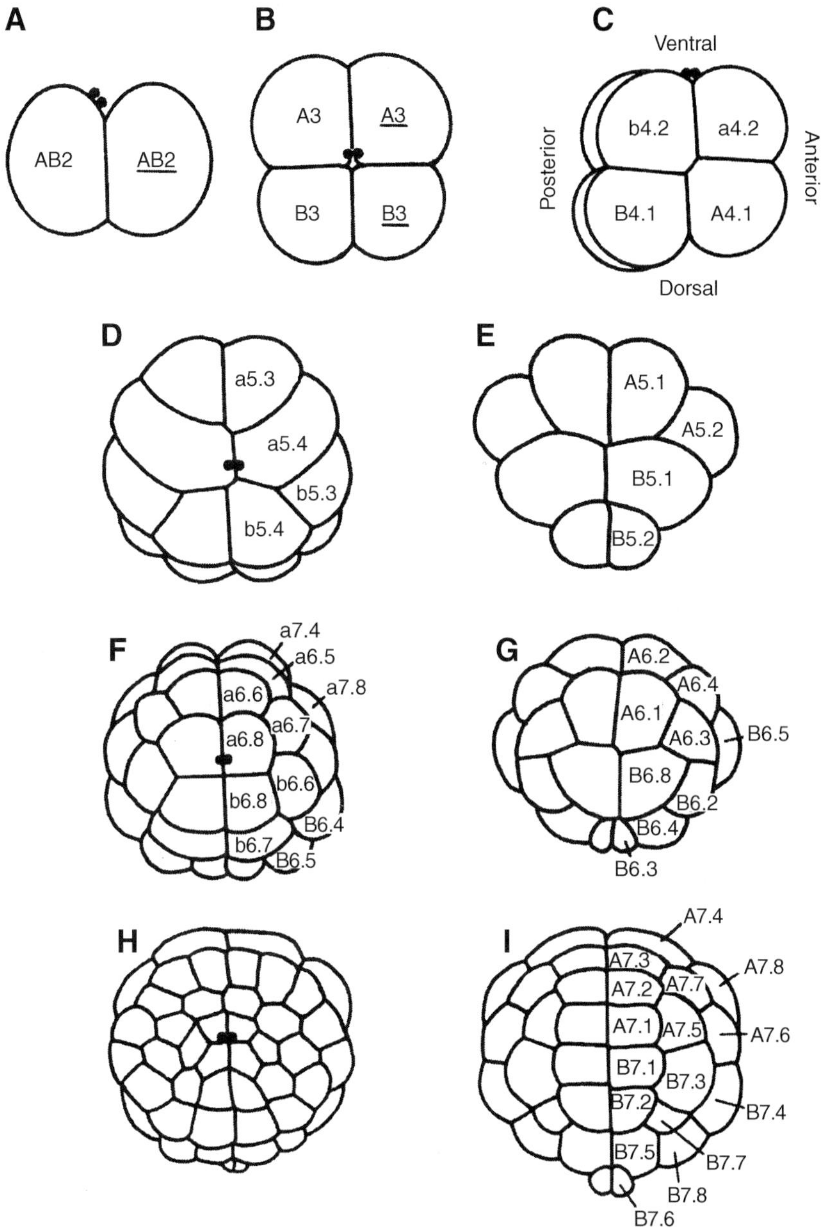

FIG. 3 Ascidian cleavage pattern. (A–C) Two-cell (A), four-cell (B), and eight-cell (C) embryos. (D–E) Sixteen-cell embryos viewed from the animal (D) and vegetal (E) poles. (F–G) Thirty-two cell embryos viewed from the animal (F) and vegetal (G) poles. H–I. Sixty-four cell embryos viewed from the animal (H) and vegetal (I) poles. In D–I blastomeres are named only on the right side of the bilaterally symmetric embryo. From Venuti and Jeffery (1989) with permission.

(Fig. 3B). The third cleavage passes through the equator of the embryo, forming four animal and four vegetal blastomeres (Fig. 3C). The animal hemisphere contains the a4.2 and a4.2 cells (or a-line cells) and the b4.2 and b.4.2 cells (or b-line cells). The vegetal hemisphere contains the A4.1 and A4.1 cells (or A-line cells) and the B4.1 and B4.1 cells (or B line cells). Slanting of the third cleavage plane results in a trapezoid-shaped eight-cell embryo, the four animal cells lying slightly anterior to the four vegetal cells. During the first three divisions, cleavage is approximately equal in the animal and vegetal hemispheres. Beginning at the fourth cleavage, however, the cleavage planes are oriented differently in the animal and vegetal hemispheres (Fig. 3D–I). As a result, a series of unequal cleavages occurs in the posterior vegetal cells, ultimately forming the small B7.6 and B7.6 blastomeres in the 64-cell embryo (Fig. 3E, G, I).

The ascidian cell lineage was first described by Conklin (1905a) and later confirmed and extended by Ortolani (1955), Nishida (1987), and Nicol and Meinertzhagen (1988). Like the cleavage pattern, the cell lineage appears to be invariant within and between species. A fate map of the uncleaved egg and eight-cell embryo is shown in Figure 4A–B. It can be seen that the ectodermal derivatives (epidermis and nervous system) originate from the animal hemisphere, the mesodermal derivatives (notochord, mesenchyme, and muscle) originate from the equatorial region, and the endodermal derivatives arise from the vegetal region. According to the fate map, most larval tissues originate from more than one quadrant of the eight-cell embryo. The epidermis is derived from the a and b line blastomeres, the nervous system from the a-, b-, and A-line blastomeres, the notochord and endoderm from the A- and B-line blastomeres, and muscle cells from the A-, b-, and B-line blastomeres (Fig. 4B). The muscle cells in the anterior and middle region of the tail (28 out of 42 muscle cells in *Halocynthia roretzi*) originate from B-line blastomeres and are called primary muscle cells. The posterior muscle cells originate from the A- and b-line blastomeres and are called secondary muscle cells. Primary and secondary lineages also exist for the notochord cells. The only tissue that arises from a single quadrant (B-line cells) is the mesenchyme.

III. Determinants

Cytoplasmic determinants are maternal factors that influence cellular or positional fates by activating differential gene expression in the embryo. Evidence for the existence of a cytoplasmic determinant is established by two operational criteria. First, removal or inactivation of the determinant should result in a change in fate related to its function. This criterion shows that the determinant is necessary for a developmental event. Second, transplantation of the determinant to an ectopic position should result in

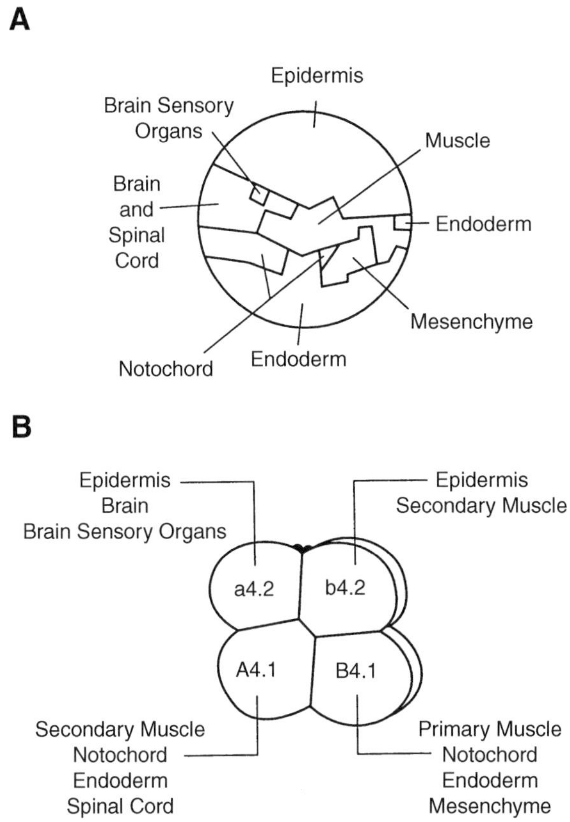

FIG. 4 Ascidian fate maps. (A) Fate map of the uncleaved zygote according to Nishida (1987). (B) Derivatives of the four quadrants of the eight-cell embryo. From Venuti and Jeffery (1989) with permission.

a change in fate comparable to the activity of the determinant. This criterion establishes that the determinant is sufficient to promote a developmental event. Classic studies of ascidian determinants usually did not satisfy both these criteria because procedures for transplantation have only recently become available. In the last decade, however, cytoplasmic determinants have been demonstrated to be both necessary and sufficient for a variety of developmental events in ascidians.

A. Classic Studies of Determinants

In 1887, L. Chabry reported the results of an experiment that launched the field of experimental embryology (Fischer, 1991). When one blastomere of an ascidian embryo was destroyed at the two-cell stage, the remaining blastomere continued to cleave as if it were part of a whole embryo, and a half-larva was formed. Chabry (1887) thus showed that bilateral symmetry is determined as early as the two-cell stage. A similar experiment at the four-cell stage showed that embryos could not compensate for missing blastomeres, thus implying a mosaic form of development. Mosaic development was in contrast to the results obtained in similar experiments with sea urchin and other regulative embryos. Studies of ascidian development carried out immediately following Chabry's work were mostly descriptive in nature, defining the cleavage pattern and the relationship between the egg and embryonic axes (Venuti and Jeffery, 1989). E. G. Conklin (1905a, b, c) inaugurated further experimental studies of ascidian embryos after his discovery of colored cytoplasmic regions in *Styela partita* eggs.

1. Ooplasmic Segregation

In the nineteeth century, experimental studies had been conducted with the colorless or opaque embryos of European ascidian species. In 1905, Conklin discovered that embryos of the American species *Styela partita* exhibit colored cytoplasmic regions (Conklin, 1905a). Conklin described six regions, corresponding to the six larval tissues, naming them the ectoplasm, myoplasm, mesenchymoplasm, (noto) chordoplasm, neuroplasm, and endoplasm. Modern researchers recognize the ectoplasm, endoplasm, and myoplasm as distinct regions. The other three regions are probably shading effects that occur at the boundaries of the three primary cytoplasmic regions. The colored cytoplasms of *Styela partita* eggs are also present in the related species *Styela plicata* and *Styela clava,* which have been the subjects of more recent studies of determinant localization (Jeffery, 1984a). The same cytoplasmic regions also are present in more distantly related species, either marked by other colors (e.g. orange in *Boltenia villosa;* Jeffery, 1982) or as opaque (e.g. *Ciona intestinalis;* Berg and Humphrey, 1960) or transparent regions (e.g., *Phallusia mammillata,* Sardet *et al.,* 1989).

The following description of colored cytoplasmic regions applies to *Styela* eggs. The ectoplasm, a transparent cytoplasmic region containing a large number of ribosomes, endoplasmic reticulum, and other cytosolic components, is derived primarily from the nucleoplasm of the germinal vesicle (GV). After GV breakdown, the ectoplasm is localized in the animal hemisphere. During cleavage, the ectoplasm is passed to the a- and b-line blastomeres, which produce the larval epidermis, the neural tissues, and some of the secondary muscle cells. Smaller quantities of ectoplasm are also present in the perinuclear region of cells in other parts of the embryo. The endoplasm, an opaque cytoplasmic region in the vegetal hemisphere of the egg, consists primarily of yolk granules and is distributed mainly to the A- and

B-line blastomeres in the vegetal hemisphere. The myoplasm, a yellow cytoplasmic region containing mitochondria, pigment granules, and associated cytoskeletal elements (Jeffery and Meier, 1983), is localized in the vegetal posterior region of the embryo. This cytoplasmic region is distributed primarily to B-line blastomeres, which are precursors of the primary muscle cells. A small portion of myoplasm is also distributed to the A-line blastomeres (Sardet *et al.,* 1989).

The colored cytoplasmic regions of *Styela* eggs are localized during a series of cytoplasmic movements known as ooplasmic segregation (Jeffery, 1984a; Fernández *et al.,* 1998). The movements of the myoplasm during ooplasmic segregation are shown in Figures 5 (see color insert) and 6. Ooplasmic segregation occurs in two phases. During the first phase of ooplasmic segregation (OS1), the cortical myoplasm is translocated into a cap surrounding the vegetal pole (Fig. 5A). As the myoplasm is translocated, the ectoplasm moves into the region immediately above the vegetal cap, and the endoplasm is displaced into the animal hemisphere. OS1 is mediated by the contraction of actin-containing cytoskeletal network in the myoplasm (Jeffery and Meier, 1983). The contraction of this network appears to be triggered by a calcium flux initiated in the vegetal cortex and propagated through the egg cytoplasm (Speksnijder *et al.,* 1989, 1990). The second phase of ooplasmic segregation (OS2) begins when the sperm aster forms near the vegetal pole. During OS2, the myoplasm moves to the future posterior pole of the egg, where it spreads laterally to form the yellow crescent (Fig. 5B-C). Yellow crescent formation is mediated by the aster microtubules, which polymerize around the male pronucleus in the vegetal hemisphere (Sawada and Schatten, 1989). During OS2, the ectoplasm forms a clear crescent above the yellow crescent and then flows into the animal hemisphere. Finally, the endoplasm is displaced into the vegetal hemisphere. The future dorsal-ventral and anterior-posterior axes of the embryo are determined during ooplasmic segregation. The dorsal pole of the embryo corresponds to the cap of myoplasm after the completion of the OS1, whereas the posterior pole of the embryo corresponds to the position of the yellow crescent after the completion of OS2.

A similar series of cytoplasmic translocations occurs during the period between fertilization and first cleavage in other ascidian species (Sawada and Osanai, 1981; Sardet *et al.,* 1989). In *Phallusia mammillata,* the transparent myoplasm forms a cap in the vegetal hemisphere during OS1. The focal point of myoplasmic segregation, however, is not necessarily localized at the vegetal pole, but appears to be directed to a point on the egg surface opposite the site of sperm entry, which usually occurs in the animal hemisphere (Speksnijder *et al.,* 1989; Roegiers *et al.,* 1995). As the myoplasm moves posteriorly, a vegetal button is protruded near the vegetal pole. Recent studies have shown that the vegetal button contains a dense accumulation of granular materials resembling ribonucleoprotein particles (Roegiers *et al.,* 1999; Roegiers, 1999). These granular materials are reminiscent of "nuage," the germ plasm-associated organelles found at the poles of other animal eggs (Eddy, 1975; Kloc *et al.,* 1996, Chap. 5). The fate and developmental function of the vegetal button materials remains to be determined, although its structure suggests that it may have affinities with the centrosome-attracting body (see below).

2. Blastomere Isolation and Cleavage Arrest

Blastomere isolation experiments conducted in numerous ascidian species suggest that cell fates are determined autonomously by segregation of cytoplasmic determinants (Conklin, 1905b, c, 1911; Berrill, 1929; Cohen and Berrill, 1932; Reverberi and Minganti, 1946, 1947; Nishida, 1992b). Additional evidence supporting the determinant model was obtained by J. R. Whittaker and colleagues via the cleavage arrest method (Whittaker, 1973, 1977; Meedel *et al.,* 1987). In this method, embryos are treated with cytochalasin to arrest cytokinesis, but nuclear division continues in the arrested blastomeres, which eventually express markers according to their normal fate. For example, the cleavage-arrested B-line blastomeres fated to produce muscle cells synthesize acetylcholinesterase (AChE), muscle actin (Fig. 6), and myosin (Whittaker, 1973; Meedel *et al.,* 1987; Jeffery, 1989b; Makabe *et al.,* 1990). Cleavage-arrested four-cell embryos with AChE expression in the B3 and B3 blastomeres are shown in Figure 10D. Likewise, cleavage-arrested A- and B-line cells fated to form endoderm produce alkaline phosphatase (AP) (Whittaker, 1977), an endoderm marker, and a-line cells fated to form the pigmented brain sensory organs produce the melanizing enzyme tyrosinase (Whittaker, 1973). Many other tissue-specific markers are expressed autonomously in cleavage arrested blastomeres ac-

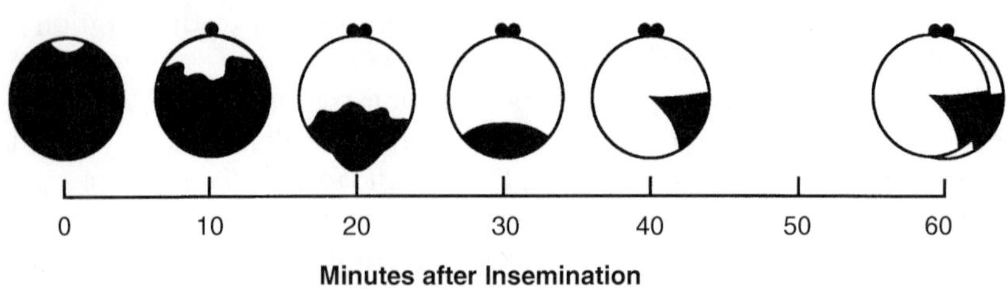

FIG. 6 Timing of myoplasmic segregation between fertilization and first cleavage in *Styela clava.* The myoplasm (shaded area) occupies the entire cortex, except for a clear area surrounding the animal pole, in unfertilized eggs. OS1 is completed 30 min after fertilization, OS2 is completed 40 min after fertilization, and first cleavage is completed 60 min after fertilization. From Jeffery (1990a).

cording to the cell lineage (Satoh, 1994). These results support the hypothesis that expression of tissue-specific markers is controlled by cytoplasmic determinants partitioned into different blastomeres during cleavage.

Cytoplasmic determinants were at one time suspected to govern the development of all six larval tissues. Additional blastomere isolation experiments showed that other developmental processes are involved in specifying some of the larval tissues (Rose, 1939). In a key experiment, Reverberi and Minganti (1946) isolated pairs of blastomeres from eight-cell *Ciona intestinalis* embryos and showed that the isolates developed into partial embryos containing only a subset of the spectrum of tissues expected from the fate map. Muscle cells were produced in isolates containing B-line blastomeres, epidermal cells in isolates containing a- and/or b-line blastomeres, and endodermal cells in isolates containing A- and/or B-line blastomeres. However, the brain sensory organs were absent from isolates containing only a-line cells, whose fate is to produce the neural derivatives. These results were later confirmed by blastomere recombination experiments (Reverberi and Minganti, 1947; Reverberi *et al.,* 1960; Nishida, 1991) and imply that neural tissues are specified by inductive processes.

An unexpected result of the blastomere isolation experiments was the development of tissues that did not correspond to Conklin's original fate map (Reverberi and Minganti, 1946). For example, a few muscle cells were found in partial embryos lacking the B-line cells. When the cell lineage was later refined by Nishida (1987) these secondary muscle cells were found to be derived from the A- and b-line blastomeres. The blastomere isolation experiments were repeated with molecular markers by N. Satoh and colleagues (Deno *et al.,* 1985) and Meedel *et al.* (1987), confirming the modern cell lineage analysis and the existence of multiple sources of muscle cells.

3. Centrifugation and Cytoplasmic Redistribution

To investigate the potential of myoplasm to control cell fate, Conklin (1931) displaced the myoplasmic organelles to ectopic regions by centrifuging *Styela partita* eggs. At low centrifugal forces, the mitochondria and yellow pigment granules were easily displaced, but the reticulate ground substance in which they are embedded was not affected. The displacement of mitochondria and yellow pigment granules from the myoplasm did not affect muscle cell development, thus indicating that these organelles do not contain muscle determinants. However, centrifugation at higher forces, which affected the mitochondria, yellow pigment granules, and the myoplasmic ground substance, caused muscle cells to develop at ectopic positions in the embryo, thereby bringing up the possibility that muscle determinants are localized in the ground substance. Jeffery and Meier (1984) obtained similar results that suggest that Conklin's myoplasmic ground substance is an actin-containing cytoskeletal matrix. The structure and composition of the myoplasmic cytoskeleton has been recently reviewed (Jeffery, 1995). Centrifugation experiments thus provided the first evidence that muscle determinants have the ability to change cell fate during embryogenesis.

Using a procedure devised by T. H. Morgan (1910), Whittaker (1980) conducted cytoplasmic redistribution experiments to further address the ability of cytoplasmic determinants to specify muscle cell fate. As shown in Figure 3C, the third cleavage of the ascidian embryo is equatorial, distributing the myoplasm to the B4.1 and $\underline{\mathrm{B4.1}}$ cells at the eight-cell stage. However, if four-cell embryos are compressed in a plane perpendicular to the animal-vegetal axis, the third cleavage passes through this axis, distributing the myoplasm to four rather than two cells in the eight-cell embryo. The application of cleavage arrest procedures to the compressed embryos allowed the fate of the blastomeres containing ectopic myoplasm to be evaluated. In cleavage-arrested eight-cell embryos as many as four cells, rather than the usual two cells, were shown to produce AChE (Whittaker, 1980). According to the cell lineage, the cells receiving ectopic myoplasm would have normally produced epidermis, and thus the presence of myoplasm changed the fate of these cells. These results suggest that muscle determinants are localized in the cytoplasm (nuclear assignments were not modified by cytoplasmic redistribution), are present in the myoplasm (the major cytoplasmic region altered), and are sufficient to produce ectopic muscle cells.

B. Recent Studies of Determinants

H. Nishida (1992a) has developed a transplantation method based on fusion of an anuclear egg fragment to a blastomere or a whole egg before cleavage. The procedure is outlined in Figure 7. The use of *Halocynthia roretzi* eggs and embryos, which are large and amenable to microsurgery, is an important feature of the fusion method. Due in large part to this new technology, there is now conclusive evidence for the existence of five types of determinants in ascidian eggs.

1. Epidermal Determinants

The epidermal determinants are thought to be localized in the ectoplasm. Conklin (1905a) noted that most of the ectoplasm is derived from the GV. When the GV ruptures at the beginning of oocyte maturation its contents are released into the animal pole region, where they mix with transparent cytoplasm to form the ectoplasm. Evidence for the existence of epidermal determinants was provided by blastomere fusion experiments. In these

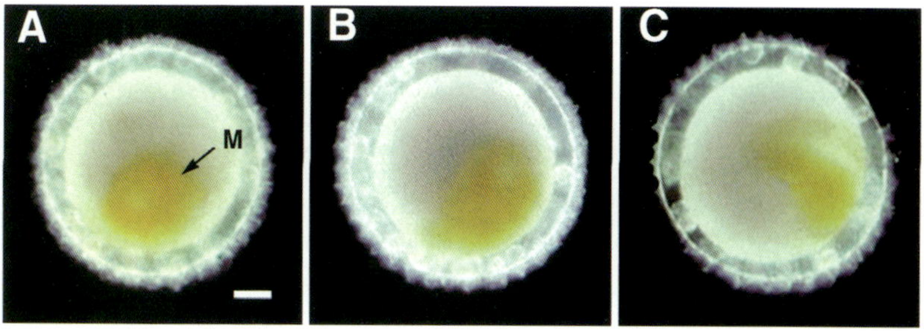

CHAPTER 1, FIG. 5 Ooplasmic segregation of the yellow myoplasm (M) in *Styela clava* eggs. (A) After the completion of OS1. (B) During OS2. (C) After the completion of OS2. The vegetal pole is at the bottom of each frame. A follicle surrounds the eggs. Scale bar: 20μm. From Localization of ribosomal protein L5 mRNA in myoplasm during ascidian development. Swalla and Jeffery, *Dev. Genet.* Copyright © 1996, Wiley-Liss, Inc., a subsidiary of John Wiley & Sons, Inc.

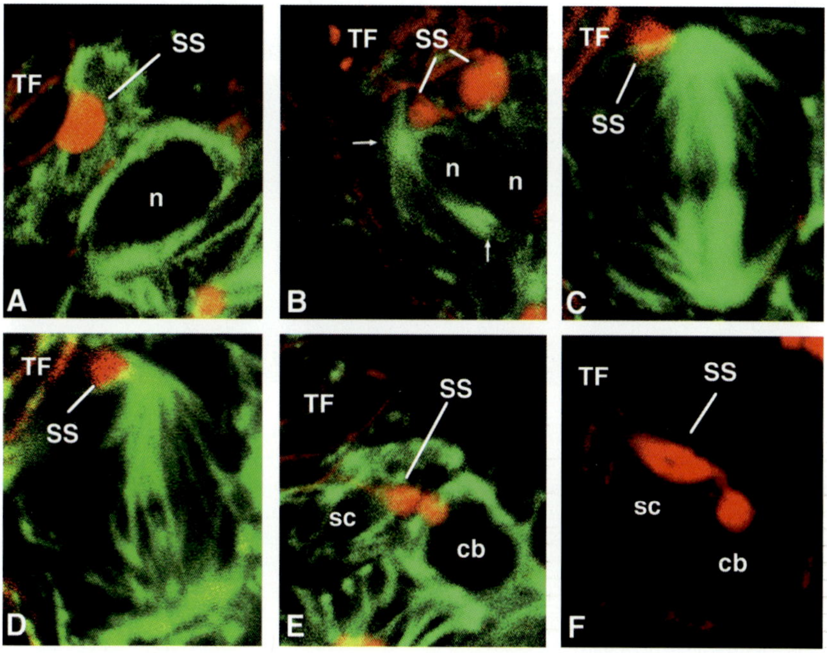

CHAPTER 3, FIG. 3 The germline stem cell (GSC) cycle, with spectrin labeled in red and tubulin in green. TF, terminal filament; SS, spectrosome; SC, germline stem cell; Cb, cystoblast; n, the nucleus. (A) An interphase GSC. (B) An 8-μm confocal stack showing a prophase SC and an interphase SC for comparison of their spectrosome size. The two segregating mitotic asters in the propase stem cell are indicated by arrows. The red spot in the lower right corner is a partial image of a fusome out of the plane of the interphase stem cell. (C) A metaphase GSC with the spectrosome associated with the terminal filament and the apical pole of the mitotic spindle. (D) An anaphase GSC as imaged in a 4-μm confocal stack. (E, F) A GSC at its telophase or the next interphase; only spectrin staining is shown. From Deng and Lin, 1997, with permission.

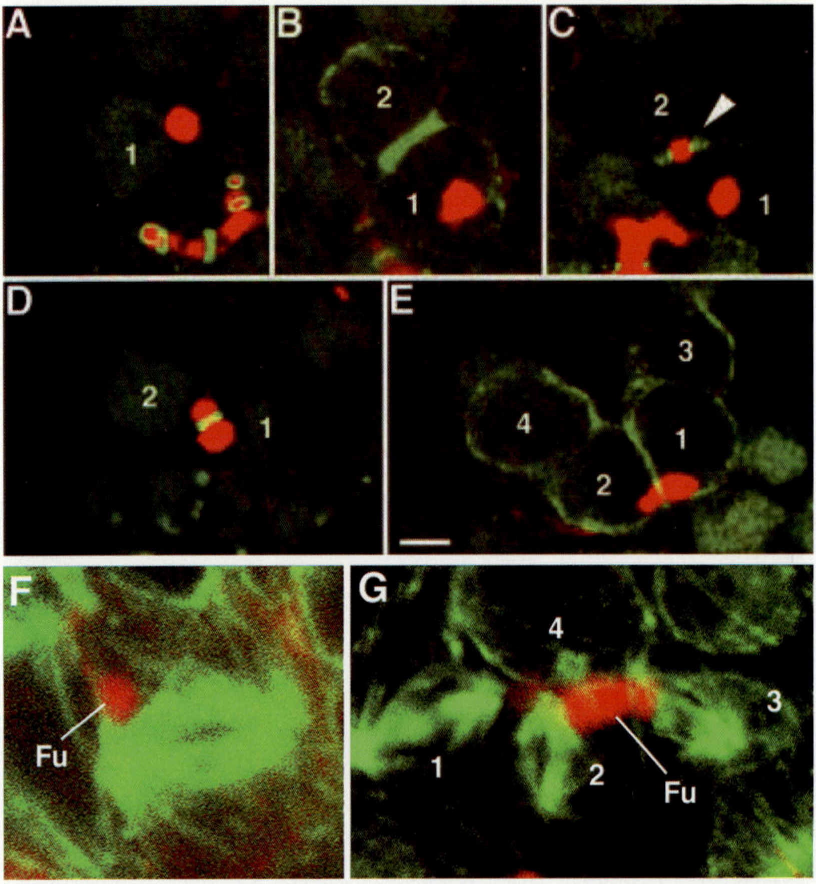

CHAPTER 3, FIG. 4 Growth of the fusome and its role in anchoring mitotic spindles during cytotocyte divisions. Images in A–E are stained for HTS (red) and for anillin (green), a marker for the ring canal and the nucleus. (A) A late interphase cytoblast with the spectrosome located in the apical region of the cytoplasm. (B) At cytokinesis, the spectrosome is often located in the mother cytoblast (cell 1), while anillin is enriched in the cleavage furrow. (C) In early interphase of a two-cell cyst, a fusome plug forms in the new ring canal (arrowhead); the preexisting fusome in the mother cytoblast (cell 1) begins to move toward it. (D) As interphase progresses, anillin continues to accumulate in the nuclei, and the old and the new fusome materials fuse in the ring canal region so that cell 1 retains a larger part of the fusome. (E) Following mitosis of the two-cell cyst, cell 1 still contains more fusomeal material than cell 2. The bar in E denotes 3 μm for A–E. Images in F–G are stained for HTS (red) and for tublin (green). (F) A metophase cytoblast with the spectrosome associated with one pole of the mitotic spindle. (G) A dividing four-cell cyst with all four mitotic spindles associated with the fusome. From de Cuevas and Spradling (1998) with permission from Company of Biologists Ltd.

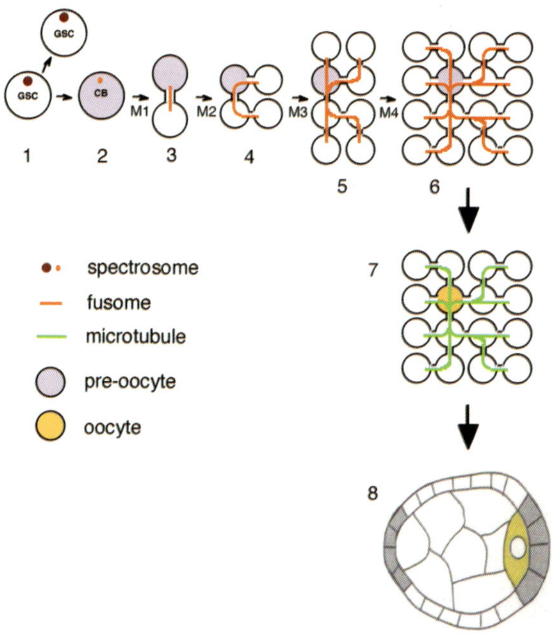

CHAPTER 3, FIG. 5 A schematic drawing of early oogenesis in *Drosophila*, illustrating oocyte determination and translocation. The asymmetric division of a GSC (1) produces a cytoblast (CB, 2), which then undergoes four rounds of incomplete divisions (M1–4) to generate a germlike cyst containing 16 interconnected cells (6). The divisions are all asymmetric, as marked by the asymmetric growth of the fusome (orange structures in 3–6) and the asymmetric segregation of the preexisting ring canals, so that the original cystoblast (pink cell in 2) is destined to become the pre-oocyte in the newly formed 16-cell cyst (pink cell in 6). In this cyst, the polarized fusome guides the formation of a polarized microtubule network (green structure in 7), which then mediates the intracyst transport that leads to oocyte differentiation (yellow cell in 7). Subsequently, the differentiated oocyte relocates to the posterior end of the cyst inside an egg chamber (yellow cell in 8).

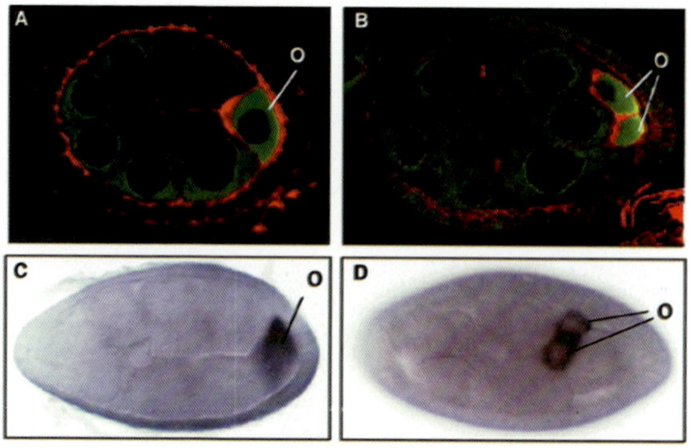

CHAPTER 3, FIG. 6 *spn* double mutant combinations produce ectopic oocytes. In wild-type egg chambers (A, C), the BicD protein (green in A) and *oskar* RNA (C) are localized to the oocyte in the posterior region. In *spn* double mutant egg chambers (B, D), the BicD protein (green in B) and oskar RNA (D) are localized to two oocytes. Data adopted from Gonzales-Reyes *et al.*, 1997, with permission from Company of Biologists Ltd.

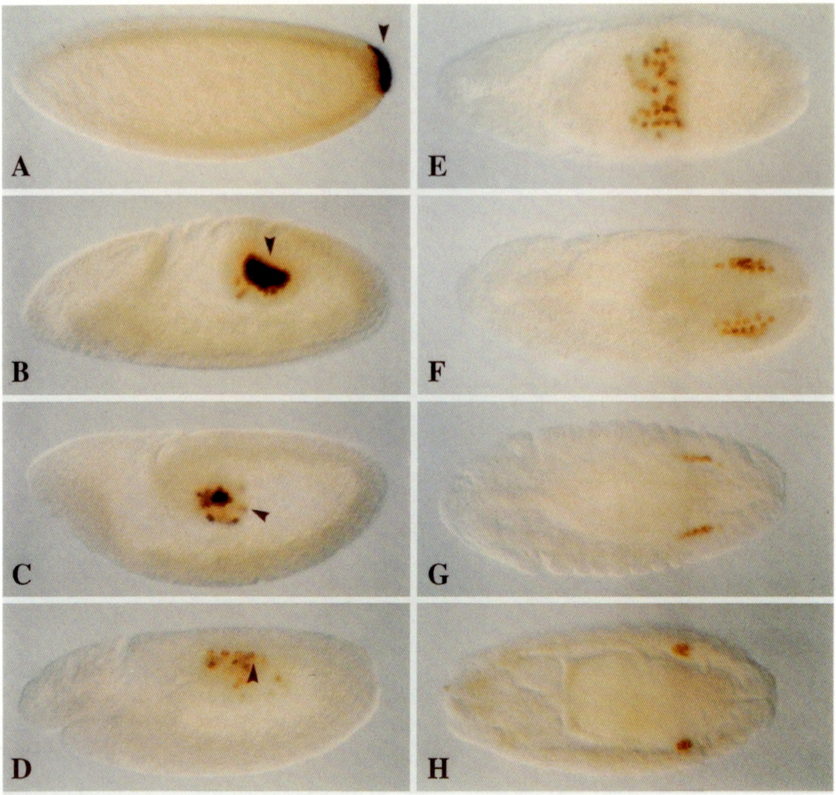

CHAPTER 5, FIG. 2 Migration of pole cells (labeled with Vasa antibody) from the posterior pole of the blastoderm (A) into the posterior midgut pocket (B); migration out of the pocket into the surrounding mesoderm (C, D, and E), followed by the aggregation of pole cells into the embryonic gonad (F, G, and H). In each figure, anterior is to the right; A to D are lateral views; E to H are dorsal views of the embryo. (Figure used courtesy of Company of Biologists and the authors, Moore *et al.*, 1998).

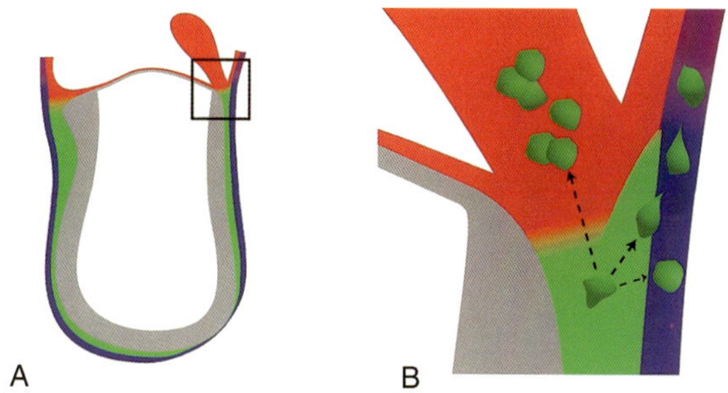

CHAPTER 6, FIG. 1 Onset of PGC migration, a schema derived from confocal microscopy of living embryos expressing green fluorescent protein in germ line cells (Anderson *et al.*, 1999a). (A) A hypothetical midsaggittal section through a mouse embryo on the eighth day of gestation. Anterior is to the left. (B) The posterior end, indicated by the box A. Germ cells (dark green) emerge from the epiblast by passing through the primitive streak (gray). They actively migrate ventrally through the forming mesoderm (green) directly into the embryonic endoderm (blue) and posteriorly into the allantois (red). Those in the endoderm become incorporated into the hindgut. It is not clear whether those in the allantois reach the gonad, or where the boundary is between those that do and those that do not reach the gonad.

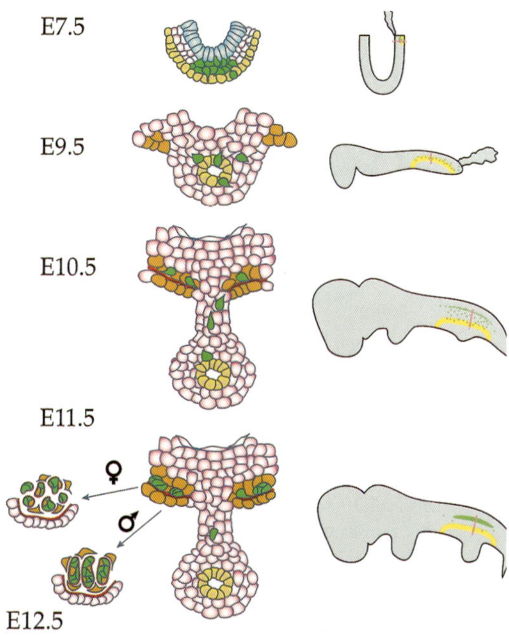

CHAPTER 6, FIG. 2 A schema for PGC migration in the mouse embryo. At each stage, a schematic cross-section is shown at the postion of the red line in the schematic saggittal section (artificially straightened out for clarity). PGCs (green) migrate from the primitive streak (blue) into the posterior embryonic endoderm (yellow) around E7.5. They remain in the hindgut until E9.5, when they migrate laterally and dorsally out of the hindgut epithelium; they enter the genital ridges between E10.5 and E11.5, where they associate with gonadal somatic cells (brown) and the forming basement membrane (red). The precise boundaries of the gonadal somatic cells are unknown.

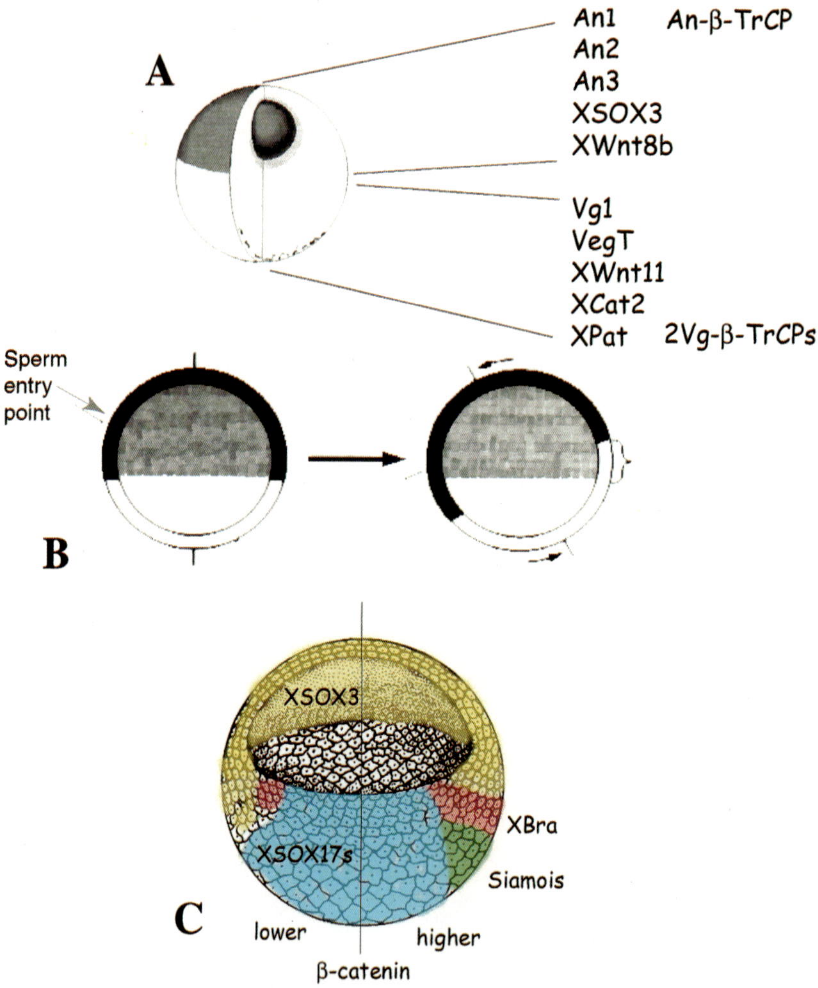

CHAPTER 8, FIG. 3 Patterning in the *Xenopus* oocytes/embryo. (A) Shown in a schematic form is the *Xenopus* oocyte, with pigmented animal hemisphere and the animally located germinal vesicle. Localized mRNAs are indicated. (B) Following sperm entry (indicated by the arrow on the left) the cortex of the egg rotates ~30° toward the side of the sperm entry, while the central region of the egg remains oriented in the earth's gravitational field. This creates an asymmetry between the future dorsal (marked by the "I") and further ventral sides of the embryo. (C) The early rotation creates an asymmetry in the level of β-catenin (and other Wnt pathway components) between the future dorsal and ventral sides of the embryo. At the late blastula stage, this leads to the dorsal activation of the *Siamois, Twin* and *Xnr-3* genes. However, the region of *Siamois* activation is not coincident with the region of β-catenin stablization, and it seems likely that modulation of β-catenin signaling by other embryonic factors is involved in delimiting the *Siamois, Twin,* and *Xnr3* expression domains. Possible candidates for these modulars include the maternal SOX, XSOX3, and the endodermally expressed SOXS, XSOX17α/β.

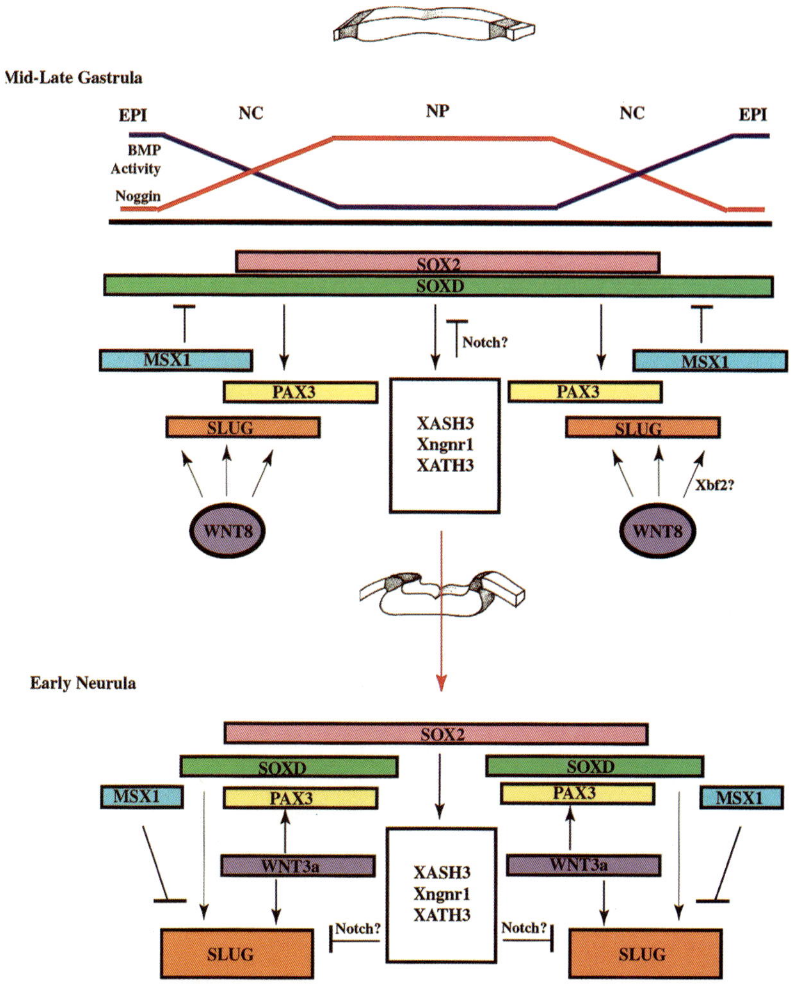

CHAPTER 8, FIG. 6 Medial-lateral patterning of the neural plate. At the top is a schematic of the neural plate in the mid-late gastrula stage embryo. A ventrodorsal BMP activity gradient is generated by the secretion of BMP-antagonist (eg. noggin) from the Spemann organizer. This BNM activity gradient divides the neuroectoderm into three domains: neural plate (NP), characterized by low BMP activity; neural crest (NC) (gray in the upper schematic), with intermediate levels of activity; and the embryonic epidermis (EPI), with high levels of BMP signaling. Noggin induces SOX2 and SOXD expression, which in turn induce the expression of the proneural genes XASH3, Xngnrl, and XATH3 in the neural plate. Proneural genes activate the expression of Delta, which acts through the Notch pathways, inhibits neural differentiation, potentially blocking the expression of SOXD and other neural markers from the medial domain of the neural plate. Msxl is known to inhibit SOXD expression, thus further refining its domain. Xwnt8, expressed in the underlying paraxial mesoderm, activates via a mediator (possibly Xbf2) to induce Pax3, Msxl, and Slug expression. At later stages, Xwnt3a begins to be expressed at the lateral edges of the neural plate and enhance the expression of Xslug and Pax3.

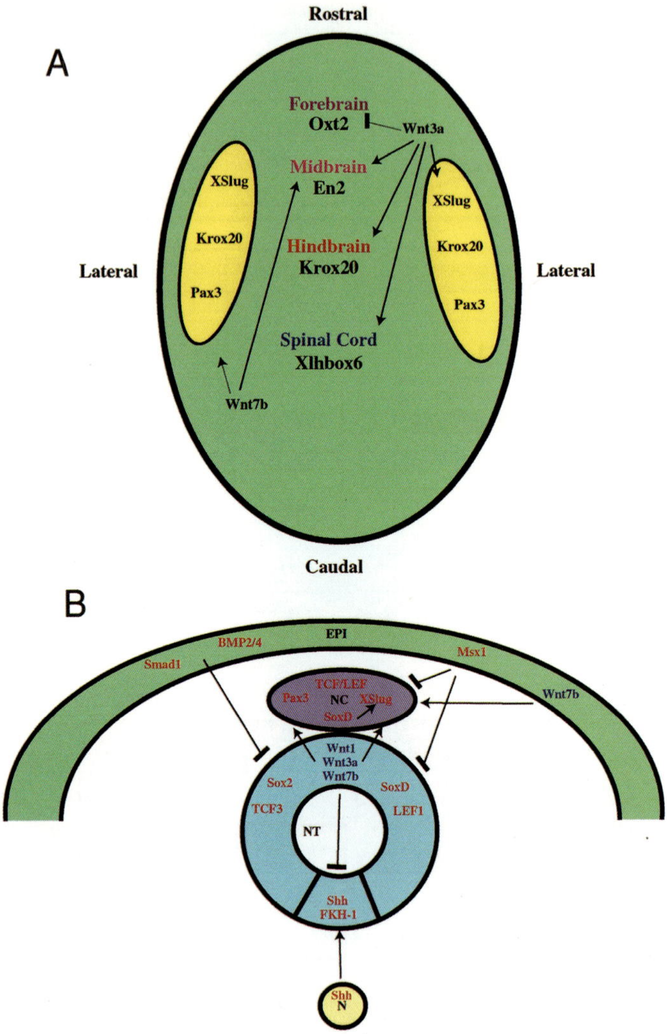

CHAPTER 8, FIG. 7 Rostral-caudal and dorsal-ventral patterning. (A) Rostal-caudal patterning: Xwnt3a and Xwnt7b are expressed at the lateral edges of the neural plate and later expressed in the dorsal neural tube. Xwnt3a is a potent caudalizing factor that can induce posterior and inhibit anterior neural genes in the developing neural tube, wheras Xwnt7b has only a slight caudalizing effect. Both Wnts appear to maintain the position of the neural crest domain, characterized by the expression of Slug, Pax3, and Krox2O. (B) Dorsal-ventral patterning: Wnt signaling plays a critical role in maintaining the expression of genes along the dorsal-ventral axis. Wnts are able to enhance the expression of dorsal genes such as the dorsal neural tube (NT) marker, Xwntl, and neural crest (NC) markers, XSlug, and Pax3. Wnts signalling effects are potentially modulated by the differential expression of TCFs and SOXs in the neural tube and crest. The ventral character of the neural tube is maintained by Shh signals emanating from the notocord (N). BMPs (via Smadl) and Msxl maintain epidermal gene expression while suppressing neural gene expression.

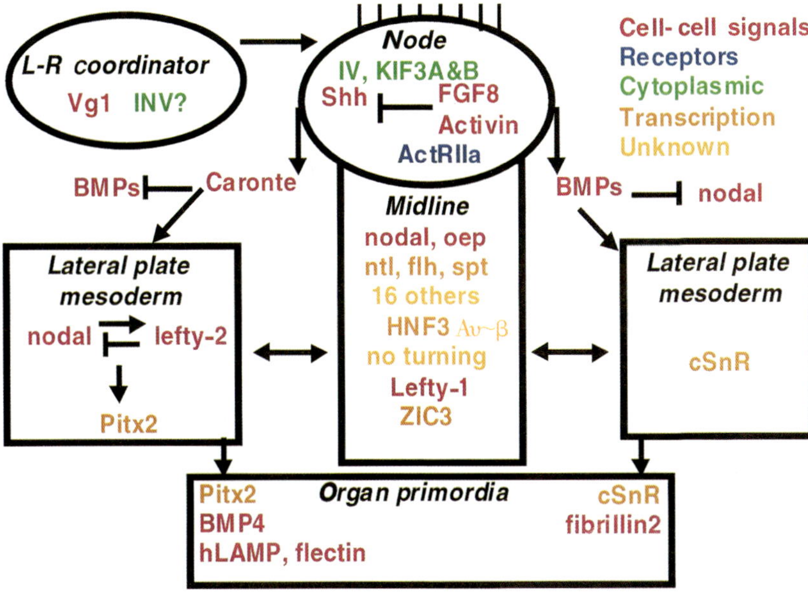

CHAPTER 9, FIG. 2 Genes and embryonic structures implicated in left-right development. Names of genes are color-coded to indicate cell-cell signals (red), signal receptors (blue), cytoplasmic proteins (green), transcription factors (brown), and genes of unknown function (orange). Genes implicated in early steps in left-right development are at the top; later genes in the pathway are at the bottom. Positions in the embryo are indicated in boxes, including the node and midline (notocord and floor-plate) in the middle of the figure, and left and right lateral plate mesoderm (LPM) on the left and right sides, respectively. Most of these early embryonic tissues contribute to bilaterally symmetric structures. The organ primordia (e.g., heart, intestine, parts of the brain) that give rise to asymmetric structures are indicated in the box at the bottom. Adapted and updated from Ramsdell and Yost, 1998.

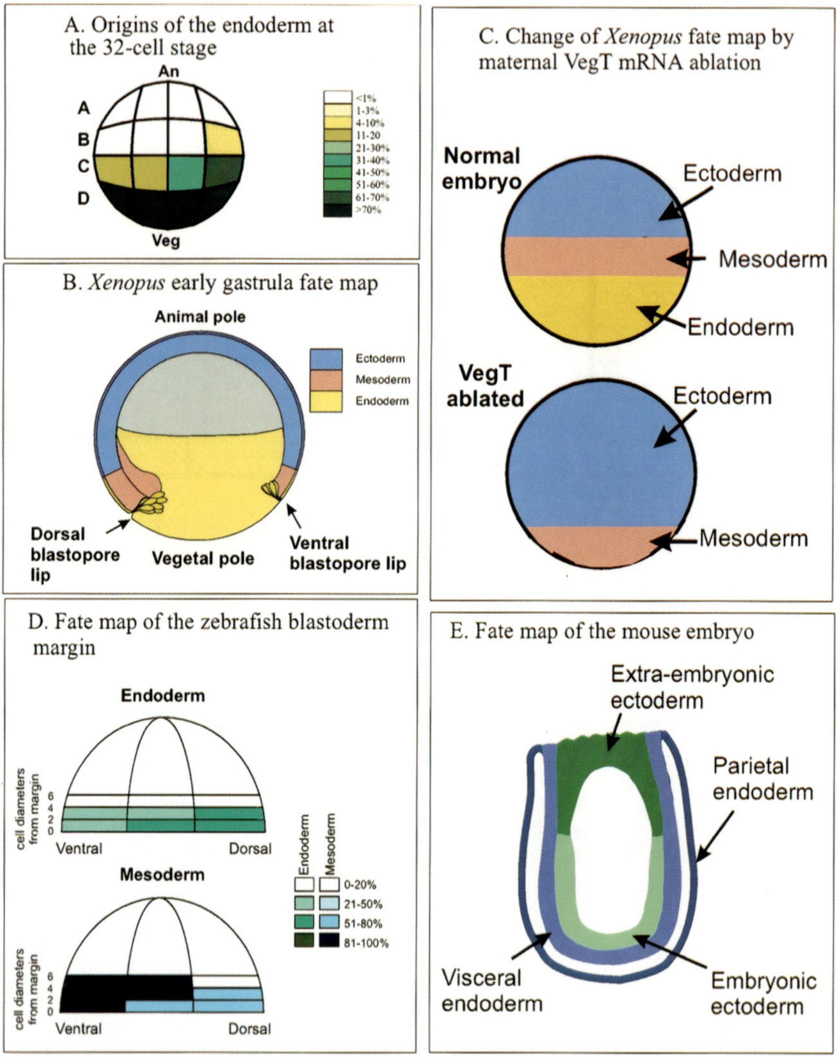

CHAPTER 10, FIG. 1 (A) Fate map of the *Xenopus* embryo at the 32-cell stage. Only the proportions of daughter cell volumes forming endoderm are shown (from Dale and Slack, 1987, with permission from Company of Biologists Ltd.). (B) Fate map of the early *Xenopus* gastrula (adapted from Keller, 1991). (C) Effect on the *Xenopus* fate map of deleting maternal VegT mRNA by oligonucleotide injection (based on findings of Zhang *et al.*, 1998a). (D) Fate map of the zebrafish blastoderm margin, showing percentage of clones giving rise to endoderm and mesoderm (Warga and Nüsslein-Volhard, 1999). (E) Fate map of the mouse embryo at the egg cylinder stage. According to the nomenclature in mouse all three germ layers of the embryo proper arise from the embryonic ectoderm.

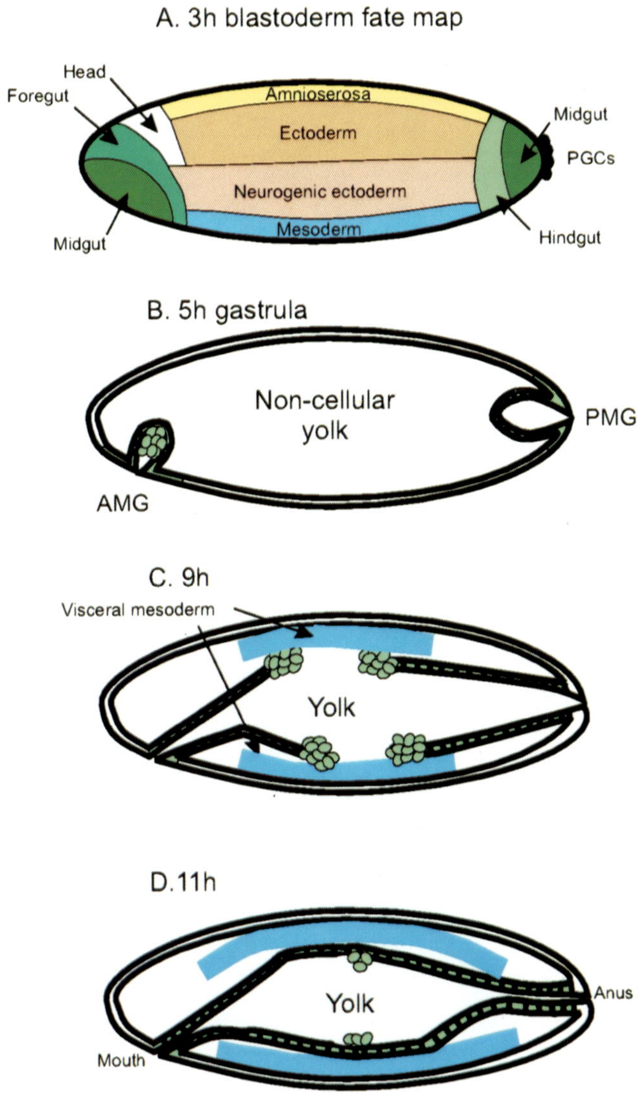

CHAPTER 10, FIG. 6 Development of the endoderm in *Drosophila* (redrawn from Gerhardt and Kirschner, 1997, with permission). (A) Fate at the blastoderm stage. (B–D) Successive stages in migration of the gut precursors. The figure is greatly simplified by ignoring the extension and retraction of the germ band around the embryo. See the text for further details.

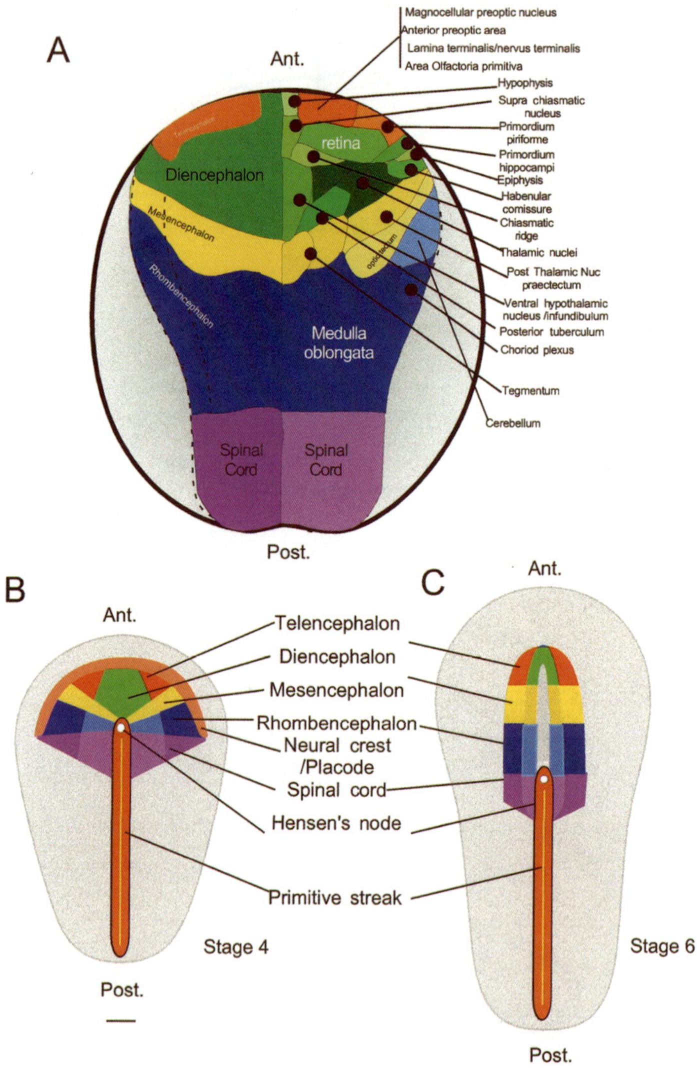

CHAPTER 11, FIG. 1 Neural plate stage fate maps. (A) *Xenopus* neural plate stage embryo (stage 15) showing the fate cells at the late tadpole stage (stage 42). On the left is shown the fate of the major divisions of the nervous system; finer brain structures are indicated on the right. Dorsal fates generally lie laterally while ventral fates lie medially. (Adapted from Eagleson and Harris, 1990). (B) Chick early neural plate stage (stage 4) with similar neural fates indicated with indentical colors as in A. Dorsal and ventral boundaries are indicated by different shades of color that approximate the separation of the alar and basal plates (ventral fates are medial and dorsal fates are more lateral). (C) Chick late neural plate stage embryo (stage 9) colored as in A and C. Elongation due to convergence and extension during the continuing movements of gastrulation. [B and C are derived from published fate maps and modified according to the suggestions of Claudio Stern (Rudnick, 1944; Spratt, 1952; Rosenquist, 1966; Bortier and Vakaet, 1992).]

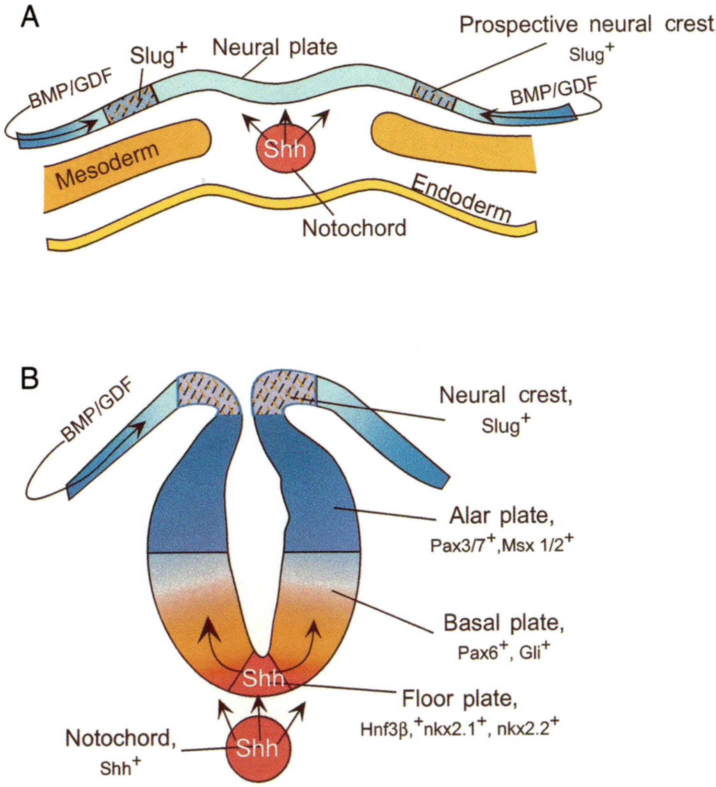

CHAPTER 11, FIG. 2 Signaling and development of the neural tube in the chick. (A) Neural plate stage of neural tube development showing the signaling by the notocord (red, SHH) and lateral ectoderm (dark blue, BMP/GDF). Prospective neural crest ($Slug^+$) is indicated at the border between the future neural plate and the ectoderm. (B) Neural Groove Stage after morphogenetic movements and early signaling. Continued SHH signaling by the notocord is indicated as well as by the developing floor plate cells (HNF3-, $Nkx2.1^+$, $Nkx2.2^+$), which generates a gradient of SHH signaling. Dorsally, BMP/GDF signaling acts to activate $Pax3/7^+$ and $Msx1/2^+$ gene expression. (C) Neural tube stage showing BMP/GDF signaling dorsally by the roof plate ($BMP4^+$, $Lmx\text{-}1^+$). Dorsal signals have established dorsal interneuron precursor cells (D_p1, D_p2) in the dorsolateral zone ($BMP7^+$, $Dsl1^+$) while neural crest cells have separated from the neural tube and have begun migrating. Medially, dorsal and ventral signaling has established domains of D1x1/2 expression. Ventrally, SHH signaling in the notocord and floor plate have initiated the development of motoneuron lineages. (D) Differentiated neural tube showing a variety of cell types as defined by molecular markers indicated in Table 1. Motoneurons have begun axon outgrowth and targeting and are indicated together with cell type specific markers for each neuronal subclass.

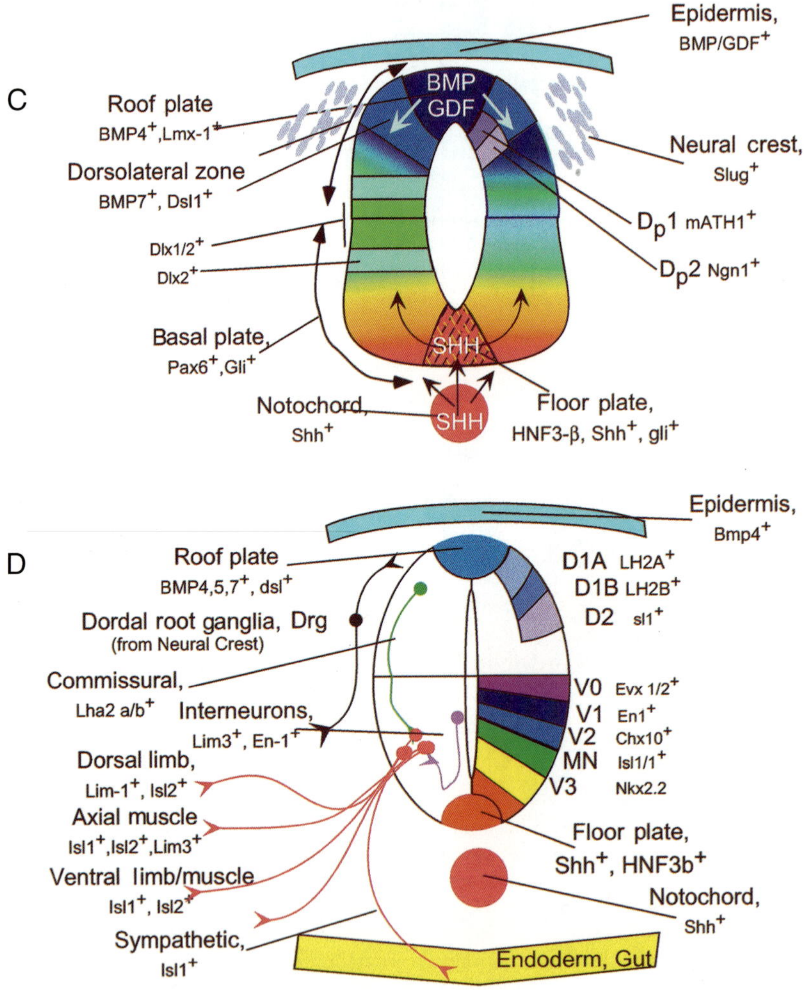

CHAPTER 11, FIG. 2 *(continued)*

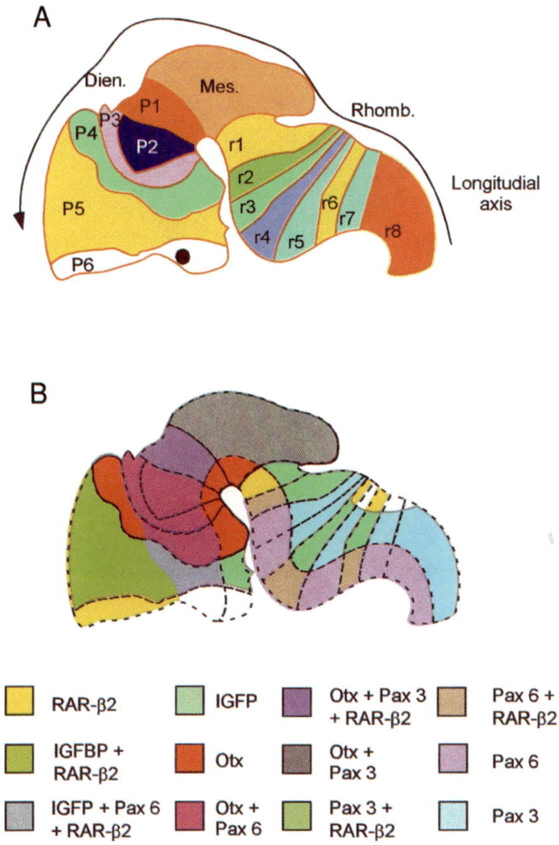

CHAPTER 11, FIG. 5 Prosomeric model of brain organization. (A) The future forebrain, midbrain, and hindbrain can be divided into segments or neuromeres based on morphological and gene expression data. A series of constrictions, called rhombomeres, are observed in the hindbrain that are barriers to cell movements and act to restrict cell lineage. As indicated in Figure 4, a number of genes are differentially expressed that correspond to the morphological constrictions and support a segmented model in this caudal region of the developing brain. In the rostral regions, the morphological evidence is less compelling, but an increasing number of genes are being indentified that appear to indicate further segmentation of forebrain and midbrain (adapted from Puelles and Rubenstein, 1993). (B) Gene expression patterns of a number of genes from mouse and rat have been prepared in separate colors and overlaid on the neuromeric model. The various expression patterns were then made partially transparent in the graphics program to allow the visualization of all the genes in a single image. The results are presented in the figure as supporting evidence of the proposed model (adpated from Puelles and Rubenstein, 1993).

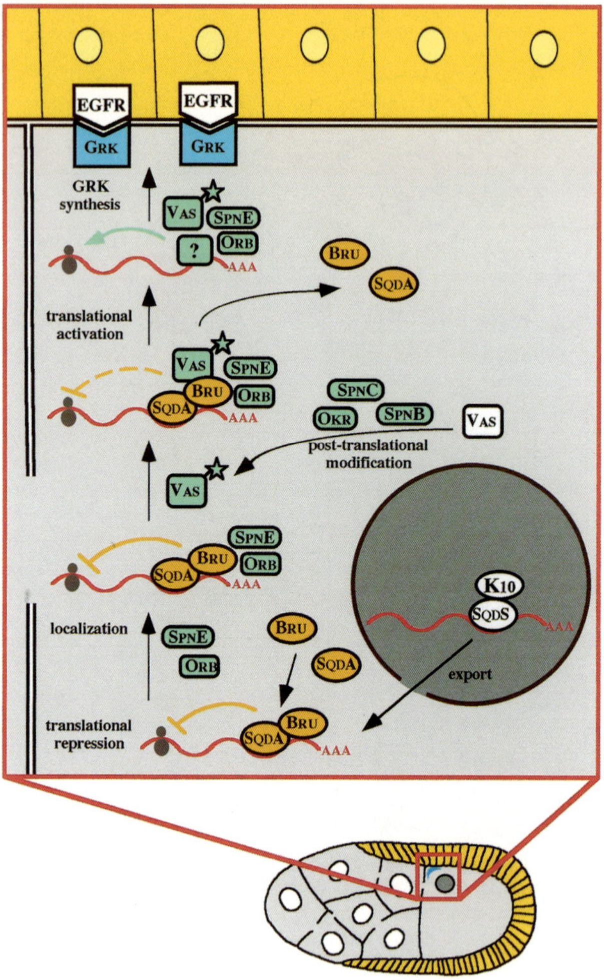

CHAPTER 14, FIG. 2 Model for RNA localization and translational control of the dorsalizing GRK signal (modified from Ghabrial and Schüpbach, 1999; Norvell *et al.*, 1999). Many of the molecular interactions shown are hypothetical. By stage 8, the oocyte nucleus has been migrated to the dorso-anterior corner of the oocyte. *grk* RNA is newly transcribed in the oocyte nucleus and associates through its 3′ UTR with SQDS. K10 is required for the nuclear retention of SQDS. Upon export to the oocytes cytoplasm, SDQA and BRU are recruited to the *grk* 3′UTR. BRU is required to translationally repress *grk* transcripts that are not localized to the dorso-anterior. *grk* RNA localization to the dorso-anterior corner of the oocyte requires ORB and SPN-E/HLS. BRU-mediated translation repression of localized *grk* transcripts is relieved through translational activation by ORB and VAS. SPN-B, SPN-C, and OKRA regulate post-translational modification of VAS. Translational activation leads to dissociation of BRU from the *grk* transcript and production of GRK protein. GRK signals to the overlying dorsal anterior follicle cells through EGFR, ultimately resulting in dorsal cell fate specification. *grk* RNA is shown in red, translational repressors in orange and translational activators in green.

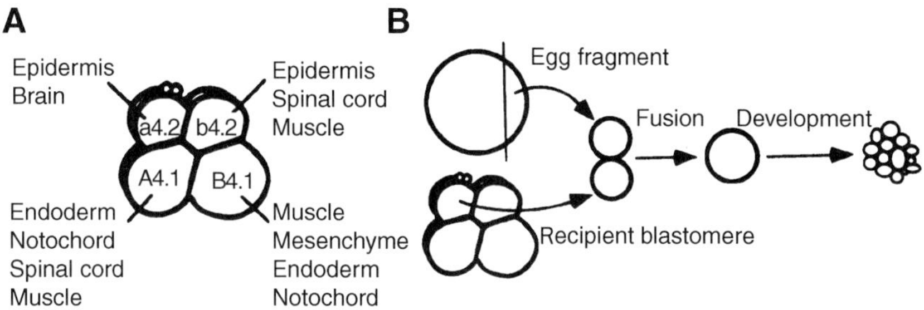

FIG. 7 The fusion method for transplantation of egg cytoplasmic regions in *Halocynthia roretzi.* (A) Derivatives of the eight-cell embryo. (B) Fusion procedure. An anucleate fragment is cut from the egg. The fragment is fused to a donor blastomere isolated from an eight-cell embryo. The fusion product develops into a partial embryo and can be assayed for tissue differentiation. In a variation of this theme, the donor can be another egg rather than a blastomere. From Yamada and Nishida (1996) with permission.

experiments, anucleate fragments from the animal hemisphere of mature *Halocynthia roretzi* eggs were fused with isolated A- and B-line blastomeres, which do not contribute to epidermis (Nishida, 1994a). The potential of the fusion products to make epidermis was assayed with a monoclonal antibody for differentiated epidermal cells (Nishikata *et al.,* 1987). The results showed that A- and B-line cells were able to produce the epidermal antigen when fused to animal cytoplasmic fragments.

Further studies were conducted to determine the origin and distribution of the ectodermal determinants. First, a series of fusion experiments was conducted as described above, except that anucleate fragments from the animal, equatorial, and vegetal portions of unfertilized eggs, as well as fertilized eggs at various times during ooplasmic segregation, were used as potential donors of epidermal determinants (Yamada and Nishida, 1996). The results showed that epidermal determinants are distributed in a gradient with a maximal activity in the animal hemisphere of unfertilized eggs. During OS1, however, the epidermal determinants were concentrated in the equatorial zone and then returned to the animal hemisphere after the completion of OS2. Hence, the distribution of epidermal determinants in *Halocynthia roretzi* eggs correlates with the ectoplasm in *Styela* eggs.

Other experiments addressed the origin of the epidermal determinants in oocytes (Bates and Nishida, 1998). In these experiments, fragments of *Halocynthia* oocytes with or without an intact GV (or released GV plasm) were produced at various times during maturation. The fragments were fertilized, and the cleavage products assayed for their capacity to form epidermis by using the epidermal antibody. The results showed that the ability to produce epidermis after fertilization and cleavage existed in oocyte fragments containing or lacking the GV/GV plasm. These results suggest that epidermal determinants are resident in the oocyte cytoplasm.

The concentration of epidermal determinants appears to be critical to the timing of tissue-specific gene expression in *Halocynthia roretzi* embryos (Ishida and Satoh, 1998). To demonstrate this point, unfertilized eggs were bisected across the equatorial plane, and embryos that developed from fertilized animal and vegetal fragments were assayed with *HrEpiC,* a zygotic gene expressed specifically in epidermal cells (Ueki *et al.,* 1991). The key feature of this experiment is that animal and vegetal fragments of unfertilized eggs will both develop after insemination (Ortolani, 1958), forming smaller embryos enriched in either animal or vegetal cytoplasm. During normal development, *HrEpiC* transcripts first appear in the epidermal progenitor cells at the 64-cell stage (Ishida *et al.,* 1996). *HrEpiC* mRNA appeared at the normal time in embryos produced from animal fragments. In contrast, *HrEpiC* transcripts appeared one division cycle later in embryos produced from vegetal fragments. These results indicate that the animal hemisphere has the highest concentration of epidermal determinants and that these control tissue-specific gene expression in a stoichiometric fashion.

The animal blastomeres undergo more cell divisions than some of their vegetal counterparts (e.g., muscle, notochord, and endoderm cells). Therefore, the animal hemisphere also may contain determinants that control the frequency of cell division.

2. Muscle and Posterior Determinants

The fusion method was applied to *Halocynthia* eggs and embryos to obtain further evidence for muscle determinants (Nishida, 1992a). A posterior (myoplasm-containing) fragment from a mature egg was fused with an a-line blastomere, the only cell type that does not normally produce primary or secondary muscle cells, and the ability to produce muscle cells was assayed by examining AChE, muscle actin, and myosin production. The posterior-vegetal cytoplasm caused the a-line progeny to express muscle cell features, thus supporting the existence of transplantable muscle determinants.

The distribution of muscle determinants was followed by fusing fragments from various egg regions to the a-line blastomeres (Yamada and Nishida, 1996). The results showed that muscle determinants are distributed in a gradient in the unfertilized egg with a maximum at the vegetal pole. During ooplasmic segregation, the muscle determinants are segregated first into the vegetal pole region and then into the posterior equatorial zone of the uncleaved zygote. The distribution of muscle determinants in the *Halocynthia* egg thus correlates with the localization of myoplasm in the *Styela* egg.

In addition to specifying muscle cell fate, the vegetal posterior region of *Halocynthia roretzi* eggs is known to harbor posterior determinants (Nishida, 1994b). The posterior region of the embryo is marked by a unique cleavage pattern (Fig. 3). Blastomeres in the anterior region of the embryo tend to undergo relatively equal cleavages, whereas the most posterior blastomeres undergo unequal cleavages (Fig. 3E, G. I). The products of unequal cleavage, the small B5.2, B6.3, and B7.6 cells, are located at the posterior pole of the embryo. The fate of the B5.2 cells is complex, with contributions to endoderm, muscle, and mesenchyme. By the 64-cell stage, however, the B7.6 cells are fate restricted to produce cells in the larval endodermal strand (Nishida, 1987).

When vegetal posterior cytoplasm was removed from eggs after the completion of OS2, the resulting embryos became radialized and failed to develop muscle cells. In place of muscle cells, these embryos developed a ring of notochord cells around the circumference of the blastopore. In another experiment, vegetal anterior fragments from a host egg were fused to the anterior region of a donor egg lacking the vegetal-posterior region. This operation resulted in the complete reversal of polarity, with unequal cleavage and muscle cells now occurring in anterior rather than posterior blastomeres. These results show that the posterior-vegetal region contains determinants that regulate the generation of unequal posterior cleavages.

The results obtained with *Halocynthia roretzi* eggs are in contrast to earlier studies in *Styela plicata* (Bates and Jeffery, 1987a). *Styela* eggs are much smaller than those of *Halocynthia,* and different methods are necessary to delete cytoplasmic regions. The procedure for producing egg fragments in *Styela* makes use of the follicle, an egg envelope that surrounds the embryo throughout development (Fig. 5). Cytoplasmic fragments are produced by puncturing the follicle, allowing cytoplasm to extrude, and then cutting across the cytoplasmic bridge connecting the two parts of the egg (Fig. 8). When small fragments were removed from the vegetal-posterior region of *Styela* eggs after the completion of OS2 there was no effect on development (Bates and Jeffery, 1987a). Unfortunately, however, larger anucleate fragments could not be produced by this method because extrusion of the vegetal-posterior cytoplasm invariably removed the zygotic nucleus. Hence, although *Styela* embryos show the same posterior cleavage pattern as *Halocynthia* eggs and must contain factors that regulate this pattern, the position of the determinants may be adjusted to accommodate the smaller egg volume.

Recently, a novel organelle, the centrosome-attracting body (CAB) (Fig. 9), has been identified in the posterior blastomeres of *Halocynthia roretzi* embryos (Hibino *et al.,* 1998). Ultrastructural studies show that the CAB consists of electron dense granules resembling the germ plasm of other animals (Iseto and Nishida, 1999). At the beginning of unequal cleavage,

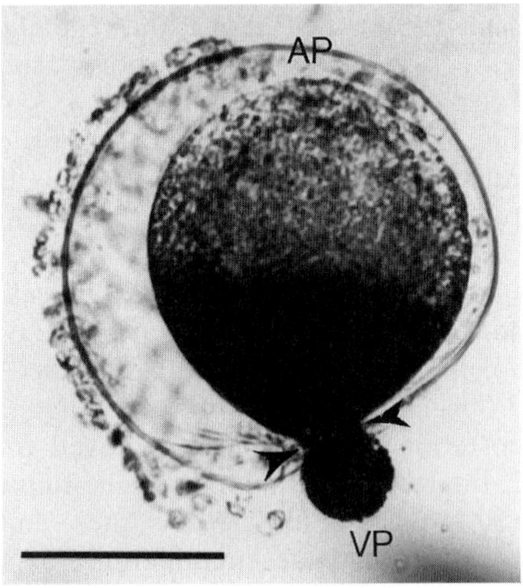
AP
VP

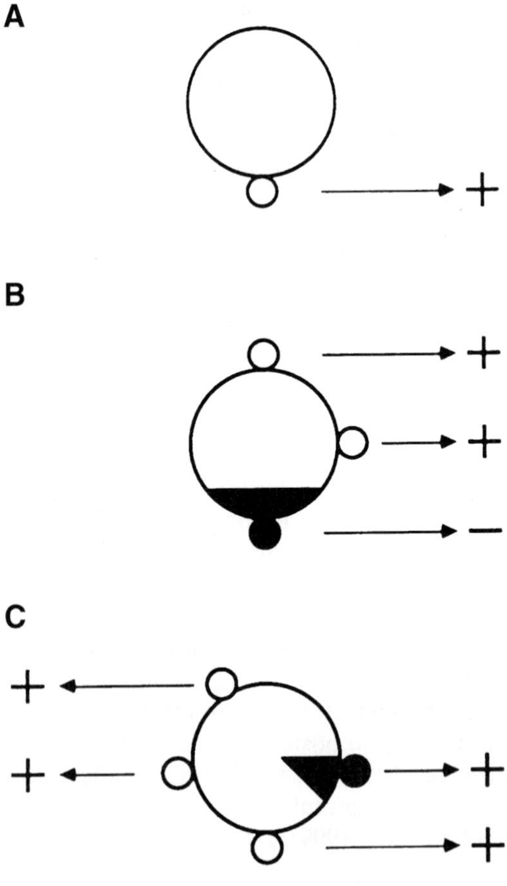
A
+
B
+
+
–
C
+
+
+
+

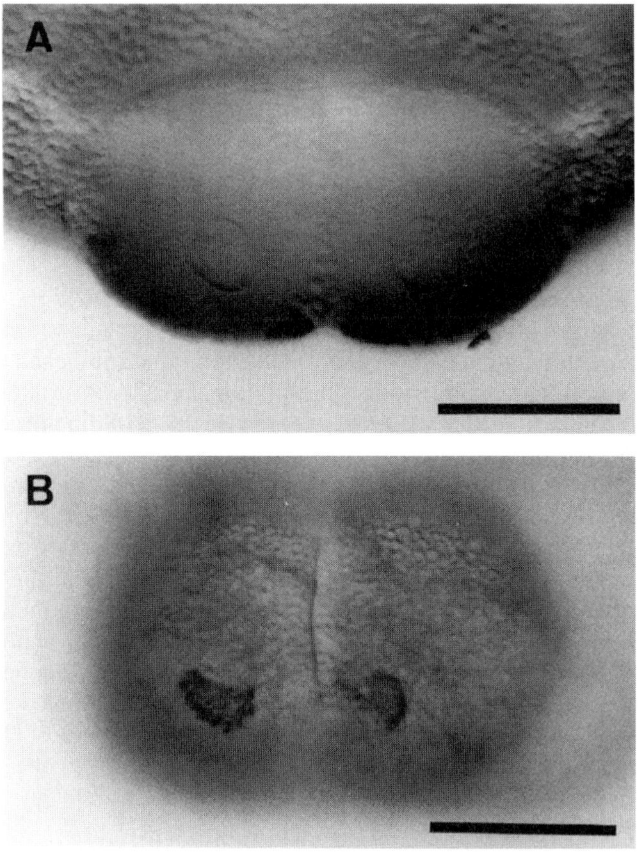

FIG. 9 The centrosome-attracting body (CAB) in *Ciona intestinalis* embryos. Vegetal (A) and posterior (B) views of the CAB stained with antikinesin antibody in the B5.2 blastomeres of 16-cell embryos. From Nishikata *et al.* (1999).

FIG. 8 The effect of removal of anucleate egg fragments on development of *Styela* embryos. Upper frame. A portion of myoplasm partially extruded from the vegetal pole (VP) after the completion of OS1. The dark area in the egg is the myoplasm. Scale bar: 100μm. From Bates and Jeffery, 1987a. Lower frame. (A) Anucleate fragments removed from the vegetal pole of unfertilized eggs have no effect on development. (B) Anucleate fragments removed from the vegetal pole but not the animal pole or the equatorial region after the completion of OS1 abolish gastrulation. (C) Anucleate fragments removed from the animal, vegetal, anterior, and posterior poles after the completion of OS2 have no effect on development. Normal development is indicated by arrows with +. Inhibition of gastrulation is indicated by arrow with–. From Jeffery (1992) with permission from Company of Biologists Ltd.

a thick bundle of microtubules appears between the CAB and the centrosome. Subsequently, the microtubule bundle shortens, drawing the centrosome and interphase nucleus toward the CAB, situated at the posterior pole of the B5.2, B6.3, and B7.6 blastomeres. The formation of an asymmetric cleavage furrow at the midpoint of the spindle apparatus then completes the unequal cleavage process. The CAB is not apparent as a visible structure before fertilization or during ooplasmic segregation, but begins to condense from numerous foci in the vegetal-posterior cytoplasm at the two-cell stage. The CAB precursors assemble during the next two cleavages, and become localized as a single entity in the posterior-vegetal region of the B4.1 and B4.1 blastomeres at the eight-cell stage. The CAB is then restricted to the smallest of each unequal-cleaving B-line blastomere during each cell cycle. The presence of the granular nuage-like CAB in the B7.6 blastomeres brings up the possibility that these cells may represent the germ line precursors in ascidians.

Two kinds of experiments were designed to determine the role of the CAB in equal posterior cleavage (Nishikata *et al.,* 1999). First, the CAB was dissolved using sodium dodecyl sulfate (SDS). The loss of the CAB in SDS-treated embryos resulted in equalization of cleavages in the posterior blastomeres. Second, the posterior-vegetal region containing the CAB was removed from donor eggs and fused to the anterior-vegetal region of host eggs. In this way, eggs were created with two CABs, one at the anterior and one at the posterior pole. Eggs with double CABs showed posterior-like unequal cleavages at both the anterior and posterior poles. Hence, the CAB appears to be a cytoplasmic organelle that regulates the posterior cleavage pattern.

3. Endodermal Determinants

The existence and localization of endodermal determinants was also studied by the fusion method (Nishida, 1993; Yamada and Nishida, 1996). Similar to the design of previous experiments, anucleate fragments were fused to a-line blastomeres, which do not contribute progeny to endoderm, and assayed for AP expression. The results showed the existence of endodermal determinants that were sufficient to promote AP expression when introduced into non-endodermal blastomeres. The endodermal determinants were distributed in a gradient with maximal activity at the vegetal pole in the unfertilized egg. They moved vegetally after fertilization and eventually occupied the vegetal hemisphere of the uncleaved zygote. The endodermal determinants in *Halocynthia* eggs thus closely parallel the localization of endoplasm in *Styela* eggs.

4. Gastrulation Determinants

Gastrulation consists of cell movements that internalize the endoderm, mesenchyme, and muscle precursor cells, and allow the epidermal cells to spread over the external surface of the ascidian embryo. These cell movements continue through neurulation and larval tail morphogenesis. Gastrulation appears to be initiated at the 110-cell stage by invagination of several large endodermal cells at or near the vegetal pole, thus creating the blastopore and archenteron (Jeffery, 1992). The invagination of the endoderm cells is followed by involution of the presumptive notochord, mesenchyme, and muscle cells over the blastopore lips in concert with epiboly of the animal hemisphere cells.

The first evidence for determinants controlling gastrulation was obtained by G. Ortolani (1958), who bisected eggs through the equatorial region. When unfertilized eggs were split in this manner and the fragments fertilized, the animal and vegetal fragments both gastrulated and developed into normal larva. However, when fertilized eggs were bisected during ooplasmic segregation, only the vegetal fragments were able to gastrulate and form complete larvae. The results suggest that gastrulation determinants reside in the vegetal hemisphere of fertilized eggs.

The egg fragmentation experiments were refined by Bates and Jeffery (1987a), who removed small cytoplasmic fragments from various parts of *Styela* eggs via the cytoplasmic deletion procedure described above (Fig. 8). These investigators showed that gastrulation was not affected by removing anucleate fragments from unfertilized eggs. However, when the same experiment was done after OS1, gastrulation could be inhibited by removal of a small cytoplasmic region from the vegetal pole. Eggs lacking this region cleave normally and develop endoderm and muscle cells, but do not gastrulate. In contrast, removal of a small region from the animal pole or the vegetal pole after OS2 has no affect on development (Bates and Jeffery, 1987a).

UV irradiation of the vegetal pole region at the completion of OS1 prevents gastrulation but not muscle and endoderm development (Jeffery, 1990a) (Fig. 10), thus mimicking the effects of cytoplasmic deletion (Fig. 11). Moreover, the effect of UV light is transient, with maximal activity during OS1 (Fig. 12A), and the UV dose required to inhibit gastrulation is less than that necessary to abolish muscle or endoderm development (Fig. 12B). The results of the cytoplasmic deletion and UV irradiation experiments suggest that gastrulation determinants are concentrated in the vegetal pole region for only a short time after the completion of OS1, and are distinct from muscle and endoderm determinants. Presumably, the gastrulation determinants are translocated to the vegetal pole with the myoplasm during OS1, but then either become insensitive to UV or spread throughout the vegetal hemisphere during OS2.

The role of determinants in gastrulation also was investigated in *Halocynthia roretzi* embryos (Nishida, 1996). Removal of a small portion of the vegetal pole cytoplasm by microsurgery after OS1 blocks gastrulation, thereby confirming the results of Bates and Jeffery (1987a). In contrast to

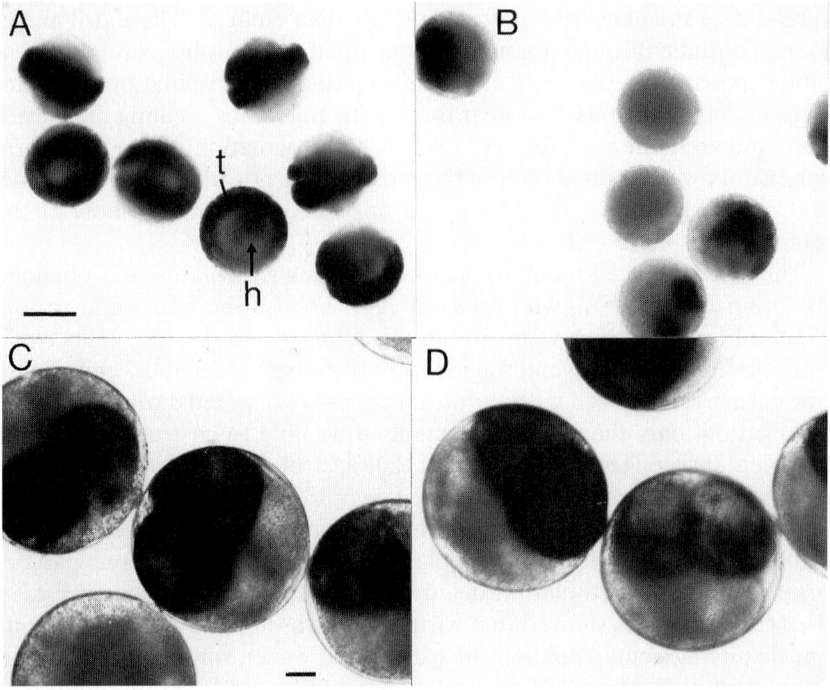

FIG. 10 Relationship between gastrulation and muscle determinants determined by UV irradiation and AChE activity. (A) Tailbud stage embryos developing from normal egg showing AChE positive muscle cells in tails (t) curling around the larval heads (h). (B) Embryos developing from eggs irradiated with a low UV dose that abolishes gastrulation but does not affect AChE production in muscle cells. (C) Embryos developing from eggs cleavage arrested at the four cell stage. (D) Embryos that developed from embryos irradiated with a low UV dose and then cleavage arrested at the four cell stage. AChE activity is observed in the two posterior cells in C and D. Embryos in B–D at the same chronological stage as A. From Jeffery (1990a).

Styela embryos, however, *Halocynthia* embryos deficient in vegetal pole cytoplasm cleave radially rather than bilaterally and lack the derivatives of the vegetal blastomeres (Nishida, 1996). Essentially, the vegetal blastomeres assume the cleavage pattern and fate of the animal blastomeres. By making larger and larger cytoplasmic deletions, it was also shown that the gastrulation determinants spread throughout the vegetal hemisphere of *Halocynthia* eggs during OS2. The vegetal pole deletion experiments indicate that there are commonalties and differences in the responses of *Styela* and *Halocynthia* eggs to deletion of the vegetal pole cytoplasm. When the techniques devised to remove anucleate fragments from *Styela* eggs (Fig. 8) were applied to *Halocynthia,* the results were unchanged (Nishida, 1996). Therefore, these

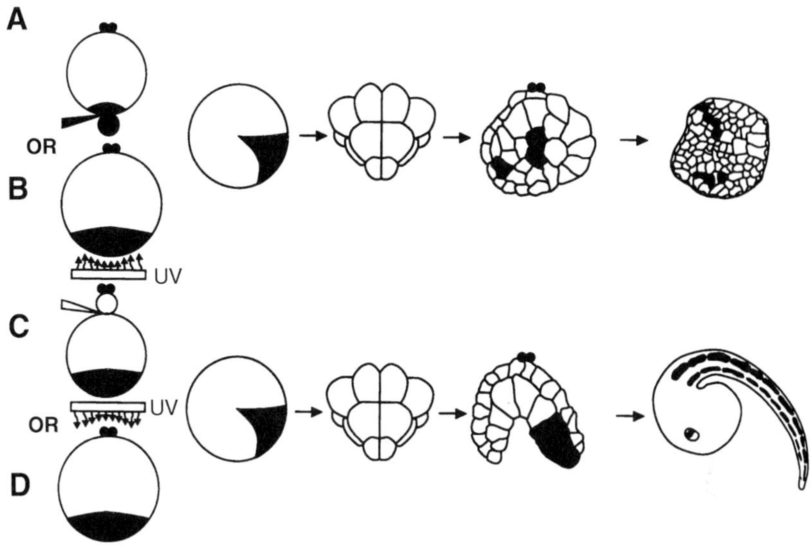

FIG. 11 A comparison of the effects of deletion (A, C) and UV irradiation (B, D) of the vegetal pole (upper row) and animal pole (lower row) regions of *Styela* eggs on gastrulation and muscle development. Dark areas represent myoplasm or AChE-positive muscle cells. From Jeffery and Swalla (1990a).

differences must be due to evolutionary variations in the distribution of cytoplasmic determinants in *Styela* and *Halocynthia* eggs.

Cytoplasmic fusion experiments have been done to test the developmental potency of the vegetal pole region in *Halocynthia* embryos (Nishida, 1996). In these experiments, a small vegetal-pole region was removed at the completion of OS1 and fused to other parts of the same egg, including the animal pole region. The results showed that gastrulation could be induced at any site in the embryo, including the animal pole, by the ectopically positioned vegetal-pole cytoplasm. In the case of animal pole transplantation, gastrulation was initiated at the animal pole, and tissues normally derived from the vegetal blastomeres were formed in the animal hemisphere. These manipulations resulted in complete reversal of animal vegetal polarity. In summary, the results of cytoplasmic deletion, UV irradiation, and cytoplasmic fusion experiments show that cytoplasmic determinants localized at the vegetal pole are both sufficient and necessary for gastrulation.

5. Determinants and Inductive Processes

Cytoplasmic determinants account for the specification of three larval tissues, epidermis, muscle, and endoderm, as well as regional fate decisions,

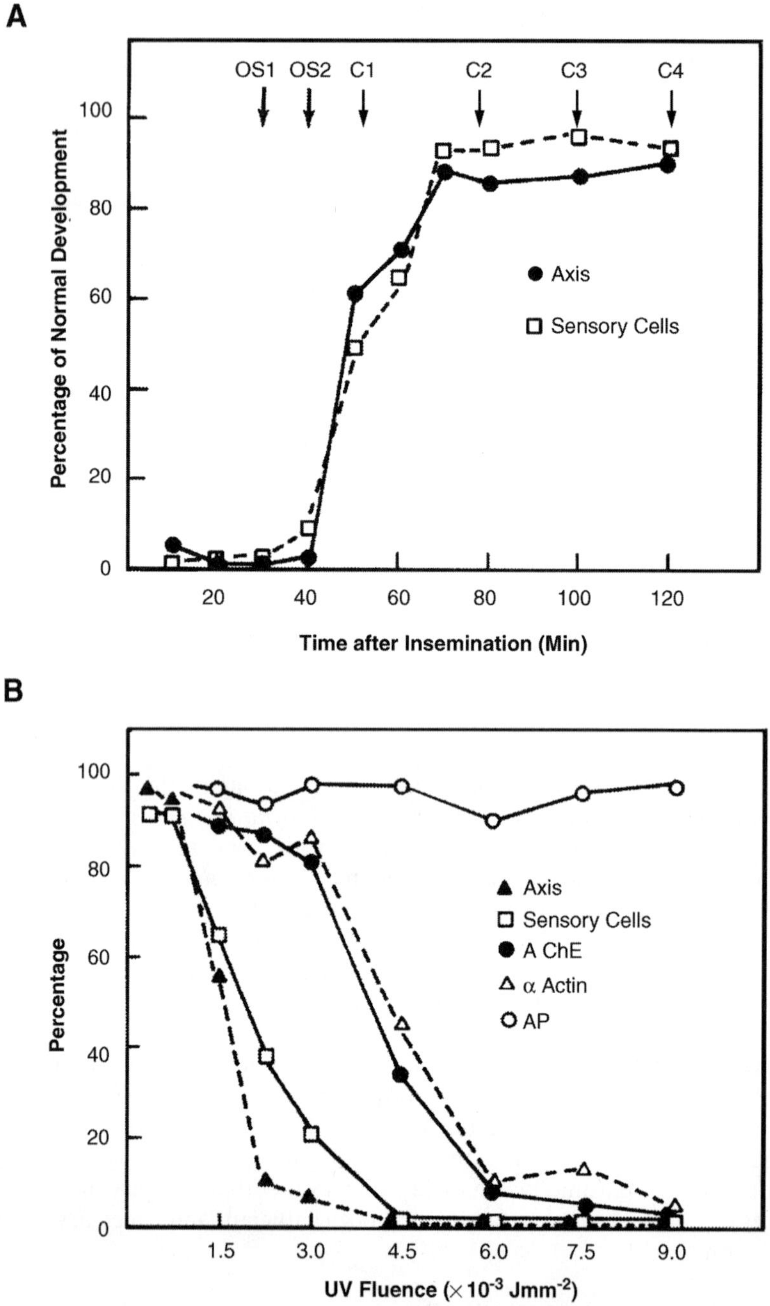

FIG. 12 Effects of UV irradiation on determinant function in *Styela clava* embryos. (A) UV sensitivity periods for axis (gastrulation) and sensory organ development. (B) UV dose response curves for gastrulation (axis and sensory organ), muscle (AChE and actin) development, and endoderm (AP) development. From Jeffery (1990a).

such as the posterior cleavage pattern and gastrulation. What processes are used to specify the other three tissues in the tadpole larva? The notochord, mesenchyme, and neural tissues appear to be specified by inductive processes (Nishida, 1997). The endoderm cells have been shown to induce notochord cells at the 32-cell stage in *Halocynthia roretzi* (Nakatani and Nishida, 1994). Likewise, the neural cells are also specified by cell-cell interactions, although the identity of the inducing cells may differ between species (Reverberi *et al.,* 1960; Nishida, 1991). Little is known about the factors responsible for mesenchyme cell specification. As described above, larval muscle cells are derived from three sources in ascidians (Nishida, 1987). The secondary muscle cells, derivatives of the A- and b-line cells, also are specified by cell interactions, but again the timing of this process and the inducing tissues remain to be resolved (Nishida, 1992b). Finally, the processes underlying the development of competence in the tissues specified by induction have not been investigated, and may be mediated by the activity of additional cytoplasmic determinants.

C. The Three-Center Model

The localization of the five types of cytoplasmic determinants during ooplasmic segregation is summarized in Figure 13. A model involving three organization centers is proposed to explain the distribution of these determinants (Fig. 14). During OS1, the determinants become attenuated into animal

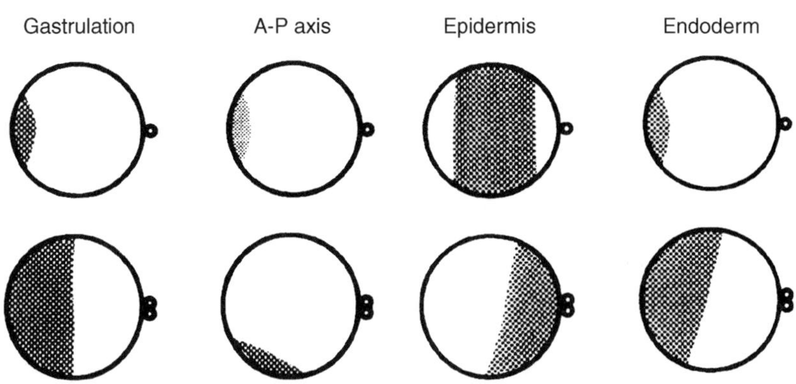

FIG. 13 Distribution of cytoplasmic determinants in *Halocynthia roretzi* eggs. (Upper row) Distribution after the completion of OS1. (Lower row) Distribution after completion of OS2. Animal poles (top). Posterior poles (right). From Nishida (1996) with permission from Company of Biologists Ltd.

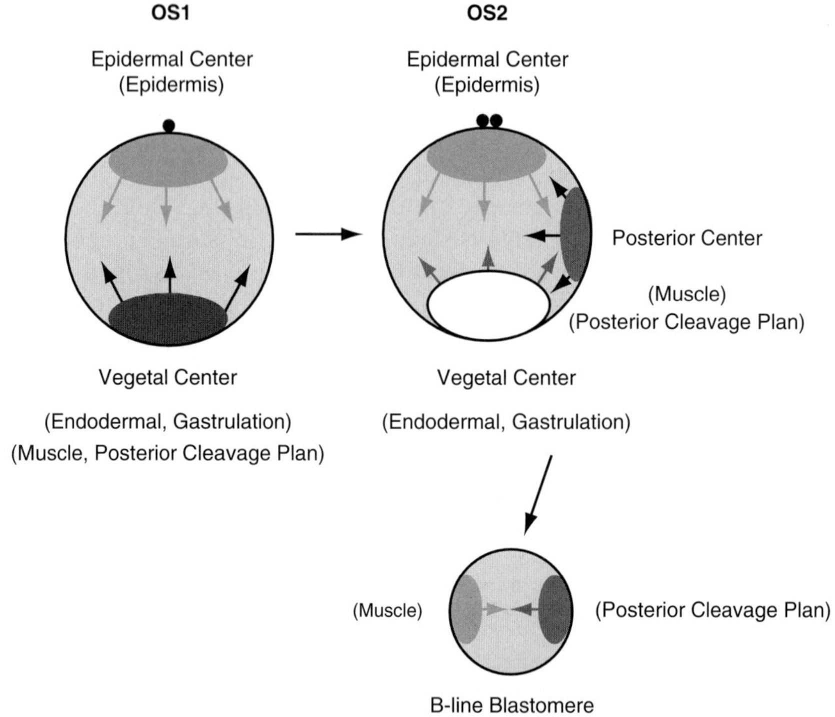

FIG. 14 Three-center model. (Left) The completion of OS1 results in epidermal and vegetal organization centers. (Right) The completion of OS2 results in the appearance of a posterior organization center. (Bottom, right) Later, in the B-line blastomeres, the muscle determinants are released into the anterior region where they will be inherited by muscle cells, and the posterior cleavage plan determinants remain at the posterior pole.

(epidermal) and vegetal organization centers (Fig. 14, left). The muscle, posterior, endoderm, and gastrulation determinants are localized in the vegetal organization center, which is consolidated around the vegetal pole. This region is called the vegetal organization center. During OS2, the vegetal organization center is modified, thus releasing the endoderm and gastrulation determinants and creating a new organization center in the myoplasmic cresent containing muscle and posterior determinants. We call this region the myoplasmic orgranization center. Later, in the B-line blastomeres, the muscle and posterior determinants separate, the latter possibly in association with the CAB. The third organization center at the posterior pole of the embryo is called the posterior organization center (or postplasm, Nishida and Makabe, 1999).

IV. Evolution of Determinants

Taxonomists have identified two orders of ascidians: the Pleurogona and the Enterogona. The Pleurogona includes *Halocynthia, Styela,* and *Molgula,* whereas the Enterogona includes *Ciona* and *Phallusia.* These five genera represent the organisms used in the majority of current studies of ascidian development. The evolutionary distance between these five genera is considerable (Wada *et al.,* 1992). Therefore, it would not be surprising to discover that changes in cytoplasmic determinants have occurred during ascidian evolution. We have already described evidence that some determinants may exhibit different localizations in *Halocynthia* and *Styela.* The conventional mode of development can evolve rapidly in ascidians, as evidenced by the existence of closely related species with radically different larval forms (Berrill, 1931; 1935; Jeffery and Swalla, 1992a). Hence, studies of the evolution of determinants have been focused on their role in changing the tadpole larva.

A. Modes of Larval Development

The tadpole larva has undergone constructive and regressive changes during evolution (Jeffery and Swalla, 1992a; Jeffery, 1997). The constructive changes involve increasing the size and complexity of the head and tail. Solitary species, such as *Ciona intestinalis, Styela plicata, Phallusia mammillata,* and *Halocynthia roretzi,* exhibit conventional tadpole larvae (Fig. 2). In these species, the differentiation of head endoderm into adult digestive organs is not initiated until late in larval life or during metamorphosis, and the tail is relatively simple, consisting of about 38 to 42 muscle cells and 40 notochord cells. In contrast, most colonial species, such as *Ecteinascidia turbinata,* have greatly enlarged heads and tails (Berrill, 1935). The tadpoles of these species have evolved a precocious differentiation of head endoderm into adult pharyngeal and gut tissues during larval development and a robust tail containing up to a thousand muscle cells. It is interesting that despite the large increase in muscle cells, the enhanced tail still contains only 40 large notochord cells. Whether evolutionary changes in the head and tail of these giant tadpole larvae are mediated by cytoplasmic determinants is unknown, but this should be a fascinating topic for future research.

In other species, the tadpole larva has lost its tail, a modification known as anural development (Berrill, 1931). The anural developers have cleavage and gastrulation patterns similar to their counterparts with tailed larvae; however, the brain sensory organs, notochord and tail muscle cells fail to differentiate (Swalla and Jeffery, 1990; Kusakabe *et al.,* 1996). The family

Molgulidae, which contains many anural developers (Jeffery and Swalla, 1990b), includes closely related species that show conventional or anural development (Hadfield *et al.,* 1995; Jeffery *et al.,* 1999). For example, *Molgula oculata* exhibits a typical tailed larva, whereas its anural sister species, *Molgula occulta,* has evolved a tailless larva. *M. occulta,* a recently derived anural species, has retained vestigial notochord and muscle cell progenitors; however, these cells do not undergo terminal differentiation.

The recent divergence of *M. oculata* and *M. occulta* permits interspecific hybridization experiments to be done in the laboratory (Swalla and Jeffery, 1990; Jeffery and Swalla, 1992b). When *M. oculata* eggs are fertilized with *M. occulta* sperm, tailed larvae identical to *M. oculata* tadpoles are formed. In contrast, when *M. occulta* eggs are fertilized by *M. oculata* sperm, tails can be restored, thus suggesting that zygotic factors are involved in evolutionary regression of the tadpole larva. Although notochord and brain sensory organ differentiation are restored in the *M. occulta* × *M. oculata* hybrids, functional muscle cells are not formed. The absence of muscle cell differentiation suggests that maternal as well as zygotic changes have occurred during the evolution of anural development.

B. Evolution of Myoplasm

Because anural developers lack functional muscle cells, current research has focused on the evolution of the myoplasm in these species. The myoplasm lacks pigmentation in *Molgula* species. Fortunately, however, the NN18 monoclonal antibody, which reacts specifically with the myoplasmic cytoskeletal protein p58 (Swalla *et al.,* 1991), allows the myoplasm to be detected in molgulid ascidians (Fig. 15). The myoplasm of the tailed species *M. oculata* is similar to that described in *Styela* and other species with tailed larva. However, the myoplasm has regressed in *M. occulta.* In some clutches of *M. occulta* eggs, the myoplasm is not detectable by NN18 staining in the egg or during embryogenesis. In other clutches, however, the myoplasm is present but attenuated to a small part of the egg and cleaving embryo (Fig. 15, right). Interestingly, hybrid embryos capable of restoring the tail after fertilization with *M. oculata* sperm could only be obtained from clutches of *M. occulta* eggs having a detectable myoplasm (Jeffery and Swalla, 1992b), suggesting that this cytoplasmic region may have a general role in tail formation.

Studies of oogenesis in *M. occulta* eggs have revealed the cellular basis for the absence of myoplasm (Swalla *et al.,* 1991). As determined by NN18 staining (Fig. 16), the myoplasm is present in young *M. oculata* and *M. occulta* oocytes (Fig. 16A, B, E, F). In *M. oculata,* as in other species with tailed larvae, the myoplasm is gradually localized in the oocyte cortex

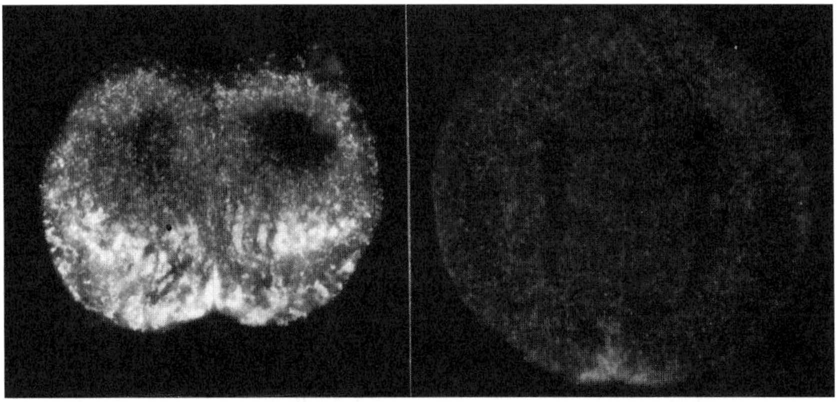

FIG. 15 Evolutionary regression of the myoplasm during anural development. Two-cell *M. oculata* (left) and *M. occulta* (right) embryos stained with NN18 antibody. The *M. occulta* embryo has a highly reduced myoplasm relative to that of its sister species *M. oculata.* From Jeffery and Swalla (1992b).

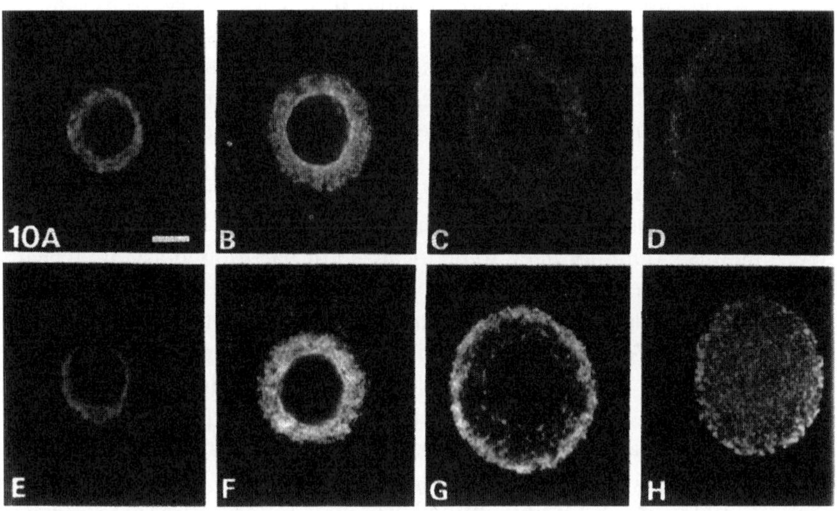

FIG. 16 Alteration of the myoplasm during oogenesis of the anural species, *M. occulta.* (Bottom row) Progressive localization of the myoplasm during oogenesis in *M. oculata,* which exhibits a tailed larva. (Top row) Initial appearance and subsequent disappearance of the myoplasm during oogenesis of *M. occulta,* which exhibits an anural larva. The myoplasm is stained with NN18 antibody. From Swalla *et al.* (1991) with permission from Company of Biologists Ltd.

during later oogenesis and persists in that region in the mature egg (Fig. 16G, H). In *M. occulta,* the myoplasm is present in young oocytes but is not localized to the cortex and disappears later in oogenesis, presumably because p58 is degraded (Fig. 16C, D). When a reduced myoplasm is present in *M. occulta* eggs (see above), p58 remains dispersed in the cytoplasm throughout oogenesis, but during ooplasmic segregation it becomes concentrated at the vegetal and posterior poles of the zygote.

The presence of myoplasm was also investigated in other anural species (Swalla *et al.,* 1991), which have evolved independently from ancestors with tailed larvae. In these species, as in *M. occulta,* the myoplasm is present in young oocytes but disappears later in oogenesis (Fig. 17). The results suggest that muscle determinants may be absent or decreased below a threshold level in eggs of anural species, thus resulting in independent episodes of muscle cell regression and anural development.

V. Identification of Determinants

A widely accepted hypothesis is that determinants are maternal mRNAs localized in the egg and partitioned into specific blastomeres during cleavage. These localized mRNAs are thought to encode transcription factors or other molecules that regulate the expression of tissue- and/or region-specific genes in the embryo. Maternal mRNAs are economical candidates for determinants because they can be translated into many copies of regulatory proteins. Moreover, there is precedent for maternal mRNA determinants in *Drosophila* (Nüsslein-Volhard, 1991) and *C. elegans* (Mello *et al.,* 1992; 1996; Draper *et al.,* 1996). However, there are alternative possibilities for the identity and function of determinants. Uniformly distributed mRNA determinants would not need to be localized or partitioned asymmetrically if they are differentially degraded or activated in specific regions of the embryo. In this scenario, the determinant would be the localized factor regulating mRNA turnover or translational activation. Likewise, localized or regionally activated maternal proteins, rather than mRNAs, could be determinants. For example, p58 (Swalla *et al.,* 1991), myoplasmin C1 (Nishikata and Wada, 1996), a heat shock protein (Bates and Bishop, 1996), ankyrin (Jeffery and Swalla, 1993), spectrin (Jeffery, 1993), Na^+K^+ATPase (Jeffery, 1993), and other unidentified proteins (Jeffery, 1985) are restricted to the myoplasm in various species. There are other possibilities as well, and different molecular identities and functions for various determinants might be expected, considering the evolutionary diversity of ascidians. Many researchers, However, have embraced the localized mRNA hypothesis, which will be reviewed here.

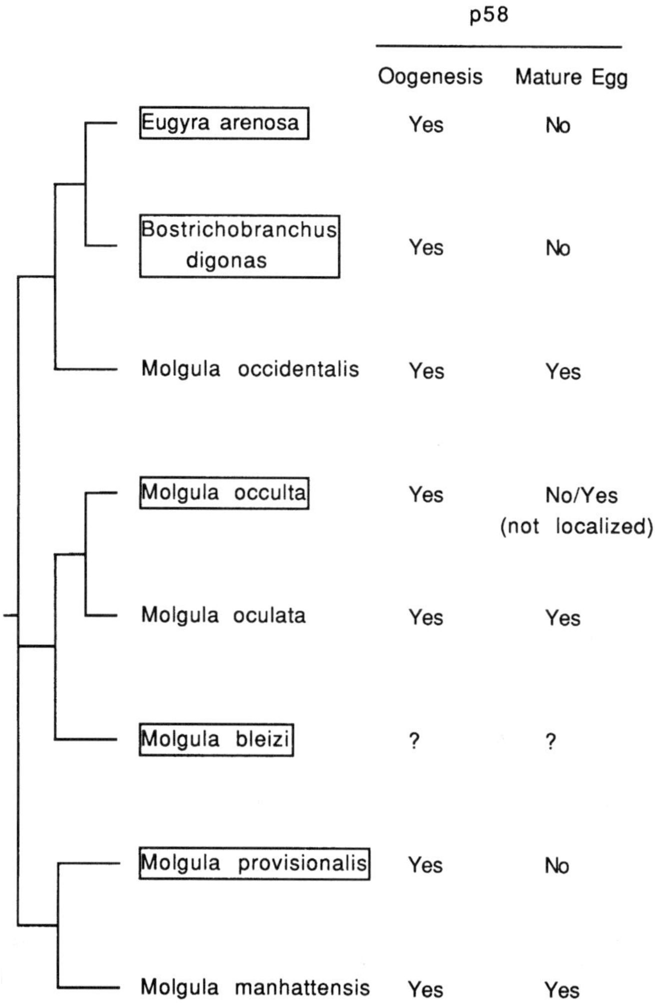

FIG. 17 Independent loss of the myoplasm during the oogenesis of anural species. (Left) Phylogeny according to Hadfield *et al.* (1995) and Jeffery *et al.* (1999). (Right) Presence of myoplasm (p58) in the oocyte or mature egg. Anural species are boxed.

A. Localized mRNAs

One of the first searches for localized mRNAs was carried out in *Styela plicata* eggs by using poly (U), which hybridizes to the poly (A) tails of mRNAs, as an *in situ* hybridization probe (Jeffery and Capco, 1978). The results suggested that poly (A) RNA is concentrated in the GV of mature oocytes (Fig. 18A). In contrast, this method showed that endoplasm and myoplasm are relatively depleted in poly (A) RNA. After oocyte maturation, the poly (A) RNA is released into the ectoplasm, segregated with this cytoplasmic region during ooplasmic segregation, and eventually partitioned into the animal blastomeres (Fig. 18B–D). This was an exciting result because it suggested for the first time that mRNA molecules are not evenly distributed in the cytoplasm and could be concentrated in a region of the egg that is distributed differentially during early development.

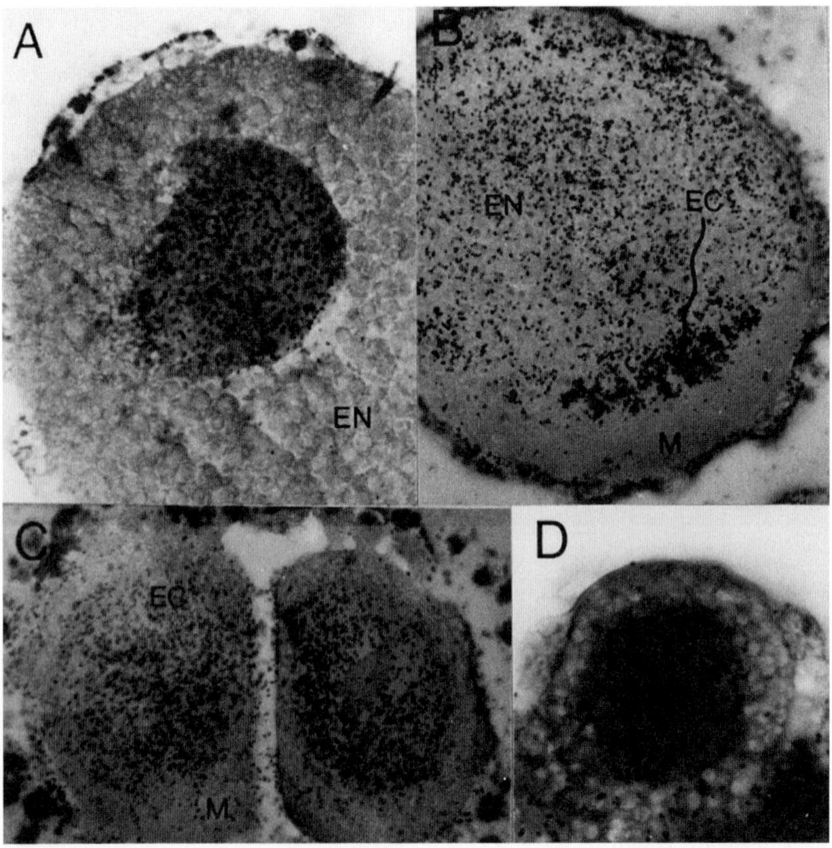

FIG. 18 Distribution of poly (A) RNA in *Styela plicata* eggs. (A) Primary oocyte showing poly (A) localization in the GV. (B) Fertilized egg after the completion of OS1 showing poly (A) concentrated in the ectoplasm (EC) and depletion in the myoplasm (M). (C) Two-cell stage showing poly (A) concentration in the ectoplasm and depletion in the myoplasm. (D) Animal blastomere of eight-cell embryo showing poly (A) concentration in the ectoplasm. En: endoplasm. From Jeffery and Capco (1978).

As *in situ* hybridization techniques improved, it was shown that actin mRNA was localized and preferentially segregated with the myoplasm in *Styela* eggs (Jeffery *et al.*, 1983). This was the first demonstration of mRNA localization in any cell type. Subsequently, localized mRNAs have been demonstrated in the eggs and embryos of numerous animals (Micklem, 1995), as well as in somatic cells. In ascidian eggs, mRNAs appear to be localized by their association with the cytoskeleton (Jeffery, 1984b; 1989b). This property was determined by treating eggs with the nonionic detergent Triton X-100 under conditions that extract soluble proteins and retain cytoskeletal elements in an insoluble matrix. The results show that mRNAs are retained in cytoskeletal residues at the same places as they are localized in intact eggs and embryos.

B. Endodermal Alkaline Phosphatase

As discussed above, the potential to develop AP segregates with the endoplasm and is restricted to the endodermal cells (Whittaker, 1977). Inhibition of protein synthesis blocks the appearance of AP, implying that enzyme synthesis is dependent on zygotic translational activity rather than activation of a preexisting maternal protein. In addition, inhibition of zygotic RNA synthesis by actinomycin D does not affect AP activity, although the appearance of tissue specific markers in other lineages is abolished. Hence, Whittaker (1977) concluded that endodermal AP is controlled by translational activation of a maternal AP mRNA.

Bates and Jeffery (1987a) examined AP expression in nucleate and anucleate fragments of fertilized *Styela plicata* eggs. If activation of a maternal mRNA is responsible for AP translation, AP activity would be expected in both nucleate and anucleate fragments. Only the nucleate fragments, however, showed the capacity to produce AP. These experiments were repeated and confirmed with *Ciona* (Whittaker, 1990) and *Halocynthia roretzi* (Bates, 1991) eggs.

The conflicting results led to a direct search for maternal AP mRNA in the *Halocynthia roretzi* egg. Kumano and Nishida (1998) have recently cloned and determined the expression of *Hr*ES-AP, a single copy gene encoding an endodermal AP (Fig. 19). Although a small amount of maternal *Hr*ES-EP mRNA was detected in eggs by northern blot analysis, it appears to be insufficient to explain the levels of AP observed during embryogenesis. The level of zygotic *Hr*ES-AP mRNA increases about 14-fold over the maternal amount by the neurula stage, and the zygotic transcripts are restricted to the endodermal cells. Therefore, at least in *Halocynthia,* the results suggest that AP synthesis is directed by zygotic mRNA.

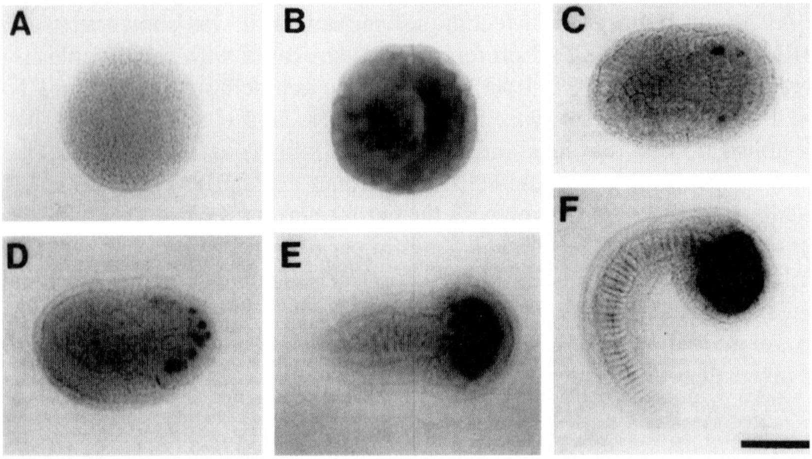

FIG. 19 Expression of AP mRNA during *Halocynthia roretzi* development. (A) Unfertilized egg. (B) Gastrula, vegetal view. (C) Neurula, dorsal view with anterior on the right. (D–F) Transitions from early to mid tailbud, anterior on right in each frame. From Kumano and Nishida (1998).

Although the *Halocynthia* results are consistent with a dependency of endodermal AP expression on zygotic rather than maternal transcripts, there are critical differences in the timing and drug dependencies of endodermal AP expression duing embryogenesis of *Halocynthia* and other ascidian species. In *Ciona* and *Styela,* AP activity begins to be detected during gastrulation and is insensitive to actinomycin D (Whittaker, 1977; 1990; Bates and Jeffery, 1987b). In *Halocynthia,* however, it begins at the mid-tailbud stage and is sensitive to actinomycin D (Nishida and Kumano, 1997; Kumano and Nishida, 1998). Hence, it is possible that endodermal AP is regulated by different mechanisms in various ascidian species.

C. UV Sensitivity

As described earlier, UV irradiation of the vegetal pole during OS1 abolishes gastrulation and larval axis formation in *Styela clava* embryos (Jeffery, 1990a). To determine the nature of the UV chromophores, Jeffery (1990a) used a UV cut-off filter that absorbs light below 270 nm (including the 260 nm maximum for nucleic acid inactivation) and transmits light above 270 nm (including the 280 nm peak for protein inactivation). The results of UV irradiation experiments with the cut-off filter are shown in Figure 20. The UV inactivation curve for gastrulation (Fig. 20C) resembles the curve for inactivation of T2 phage DNA (Fig. 20A) rather than DNase I (Fig. 20B), thus suggesting that the chromophores are nucleic acids (presumably RNA molecules). Similar results were obtained for muscle cell development (Fig. 20D), although the inactivation curve was shifted toward the DNase I curve, thereby suggesting that muscle determinants may have a protein component. The UV cut-off filter results provided the first evidence that maternal RNAs may be determinants in ascidian embryos.

UV irradiation experiments have also been performed with *Ciona savignyi* eggs (Marikawa *et al.*, 1995; Marikawa and Satoh, 1995). Centrifugation of unfertilized eggs in the presence of cytochalasin yielded four fragments of different colors: large nucleated red fragments and small anucleate clear,

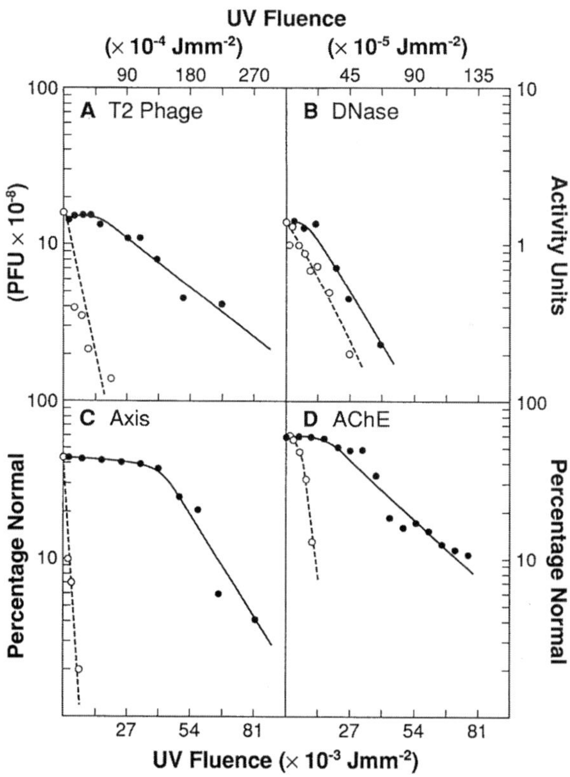

FIG. 20 Absorption characteristics of gastrulation and muscle determinants studied by irradiation of *Styela clava* eggs with a UV cut-off filter. (A) Inactivation curve for T2 phage. (B) UV inactivation curve for DNase 1 activity. (C) Inactivation curve for gastrulation (axis). (D) Inactivation curve for muscle (AChE) development. The UV source use in these experiments emits maximally at 260 nm. (Filled circles) UV with cut-off filter. (Open circles) UV without cut-off filter. A–C from Jeffery (1991) with permission.

black, and brown fragments (Marikawa *et al.,* 1994). When these fragments were inseminated, only the red fragments were able to cleave, and partial embryos developed containing epidermal cells but no muscle cells (Fig. 21A). However, when red fragments were fused with the black fragments (but not clear or brown fragments) and inseminated, the former were able to cleave and form partial embryos with muscle cells (Fig. 21B), thus showing that muscle determinants are localized in the black fragments. The black fragments also contain p58 and thus the myoplasmic cytoskeletal domain (Marikawa, 1995). Marikawa *et al.* (1995) showed that UV irradiation of black fragments abolished their ability to promote muscle cell development after fusion with red fragments (Fig. 21C). The ability to promote muscle cell development could be rescued by injecting poly (A) RNA extracted from nonirradiated black fragments into the red fragments (Fig. 21D). However, the recovery of muscle-forming capacity required fusion with UV irradiated black fragments. The results show that maternal mRNAs, along with other unidentified components of the black fragments, are able to promote muscle formation in partial embryos that developed from the red fragments.

Similar experiments have been conducted to identify UV-sensitive endodermal determinants in *Ciona savignyi* embryos (Marikawa and Satoh, 1995). Like the muscle determinants, the endodermal determinants are localized in the black fragments, and the latter have the ability to rescue AP expression when fused to the red fragments. It was known from experiments with *Styela clava* eggs that the endodermal determinants are more resistant to UV irradiation than the gastrulation determinants (Jeffery, 1990a; see Fig. 12B). In *Ciona* black fragments, endodermal determinants are also more resistant to UV irradiation than muscle determinants (Marikawa and Satoh, 1995). Moreover, experiments with the UV cut-off filter suggested that the endodermal determinants, like the gastrulation determinants, are likely to be RNA molecules.

D. Organized Searches for Determinants

In the context of hierarchies of developmental regulatory genes, two ways can be envisioned to search for determinants. The first way is to develop strategies to isolate the determinants themselves, which presumably lie at the base of gene hierarchies in the egg. This "bottom-up" approach presumes a prior understanding of the molecular nature of determinants. This is justified in ascidians by the UV irradiation results, which support the possibility that cytoplasmic determinants are maternal mRNAs.

The second way to search for determinants is to begin with the downstream genes they regulate in the embryo and then characterize genes upstream in the gene hierarchy until the determinant is reached in the egg.

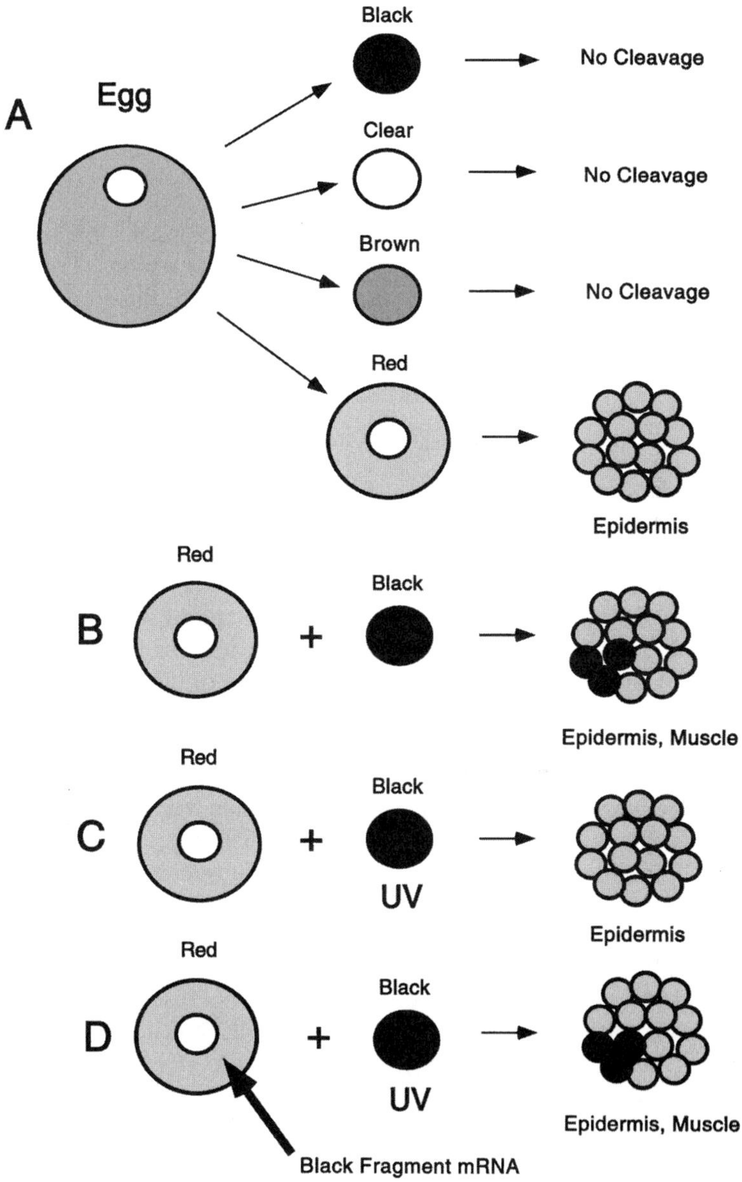

FIG. 21 Analysis of muscle determinants in *Ciona savignyi* by egg fragmentation. (A) Production and development of egg fragments. (Left) Intact unfertilized egg with nucleus. (Middle) Centrifugation of unfertilized eggs in the presence of cytochalasin results in the production of anucleate black, clear, and brown fragments and nucleate red fragments. (Right) After insemination only the red fragments cleave, and the embryos formed contain only epidermal cells. (B) Fusion of red fragments with black fragments and insemination results in the development of embryos with muscle cells. (C) UV irradiation of black fragments prior to fusion with red fragments blocks muscle cell development. (D) Injection of black fragment mRNA into red fragments prior to fusion with UV-irradiated black fragments results in the rescue of muscle cell development.

This "top-down" strategy can only be tactical if the gene cascade is rather short and simple. This approach appears to be justified in ascidians by their small genomes and early expression of tissue-specific genes.

1. Searching from "Bottom Up"

The strategies being used to search for cytoplasmic determinants from "bottom up" include fractionation of egg cytoplasmic regions, UV irradiation, and production of egg fragments, combined with differential cloning procedures and identification of expressed sequence tags (EST). Some approaches also capitalize on evolutionary differences in ascidian larvae, subtractive hybridization of maternal mRNAs, and random screens of cDNA libraries.

a. Fractionation of Myoplasm. *Styela* eggs are advantageous in the quest for muscle determinants because of their colored cytoplasmic regions. A method has been devised to fractionate myoplasm from other cytoplasmic regions at the yellow crescent stage (Jeffery, 1985). Using this method, myoplasmic mRNAs and proteins were compared to those present in other regions of the egg. Two-dimensional gel electrophoresis revealed that 15 of 133 detectable egg polypeptides were restricted to the myoplasm. Some of the myoplasmic proteins are identifiable; they include several isoforms of actin and p58, the myoplasmic protein recognized by the NN18 antibody. Isolation of yellow crescents in the presence of an RNase inhibitor allowed intact mRNAs to be extracted from the myoplasmic fraction. *In vitro* translation and 2D gel electrophoresis distinguished about 125 messages, but their encoded proteins were not qualitatively different from those directed by mRNA isolated from the nonmyoplasmic regions of the egg (Jeffery, 1985). However, this approach would only detect the most abundant messages encoding acidic polypeptides.

In another approach, also based on yellow crescent isolation, Swalla and Jeffery (1995) differentially screened a *Styela* egg cDNA library with probes made from myoplasmic and nonmyoplasmic poly (A) RNA. The differential screen yielded several clones encoding a 1.2 kb RNA, which was designated yellow crescent (YC) RNA. *In situ* hybridization showed that YC RNA is maternal and localized in the myoplasm of oocytes, eggs, and early cleaving embryos (Fig. 22; Fig. 25E, F). At the eight-cell stage, most of the YC RNA is distributed to the B4.1 blastomeres (Fig. 23), and is later present in the tail muscle cells. The YC transcripts appear to be tenaciously associated with the myoplasmic cytoskeleton, as shown by their retention in the cytoskeletal matrix following Triton X-100 extraction (Swalla and Jeffery, 1995). YC RNA reaches the myoplasm during oogenesis by a highly ordered process (Fig. 22). Transcripts first appear in the perinuclear region of young

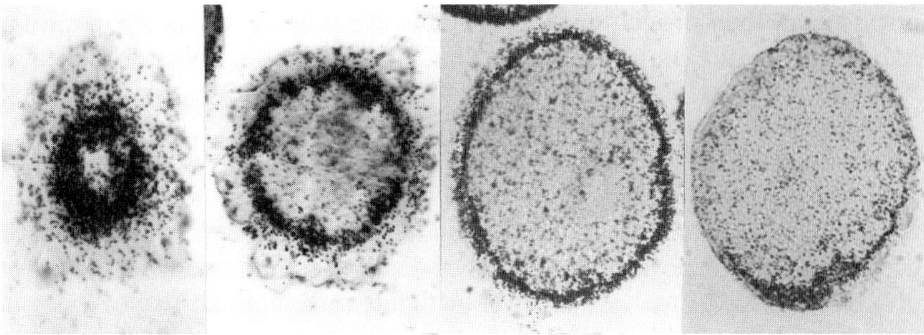

FIG. 22 Localization of YC RNA during oogenesis and ooplasmic segregation in *Styela clava*. From left to right: previtellogenic oocyte, vitellogenic oocyte, unfertilized egg, fertilized egg after completion of OS1. From Swalla and Jeffery (1996a).

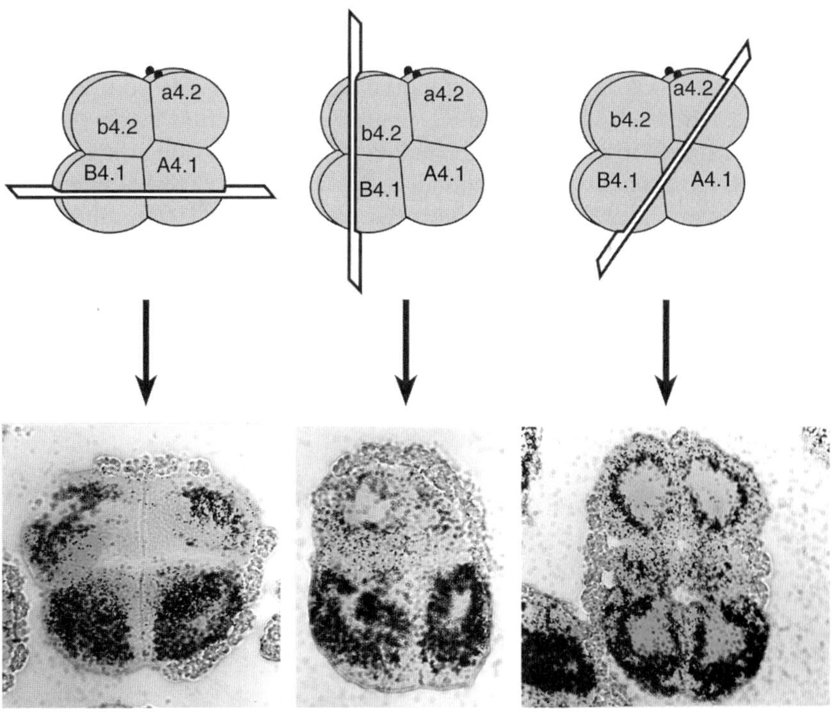

FIG. 23 Distribution of YC RNA in the quadrants of an eight-cell *Styela clava* embryo. Top row, from left to right: eight-cell embryos showing planes of sectioning through A- and B-line blastomeres, b- and B-line blastomeres, and b-, A-, and B-line blastomeres. Bottom row, from left to right: distribution of YC RNA in sections of eight-cell embryos cut in the planes shown in the top row.

oocytes and then progressively move outward toward the cortex in a ring-like structure. It may be significant that *Vg1* mRNA, which is localized in the vegetal cortex of *Xenopus* eggs, is translocated through the cytoplasm in a similar manner during oogenesis (Melton, 1987).

Although YC RNA has a poly (A) region at its 3′ terminus, it contains only a short protein coding region (Swalla and Jeffery, 1995). The open reading frame region is sufficient to encode a polypeptide of only 49 amino acids. Hence, it is possible that YC RNA may be a member of a family of noncoding RNAs involved in cell growth and development (e.g., Askew *et al.,* 1990; Crespi *et al.,* 1994). However, it should also be noted that YC RNA has sequence motifs in common with the mitochondrial large-ribosomal RNA (mlrRNA). *Halocynthia roretzi* mlrRNA has been cloned and shown to be localized in the myoplasm and segregated to the muscle cells (Oka *et al.,* 1999). Based on cytoplasmic stratification by centrifugation, most of the YC RNA appears to be present in the cytoplasm rather than in mitochondria. *Styela clava* eggs can be stratified into four layers by centrifugation (Jeffery and Meier, 1984). From centripetal to centrifugal pole, these layers consist of myoplasmic cytoskeletal matrix and pigment granules, clear cytoplasm, mitochondria, and endoplasmic yolk granules (Fig. 24). Although some of the YC is present in the third layer, the bulk of the YC transcripts stratify in the first layer, thus supporting the presence of YC RNA outside the mitochondria. In *Drosophila* (Kobayashi *et al.,* 1993; Ding *et al.,* 1994) and *Xenopus* (Kobayashi *et al.,* 1994), mlrRNA is localized outside the mitochondria in the egg cytoplasm, but its role in development is not understood. Functional studies will be required to determine the developmental significance of YC RNA.

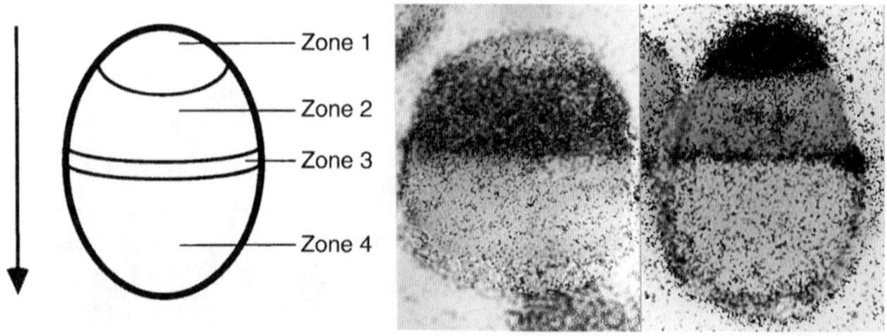

FIG. 24 Stratification of PCNA mRNA and YC RNA in unfertilized *Styela clava* eggs by centrifugation. (Arrow) Direction of centrifugation. (Left) Diagram showing distribution of cytoplasmic organelles in four zones. (Zone 1) Myoplasmic cytoskeletal domain and yellow pigment granules. (Zone 2) Ectoplasmic zone. (Zone 3) Mitochondria. (Zone 4) Endoplasmic yolk granules. (Middle) PCNA mRNA. (Right) YC RNA.

The identification of YC RNA inadvertently led to the isolation and characterization of two other localized mRNAs in *Styela clava* eggs (Swalla and Jeffery, 1996a, b). These mRNAs were isolated by screening an oocyte cDNA library with a double-stranded YC cDNA insert and thus may exhibit regions related to YC RNA. The first mRNA, which encodes the L5 ribosomal protein, is concentrated in the myoplasm and distributed preferentially to the muscle cells during embryogenesis (Swalla and Jeffery, 1996a). It is unknown whether the L5 protein is localized in the myoplasm. However, L5 protein is known to be a critical factor in translational regulation and ribosome biogenesis (Deshmukh *et al.,* 1993). Therefore, a possible function of L5 mRNA localization may be to promote ribosome biosynthesis in the muscle cells, which devote a large part of their metabolic activity to synthesis of contractile proteins. The second mRNA encodes the proliferating cell nuclear antigen (PCNA) (Swalla and Jeffery, 1996b), a cofactor of DNA polymerase δ that is essential for DNA replication and cell division (Bravo *et al.,* 1987). The maternal PCNA mRNA is concentrated in the ectoplasm and distributed preferentially to the epidermal cells during embryogenesis (Fig. 25). In egg centrifugation experiments, PCNA mRNA stratifies to the second layer (see above), which consists of ectoplasmic components (Fig. 24). PCNA protein is also localized in the ectoplasm (Swalla and Jeffery, 1996b). It is possible that PCNA is localized in the ectoplasm in order to concentrate maximal DNA replication capacity in the animal hemisphere cells, which undergo the most cell divisions during embryogenesis.

b. UV Irradiation. As described above, the gastrulation determinants appear to be UV-sensitive RNA molecules localized at the vegetal pole. In an attempt to identify candidate mRNAs, Jeffery (1990b) compared the proteins encoded by mRNAs isolated from UV-irradiated and normal *Styela* eggs by *in vitro* translation. Of 433 polypeptides detectable in 2D gels, 21 of these were missing in the UV irradiated embryos (Fig. 26). The most prominent is a 30 kDa polypeptide (p30), which appears to be associated with the cytoskeleton. Several lines of evidence suggest that p30 may be involved in gastrulation (Jeffery, 1990b). First, p30 synthesis peaks during gastrulation. Second, p30 and gastrulation are abolished by the same threshold dose of UV irradiation. Third, the UV sensitivity period for inhibiting p30 synthesis and gastrulation coincide during OS1. Further investigation of the role of this factor in gastrulation awaits the isolation and characterization of p30.

c. Egg Fragmentation. The experimental system involving the production of egg fragments and UV irradiation has been combined with subtractive cloning to search for determinants in *Ciona savignyi* (Yoshida *et al.,*

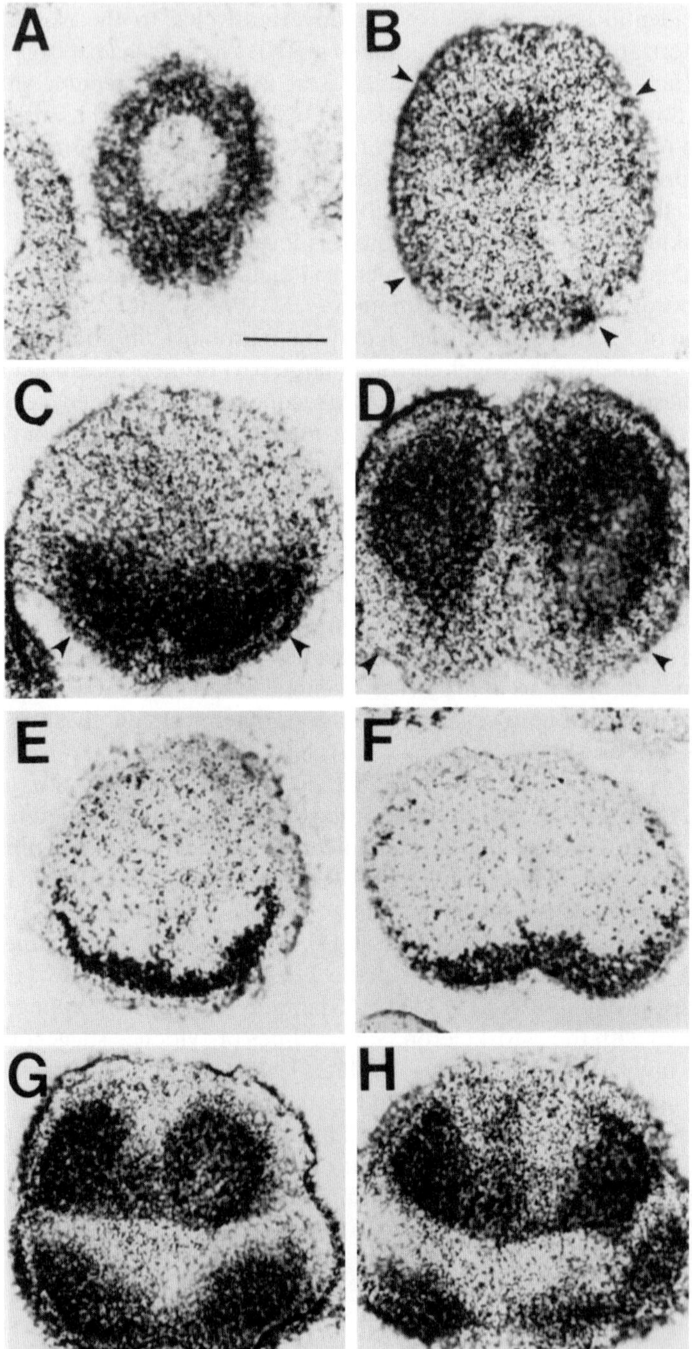
A
B
C
D
E
F
G
H

1996). In these experiments, red fragment mRNA was subtracted from black fragment mRNA, and the subtracted probe was used to screen a cDNA library. The screen resulted in the isolation of *posterior end mark* (*pem*), a black fragment-enriched mRNA that is localized in the myoplasm. In striking contrast to other RNAs localized in the myoplasm (see above), *pem* mRNA is not eventually distributed to the muscle cells. Instead, it becomes progressively localized in the most posterior embryonic cells (B5.2, B.6.3, and B7.6 cells) and is deposited in only a few endodermal strand cells of the tadpole larva (Fig. 27). The CAB may be the agent that localizes *pem* mRNA in the posterior blastomeres.

The deduced amino acid sequence of the pem protein does not contain any motifs suggestive of its function. However, overexpression experiments, in which synthetic *pem* mRNA was injected into fertilized eggs, resulted in tadpole larvae with anterior and dorsal deficiencies (Yoshida *et al.,* 1996). It is interesting that treatment with lithium chloride, a dorsalizing agent in vertebrate embryos, rescues the development of anterior and dorsal structures in larvae overexpressing pem (Yoshida *et al.,* 1998). Hence, the pem protein may be involved in posterior patterning via a signaling cascade affected by lithium.

d. An Evolutionary Approach. Another strategy being used to seek determinants capitalizes on the novel evolutionary system consisting of ascidian species with different modes of larval development (Jeffery, 1997). As described above, the myoplasm is absent or reduced in oocytes and eggs of the tailless species *M. occulta,* relative to its sister species, *M. oculata.* When mRNA of *M. occulta* oocytes was subtracted from its counterpart in *M. oculata,* three cDNA clones (designed as Uro-1, 2, and 11) were isolated with messages that are absent or downregulated in *M. occulta* eggs (Swalla *et al.,* 1993). The three cDNA clones encode single copy genes designated *Cymric* (Uro-1), *lynx* (Uro-2), and *Manx* (Uro-11), after varieties of tailless cats. The putative proteins encoded by these genes contain

---

FIG. 25 Distribution of PCNA mRNA during *Styela clava* development. (A–D, G–H) PCNA mRNA. (A) Previtellogenic oocyte. (B) Unfertilized egg; section through animal-vegetal axis. Arrowheads indicate the cortical myoplasm. (C) Fertilized egg after OS1; section through animal-vegetal axis. PCNA mRNA localized in the ectoplasm transiently diffuses into the myoplasm. Arrowheads indicate approximate border between ectoplasm (top) and myoplasm (bottom) in PCNA mRNA localization domain. (D) Two-cell embryo. PCNA mRNA is localized in the ectoplasm. Arrowheads indicate domain of the myoplasm. (G) Eight-cell embryo; section through vegetal hemisphere. (H) Sixteen-cell embryo; section through vegetal hemisphere. (E–F) YC RNA. (E) Fertilized egg after OS1; section through animal-vegetal axis. (D) Two-cell embryo, posterior on bottom.

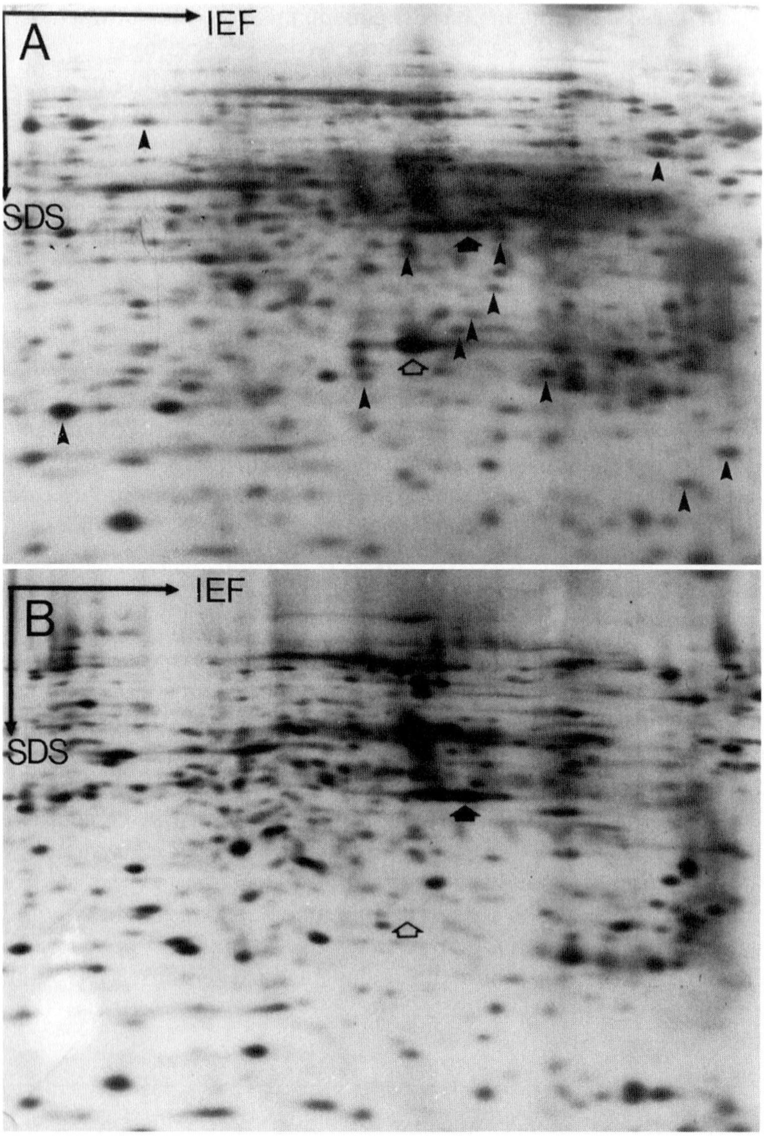

FIG. 26 Autoradiographs of 2D gels containing radioactive proteins translated *in vitro.* by mRNA extracted from normal (A) and UV-irradiated (B) *Styela clava* embryos. The synthesis of p30 (large open arrowhead) is abolished at UV doses that affect gastrulation. Other polypeptides that are eliminated or reduced in this experiment are indicated by small solid arrowheads. Large solid arrowheads indicate the position of muscle actin. From Jeffery (1990b).

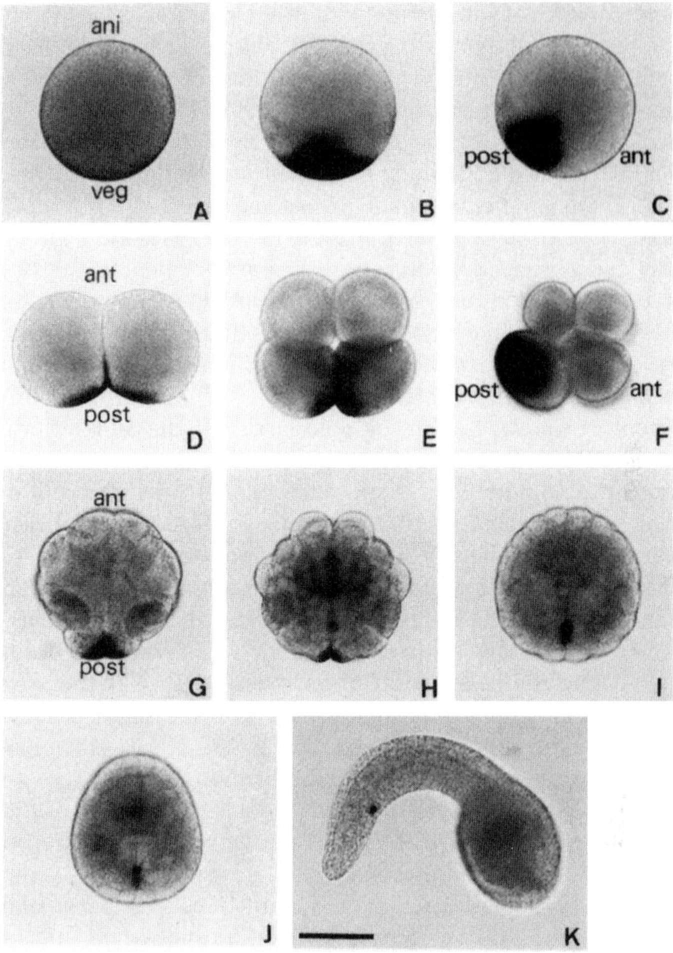

FIG. 27 Distribution of *pem* mRNA during *Ciona savignyi* development. (A) Unfertilized egg. (B) Fertilized egg after completion of OS1. (C) Fertilized egg after completion of OS2. (D) Two-cell embryo. (E) Four-cell embryo. (F) Eight-cell embryo. (G) Sixteen-cell embryo. (H) Thirty-two-cell embryo. (I) Sixty-four-cell embryo. (J) Neurula. (K) Tailbud. Scale bar: 100μm. From Yoshida *et al.* (1996) with permission from Company of Biologists Ltd.

sequence motifs that provide clues to their functions (Swalla *et al.,* 1993, 1995). The Cymric protein is a Shark family, nonreceptor tyrosine kinase (Chan *et al.,* 1994; Ferrante *et al.,* 1995) with SH2 domains and ankyrin repeats at its N-terminus, implying that it could be involved in a signaling cascade in the egg. The lynx protein contains a transmembrane domain and a leucine zipper, suggesting that it may be an integral membrane component capable of interacting with other proteins. The Manx protein contains a putative DNA binding site resembling the bacterial histone-like protein and a zinc finger, suggesting that it may be a transcription factor.

In cloning the *Manx* gene, Swalla *et al.* (1999) identified *bobcat,* a gene embedded in an intron of the *Manx* transcription unit. The *bobcat* gene encodes a DEAD-box containing RNA helicase similar in structure to p68 (Ford *et al.,* 1988) and may be involved in some aspect of RNA metabolism, translation, or localization. The *Manx* and *bobcat* genes are expressed both maternally and zygotically. Antisense procedures have been developed to investigate the function of zygotic genes in ascidian embryos (Swalla and Jeffery, 1996c; Olsen and Jeffery, 1997; Olsen *et al.,* 1999; Satou, 1999). Antisense experments have shown that zygotic *Manx* and *bobcat* expression is required to rescue the tail in *M. occulta* × *M. oculata* hybrids and for development of the *M. oculata* tadpole larva (Swalla and Jeffery, 1996c; Swalla *et al.,* 1999). The role of the maternal mRNAs, which are of low abundance in the mature egg, is still uncertain. However, recent evidence shows that some of the Uro transcripts are localized in a spherical mass on the vegetal side of the GV in *M. oculata* oocytes. This mass resembles the METRO, a localization of mRNAs and other components destined for delivery into the germ plasm in *Xenopus* oocytes (Kloc *et al.,* 1996). Therefore, like *Xenopus* oocytes (Kloc and Etkin, 1995), ascidian oocytes may exhibit two distinct pathways for transportation of RNA to the cortex. One pathway, which may transport some of the Uro messages, resembles the METRO, while the other, which transports YC RNA, is similar to the pathway responsible for *Vg-1* mRNA localization.

e. Subtraction of Maternal mRNAs. Subtractive hybridization of fertilized egg mRNAs (maternal messages) with gastrula mRNAs (enriched in zygotic mRNAs) is being carried out to identify localized maternal mRNAs in *Ciona savignyi* (Satou and Satoh, 1997). The subtracted mRNAs have been used to make a library of about 1000 different clones. To screen for localized messages, clones were isolated randomly from this library and their distribution in eggs and embryos was determined by *in situ* hybridization. This procedure resulted in the isolation of five clones encoding different mRNAs localized in the myoplasm of *Ciona savignyi* eggs (Satou and Satoh, 1997; Satou, 1999).

Because the distribution of these mRNAs is identical to that of *pem* mRNA (e.g., initial localization in the myoplasm followed by restriction to the most posterior blastomeres of the embryo), they were named *pem-2, pem-3, pem-4, pem-5,* and *pem-6* (*pem* itself was not among the identified clones). It may be significant that all the *pem* RNAs show at least one copy of the short conserved sequence UUAUUU in their 3′ noncoding regions, which may be a localization signal. The putative amino acid sequences of the pem-2, pem-3, pem-4, and pem 5 proteins suggest that they may have regulatory functions related to the determination of posterior fates in the ascidian embryo.

The pem 2 protein has an SH3 domain and other motifs that are indicative of members of the Dbl family of signal transducing molecules. The Dbl protein family consists of guanine nucleotide exchange factors that induce changes in the actin cytoskeleton through activation of Rho-like GTPases. hPEM-2, a human homologue of pem-2, has been identified and functionally characterized (Reid *et al.,* 1999). Biochemical activity assays indicate that hPEM-2 specifically activates the cell cycle protein cdc42, whereas ectopic expression of hPEM-2 in fibroblasts shows that this protein controls cell shape through polymerization of the cortical actin cytoskeleton. Hence, ascidian pem-2 may be involved in transducing an intracellular signal related to contractile ring formation during cytokinesis.

The pem-3 protein contains KH domains (Satou, 1999), thus suggesting that it may be an RNA binding protein (Siomi *et al.,* 1993). The pem-3 KH domains show extensive similarity to those of MEX-3, an RNA binding protein encoded by a localized mRNA in *C. elegans* embryos (Draper *et al.,* 1996). The *mex-3* gene specifies anterior blastomere identity in *C. elegans* but is also localized in the P granules, which are involved in nematode germ cell determination. If the *pem-3* gene is the orthologue of *mex-3,* this provides support for the idea that the CAB-containing B7.6 blastomeres may be germ cell precursors in ascidians.

The pem-4 protein has three zinc fingers and a nuclear localization sequence, thus suggesting that it may function as a transcription factor.

The pem-5 protein is apparently related to the protein product of the *Drosophila* maternal effect gene *spire* (Wellington *et al.,* 1999). The spire protein contains actin binding domains and is required for localization of determinants to the posterior pole and the dorsal anterior corner of the developing oocyte. Unlike *pem-5* mRNA in *Ciona,* however, *spire* mRNA is not localized to the posterior pole in *Drosophila.* In this light, it would be interesting to know whether the ascidian pem-5 protein is localized at the posterior pole. The results of the *pem-5* mRNA localization studies bring up the possibility that pem-5 is involved in localizing determinants to the posterior pole, possibily via interactions with the actin cytoskeleton.

The *Ciona* screen also identified three maternal genes (*CsEndo-I, C-Endo-2,* and *C-Endo*-3) with distribution patterns suggesting localization in the endoplasm (Imai *et al.,* 1999). Although these mRNAs could not be detected in the egg, they were concentrated in the endodermal cells by the 8- to 16-cell stage. Perhaps the *C-Endo* transcripts are initially spread throughout the egg cytoplasm and progressively localized in the posterior region during cleavage (see below). CsEndo-1 encodes a putative nuclear protein. The CsEndo-3 protein exhibits a transmembrane region and a PDZ domain, which suggest it is an integral membrane protein. CsEndo-2 has no similarities to known proteins. Hence, the subtractive hybridization approach has yielded maternal mRNAs that may function as determinants in the myoplasm and endoplasm of ascidian eggs and embryos.

f. Mass Screening of Maternal mRNAs. Randomly isolated clones from an arrayed egg cDNA library are being sequenced and screened by *in situ* hybridization to develop an EST database in *Halocynthia roretzi* (Nishida and Makabe, 1999; Kawashima *et al.,* 2000; Sasakura *et al.,* 2000). Thus far, 150 of more than 2000 partially sequenced clones have been assayed for localization, and some of these ESTs represent mRNAs that are localized in the eight-cell embryo. The *Hrsmad1/5* mRNA is localized in the ectoplasm and encodes a member of the smad gene family (Kobayashi *et al.,* 1999). The smad proteins have the ability to bind TGF-β receptors and are thought to mediate intracellular signaling by the TGF-β class of growth factors (Heldin *et al.,* 1997). The other seven maternal mRNAs show posterior localization patterns resembling the *Ciona savignyi pem* mRNAs (Nishida and Makabe, 1999). Two distinct localization patterns were observed within this group of seven posterior-localized mRNAs.

Three of the posteriorly localized maternal mRNAs exhibited localization patterns identical to the *pem* mRNAs. These mRNAs are designated type I postplasmic mRNAs. The type I transcripts are localized in the cortical myoplasm of unfertilized eggs, retain their association with the myoplasm during ooplasmic segregation and the early cleavages, and are gradually partitioned to the most posterior blastomeres after the eight-cell stage. One of these messages, *HrPOPK-I,* encodes a putative protein serine/threonine kinase, thereby suggesting a role in signal transduction (Sasakura *et al.,* 1998a). Another mRNA encodes the ascidian *Wnt-5* protein (Sasakura *et al.,* 1998b) (Fig. 28). A maternal *Wnt-5* mRNA showing the same posterior localization pattern (but with a different zygotic expression pattern) has also been described in *Ciona intestinalis* (Davis and Smith, 1999). The identification of a posteriorly localized *Wnt-5* mRNA is intriguing because Wnt proteins encode secreted factors involved in many different signaling processes (Moon *et al.,* 1997). The third type I mRNA (*HrZF-1*) encodes a novel zinc finger protein (Sasakura *et al.,* 2000), which may be a transcription factor.

Another class of posteriorly localized mRNAs is distributed evenly in the unfertilized egg and then becomes translocated after fertilization. These mRNAs are designated type II postplasmic mRNAs. The *C-Endo* mRNAs described above may be examples of type II postplasmic mRNAs in *Ciona savignyi.* Three ESTs encoding type II mRNAs have been recently described in *Halocynthia* and designated *HrPet-1-3* (Sasakura *et al.,* 2000). The *HrPet-1* mRNA encodes a protein with homology to the yeast MIC1

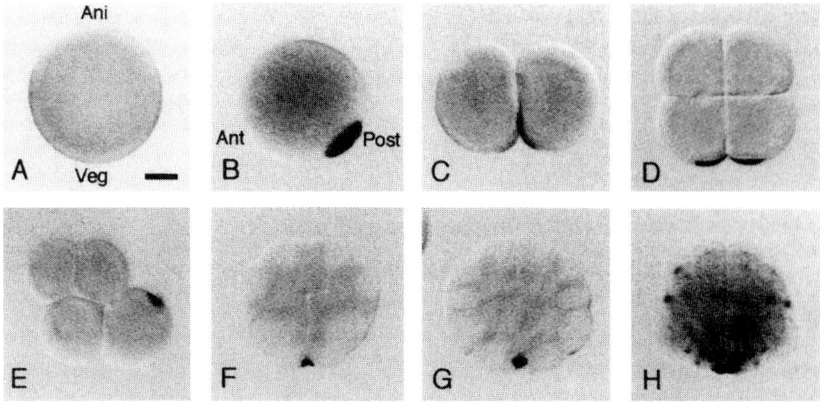

FIG. 28 Distribution of *Wnt-5* mRNA during early development of *Halocynthia roretzi.* (A) Unfertilized egg. (B) Fertilized egg after completion of OS2. (C) Four-cell embryo; posterior is at bottom. (D) Eight-cell embryo; posterior is at the right. (F) Sixteen-cell embryo; posterior is at the bottom. G. Thirty-two-cell embryo; posterior at the bottom. H. Sixty-four-cell embryo; animal view. *Wnt-5* mRNA expression is maternal (A-G) until the 64-cell stage (H), when zygotic expression begins. From Sasakura *et al.* (1998b).

transcription-factor interacting protein. The other HrPet proteins have no correlates within the current protein database.

Sasakura *et al.* (2000) have shown that type I and II postplasmic mRNAs are associated with the cytoskeleton. However, the mRNA-cytoskeletal associations differ in their sensitivity to cytochalasin B and nocodazole, inhibitors of microfilament and microtubule function, respectively. The translocation of type I mRNAs is sensitive to cytochalasin B during OS1 and to nocodazole during OS2, whereas type II mRNA localization is sensitive to cytochalasin B (and not to nocodazole) throughout ooplasmic segregation. In addition, based on egg stratification, type I mRNAs but not type II mRNAs, are tightly associated with the unfertilized egg cortex. Therefore, two cytoskeletal systems appear to be involved in translocating mRNAs to the vegetal posterior region during ooplasmic segregation. Sasakura *et al.* (2000) argue that these two pathways for type I and II mRNA localization in *Halocynthia* are not homologous to the dual RNA localization pathways in *Xenopus* (Kloc and Etkin, 1995), which operate during oogenesis rather than after fertilization.

A striking result of the EST analysis is that about 5% of the sampled cDNA clones encode messages that are localized in the egg and/or embryo. It can be concluded from these studies that the proportion of localized mRNAs is unexpectedly high and that many more of these messages will soon be discovered in the posterior organization center of ascidian eggs.

It is also significant that most of the localized mRNAs show distribution patterns resembling the original *pem* mRNAs: initial localization in the myoplasm and eventual restriction to the most posterior blastomeres of the embryo. The localization of these interesting mRNAs implies a role in determining posterior developmental features.

2. Searching from "Top Down"

The strategy being used to search for determinants from "top down" involves four steps: (1) identifying a tissue-specific gene, (2) characterizing the *cis-* and *trans*-acting elements that regulate this gene, (3) proceeding upstream to a gene that controls the tissue-specific gene by cloning/identifying the *trans*-acting regulatory proteins, and (4) repeating steps 1 to 3 until an upstream gene encoding a factor fitting the criteria for a determinant is reached. Many tissue-specific genes have been cloned and are suitable as a starting point for this analysis (Satoh *et al,* 1996; Corbo *et al.,* 1997, 1998; Olsen and Jeffery, 1997; Takahashi *et al.,* 1997, 1999). Below I review progress on identifying muscle determinants using the "top-down" strategy.

a. Navigating Upstream in the Muscle Pathway. Muscle cell determination, as measured by tissue-specific gene transcription, begins early during the cleavage period. The expression of muscle actin and myosin genes is first detected at the 32-cell stage in *Halocynthia roretzi* (Satou *et al.,* 1995). At this time, the B-line cells, in which the zygotic transcripts accumulate, are not fate restricted and will eventually produce endoderm, mesenchyme, and muscle cells. The muscle-specific mRNAs accumulate exclusively in the muscle cell progeny by an unknown process (differential segregation or degradation?). It is surprising that the expression of these muscle "housekeeping" genes appears to precede that of a single-copy *MyoD*-related gene, which has been cloned in *Halocynthia roretzi* (Araki *et al.,* 1994) and *Ciona intestinalis* (Meedel *et al.,* 1997). The *MyoD* mRNA is zygotic and hence not a candidate for a muscle determinant (Araki *et al.,* 1994; Meedel *et al.,* 1997). (A maternal MyoD protein, however, is still a possibility.)

Because of early zygotic activation, the upstream factors regulating the actin and myosin genes are being studied to seek the shortest route to muscle determinants. The upstream regulatory elements sufficient for tissue-specific expression of the muscle actin genes consist of only 38 base pairs (Satou and Satoh, 1996), a finding that suggests that the *trans*-acting factors regulating these genes may not be complex. Erives *et al.* (1998) have recently cloned the *snail* gene, which is involved in mesoderm determination in *Drosophila* (Leptin, 1991), and have shown that it is first expressed in B-line cells at the 32-cell stage in *Ciona intestinalis* embryos (Fig. 29). A *Halocynthia roretzi snail* gene has also been cloned, although it does not

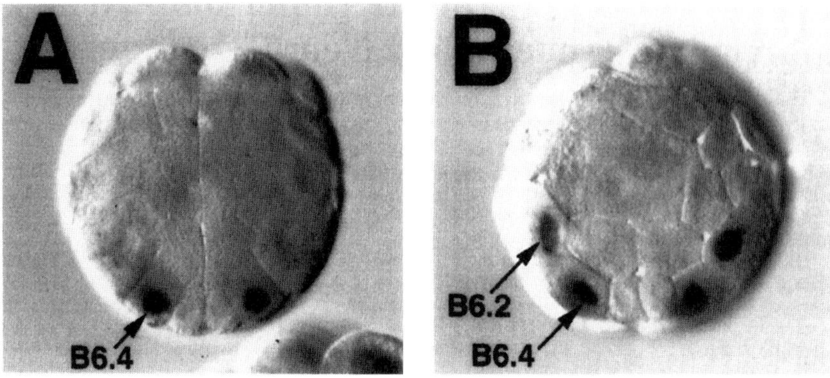

FIG. 29 Expression of zygotic *snail* mRNA in *Ciona intestinalis* embryos. (A) Transcripts are initially detected in nuclei of the B6.4 blastomeres in 32-cell embryos. (B) Later during the 32-cell stage transcripts also appear in the nuclei of the B6.2 blastomeres. From Corbo *et al.* (1998) with permission from Company of Biologists Ltd.

appear to be expressed until one cleavage cycle later (Wada and Saiga, 1999). The *Ciona snail* gene enhancer contains seven E-box sequences, including three sequences containing a nonconventional AC-core motif. Mutation of the AC-core sequence results in loss of *snail* expression (Erives *et al.*, 1998). It is significant that E boxes with the same unusual AC-core motifs occur in the short promoter regions of *Halocynthia* muscle actin genes activated at the 32-cell stage (Kusakabe *et al.*, 1995). Hence, muscle determinants may correspond to *trans*-acting factors that activate the *snail* and muscle actin genes by binding to these E-box motifs.

In *Drosophila,* the *twist* gene is partially responsible for activating *snail* expression in the mesoderm (Ip *et al.*, 1992). By analogy, the ascidian *twist* homologue has been proposed as a candidate for a muscle determinant (Erives *et al.*, 1998). This proposal implies that *twist* is encoded by a localized maternal message, or that the twist protein itself is maternal and localized in the ascidian egg. Cloning and characterization of ascidian *twist* gene will be necessary to test this hypothesis.

3. Signaling Cascade at the Posterior Pole?

Table I summarizes the twenty localized mRNAs currently identified in ascidian eggs. The majority of these mRNAs are localized in the myoplasm, at least in the egg and early cleaving embryo, and most of these transcripts later exhibit a posterior localization pattern resembling *pem* mRNA. The sub-myoplasmic region containing these messages, which includes the CAB, corresponds to the posterior organization center (Fig. 14), and has been

TABLE I

Summary of Localized RNAs in Ascidian Eggs and Embryos

RNA	Species	Localization	Putative Function
PCNA	*Styela clava*	Ectoplasm	DNA replication
Hrsmad1/5	*Halocynthia roretzi*	Ectoplasm	Signal transduction
YC	*Styela clava*	Myoplasm (muscle)	Unknown; may be related to or same as mlrRNA
mlrRNA	*Halocynthia roretzi*	Myoplasm (muscle)	Mitochondrial ribosome component
L5	*Styela clava*	Myoplasm (muscle)	Cytoplasmic ribosome component
pem	*Ciona savignyi*	Myoplasm (postplasm)	Unknown
pem-2	*Ciona savignyi*	Myoplasm (postplasm)	Signal transduction
pem-3	*Ciona savignyi*	Myoplasm (postplasm)	RNA binding protein
pem-4	*Ciona savignyi*	Myoplasm (postplasm)	Transcription factor
pem-5	*Ciona savignyi*	Myoplasm (postplasm)	Determinant localization
pem-6	*Ciona savignyi*	Myoplasm (postplasm)	Unknown
HrPOPK-1	*Halocynthia roretzi*	Myoplasm (postplasm)	Signal transduction
Wnt-5A	*Halocynthia roretzi* *Ciona intestinalis*	Myoplasm (postplasm)	Signal transduction
HrZF-1	*Halocynthia roretzi*	Myoplasm (postplasm)	Transcription factor
HrPet-1	*Halocynthia roretzi*	Myoplasm (postplasm)	Transcription factor binding protein
HrPet-2	*Halocynthia roretzi*	Myoplasm (postplasm)	Unknown
HrPet-2	*Halocynthia roretzi*	Myoplasm (postplasm)	Transcription factor binding protein
CsEndo-1	*Ciona savignyi*	Endoplasm	Nuclear protein
CsEndo-2	*Ciona savignyi*	Endoplasm	Unknown
CsEndo-3	*Ciona savignyi*	Endoplasm	Transmembrane protein

Relevant references are found in the text. Some Uro mRNAs are localized in oocytes, but because their localization has yet to be described in the egg they are not included in this table.

designated the postplasm (Nishida and Makabe, 1999). The posterior organization center/postplasm may represent an epicenter for mRNA localization in the ascidian embryo. It is impressive that the posteriorly localized messages code for a variety of potential developmental regulatory factors including putative transcription factors and the components of signal transducing systems. Hence, the posterior determinants may function by activating a signal transduction cascade(s) leading to zygotic gene expression in the posterior region of the embryo.

VI. Concluding Remarks

After more than a hundred years of research, we now have considerable information about the biology of cytoplasmic determinants in ascidians. We know that these factors are localized in distinct cytoplasmic regions in the egg and embryo, that their localization is mediated by association with the cytoskeleton, and that these factors are both necessary and sufficient to specify cell and positional fates during embryogenesis. In addition, we know that there are at least five kinds of cytoplasmic determinants, whose localization and function are distinguishable by specific assays. In addition, some of determinants localized in the posterior region of the embryo may be responsible for specifying the prospective germ line. Finally, we have evidence suggesting that cytoplasmic determinants are localized maternal mRNAs. However, in no case have the determinants been identified and characterized with certainty.

In the future, it is likely that recent advances in molecular technology, including overexpression of genes via injection of synthetic mRNAs (Yoshida *et al.,* 1996), suppression of gene expression by antisense oligonucleotides (Swalla and Jeffery, 1996c), and transformation of genes by electroporation (Corbo *et al.,* 1997) or microinjection (Hikosaka *et al.* 1992), will be instrumental in investigating the function of localized mRNAs as determinants. In addition, as ascidian genetics matures, this approach will be a powerful new tool in the quest for cytoplasmic determinants.

I have reviewed several approaches that are being used to identify potential cytoplasmic determinants. Some of these approaches begin at the bottom of a gene hierarchy, and are designed to identify determinants directly, whereas others begin at the top of a gene hierarchy and are designed to identify tissue-specific gene regulators and work upstream toward the determinants. It is anticipated that these opposing strategies will eventually converge on the same factors.

Once cytoplasmic determinants are identified and characterized many exciting questions can be envisioned in the future. It will be important to know the molecular composition of the determinants. Are they single uniform entities or factors composed of many parts with different functions? It will also be critical to know how the determinants are assembled with the cytoskeleton, how the determinants function to activate (or repress?) batteries of tissue-specific genes during embryogenesis, and how they intersect with the signaling pathways involved in inductive processes. Finally, it will be important to determine how the functions of cytoplasmic determinants have evolved in different ascidian species. The new century that looms before us is certain to be an exciting time for research on cytoplasmic determinants in ascidians.

Acknowledgments

Research in my laboratory on ascidian development is supported by NIH (HD-13970) and NSF (IBN-9996146) grants. I thank Professor T. S. Painter for the cut-off filters used in the UV irradiation experiments, and possibly as coverslips in his original polytene chromosome squashes. This article is dedicated to my former graduate and postdoctoral students W. R. Bates, R. L. Beach, R. D. Brodeur, D. G. Capco, F. Harrington, T. Kusakabe, C. L. Olsen, E. L. Pederson, B. J. Swalla, C. R. Tomlinson, J. A. Uzman, J. M. Venuti, M. E. White, M. Wilson, and J. Zhou, who have collaborated with me in studies of ascidian development and evolution. Our studies on ascidians have been carried out at the University of Texas at Austin; Marine Biological Laboratory, Woods Hole; University of California, Davis; and Station Biologique de Roscoff, France.

References

Araki, I., Saiga, H., Makabe, K. W., and Satoh, N. (1994). Expression of AMD1, a gene for a major MyoD1-related factor in the ascidian *Halocynthia roretzi. W. Roux's Arch Dev. Biol.* **203,** 320–327.

Askew, D. D., Li, J., and Ihle, J. N. (1990). Retroviral insertions in the murine His-1 locus activate the expression of a novel RNA that lacks an extensive open reading frame. *Mol. Cell. Biol.* **14,** 1743–1751.

Bates, W. R. (1991). Analysis of morphogenetic determinants in ascidian eggs and embryos by cytoplasmic deletion and microinjection. *Can. J. Zool.* **69,** 2811–2818.

Bates, W. R., and Bishop, C. (1996). Localization of constitutive heat shock proteins in developing ascidians. *Dev. Growth Differen.* **38,** 307–314.

Bates, W. R., and Jeffery, W. R. (1987a). Localization of axial determinants in the vegetal pole region of ascidian eggs. *Dev. Biol.* **130,** 98–107.

Bates, W. R., and Jeffery, W. R. (1987b). Alkaline phosphatase expression in ascidian egg fragments and andromerogons. *Dev. Biol.* **119,** 382–389.

Bates, W. R., and Nishida, H. (1998). Developmental roles of nuclear complex factors released during oocyte maturation in the ascidians *Halocynthia roretzi* and *Boltenia villosa. Zool. Sci.* **15,** 69–76.

Berg, W. E., and Humphrey, W. J. (1960). Electron microscopy of four-cell stages of the ascidians *Ciona* and *Styela. Dev. Biol.* **2,** 42–60.

Berrill, N. J. (1929). Studies in tunicate development. I. General physiology of development of simple ascidians. *Philos. Trans. R. Soc. London* (Series B) **218,** 37–78.

Berrill, N. J. (1931). Studies in tunicate development. II. Abbreviation of development in the Molgulidae. *Philos. Trans. R. Soc. London* (Series B) **219,** 281–346.

Berrill, N. J. (1935). Studies in tunicate development. III. Differential retardation and acceleration. *Philos. Trans. R. Soc. London* (Series B) **225,** 255–326.

Bravo, R., Frank, R., Blundell, P. A., and MacDonald-Bravo, H. (1987). Cyclin/PCNA is the auxiliary protein of DNA polymerase δ. *Nature* **326,** 515–517.

Chabry, L. (1887). Contribution á embryologie normale et térotologique des Ascidies simples. *J. Anat. Physiol.* **23,** 167–319.

Chan, T. A., Chu, C. A., Rauen, K. A., Kroiher, M., Tatarewicz, S. M., and Steele, R. E. (1994). Identification of a gene encoding a novel protein-tyrosinse kinase containing SH2 domains and ankyrin-like repeats. *Oncogene* **9,** 1253–1259.

Cohen, A., and Berrill, N. J. (1932). The development of isolated blastomeres of the ascidian egg. *J. Exp. Zool.* **74,** 91–117.

Conklin, E. G. (1905a). The organization and cell lineage of the ascidian egg. *J. Acad. Nat. Sci. Philadelphia* **13,** 1–119.

Conklin, E. G. (1905b). Organ forming substances in the eggs of ascidians. *Biol. Bull.* **8,** 205–230.

Conklin, E. G. (1905c). Mosaic development in ascidian eggs. *J. Exp. Zool.* **2,** 145–233.

Conklin, E. G. (1911). The organization of the egg and development of single blastomeres of *Phallusia mammillata. J. Exp. Zool.* **60,** 1–120.

Conklin, E. G. (1931). The development of centrifuged eggs of ascidians. *J. Exp. Zool.* **60,** 1–119.

Corbo, J. C., Erives, A., Di Gregorio, A., Chang, A., and Levine, M. (1997). Dorsoventral patterning of the vertebrate neural tube is conserved in a protochordate. *Development* **124,** 2335–2344.

Corbo, J. C., Fujiwara, S., Levine, M., and Di Gregorio, A. (1998). Suppressor of hairless activitates *Brachyury* expression in the *Ciona* embryo. *Dev. Biol.* **203,** 358–368.

Crespi, M. D., Jurkevitch, E., Poiret, M., d'Aubenton-Carafa, Y., Petrovics, A., Kondorosi, E., and Kondorosi, A. (1994). *enod40,* a gene expressed during nodule organogenesis, codes for a nontranslatable RNA involved in plant growth and development. *EMBO J.* **13,** 5099–5112.

Davis, S. W., and Smith, W. C. (1999). Conserved features of *wnt-5* expression in vertebrates and the ascidian *Ciona intestinalis.* Submitted.

Deno, T., Nishida, H., and Satoh, N. (1985). Autonomous muscle cell differentiation in partial ascidian embryos according to the newly verified cell lineages. *Dev. Biol.* **104,** 322–328.

Deshmukh, M., Tsay, Y.-F., Paulovich, A. G., and Woolford, J. R. Jr. (1993). Yeast ribosomal protein L5 is required for stability of newly-synthesized 5S rRNA and the assembly of 60S ribosomal subunits. *Mol. Cell Biol.* **13,** 2835–2845.

Di Gregorio, A., and Levine, M. (1998). Ascidian embryogensis and origins of the chordate body plan. *Curr. Opin. Genet. Dev.* **8,** 457–463.

Ding, D., Whittaker, K. L., and Lipshitz, H. D. (1994). Mitochrondrially encoded 16S large ribosomal RNA is concentrated in the posterior polar plasm of early *Drosophila* embryos but it is not required for pole cell formation. *Dev. Biol.* **163,** 503–515.

Draper, B. W., Mello, C. C., Bowerman, B., Hardin, J., and Priess, J. R. (1996). MEX-3 is a KH domain protein that regulates blastomere identity in early C. elegans embryos. *Cell* **87,** 205–216.

Eddy, E. M. (1975). Germ plasm and the differentiation of the germ cell line. *Int. Rev. Cytol.* **43,** 229–280.

Erives, A., Corbo, J., and Levine, M. (1998). Lineage-specific regulation of the *Ciona snail* gene in the embryonic mesoderm and neuroectoderm. *Dev. Biol.* **194,** 213–225.

Fernández, J., Roegiers, F., Cantillana, V., and Sardet, C. (1998). Formation and localization of cytoplasmic domains in leech and ascidian zygotes. *Int. J. Dev. Biol.* **42,** 1075–1084.

Ferrante, A. W. Jr., Reinke, R., and Stanley, E. R. (1995). Shark, A Src homology 2, ankyrin repeat, tyrosine kinase, is expressed on the apical surfaces of ectodermal epithelia. *Proc. Natl. Acad. Sci. USA* **92,** 1911–1915.

Fischer, J.-L. (1991). Laurent Chabry and the beginnings of experimental embryology in France. In "A Conceptual History of Modern Embryology" (S. Gilbert, Ed.), pp. 31–41. Plenum Press, New York.

Ford, M. J., Anton, I. A., and Lane, D. R. (1988). Nuclear protein with sequence homology to translation intiation factor eIF-4A. *Nature* **322,** 736–738.

Hadfield, K. A., Swalla, B. J., and Jeffery, W. R. (1995). Multiple origins of anural development in ascidians inferred from rDNA sequences. *J. Mol. Evol.* **40,** 413–427.

Heldin, C.-H., Miyazono, K., and ten Dijke, P. (1997). TGF-β signaling from membrane to nucleus through SMAD proteins. *Nature* **390,** 465–471.

Hibino, T., Nishikata, T., and Nishida, H. (1998). Centrosome-attracting body: A novel structure closely related to unequal cleavages in the ascidian embryo. *Dev. Growth Differen.* **40,** 85–95.

Hikosaka, A., Kusakabe, T., Satoh, N., and Makabe, K. W. (1992). Introduction and expression of recombinant genes in ascidian embryos. *Dev. Growth Differen.* **34,** 627–634.

Imai, K., Satoh, N., and Satou, Y. (1999). Identification and characterization of maternally expressed genes with mRNAs that are segregated with the endoplasm of early ascidian embryos. *Int. J. Dev. Biol.* **43,** 125–133.

Ip, Y. T., Park, R., Kosman, D., Yazdanbakhsh, K., and Levine, M. (1992). Dorsal-twist interactions establish *snail* expression in the presumptive mesoderm of the *Drosophila* embryo. *Genes Dev.* **6,** 1518–1530.

Iseto, T., and Nishida, H. (1999). Ultrastructural studies on the centrosome-attracting body: Electron dense matrix and its role in unequal cleavages in ascidian embryos. *Dev. Growth Differen.* **41,** 601–609.

Ishida, K., and Satoh, N. (1998). Quantity of prelocalized maternal factor is associated with the timing of initiation of an epidermis-specific gene expression of the ascidian embryo. *Dev. Genes Evol.* **208,** 151–156.

Ishida, K., Ueki, T., and Satoh, N. (1996). Spatio-temporal expression patterns of eight epidermis-specific genes in the ascidian embryo. *Zool. Sci.* **13,** 699–709.

Jeffery, W. R. (1982). Calcium ionophore polarizes ooplasmic segregation in ascidian eggs. *Science* **216,** 545–547.

Jeffery, W. R. (1984a). Pattern formation by ooplasmic segregation in the ascidian egg. *Biol. Bull.* **166,** 177–198.

Jeffery, W. R. (1984b). Spatial distribution of messenger RNA in the cytoskeletal framework of ascidian eggs. *Dev. Biol.* **103,** 482–492.

Jeffery, W. R. (1985). Identification of proteins and mRNAs in isolated yellow crescents of ascidian eggs. *J. Embryol. Exp. Morph.* **89,** 275–287.

Jeffery, W. R. (1988). The role of cytoplasmic determinants in embryonic development. In "Developmental Biology: A Comprehensive Synthesis" (L. Browder, Ed.), pp. 1–53. Plenum Press, New York.

Jeffery, W. R. (1989a). Requirement of cell division for muscle actin expression in the primary muscle lineages of ascidian embryos. *Development* **105,** 75–84.

Jeffery, W. R. (1989b). Localized mRNA and the egg cytoskeleton. *Int. Rev. Cytol.* **118,** 150–195.

Jeffery, W. R. (1990a). Ultraviolet irradiation during ooplasmic segregation prevents gastrulation, sensory cell induction, and axis formation in the ascidian embryo. *Dev. Biol.* **140,** 388–400.

Jeffery, W. R. (1990b). An ultraviolet-sensitive maternal mRNA encoding a cytoskeletal protein may be involved in axis formation in the ascidian embryo. *Dev. Biol.* **141,** 141–148.

Jeffery, W. R. (1991). Ultraviolet-sensitive determinants of gastrulation and axis development in the ascidian embryo. In "Gastrulation: Movements, Patterns, and Molecules" (R. Keller, W. H. Clark, Jr., F. Griffin, Eds.), pp. 225–250. Plenum Press, New York.

Jeffery, W. R. (1992). A gastrulation center in the ascidian egg. *Development* **Suppl. 1992,** 53–63.

Jeffery, W. R. (1993). The myoplasm of ascidian eggs: A plasma membrane skeleton which is modified during evolution. *Biol. Res.* **26,** 481–490.

Jeffery, W. R. (1994). Actin as a tissue-specific marker in studies of ascidian development and evolution. *Adv. Dev. Biol.* **3,** 137–183.

Jeffery, W. R. (1995). Development and evolution of an egg cytoskeletal domain in ascidians. *Curr. Top. Dev. Biol.* **31,** 243–276.

Jeffery, W. R. (1997). Evolution of ascidian development. *BioEssays* **47,** 417–425.

Jeffery, W. R. (2000). Tunicates: Models for chordate evolution and development at low genomic complexity. In "Comparative Genomics" (M. Clark, Ed.). Kluwer Press, Boston (in press).

Jeffery, W. R., and Capco, D. G. (1978). Differential accumulation and localization of maternal poly (A)-containing RNA during early development of the ascidian, *Styela. Dev. Biol.* **67,** 152–166.

Jeffery, W. R., and Meier, S. (1983). A yellow crescent cytoskeletal domain in ascidian eggs and its role in early development. *Dev. Biol.* **96,** 125–143.

Jeffery, W. R., and Meier, S. (1984). Ooplasmic segregation of the myoplasmic actin network in stratified ascidian eggs. *W. Roux's Arch. Dev. Biol.* **193,** 257–262.

Jeffery, W. R., and Swalla, B. J. (1990a). The myoplasm of ascidian eggs: A localized cytoskeletal domain with multiple roles in embryonic development. *Sem. Cell Biol.* **1,** 373–381.

Jeffery, W. R., and Swalla, B. J. (1990b). Anural development in ascidians: Evolutionary modification and elimination of the tadpole larva. *Sem. Dev. Biol.* **1,** 253–261.

Jeffery, W. R., and Swalla, B. J. (1992a). Evolution of alternate modes of development in ascidians. *Bioessays* **14,** 219–226.

Jeffery, W. R., and Swalla, B. J. (1992b). Factors necessary for restoring an evolutionary change in an anural ascidian embryo. *Dev. Biol.* **153,** 194–205.

Jeffery, W. R., and Swalla, B. J. (1993). An ankyrin-like protein in ascidian eggs and its role in the evolution of direct development. *Zygote* **1,** 197–208.

Jeffery, W. R., Tomlinson, C. R., and Brodeur, R. D. (1983). Localization of actin messenger RNA during early ascidian development. *Dev. Biol.* **99,** 408–417.

Jeffery, W. R., Swalla, B. J., Ewing, N., and Kusakabe, T. (1999). Evolution of the ascidian anural larva: Evidence from embryos and molecules. *Mol. Biol. Evol.* **16,** 646–654.

Katz, M. C. (1983). The comparative anatomy of the tunicate tadpole *Ciona intestinalis. Biol. Bull.* **164,** 1–27.

Kawashima, T., Kawashima, S., Kanehisa, M., Nishida, H., and Makabe, K. W. (2000). MAGEST: Maboya gene expression patterns and sequence tags. *Nuc. Acid Res.* **28,** 133–135.

Kloc, M., and Etkin, L. D. (1995). Two distinct pathways for the localization of mRNAs at the vegetal cortex in *Xenopus oocytes. Development* **121,** 287–297.

Kloc, M., Larabell, C., and Etkin, L. D. (1996). Elaboration of the messenger transport organizer pathway for localization of RNA to the vegetal cortex of *Xenopus oocytes. Dev. Biol.* **180,** 119–130.

Kobayashi, A., Sasakura, Y., Ogasawara, M., and Makabe, K. W. (1999). A maternal RNA encoding *smad1/5* is segregated to animals blastomeres during ascidian development. *Dev. Growth Differen.* **41,** 419–427.

Kobayashi, S., Amikura, R., and Okada, M. (1993). Presence of mitochondrial large ribosomal RNA outside mitochondria in germ plasm of *Drosophila melanogaster. Science* **260,** 1521–1524.

Kobayashi, S., Amikura, R., and Okada, M. (1994). Localization of mitochondrial large RNA in germinal granules and the consequent segregation of the germ line. *Int. J. Dev. Biol.* **38,** 193–199.

Kowalevsky, A. O. (1866). Entwicklungsgeschichte der einfachen Ascidien. *Mem. Acad. Sci. St. Petersberg* **10,** 1–19.

Kumano, G., and Nishida, H. (1998). Maternal and zygotic expression of the endoderm-specific alkaline phosphatase gene in embryos of the ascidian, *Halocynthia roretzi. Dev. Biol.* **198,** 245–252.

Kusakabe, T., Hikosaka, A., and Satoh, N. (1995). Coexpression and promoter function in two muscle actin gene complexes of differential structural organization in the ascidian *Halocynthia roretzi. Dev. Biol.* **169,** 461–472.

Kusakabe, T., Swalla, B. J., Satoh, N., and Jeffery, W. R. (1996). Mechanism of an evolutionary change in muscle cell differentiation in ascidians with different modes of development. *Dev. Biol.* **174,** 379–392.

Lambert, C. C., and Laird, C. (1971). Molecular properties of tunicate DNA. *Biochim. Biophys. Acta* **240,** 39–45.

Leptin, M. (1991). *twist* and *snail* as positive and negative regulators during *Drosophila* mesoderm development. *Genes Dev.* **5,** 1568–1576.

Makabe, K. W., Fujiwara, S., Saiga, H., and Satoh, N. (1990). Specific expression of myosin heavy chain gene in muscle lineage cells of the ascidian embryo. *W. Roux's Arch. Dev. Biol.* **199,** 307–313.

Marikawa, Y. (1995). Distribution of myoplasmic cytoskeletal domains among egg fragments of the ascidian *Ciona savignyi:* The concentration of the deep filament lattice in the fragment enriched in muscle determinants. *J. Exp. Zool.* **271,** 348–355.

Marikawa, Y., and Satoh, N. (1995). Ultraviolet-sensitive ooplasmic factors are responsible for the development of endoderm-specific alkaline phosphatase in the ascidian embryos. *Dev. Growth Differen.* **38,** 167–173.

Marikawa, Y., Yoshida, S., and Satoh, N. (1994). Development of egg fragments of the ascidian *Ciona savignyi:* The cytoplasmic factors responsible for muscle differentiation are separated into a specific fragment. *Dev. Biol.* **162,** 134–142.

Marikawa, Y., Yoshida, S., and Satoh, N. (1995). Muscle determinants in the ascidian egg are inactivated by UV irradiation and the inactivation is partially rescued by injection of maternal mRNA. *Dev. Biol.* **204,** 180–186.

Meedel, T. H., Crowther, R. J., and Whittaker, J. R. (1987). Determinative properties of muscle lineages in ascidian embryos. *Development* **100,** 245–260.

Meedel, T. H., Farmer, S. C., and Lee, J. J. (1997). The single *MyoD* family gene of *Ciona intestinalis* encodes two differentially expressed proteins: Implications for the evolution of chordate muscle gene regulation. *Development* **124,** 1711–1721.

Mello, C. C., Draper, B. W., Krause, M., Weintraub, H., and Priess, J. R. (1992). The pie-1 and mex-1 genes and maternal control of blastomere identity in early *C. elegans* embryos. *Cell* **70,** 163–176.

Mello, C. C., Schubert, C., Draper, B., Zhang, W., Lobel, R., and Priess, J. R. (1996). The PIE-1 protein and germline specification in *C. elegans* embryos. *Nature* **382,** 710–712.

Melton, D. A. (1987). Translocation of a localized maternal mRNA to the vegetal pole of *Xenopus oocytes. Nature* **328,** 80–82.

Micklem, D. R. (1995). mRNA localization during development. *Dev. Biol.* **171,** 377–395.

Moody, R., Davis, S. W., Cubas, F., and Smith, W. C. (1999). Isolation of developmental mutants of the ascidian *Ciona savignyi. Mol. Gen. Genet.* **262,** 199–206.

Moon, R. T., Brown, J. D., and Torres, M. (1997). *Wnts* modulate cell fate and behavior during vertebrate development. *Trends Genet.* **13,** 157–162.

Morgan, T. H. (1910). The effects of altering the position of the cleavage planes in eggs with precocious specification. *Arch. Entw. Zool.* **227,** 155–158.

Nakatani, Y., Moody, R., and Smith, W. C. (1999). Mutations affecting tail and notochord development in the ascidian *Ciona savignyi. Development* **126,** 3293–3301.

Nakatani, Y., and Nishida, H. (1994). Induction of notochord during ascidian embryogenesis. *Dev. Biol.* **166,** 289–299.

Nicol, D. R., and Meinertzhagen, I. A. (1988). Development of the central nervous system of the larva of the ascidian. *Ciona intestinalis* L. I. The early lineages of the neural plate. *Dev. Biol.* **130,** 721–736.

Nishida, H. (1987). Cell lineage analysis in ascidian embryos by intracellular injection of a tracer enzyme. III. Up to the tissue restricted stage. *Dev. Biol.* **99,** 382–394.

Nishida, H. (1991). Induction of brain and sensory pigment cells in the ascidian embryo analyzed by experiments with isolated blastomeres. *Development* **112,** 389–395.

Nishida, H. (1992a). Regionality of egg cytoplasm that promotes muscle differentiation in embryo of the ascidian, *Halocynthia roretzi. Development* **116,** 521–529.

Nishida, H. (1992b). Developmental potential for tissue differentiation of fully dissociated cells of the ascidian embryo. *W. Roux's Arch. Dev. Biol.* **201,** 81–87.

Nishida, H. (1993). Localized regions of the egg cytoplasm that promote expression of endoderm-specific alkaline phosphatase in embryos of the ascidian *Halocynthia roretzi. Development* **118,** 1–7.

Nishida, H. (1994a). Localization of egg cytoplasm that promotes differentiation to epidermis in embryos of the ascidian *Halocynthia roretzi. Development* **120,** 235–243.

Nishida, H. (1994b). Localization of determinants for the formation of the anterior-posterior axis in the eggs of the ascidian *Halocynthia roretzi. Development* **120,** 3093–3104.

Nishida, H. (1996). Vegetal egg cytoplasm promotes gastrulation and is responsible for specification of vegetal blastomeres in embryos of the ascidian *Halocynthia roretzi. Development* **122,** 1271–1279.

Nishida, H. (1997). Cell fate specification by localized cytoplasmic determinants and cell interactions in ascidian embryos. *Int. Rev. Cytol.* **176,** 245–306.

Nishida, H., and Kumano, G. (1997). Analysis of the temporal expression of endoderm-specific alkaline phosphatase during development of the ascidian *Halocynthia roretzi. Dev. Growth Differen.* **39,** 199–205.

Nishida, H., and Makabe, K. W. (1999). Maternal information and localized maternal mRNA in eggs and early embryos of the ascidian *Halocynthia roretzi. Invert. Reprod. Dev.* **36,** 1–3.

Nishikata, T., Hibino, T., and Nishida, H. (1999). The centrosome-attracting body, microtubule system, and posterior egg cytoplasm are involved in positioning of cleavage planes in the ascidian embryo. *Dev. Biol.* **209,** 72–85.

Nishikata, T., Mita-Miyazawa, 1., Deno, T., Takamura, K., and Satoh, N. (1987). Expression of epidermis-specific antigens during embrogenesis of the ascidian *Halocynthia roretzi. Dev. Biol.* **121,** 408–416.

Nishikata, T., and Wada, M. (1996). Molecular characterization of myoplasmin-C1: A cytoskeletal component localized in the myoplasm of the ascidian egg. *Dev. Genes Evol.* **206,** 72–79.

Nüsslein-Volhard, C. (1991). Determination of the embryonic axis of *Drosophila. Development* **Suppl. 1991,** 1–10.

Oka, T., Amikura, R., Kobayashi, S., Yamamoto, H., and Nishida, H. (1999). Localization of mitochondrial large ribosomal RNA in the myoplasm of the early ascidian embryo. *Dev. Growth Differen.* **41,** 1–8.

Olsen, C. L., and Jeffery, W. R. (1997). A forkhead gene related to *HNF-3β* is required for gastrulation and axis formation in the ascidian embryo. *Development* **124,** 3609–3619.

Olsen, C. L., Natzle, J. E., and Jeffery, W. R. (1999). The *forkhead* gene *FH1* is involved in evolutionary modification of the ascidian tadpole larva. *Mech. Dev.* **85,** 49–58.

Ortolani, G. (1955). The presumptive territory of the mesoderm in the ascidian germ. *Experientia* **11,** 445–446.

Ortolani, G. (1958). Cleavage and development of egg fragments in ascidians. *Acta Embryol. Morphol. Exp.* **1,** 247–272.

Reid, T., Balhoorn, A., Ahmadian, M. R., and Collard, J. G. (1999). Identification and characterization of hPEM-2, a guanine nucleotide exchange factor specific for cdc42. *J. Biol. Chem.* **274,** 33587–33593.

Reverberi, G., and Minganti, A. (1946). Fenomeni di evocazione nello sviluppo dell' uovo di Ascide. Rsultati dell'indagine sperimentale sull' uovo di *Ascidiella aspersa* e di *Ascidia malaca* all stadio di 8 blastomeri. *Pubbl. Stn. Zool. Napoli* **20,** 199–152.

Reverberi, G., and Minganti, A. (1947). La distribuzione delle potentze neiz germ di Ascidiealle stadio de otto blastomeri, analizzata mediante le combinazione e I trpiants di blastomeri. *Pubbl. Stn. Zool. Napoli* **21,** 1–35.

Reverberi, G., Ortolani, G., and Farinella-Ferruzza, N. (1960). The causal formation of the brain in the ascidian larva. *Acta Embryol. Morphol. Exp.* **3,** 296–336.

Roegiers, F. (1999). Reorganisations corticales et cytoplasmiques du zygote d'Ascidie. Ph.D. Thesis, Universite de Nice-Sophia Antipolis.

Roegiers, F., McDougall, A., and Sardet, C. (1995). The sperm entry point defines the orientation of the calcium induced contraction wave that directs the first phase of cytoplasmic reorganization in the ascidian egg. *Development* **121,** 3457–3466.

Roegiers, F., Djediat, C., Dumollard, R., Rouviere, C., and Sardet, C. (1999). Phases of cytoplasmic and cortical reorganizations of the ascidian zygote between fertilization and first division. *Development* **126,** 3101–3117.

Rose, S. M. (1939). Embryonic induction in the ascidia. *Biol. Bull.* **76,** 216–232.

Sardet, C., Speksnijder, J. E., Inouye, S., and Jaffe, L. (1989). Fertilization and ooplasmic movements in the ascidian egg. *Development* **105,** 237–249.

Sasakura, Y., Ogasawara, M., and Makabe, K. W. (1998a). Maternally localized RNA encoding a serine/threonine protein kinase in the ascidian, *Halocynthia roretzi. Mech. Dev.* **76,** 161–163.

Sasakura, Y., Ogasawara, M., and Makabe, K. W. (1998b). *HrWnt-5:* A maternally expressed ascidian *Wnt* gene with posterior localization in early embryos. *Int. J. Dev. Biol.* **42,** 573–579.

Sasakura, Y., Ogasawara, M., and Makabe, K. W. (2000). Two pathways of maternal RNA localization at the posterior-vegetal cytoplasm in early ascidian embryos. *Dev. Biol.* **220,** 365–378.

Satoh, N. (1994). "Development Biology of Ascidians." Cambridge University Press, Cambridge, U.K.

Satoh, N., and Jeffery, W. R. (1995). Chasing tails in ascidians: Developmental insights into the origin and evolution of chordates. *Trends Genet.* **11,** 354–357.

Satoh, N., Makabe, K. W., Katsuyama, Y., Wada, S., and Saiga, H. (1996). The ascidian embryo: An experimental system for studying genetic circuitry for embryonic cell specification and morphogenesis. *Dev. Growth Differen.* **38,** 315–340.

Satou, Y. (1999). *posterior end mark 3* (*pem-3*), an ascidian maternally expressed gene with localized mRNA encodes a protein with *Caenorhabditis elegans.* MEX-3-like KD domains. *Dev. Biol.* **212,** 337–350.

Satou, Y., Kusakabe, T., Araki, I., and Satoh, N. (1995). Timing of initiation of muscle gene expression in the ascidian embryo precedes that of developmental fate restriction in muscle lineage cells. *Dev. Growth Differen.* **37,** 319–327.

Satou, Y., and Satoh, N. (1996). Two cis-regulatory elements are essential for the muscle-specific expression of an actin gene in the ascidian embryo. *Dev. Growth Differen.* **38,** 565–573.

Satou, Y., and Satoh, N. (1997). *posterior end mark 2* (*pem-2*), *pem-4, pem-5,* and *pem-6*: Maternal genes with localized mRNA in the ascidian embryo. *Dev. Biol.* **192,** 467–481.

Sawada, T., and Osanai, K. (1981). The cortical contraction related to the ooplasmic segregation in *Ciona intestinalis* eggs. *W. Roux's Arch. Dev. Biol.* **190,** 208–214.

Sawada, T., and Schatten, G. (1989). Effects of cytoskeletal inhibitors on ooplasmic segregation and microtubule organization during fertilization and early development in the ascidian *Molgula occidentalis. Dev. Biol.* **79,** 132, 331–342.

Siomi, H., Matunis, M. J., Michael, W. M., and Dreyfuss, G. (1993). The pre-mRNA binding K protein contains a novel evolutionary conserved motif. *Nuc. Acid. Res.* **21,** 1193–1198.

Speksnijder, J. E., Jaffe, L. F., and Sardet, C. (1989). Polarity of sperm entry in the ascidian egg. *Dev. Biol.* **133,** 180–184.

Speksnijder, J. E., Sardet, C., and Jaffe, L. F. (1990). The activation wave of calcium in the ascidian egg and its role in ooplasmic segregation. *J. Cell. Biol.* **110,** 1589–1598.

Swalla, B. J., and Jeffery, W. R. (1990). Interspecific hybridization between an anural and urodele ascidian: Differential expression of urodele features suggests multiple mechanisms control anural development *Dev. Biol.* **142,** 319–334.

Swalla, B. J., and Jeffery, W. R. (1995). A maternal RNA localized in the yellow crescent is segregated to the larval muscle cells during ascidian development *Dev. Biol.* **170,** 353–364.

Swalla, B. J., and Jeffery, W. R. (1996a). Localization of ribosomal protein L5 mRNA in myoplasm during ascidian development. *Dev. Genet.* **19,** 258–267.

Swalla, B. J., and Jeffery, W. R. (1996b). PCNA mRNA has a 3' UTR antisense to yellow crescent RNA and is localized in ascidian eggs and embryos. *Dev. Biol.* **178,** 23–34.

Swalla, B. J., and Jeffery, W. R. (1996c). Requirement of the *Manx* gene for expression of chordate features in a tailless ascidian larva. *Science* **274,** 1205–1208.

Swalla, B. J., Badgett, M. R., and Jeffery, W. R. (1991). Identification of a cytoskeletal protein localized in the myoplasm of ascidian eggs: Localization is modified during anural development. *Development* **111,** 425–436.

Swalla, B. J., Makabe, K. W., Satoh, N., and Jeffery, W. R. (1993). Novel genes expressed differentially in ascidians with alternate modes of development. *Development* **119,** 307–318.

Swalla, B. J., Makabe, K. W., Satoh, N., and Jeffery, W. R. (1995). *Cymric,* a new member of the shark family of non-receptor tyrosine kinases, is expressed differentially in ascidians with alternate modes of development. *Mol. Biol. Cell* **6,** 81a.

Swalla, B. J., Just, M. A., Pederson, E. L., and Jeffery, W. R. (1999). A multigene locus containing the *Manx* and *bobcat* genes is required for development of chordate features in the ascidian tadpole larva. *Development* **126,** 1643–1653.

Takahashi, H., Ishida, K., Makabe, K. W., and Satoh, N. (1997). Isolation of cDNA clones for genes that are expressed in the tail region of the ascidian tailbud embryo. *Int. J. Dev. Biol.* **41,** 691–698.

Takahashi, H., Hotta, K., Erives, A., Di Gregorio, A., Zeller, R. W., Levine, M., and Satoh, N. (1999). *Brachyury* downstream differentiation in the ascidian embryo. *Genes Dev.* **13,** 1519–1523.

Ueki, T., Makabe, K. W., and Satoh, N. (1991). Isolation of cDNA clones for epidermis-specific genes of the ascidian embryo. *Dev. Growth Differen.* **33,** 319–327.

Venuti, J. M., and Jeffery, W. R. (1989). Cell lineage and determination of cell fate in ascidian embryos. *Int. J. Dev. Biol.* **33,** 197–212.

Wada, H., Makabe, K. W., Nakauchi, M., and Satoh, N. (1992). Phylogenetic relationships between solitary and colonial ascidians, as inferred from the sequence of the central region of their respective 18S rDNA. *Biol. Bull.* **183,** 448–455.

Wada, S., and Saiga, H. (1999). Cloning and early expression of *Hrsna,* a *snail* family gene, of the ascidian *Halocynthia roretzi*: Implication in the origin of mechanisms for mesoderm specification and body axis formation in chordates. *Dev. Growth Differen.* **41,** 9–18.

Wellington, A., Emmons, S., James, B., Calley, J., Grover, M., Tolias, P. and Manseau, L. (1999). Spire contains actin binding domains and is related to ascidian posterior end mark-5. *Development* **126,** 5267–5274.

Whittaker, J. R. (1973). Segregation during ascidian embryogenesis of egg cytoplasmic information for tissue specific enzyme development. *Proc. Nat. Acad. Sci. USA* **70,** 2096–2100.

Whittaker, J. R. (1977). Segregation during cleavage of a factor determining endodermal alkaline phosphatase development in ascidian embryos. *J. Exp. Zool.* **202,** 139–154.

Whittaker, J. R. (1980). Acetylcholinesterase development in extra cells caused by changing the distribution of myoplasm in ascidian embryos. *J. Embryol. Exp. Morph.* **55,** 343–354.

Whittaker, J. R. (1990). Determination of alkaline phosphatase expression in endodermal cell lineages of an ascidian embryo. *Biol. Bull.* **178,** 222–230.

Yamada, A., and Nishida, H. (1996). Distribution of cytoplasmic determinants in unfertilized eggs of the ascidian *Halocynthia roretzi. Dev. Genes Evol.* **206,** 297–304.

Yoshida, S., Marikawa, Y., and Satoh, N. (1996). *posterior end mark,* a novel maternal gene encoding a localized factor in the ascidian embryo. *Development* **122,** 2005–2012.

Yoshida, S., Marikawa, Y., and Satoh, N. (1998). Regulation of the trunk-tail patterning in the ascidian embryo: A possible interaction of cascades between lithium/β-catenin and localized maternal factor pem. *Dev. Biol.* **202,** 264–279.

RNA Localization and Germ Cell Determination in *Xenopus*

M. Kloc, S. Bilinski*, A. P. Chan, L. H. Allen, N. R. Zearfoss, and L. D. Etkin
Department of Molecular Genetics, University of Texas M. D. Anderson Cancer Center, Houston, Texas 77030;* Institut of Zoology, Jagiellonian University, Krakow, Poland

---

In many organisms the proper development of the embryo depends on the asymmetrical distribution of maternal RNAs and proteins in the egg. Although the *Xenopus* oocyte is radially symmetrical it contains distinct populations of maternal RNAs that are localized either in the animal or vegetal pole. The process of localization of RNAs in *Xenopus* oocytes occurs during the long period of oocyte differentiation and growth that is accompanied by the elaboration of oocyte polarity. Some of the vegetally localized RNAs, such as Vg1, VegT, and Xwnt11, are involved in axial patterning and germ layer specification. Others, such as Xdazl and Xcat2, which are located in the germ plasm, are likely to play a role in the specification of germ cell fate. We will discuss the different aspects of RNA localization in *Xenopus* in the context of the differentiation of the germ cells and the development of the oocyte polarity.

KEY WORDS: RNA localization, Oogenesis, Germ cell, Differentiation, Maternal information.

I. Introduction

From before the turn of the last century embryologists have asked how the single cell egg can generate a complex organism with different tissue types, organs, and a defined axial polarity. Although a great deal of information has been gathered regarding this process, the fundamental issue is still not resolved. Observations of the developmental fate of embryos with specific blastomeres removed showed that there were two basic types of development: regulative and mosaic (Jeffery, Chapter 1). In the mosaic embryo each blastomere (or group of blastomeres) played a distinct and irreplaceable role in the formation of a normal adult organism. The discovery of the mosaic type of development led to the concept of "localized cytoplasmic determinants." The nature of these determinants was unknown, but it was believed that they were maternal in origin and that they were responsible for the specification of the cell fates in developing embryo. With the recent discoveries of different species of RNAs localized in the specific regions of egg cytoplasm, the concept of localized maternal determinants has become a paradigm in modern biology.

Many articles were published in the 1990s describing the phenomenon of RNA localization in the cytoplasm of somatic cells and oocytes and embryos of different animal and plant species (St. Johnston, 1995; Bashirullah *et al.,* 1998; Hazelrigg, 1998; Gavis, 1997; King *et al.,* 1999; Mowry and Cote, 1999). In fact, it is now known that the phenomenon of localized RNA is common for evolutionarily distant organisms such as budding yeast and vertebrates. This suggests that RNA localization represents a highly conserved and evolutionarily ancient strategy that allows cells to produce local asymmetries in the certain subdomains of the cytoplasm.

In the majority of invertebrates and vertebrates the proper development of the embryo depends on asymmetrical distribution of the maternal RNAs and proteins in the cytoplasm of the egg. One of the most spectacular examples of molecules that are asymmetrically deposited during oogenesis is the localization of different species of RNAs during oogenesis in *Drosophila* (Cooperstock and Lipshitz, Chapter 14) and *Xenopus* (Yisraeli and Yanis, Chapter 13). Although the *Xenopus* oocyte is radially symmetrical, it contains distinct populations of maternal RNAs that are localized either in the animal or the vegetal pole (Rebegliati *et al.,* 1985; Melton, 1987; Yisraeli *et al.,* 1990; Mowry and Cote, 1999; Mosquera *et al.,* 1993; Kloc *et al.,* 1993a; Weeks and Melton, 1987a,b). The vegetally localized mRNAs such as Vg1 and Xwnt11 are probably involved in the axial patterning of the embryo (Weeks and Melton, 1987b; Ku and Melton, 1993; Thomsen and Melton, 1993). Another vegetally localized mRNA, Veg-T, which is a transcription factor of the brachyury family, plays a role in the specification of the mesodermal and endodermal cell fates and is essential for the embryonic endoderm formation (Horb and Thomsen, 1997; Stennard *et al.,* 1996; Lustig *et al.,* 1996; Clements *et al.,* 1999; Casey *et al.,* 1999; Tada *et al.,* 1998; Kofron *et al.,* 1999; Zhang *et al.,* 1998). Some of the vegetally localized RNAs such as Xcat2, Xlsirts, Xdazl, and Xwnt11 are likely to be involved in the specification of the germ line fate (Mosquera *et al.,* 1993; Kloc *et al.,* 1993a, 1998; Houston *et al.,* 1998; Houston and King, 2000). These RNAs are localized in the specialized subdomain of the vegetal ooplasm called the germ plasm. During cleavages the germ plasm, together with localized RNAs, becomes acquired by the primordial germ cells of the embryo. This indicates that the mechanisms responsible for the proper localization of different RNAs in the oocyte play an extremely important role in axial patterning and specification of cell fates and embryonic germ layers of the future embryo.

Studies showed that vegetally localized RNAs made *in vitro* and injected into *Xenopus* oocytes can localize and anchor at the vegetal pole very efficiently within 24 to 48 hours (Kloc *et al.,* 1993a; Kloc *et al.,* 1996; Mowry and Melton, 1992; Yisraeli and Melton, 1988; Yisraeli *et al.,* 1990; Zhou and King, 1996a, b). However, in reality the process of differential localization of RNAs in full-grown oocytes is not instantaneous but occurs during the long period of oocyte differentiation and growth that is accompanied by the elaboration of oocyte polarity, and in *Xenopus* it can last for several years. In this chapter we will discuss the different aspects of RNA localization in *Xenopus* in the context of differentiation of the germ cells and development of oocyte polarity.

II. Origin and Development of Oocyte Polarity

In *Xenopus* the major period of oocyte growth is divided into six consecutive stages (stage 1 to stage 6; Dumont, 1972). The stage 1 oocyte is transparent, has a centrally located nucleus, and, since it is devoid of pigment and yolk, does not exhibit any signs of external polarity. The primary sign of internal polarity in the stage 1 oocyte is the location of the prominent cytoplasmic organelle called the mitochondrial cloud or Balbiani body (Guraya, 1979; Hausen and Riebesell, 1991). The mitochondrial cloud is juxtaposed to the nucleus and always faces the future vegetal pole of the oocyte. Electron microscopy revealed that the mitochondrial cloud contains numerous mitochondria, membranes of endoplasmic reticulum, and the germ plasm (Al-Mukhtar and Webb, 1971; Coggins, 1973; Billett and Adam, 1976; Tourte *et al.,* 1984; Heasman *et al.,* 1984; Wylie *et al.,* 1985; Kloc *et al.,* 1998). Light and electron microscopy immunostaining and *in situ* hybridization showed that spectrin, γ tubulin, EF1α, DEAD box proteins, and several localized RNAs are also present in the mitochondrial cloud (Viel *et al.,* 1990; Kloc *et al.,* 1998; Kloc and Etkin, 1998; St. Johnston, 1995; Bashirullah *et al.,* 1998; Watanabe *et al.,* 1992; Komiya *et al.,* 1994; Kobayashi *et al.,* 1994, 1998). Starting from late stage 2 of oogenesis, oocytes begin to accumulate yolk and pigment, and the mitochondrial cloud fragments into many islands, which migrate preferentially toward, and anchor at, the vegetal cortex of the oocyte (Figure 1).

Full-grown stage 6 oocytes of wild type *Xenopus* show a striking external polarity with the dark, pigmented animal hemisphere and light, only slightly pigmented, vegetal half. The polarity of pigmentation is the extrinsic reflec-

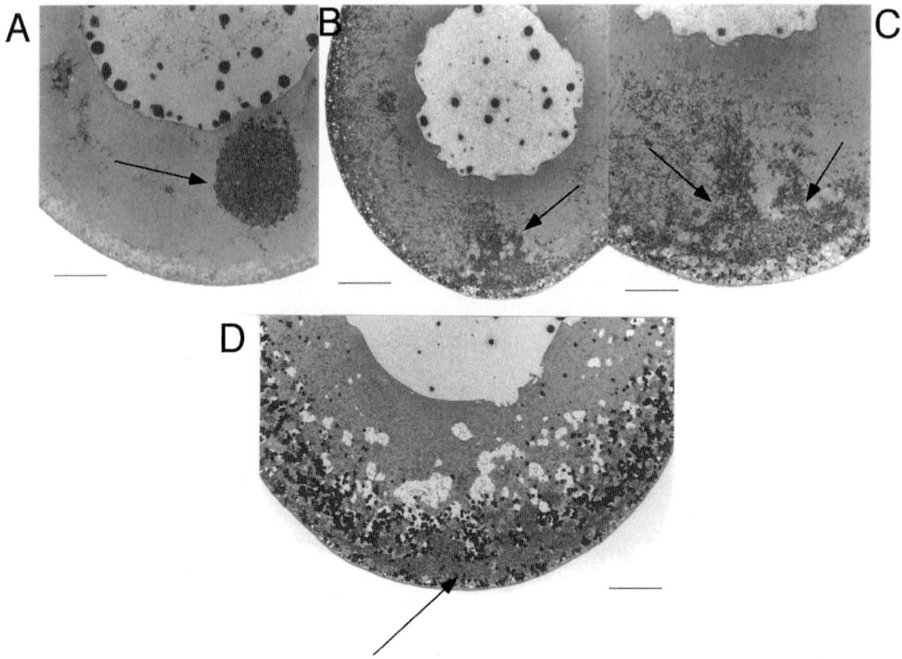

FIG. 1 Dispersion of mitochondrial cloud in growing *Xenopus* oocytes. Semithin sections of different stage oocytes stained with methylene blue showing the changes in the mitochondrial cloud (arrows) during oogenesis. (A) Stage 1 oocyte, bar = 25 μm. (B) Late stage 1 oocyte, bar = 50 μm. (C) Late stage 1 oocyte, bar = 25 μm. (D) Stage 2 oocyte, bar = 50 μm.

tion of a much more complex internal polarity at the subcellular and molecular levels. There is polarization along the animal-vegetal axis in the distribution of oocyte organelles; the nucleus is shifted toward the animal half, which contains small yolk platelets. The Golgi apparatus, large yolk platelets, specialized regions of endoplasmic reticulum, and germ plasm are located in the vegetal half and/or vegetal cortex of the oocyte (Gerhart *et al.*, 1986; Hausen and Riebesell, 1991). The polarity at the molecular level is exemplified by differential distribution of animally and vegetally localized RNAs as well as various proteins.

Since there is no indication that external signals from follicular cells, tissues of the ovarian wall, or a gravitational force play any role in determination of oocyte polarity, the development of polarity is likely to be autonomous and an intrinsic property of the oocyte itself. However, there is a very strong indication that the elaboration of oocyte polarity that takes place between stage 1 and stage 6 of oogenesis is based on the antecedent polarity that developed in germ cells long before the sexual differentiation of the gonad took place. Already the organization of primary oogonium shows some signs of polarity in the distribution of different organelles.

Primary oogonia are large, roundish, or oval shaped cells with lobulated nuclei. The Golgi apparatus, the majority of the endoplasmic reticulum, and a very prominent mitochondrial mass are positioned to the one side of the nucleus (Al-Mukhtar and Webb, 1971; Coggins, 1973; Kalt and Gall, 1974). The axial polarity is readily visible in secondary oogonia of 16 cell nests. A secondary oogonium is pear-shaped, and its round nucleus is shifted toward the dilated part of the cell. The mitochondrial mass, endoplasmic reticulum, and Golgi complex are located in the narrowed part of the cell. Ultrastructural analysis of the mitochondrial mass in 16-cell-nest oogonia showed that it contains numerous mitochondria. The space between mitochondria is filled with an electron dense mitochondrial cement, which is probably the precursor of granulofibrillar material and germinal granules that are the components of germ plasm. The mitochondrial mass also contains Xlsirts RNA and spectrin (Figure 2). The central part of the mitochon-

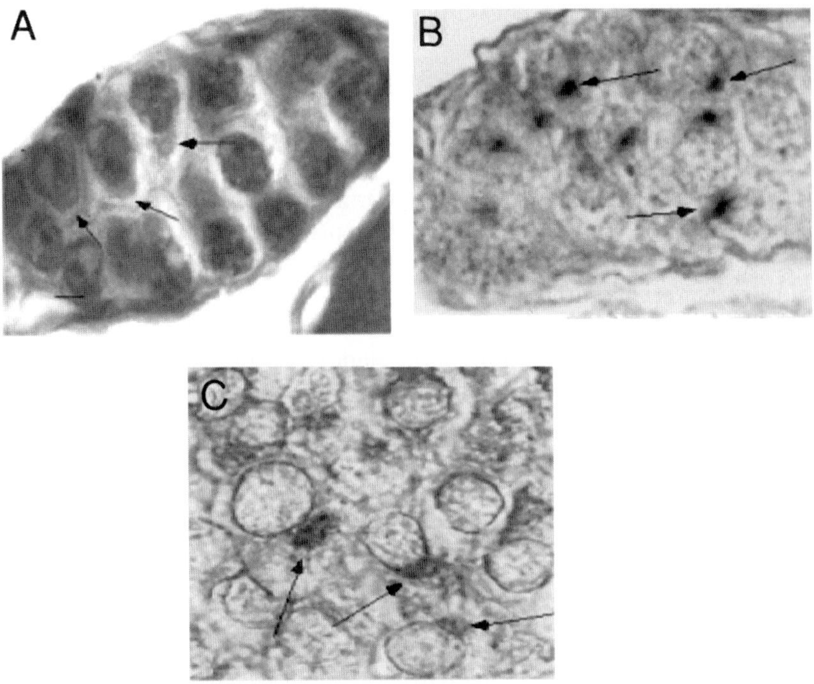

FIG. 2 Composition of mitochondial aggregates in 16-cell nests. Histological sections through an ovary of young froglet stained with hematoxylin and eosin (A), immunostained with spectrin (B), and *in situ* hybridization with xlsirts (C). Arrows point to the mitochondrial mass.

drial mass is occupied by a centrosome that contains γ tubulin and a pair of centrioles (Al-Mukhtar and Webb, 1971; Coggins, 1973; Kalt and Gall, 1974; Kloc and Etkin, 1998; Kloc *et al.,* 1998; Pepling *et al.,* 1999).

Late pachytene and early diplotene oocytes are round and surrounded by follicular cells. The oocyte nucleus is centrally located, and there are several small aggregates of mitochondria adjacent to the nuclear membrane. Electron microscopy shows that all these aggregates contain mitochondria and mitochondrial cement and all contain similar amounts of Xlsirts, Xcat2, and Xwnt11 RNAs (M. Kloc and L. D. Etkin, unpublished observations). However, only one of these aggregates contains a centrosome and centrioles (Al-Mukhtar and Webb, 1971; Coggins, 1973; Heasman *et al.,* 1984; Klymkowsky and Karnovsky, 1994; Kloc *et al.,* 1996; Kloc and Etkin, 1998). In prestage 1 oocytes (diplotene oocytes just before entering the major period of growth), one of the mitochondrial aggregates (the one which contains a centrosome, a centriole, and faces the future vegetal pole of the oocyte) enlarges, becoming a pre-mitochondrial cloud, which is the precursor of the mitochondrial cloud in stage 1 oocytes. At the same time, Xlsirts, Xcat2, and Xwnt11 RNAs gradually disappear from the smaller mitochondrial aggregates but remain in the premitochondrial cloud (Kloc *et al.,* 1996). The other mitochondrial aggregates will disperse in later stages of oogenesis. Ultrastructural analysis shows that mitochondrial cement in the premitochondrial cloud is gradually replaced by granulofibrillar material and eventually by germinal granules (Kloc *et al.,* 2000). There is a very strong indication that the premitochondrial cloud of a stage 1 oocyte originates from the mitochondrial mass present in primary and secondary oogonium. All these structures share common components such as centrioles, γ tubulin, spectrin, Xlsirts RNA, mitochondria, mitochondrial cement, and/or granulofibrillar material characteristic of germ plasm. The polarity of a stage 6 oocyte, and ultimately the egg, is critical for the proper outcome of embryogenesis. The observations of the elaboration of oocyte polarity show how the earliest events in gametogenesis are an integral and inseparable part of the development of the new individual.

III. Pathways of RNA Localization

One of the signs of intrinsic molecular polarity of the *Xenopus* oocyte is the differential distribution of many types of maternal RNAs that are preferentially localized to either the animal or the vegetal pole of the oocyte (Table I). The proper localization of RNAs is a critical component in the differentiation of the oocyte and subsequent patterning and lineage

TABLE I
RNAs Localized to the Animal Pole

Name	Function	References
An1 and b	Ubiquitin-like	Linnen *et al.*, 1993 Rebagliati *et al.*, 1985
An2	Subunit of mt ATPase	Rebagliati *et al.*, 1985 Weeks and Melton, 1987a
An3	RNA binding protein/RNA helicase	Gururajan *et al.*, 1991 Rebagliati *et al.*, 1985
An4 a and b	Unknown	Hudson *et al.*, 1996
βTrCP	βtransducin signaling protein	Hudson *et al.*, 1996
G protein	Signaling molecule	Otte *et al.*, 1992
oct60	Transcription factor	Hinkley *et al.*,1992
PABP	Poly (A) binding protein	Schroeder and Yost, 1996
PKCα	Signaling molecule	Otte and Moon, 1992
Xl-21	Transcription factor?	Kloc *et al.*, 1991a
Xlan-4	Contains PEST sequences	Reddy *et al.*, 1992
Xlcaax-1	Membrane and ATP binding protein	Kloc *et al.*, 1989 Kloc *et al.*, 1991b

determination in the embryo. Therefore, we will consider how RNAs are localized during oogenesis and their potential role in these processes.

A. Animal Localization

Some of the maternal RNA species in *Xenopus* are not equally distributed between the animal and vegetal half, but are twice as abundant in the animal pole than the vegetal pole (Rebagliati *et al.*, 1985). This can probably be attributed to the large amount of yolk in the vegetal cytoplasm versus the animal cytoplasm, and not to the active transport of these RNAs to the animal pole. However, there are examples of RNAs that are more than four-fold enriched in the animal half, such as Anla and b, An2, An3, An4a, b, X121, Xlan-4, Xlcaax-1, βTrCP, PKC α, oct60, G protein, and PABP (Table I) (Rebagliati *et al.*, 1985; Weeks and Melton, 1987a; Kloc *et al.*, 1989, 1991b, Kloc 1993b; Otte *et al.*, 1992; Otte and Moon, 1992; Hinkley *et al.*, 1992; Reddy *et al.*, 1992; Linnen *et al.*, 1993; Hudson *et al.*, 1996; Schroeder and Yost, 1996). Unlike the other RNAs that are in a gradient, PABP is localized to a thin rim beneath the animal cortex (Schroeder and Yost, 1996). Although, with the exception of PABP, there is no detailed information on the pattern of distribution of RNAs in the animal pole, it seems that none of these RNAs is tightly anchored in the animal cortex. The enrichment of certain species of maternal RNAs in the animal hemisphere of the oocyte suggests the existence of preferential transport or/ and preferential retention in the animal pole. However, so far there is no indication of any mechanism such as cis and trans-acting factors and cytoskeletal components for active transport of these RNAs to the animal hemisphere.

B. Vegetal Localization

There are two major vegetal pathways of RNA localization in *Xenopus* oogenesis, and there is possibly a third intermediate pathway that may utilize aspects of the other two pathways.

1. Early or METRO Pathway

Several different RNAs are known to use this localization pathway (Table II). These include Xcat2, Xcat3, Xdazl, Xwnt11, Xlsirts, B6, B7, C10, Xpat and Xotx1 (Elinson *et al.,* 1993; Kloc *et al.,* 1993a; Ku and Melton, 1993; Mosquera *et al.,* 1993; Forristall *et al.,* 1995; Zhou and King, 1996a; Hudson and Woodland, 1998; Houston *et al.,* 1998; King, 1995; Simeone *et al.,* 1993;

TABLE II
RNAs Localized to the Vegetal Cortex Through the METRO (Early) Pathway

Name	Function	Sublocalization in the Germ Plasm	References
Xcat2	Zinc finger RNA binding protein, translational repression	GFM and germinal granules	Mosquera *et al.,* 1993; Zhou and King, 1996a; Kloc *et al.,*1998; MacArthur *et al.,* 1999
Xcat3	RNA helicase	Unknown	Elinson *et al.,* 1993; Mosquera *et al.,* 1993
Xdazl	RNA binding, promotion of meiosis	Unknown	Houston *et al.,* 1998
B6, B7, C10	Unknown	Unknown	King, 1995
Xpat	Unknown	Unknown	Hudson and Woodland, 1998
Xlotax1	Homeobox gene	Unknown	Pannese *et al.,* 2000; Simeone *et al.,* 1993
Xlsirts	Noncoding RNA, Vg1 anchoring	Fibrillar network between mitochondria	Kloc *et al.,* 1993a; Kloc and Etkin, 1994; Kloc *et al.,* 1998
Xwnt11	Secreted signaling protein	Fibrillar network between mitochondria	Ku and Melton, 1993; Kloc *et al.,* 1998

Pannese *et al.,* 2000). Light and electron microscopy *in situ* hybridization with Xcat2, Xwnt11, Xlsirts probes showed that in oocytes at the nest stage of development only Xlsirt is present in the mitochondrial mass; it is localized on the periphery of the mitochondrial cement between mitochondria. In late pachytene and early diplotene oocytes Xcat2, Xwnt11 and Xlsirts RNAs are more or less uniformly distributed between mitochondrial aggregates surrounding the oocyte nucleus (Kloc *et al.,* 1996, 1998). In prestage 1 oocytes Xcat2, Xwnt11, and Xlsirts RNAs remain only in the premitochondrial cloud, and they are a component of germ plasm. Xcat2 is present on GFM and on the periphery of forming germinal granules, and Xlsirts and Xwnt11 RNAs are limited to the fibrillar intermitochondrial matrix (Kloc *et al.,* 1998, 2000). In stage 1 oocytes these RNAs are localized at the tip of the mitochondrial cloud in the area called the METRO (message transport organizer). Electron microscopy *in situ* hybridization showed that the METRO region contains germ plasm and that Xcat2 RNA is present in the germinal granules. However, Xwnt11 and Xlsirts RNAs are located on the fibrillar matrix between mitochondria (Kloc *et al.,* 1998, 2000). In late stage 1 and early stage 2 oocytes, when the mitochondrial cloud starts fragmenting and moving toward the vegetal pole, Xcat2, Xwnt11, Xlsirts RNAs are localized in a disc at the apex of the vegetal pole. Individual RNAs are sorted inside the disc of the METRO; Xcat2 is on the disc's periphery, Xwnt11 in its center, and Xlsirts in the entire disc (Kloc and Etkin, 1995, 2000). From stage 2 to stage 4 of oogenesis, all these RNAs become localized together with the migrating islands of the mitochondrial cloud at the apex of the vegetal pole. Electron microscopy *in situ* hybridization showed that at stages 2 to 4, Xcat2 RNA is still present in germinal granules, which are now localized in the vegetal cortex of the oocyte. Xwnt11 and Xlsirt also remain associated on the fibrillar matrix between mitochondria (Kloc *et al.,* 1998). Light microscopy *in situ* hybridization showed that Xdazl and Xotx1 RNAs localize in the pattern very similar to Xcat2 RNA (Houston *et al.,* 1998; Pannese *et al.,* 2000). In unfertilized and fertilized eggs Xcat2, Xlsirts, Xwnt11, Xdazl, and Xpat RNAs are present in the patches of germ plasm localized in the vegetal cortex. During cleavages these RNAs segregate together with the germ plasm to the vegetal blastomeres of the embryo (Hudson and Woodland, 1998; Houston *et al.,* 1998; Kloc *et al.,* 1998). However, it would be inaccurate to conclude that all RNAs localized through the early pathway are involved exclusively in the specification of germ cell lineage, since Xwnt11 is known to be involved in axial patterning (Ku and Melton, 1993).

2. Late or Vg-like Pathway

This pathway of vegetal localization was first described for Vg1 RNA (Rebagliati *et al.,* 1985; Melton, 1987) and subsequently for Xcat 4, B9,

B12 and VegT/XBrat/Xombi/Antipodean (King, 1995; Lustig *et al.*, 1996; Stennard *et al.*, 1996; Zhang and King, 1996; Zhang *et al.*, 1998; Horb and Thomsen, 1997) (Table III). These RNAs are uniformly distributed in stage 1 oocytes and are excluded from the mitochondrial cloud. In late stage 2 oocytes Vg1 RNA accumulates in a conspicuous, crescent-like structure located between the nucleus and the vegetal pole. Immunostaining and electron microscopy analysis show that the crescent contains spectrin and tubulin and consists of a concentration of mitochondria and vesicles of endoplasmic reticulum (Palecek and Ubbels, 1997; Chan *et al.*, 1999; M. Kloc, A. P. Chan, and L. D. Etkin, unpublished). During stage 2 and stage 3 of oogenesis Vg1 RNA accumulates in the wedge-shaped structure that forms between the crescent and the vegetal cortex. The formation of the wedge overlaps temporally and spatially with the migration of the islands of the mitochondrial cloud and the formation of the disc of the METRO pathway (Kloc and Etkin, 1994, 1995, 1998; Forristall *et al.*, 1995; Yisraeli and Melton, 1988). Finally, from stage 4 of oogenesis onward, Vg1 RNA translocates and anchors at the vegetal cortex of the oocyte. First the localization of Vg1 RNA is limited to the apex of the vegetal pole. Eventually it spreads toward the equator and becomes localized to the whole vegetal cortex of the oocyte. It seems that other RNAs from this group behave either identically or very similarly to Vg1 RNA (Melton, 1987; Yisraeli *et al.*, 1990; Elinson *et al.*, 1993; Pondel and King, 1988). Light and electron microscopy *in situ* hybridization showed that RNAs that localize by the late pathway are excluded from the germ plasm, not only in the oocytes but also in the embryos (Kloc *et al.*, 1998, 2000). During oocyte maturation Vg1 is released from the cortex but remains enriched in the vegetal cytoplasm. Recent studies showed that the proper distribution of Vg1 protein is important in the establishment of dorsal identity and the right-left symmetry of the embryo (Thomsen and Melton, 1993; Hyatt *et al.*, 1996).

It is very interesting that injected RNAs, which normally localize through the METRO pathway when their endogenous counterparts have already completed their translocation, can use the late pathway to localize to the

TABLE III
RNAs Localized to the Vegetal Cortex Through the Late Pathway

Name	Function	References
B9, B12	Unknown	King, 1995
Vg1	TGFβ	Rebagliati *et al.*, 1985
VegT	Transcription factor	Zhang and King, 1996; Zhang *et al.*, 1998; Stennard *et al.*, 1996; Lustig *et al.*, 1996; Horb and Thomsen, 1997
Xcat4	Unknown	King, 1995

vegetal cortex (Kloc *et al.,* 1993a; Zhou and King, 1996b; Kloc *et al.,* 1996). This indicates the presence of cis- and trans-acting factors common for both early and late pathways. In fact, the recently identified 25 nt element of fatvg (Chan *et al.,* 1999) is a candidate for such a late element, since it directs the localization of fatvg through the late pathway and similar sequences are also found in many of RNAs of the early pathway.

3. Intermediate Pathway

This pathway is used by several different RNAs, including fatvg (Chan *et al.,* 1999) (Table IV). In stage 1 oocytes these RNAs are uniformly distributed throughout the cytoplasm and are excluded from the mitochondrial cloud, which is typical for RNAs using the late pathway. However, in stage 2 oocytes fatvg RNA is present in the mitochondrial cloud in the pattern reminiscent of METRO pathway localized RNAs. Electron microscopy *in situ* hybridization showed that fatvg RNA is localized in the germ plasm between mitochondria but is excluded from the germinal granules (A. Chan, M. Kloc, and L. Etkin, unpublished). After the mitochondrial cloud reached the vegetal cortex this RNA was first present in the crescent and subsequently in the wedge—typical of the late pathway-localizing RNAs (Chan *et al.,* 1999). However, contrary to the behavior of late pathway RNAs, fatvg RNA is present in the germ plasm in oocytes and embryos (Chan *et al.,* 1999).

IV. Role of Cytoskeleton and ER in RNA Localization

We do not know how RNAs of the METRO pathway enter the premitochondrial aggregates and the mitochondrial cloud. Plausible mechanisms include the direct and targeted transport of RNAs from the nucleus or from the cytoplasm and/or the selective entrapment of RNAs in these structures. We also do not know the mechanism responsible for the move-

TABLE IV
RNAs Localized to the Vegetal Cortex Through the Intermediary Pathway

Name	Function	References
fatvg	ADRP-related	Chan *et al.,* 1999
XRMC/hermes	RNA binding protein	Gerber *et al.,* 1999; N. R. Zearfoss and L. D. Etkin, unpublished

ment of the mitochondrial cloud and cloud fragments to the vegetal pole of the oocyte. The results of experiments with microtubule-depolymerizing drugs (nocodazole and colchicine) and microfilament-disrupting agents (cytocholasin B) showed that the transport of RNAs from the nucleus to the mitochondrial cloud, as well as the association of these RNAs with the cloud, does not depend on the presence of intact microtubules or microfilaments (Kloc *et al.*, 1996). However, we do not know the mechanism responsible for the movement of mitochondrial cloud and its fragments to the vegetal cortex. It is possible that this movement is simply a passive displacement caused by the vegetally directed streaming of oocyte cytoplasm. Since the presence of vitellogenin in culture medium is necessary for the localization of injected RNAs, it is possible that a driving force of localization may be the asymmetric accumulation of yolk platelets that occurs from an animal to vegetal direction. The answer to this question will await the technical development of an *in vitro* system that can recapitulate the movement of the cloud. However, anchoring of METRO pathway-localizing RNAs in the disc and in the vegetal cortex depends on the presence of intact microfilaments, and it becomes disrupted in the presence of cytocholasin B (Kloc and Etkin, 1995).

However, localization of RNAs by the late pathway is microtubule dependent, and anchoring of Vg1 RNA at the vegetal cortex depends on the presence of intact microfilaments (Yisraeli *et al.,* 1990) and Xlsirts RNA (Kloc and Etkin, 1994). There is no information available yet concerning the involvement of cytoskeleton in the localization and anchoring of the RNAs of the intermediate pathway.

Confocal microscopy studies of the organization of the cytoskeleton during *Xenopus* oogenesis showed that stage 1 oocytes contain a poorly ordered microtubule array (Gard, 1991). Starting from stage 3 of oogenesis, the microtubule array becomes more complex and polarized. In stage 6 oocytes, microtubules radiate from the germinal vesicle. It was suggested that the germinal vesicle acts as a microtubule organizing center (MTOC), and that oocyte microtubules had their plus ends directed toward the oocyte cortex and the minus ends directed towards the germinal vesicle (Gard, 1991). However, the latest electron microscope observation of hook-decorated microtubules showed that from stage 3 of oogenesis onward, the majority of microtubules have their minus ends directed toward the oocyte cortex (Pfeiffer and Gard, 1999). These observations do not give any clues as to why certain molecules such as the late pathway localized RNAs are transported preferentially to the vegetal cortex of the oocyte. One possible explanation is that the vegetal hemisphere of the oocyte contains a subpopulation of specialized microtubules and/or specialized molecular motors that are able to direct selected molecules preferentially toward the vegetal pole. We postulated that the γ tubulin-positive centrosome found in the mitochondrial cloud of stage 1 oocytes could act as a MTOC that establishes distinct microtubule tracks for the localization of Vg1 and other late pathway RNAs (Kloc and Etkin, 1998) (Figure 3).

Another question concerns how the RNA molecules move on the microtubules. Studies on the localization of RNA in somatic cells have shown that the RNA-multiprotein particles attach directly to and translocate along microtubules (Knowles *et al.*, 1996; Ainger *et al.*, 1993). It is interesting that in insect cells the granules of nuage material show in electron microscope a tight association with the microtubule tracks (Bilinski and Jaglarz, 1999). It seems that in *Xenopus* oocytes, the movement of the late pathway localized RNAs is assisted by a specialized subpopulation of endoplasmic reticulum vesicles (Deshler *et al.,* 1997, Etkin, 1997; Kloc and Etkin, 1998). The combination of *in situ* hybridization and immunostaining showed that in late stage 1 oocytes, while Vg1 RNA is still uniformly dispersed in the cytoplasm, a portion of endoplasmic reticulum (ER) forms a cap on the top of the migrating mitochondrial cloud. In late stage 2 and early stage 3

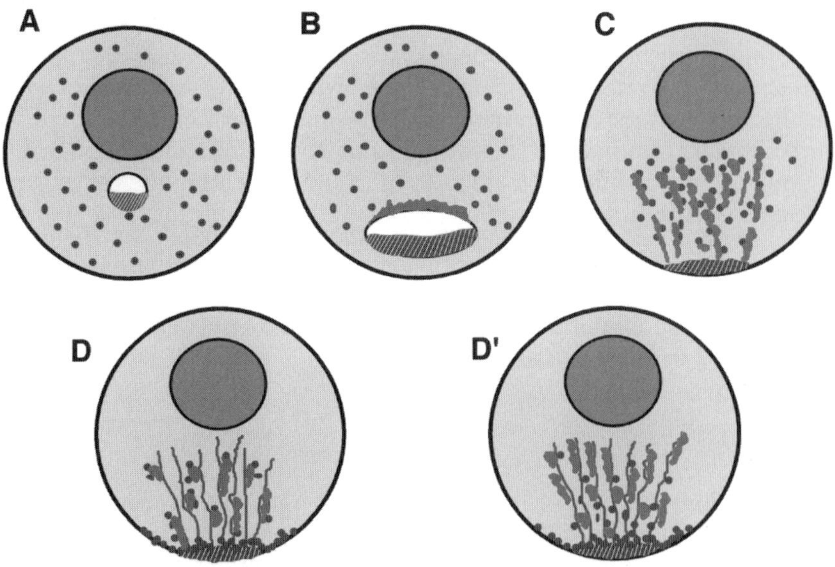

FIG. 3 Model for establishment of the late pathway in stage 2–3 oocytes, showing interaction between the METRO and late pathways during oogenesis. (A) Vg1 (black dots) is distributed throughout the cytoplasm while other RNAs are within the METRO region of the mitochondrial cloud (hatched). (B) By stage 2 there is a cap of endoplasmic reticulum (ER) associated with the cloud (gray dots). (C) The ER forms a wedge-shaped structure on which the vg1 mRNA is associated. (D) The Vg1-containing vesicles move along microtubules translocating to the vegetal cortex, or (D′) the ER serves as a matrix for the microtubules on which the Vg1 translocates to the vegetal cortex.

oocytes Vg1 RNA is tightly associated first with a crescent-shaped region of ER, and subsequently with a wedge-shaped subdomain of ER (Figure 3). In late stage oocytes Vg1 RNA and ER are present in vesicle-like structures near the vegetal cortex (Kloc and Etkin, 1998). Recent studies demonstrated that in fractionated *Xenopus* oocytes, the ER-containing fraction is also enriched in the Vera/Vg1 RBP protein that binds to the Vg1 3′-UTR. It is possible that this protein could play a role in the association of Vg1 RNA with the crescent and the wedge-shaped subdomains of oocyte ER and the microtubles (Deshler *et al.,* 1997; Kloc and Etkin, 1998). A close association between the ER and microtubules have been observed in many other systems (Lee and Chen, 1988; Allan and Vale, 1994; Dabora and Sheetz, 1988). Recently, it has been shown that ER vesicles are able to use plus and minus end-directed motors to move on the microtubule tracks in *Xenopus* eggs, embryos, and XTC cells (Lane and Allan, 1999). On the basis of above information we proposed the following model of localization of late pathway RNAs. We postulated that there is a continuity between the early and late localization pathways in which the mitochondrial cloud serves as the focal point for the organization of the unique subdomain of ER that piggybacks on the migrating mitochondrial cloud. At the same time, the centrosome of mitochondrial cloud organizes the tracks of microtubules in the vegetal hemisphere of the oocyte. Vg1 RNA associates with the ER through the vera protein and ER vesicles are moved along the microtubules. Alternatively, the wedged-shaped structure of the ER serves as a matrix for the formation of tracks of microtubules (Figure 3; Kloc and Etkin, 1998; Etkin, 1997).

V. The Role of Cis-acting Elements and Trans-acting Factors in RNA Localization

A. Cis-acting Elements

For the majority of localized mRNAs studied to date, the sequences that are necessary and usually sufficient for localization map to the 3′-untranslated region (3′-UTR, for review see Bashirullah *et al.,* 1998; Yisraeli and Yanov, Chapter 13). In *Xenopus,* the localization elements of four vegetally localized mRNAs, Vg1, fatvg, and Xcat2 (Mowry and Melton, 1992; Deshler *et al.,* 1997; Havin *et al.,* 1998; Gautreau *et al.,* 1977; Chan *et al.,* 1999; Zhou and King, 1996a,b), and the noncoding Xlsirts have been characterized (Kloc *et al.,* 1993a). The vegetal localization of Vg1 is mediated by a 340 nt element (VgLE) in the 3′-UTR. The VgLE contains several redundant elements such as two copies of E1, four copies of E2, one E2-like element, two copies of E3, three copies of E4, and VM1/hexanucleotide elements. The E2 element (UUCAC) is critical for the proper vegetal localization of Vg1, since the deletion of all E2 elements abolished the ability of injected Vg1 to localize. However, the other elements also contribute to Vg1 localization in a quantitative and redundant fashion (Deshler *et al.,* 1997).

The fatvg 3′ UTR (356 nt), which is responsible for the localization of fatvg mRNA to the vegetal cortex, also consists of several reiterated localization elements. Sequences between 1 and 102 and 188 and 356 nt are necessary for efficient localization to the vegetal cortex. The first 102 nt of fatvg 3′ UTR contains a short 25-nt localization element called FVLE1, which, as a single copy, can direct vegetal cortical localization. However, this element is insufficient to target fatvg RNA to the mitochondrial cloud (Chan *et al.,* 1999) and therefore should be considered a late localization element. This 25nt FVLE1 is different from any Vg1 localization elements reported. In fact, it is interesting that similar sequences are also found in a variety of RNAs that localize via the early pathway such as Xlsirts and Xcat2. The 3′ UTR of fatvg also contains three E2 elements, but none of them are necessary for proper localization (Chan *et al.,* 1999).

Analysis of localization elements in the 3′ UTR of Xcat2 RNA, an RNA that utilizes the METRO pathway of localization, showed that it contains two presumably independent localization signals; one for localization to the mitochondrial cloud and a second signal for localization to the vegetal cortex. The 3′ UTR of Xcat2 consists of 392 nucleotides. The first 227 nucleotides of the 3′ UTR are necessary for localization of Xcat2 into the mitochondrial cloud. The second bipartite signal consists of 150 nt immediately adjacent to the ORF, as well as an additional sequence of 120 nt at the 3′ end of 3′ UTR, both of which are necessary and sufficient for vegetal cortical localization (Zhou and King, 1966 a,b).

The injection of mutated labeled Xcat2 mRNA into oocytes combined with electron microscopy analysis gave new insights into the role of the 3′UTR with regard to the targeting of localized RNA to subcellular structures. An Xcat2 mutant mRNA that had a 164 nt deletion in the 3′ end of 3′ UTR but had an intact MCLE (mitochondrial cloud localization element) that consisted of proximal 227 nucleotides of 3′ UTR clearly targeted the cloud but did not associate with the germinal granules. Therefore, we identified the distal 164 nt of the 3′ UTR as the GGLE (germinal granules localization element). The GGLE was also sufficient to target a heterologous RNA that contained its own MCLE to the germinal granules (Kloc *et al.,* 2000). This is the first example of a localization element involved in targeting an mRNA to the germinal granules and demonstrates the importance of cis-acting RNA elements in the assembly of germinal granules.

One of the early pathway-localizing RNAs is Xlsirts. Xlsirts are a family of noncoding RNAs that contain a variable number of 79 to 85 nt tandem repeats. We have shown that two tandem repeats are necessary for the mitochondrial cloud and vegetal cortical localization of Xlsirts RNA (Kloc *et al.,* 1993a). More detailed analysis of this sequence indicates that it contains three known localization subelements identical to Vg1 localization elements: two VM1 and one E4 element, as well as one element with high homology to the fatvg localization element FVLE1 (Chan *et al.,* 1999). The significance of this finding is still under study.

B. Trans-acting Factors

So far there is no information available on the identity of any trans-acting factors involved in the localization of RNAs by the early pathway. However, three laboratories working on the late pathway utilizing RNA Vg1 were able to identify several RNA-binding proteins that form a complex with Vg1 localization elements and recognize the cis-acting elements within the Vg1 3′ UTR. *In vitro* cross-linking experiments and mobility shift assays identified a set of six proteins (33, 36, 40, 69, and 78 kDa) that bind the VgLE (Mowry, 1996). One of these proteins was later identified independently by two laboratories as both the 75 kDa Vera (*V*g1-*ER*-*A*ssociated) protein (Deshler *et al.,* 1997) and the 69 kDa Vg1RBP, which is believed to be a microtubule-associated protein (Elisha *et al.,* 1995; Schwartz *et al.,* 1992). These two proteins are identical to each other as well as to the chicken zipcode-binding protein 1 (ZBP-1; Ross *et al.,* 1997). In chick fibroblasts the ZBP-1 binds to the 54 nt zipcode of β-actin RNA, which targets the RNA to the leading edge of fibroblasts (Kislauskis *et al.,* 1994). This homology points to the common aspects of RNA localization pathways in different cell types and different organisms.

Our understanding of mechanisms responsible for the localization of different RNAs is still in its infancy. Although we do not know the exact role of different RNAs localized in *Xenopus* oocyte, it is plausible that at least those that are components of germ plasm play a role as cytoplasmic determinants responsible for the specification of the germ cell fate. In the next section we will discuss the specification of the germ line and germ cell development during *Xenopus* embryogenesis.

VI. Specification of Germ Line

The nature of the mechanisms underlying the differentiation of germ line versus soma in developing animal embryos has occupied biologists for over a century. In the 1880s Wilhelm Roux and August Weismann were working on the theory of germ cell determination. In 1885 and 1892 Weismann refined the germ plasm theory. He postulated the existence of the "uninterrupted germ line" throughout the entire animal life cycle as well as the presence of nuclear determinants that produce cytoplasmic substances that dictate the cell's destiny (Weismann 1885, 1892). The concept of nuclear and cytoplasmic determinants agreed very well with the earlier observations of von Baer (1837) and subsequently others (Whitman, 1878; Boveri, 1892; Driesch and Morgan, 1895; Conklin, 1897, 1905; Wilson, 1903) on animal eggs of different animal phyla that very often showed striking polarity and stratification of cytoplasmic components. After fertilization, different regions of egg cytoplasm were differentially distributed among forming blastomeres whose descendants gave rise to specific cell lineages. One of the most striking examples of a structure characteristic of the germ cell lineage is the "germ plasm" present in oocytes and embryos of Anuran amphibians. In *Xenopus,* early previtellogenic oocytes contain germ plasm that is part of the mitochondrial cloud (Balbiani body). During vitellogenesis the mitochondrial cloud disperses and fragments of the cloud containing germ plasm migrate toward the vegetal pole of the oocyte, where they become anchored at the vegetal cortex. After fertilization, during consecutive cleavages, the germ plasm segregates to vegetal blastomeres. Later in development the descendants of these blastomeres become the primordial germ cells of the embryo (Balbiani, 1864; Billett and Adam, 1976; Guraya, 1979; Hausen and Riebesell, 1991; Saffman and Lasko, 1999).

A. What Is the "Germ Plasm"?

The germinal plasm in amphibians was first described by Bounoure in 1939. In the vegetal cortical region of eggs of *Rana temporaria* he discovered small patches of specialized cytoplasm, which he collectively called the "germinal plasm." He followed the fate of the germinal plasm during embryogenesis and showed that during the first two cleavages it stayed in the vegetal region and was uniformly distributed between two, and later four, blastomeres. During subsequent cleavages the germinal plasm underwent differential segregation in that it was distributed into only one of the two daughter cells. This resulted in the formation of a limited number of germinal plasm-containing vegetal blastomeres. In late blastula and gastrula stages germinal plasm-containing cells were displaced toward the endodermal mass of the embryo. The germinal plasm in these cells occupied a juxtanuclear position. Further analysis of the fate of these cells showed that they eventually became primordial germ cells (PGCs; Eddy, 1975; Whitington and Dixon, 1975; Nieuwkoop and Sutasurya, 1979).

Experimental evidence showing that only blastomeres that contain a sufficient amount of germ plasm will develop into germ cells indicates a crucial role of germ plasm in the development of the germ line in *Xenopus* and other Anuran amphibians (Nieuwkoop and Sutasurya, 1979; Buehr and Blackler, 1970; Tanabe and Kotani, 1974). Mahowald and Hennen (1971) showed that the patches of germinal plasm present in the unfertilized egg consisted of components nearly identical to the germinal plasm in vegetal blastomeres and PGCs of the frog embryo.

Analysis at the ultrastructural level showed that germinal plasm had two major components: mitochondria and membraneless electron-dense bodies. These electron-dense bodies were described as having a fibrillar, granular, or fibrillo-granular appearance and have been collectively termed "germinal granules" by some authors and "nuage" by others (Al-Muhtar and Webb, 1971; Billett and Adam, 1976; Heasman *et al.,* 1984, Wylie *et al.,* 1985; Saffman and Lasko, 1999). However, the term "nuage" was usually reserved for electron-dense bodies that were found in close contact with the nuclear membrane (Coggins, 1973). However, electron-dense material that was found in close contact with mitochondrial membranes in nest stage oocytes and in spermatocytes was usually termed "mitochondrial cement" or "intermitochondrial cement" (Andre, 1962; Clerot, 1968). Heasman *et al.* (1984) described two categories of electron dense material in the germinal plasm of *Xenopus* oocytes: electron-dense material in the form of elongated and branching structures called granulo-fibrillar material (GFM), and round compact electron dense material called germinal granules. They suggested that mitochondrial cement was probably the precursor of GFM, which gives rise to germinal granules. Using Xcat2 RNA as a molecular marker and electron microscopy *in situ* hybridization, we were able to follow the relationship between structures within the germ plasm during oogenesis in *Xenopus* (Kloc *et al.,* 2000). Our data suggested that the germinal granules were indeed derived from the GFM. Recent studies on the role of Xcat2 suggest that Xcat2 RNA is translationally repressed during oogenesis (MacArthur *et al.,* 1999), and during embryogenesis Xcat2 protein may be involved in the translational regulation of other RNAs specific to the primordial germ cells. However, the roles of most of the individual components of germ plasm remain a complete mystery, although it was shown that Xdazl functions in germ cell migration (Houston and King, 2000).

There is a great deal of confusion in old and new literature concerning the terminology used to describe germ plasm and its components in amphibian oocytes and embryos. Terms such as germ(inal) plasm, nuage, electron-dense material, mitochondrial cement, granulo-fibrillar material, and germ(inal) granules have been, and are, very often used interchangeably to describe the same or different structures. On the basis of a vast body of literature and our own results (see below), we feel that the following terminology most accurately reflects the current status of structures within this region of the cytoplasm:

Germ(inal) plasm: the region of specialized cytoplasm containing mitochondria and mitochondrial cement, GFM, and/or germ(inal) granules. Germinal plasm contains spectrin and Xlsirts and Xwnt11 RNAs localized on a fibrillar matrix between mitochondria and Xcat2 RNA localized in GFM and germinal granules. In stage 1 oocytes the germ plasm occupies the vegetal tip of the mitochondrial cloud. In stage 2 oocytes the germ plasm is located in the migrating islands of the mitochondrial cloud. When the islands approach the vegetal cortex of the oocyte, the germ plasm becomes localized first as a discrete disc at the apex of the vegetal cortex (stage 2 and stage 3 oocytes) and subsequently as small patches in the vegetal cortex (stage 4 to stage 6 oocytes). After maturation the patches of germ plasm aggregate, and during cleavages they segregate into vegetal blastomeres and ultimately into primordial germ cells in the developing embryo.

Nuage: electron-dense material of unknown composition closely apposing the nuclear membrane. This material may be a precursor of the mitochondrial cement.

Mitochondrial cement: electron-dense material of unknown composition in close contact with mitochondrial membranes, probably a precursor of GFM.

Granulo-fibrillar material (GFM): branching or elongated electron-dense strands of granulo-fibrillar material containing Xcat2 RNA; a precursor of germinal granules.

Germ(inal) granules: electron-dense bodies of granulo-fibrillar or granular material containing Xcat2 and the large and small subunits of mitochondrial ribosomal RNA. Germinal granules change their appearance between stages 1 and 6 of oogenesis and then again during embryogenesis. In stage 1 oocytes germinal granules are round or roundish and very compact; between stage 2 and 6 of oogenesis they become irregular in shape and contain areas of low electron density. After fertilization germinal granules form irregularly shaped branching structures (M. Kloc and L. D. Etkin, unpublished). Between late neurula (stage 18) and early tail-bud (stage 25) they change into irregular string-like bodies, and finally into round granular bodies at the feeding tadpole stage (stage 46; Ikenishi and Kotani, 1975).

B. Germ Cell Development during Embryogenesis

In *Xenopus,* as in the majority of vertebrates, functional gametes develop from stem cells located in the gonads. The stem cells, also called gonocytes or primordial germ cells, are part of the reproductive cell line that is also called the germ line. The origin of primordial germ cells predates the development of the gonad and can be traced far back to the events that take place before and after fertilization of the egg. In immature but fully grown stage 6 oocytes the vegetal cortical region of the cytoplasm contains small patches of yolk-free cytoplasm with unique staining properties, which are collectively called the germ(inal) plasm (Czolowska, 1969, 1972; Nieuwkoop and Sutasurya, 1979; Saffman and Lasko, 1999). During oocyte maturation the germ plasm patches aggregate and fuse into large islands of germ plasm situated at the apex of the vegetal cortex of the egg. Recent studies (Robb *et al.,* 1996) showed that the process of aggregation and fusion of germ plasm depends on the protein Xklp1, which belongs to the family of ATP dependent kinesin-like motor proteins. After fertilization, during the first and several consecutive cleavages, the germ plasm segregates to, and stays in, the vegetal blastomeres. At the beginning of gastrulation the blastomeres, which contain germ plasm and can now be called primordial germ cells (PGCs), migrate upward in the direction of the floor of blastocoel. This is the first visible sign of segregation of cells of the germ line from the soma of the embryo. Some authors call these cells presumptive primordial germ cells (pPGCs) and they limit the term "PGCs" to the germ cells already settled in the gonad. In gastrula and neurula stages of development the PGCs become located deep inside the endoderm. At stage 10 of gastrula development (staging, according to Nieuwkoop and Faber, 1994) the PGCs are usually embedded in the lower part of the endoderm between the floor of blastocoel and the vegetal pole. Between late neural fold stage and tailbud (stages 17 to 28) all PGCs usually become aggregated into one or two clusters of cells located in the endoderm below the cavity of archenteron. At this stage each PGC contains a single island of germ plasm in close proximity to the nucleus. In tailbud embryos (between stages 28 and 31) PGCs initiate migration toward the lateral part of the endodermal mass and toward the splanchnopleures (the splanchnopleure is the visceral layer of lateral plate mesoderm). It seems plausible that this is an active process of migration, since there is no extensive endodermal or mesodermal movement at this stage of development. Also, the roundish or ameboidal shape of PGCs and the large intercellular spaces around them suggest active migration (Kamimura *et al.,* 1976). Between stages 31 and 36 of development (hatching stage) almost all PGCs gather in the dorsal region of the endoderm around the archenteron and close to the splanchnopleures. In stage 41 tadpoles the PGCs accumulate between the two approaching splanchnopleures that form dorsal mesentery at stage 42. In stage 43 tadpoles PGCs occupy dorsal mesentery, where they form an unpaired, median genital ridge. From stage 44 tadpoles to stage 46 tadpoles the PGCs start to migrate laterally from the unpaired genital ridge toward the coelomic wall along both sides of dorsal mesentery to the sites of the presumptive paired gonadal ridges. The PGCs are easily distinguishable from somatic cells by their large size, huge lobulated nucleus, the presence of yolk platelets, and the presence of cytoplasmic electron-dense bodies, which are often attached to mitochondria (Eddy and Ito, 1971; Wylie and Heasman, 1976). Observations by electron microscopy suggest that these bodies represent remnants of the germ plasm observed in the PGCs of earlier developmental stages. Recent studies showed the presence of Xcat2 protein in the germ plasm of PGC during blastula and gastrula stages of development (MacArthur *et al.,* 1999). Kamimura *et al.* (1976) counted the number of PGCs in *Xenopus* embryos between stages 12 and 46 of development and found that during this period the number of PGCs increases about four-fold (from 15 to 60 on average), thus suggesting that each initial PGC in the gastrula probably divided twice. The lateral migration of PGCs from the median genital ridge toward the coelomic walls is completed at stage 46, and in stage 48 tadpoles paired genital ridges form folds that visibly protrude into the coelomic cavity. The presence of cytoplasmic processes on the surface of migrating PGCs suggests that their migration from the endoderm to the gonadal anlagen is also, at least partially, an active process (Wylie and Heasman, 1976). Recent experiments with antisense oligonucleotide depletion of Xdazl RNA indicated that its gene product is necessary for formation and migration of PGCs from the endoderm (Houston and King, 2000).

In stage 49 tadpoles the somatic cells from the mesonephric region of the embryo enter the genital ridges and form the medullary tissue of the future gonad. The PGCs are at first located exclusively in the "cortical region" of the gonadal rudiment. When the formation of the gonad has been completed the anterior part of the former genital ridge devoid of germ cells will form the fat body (Nieuwkoop and Faber, 1994). There are usually between 25 and 30 PGCs in a single gonadal anlage. Until stage 52 of development (three weeks of age) the PGCs remain quiescent, and young gonads remain sexually undifferentiated. At this stage of development PGCs can differentiate into female or male gametes, depending on the sexual identity of the surrounding somatic tissue. One can experimentally transplant PGCs from a female to a male and they will form male gametes containing a female set of chromosomes; PGCs transplanted from the male to the female will form female gametes with a male set of chromosomes. At stage 52 when sexual differentiation begins, the male PGCs will leave the cortical region of the gonad and enter the medullary tissue. From this moment they will resume mitotic divisions and will be called primary spermatogonia and the gonad will become testis. For further description of testes development and spermatogenesis, see Reed and Stanley (1972), Kalt (1973), and Nieuwkoop and Faber (1994). In the female, the PGCs will stay exclusively in the cortical region of the gonad, where they will resume mitoses and will be called primary oogonia and the gonad will become the ovary. Primary oogonia can probably multiply for some time by asynchronous divisions before entering the phase of synchronous and incomplete mitoses, which result in the formation of "secondary oogonia." It is unclear how many times and how long the primary oogonia can multiply by asynchronous divisions. Some textbooks indicate that there is a phase of slow and asynchronous division of primary oogonia between three and four weeks of *Xenopus* development, but we have never found any original studies that would support or repudiate this statement (Hausen and Riebesell, 1991). However, it is well documented that by four weeks of development (stages 54 to 55) the *Xenopus* ovary contains pairs of secondary oogonia connected by intercellular bridges (Coggins, 1973; Hausen and Riebesell, 1991). Individual primary oogonium undergoes four synchronous and incomplete mitoses, which result in the formation of 16 pear-shaped sister oogonial cells connected at their bases by short intercellular bridges. Individual groups of 16 interconnected oogonia are called the cysts or nests (Hausen and Riebesell, 1991; Pepling *et al.,* 1999). At stage 55 of tadpole development, 16 cells of the nest enter, in precise synchrony, the prophase of first meiotic division. At this moment they will become the oocytes. The synchrony persists throughout leptotene, zygotene, and early pachytene stages of first meiotic prophase. In late pachytene the oocytes in individual nests become separated by invading follicular cells; intercellular bridges between the oocytes break, and around stage 57 of tadpole development young oocytes, surrounded by follicular cells, will asynchronously enter the diplotene phase of first meiotic prophase. Oocytes arrest in diplotene and enter their major growth period, which, depending on the hormonal condition of the frog, can last for several years. After natural or *in vitro* stimulation by progesterone, fully grown stage 6 oocytes will resume meiosis in the process of oocyte maturation. During maturation the oocyte will complete the first meiotic division and will arrest in metaphase of the second meiotic division, awaiting fertilization (Hausen and Riebesell, 1991; Nieuwkoop and Faber, 1994). It seems that at this point of oocyte development all vegetally localized RNAs are already at their final destination at the vegetal cortex, awaiting to perform their functions in germ cell specification or specification of the body axes.

VII. Conclusions and Future Directions

The discovery of new localized RNAs suggests that the phenomenon and a need for RNA localization during *Xenopus* oogenesis is a much more widespread phenomenon than previously thought. A most important challenge in the future will be to define the functions of all known localized RNAs. The latest developments in transgenic *Xenopus* technologies, as well as antisense oligo nucleotide injections or double stranded RNA interference of gene function, are very promising techniques through which we can begin to analyze function. Another important challenge is to unravel the complex machinery and the cis- and trans-acting factors involved in the localization of RNAs. At present, besides Xlklp1, which participates in germ plasm aggregation, no single molecular motor has been identified that is responsible for the translocation of RNAs to the animal or vegetal pole. Although a number of different cis-acting elements have been identified, a limited number of factors have been found that interact with these signals. The most characterized factor is the zipcode-binding protein that is believed to be involved in RNA transport in both oocytes and somatic cells. It is obvious that this protein alone cannot be responsible for the specificity of the movement and targeting of the RNA to the vegetal cortex and to specific destinations within the oocyte germ plasm. The recent discovery of the GGLE signal targeting Xcat2 RNA to germinal granules indicates that the cis-acting elements can be responsible for very precise targeting of RNAs, not only to the particular region of the oocyte cortex, but also to specific subcellular structures. The discoveries of new localized RNAs and the growing number of laboratories involved in the studies on localized RNAs promise rapid progress and the answers at least to some of these questions in the very near future.

Acknowledgments

The work from the Etkin laboratory was supported by grants from the National Science Foundation and NIH (L. D. E.). R. Z. and L. A. were supported by a NIH training grant.

References

Ainger, K., Avossa, D., Morgan, F., Hill, S. J., Barry, C., Barbarese, E., and Carson, J. H. (1993). Transport and localization of exogenous myelin basic protein mRNA microinjected into oligodendrocytes. *J. Cell Biol.* **123,** 431–441.

Allan, V., and Vale, R. (1994). Movement of membrane tubules along microtubules *in vitro:* Evidence for specialized sites of motor attachment. *J. Cell Sci.* **107,** 1885–1897.

Al-Mukhtar, K. A. K., and Webb, A. C. (1971). An ultrastructural study of primordial germ cells, oogonia, and early oocytes in *Xenopus laevis. J. Embryol. Exp. Morphol.* **26,** 195–217.

Andre, L. (1962). Contribution à la connaissance du chondriome. Etude de ses modifications ultrastructurales pendant la spermatogenèse. *J. Ultr. Res. Suppl.* **3,** 1–185.

Balbiani, E. G. (1864). Sur la constitution du germe dans l'oeuf animal avant la fècondation. C.R. Hebd. Seances Acad Sci **58,** 584.

Bashirullah, A., Cooperstock, R. L., and Lipshitz, H. D. (1998). RNA localization in development. *Annu. Rev. Biochem.* **67,** 335–394.

Bilinski, S. M., and Jaglarz, M. K. (1999). Organization and possible functions of microtubule cytoskeleton in Hymenopteran nurse cells. *Cell Motil.* and *Cytoskel.* **43,** 213–220.

Billett, F. S., and Adam, E. (1976). The structure of the mitochondrial cloud of *Xenopus laevis* oocytes. *J. Embryol. Exp. Morphol.* **36,** 697–710.

Bounoure, L. (1939). L'origine des cellules reproductrices et le probleme de la lignee germinale. pp. 271 Gauthier-Villars, Paris.

Boveri, T. (1892). Ueber die Entstehung des Gegensatzes zwischen den Geschlechtszellen und den somotishen Zellen bei Ascaris megalocephala. *Sitzungsber. Ges. Morph. Physiol. München* **8,** 114–125.

Buehr, M. L., and Blackler, A. W. (1970). Sterility and partial sterility in the South African clawed toad following the pricking of the egg. *J. Embryol. Exp. Morphol.* **23,** 375–384.

Casey, E. S., Tada, M., Fairclough, L., Wylie, C. C., Heasman, J., and Smith, J. C. (1999). Bix4 is activated directly by VegT and mediates endoderm formation in *Xenopus* development. *Development* **126,** 4193–4200.

Chan, A. P., Kloc, M., and Etkin, L. D. (1999). Fatvg encodes a new localized RNA that uses a 25-nucleotide element (FVLE1) to localize to the vegetal cortex of *Xenopus* oocytes. *Development* **126,** 4943–4953.

Clements, D., Friday, R. V., and Woodland, H. R. (1999). Mode of action of VegT in mesoderm and endoderm formation. *Development* **126,** 4903–4911.

Clerot, J. C. (1968). Mise en evidence par cytochimie ultrastructurale de l'emission de proteines par le noyeau d'auxocytes de Batracines. *J. Microsc.* **7,** 973–992.

Coggins, L. W. (1973). An ultrastructural and radioautographic study of early oogenesis in the toad *Xenopus laevis. J. Cell Sci.* **12,** 71–93.

Conklin, E. G. (1897). The embryology of Crepidula. A contribution to the cell lineage and early development of some marine gasteropods. *J. Morphol.* **13,** 1–226.

Conklin, E. G. (1905). The organization and cell-lineage of the ascidian egg. *J. Acad. Natural Sci. Philadelphia* **13,** 1–119.

Czolowska, R. (1969). Observations on the origin of the "germinal cytoplasm" in *Xenopus laevis. J. Embryol. Exp. Morphol.* **22,** 229–251.

Czolowska, R. (1972). The fine structure of the "germinal plasm" in the egg of *Xenopus laevis. Wilhelm Roux's Arch.* **169,** 335–344.

Dabora, S., and Sheetz, M. (1988). The microtubule-dependent formation of tubulovesicular network with characteristics of ER from cultured cell extracts. *Cell* **54,** 27–35.

Deshler, J. O., Highett, M. I., and Schnapp, B. J. (1997). Localization of *Xenopus* Vg1 mRNA by Vera protein and the endoplasmic reticulum. *Science* **276,** 1128–1131.

Driesch, H., and Morgan, T. H. (1895). Zur Analyse der ersten Entwicklungsstadien des Ctenophoroneies. I. Von der Entwicklung einzelner Ctenophorenblastomeren. *Arch Entwicklungsmech Org.* **2,** 204–215.

Dumont, J. N. (1972). Oogenesis in *Xenopus laevis* (Daudin). I. Stages of oocyte development in laboratory maintained animals. *J. Morphol.* **136,** 153–179.

Eddy, E. M. (1975). Germ plasm and the differentiation of the germ cell line. *Int. Rev. Cytol.* **43,** 229–280.

Eddy, E. M., and Ito, S. (1971). Fine structural and radioautographic observations of dense perinuclear cytoplasmic material in tadpoles oocytes. *J. Cell Biol.* **49,** 90–108.

Elinson, R. P., King, M. L., and Forristall, C. (1993). Isolated vegetal cortex from *Xenopus* oocytes selectively retains localized mRNAs. *Dev. Biol.* **160,** 554–562.

Elisha, Z., Havi, L., Ringel, I., and Yisraeli, J. K. (1995). Vg1 RNA binding protein mediates the association of Vg1 RNA with microtubules in *Xenopus* oocytes. *EMBO J.* **14,** 5109–5144.

Etkin, L. D. (1997). A new face for the endoplasmic reticulum: RNA localization. *Science* **276,** 1092–1093.

Forristall, C., Pondel, M., and King, M. L. (1995). Patterns of localisation and cytoskeletal association of two vegetally localized RNAs, Vg1 and Xcat2. *Development* **121,** 201–208.

Gard, D. L. (1991). Organization, nucleation, and acetylation of microtubules in *Xenopus laevis* oocytes: A study by confocal immunofluorescence microscopy. *Dev. Biol.* **143,** 346–362.

Gavis, L. (1997). Expeditions to the pole: RNA localization in *Xenopus* and *Drosophila. Trends Cell Biol.* **7,** 485–492.

Gautreau, D., Cote, C. A., and Mowry, K. L. (1997). Two copies of a subelement from the Vg1 RNA localization sequence are sufficient to direct vegetal localization in *Xenopus* oocytes. *Development* **124,** 5013–5020.

Gerber, W. V., Tatiana, A., Yatskievych, P., Antin, K., Correia, C., Conlon, R., Krieg, P. (1999). The RNA-binding protein gene, hermes, is expressed at high levels in the developing heart. *Mech. Dev.* **80,** 77–86.

Gerhart, J., Danilchik, M., Roberts, J., Rowning, B., and Vincent, J. P. (1986). "Gametogenesis and the Early Embryo," pp. 305–319. Alan R. Liss, Inc., New York.

Guraya, S. S. (1979). Recent advances in the morphology, cytochemistry, and function of Balbiani's vitelline body in animal oocytes. *Int. Rev. Cytol.* **59,** 249–321.

Gururajan, R., Perry, O. K. H., Melton, D. A., and Weeks, D. L. (1991). The *Xenopus* localized messenger RNA An3 may encode an ATP-dependent RNA helicase. *Nature* **349,** 717–719.

Hausen, P., and Riebesell, M. (1991). "The Early Development of *Xenopus laevis.* An Atlas of the Histology." Springer-Verlag, New York.

Havin, L., Git, A., Elisha, Z., Oberman, F., Yaniv, K., Schwartz, S. P., Standart, N., and Yisraeli, J. K. (1998). RNA-binding protein conserved in both microtubule- and microfilament-based RNA localization. *Genes Dev.* **12,** 1593–1598.

Hazelrigg, T. (1998). The destinies and destinations of RNAs. *Cell* **95,** 451–460.

Heasman, J., Quarmby, J., and Wylie, C. C. (1984). The mitochondrial cloud of *Xenopus* oocytes: The source of germinal granule material. *Dev. Biol.* **105,** 458–469.

Hinkley, C. S., Martin, J. F., Leibham, D., and Perry, M. (1992). Sequential expression of multiple POU proteins during amphibian early development. *Mol. Cell. Biol.* **12,** 638–649.

Horb, M. E., and Thomsen, G. H. (1997). A vegetally localized T-box transcription factor in *Xenopus* eggs specifies mesoderm and endoderm and is essential for embryonic mesoderm formation. *Development* **124,** 1689–1698.

Houston, D. W., and King, M. L. (2000). A critical role for Xdazl, a germ plasm-localized RNA, in the differentiation of primordial germ cells in *Xenopus. Development* **127,** 447–456.

Houston, D. W., Zhang, J., Maines, J. Z., Wasserman, S. A., and King, M. L. (1998). A *Xenopus* DAZ-like gene encodes an RNA component of germ plasm and is a functional homologue of Drosophila boule. *Development* **125,** 171–180.

Hudson, C., and Woodland, H. R. (1998). Xpat, a gene expressed specifically in germ plasm and primordial germ cells of *Xenopus laevis. Mech. Dev.* **73,** 159–168.

Hudson, J. W., Alarcon, V. B., and Elinson, R. P. (1996). Identification of new localized RNAs in the *Xenopus* oocyte by differential display PCR. *Dev. Genet.* **19,** 190–198.

Hyatt, B., Lohr, J., and Yost, H. J. (1996). Initiation of vertebrate left-right axis formation by maternal Vg1. *Nature* **384,** 62–65.

Ikenishi, K., and Kotani, M. (1975). Ultrastructure of the "germinal plasm" in *Xenopus* embryos after cleavage. *Dev. Growth Differen.* **17,** 101–110.

Kalt, M. R. (1973). Ultrastructural observations on the germ line of *Xenopus laevis. Z. Zellforsch. Mikrosk. Anat.* **138,** 41–62.

Kalt, M. R., and Gall, J. G. (1974). Observations on early germ cell development and premeiotic ribosomal DNA amplification in *Xenopus laevis. J. Cell Biol.* **62,** 460–472.

Kamimura, M., Ikenishi, K., Kotani, M., and Matsuno, T. (1976). Observations on the migration and proliferation of gonocytes in *Xenopus laevis. J. Embryol. Exp. Morphol.* **36,** 197–207.

King, M. L. (1995). mRNA localization during frog oogenesis. In "Localized RNAs." (H. D. Lipshitz, Ed.) pp. 137–148. Landes Publishing, Austin, TX.

King, M. L., Zhou, Y., and Bubunenko, M. (1999). Polarizing genetic information in the egg: RNA localization in the frog oocyte. *BioEssays* **21,** 546–557.

Kislauskis, E. H., Zhu, X., and Singer, R. H. (1994). Sequences responsible for intracellular localization of beta-actin messenger RNA also affect cell phenotype. *J. Cell Biol.* **127,** 441–451.

Kloc, M., and Etkin, L. D. (1994). Delocalization of Vg1 mRNA from the vegetal cortex in *Xenopus* oocytes after destruction of Xlsirt RNA. *Science* **265,** 1101–1103.

Kloc, M., and Etkin, L. D. (1995). Two distinct pathways for the localization of RNAs at the vegetal cortex in *Xenopus* oocytes. *Development* **121,** 287–297.

Kloc, M., and Etkin, L. D. (1998). Apparent continuity between the messenger transport organizer and late RNA localization pathways during oogenesis in *Xenopus. Mech. Dev.* **73,** 95–106.

Kloc, M., Miller, M., Carrasco, A. E., Eastman, E., and Etkin, L. (1989). The maternal store of the xlgv7 mRNA in full-grown oocytes is not required for normal development in *Xenopus. Development* **107,** 899–907.

Kloc, M., Reddy, B. A., Miller, M., Eastman, E., and Etkin, L. D. (1991a). x121: A localized maternal transcript in *Xenopus laevis. Mol. Reprod. Dev.* **28,** 341-345.

Kloc, M., Reddy, B., Crawford, S., and Etkin, L. D. (1991b). A novel 110-kDa maternal CAAX box-containing protein from *Xenopus* is palmitoylated and isoprenylated when expressed in baculovirus. *J. Biol. Chem.* **266,** 8206–8212.

Kloc, M., Spohr, G., and Etkin, L. D. (1993a). Translocation of repetitive RNA sequences with the germ plasm in *Xenopus* oocytes. *Science* **262,** 1712–1714.

Kloc, M., Li, X. X., and Etkin, L. D. (1993b). Two upstream cysteines and the CAAX motif but not the polybasic domain are required for membrane association of Xlcaax in *Xenopus* oocytes. *Biochemistry* **32,** 8207–8212.

Kloc, M., Larabell, C., and Etkin, L. D. (1996). Elaboration of the messenger transport organizer pathway for localization of RNA to the vegetal cortex of *Xenopus* oocytes. *Dev. Biol.* **180,** 119–130.

Kloc, M., Larabell, C., Chan, P.-Y. A., and Etkin, L. D. (1998). Contribution of METRO pathway localized molecules to the organization of the germ cell lineage. *Mech. Dev.* **75,** 81–93.

Kloc, M., Bilinski, S., Chan, A. P.-Y., and Etkin, L. D. (2000). The targeting of Xcat2 mRNA to the germinal granules depends on a cis-acting germinal granule localization element within the 3′UTR. *Dev. Biol.* **217,** 221–229.

Klymkowsky, M. W., and Karnovsky, A. (1994). Morphogenesis and the cytoskeleton: Studies of the *Xenopus* embryo. *Dev. Biol.* **165,** 372–384.

Knowles, R. B., Sabry, J. H., Martone, M. E., Deerinck, T. J., Ellisman, M. H., Bassell, G. J., and Kosik, K. S. (1996). Translocation of RNA granules in living neurons. *J. Neurosci.* **16,** 7812–7820.

Kobayashi, S., Amikura, R., and Mukai, M. (1998). Localization of mitochondrial large ribosomal RNA in germ plasm of *Xenopus* embryos. *Curr. Biol.* **8,** 1117–1120.

Kobayashi, S., Amikura, R., and Okada, M. (1994). Localization of mitochondrial large rRNA in germinal granules and the consequent segregation of germ line. *Int. J. Dev. Biol.* **38,** 193–199.

Kofron, M., Demel, T., Xanthos, J., Lohr, J., Sun, B., Sive, H., Osada, S.-I., Wright, C., Wylie, C., and Heasman, J. (1999). Mesoderm induction in *Xenopus* is a zygotic event regulated by maternal VegT via TGF(beta) growth factors. *Development* **126,** 5759–5770.

Komiya, T., Itoh, K., Ikenishi, K., and Furusawa, M. (1994). Isolation and characterization of a novel gene of the DEAD box protein family which is specifically expressed in germ cells of *Xenopus laevis. Dev. Biol.* **162,** 354–363.

Ku, M., and Melton, D. A. (1993). Xwnt-11: A maternally expressed *Xenopus* wnt gene. *Development* **119,** 1161–1173.

Lane, J. D., and Allan, V. J. (1999). Microtubule-based endoplasmic reticulum motility in *Xenopus laevis:* Activation of membrane-associated kinesin during development. *Mol. Biol. Cell* **10,** 1909–1922.

Lee, C., and Chen, L. B. (1988). Dynamic behavior of endoplasmic reticulum in living cells. *Cell* **54,** 37–46.

Linnen, J. M., Bailey, C. P., and Weeks, D. L. (1993). Two related localized mRNAs from *Xenopus laevis* encode ubiquitin-like fusion proteins. *Gene* **128,** 181–188.

Lustig, K. D., Kroll, K. L., Sun, E. E., and Kirschner, M. W. (1996). Expression cloning of a *Xenopus* T-related gene (Xombi) involved in mesodermal patterning and blastopore lip formation. *Development* **122,** 4001–4012.

MacArthur, H., Bubunenko, M., Houston, D. W., and King, M. L. (1999). Xcat RNA is a translationally sequestered germ plasm component in *Xenopus. Mech. Dev.* **84,** 75–88.

Mahowald, A. P., and Hennen, S. (1971). Ultrastructure of the "germ plasm" in eggs and embryos of Rana pipiens. *Dev. Biol.* **24,** 37–53.

Melton, D. A. (1987). Translocation of a localized maternal mRNA to the vegetal pole of *Xenopus* oocytes. *Nature* **328,** 80–82.

Mosquera, L., Forristall, C., Zhou, Y., and King, M. L. (1993). A mRNA localized to the vegetal cortex of *Xenopus* oocytes encodes a protein with a nanos-like zinc finger domain. *Development* **117,** 377–386.

Mowry, K. L. (1996). Complex formation between stage-specific oocyte factors and a *Xenopus* mRNA localization element. *Proc. Natl. Acad. Sci. USA* **93,** 14608–14613.

Mowry, K. L., and Cote, C. A. (1999). RNA sorting in *Xenopus* oocytes and embryos. *FASEB J.* **13,** 435–445.

Mowry, K. L., and Melton, D. A. (1992). Vegetal messenger RNA localization directed by a 340-nt RNA sequence element in *Xenopus* oocytes. *Science* **255,** 991–994.

Nieuwkoop, P. D., and Faber, J. (1994). "Normal Table of *Xenopus laevis* (Daudin)." Garland Publishing, Inc., New York, London.

Nieuwkoop, P. D., and Sutasurya, L. A. (1979). "Primordial Germ Cells in the Chordates." Cambridge University Press, Cambridge.

Otte, A. P., and Moon, R. T. (1992). Protein kinase C isozymes have distinct roles in neural induction and competence in *Xenopus. Cell* **68,** 1021–1029.

Otte, A. P., McGrew, L. L., Olate, J., Nathanson, N. M., and Moon, R. T. (1992). Expression and potential functions of G-protein alpha subunits in embryos of *Xenopus laevis. Development* **116,** 141–146.

Palecek, J., and Ubbels, G. A. (1997). Dynamic changes in the tubulin cytoskeleton during oogenesis and early development in the anuran amphibian *Xenopus laevis* (Daudin). *Folia Hist. Cytobiol.* **35,** 3–18.

Pannese, M., Cagliani, R., Pardini, C. L., and Boncinelli, E. (2000). Xotx1 maternal transcripts are vegetally localized in *Xenopus laevis* oocytes. *Mech. Dev.* **90,** 111–114.

Pepling, M. E., de Cuevas, M., and Spradling, A. C. (1999). Germline cysts: A conserved phase of germ cell development. *Trends Cell Biol.* **9,** 257–262.

Pfeiffer, D. C., and Gard, D. L. (1999). Microtubules in *Xenopus* oocytes are oriented with their minus-ends toward the cortex. *Motil. Cytoskel.* **44,** 34–43.

Pondel, M. D., and King, M. L. (1988). Localized maternal mRNA related to transforming growth factor beta mRNA is concentrated in a cytokeratin-enriched fraction from *Xenopus* oocytes. *Proc. Natl. Acad. Sci. USA* **85,** 7612–7616.

Rebagliati, M. R., Weeks, D. L., Harvey, R. P., and Melton, D. A. (1985). Identification and cloning of localized maternal RNAs from *Xenopus* eggs. *Cell* **42,** 769–777.

Reddy, B. A., Kloc, M., and Etkin, L. D. (1992). The cloning and characterization of a localized maternal transcript in *Xenopus laevis* whose zygotic counterpart is detected in the CNS. *Mech. Dev.* **39,** 143–150.

Reed, S. C., and Stanley, H. P. (1972). Fine structure of spermatogenesis in the South African clawed toad *Xenopus laevis* Daudin. *J. Ultrastruct. Res.* **41,** 277–295.

Robb, D. L., Heasman, J., Raats, J., and Wylie, C. (1996). A kinesin-like protein is required for germ plasm aggregation in *Xenopus. Cell* **87,** 823–831.

Ross, A. F., Oleynikov, Y., Kislauskis. E. H., Taneja, K. L., and Singer, R. H. (1997). Characterization of a beta-actin mRNA zipcode-binding protein. *Mol. Cell Biol.* **17,** 2158–2165.

Saffman, E. E., and Lasko, P. (1999). Germline development in vertebrates and invertebrates. CMLS, *Cell. Mol. Life. Sci.* **55,** 1141–1163.

Schroeder, K. E., and Yost, H. J. (1996). *Xenopus* poly (A) binding protein maternal RNA is localized during oogenesis and associated with large complexes in blastula. *Dev. Genet.* **19,** 268–276.

Schwartz, S. P., Aisenthal, L., Elisha, Z., Oberman, F., and Yisraeli, J. K. (1992). A 69-kDa RNA-binding protein from *Xenopus* oocytes recognizes a common motif in two vegetally localized maternal mRNAs. *Proc. Natl. Acad. Sci. USA* **89,** 11895–11899.

Simeone, A., Acampora, D., Mallamaci, A., Stornaiuolo, A., D'Apice, M. R., Nigro, V., and Boncinelli, E. (1993). A vertebrate gene related to orthodenticle contains a homeodomain of the bicoid class and demarcates anterior neuroectoderm of gastrulatimng mouse embryo. *EMBO J.* **12,** 2735–2747.

St. Johnston, D. (1995). The intracellular localization of messenger RNAs. *Cell* **81,** 161–170.

Stennard, F., Carnac, G., and Gurdon, J. B. (1996). The *Xenopus* T-box gene, Antipodean, encodes a vegetally localised maternal mRNA and can trigger mesoderm formation. *Development* **122,** 4179–4188.

Tada, M., Casey, E. S., and Smith, J. C. (1998). Bix1, a direct target of *Xenopus* T-box genes, causes formation of ventral mesoderm and endoderm. *Development* **125,** 3997–4006.

Tanabe, K., and Kotani, M. (1974). Relationship between the amount of the "germinal plasm" and the number of primordial germ cells in *Xenopus laevis. J. Embryol. Exp. Morphol.* **31,** 89–98.

Thomsen, G. H., and Melton, D. (1993). Processed Vg1 protein is an axial mesoderm inducer in *Xenopus. Cell* **74,** 433–441.

Tourte, M., Mignotte, F., and Mounolou, J.-C. (1984). Heterogeneous distribution and replication activity of mitochondria in *Xenopus laevis* oocytes. *Eur. J. Cell. Biol.* **34,** 171–178.

Viel, A., Armand, M. J., Callen, J. C., Gomez De Gracia, A., Denis, H., and le Maire, M. (1990). Elongation factor 1 alpha (EF-1 alpha) is concentrated in the Balbiani body and accumulates coordinately with the ribosomes during oogenesis of *Xenopus laevis. Dev. Biol.* **141,** 270–278.

von Baer, K. E. (1837). "Uber Entwicklongsgesichte der Tiere: Beobachtung und reflexion." part 2. Konigsberg.

Watanabe, M., Itoh, K., Abe, K., Akizawa, T., Ikenishi, K., and Furusawa, M. (1992). Immunolocalization of DEAD family proteins in germ line cells of *Xenopus* embryos. *Dev. Growth Differen.* **34,** 223–231.

Weeks, D. L., and Melton, D. A. (1987a). A maternal mRNA localized to the animal pole of *Xenopus* eggs encodes a subunit of mitochondrial ATPase. *Proc. Natl. Acad. Sci. USA* **84,** 2798–2802.

Weeks, D. L., and Melton, D. A. (1987b). A maternal mRNA localized to the vegetal hemisphere in *Xenopus* eggs codes for a growth factor related to TGF-beta. *Cell* **51,** 861–867.

Weismann, A. (1885). "Die Continuitat des Keimplasmas als Grundlage einer Theorie der Vererbung," pp. 122. Ed. Fischer, Jena.

Weismann, A. (1892). "Das keimplasma. Eine Theorie der Verebung," pp. 628. Ed. Fisher, Jena.

Weismann, A. (1903). Die Entwickelung der Dipteren im Ei. *Zeitschr. wiss Zool.* **13,** 159–200.

Whitington, P. M., and Dixon, K. E. (1975). Quantitative studies of germ plasm and germ cells during early embryogenesis of *Xenopus laevis. J. Embryol. Exp. Morphol.* **33,** 57–74.

Whitman, C. O. (1878). The embryology of Clepsine. *Q. J. Microsc. Sci.* **18,** 215–315.

Wilson, E. B. (1903). Experiments on cleavage and localization in the nemertine egg. *Arch. Entwicklungsmech Org.* **16,** 411–460.

Wylie, C. C., and Heasman, J. (1976). The formation of the gonadal ridge in *Xenopus laevis.* I. A light and transmission electron microscope study. *J. Embryol. Exp. Morphol.* **35,** 125–138.

Wylie, C. C., Holwill, S., O'Driscoll, M., Snape, A., and Heasman, J. (1985). Germ plasm and germ cell determination in *Xenopus laevis* as studied by cell transplantation analysis. *Cold Spring Harbor Symp. Quant. Biol.* **50,** 37–43.

Yisraeli, J. K., and Melton, D. A. (1988). The maternal RNA Vg1 is correctly localized following injection into *Xenopus* oocytes. *Nature* **336,** 592–595.

Yisraeli, J. K., Sokol, S., and Melton, D. A. (1990). A two-step model for the localization of maternal mRNA in *Xenopus* oocytes: Involvement of microtubules and microfilaments in the translocation and anchoring of Vg1 mRNA. *Development* **108,** 289–298.

Zhang, J., Houston, D. W., King, M. L., Payne, C., Wylie, C., and Heasman, J. (1998). The role of maternal VegT in establishing the primary germ layers in *Xenopus* embryos. *Cell* **94,** 515–524.

Zhang, J., and King, M. L. (1996). *Xenopus* VegT RNA is localized to the vegetal cortex during oogenesis and encodes a novel T-box transcription factor involved in mesodermal patterning. *Development* **122,** 4119–4129.

Zhou, Y., and King, M. L. (1996a). Localization of Xcat-2 RNA, a putative germ plasm component, to the mitochondrial cloud in *Xenopus* stage I oocytes. *Development* **122,** 2947–2953.

Zhou, Y., and King, M. L. (1996b). RNA transport to the vegetal cortex of *Xenopus* oocytes. *Dev. Biol.* **179,** 173–183.

Asymmetric Germ Cell Division and Oocyte Determination during *Drosophila* Oogenesis

Wei Deng and Haifan Lin
Department of Cell Biology, Duke University Medical School
Durham, North Carolina 27710

Early oogenesis in *Drosophila,* with a stereotypic pattern of germ cell division and differentiation, provides an attractive model for studying cell lineage and patterning. *Drosophila* oogenesis is initiated when a germline stem cell divides asymmetrically to produce a daughter stem cell and a cytoblast. The cystoblast then undergoes four rounds of incomplete mitoses to form a 16-cell cyst, accompanied by the formation of the fusome. Within the cyst, one of the two cells with four intercellular bridges differentiates into an oocyte while the rest become nurse cells. The oocyte then translocates within the cyst to a posterior position, which defines the anterio-posterior axis of the future embryo. Recent studies have shown that the asymmetric germline stem cell division is controlled by somatic signaling involving *piwi, fs(1)Yb,* and the *dpp* pathway as well as by intrinsic mechanisms involving *pumilio, nanos, arrest, bag-of-marbles,* and the spectrosome—the fusome precursor in the stem cells and the cystoblast. The spectrosome in the cystoblast appears to play an important role in polarized fusome growth during cyst formation. The fusome may guide the formation of a polarized microtubule network for the intracyst transport of certain RNAs and proteins to the cystocyte destined to become the oocyte. Genes such as *egalitarian, Bicaudal D, stonewall,* and *encore* are important for oocyte determination, while differential adhesion between the oocyte and its surrounding pre-follicle cells, as mediated by *armadillo,* α-catenin, *shotgun,* and the *spindle* genes, is crucial for oocyte translocation. Early oogenesis shares many parallel features to early spermatogenesis, although distinct differences are also observed at both the phenomenological and mechanistic levels. The study of oogenesis, progressing at an exciting rate, contributes significantly to our understanding of the mechanisms underlying proliferation, differentiation, and patterning.

KEY WORDS: Stem cells, germline, asymmetric division, oogenesis, oocytes, spermatogenesis, cell fate, *Drosophila.*

I. Introduction

Oogenesis in *Drosophila,* with its invariant pattern of cell proliferation and differentiation, provides an excellent opportunity to study lineage and patterning mechanisms during development. In particular, early oogenesis in *Drosophila* is characterized by a series of asymmetric divisions in the germline. These precise asymmetric divisions lead to the development of the ovarian germline lineage and a defined differentiation pattern within the lineage. It thus offers a valuable model for investigating the role of asymmetric division in cell proliferation, differentiation and patterning within a cell lineage.

Drosophila oogenesis occurs in ovarioles, the functional units of the ovary (Fig. 1). In each ovariole, two to three germline stem cells are present at its apical tip in a specialized structure called the germarium (Brown and King, 1962, 1964; Wieschaus and Szabad, 1979; Lin and Spradling, 1993). The germarium can be divided into three regions (Fig. 1). Region 1 contains mitotically active germline cells, region 2 contains differentiating germline cysts, and region 3 contains a single newly formed egg chamber. Germline stem cells are located in region 1 in direct contact with the somatic terminal

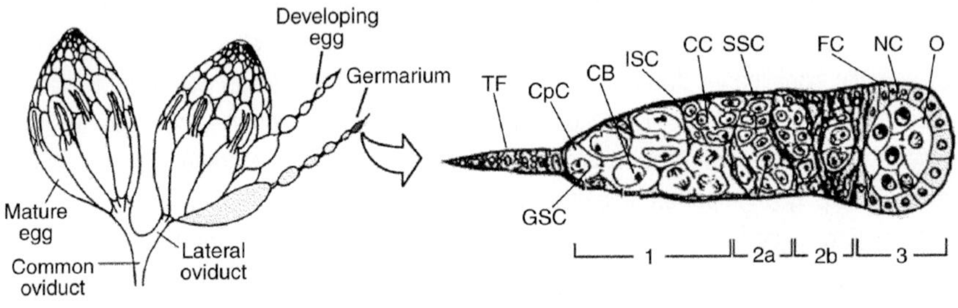

FIG. 1 The *Drosophila* ovary and germarium. The *Drosophila* ovary (left) is composed of 16 to 18 beads-on-a-string structures called ovarioles. An ovariole is a functional unit of oogenesis, starting with the germarium and followed by a string of successively developed egg chambers. The germarium (right), including regions 1 to 3. The germline stem cell (GSC) basal to the terminal filament (TF) divides asymmetrically to process a cystoblast (CB), which then further divides with incomplete cytokinesis to produce a germline cyst containing interconnected cystocytes (CC). These mitotic events occur in region 1. In region 2, intracyst transport of certain RNA, proteins, and organelles occurs, leading to the differentiation of 15 nurse cells (NC) and an oocyte (O) within each cyst. In region 3, a differentiated germline cyst becomes completely enveloped by a monolayer of follicle cells (FC) to form an egg chamber. CpC, somatic cap cells, are the basal cells of the terminal filament; ISC, nonmitotic inner sheath cells; SSC, somatic stem cells. Modified from King, 1970, with permission.

filament cells at the apex. Germline stem cells undergo oriented asymmetric divisions to produce a daughter stem cell, which remains associated with the terminal filament, and a differentiated daughter, the cystoblast, which becomes displaced one cell away from the terminal filament (Lin and Spradling, 1997; Deng and Lin, 1997). The cystoblast then undergoes four rounds of asymmetric divisions with incomplete cytokinesis to form a germline cyst containing 16 cells called cystocytes. Cystocytes within a cyst are interconnected in an invariant pattern by cytoplasmic bridges called ring canals. As the cyst moves to region 2 of the germarium, it becomes enveloped by follicle cells to form an egg chamber. Meanwhile, within the cyst, one of the two cystocytes with four ring canals differentiates into the oocyte and the other 15 cystocytes differentiate into nurse cells. The oocyte then relocates to the posterior region of the cyst, which defines the anterio-posterior axis of the egg chamber and the future embryo. The polarized egg chamber eventually buds off the germarium, joins preexisting egg chambers in a linear array to form an ovariole, and develops into a mature egg (King, 1970; Spradling, 1993). An ovariole thus resembles an assembly line in which each egg chamber represents a differentiated and precisely patterned structure derived from a single germline stem cell division. Moreover, the position of the egg chamber along the ovariole corresponds to its birth order and developmental stage. These attributes render *Drosophila* oogenesis a unique model to study lineage development.

Several questions of general significance can be readily addressed by using early oogenesis in *Drosophila* as a model. For example, what is the mechanism that controls the self-renewing division of germline stem cells? How are the asymmetric fates of the two stem cell daughters established? These two questions are central to the understanding of stem cell biology. For instance, what controls the four synchronous divisions of the cystoblast that lead to the formation of a 16-cell cyst? What leads to the invariant differentiation of the 16 cystocytes into one oocyte and 15 nurse cells? These two questions probe into the essential concepts of proliferation versus differentiation. Finally, what leads the inevitable repositioning of the oocyte within the chamber to define the anterio-posterior axis of the egg and the embryo? This question bears the general significance toward understanding how morphogenic movement leads to patterning.

Rapid progress has been made toward understanding these important questions on *Drosophila* early oogenesis. This is attributed to the effective application of combined genetic, cell biological, and molecular approaches. In this chapter, we intend to review the latest progress on *Drosophila* early oogenesis research and to evaluate its significance to our understanding of cell lineage and patterning in general.

II. Self-Renewing Asymmetric Division of Germline Stem Cells as a Continuous Source of Germ Cells for Oogenesis

In *Drosophila,* oogenesis is initiated when a germline stem cell divides asymmetrically to produce a daughter stem cell and a cystoblast. Defects in germline stem cell division affect not only the maintenance of germline stem cells but also the further development of the cystoblast. Studies in the past several years have revealed that germline stem cell division is controlled by both extrinsic signals from the apical somatic cells and intrinsic factors that act within the germline stem cell itself. This section will review the key features of such a dual regulatory mechanism and discuss its general implication in understanding stem cell biology.

A. Somatic Signaling Mechanism

Signals from the terminal filament cells are essential for the self-renewal of germline stem cells (Figs. 1, 2). The involvement of the terminal filament was first observed when laser ablation of the apical half of the terminal filament cells resulted in the abnormal rate of germline stem cell division (Lin and Spradling, 1993). The essential function of the terminal filament, including its basal cells, called cap cells, in maintaining germline stem cells was further revealed by the discovery of two genes, *piwi* and *fs(1)Yb* (*Yb*), whose expression in these cells is required for the self-renewal of germline stem cells (Lin and Spradling, 1997; Cox *et al.,* 1998; 2000; King and Lin, 1999). These findings demonstrate the crucial role of the somatic signaling mechanism in maintaining germline stem cells. Molecular genetic analyses indicate that *piwi* and *Yb* are unlikely somatic signals. However, the gene product of *depacantaplegic* (*dpp*), a TGF-β homolog in *Drosophila,* is a signal that is probably emanated from the apical somatic cells to control the proliferation of germline stem cells as well (Xie and Spradling, 1998). The apical somatic cells might form a microenvironment or "niche" to regulate the proliferation of germline stem cells. A brief treatment of each of these genes is given below.

1. *piwi*

piwi is defined by a group of noncomplementing mutations in the 32C region of the genome that affect germline stem cell divisions (Lin and Spradling, 1997). *piwi* mutant ovarioles often contain a germlineless germarium and only two to three egg chambers. Phenotypic analysis suggests that these two to three egg chambers are derived from the self-differentiation

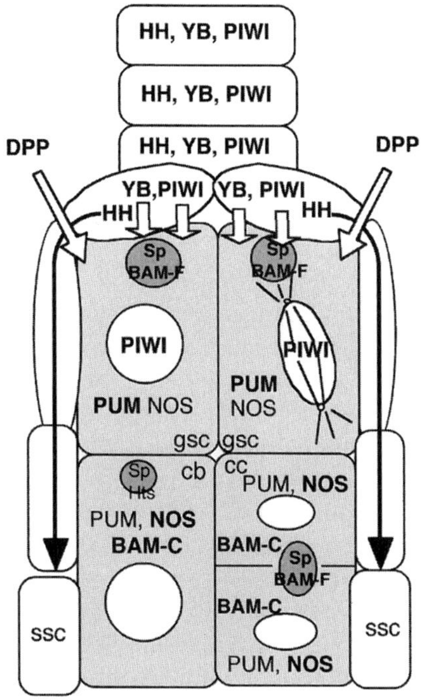

FIG. 2 A magnified view of the apical region of the germarium showing the topology of the GSC region and the gene expression pattern in that region. There are usually two GSCs (gsc) in the germarium in contact with the terminal filament cells, including cap cells, expressing *Yb, piwi* and *hh*. Although *Yb*- and *piwi*-mediated signaling are required for GSC maintenance, *hh* is only known to regulate the division and differentiation of somatic stem cells (SSC) 2 to 5 cells away from TF. In GSCs, spectrosomes (shaded spheres) containing spectrin (Sp) HTS (not shown) and BAM-F reside in the apical region of the cytoplasm both at interphase and during mitosis, apposed to the signaling somatic cells. During mitosis, the spectrosome anchors one pole of the spindle so that the divisional plan is approximately perpendicular to the apico-basal axis of the germarium. As a result, the daughter GSC remains in contact with TF while the cystoblast (cb) becomes one cell away from the somatic cells. The PUM and PIWI proteins are present at a high level in GSCs and at a low level in cystoblasts and cysts. NOS protein shows a complementary distribution profile, while BAM-C is only present in cystoblasts and early cysts. The source of DPP has not been identified. Modified from Lin (1998) *Curr. Op. Cell Biol.* **10,** 687–690 with permission from Company of Biologists Ltd.

of the two to three germline stem cells in the germarium (Lin and Spradling, 1997; Cox *et al.,* 1998). *piwi* is thus required for the self-renewing division of germline stem cells. Germline and somatic clonal analyses further suggest that the expression of *piwi* in the terminal filament cells is essential for maintaining germline stem cells and for promoting their division, while *piwi* expression in germline stem cells also facilitates stem cell division in a cell-autonomous manner (Cox *et al.*, 1998, 2000; Fig. 2). The PIWI protein is present in the nucleoplasm in all types of ovarian cells. Among somatic cells, PIWI is abundantly present in the basal terminal filament cells (also known as cap cells) and inner sheath cells, consistent with its somatic signaling function (Cox *et al.*, 2000). In the germline, PIWI is present at a high level in the stem cells but is sharply reduced in the cystoblast, consistent with its cell-autonomous function in promoting stem cell division. These observations revealed the dual role of *piwi* in regulating germline stem cell maintenance and division (Cox *et al.*, 2000).

An exciting aspect of PIWI molecular biology is that it represents a novel class of stem cell genes evolutionarily conserved from animal to plant kingdoms (Cox *et al.*, 1998; Benfey, 1999). In *C. elegans,* the two *piwi* homologs, *piwi* related gene-1 (*prg-1*) and *prg-2,* are required for germline proliferation and maintenance (Cox *et al.*, 1998). Furthermore, two *piwi*-like genes in *Arabidopsis, zwille* and *argonaut,* are required for maintaining apical meristem cells (Bohmert *et al.*, 1998; Moussian *et al.*, 1998). In mammalian systems, including humans, *piwi* homologs are expressed in the early spermatogenic region in the seminiferous tubules (W. Deng, Dan Qiao, and H. Lin, unpublished data). Removing a murine *piwi* homolog, *miwi,* from mice causes azoospermia, thus suggesting its essential role in spermatogenesis, possibly in germline stem cell division (W. Deng and H. Lin, unpublished data). Therefore, *piwi* family genes appear to be the first class of genes with an important stem cell function that is conserved during evolution.

The *piwi* family genes encode highly basic novel proteins. These proteins share weak homology to eIF2C, which is a major component of the cytoplasmic protein fraction that stimulates the formation of a ternary complex between the met-tRNA, GTP, and the eIF2 (Roy *et al.*, 1988; Zou *et al.*, 1998). A member of the *piwi* family genes in *Drosophila, aubergine* (also known as *sting),* functions to regulate mRNA processing and translation (Wilson *et al.*, 1996; Schmidt *et al.*, 1999). A *piwi*-like gene in C. *elegans, rde-1,* is involved in dsRNA-mediated interference (Tabara *et al.*, 1999). All of these observations point to the potential biochemical activity of the PIWI family proteins in regulating RNA-related processes.

2. *Yb*

The importance of *Yb* in germline stem cell regulation is indicated by its mutant phenotype (King and Lin, 1999). Like *piwi, Yb* mutant ovarioles often contain a rudimentary germarium and only two to three egg chambers. The rudimentary germaria often lack germline cells or contain a small number of ill-differentiated germline cells. *Yb* is specifically expressed in the terminal filament to regulate germline stem cell division (Fig. 2); *Yb* mutant germline clones develop normally (King and Lin, 1999).

In addition to having a role in regulating germline stem cell division, *Yb* also controls the proliferation of ovarian somatic stem cells residing midway in the somatic shell of the germarium (J. King and H. Lin, unpublished data; Figs. 1,2). These stem cells are responsible for producing follicle cells that envelop the germline cyst to form an egg chamber. Loss of *Yb* function leads to a drastic reduction of somatic stem cell proliferation that matches the reduction in germline cyst production, while overexpression of *Yb* leads to the overproliferation of follicle cells to exceed the need for egg chamber formation (J. King, D. Cox, and H. Lin, unpublished data). This function of *Yb* is very similar to the role of the *hedgehog* (*hh*) signaling pathway in somatic stem cell division (Forbes *et al.,* 1996). Thus, *Yb* is the first gene known to mediate a signaling pathway that regulates the coordinated division of two distinctive populations of stem cells within a common structure.

Yb encodes a novel protein with a putative adenosine triphosphate (ATP) binding site (King and Lin, 1999). All existing *Yb* mutations are genetically null. Molecular analysis of these mutations reveals that they are missense mutations clustered within the carboxyl half of the protein, thus suggesting the importance of this region for YB function (King and Lin, 1999). There is no structural feature that suggests YB is a signaling protein itself. Instead, genetic and molecular analyses indicate that YB is required for the expression of *piwi* and *hh* in the terminal filament to control the division of germline and somatic stem cells, respectively (J. King, D. Cox, and H. Lin, unpublished data). This upstream status of *Yb* with respect to *piwi* is confirmed by epistasis analysis between *Yb* and *piwi.* The upstream status of *Yb* with respect to the *hh* signaling pathway is also reflected by the overexpression of *hh* in *Yb* mutants, which bypasses the *Yb* requirement for somatic stem cell division, resulting in overproduction of follicle cells. This effect is very similar to the overexpression of *hh* in the Yb^+ background. In addition, the overexpression of *hh* in *Yb* mutants partially rescues the self-renewing ability of germline stem cells. This, together with the low penetrant phenotype of the germline stem cell defects observed in *hh* mutants, suggests that the *hh* signaling pathway is also involved in maintaining germline stem cells, probably as a redundant safeguard mechanism (J. King, D. Cox, and H. Lin, unpublished data). Hence, the study of the regulatory relationship among *Yb-, piwi-,* and *hh*-mediated mechanisms reveals a bifurcating pathway in which *Yb* controls germline and somatic stem cell division through the *piwi-* and *hh*-mediated mechanisms, respectively. This pathway for the first time reveals how the division of two distinct populations of stem cells within a common tissue or organ is coordinated by a common signaling mechanism—a question essential for the understanding of the development and homeostasis of complex tissues.

3. *dpp*

As a member of the TGF-β superfamily, *dpp* encodes a vertebrate BMP2/4 homolog and functions as both a local signal and a long distance morphogen in the patterning of *Drosophila* embryos and imaginal discs (Lawrence and Struhl, 1996). In adult ovaries, *dpp* also plays a signaling role in controlling germline stem cell division (Xie and Spradling, 1998). Overexpression of *dpp* induces hyperproliferation of germline stem cells without differentiation, resulting in germaria filled with numerous tumorous germline stem cells. Consistent with a positive role of *dpp* in regulating germline stem cell division, germline clones mutant for *thick vein* (*tkv*) and *punt,* which encode the receptors for the DPP ligand, fail to maintain germline stem cells and significantly reduce their division rate (Xie and Spradling, 1998). Similar defects are also seen in germline clones mutant for the downstream effectors of the DPP signal, such as *mad* and *medea* (Xie and Spradling, 1998). These observations suggest that the DPP signal acts directly on germline stem cells to regulate their division (Fig. 2). It remains unknown whether the DPP signal is generated by the apical somatic cells or is an autocrine or paracrine in the germline. However, given that *dpp* was overexpressed in the Gal4/UAST system, which only works efficiently in somatic cells (Rorth, 1996, 1998), it is likely that the DPP signal is generated in somatic cells. If so, DPP is the first identified somatic signal that regulates the division of stem cells in the germline. Further study of the DPP-mediated mechanism in germline stem cell division should significantly advance our understanding of the cell-cell interaction mechanisms involved in controlling stem cell division.

B. Intrinsic Mechanisms

The division of germline stem cells is intrinsically polarized, as marked by the asymmetric localization of a novel cytoplasmic organelle called the spectrosome during the cell cycle (Lin and Spradling, 1997; Deng and Lin, 1997). The spectrosome is part of the intracellular machinery that controls the asymmetric division of germline stem cells (Deng and Lin, 1997; de Cuevas and Spradling, 1998). Moreover, recent studies have shown that the PUMILIO (PUM), PIWI, NANOS (NOS), and BAM-C proteins [encoded by *bag of marbles* (*bam*)] are differentially expressed between the germline stem cell and the cystoblast. These proteins are involved in determining the cell fate between the stem cell and the cystoblast (Lin and Spradling, 1997; Forbes and Lehmann, 1998; Parisi and Lin, 1999; Bhat, 1999; Cox *et al.,* 2000; McKearin and Ohlstein, 1995; Ohlstein and McKearin, 1997).

1. The Spectrosome

The spectrosome is a spherical cytoplasmic organelle in germline stem cells and cystoblasts that contains small membranous vesicles but excludes most ribosomes and mitochondria, as revealed by electron microscopy (Lin *et al.*, 1994). It contains membrane skeleton proteins such as α-spectrin, β-spectrin, ankyrin, and an adducin-like protein encoded by the *hu-li tai shao* (*hts*) gene (Lin *et al.*, 1994; de Cuevas *et al.*, 1996), and is closely associated with a centriole both at interphase and during mitosis (Lin *et al.*, 1994; Lin and Spradling, 1995). The spectrosome exhibits a very striking asymmetric cell cycle behavior (Deng and Lin, 1997; de Cuevas and Spradling, 1998; see Fig. 3 on color insert). At interphase, it is always located at the apical cytoplasmic region of the germline stem cell, adjacent to the terminal filament cells. At prometaphase, it is associated with one of the two emerging mitotic asters, which leads to its association with one pole of the mitotic spindle upon its formation at metaphase (Deng and Lin, 1997). Subsequently, at anaphase, the spectrosome remains adjacent to the terminal filament and associated with the spindle pole. However, at telophase, when the mitotic spindle is disassembled, the spectrosome grows and elongates toward the distal daughter cell, forming a bar-shaped structure traversing the cleavage furrow (Deng and Lin, 1997). It is interesting that the stem cell cleavage furrow, rich in anillin, persists into the G2 phase of the next cell cycle (de Cuevas and Spradling, 1998). The bar-shaped spectrosome is maintained during this time. There is a newly formed plug composed of spectrosomal material within the cleavage furrow, which eventually fuses with the migrating spectrosome and becomes part of the bar-shaped spectrosome spanning the cleavage furrow (de Cuevas and Spradling, 1998). Eventually at G2 phase, the closure of the cleavage furrow partitions the spectrosome asymmetrically, so that two-thirds remains in the daughter stem cell while one-third is segregated to the newly formed cystoblast (Deng and Lin, 1997; de Cuevas and Spradling, 1998).

The asymmetric behavior of the spectrosome during the stem cell cycle may serve two functions in setting up the diverging cell fates of the two daughter cells. The first potential function is to anchor one pole of the mitotic spindle to the somatic signaling cells. As a result, after each division, one daughter remains in contact with the signaling cells and remains as a stem cell, while the other daughter cell is displaced one cell away from the signaling cells and differentiates into a cystoblast (Lin and Spradling, 1997; Deng and Lin, 1997). The second potential function is to asymmetrically localize and segregate factors important for the division (Lin *et al.*, 1994). Indeed, the spectrosome also contains the regulatory molecules such as cyclin A and the fusomal form of BAM protein (McKearin and Ohlstein, 1995).

The first function has been shown to be true (Deng and Lin, 1997). Eliminating the adducin-like HTS protein in the germline by the *hts*[1] mutation abolishes the spectrosome. This demonstrates that the HTS protein is essential for the integrity of the structure (Yue and Spradling, 1992; Lin *et al.*, 1994). The spectrosome-free germline stem cells divide at a normal rate, suggesting the spectrosome does not play a role in controlling the rate of stem cell division (Lin and Spradling, 1997). However, in these spectrosome-free stem cells, the mitotic spindle orientation is completely randomized (Deng and Lin, 1997). This observation suggests that the spectrosome anchors one pole of the mitotic spindle along the apical-basal axis of the germarium to define the proper orientation for stem cell division. As a result of this randomly oriented division, the further differentiation of the cystoblast into a germline cyst is severely compromised. This further suggests the possibility that the spectrosome-defined divisional orientation is important to guarantee the proper asymmetry of each stem cell division to generate a functional cystoblast. It is interesting that the germline stem cell population is still maintained in the *hts*[1] mutants (Lin and Spradling, 1997), presumably because 50% of the stem cell daughters in a population will still contact the somatic signaling cells in a stochastic manner due to the topology of the germarium. Hence, the spectrosome is probably not an essential mechanism for stem cell maintenance but is an important mechanism to achieve the precise asymmetry of each stem cell division to produce a functional cystoblast for oogenesis in an efficient manner.

Studies have also started to shed light onto how the spectrosome anchors the mitotic spindle. McGrail and Hays (1997) reported that the minus-end microtubule motor dynein is localized to the region of the cytoplasm where the spectrosome is associated with the mitotic spindle. Moreover, they showed that in cytoplasmic dynein mutants, a derivative structure of the spectrosome called the fusome in the germline cyst is no longer associated with the mitotic spindle (McGrail and Hays, 1997; see below). Therefore, it is likely that the dynein-mediated machinery is essential for the association of the spindle pole to the spectrosome.

2. *pum* and *nos*

The involvement of *pum,* a maternal effect gene, in germline stem cell regulation was first identified by the discovery of a novel class of *pum* mutations called *ovarette* (*ovt*) alleles (Lin and Spradling, 1997). Ovarioles from *pum*^{ovt} mutant ovaries often contain two to three clusters of undifferentiated cells, and occasionally two to three egg chambers, even though the initial number of germline stem cells at the onset of oogenesis is basically normal. This phenotype suggests that germline stem cells in *pum*^{ovt} mutants have undergone symmetric divisions without differentiation into cystoblasts. Hence, *pum* is required for the proper asymmetric divisions that maintain functional germline stem cells and cystoblasts (Lin and Spradling, 1997). When pum^{ovt} mutant alleles are placed *in trans* to the stronger maternal effect *pum* alleles, region 1 of the mutant germarium is often devoid of germ cells, thus suggesting a more severe failure in germline stem cell maintenance (Forbes and Lehmann, 1998; Parisi and Lin, 1999). *pum* is required cell-autonomously for germline stem cell function (Forbes and Lehmann, 1998; Parisi and Lin, 1999). Moreover, the requirement of *pum* for germline stem cell function is independent of its earlier cell-autonomous function in stem cell precursors but is directly involved in germline stem cell division during oogenesis (Parisi and Lin, 1999).

Consistent with its role in asymmetric germline stem cell division, PUM is a translational suppressor that is present at a high level in the germline stem cell but at a low level in the cystoblast (Forbes and Lehmann, 1998; Parisi and Lin, 1999; Fig. 2). Systematic genetic and molecular analyses reveal the existence of two PUM isoforms that are required for oogenesis, the 156 kDa and 130 kDa isoforms, even though either isoform alone at a haploid dose is sufficient for embryogenesis (Parisi and Lin, 1999). The two isoforms appear to have similar or identical biochemical activity, since increasing the dose of one isoform can compensate for the lack of the other isoform to rescue germline stem cell divisions (Parisi and Lin, 1999). These results suggest that the dose of PUM determines whether a germline stem cell can undergo self-renewing asymmetric divisions.

nos mutant ovaries also fail to maintain germline stem cells, apparently owing to the premature death of these cells (Forbes and Lehmann, 1998). Most germline-containing ovarioles in newly eclosed *nos* mutant females initially have spectrosome-bearing germline stem cells in contact with the apical somatic cells. However, these cells apparently do not actively divide and eventually disappear in approximately three weeks. This indicates that *nos* is important for the survival of germline stem cells. It is interesting that *nos* appears to have an important function in cyst development as well. NOS is detected at a low level in germline stem cells and developing cysts but increases to a much higher level in germline cysts in region 2 (Fig. 2). In *nos* mutants, dying cysts can often be found in region 2 of the germarium. These observations suggest that *nos* is required for the survival of germline cysts as well.

It remains an open question how *pum* and *nos* regulate the asymmetric division of germline stem cells. Presumably, they do so by differentially suppressing the translation of certain mRNAs in the stem cell and the cystoblast. During embryogenesis, PUM binds to the *nos* response element (NRE) in the 3′-UTR of the *hunchback* mRNA, followed by NOS binding to PUM and the RNA, thus forming a ternary complex to suppress translation (reviewed in Parisi and Lin, 2000). Given that this interaction appears to be well conserved during evolution in diverse developmental processes (reviewed in Parisi and Lin, 2000), one would expect that PUM and NOS act as binding partners during early oogenesis as well. However, the oogenic phenotypes of *pum* and *nos* are apparently different. Since at least two isoforms of PUM are required for oogenesis, and since PUM and NOS have different expression patterns, it is possible that PUM and NOS interact with different partners (e.g., different PUM- or NOS-like molecules) to regulate different target mRNAs. Searching for these partners and targets will be the focus of future research.

3. *piwi*

As mentioned above, the PIWI protein is present at a high level in the germline stem cell but at a low level in the cystoblast (Cox *et al.,* 2000). Although PIWI in the germline is not essential for maintaining the germline stem cell, removing PIWI from the germline decreases germline stem cell division four- to five-fold, as shown by single germline stem cell clonal analysis (Cox *et al.,* 2000). It is interesting to note that the cell autonomous function of PIWI is not limited to germline stem cell division; it may also function in cystoblast development and subsequent stages of oogenesis (see below).

4. *bag-of-marbles* (*bam*)

Among all the genes involved in the asymmetric division of the germline stem cell and the differentiation of the cystoblast, *bam* is the best candidate for promoting cystoblast differentiation (McKearin and Spradling, 1990; McKearin and Ohlstein, 1995; Ohlstein and McKearin, 1997). The *bam* RNA is detectable only in cystoblasts and in two- and four-cell cysts but not in germline stem cells or later cysts (McKearin and Spradling, 1990). Likewise, the cytoplasmic form of the BAM protein, BAM-C, is only synthesized in cystoblasts, two-, four-, and eight-cell developing cysts (McKearin and Ohlstein, 1995; Fig. 2). *bam* mutant germaria contain tumorous germline cells that behave like germline stem cells (McKearin and Spradling, 1990; McKearin and Ohlstein, 1995, D. Cox and H. Lin, unpublished data); that is, they undergo complete cytokinesis and contain spectrosome-like structure and a high level of PIWI. This phenotype suggests that *bam* is required for the differentiation of stem cell daughters into cystoblasts. Consistent with this, ectopic expression of *bam* eliminates the germline stem cell population, thus suggesting that ectopic expression of *bam* in germline stem cells is sufficient to promote their differentiation into cystoblasts (Ohlstein and McKearin, 1997).

The BAM protein also exists as the fusomal isoform (BAM-F; McKearin and Ohlstein, 1995; Fig. 2). Given that BAM-F is present in both germline stem cells and cystoblasts whereas BAM-C is present only in cystoblasts and mitotic cysts, it is postulated that BAM-C is the active form of BAM required for the cystoblast differentiation and cyst formation during early oogenesis (Ohlstein and McKearin, 1997).

5. Interaction between *piwi* and *bam*

One important turning point in understanding germline stem cell proliferation versus differentiation comes from a recent study on the regulatory relationship between *piwi* and *bam* (D. Cox and H. Lin, unpublished results). Studies reviewed above suggest that *piwi* is required for germline stem cell (GSC) self-renewal, whereas *bam* is required for cystoblast differentiation. Consistent with this, PIWI and BAM have complementary expression patterns in GSCs and cystoblasts (D. Cox and H. Lin, unpublished data). Despite this, a recent study shows that increasing the *piwi* gene dosage to three copies rescues the *bam* mutant phenotype in both females and males, while overexpression of BAM in a *piwi* mutant also partially rescues the *piwi* mutant phenotype to maintain germline stem cells (D. Cox and H. Lin, unpublished data). Hence, *piwi* and *bam* can compensate for each other's activities in germline stem cell division and cystoblast differentiation. The ability of *bam* to substitute for *piwi* in germline stem cell division suggests that BAM is a mitotic factor instead of a differentiation factor. As a mitotic factor, BAM may promote cystoblast differentiation by promoting its four synchronous divisions to form a 16-cell cyst (McKearin and Christerson, 1994; D. Cox and H. Lin, unpublished data).

The mitosis-promoting activity of *bam* is consistent with the analyses of *encore* (*enc*) and *Rbp9* mutants (Hawkins *et al.*, 1996; Kim-Ha *et al.*, 1999). *enc* is involved in regulating the number of germline mitoses following cystoblast differentiation. Mutations in *enc* result in exactly one extra round of mitosis in germline cysts, leading to five rounds of mitoses (Hawkins *et al.*, 1996). Analysis of *bam* mRNA expression in *enc* mutant germaria reveals a corresponding expansion of *bam* expression (Hawkins *et al.*, 1996). Furthermore, the *bam* mutation dominantly suppresses the *enc* mutant phenotype, suggesting that the additional round of mitosis observed in *enc* mutants is dependent on bam^{+} activity (Hawkins *et al.*, 1996). Similarly, *Rbp9* function is required for the regulation of cystocyte differentiation (Kim-Ha *et al.*, 1999). *Rbp9* mutant germaria are filled with undifferentiated germ cells, causing a germline tumor phenotype similar to that observed in *bam* mutants. Analysis of *Rbp9* mutant germaria with antibodies to BAM reveals a 5- to 10-fold increase in the number of BAM-expressing cells (Kim-Ha *et al.*, 1999). Furthermore, RBP9 has been shown to bind specifically to the 3′ untranslated region of *bam* mRNA *in vitro* (Kim-Ha *et al.,* 1999). This suggests that RBP9 functions by binding to *bam* mRNA to downregulate BAM protein expression, which allows the cystocyte to exit the mitotic cell cycle and differentiate (Kim-Ha *et al.,* 1999). Taken together, these results implicate *bam* as a mitotic promoter. The ability of *bam* to maintain *piwi*⁻ germline stem cells provides the most definitive evidence in supporting the mitotic role of *bam.* These results suggest that *piwi* and *bam* act as mitotic promoters in a common pathway and therefore their activities are functionally interchangeable in promoting germline stem cell self-renewal and cystoblast division.

The interaction between *piwi* and *bam* appears to be specific only between these two genes, since increasing *piwi* gene dosage cannot compensate for the activity of *ovarian tumor* (*otu*) or *Sex lethal* (*Sxl*), two other genes required for germ cell differentiation (D. Cox and H. Lin, unpublished data). Furthermore, this recent study suggests that *piwi* and *bam* may have novel functions in the transition from primordial germ cells (PGCs) to germline stem cells, in which *piwi* is required for BAM-C expression. These novel discoveries led to a hypothesis for the interaction of *piwi* and *bam* before and during oogenesis in which both genes act to promote mitosis and therefore can function interchangeably in either germline stem cell division or cystoblast differentiation. Our results suggest that the stem cell and cystoblast fates may not be determined by cell-fate-specific factors. Instead, these two opposing cell fates are determined by the uniquely combined doses of regulatory molecules in each cell type.

6. *benign gonial cell neoplasm* (*bgcn*)

The *bgcn* mutant shows *bam*-like phenotype, thus suggesting the requirement of *bgcn* for cystoblast differentiation (Lin *et al.,* 1994; Lavoie *et al.,* 1999). *bgcn* mutant ovarioles are filled with mitotically active hyperproliferated germline cells that contain spectrosome-like structures (Lin *et al.,* 1994). It has been found that *bam* is expressed normally in *bgcn* mutants but that the BAM protein fails to be localized to the fusome (Lavioe *et al.,* 1999). Furthermore, overexpression of *bam* is not able to promote the aberrant germline stem cell differentiation in *bgcn* mutant background. These observations confirm that *bgcn* is another factor important for cystoblast differentiation and suggest that *bgcn* may promote cystoblast differentiation by regulating the cellular localization of the BAM protein.

III. Germline Cyst Formation via Asymmetric Cell Divisions

Once the cystoblast is generated, it immediately embarks on a differentiation process to form a germline cyst. In *Drosophila,* a cystoblast undergoes four rounds of mitotic divisions to generate a 16-cell cyst (see Introduction). These divisions have the following characteristics: they are synchronized, asymmetric, and have incomplete cytokinesis. They are also accompanied by the formation and development of an intercellular structure called the fusome that connects cystocytes through ring canals (Storto and King, 1989; Lin and Spradling, 1995). The mature fusome in the resulting 16-cell cyst is a highly branched structure that reflects the pattern of asymmetric cystocyte divisions. The fusome appears to be derived from the spectrosome in the cystoblast and is intimately related to germline cyst development and oocyte determination (Lin and Spradling, 1995; Deng and Lin, 1997; de Cuevas and Spradling, 1998). Therefore, a discussion of the germline cyst formation mechanism should start with a review of fusome biology.

A. The Role of the Fusome in Germline Cyst Formation

The fusome was discovered in the ovary of the greater diving beetle *Dytiscus marginalis* at the turn of the century by Italian biologist Giardina (1901), who thought that the fusome arose from a remnant of the mitotic spindles during the cystocyte divisions. This structure was later observed in the gonads of other insect species. During the past three decades, light and electron micropy studies reveal a detailed picture of the fusome structure (Telfer, 1975). Recent studies further identified several key molecular components of the fusome (Lin *et al.,* 1994; McKearin and Ohlstein, 1995; de Cuevas and Spradling, 1998; Leon and McKearin, 1999; Lilly *et al.,* 2000). Like the spectrosome, the fusome comprises numerous membranous vesicles and membrane skeleton proteins such as spectrin, the adducin-like HTS protein, and ankyrin, as well as a fly homologue of the transitional endoplasmic reticulum ATPase (TER94). The fusome also localizes regulatory factors such as cyclin A and BAM. During cystocyte divisions, the fusome is always associated with one pole of the mitotic spindle in every dividing cystocyte, thus anchoring the asymmetric orientation of the division to define the pattern of cyst formation and the polarity of the subsequent oocyte differentiation (Storto and King, 1989; Lin and Spradling, 1995; Deng and Lin, 1997). The fusome thus plays a central role in germline cyst formation and oocyte determination.

1. Cyst Formation Is Accompanied by Fusome Biogenesis

The morphogenesis of the fusome during cyst formation has been systematically analyzed (Lin and Spradling, 1995; de Cuevas and Spradling, 1998; Fig. 3). The presence of the fusome at each stage of cyst development and its asymmetric association with the mitotic spindles was first visualized by using antispectrin and antitubulin antibodies as markers (Lin and Spradling,

1995). In the newly formed cystoblast, the spectrosome is usually located at the apical region of the cell, opposite the future site of cytokinesis (Lin and Spradling, 1995). During mitosis, it keeps its shape and position. At metaphase and anaphase, one pole of the mitotic spindle is always associated with the cystoblast spectrosome (Lin and Spradling, 1995). At cytokinesis, as visualized by antianillin antibody staining, a ring canal forms at the cleavage furrow to prevent complete cytokinesis (de Cuevas and Spradling, 1998). Meanwhile, the spectrosome is relocated toward the forming ring canal. By the end of the mitosis, in the newly formed two-cell cyst, only the mother cyst cell (cystoblast) contains the original spectrosome. In the interphase of the next cell cycle, a small plug of the fusomal material is formed in the new ring canal. As interphase progresses, the fusome and the plug migrate toward each other and eventually fuse together to form a two-cell stage fusome. During the fusion, there is also addition of new fusomal material formed between the original spectrosome and the plug as they fuse. The resulting fusome now spans the ring canal in the two-cell cyst (de Cuevas and Spradling, 1998). The cell with the spectrosome at cytokinesis (i.e., the cystoblast) contains more fusome material than its sister cell. This asymmetric rule is observed also in the next three mitoses during the formation of four-cell, eight-cell and 16-cell cysts. After each round of mitosis, the fusomal plugs form in the nascent ring canals, move toward the original fusome, and fuse with it. As the fusomal plugs migrate toward the original fusome, the nascent ring canals associated with them move together with them. This centripetal ring canal movement changes the relative spatial position of the cystocytes and is responsible for the rosette pattern of cystocyte interconnections within a cyst. The polarized growth of the fusome suggests that fusome morphogenesis is an intrinsically polarized process.

2. Molecular Mechanisms of Fusome Biogenesis

Although fusome biogenesis has been systematically documented, the molecular mechanisms regulating this process remain a mystery. Several molecules have been identified as important for fusome formation and/or integrity. For example, the adducin-like HTS protein and spectrin are essential for the integrity of the spectrosome and fusome (Lin *et al.,* 1994; de Cuevas and Spradling, 1998). In the absence of either molecule, the spectrosome and fusome do not form; their components become dispersed throughout the cytoplasm. Hence, it is likely that these membrane skeletal proteins may form a mesh around the membranous vesicles in the fusome similar to their association with the cell membrane. This "fusomal skeleton" is essential for maintaining the fusome structure.

Other candidate molecules for fusome biogenesis include BAM, BGCN, and the cytoplasmic dynein. BAM-F is a component of the fusome. The BAM-F is not localized to the fusome in the *bgcn* mutant, indicating that *bgcn* is required for the fusomal localization of BAM-F (Lavoie *et al.*, 1999). Moreover, in *bam* mutant ovaries, the number of vesicles in the fusome is greatly reduced and the spectrosome structure is severely compromised (McKearin and Ohlstein, 1995). Furthermore, the BAM protein interacts with the *Drosophila* homolog of a protein (TER94) involved in vesicle fusion in yeast and vertebrate cells (Leon and McKearin, 1999). These observations together suggest that BAM is involved in recruiting vesicles to the spectrosome and fusome.

Cytoplasmic dynein and the dynectin complex may also play a role in vesicle recruitment and/or organization (McGrail and Hays, 1997). Cytoplasmic dynein is a minus-end directed microtubule motor that is localized to the region where the fusome is associated with the mitotic spindle. In cytoplasmic dynein heavy chain (*Dhc64C*) mutants, the fusome is disorganized and less branched. In addition to dynein, the Glued protein, a component in the dynactin complex, has been shown to colocalize with dynein throughout oogenesis and physically interact with the dynein heavy chain (McGrail *et al.*, 1995). It has been found that one of the dynactin components can associate with Golgi-spectrin in mammals (Holleran *et al.*, 1996). Therefore, dynein may be associated with the fusomal vesicles via dynactin-spectrin mediated interactions and bring the vesicles to the fusome along the microtubule via its motor function.

3. Functions of the Fusome in Cyst Formation

During germline cyst formation, the fusome plays an essential role in synchronizing cystocyte divisions, defining the cystocyte interconnecting pattern, and ensuring four rounds of cystoblast divisions (Deng and Lin, 1997). These functions were first implicated by the interconnecting pattern of the fusome in the cyst as well as by the association of the fusome with one pole of the mitotic spindles. Definitive evidence for the aforementioned function of the fusome came from the study of the *hts*[1] mutant cysts (Deng and Lin, 1997), in which the lack of the adducin-like HTS protein leads to the complete elimination of the fusome (Lin *et al.*, 1994). Since the HTS protein is only detectable in the fusome in wild-type germline cysts, it is reasonable to assume that the *hts* mutation specifically eliminates fusomes in germline cysts. In *hts* mutants, the orientation of the cystocyte division becomes randomized, the division of individual cystocytes becomes asynchronous, and their division number is significantly reduced so that the cysts often contain only two to four cystocytes (Deng and Lin, 1997). Similar, though less severe, defccts are found in germline clones deficient for

α-spectrin (de Ceuvas and Spradling, 1998). These defects are not caused by lapses in communication between the cystocytes through aberrant ring canals, since ring canals are formed normally in *hts*[1] and α-spectrin mutant cysts (Lin *et al.,* 1994). Therefore, the fusome plays a key role in facilitating and synchronizing cystocyte divisions and in defining their interconnecting pattern.

The function of the fusome in synchronizing and coordinating the cystocyte cell cycle may lie in its ability to recruit cell cycle regulatory factors such as cyclin A, which is associated with the spectrosome in germline stem cells and cystoblast and with the fusome in germline cysts (Lilly *et al.,* 2000). In somatic cells, cyclin A, together with its associated Cdk1, is known to be essential for G2/M transition during the cell cycle. The accumulation of cyclin A in the fusome in S and G2 phases of the cystocyte cell cycle suggests that cyclin A may promote the cystocyte mitosis in a similar manner. In ovaries where cyclin A is overexpressed, a high percentage of egg chambers contain 32 cells, indicating that the germline cysts undergo an additional round of division. This activity of cyclin A requires an intact fusome, since no extra division occurs in *hts* mutant ovaries (Lilly *et al.,* 2000). Female sterile mutations in *UbcD1,* which encodes an E2 ubiquitin-conjugating enzyme, also causes the 32-cell cyst phenotype (Lilly *et al.,* 2000). It is therefore possible that the fusome prevents cyclin A degradation by the ubiquitin-mediated pathway.

In addition to cyclin A, *bam* is required for cystocyte divisions, since the null allele of the *bam* gene can dominantly suppress the effect of cyclin A overexpression (Lilly *et al.,* 2000). Moreover, as mentioned above, *encore* is also involved in regulating cystocyte divisions. It does so possibly by regulating *bam* expression during the cyst formation at the posttranscriptional level, since *encore* regulates *gurken* expression posttranscriptionally to determine the axis of the oocyte in later stages of oogenesis (Hawkins *et al.,* 1997). Another gene involved in regulating *bam* expression is RBP9, a *Drosophila* RNA-binding protein and a member of the Hu family proteins (see above; Kim-Ha *et al.,* 1999).

The *encore*-like phenotype is also observed at a low frequency in *stonewall* mutant ovaries (Clark and McKearin, 1996; see below). However, *bam* is expressed normally in *stonewall* mutant germarium (Clark and McKearin, 1996), a finding that might explain the low expressivity of the 32-cell egg chamber phenotype.

Recent studies have shown that cytoplasmic dynein not only plays a role in fusome biogenesis, but also in anchoring mitotic spindles to the fusome (McGrail and Hays, 1997). In dynein mutants, the mitotic spindles become randomly oriented and dissociated from the defective fusome.

B. RNA Binding Proteins in Cyst Formation

Null or strong hypomorphic mutations in genes such as *orb, vasa,* and *arrest/Bruno* lead to pleiotropic defects in germline cyst formation and eventual germline cyst degeneration (Lantz *et al.,* 1994; Styhler *et al.,* 1998; M. Parisi, W. Deng, and H. Lin, unpublished data). These mutant flies have rudimentary ovaries, where each ovariole contains a small germarium followed by an interfollicular stalk-like structure. In addition, the *arrest* mutant ovarioles also contain two to three pseudo-egg chambers, where some mitotically active germline cells are surrounded by a layer of morphologically normal follicle cells (M. Parisi, W. Deng and H. Lin, unpublished data). Molecular marker analyses reveal that the partial germaria contain germline stem cells at the apical tip and several developing germline cysts. However, cells within these "cysts" are not interconnected by ring canals (M. Parisi, W. Deng, and H. Lin, unpublished data). Likewise, in *orb* and *vasa* mutants, germline cysts rarely reach the 16-cell stage before they degenerate (Lantz *et al.,* 1994; Styhler *et al.,* 1998). These observations suggest that these mutations either affect the asymmetry of the germline stem cell division required for the formation of a functional cystoblast or directly affect cyst development. Therefore, *orb, vasa,* and *arrest* (*Bruno*) play essential roles in early germline cyst development.

It is interesting that *orb, vasa,* and *arrest* (*Bruno*) all encode RNA binding proteins (Lantz *et al.,* 1992; Hay *et al.,* 1988; Lasko and Ashburner, 1988; Kim-Ha *et al.,* 1995; Webster *et al.,* 1997). It has been shown that VASA and BRUNO interact with each other to regulate *oskar* translation (Webster *et al.,* 1997), whereas ORB is required for translational control of *oskar* by regulating the polyadenylation of its messenger RNA (Chang *et al.,* 1999). It is possible that these RNA-binding proteins might be involved in germline cyst development by regulating the translation of *oskar* or other oogenic genes in a similar manner.

IV. Oocyte Determination

After a 16-cell germline cyst is formed, it moves to region 2 of the germarium, where the fusome within the cyst degenerates and cystocytes cease mitosis. Subsequently, one of the two cystocytes with four ring canals in the cyst will differentiate into an oocyte while the remaining cells become nurse cells. The first definitive sign of oocyte differentiation is its entry into meiosis and arrest at prophase I, as marked by the formation of the synaptonemal complex (Carpenter, 1975). The nurse cells, however, enter the endo-replication cell cycle, replicating their DNA without division, forming a polyploid genome. The mechanism of oocyte determination has been a focus of research because of its importance in oogenesis. In addition, it provides an ideal model to study how asymmetric transport and localization of cytoplasmic components lead to differentiation.

From 1966 to 1970, three models were proposed to explain the oocyte-determination mechanism in *Drosophila* (Suter and Steward, 1991). The mosaic model (Koch *et al.*, 1967), based on the asymmetric division pattern during cyst formation, proposes that the original cystoblast contains oocyte-determination factors and retains them during four rounds of asymmetric divisions. The transportation model is based on the observation that certain factors are specifically accumulated into the oocyte during oocyte differentiation (Mahowald and Strassheim, 1970). It proposes that the occyte is specified because its determinants are transported into one of the two four-ring-canal cystocytes after cyst formation. Finally, the induction model is based on the observation that the oocyte is always localized in the posterior of the egg chamber in contact with the follicle cell. This led to the hypothesis that the posterior follicle cell induces oocyte differentiation through cell-cell interaction (Koch and King, 1966; King, 1970). Various cellular and molecular events have been discovered to substantiate the first two models, while suggesting that the third model is unlikely to be true. Particularly, the fusome and the microtubule network have been discovered to play important roles in oocyte determination (Lin *et al.*, 1994; Cooley and Theurkauf, 1994). These new findings led to the fusome polarity hypothesis (Lin and Spradling, 1995). This hypothesis, similar to certain aspects of the mosaic and transportation models, as detailed below, is the first systematic attempt to assign the role of specific cytoskeletal components in oocyte determination.

A. Intrinsic Polarity of the Fusome and Oocyte Specification

The asymmetric segregation of oocyte components during cystoblast and cyst formation was first observed in *Dytiscus marginalis* (Giardina, 1901). However, it was thought that the fusome was derived from the mitotic spindle residues, first formed as a bar-shaped symmetric structure spanning the ring canal of the two-cell cyst following cystoblast division (Telfer, 1975). The cystopblast division was assumed to be symmetric. These assumptions, however, are not true, since the cystoblast was found to contain the spectrosome, which is associated with one pole of the mitotic spindle during cystoblast division (Lin and Spradling, 1995; see Fig. 4 on color insert). As a result, the spectrosome is retained in the mother cystoblast. Hence, the cystoblast division is an asymmetric division that resembles yeast budding, with the spectrosome-containing cystoblast equivalent to the mother cell. The subsequent three cystocyte divisions are also asymmetric, with the preexisting fusome and ring canals invariantly segregated to the old cystocyte (Lin and Spradling, 1995; Fig. 4). Each division is followed by the fusome growth toward the newly formed ring canal. In hts^1 and α-spectrin mutant germline cysts, where the fusome is abolished, oocytes usually fail to be specified and all cystocytes in the cysts develop into nurse cells (Lin *et al.*, 1994; de Cuevas *et al.*, 1996). These observations reveal the importance of the spectrosome and/or fusome in oocyte determination.

Based on the morphogenesis of the fusome and the defect of the fusome-free hts^1 mutant in oocyte determination, the fusome polarity hypothesis was proposed (Fig. 5, see color insert). According to this hypothesis, the fusome is formed by the migration of the spectrosome in the cystoblast to the newly formed ring canal in the two-cell cyst followed by the addition of new material to the spectrosome from the daughter cystocyte. A fusome formed this way is a polarized structure; the origin of the polarity resides at the original spectrosomal end in the cystoblast. During the subsequent three rounds of cystocyte divisions, new fusomal materials are added to the branching ends, forming a radially polarized structure with the origin of the polarity still retained in the original cystoblast. This polarized fusome serves as a template to guide the formation of a polarized microtubule network, which leads to polarized intracyst transport of mRNA, proteins, and other molecules and subcellular structures, thus resulting in oocyte differentiation. Hence, this model suggests that the oocyte is the direct decent of the original cystoblast.

Although the polarity of the fusome remains an abstract concept, the hypothesized biogenesis of the fusome has been shown to be true (de Cuevas and Spradling, 1998). Moreover, the absence of the fusome in hts^1 mutant is correlated with the disruption of the polarized microtubule array in the germline cyst and the delocalization of the oocyte-specific RNAs (Deng and Lin, 1997).

The polarized fusome model represents the current view of the oocyte-determination mechanism. The nature of the fusome polarity and how it guides the microtubule organization in the cyst remain to be investigated. It is possible that the fusome polarity is reflected in its radial growth following each round of cystocyte division, which may guide microtubule organization. However, given the known ability of the fusome to associate with the centrosome, it is more likely that the origin (i.e., the spectrosomal region) of the fusome retains certain factors that keep its associated centrosome (the orignal centrosome in the cystoblast) active, which then serves as a nucleation center to form a radial array of microtubules that extends into every cystocyte through ring canals. This hypothesis is supported by the fact that, even though there are 32 centrosomes in 16 cystocytes in a newly formed cyst, only one microtubule array is observed (Theurkauf *et al.,* 1993). This microtubule array originates from one of the two cells with four ring canals, suggesting that only the centrosome(s) in that cystocyte is active. The active centrosome may interact with microtubules via cytoplasmic dynein, since female sterile mutations in the dynein heavy chain gene (*Dhc64C*) cause the defect in fusome organization and the failure of oocyte determination (McGrail and Hays, 1997).

B. Polarized Microtubule Network in Oocyte Determination

The first evidence for the polarized transportation during oocyte specification comes from ultrastructural studies (Mahowald and Strassheim, 1970). In the newly formed germline cyst, all 16 cells contain centrioles, which are located in a juxtanuclear position. These centrioles begin to move toward the future oocyte in region 2b; some degenerate and others translocate into the oocyte in the stage-1 egg chamber. In addition to the centriole, molecular genetic studies in the 1990s have identified a number of gene products that are specifically localized to the future oocyte following germline cyst formation (Table I). Some of these molecules, such as the BicD and EGL proteins, are essential for specifying the oocyte fate (see below), whereas other molecules are dispensable for the initial oocyte determination. Nevertheless, these molecules are critical for oocyte development in later stages of the oogenesis.

The differential localization of these molecules to the oocyte is mediated by a microtubule-based transport system (Koch and Spitzer, 1983; Theurkauf *et al.,* 1993). Although microtubules appear to be uniformly organized in germarial region 1, where dividing germline cysts reside, they become clearly organized into a radial array with a single microtubule organization center (MTOC) in 16-cell cysts, starting in region 2a and persisting in region 2b. The localization of the MTOC in the 16-cell cyst coincides with the position of the pro-oocyte in the cyst, thus suggesting that the differentiating oocyte could be the center of this polarized microtubule network. This network is important in oocyte differentiation. When microtubule dynamics are disrupted by drug treatment, the polarized microtubule network can no longer be detected, and oocyte-specific markers, such as *oskar* and cyclin B mRNAs, are no longer concentrated in one cell. As a result, all 16 cystocytes develop into nurse cells.

Microtubules originating from the MTOC within the pro-oocyte pass through the ring canals into the adjacent cells, suggesting that they may serve as tracks for transporting various RNAs, proteins, and organelles into the future oocyte. Indeed, the localized RNAs, such as *oskar* and *fs(1)K10* mRNAs, are enriched in cytoskeleton extracts (Pokrywka and

TABLE 1

Molecules Accumulated in the Oocyte During Oocyte Specification

Gene Products	Functions in Oogenesis and Germline Development	References
oskar mRNA	Germ plasm assembly, pole cell formation	Kim-Ha *et al.,* 1991 Ephrussi *et al.,* 1991 Ephrussi and Lehmann, 1992
Bicaudal D mRNA and protein	Oocyte determination, antero-posterior, and dorsal-ventral patterning of the oocyte	Suter *et al.,* 1989 Wharton and Struhl, 1989 Suter and Steward, 1991 Ran *et al.,* 1994 Swan and Suter, 1996
egalitarian mRNA and protein	Oocyte determination, dorsal-ventral patterning of the oocyte	Mach and Lehmann, 1997
orb mRNA and protein	Cyst formation, antero-posterior and dorsal-ventral patterning of the oocyte	Lantz *et al.,* 1992, 1994 Christerson and McKearin, 1994
nanos mRNA	Germline stem cell survival, cystoblast differentiation	Wang *et al.,* 1994 Forbes and Lehmann, 1998
fs (1)K10 mRNA	Dorsal-ventral patterning in the oocyte	Schubpach, 1987 Cheung *et al.,* 1992
hts mRNA	Cyst formation and oocyte determination	Yue and Spradling, 1992 Lin *et al.,* 1994
cyclin B short mRNA	A component of the germ plasm	Dalby and Glover, 1992, 1993
Bicaudal C mRNA	Anterior follicle cell patterning through germline-soma interaction	Mahone *et al.,* 1995
tosca mRNA	Unknown, encode a *Drosophila* homolog of RAD2 family protein	Digilio *et al.,* 1996
yemanuclein-alpha mRNA and protein	Unknown, encode oocyte-specific nuclear protein with DNA binding properties	Ait-Ahmed *et al.,* 1987, 1992 Capri *et al.,* 1997
Dhc64C protein	Cyst formation and oocyte determination, polarized transportation in later stage oocytes	Li *et al.,* 1994 McGrail and Hays, 1997
p150^{Glued} protein	A component of the dynactin complex, colocalization with dynein	McGrail *et al.,* 1995
fs(2)cup protein	Nurse cell chromosome structure, interacting with otu in a variety of processes	Keyes and Spradling, 1997
otu p104 protein	Germ cell proliferation, cyst formation and differentiation, nurse cell chromosome structure germline sex determination	Storto and King, 1988 Pauli *et al.,* 1993 Sass *et al.,* 1995 Mal'ceva *et al.,* 1997

Stephenson, 1994). The localization of these RNAs is disrupted in ovaries treated with drugs interfering with microtubule dynamics (Theurkauf *et al.,* 1993; Pokrywka and Stephenson, 1995). The cytoplasmic dynein may be a candidate motor required for this microtubule-mediated transportation. Dynein heavy chain begins to concentrate in the presumptive pro-oocytes of the late region 2a cysts immediately following the formation of the polarized microtubule network (Li *et al.,* 1994). Female sterile mutations of *Dhc64C* affect oocyte determination in the germline cyst (McGrail and Hays, 1997). It is not surprising that the activity of cytoplasmic dynein in oocyte determination is manifested in the germline, since oocyte determination fails in dyenin mutant germline clones (McGrail and Hays, 1997).

It is possible that ring canals are also involved in guiding the formation of the polarized microtubule array. An obvious correlation for this is that the oocyte always differentiates from one of the two cystocytes with four ring canals. Furthermore, the *hts* gene encodes both a fusome isoform (HTS-F) and a ring canal isoform (HTS-RC). In hts^1 mutant ovaries, both HTS-F and HTS-RC are affected, resulting in the disruption of the formation of both the fusome and the ring canal. Therefore, the failure of oocyte determination in hts^1 mutant cysts is possibly due to the combined defects in the fusome and ring canals. The HTS-RC minigene is able to partially restore oocyte specification in hts^1 mutant, thus suggesting the involvement of ring canals in oocyte determination (Smith-Leiker *et al.,* 1997).

C. Genes Important for Oocyte Determination

Genetic analyses have identified a number of genes critical for oocyte determination. These genes include *Bicaudal D* (*BicD,* Wharton and Struhl, 1989; Suter *et al.,* 1989; Suter and Steward, 1991; Ran *et al.,* 1994), *egalitarian* (*egl,* Carpenter, 1994; Mach and Lehmann, 1997), *stonewall* (*stwl,* Clark and McKearin, 1996, *encore* (*enc,* Hawkins *et al.,* 1996), *orb* (Lantz *et al.,* 1992, 1994, Chang *et al.,* 1999), and *vasa* (Styhler *et al.,* 1998). Among them, *BicD, egl,* and *stwl* are required for oocyte determination but not cyst formation, while *enc, orb,* and *vasa* affect both cyst formation (see above) and oocyte determination. Some of these factors, such as *BicD, egl* and *orb,* are specifically located to the oocyte and are more directly involved in oocyte determination, while others are evenly distributed among the cystocytes to regulate the expression of the oocyte determination factors. Analyses of these genes have started to reveal molecular processes involved in oocyte determination.

1. *Bicaudal D (BicD)*

The *BicD* gene is first identified by bicaudal embryonic phenotype caused by the gain-of-function mutation of the gene (Mohler and Wieschaus, 1986).

The analysis of two loss-of-function mutations and null mutations in the gene underscore the importance of *BicD* in oocyte determination (Suter *et al.,* 1989; Suter and Steward, 1991; Ran *et al.,* 1994). In these mutant flies, germline cysts are formed, with normal fusomes and a normal pattern of intercystocyte connections (Theurkauf *et al.,* 1993; Lin *et al.,* 1994). However, no MTOC is detectable in mutant 16-cell cysts (Theurkauf *et al.,* 1993). Moreover, oocyte-specific transcripts, such as *oskar, orb,* and *fs(1)K10* mRNA, fail to be transported into the future oocyte (Suter and Steward, 1991; Lantz *et al.,* 1992; Ran *et al.,* 1994). Eventually, all 16 cystocytes differentiate into nurse cells (Suter and Steward, 1991; Ran *et al.,* 1994). Hence, *BicD* is required for establishing the polarized microtubule transport network in the cyst for oocyte specification. This is further supported by the following observations. The BicD protein is differentially localized to the pro-oocyte, beginning in region 2a and becoming clearly detectable at region 2b. $BicD^{PA66}$ mutation affects the accumulation of BicD protein in the pro-oocyte and also leads to failure of oocyte specification (Theurkauf *et al.,* 1993; Suter and Steward, 1991). The BicD accumulation is independent of the microtubule transport network but requires phosphorylation (Suter and Steward, 1991). Therefore, the localization of the BicD protein in the pro-oocyte may be necessary for the microtubule network formation.

BicD is also necessary for the posterior positioning of the oocyte during the follicle formation (Ran *et al.,* 1994). However, the posterior positioning is not correlated to oocyte determination. In hypomorphic *BicD* mutant, the presumptive oocytes locate at the posterior and show signs of initial oocyte differentiation but are eventually completely reverted to nurse cells.

BicD contains a coiled-coil domain, which has homology to the tail domain of the myosin heavy chain (Suter and Steward, 1991). It has been shown that this domain is essential for the interaction between BicD and lamin Dm_0 in the yeast two-hybrid system, suggesting the coiled-coil domain is capable of mediating the interaction between BicD and other proteins (Stuurman *et al.,* 1999). However, the functional significance of this domain in oocyte determination remains unknown.

2. *egalitarian* (*egl*)

The *egl* mutant phenotype is very similar to that of *BicD* in that cystocyte divisions occur normally (Schupbach and Wieschaus, 1991; Lin *et al.,* 1994; Mach and Lehmann, 1997). However, all of the 16 cells within the cyst differentiate into nurse cells. Like *BicD,* the activity of *egl* is required in the germline for oocyte determination, as indicated by pole cell transplantation experiment (Mach and Lehmann, 1997). The *egl* transcript is enriched in the presumptive oocyte when the newly formed cyst moves to germarial region 2a (Mach and Lehmann, 1997). Although EGL protein is initially evenly distributed in the newly formed 16-cell cysts, it begins to accumulate in the pro-oocytes in region 2b and eventually becomes localized in the oocyte in stage-1 egg chambers. Unlike BicD, the differential localization of EGL appears to depend on the polarized microtubule network, since its localization in later stages of oogenesis is disrupted in colchicine-treated ovaries (Mach and Lehmann, 1997). EGL is not necessary for the formation of the microtubule network. In *egl* mutant, the MTOC is formed in region 2a cysts but is no longer detectable in region 2b and 3, thus suggesting that *egl* is required for the maintenance of the microtubule array (Theurkauf *et al.,* 1993). Throughout oogenesis, EGL protein and BicD protein colocalize with each other. Moreover, the oocyte localization of the two proteins is interdependent: in *egl* mutant ovaries where EGL protein is not localized or is not produced, BicD protein fails to localize into the pro-oocyte. Nevertheless, the microtubule network still forms initially. This suggests that the localization of BicD to the presumptive oocyte is not a prerequisite for BicD function in initiating microtubule net work formation. In *BicD* mutant ovaries where BicD protein is evenly distributed among the cystocytes, the pro-oocyte localization of the EGL protein is disrupted, suggesting a second role of BicD in maintaining the microtubule network through interaction with EGL. Consistent with this interdependence, EGL protein and BicD protein physically interact with each other in co-immunoprecipitation assays. In addition, *egl* mutation can suppress the bicaudal phenotype of the dominant *BicD* mutant. These results provide molecular and genetic evidence for the interaction between the two proteins (Mach and Lehmann, 1997).

3. *stonewall* (*stwl*)

Similar to *BicD* and *egl* mutants, germline cyst development appears normal in *stwl* mutants, and *stwl* mutant ovaries often contain egg chambers with 16 nurse cell but no oocyte. However, a small percentage of cysts undergo an extra round of division (Clark and McKearin, 1996; see above). In the mutant egg chambers that do contain an oocyte, the karyosome in the oocyte is larger than wild type and appears to have partial polyploidy (Clark and McKearin, 1996). These defects suggest that *stwl* plays an important role in oocyte determination. In *stwl* mutant cysts, the BicD and Orb proteins are no longer localized to the presumptive oocyte (Clark and McKearin, 1996), thus indicating a requirement of *stwl* for the proper localization of these critical molecules in oocyte specification.

Stwl encodes a basic protein containing a helix-turn-helix (HTH) motif, which is a signature of a DNA-binding domain (Clark and McKearin, 1996). Therefore, *stwl* is probably a transcription factor that regulates the expression of genes important for oocyte determination but is not directly involved in the localization of these gene products. Consistent with this hypothesis, STWL, unlike BicD and EGL, is evenly distributed within the germline cysts in the nucleus of every cystocyte (Clark and McKearin, 1996).

4. *encore* (*enc*)

The role of *enc* in oocyte differentiation was revealed by Hawkins and her colleagues (Hawkins *et al.,* 1996), who analyzed the phenotype of temperature-sensitive alleles of the *enc* gene. At 25°C, these mutations cause an extra round of cystocyte division, generating 32-cell cysts (see above); at 18°C, the oocyte fails to differentiate normally in an otherwise normal 16-cell cyst. The oocyte nuclei in most mutant egg chambers become polyploid despite their significantly smaller size. In addition, *oskar* mRNA, though initially localized normally in the presumptive oocyte of germline cysts in regions 2 and 3, becomes evenly distributed among all cystocytes in postgermarial egg chambers. These observations suggest that the *enc* gene is important for maintaining the oocyte identity following its initial establishment. Alternatively, *enc* mutation could cause the delayed differentiation of cystocytes into nurse cells and oocytes, which compromises their proper development.

D. From Ectopic Oocytes to Oocyte Determination Mechanism

Recently, mutations in several genes have been found to induce ectopic oocyte formation. Analyses of the phenotype of these mutants provide valuable information on mechanisms underlying oocyte differentiation, especially with regard to the role of cell cycle progression.

1. *cyclin E*

Cyclin E is a cell cycle regulator important for the G1-S transition during mitosis. Interestingly, hypomorphic cyclin E mutations frequently cause the formation of two to three germinal vesicles (i.e., oocyte nuclei) in 16-cell cysts (Lilly and Spradling, 1996). This defect can be detected as early as stage 3 of oogenesis. Germ cells with ectopically formed germinal vesicles often contain three or four ring canals and lay next to the endogenous oocyte. Although their nuclei undergo oocyte-like differentiation, these germ cells are not enriched in oocyte-specific markers such as BicD and Orb proteins (Gonzalez-Reyes *et al.,* 1997), nor do they accumulate yolk or show any other signs of oocyte-like cytoplasmic differentiation (Lilly and Spradling, 1996). Furthermore, morphological defects are observed in both ectopic and endogenous oocyte nuclei (Lilly and Spradling, 1996). These observations indicate that the cell cycle machinery plays an important role in the nuclear but not cytoplasmic aspect of oocyte differentiation.

How can a reduction in cyclin E expression induce ectopic germinal vesicle formation? Lilly and Spradling (1996) proposed two models to explain this phenotype. Both models suggest that cell cycle regulation and oocyte specification factors interact with each other to direct the choice between oocyte and nurse cell nuclear fate. It is known that after 16-cell cyst formation, all cystocytes enter premeiotic S phase asynchronously, and the two cystocytes with four ring canals progress the furthest into meiosis. Electron microscopic analysis by Carpenter (1975) shows the presence of a gradient in the premeiotic progression within the cyst. The two cells with four ring canals form long synaptonemal complexes and start chromosome condensation, whereas some cells with three ring canals assemble short synaptonemal complexes. However, only one of the four-ring-canal cells will progress further into meiosis and differentiate into the oocyte; the other 15 cells eventually regress from meiosis, enter the endocycle, and become polyploid nurse cells. Lilly and Spradling propose that since cyclin E is a G1-S promoter, the reduction in cyclin E expression makes the cells more reluctant to regress from meiosis and promote their nuclei to develop into germinal vesicles. Alternatively, the cyclin E mutation might increase the sensitivity of the cystocyte to oocyte specification factors, which are distributed within the cyst in a radial gradient with the highest concentration in the future oocyte. Hence, the cyclin E mutation can promote the cystocyte adjacent to the true oocyte to develop into an ectopic oocyte.

2. *spindle* Genes

Studies on the *spindle* (*spn*) genes also relate cell cycle mechanism to oocyte determination. Five *spn* genes, *spn-A, spn-B, spn-C, spn-D,* and *spn-E,* were originally isolated from a screen searching for maternal effect mutations affecting the dorsal-ventral polarity of the embryo (Tearle and Nusslein-Volhard, 1987). The involvement of these genes in oocyte differentiation was first revealed by the observation that the oocyte is frequently mislocalized in *spn* mutant egg chambers (Gonzalez-Reyes and St. Johnston, 1994, Gillespie and Berg, 1995; Fig. 6, on color insert; see below). Unexpectedly, early stage egg chambers from *spn* double mutant flies often contain 14 nurse cells and two oocytes (Gonzalez-Reyes and St. Johnston, 1997). Both oocytes have four ring canals, show high levels of F-actin in the membrane cytoskeleton, accumulate oocyte-specific markers such as BicD protein and *oskar* mRNA, and fail to endoreplicate their DNA, just like wild type oocytes (Fig. 6). Therefore, unlike the ectopic oocyte in the *cyclin E* mutant cyst, the ectopic oocyte induced by the *spn* mutations differentiates like an oocyte in both cytoplasm and nucleus. However, similar to the *cyclin E* mutant, karyosomes in *spn* mutant oocytes are defective. The ectopic oocyte fails to maintain its identity and eventually reverts to the nurse cell fate by state 8 of oogenesis. These observations indicate that *spn* genes are involved in the oocyte specification probably by temporarily preventing one of the four-ring-canal cells to regress from meiosis. Recent molecular analysis reveals that *spn-B* encodes a Rad51-like protein and is necessary for double-strand break (DSB) repair in meiosis (Ghabrial *et al.,* 1998). Failure in DSB repair caused by *spn* mutations could trigger the cell cycle checkpoint machinery to affect cell cycle progression or regression.

The *spn* genes are also proposed to delay oocyte specification through their interaction with *BicD* and *egl,* given that mutations in these genes also lead to the same ventralized phenotype in late oogenesis (Gonzalez-Reyes *et al.,* 1997; Swan and Suter, 1996; Mach and Lehmann, 1997). However, the nature of these interactions is unknown.

3. *egghead*

egghead was originally isolated as a gene involved in follicle cell development and patterning during oogenesis (Goode *et al.,* 1996). Recent isolation of a female sterile allele of this gene (egh^{cm}) reveals striking oocyte determination and positioning defects in early oogenesis (Rubsam *et al.,* 1998). egh^{cm} mutant egg chambers contain either too many or too few germline cells, which are often ordered in a way such that the numbers of the germ cells in the neighboring egg chambers adds up to 16 or 32. This suggests that the abnormal germ cell numbers are mainly the result of the aberrant splitting or fusion of 16-cell cysts by follicle cells. In these egg chambers, ectopic oocytes, forming at a frequency of more than one oocyte per 16 cells, show both nuclear and cytoplasmic signatures of the oocyte. They are randomly located in egg chambers with three or more germ cells, but often contain fewer than four ring canals. It seems as though these ectopic oocytes can derive from any cystocytes within the egg chamber. These observations suggest that the ring canal number or the position of the cell within the cyst is dispensable for oocyte differentiation in split cysts. Instead, they may be induced by the aberrant contact between the germ cells and the somatic cells in the split egg chamber.

E. Revisiting Hypotheses on Oocyte Determination

As discussed above, the process of oocyte determination is not controlled by a single master switch but involves multiple steps of specification (Fig.

5). The intrinsic asymmetry of the cystoblast division may provide the initial cue to oocyte specification. Immediately after the formation of the 16-cell cyst, the polarized fusome may then guide the formation of a polarized microtubule network within the cyst. This directs a polarized intracyst transport of specific oocyte-promoting factors to the direct descendent of the cystoblast, which leads to its differentiation into the oocyte. The microtubule network in the presumptive oocyte might be involved in the retention of certain oocyte-specification factors. In addition, cell cycle regulators are involved in oocyte specification either by promoting the presumptive nurse cells to regress from meiosis and/or by keeping the presumptive oocyte in the meiotic cycle. Subsequent to the polarized transport, the differentiating oocyte is translocated into the posterior end of the newly formed egg chamber (see below). The interaction between the oocyte and the follicle cells may then help maintain the identity of the oocyte. It will be interesting to find out what factors are involved in oocyte differentiation and how these factors interplay with the cell-cycle machinery to achieve this process.

V. Oocyte Translocation and the Establishment of the Anterior-Posterior Axis of the Future Embryo

It has long been noted that the oocyte is always located at the most posterior position in egg chambers and that this location defines the posterior pole of the future embryo (King, 1970; see Fig. 1). This positioning of the oocyte is a result of the posterior translocation of the presumptive oocyte in the 16-cell cyst in region 2b, where the germline cyst is flatterned into the lens shape and spans the whole width of the germarium. Concurrently, somatic cells envelope the cyst to form an egg chamber. The egg chamber polarity becomes evident in region 3 when a spherical stage-1 egg chamber is formed.

As the stage-1 egg chamber buds off the germarium, the oocyte at its posterior position induces the contacting terminal follicle cells to adopt a posterior fate through the Gurken-Torpedo (Egfr) signaling pathway (Gonzalez-Reyes *et al.* 1995; Roth *et al.*, 1995). Subsequently, during stages 6 to 9 of oogenesis, these posterior follicle cells send an unknown signal back to the oocyte to induce microtubule reorganization (Lane and Kalderon, 1994). The reorganized microtubules form an array in the oocyte with their minus end at the anterior and their plus end at the posterior (Theurkauf *et al.*, 1992; Clark *et al.*, 1994, 1997). This polarized microtubule array is responsible for localizing the *bicoid* mRNA to the anterior of the oocyte and the *oskar* mRNA to the posterior (Kim-Ha *et al.*, 1991; Ephrussi *et al.*, 1991; Pokrywka and Stephenson, 1991, 1995; Clark *et al.*, 1994), thereby establishing the anterior-posterior axis for the future embryo.

How the oocyte is translocated to the posterior position following its specification has intrigued researchers for many years. Mutations in genes such as *dicephalic*, the *spn* group, *hold up*, *shotgun*, and *armadillo* cause the oocyte to localize randomly in the egg chambers, thereby leading to various defects in axis patterning (Lohs-Schardin, 1982; Gonzalez-Reyes and St. Johnston, 1994; Gillespie and Berg, 1995; Gonzalez-Reyes *et al.*, 1997; Rotoli *et al.*, 1998; Oda *et al.*, 1997; Godt and Tepass, 1998; Gonzalez-Reyes *et al.*, 1998). Molecular genetic analyses of these genes are starting to reveal that cell-cell adhesion plays a major role in oocyte positioning.

A. *shotgun*, *armadillo*, and α-catenin

The *shotgun* gene encodes a *Drosophila* homolog of E-cadherin, a transmembrane glycoprotein that mediates cell-cell adhesion though calcium-dependent homophilic interactions (Tepass *et al.*, 1996). The intracellular side of cadherin is linked to the actin cytoskeleton though the catenin complex, which includes α-catenin and β-catenin [Armadillo (ARM) in *Drosophila*] (Peifer *et al.*, 1993; Oda *et al.*, 1994). This cell adhesion complex has important functions in the morphogenesis of epithelial cells in various developmental processes (Oda *et al.*, 1994, 1997; Uemura *et al.*, 1996; Niewiadomska *et al.*, 1999).

In *shotgun* mutants, the oocyte is often mislocalized in egg chambers. This leads to the hypothesis that cell-cell adhesion plays an important role in oocyte positioning (Godt and Tepass, 1998; Gonzalez-Reyes and St. Johnston, 1998). This hypothesis is further supported by several other lines of evidence from the molecular genetic analysis of *shotgun*, *arm*, and α-catenin. First, this hypothesis is consistent with the localization pattern of DE-cadherin, ARM, and α-catenin in the germarium. DE-cadherin is produced in every germline and follicle cell. Starting from region 2b immediately before follicle formation, a higher concentration of the DE-cadherin is found in the presumptive oocyte (Godt and Tepass, 1998; Gonzalez-Reyes and St. Johnston, 1998). After the egg chamber is formed, a higher level of DE-cadherin is also observed in the terminal follicle cells, with the highest concentration found at the boundary between the oocyte and the posterior follicle cells (Godt and Tepass, 1998). α-catenin and ARM are distributed in a similar pattern (Godt and Tepass, 1998; Gonzalez-Reyes and St. Johnston, 1998). Second, in egg chambers containing either a germline or a follicle cell clone of *shotgun*, oocytes are mislocalized in a similar way to what is observed in *shotgun* mutants. Similar defects are also found in germline clones of *arm* mutations (Godt and Tepass, 1998; Gonzalez-Reyes and St. Johnston, 1998). Thus, *shotgun* and *arm* are required in both the germline cyst and its surrounding follicle cells for proper oocyte localization. Third, when the *shotgun* expression is abolished in the posterior follicle cells by generating mutant clones of these cells, the oocyte is always apposed to the follicle cells with the highest level of DE-cadherin (Godt and Tepass, 1998). Hence, the differential concentration of DE-cadherin in the developing follicle is sufficient to position the oocyte to the posterior position. Finally, it is also observed that *shotgun* mutant germline cysts fail to acquire a lens shape in region 2b (Gonzalez-Reyes and St. Johnston, 1998), indicating that DE-cadherin plays a role in the morphogenesis of the germline cyst. In fact, *shotgun* and *arm* germline clones exhibit pleiotrophic defects in morphology and cell integrity (Peifer *et al.,* 1993; White *et al.,* 1998; Godt and Tepass, 1998).

Despite the above evidence, it is puzzling to find that both anterior and posterior follicle cells accumulate a higher level of DE-cadherin, yet the oocyte is only localized to the posterior pole. It is proposed that the posterior follicle cells form first and are in contact with the oocyte before anterior follicle cell formation (Godt and Tepass, 1998; Gonzalez-Reyes and St. Johnston, 1998). The cues for establishing the posterior follicle cells and for their high level of DE-cadherin expression have not been identified. In the germline, oocyte specification factors may induce the high-level expression of DE-cadherin in the oocyte, since DE-cadherin in the expressed at an equally low level in all 16 germline cells in *egl* mutants (Godt and Tepass, 1998). Therefore, oocyte specification seems to be an essential prerequisite for oocyte translocation. This is also consistent with the observation that the oocyte can be induced at ectopic positions in the egg chamber (see above).

B. The *spindle* Genes

The *spn* genes are required for many aspects of oocyte development, such as the formation of karyosome in the oocyte and the establishment of dorsal-ventral polarity of the oocyte (Gonzalez-Reyes *et al.,* 1997). The mislocalization of the oocyte is also observed in *spn* egg chambers at low frequencies (Gonzalez-Reyes and St. Johnston, 1994; Gonzalez-Reyes *et al.,* 1997). These mislocalized oocytes fail to form the normal anterior-posterior pattern. For example, when the oocyte is localized to the middle of the egg chamber with nurse cells at both ends, it becomes dicephalic; anterior markers, such as *bicoid* mRNA, are localized at both ends of the oocyte whereas the posterior marker is in the middle (Gonzalez-Reyes and St. Johnston, 1994).

It is not clear how the *spn* genes direct the positioning of the oocyte within the egg chamber. It is speculated that the *spn* mutations delay the choice of the oocyte between the two pro-oocytes (see above). This delay may disrupt the positioning mechanism (Gonzalez-Reyes and St. Johnston, 1998). Among the five *spn* genes, *spnE* and *spnB* have been cloned (Gillespie and Berg, 1995; Ghabrail *et al.*, 1998). *spnE* encodes an ATP-dependent RNA helicase belonging to the DE-H protein family. The DEAD/DE-H family of proteins is important for RNA processing in a variety of organisms (Gillespie and Berg, 1995). In *Drosophila*, another member of the DEAD protein family, Vasa, is necessary for the localization of *nanos* mRNA during embryogenesis (Gavis *et al.*, 1996; Dahanukar and Wharton, 1996). Hence, it is possible that *spnE* may play a role in mRNA transport and/or localization during oocyte determination. Disruption of this process may delay oocyte determination, leading to oocyte mispositioning. Consistent with this, *vasa* mutant egg chambers also contain mislocalized oocyte at a low frequency. The *spnB* gene is involved in meiotic cell cycle regulation and oocyte specification (see above); defective *spnB* gene may thus delay oocyte determination as well.

VI. A Comparative Analysis of Early Oogenesis and Early Spermatogenesis

Oogenesis and spermatogenesis in *Drosophila* share many parallel features, despite their distinctly different missions. For example, testicular germ cells proliferate in a very similar manner to ovarian germ cells (Fuller, 1993). At the apical tip of the testis, there are five to nine spermatogenic germline stem cells anchored around approximately 12 somatic hub cells. Similar to oogenesis, spermatogenesis starts with the asymmetric division of the germline stem cell, which generates one daughter stem cell that remains in contact with the hub and one differentiated daughter cell, the primary spermatogonial cell, that is displaced one cell away from the hub. The primary spermatogonial cell then undergoes four rounds of synchronized, incomplete divisions, thus giving rise to a cyst of 16 spermatogonial cells. Spermatogenic germline cyst formation, like oogenic germline cyst formation, is accompanied by morphogenesis of the fusome. Subsequently, unlike cystocytes in oogenic germline cysts, the 16 spermatogonial cells synchronously differentiate into primary spermatocytes in the premeiotic S phase, enter a prolonged G2 growth phase, undergo two meiotic divisions with incomplete cytokinesis, and generate 64 interconnected spermatids. Eventually, the 64 spermatids synchronously undergo spermiogenesis to become mature sperm. During spermatogenesis, germline stem cell division is also accompanied by somatic stem cell division in an even more strictly coupled manner than oogenesis. During oogenesis, two to three ovarian somatic stem cells are located in the region 2b and are responsible for generating ~16 prefollicle cells that will envelop each differentiating germline cyst. This coordination is achieved by the collective action of the somatic stem cells. During spermatogenesis, each germline stem cell is flanked by a spermatogenic somatic stem cell on each side. These somatic stem cells, called cyst progenitor cells, are also anchored around the hub. Accompanying each germline stem cell division, the cyst progenitor cells divide once to produce two cyst cells that envelop the resulting primary spermatogonial cell. The two cyst cells are nonmitotic but they grow in size during germline cyst development to continuously envelop the germline cyst. This section compares the two processes in light of the recent progress in the field.

A. The Testicular Fusome vs. the Ovarian Fusome

Although fusomes and ring canals exist in the germline cyst in both testes and ovaries, they contain distinct components and behave differently during spermatogenesis and oogenesis (Hime *et al.,* 1996). The testicular fusome contains F-actin in addition to membrane skeletal proteins such as α-spectrin and adducin-like protein HTS. Although Bam and cyclin A are located in the ovarian fusome, it is not known whether they are present in the testicular fusome. Unlike the ovarian fusome, the testicular fusome does not breakdown after germline cyst formation. Instead, the testicular fusome persists and enlarges during the prolonged growth period of the 16-cell cyst. During meiosis, the testicular fusome further branches and extends to new daughter cells and rearranges itself to form globular endings in the meiotic cells. These endings are connected by thread-like structures penetrating the meiotic ring canals to form a fusome. Upon the formation of 64 spermatids in a cyst, the testicular fusome appears more uniform in diameter and overall becomes thinner. Subsequently, during the spermatid elongation phase, the testicular fusome moves together with ring canals toward the distal end, loses F-actin, and eventually arranges into hexagonal structures. Hence, the testicular fusome persists and may coordinate the synchrony of the male germline cyst division and differentiation throughout the mitotic, meiotic, and postmeiotic phases of spermatogenesis. During these processes, the geometric arrangement of the ring canals correlates with the morphological changes of the fusome, suggesting that the testicular fusome may play a role in the spatial organization of germline cysts during spermatogenesis, similar to the role of ovarian fusome during oogenesis. Further experiments will reveal these potential functions of the testicular fusome.

B. The Testicular Ring Canal vs. the Ovarian Ring Canal

Both testicular and ovarian ring canals are derived from the arrested cleavage furrow as a result of incomplete cytokinesis. In *Drosophila* and other animal cells, the cleavage furrow represents the constriction of the actinomyosin-mediated contractile ring (Satterwhite and Pollard, 1992). Other proteins in the ring canal include anillin and septin family members, such as Peanut, septin 1, and septin 2 (Miller and Kiehart, 1995; Neufeld and Rubin, 1994; Fares *et al.,* 1995; Field and Alberts, 1995; Field *et al.,* 1996).

During oogenesis, 15 ring canals are formed within a germline cyst. The earliest component detected on the ovarian ring canals is anillin, which is present in the temporary ring canal between the germline stem cell and the cystoblast as well as on the ring canals in the developing cysts (de Cuevas and Spradling, 1998). The localization of anillin to the ring canals is transient and is only detected on nascent ring canals. The anillin staining on ring canal is greatly reduced in stage-1 egg chamber and is barely detectable by stage 2 (Field and Alberts, 1995; de Cuevas and Spradling, 1998). After cleavage furrow arrest, the ring canal matures by recruiting several proteins sequentially. The first detectable component on the nascent ring canal is the phosphotyrosine epitope (Robinson *et al.,* 1994). It is has been proposed that the tyrosine phosphorylation of the cleavage furrow components helps the arrest of the cleavage furrow and/or initiate ring canal formation (Robinson *et al.,* 1994). The phosphotyrosine epitopes become more prominent after the formation of the female germline cyst and can be detected on the ring canals throughout the rest of oogenesis (Robinson *et al.,* 1994). At least some of these epitopes are the substrates of nonreceptor tyrosine kinase Tec29 and Src64, which are both recruited to the ring canal (Roulier *et al.,* 1998; Guarnieri *et al.,* 1998; Dodson *et al.,* 1998). Mutations in Tec29 and Src64 genes cause a great reduction of the phosphotyrosine epitopes on ring canals and failure in ring canal growth, thus suggesting that tyrosine phosphorylation is required for the growth and maintenance of ring canals (Roulier *et al.,* 1998; Guarnieri *et al.,* 1998; Dodson *et al.,* 1998). When the ovarian germline cyst moves to region 2 of the germarium, F-actin and HTS-RC, the ring canal form of the HTS protein, are recruited to the ring canal (Robinson *et al.,* 1994; Yue and Spradling, 1992). Afterwards, when the stage-1 egg chamber is formed in region 3 of the germarium, Kelch protein can be detected on the ring canal (Xue and Cooley, 1993; Robinson *et al.,* 1994). During later stages of oogenesis, the ring canal diameter increases approximately 10-fold to reach approximately 10 μm in stage-10 egg chambers (Robinson *et al.,* 1994). Although the molecular nature of this growth remains unclear, it has been shown that both Kelch and Tec29/Src64 are required for this process, probably for maintaining the proper actin organization within ring canals (Xue and Cooley, 1993; Robinson and Cooley, 1997; Roulier *et al.,* 1998; Guarnieri *et al.,* 1998; Dodson *et al.,* 1998).

Testicular ring canals are different from ovarian ring canals in ontogeny. In the testis, four rounds of mitotic divisions are followed by two rounds of meiotic divisions, all with incomplete cytokinesis. Consequently, the 64 spermatids in a postmeiotic germline cyst are interconnected by 63 ring canals. The mitotic ring canals and the meiotic ring canals in testis are assembled in a similar manner.

Testicular ring canal formation resembles the cleavage furrow formation in other cell types (Hime *et al.,* 1996). At the late anaphase of mitotic and meiotic divisions, a contractile ring is formed at the cell periphery between the separating daughter nuclei. The contractile ring contains F-actin, the cytoplasmic myosin II, anillin, Peanut, septin 1, and septin 2. As the cell cycle progresses, the contractile ring constricts to become the cleavage furrow, which retains all the existing molecular components in the ring. Upon the arrest of the cleavage furrow at late telophase, the phosphotyrosine epitopes are accumulated onto the nascent ring canal, analogous to ovarian ring canal assembly. Whether these epitopes in males reflect the same proteins found in females is unknown. Also unknown is whether Tec29 and Src64 are involved in the phosphorylation of the male phosphotyrosine epitopes and in the formation and maintenance of the testicular ring canals. The nascent ring canals subsequently lose F-actin and myosin II staining but still retain anillin, Peanut, septin 1, and septin 2 staining. In contrast, the major components of the ovarian ring canals, F-actin, HTS-RC, and Kelch, cannot be detected on testicular ring canals. During spermatid elongation, all the 63 ring canals move toward the distal end along with the fusome. Therefore, there are striking structural and behavioral differences between testicular and ovarian ring canals, presumably reflecting their functional difference. In the testis, the major function of the ring canal is to coordinate cell division and differentiation by allowing for cell-cell communication. In contrast, ovarian ring canals not only serve as cell-cell communication tunnels but also have to sustain the massive translocation of cytoplasmic components from nurse cells to the oocyte in a process known as dumping. It is proposed that components, such as F-actin, HTS-RC, and Kelch, are essential for the formation of robust ovarian ring canals that can maintain its structure during the dumping process.

C. Difference in Genetic Regulation of Early Oogenesis and Early Spermatogenesis

Germline stem cell division and germline cyst formation occur in a very similar way in males and females. Moreover, these processes are controlled by intrinsic mechanisms and somatic signaling in both sexes. Despite these general similarities, the molecular events involved in germline stem cell division and cyst formation during oogenesis are different from these processes in spermatogenesis. This section compares and contrasts the molecular mechanisms involved in these early oogenic and spermatogenic processes.

1. Intrinsic Mechanism—*bam* and *bgcn*

bam and *bgcn* are required for the differentiation of the germline stem cell daughter in both sexes (McKearin and Spradling, 1990; Gateff, 1982; Gonczy *et al.,* 1997). Loss of *bam* or *bgcn* cause the overproliferation of germline cells, generating a massive number of undifferentiated germline cells in both ovaries and testes. These germline cells are stem cell-like in ovaries but spermatogonial cell-like in testes. The spermatogonial features of the *bam* and *bgcn* mutant cells are reflected as follows: first, these cells express enhancer trap markers for spermatogonial cells, although some enhancer trap markers for the stem cells are also expressed. Second, the division of these cells is incomplete as marked by the ring canals and the fusome expanding through the intercellular bridges. Third, these cells divide synchronously, which is characteristic of spermatogonial cells but not germline stem cells. Finally, the *bgcn* mutant germ cells produce BamC, which is only present in spermatogonial cells but not in stem cells. These features contrast *bam* and *bgcn* mutant germ cells in the ovary, and suggest that although *bam* and *bgcn* function for stem cell daughter differentiation during oogenesis, they restrict the proliferation of the spermatogonial cells during spermatogenesis. In both cases, *bam* and *bgcn* regulate the spermatogonial proliferation in a cell autonomous manner (Gonczy *et al.,* 1997).

2. Extrinsic Regulation—the TGF-β Signaling Pathway

Although the *dpp*-signaling pathway is important for germline stem cell division during oogenesis (Xie and Spradling, 1998), *dpp* and other players in the pathway, except for *punt* and *schnurri,* are not involved in the regulation of the male germ cell proliferation (Matunis *et al.,* 1997). *punt* encodes type II TGF-β receptor while *schnurri* encodes a zinc finger transcription factor, which is a known downstream effector of the *dpp* signaling pathway during *Drosophila* embryogenesis (Arora *et al.,* 1995; Grieder *et al.,* 1995; Staehling-Hampton *et al.,* 1995). Mutations in *punt* and *schnurri* result in germ cell overproliferation at a low and high frequency, respectively (Matunis *et al.,* 1997). However, the overproliferated germline cells in *punt* and *schnurri* mutants testes are not stem cell-like but spermatoginal cell-like. The overproliferated germ cells in *punt* and *schnurri* mutants undergo synchronized incomplete cytokinesis like spermatogonial cells. Moreover, they express spermatogonial markers such as BAM-C. In addition to the difference in the overproliferated germ cells, *punt* and *schnurri* are required in the somatic cyst cells to regulate the proliferation of the primary spermatogonial cells surrounded by the somatic cyst cells. This is also different from the *dpp* effectors in oogenesis, which are involved in ovarian germline stem cell regulation in a cell autonomous manner. Hence, the function of the *dpp* signaling pathway in early gametogenesis is not conserved between the two sexes.

Given that other TGF-β family members have been identified in *Drosophila* recently (Arora *et al.,* 1995; Wharton *et al.,* 1993), it is proposed that, in the testis, *punt* and *schnurri* are probably the effectors of these new TGF-βs other than DPP (Matunis *et al.,* 1997). The germ cell intrinsic factors of the *punt/schnurri*-mediated pathway have not been identified yet. It will be interesting to see whether *bam* and *bgcn* are involved in this pathway.

3. Germline Regulation of Somatic Cell Division in the Testis

An interesting aspect of early spermatogenesis that appears to be different from oogenesis is the germline regulation of somatic cell fate (Gonzcy and DiNardo, 1996). Although the overall somatic cell organization is initially normal in the absence of the germ cells, the hub cells become larger in size. BrdU pulse chase experiments showed that some of the proliferating somatic cyst progenitor cells were recruited to the hub and apparently converted their fate to hub cells (Gonzcy and DiNardo, 1996). These observations suggest that germ cells are required to prevent the cyst cell from adopting the hub cell fate. In more than 50% of agamatic testis, the cyst progenitor cells are missing, indicating that germline is required for maintaining these somatic stem cells. Most strikingly, without germ cells, the cyst progenitor cells continue to proliferate. Therefore, during early spermatogenesis, germ cells are not only regulated by the surrounding somatic cells but also required for the maintenance and proliferation restriction of these somatic cells. These features of germline-somatic interaction are not found during oogenesis (Margolis and Spradling, 1995).

VII. Concluding Remarks

Early oogenesis in *Drosophila* provides an excellent opportunity for studying mechanisms underlying cell proliferation, cell fate specification, and patterning. Recent progress has clearly revealed the requirement of both intra- and intercellular mechanisms for germline stem cell division, while germline cyst formation and oocyte determination appear to be predominantly governed by the intrinsic mechanism. Following oocyte determination the translocation of the oocyte depends on the differential expression of cell adhesion molecules in the oocyte and its surrounding prefollicular cells. Some important intrinsic factors and cell-cell signaling molecules have been identified, and their regulatory relationship has begun to be explored in the context of gene expression, cell cycle progression, cytoskeletal organization, and mitotic behavior of the germline cells. Current research supports the idea that the formation and self-renewing division of GSCs are largely under the control of extrinsic signals, which induce the intrinsic asymmetry of germline stem cells and cystoblasts. This asymmetry then provides an initial lineage cue that guides the patterning of germline cyst formation, determination of the oocyte, and translocation of the determined oocyte to the posterior pole of the germline cyst.

Elucidation of the mechanisms for early oogenesis is now occurring at an exciting rate. New genetic screens will identify more genes required for various events during early oogenesis. Meanwhile, further analysis of known inter- and intracellular mechanisms will reveal exactly how cell-cell signaling and intracellular processes occur in an orchestrated manner to ensure the progression of early oogenesis. The knowledge gained from these studies should provide valuable insight not only into the oogenic mechanism but also the mechanism of cell proliferation, differentiation, and patterning in general.

Acknowledgments

We thank Drs. Anna Chao, Jacqueline Swain, and Michael Parisi for critical reading of the manuscript. The writing of this review and the stem cell work in the Lin laboratory are supported by a NIH grant (R01HD33760) and a David and Lucile Packard Fellowship to H. L.

References

Ait-Ahmed, O., Bellon, B., Capri, M., Joblet, C., and Thomas-Delaage, M. (1992). The yemanuclein-alpha: A new *Drosophila* DNA binding protein specific for the oocyte nucleus. *Mech. Dev.* **37,** 69–80.

Ait-Ahmed, O., Thomas-Cavallin, M., and Rosset, R. (1987). Isolation and characterization of a region of the *Drosophila* genome which contains a cluster of differentially expressed maternal genes (*yema* gene region). *Dev. Biol.* **122,** 153–162.

Arora, K., Dai, H., Kazuko, S. G., Jamal, J., O'Connor, M. B., Letsou, A., and Warrior, R. (1995). The *Drosophila schnurri* gene acts in the Dpp/TGF beta signaling pathway and encodes a transcription factor homologous to the human MBP family. *Cell* **81,** 781–790.

Benfey, P. N. (1999). Stem cells: A tale of two kingdoms. *Curr. Biol.* **9,** R171–172.

Bhat, K. M. (1999). The posterior determinant gene *nanos* is required for the maintenance of the adult germline stem cells during *Drosophila* oogenesis. *Genetics* **151,** 1479–1492.

Bohmert, K., Camus, I., Bellini, C., Bouchez, D., Caboche, M., and Benning, C. (1998). *AGO*1 defines a novel locus of *Arabidopsis* controlling leaf development. *EMBO J.* **17,** 170–180.

Brown, E. H., and King, R. C. (1962). Oogonial and spermatogonial differentiation within a mosaic gonad of *Drosophila melanogaster. Growth* **26,** 53–70.

Brown, E. H., and King, R. C. (1964). Studies on the events resulting in the formation of an egg chamber in *Drosophila melanogaster. Growth* **28,** 41–81.

Capri, M., Santoni, M. J., Thomas-Delaage, M., and Ait-Ahmed, O. (1997). Implication of a 5′ coding sequence in targeting maternal mRNA to the *Drosophila* oocyte. *Mech. Dev.* **68,** 91–100.

Carpenter, A. T. (1994). *Egalitarian* and the choice of cell fates in *Drosophila melanogaster* oogenesis. *Ciba Found. Symp.* **182,** 223–246.

Carpenter, A. T. (1975). Electron microscopy of meiosis in *Drosophila melanogaster* females. I. Structure, arrangement, and temporal change of the synaptonemal complex in wild-type. *Chromosoma* **51,** 157–182.

Chang, J. S., Tan, L., and Schedl, P. (1999). The *Drosophila* CPEB homolog, *orb,* is required for Oskar protein expression in oocytes. *Dev. Biol.* **215,** 91–106.

Cheung, H. K., Serano, T. L., and Cohen, R. S. (1992). Evidence for a highly selective RNA transport system and its role in establishing the dorsoventral axis of the *Drosophila* egg. *Development* **114,** 653–661.

Christerson, L. B., and McKearin, D. M. (1994). *orb* is required for anteroposterior and dorsoventral patterning during *Drosophila* oogenesis. *Genes Dev.* **8,** 614–628.

Clark, I., Giniger, E., Ruohola-Baker, H., Jan, L. Y., and Jan, Y. N. (1994). Transient posterior localization of a kinesin fusion protein reflects anteroposterior polarity of the *Drosophila* oocyte. *Curr. Biol.* **4,** 289–300.

Clark, I. E., Jan, L. Y., and Jan, Y. N. (1997). Reciprocal localization of Nod and kinesin fusion proteins indicates microtubule polarity in the *Drosophila* oocyte, epithelium, neuron and muscle. *Development* **124,** 461–470.

Clark, K. A., and McKearin, D. M. (1996). The *Drosophila* stonewall gene encodes a putative transcriptional factor essential for germ cell development. *Development* **122,** 937–950.

Cooley, L., and Theurkauf, W. E. (1994). Cytoskeletal functions during *Drosophila* oogenesis. *Science* **266,** 590–596.

Cox, D. N., Chao, A., Baker, J., Chang, L., Qiao, D., and Lin, H. (1998). A novel class of evolutionarily conserved genes defined by *piwi* are essential for stem cell self-renewal. *Genes Dev.* **12,** 3715–3727.

Cox, D. N., Chao, A., and Lin, H. (2000). *piwi* encodes a nucleoplasmic factor whose activity modulates the number and division rate of germline stem cells. *Development.* **127,** 503–514.

Dahanukar, A., and Wharton, R. P. (1996). The Nanos gradient in *Drosophila* embryos is generated by translational regulation. *Genes Dev.* **10,** 2610–2620.

Dalby, B., and Glover, D. M. (1992). 3′ Non-Translated sequences in *Drosophila* cyclin B transcripts direct posterior pole accumulation late in oogenesis and peri-nuclear association in syncytial embryos. *Development* **115,** 989–997.

Dalby, B., and Glover, D. M. (1993). Discrete sequence elements control posterior pole accumulation and translational repression of maternal cyclin B RNA in *Drosophila. EMBO J.* **12,** 1219–1227.

de Cuevas, M., Lee, J. K., and Spradling, A. C. (1996). alpha-spectrin is required for germline cell division and differentiation in the *Drosophila* ovary. *Development* **122,** 3959–3968.

de Cuevas, M., and Spradling, A. C. (1998). Morphogenesis of the *Drosophila* fusome and its implications for oocyte specification. *Development* **125,** 2781–2789.

Deng, W., and Lin, H. (1997). Spectrosomes and fusomes anchor mitotic spindles during asymmetric germ cell divisions and facilitate the formation of a polarized microtubule array for oocyte specification in *Drosophila. Dev. Biol.* **189,** 79–94.

Digilio, F. A., Pannuti; A., Lucchesi, J. C., Furia, M., and Polito, L. C. (1996). *Tosca:* A *Drosophila* gene encoding a nuclease specifically expressed in the female germline. *Dev. Biol.* **178,** 90–100.

Dodson, G. S., Guarnieri, D. J., and Simon, M. A. (1998). Src64 is required for ovarian ring canal morphogenesis during *Drosophila* oogenesis. *Development* **125,** 2883–2892.

Ephrussi, A., Dickinson, L. K., and Lehmann, R. (1991). Oskar organizes the germ plasm and directs localization of the posterior determinant Nanos. *Cell* **66,** 37–50.

Ephrussi, A., and Lehmann, R. (1992). Induction of germ cell formation by *oskar. Nature* **358,** 387–392.

Fares, H., Peifer, M., and Pringle, J. R. (1995). Localization and possible functions of *Drosophila* septins. *Mol. Biol. Cell* **6,** 1843–1859.

Field, C. M., al-Awar, O., Rosenblatt, J., Wong, M. L., Alberts, B., and Mitchison, T. J. (1996). A purified *Drosophila* septin complex forms filaments and exhibits GTPase activity. *J. Cell Biol.* **133,** 605–616.

Field, C. M., and Alberts, B. M. (1995). Anillin, a contractile ring protein that cycles from the nucleus to the cell cortex. *J. Cell Biol.* **131,** 165–178.

Forbes, A., and Lehmann, R. (1998). *Nanos* and *Pumilio* have critical roles in the development and function of *Drosophila* germline stem cells. *Development* **125,** 679–690.

Forbes, A. J., Lin, H., Ingham, P. W., and Spradling, A. C. (1996). *hedgehog* is required for the proliferation and specification of ovarian somatic cells prior to egg chamber formation in *Drosophila. Development* **122,** 1125–1135.

Fuller, M. T. (1993). Spermatogenesis. *In* "The Development of *Drosophila melanogaster*" (A. Martinez-Arias and M. Bates, Eds.), Vol. 1, pp. 71–147. Cold Spring Harbor Laboratory Press. Cold Spring Harbor, New York.

Gateff, E. (1982). Gonial cell neoplasm of genetic origin affecting both sexes of *Drosophila melanogaster. Prog. Clin. Biol. Res.* **85,** 621–632.

Gavis, E. R., Lunsford, L., Bergsten, S. E., and Lehmann, R. (1996). A conserved 90 nucleotide element mediates translational repression of *nanos* RNA. *Development* **122,** 2791–2800.

Ghabrial, A., Ray, R. P., and Schupbach, T. (1998). *okra* and *spindle-B* encode components of the RAD52 DNA repair pathway and affect meiosis and patterning in *Drosophila* oogenesis. *Genes Dev.* **12,** 2711–2723.

Giardina, A. (1901). Origine dell' oocite e dello cellule nutrici nel *Dytiscus. Int. Monatsschrift Anat. Physiol.* **18,** 417–484.

Gillespie, D. E., and Berg, C. A. (1995). *Homeless* is required for RNA localization in *Drosophila* oogenesis and encodes a new member of the DE-H family of RNA-dependent ATPases. *Genes Dev.* **9,** 2495–2508.

Godt, D., and Tepass, U. (1998). *Drosophila* oocyte localization is mediated by differential cadherin- based adhesion. *Nature* **395,** 387–391.

Gonczy, P., and DiNardo, S. (1996). The germ line regulates somatic cyst cell proliferation and fate during *Drosophila* spermatogenesis. *Development* **122,** 2437–2447.

Gonczy, P., Matunis, E., and DiNardo, S. (1997). *bag-of-marbles* and *benign gonial cell neoplasm* act in the germline to restrict proliferation during *Drosophila* spermatogenesis. *Development* **124,** 4361–4371.

Gonzalez-Reyes, A., Elliott, H., and St. Johnston, D. (1997). Oocyte determination and the origin of polarity in *Drosophila:* The role of the *spindle* genes. *Development* **124,** 4927–4937.

Gonzalez-Reyes, A., Elliott, H., and St. Johnston, D. (1995). Polarization of both major body axes in *Drosophila* by *gurken-torpedo* signaling. *Nature* **375,** 654–658.

Gonzalez-Reyes, A., and St. Johnston, D. (1994). Role of oocyte position in establishment of anterior-posterior polarity in *Drosophila. Science* **266,** 639–42.

Gonzalez-Reyes, A., and St. Johnston, D. (1998). The *Drosophila* AP axis is polarised by the cadherin-mediated positioning of the oocyte. *Development* **125,** 3635–3644.

Goode, S., Melnick, M., Chou, T. B., and Perrimon, N. (1996). The neurogenic genes *egghead* and *brainiac* define a novel signaling pathway essential for epithelial morphogenesis during *Drosophila* oogenesis. *Development* **122,** 3863–3879.

Grieder, N. C., Nellen, D., Burke, R., Basler, K., and Affolter, M. (1995). *Schnurri* is required for *Drosophila* Dpp signaling and encodes a zinc finger protein similar to the mammalian transcription factor PRDII-BF1. *Cell* **81,** 791–800.

Guarnieri, D. J., Dodson, G. S., and Simon, M. A. (1998). SRC64 regulates the localization of a Tec-family kinase required for *Drosophila* ring canal growth. *Mol. Cell* **1,** 831–840.

Hawkins, N. C., Thorpe, J., and Schupbach, T. (1996). *Encore,* a gene required for the regulation of germ line mitosis and oocyte differentiation during *Drosophila* oogenesis. *Development* **122,** 281–290.

Hawkins, N. C., Van Buskirk, C., Grossniklaus, U., and Schupbach, T. (1997). Post-transcriptional regulation of *gurken* by *encore* is required for axis determination in *Drosophila. Development* **124,** 4801–4810.

Hay, B., Jan, L. Y., and Jan, Y. N. (1988). A protein component of *Drosophila* polar granules is encoded by *vasa* and has extensive sequence similarity to ATP-dependent helicases. *Cell* **55,** 577–587.

Hime, G. R., Brill, J. A., and Fuller, M. T. (1996). Assembly of ring canals in the male germ line from structural components of the contractile ring. *J. Cell Sci.* **109,** 2779–2788.

Holleran, E. A., Tokito, M. K., Karki, S., and Holzbaur, E. L. (1996). Centractin (ARP1) associates with spectrin revealing a potential mechanism to link dynactin to intracellular organelles. *J. Cell Biol.* **135,** 1815–1829.

Keyes, L. N., and Spradling, A. C. (1997). The *Drosophila* gene *fs(2)cup* interacts with otu to define a cytoplasmic pathway required for the structure and function of germ-line chromosomes. *Development* **124,** 1419–1431.

Kim-Ha, J., Kerr, K., and Macdonald, P. M. (1995). Translational regulation of *oskar* mRNA by Bruno, an ovarian RNA-binding protein, is essential. *Cell* **81,** 403–412.

Kim-Ha, J., Kim, J., and Kim, Y. J. (1999). Requirement of RBP9, a *Drosophila* Hu homolog, for regulation of cystocyte differentiation and oocyte determination during oogenesis. *Mol. Cell Biol.* **19,** 2505–2514.

Kim-Ha, J., Smith, J. L., and Macdonald, P. M. (1991). *oskar* mRNA is localized to the posterior pole of the *Drosophila* oocyte. *Cell* **66,** 23–35.

King, F. J., and Lin, H. (1999). Somatic signaling mediated by *fs(1)Yb* is essential for germline stem cell maintenance during *Drosophila* oogenesis. *Development* **126,** 1833–1844.

King, R. C. (1970). "*Ovarian Development in Drosophila melanogaster.*" Academic Press, New York.

Koch, E. A., and King, R. C. (1966). The origin and early differentiation of the egg chamber of *Drosophila melanogaster. J. Morphol.* **119,** 283–303.

Koch, E. A., Smith, P. A., and King, R. C. (1967). The division and differentiation of *Drosophila* cystocytes. *J. Morphol.* **121,** 55–70.

Koch, E. A., and Spitzer, R. H. (1983). Multiple effects of colchicine on oogenesis in *Drosophila:* Induced sterility and switch of potential oocyte to nurse-cell developmental pathway. *Cell Tissue Res.* **228,** 21–32.

Lane, M. E., and Kalderon, D. (1994). RNA localization along the anteroposterior axis of the *Drosophila* oocyte requires PKA-mediated signal transduction to direct normal microtubule organization. *Genes Dev.* **8,** 2986–2995.

Lantz, V., Ambrosio, L., and Schedl, P. (1992). The *Drosophila orb* gene is predicted to encode sex-specific germline RNA-binding proteins and has localized transcripts in ovaries and early embryos. *Development* **115,** 75–88.

Lantz, V., Chang, J. S., Horabin, J. I., Bopp, D., and Schedl, P. (1994). The *Drosophila orb* RNA-binding protein is required for the formation of the egg chamber and establishment of polarity. *Genes Dev.* **8,** 598–613.

Lasko, P. F., and Ashburner, M. (1988). The product of the *Drosophila* gene *vasa* is very similar to eukaryotic initiation factor-4A. *Nature* **335,** 611–617.

Lavoie, C. A., Ohlstein, B., and McKearin, D. M. (1999). Localization and function of Bam protein require the *benign gonial cell neoplasm* gene product. *Dev. Biol.* **212,** 405–413.

Lawrence, P. A., and Struhl, G. (1996). Morphogens, compartments, and pattern: Lessons from *Drosophila? Cell* **85,** 951–961.

Leon, A., and McKearin, D. (1999). Identification of TER94, an AAA ATPase protein, as a Bam-dependent component of the *Drosophila* fusome. *Mol. Biol. Cell* **10,** 3825–3834.

Li, M., McGrail, M., Serr, M., and Hays, T. S. (1994). *Drosophila* cytoplasmic dynein, a microtubule motor that is asymmetrically localized in the oocyte. *J. Cell Biol.* **126,** 1475–1494.

Lilly, M. A., de Cuevas, M., and Spradling, A. C. (2000). Cyclin A associates with the fusome during germline cyst formation in the *Drosophila* ovary. *Dev. Biol.* **218,** 53–63.

Lilly, M. A., and Spradling, A. C. (1996). The *Drosophila* endocycle is controlled by cyclin E and lacks a checkpoint ensuring S-phase completion. *Genes Dev.* **10,** 2514–2526.

Lin, H., and Spradling, A. C. (1993). Germline stem cell division and egg chamber development in transplanted *Drosophila* germaria. *Dev. Biol.* **159,** 140–152.

Lin, H., and Spradling, A. C. (1995). Fusome asymmetry and oocyte determination in *Drosophila. Dev. Genet.* **16,** 6–12.

Lin, H., and Spradling, A. C. (1997). A novel group of *pumilio* mutations affects the asymmetric division of germline stem cells in the *Drosophila* ovary. *Development* **124,** 2463–2476.

Lin, H., Yue, L., and Spradling, A. C. (1994). The *Drosophila* fusome, a germline-specific organelle, contains membrane skeletal proteins and functions in cyst formation. *Development* **120,** 947–956.

Lohs-Schardin, M. (1982). *Dicephalic*—a *Drosophila* mutant affecting polarity in follicle organization and embryonic patterning. *J. Roux Arch. Dev. Biol.* **191,** 28–36.

Mach, J. M., and Lehmann, R. (1997). An Egalitarian-BicaudalD complex is essential for oocyte specification and axis determination in *Drosophila. Genes Dev.* **11,** 423–435.

Mahone, M., Saffman, E. E., and Lasko, P. F. (1995). Localized *Bicaudal-C* RNA encodes a protein containing a KH domain, the RNA binding motif of FMR1. *EMBO J.* **14,** 2043–2055.

Mahowald, A. P., and Strassheim, J. M. (1970). Intercellular migration of centrioles in the germarium of *Drosophila melanogaster.* An electron microscopic study. *J. Cell Biol.* **45,** 306–320.

Mal'ceva, N. I., Belyaeva, E. S., King, R. C., and Zhimulev, I. F. (1997). Nurse cell polytene chromosomes of *Drosophila melanogaster otu* mutants: Morphological changes accompanying interallelic complementation and position effect variegation. *Dev. Genet.* **20,** 163–174.

Margolis, J., and Spradling, A. (1995). Identification and behavior of epithelial stem cells in the *Drosophila* ovary. *Development* **121,** 3797–3807.

Matunis, E., Tran, J., Gonczy, P., Caldwell, K., and DiNardo, S. (1997). *punt* and *schnurri* regulate a somatically derived signal that restricts proliferation of committed progenitors in the germline. *Development* **124,** 4383–4391.

McGrail, M., Gepner, J., Silvanovich, A., Ludmann, S., Serr, M., and Hays, T. S. (1995). Regulation of cytoplasmic dynein function *in vivo* by the *Drosophila* Glued complex. *J. Cell Biol.* **131,** 411–425.

McGrail, M., and Hays, T. S. (1997). The microtubule motor cytoplasmic dynein is required for spindle orientation during germline cell divisions and oocyte differentiation in *Drosophila. Development* **124,** 2409–2419.

McKearin, D., and Christerson, L. (1994). Molecular genetics of the early stages of germ cell differentiation during *Drosophila* oogenesis. *Ciba Found. Symp.* **182,** 210–219.

McKearin, D., and Ohlstein, B. (1995). A role for the *Drosophila bag-of-marbles* protein in the differentiation of cystoblasts from germline stem cells. *Development* **121,** 2937–2947.

McKearin, D. M., and Spradling, A. C. (1990). *bag-of-marbles:* A *Drosophila* gene required to initiate both male and female gametogenesis. *Genes Dev.* **4,** 2242–2251.

Miller, K. G., and Kiehart, D. P. (1995). Fly division. *J. Cell Biol.* **131,** 1–5.

Mohler, J., and Wieschaus, E. F. (1986). Dominant maternal-effect mutations of *Drosophila melanogaster* causing the production of double-abdomen embryos. *Genetics* **112,** 803–822.

Moussian, B., Schoof, H., Haecker, A., Jurgens, G., and Laux, T. (1998). Role of the *ZWILLE* gene in the regulation of central shoot meristem cell fate during *Arabidopsis* embryogenesis. *EMBO J.* **17,** 1799–1809.

Neufeld, T. P., and Rubin, G. M. (1994). The *Drosophila peanut* gene is required for cytokinesis and encodes a protein similar to yeast putative bud neck filament proteins. *Cell* **77,** 371–379.

Niewiadomska, P., Godt, D., and Tepass, U. (1999). DE-Cadherin is required for intercellular motility during *Drosophila* oogenesis. *J. Cell Biol.* **144,** 533–547.

Oda, H., Uemura, T., Harada, Y., Iwai, Y., and Takeichi, M. (1994). A *Drosophila* homolog of cadherin associated with armadillo and essential for embryonic cell-cell adhesion. *Dev. Biol.* **165,** 716–726.

Oda, H., Uemura, T., and Takeichi, M. (1997). Phenotypic analysis of null mutants for DE-cadherin and Armadillo in *Drosophila* ovaries reveals distinct aspects of their functions in cell adhesion and cytoskeletal organization. *Genes Cells* **2,** 29–40.

Ohlstein, B., and McKearin, D. (1997). Ectopic expression of the *Drosophila* Bam protein eliminates oogenic germline stem cell. *Development* **124,** 3651–3662.

Parisi, M., and Lin, H. (1999). The *Drosophila pumilio* gene encodes two functional protein isoforms that play multiple roles in germline development, gonadogenesis, oogenesis and embryogenesis. *Genetics* **153,** 235–250.

Parisi, M., and Lin, H. (2000). Translational repression: A duet of Nanos and Pumilio. *Curr. Biol.* **10,** R81–R83.

Pauli, D., Oliver, B., and Mahowald, A. P. (1993). The role of the *ovarian tumor* locus in Drosophila melanogaster germ line sex determination. *Development* **119,** 123–134.

Peifer, M. (1993). The product of the *Drosophila* segment polarity gene *armadillo* is part of a multi-protein complex resembling the vertebrate adherens junction. *J. Cell Sci.* **105,** 993–1000.

Pokrywka, N. J., and Stephenson, E. C. (1991). Microtubules mediate the localization of *bicoid* RNA during *Drosophila* oogenesis. *Development* **113,** 55–66.

Pokrywka, N. J., and Stephenson, E. C. (1994). Localized RNAs are enriched in cytoskeletal extracts of *Drosophila* oocytes. *Dev. Biol.* **166,** 210–219.

Pokrywaka, N. J., and Stephenson, E. C. (1995). Microtubules are a general component of mRNA localization systems in *Drosophila* oocytes. *Dev. Biol.* **167,** 363–370.

Ran, B., Bopp, R., and Suter, B. (1994). Null alleles reveal novel requirements for *Bic-D* during *Drosophila* oogenesis and zygotic development. *Development* **120,** 1233–1242.

Robinson, D. N., Cant, K., and Cooley, L. (1994). Morphogenesis of *Drosophila* ovarian ring canals. *Development* **120,** 2015–2025.

Robinson, D. N., and Cooley, L. (1997). *Drosophila* Kelch is an oligomeric ring canal actin organizer. *J. Cell Biol.* **138,** 799–810.

Rorth, P. (1996). A modular misexpression screen in *Drosophila* detecting tissue-specific phenotypes. *Proc. Natl. Acad. Sci. USA* **93,** 12418–12422.

Rorth, P. (1998). Gal4 in the *Drosophila* female germline. *Mech. Dev.* **78,** 113–118.

Roth, S., Neuman-Silberberg, F. S., Barcelo, G., and Schupbach, T. (1995). *cornichon* and the EGF receptor signaling process are necessary for both anterior-posterior and dorsal-ventral pattern formation in *Drosophila. Cell* **81,** 967–978.

Rotoli, D., Andone, S., Tortiglione, C., Manzi, A., Malva, C., and Graziani, F. (1998). *hold-up* is required for establishment of oocyte positioning, follicle cell fate and egg polarity and cooperates with *Egfr* during *Drosophila* oogenesis. *Genetics* **148,** 767–773.

Roulier, E. M., Panzer, S., and Beckendorf, S. K. (1998). The Tec29 tyrosine kinase is required during *Drosophila* embryogenesis and interacts with Src64 in ring canal development. *Mol. Cell* **1,** 819–829.

Roy, A. L., Chakrabarti, D., Datta, B., Hileman, R. E., and Gupta, N. K. (1988). Natural mRNA is required for directing Met-tRNA(f) binding to 40S ribosomal subunits in animal cells: Involvement of Co-eIF-2A in natural mRNA-directed initiation complex formation. *Biochemistry* **27,** 8203–8209.

Rubsam, R., Hollmann, M., Simmerl, E., Lammermann, U., Schafer, M. A., Buning, J., and Schafer, U. (1998). The egghead gene product influences oocyte differentiation by follicle cell-germ cell interactions in *Drosophila* melanogaster. *Mech. Dev.* **72,** 131–140.

Sass, G. L., Comer, A. R., and Searles, L. L. (1995). The ovarian tumor protein isoforms of *Drosophila melanogaster* exhibit differences in function, expression, and localization. *Dev. Biol.* **167,** 201–212.

Satterwhite, L. L., and Pollard, T. D. (1992). Cytokinesis. *Curr. Opin. Cell Biol.* **4,** 43–52.

Schmidt, A., Palumbo, G., Bozzetti, M. P., Tritto, P., Pimpinelli, S., and Schafer, U. (1999). Genetic and molecular characterization of sting, a gene involved in crystal formation and meiotic drive in the male germ line of *Drosophila melanogaster. Genetics* **151,** 749–760.

Schupbach, T. (1987). Germ line and soma cooperate during oogenesis to establish the dorso-ventral pattern of egg shell and embryo in *Drosophila melanogaster. Cell* **49,** 699–707.

Schupbach, T., and Wieschaus, E. (1991). Female sterile mutations on the second chromosome of *Drosophila melanogaster.* II. Mutations blocking oogenesis or altering egg morphology. *Genetics* **129,** 1119–1136.

Smith-Leiker, T. N., Sokol, N., and Cooley, L. (1997). Functional Analysis of *hu-li tai shao* (*hts*) in the germline. *38th Annual Drosophila Research Conference,* Chicago, Illinois. April 16–20, 1997.

Spradling, A. C. S. (1993). Development genetics of oogenesis. *In* "The Development of Drosophila melanogaster" (A. Martinez-Arias and M. Bates, Eds.), Cold Spring Harbor Laboratory Press. Cold Spring Harbor, New York. Vol. 1, pp. 1–70.

Staehling-Hampton, K., Laughon, A. S., and Hoffmann, F. M. (1995). A *Drosophila* protein related to the human zinc finger transcription factor PRDII/MBPI/HIV-EP1 is required for dpp signaling. *Development* **121,** 3393–3403.

Storto, P. D., and King, R. C. (1988). Multiplicity of functions for the otu gene products during *Drosophila* oogenesis. *Dev. Genet.* **9,** 91–120.

Storto, P. D., and King, R. C. (1989). The role of polyfusomes in generating branched chains of cystocytes during *Drosophila* oogenesis. *Dev. Genet.* **10,** 70–86.

Stuurman, N., Haner, M., Sasse, B., Hubner, W., Suter, B., and Aebi, U. (1999). Interactions between coiled-coil proteins: *Drosophila* lamin Dm0 binds to the bicaudal-D protein. *Eur. J. Cell Biol.* **78,** 278–287.

Styhler, S., Nakamura, A., Swan, A., Suter, B., and Lasko, P. (1998). vasa is required for GURKEN accumulation in the oocyte, and is involved in oocyte differentiation and germline cyst development. *Development* **125,** 1569–1578.

Suter, B., Romberg, L. M., and Steward, R. (1989). Bicaudal-D, a *Drosophila* gene involved in developmental asymmetry: Localized transcript accumulation in ovaries and sequence similarity to myosin heavy chain tail domains. *Genes Dev.* **3,** 1957–1968.

Suter, B., and Steward, R. (1991). Requirement for phosphorylation and localization of the Bicaudal-D protein in *Drosophila* oocyte differentiation. *Cell* **67,** 917–926.

Swan, A., and Suter, B. (1996). Role of Bicaudal-D in Patterning the *Drosophila* egg chamber in mid- oogenesis. *Development* **122,** 3577–3586.

Tabara, H., Sarkissian, M., Kelly, W. G., Fleenor, J., Grishok, A., Timmons, L., Fire, A., and Mello, C. C. (1999). The *rde-1* gene, RNA interference, and transposon silencing in C. elegans. *Cell* **99,** 123–132.

Tearle, R., and Nusslein-Volhard, C. (1987). Tubingen mutants and stocklist. *Drosophila Inform. Serv.* **66,** 209–269.

Telfer, W. H. (1975). Development and physiology of the oocyte-nurse cell syncytium. *Adv. Insect Physiol.* **11,** 223–319.

Tepass, U., Gruszynski-DeFeo, E., Haag, T. A., Omatyar, L., Torok, T., and Hartenstein, V. (1996). *shotgun* encodes *Drosophila* E-cadherin and is preferentially required during cell rearrangement in the neurectoderm and other morphogenetically active epithelia. *Genes Dev.* **10,** 672–685.

Theurkauf, W. E., Alberts, B. M., Jan, Y. N., and Jongens, T. A. (1993). A central role for microtubules in the differentiation of *Drosophila* oocytes. *Development* **118,** 1169–1180.

Theurkauf, W. E., Smiley, S., Wong, M. L., and Alberts, B. M. (1992). Reorganization of the cytoskeleton during *Drosophila* oogenesis: Implications for axis specification and intercellular transport. *Development* **115,** 923–936.

Uemura, T., Oda, H., Kraut, R., Hayashi, S., Kotaoka, Y., and Takeichi, M. (1996). Zygotic *Drosophila* E-cadherin expression is required for processes of dynamic epithelial cell rearrangement in the *Drosophila* embryo. *Genes Dev.* **10,** 659–671.

Wang, C., Dickinson, L. K., and Lehmann, R. (1994). Genetics of nanos localization in *Drosophila. Dev. Dyn.* **199,** 103–115.

Webster, P. J., Liang, L., Berg, C. A., Lasko, P., and Macdonald, P. M. (1997). Translational repressor Bruno plays multiple roles in development and is widely conserved. *Genes Dev.* **11,** 2510–2521.

Wharton, K. A., Ray, R. P., and Gelbart, W. M. (1993). An activity gradient of *decapentaplegic* is necessary for the specification of dorsal pattern elements in the *Drosophila* embryo. *Development* **117,** 807–822.

Wharton, R. P., and Struhl, G. (1989). Structure of the *Drosophila* BicaudalD protein and its role in localizing the the posterior determinant nanos. *Cell* **59,** 881–892.

White, P., Aberle, H., and Vincent, J. P. (1998). Signaling and adhesion activities of mammalian beta-catenin and plakoglobin in *Drosophila. J. Cell Biol.* **140,** 183–195.

Wieschaus, E., and Szabad, J. (1979). The Development and function of the female germ line in *Drosophila melanogaster:* a cell lineage study. *Dev. Biol.* **68,** 29–46.

Wilson, J. E., Connell, J. E., and Macdonald, P. M. (1996). *aubergine* enhances *oskar* translation in the *Drosophila* ovary. *Development* **122,** 1631–1639.

Xie, T., and Spradling, A. C. (1998). *decapentaplegic* is essential for the maintenance and division of germline stem cells in the *Drosophila* ovary. *Cell* **94,** 251–260.

Xue, F., and Cooley, L. (1993). *kelch* encodes a component of intercellular bridges in *Drosophila* egg chambers. *Cell* **72,** 681–693.

Yue, L., and Spradling, A. C. (1992). *hu-li tai shao,* a gene required for ring canal formation during *Drosophila* oogenesis, encodes a homolog of adducin. *Genes Dev.* **6,** 2443–2454.

Zou, C., Zhang, Z., Wu, S., and Osterman, J. C. (1998). Molecular cloning and characterization of a rabbit eIF2C protein. *Gene* **211,** 187–194.

The Germline in *C. elegans*: Origins, Proliferation, and Silencing

Geraldine Seydoux[1] and Tim Schedl
Department of Molecular Biology and Genetics, Johns Hopkins University School of Medicine, Baltimore, Maryland 21205; Department of Genetics, Washington University School of Medicine, St. Louis, Missouri 63110

Germ cells are essential for reproduction, yet the molecular mechanisms that underlie their unique development are only beginning to be understood. Here we review important events that lead to the establishment of the germline and the initiation of meiotic development in *C. elegans*. Formation of the germline begins in the pregastrulation embryo, where it depends on polarization along the anterior/posterior axis and on the asymmetric segregation of P granules and associated factors. During postembryonic development, the germline expands using the GLP-1/Notch signaling pathway to promote proliferation and regulate entry into meiosis. Throughout their development, germ cells also employ unique "silencing" mechanisms to regulate their genome and protect themselves against unwanted expression from repetitive sequences including transposable elements. Together these mechanisms preserve the health and reproductive potential of the germline.

KEY WORDS: Germ cells, *C. elegans,* P granules, Meiosis, GLP-1/Notch signaling, Gene silencing, Transposable elements, RNA-mediated interference.

I. Introduction

Reproduction in animals depends on unique cells called germ cells. From their initial allocation during embryogenesis to their differentiation into highly specialized gametes in later development, germ cells follow a developmental path that sets them apart from the cells that form the soma, or body, of the organism. Germ cells are essential for reproduction and the perpetuation of species, yet the molecular mechanisms that guide their exceptional development are only now beginning to be elucidated. Here we review research in the nematode *Caenorhabditis elegans* that is contributing to our understanding of the first steps in germ cell development. In

[1] gseydoux@jhmi.edu and ts@genetics.wvstl.edu

particular, we focus on three general areas: the embryonic origin of the germline, regulation of the decision to proliferate or initiate meiotic development, and the unique "silencing" mechanisms that germ cells use to regulate their genome. Other aspects of germline development have recently been reviewed: sex determination (Hansen and Pilgrim, 1999; Kuwabara, 1999; Ellis, 1998); spermatogenesis (L'Hernault and Singson, 2000), and gametogenesis (Hubbard and Greenstein, 2000).

Several attributes make *C. elegans* uniquely suited for studying germline development. First, the relative transparency of *C. elegans* tissues make it possible to identify and observe germ cells directly in live animals throughout all stages of development. This property allows for mutant phenotypes to be described with cellular precision.

Second, germ cells in *C. elegans* exhibit numerous characteristics common to germ cells in vertebrates and other invertebrates (Kloc *et al.,* Chapter 2; Mahowald, Chapter 5; Anderson *et al.,* Chapter 6). These characteristics include germ granules (Strome and Wood, 1982); a transient association with gut cells before joining the somatic gonad during embryogenesis (Sulston *et al.,* 1983); the formation of a population of proliferating stem cells from which germ cells that undergo meiosis are supplied (Kimble and White, 1981); and many events in meiotic prophase (e.g., meiotic double strand breaks; Dernberg *et al.,* 1998) and gametogenesis (e.g., storage of maternal RNAs; Anderson and Kimble, 1997).

Third, the tractability of *C. elegans* genetics makes this nematode an excellent model system to study gene function *in vivo.* New reverse genetic methods [RNA-mediated interference (Fire *et al.,* 1998) and chemically induced intragenic deletions detected by polymerase-chain reaction (PCR) (Dernberg *et al.,* 1998)] fueled by the near complete genome sequence (*C. elegans* Sequencing Consortium, 1998) have significantly increased the pace of gene discovery. Additional candidate germline developmental genes are now being identified using DNA microarray hybridization technology where germline or germ cell-type enriched transcripts are enumerated (V. Reinke, H. Smith, J. Nance, S. Ward, and S. Kim, personal communication). Finally, new transformation methods have been developed that substantially improve the ability to express transgenes in the germline (Kelly *et al.,* 1997).

II. Formation of the Germ Lineage: Establishment of Embryonic Polarity and Segregation of Germline Factors

A. Establishment of Embryonic Polarity—An Essential First Step in the Formation of the Germline

Formation of the germ lineage begins with the asymmetric divisions of the zygote P_0 and of the germline blastomeres P_1, P_2, and P_3 (Fig. 1). These

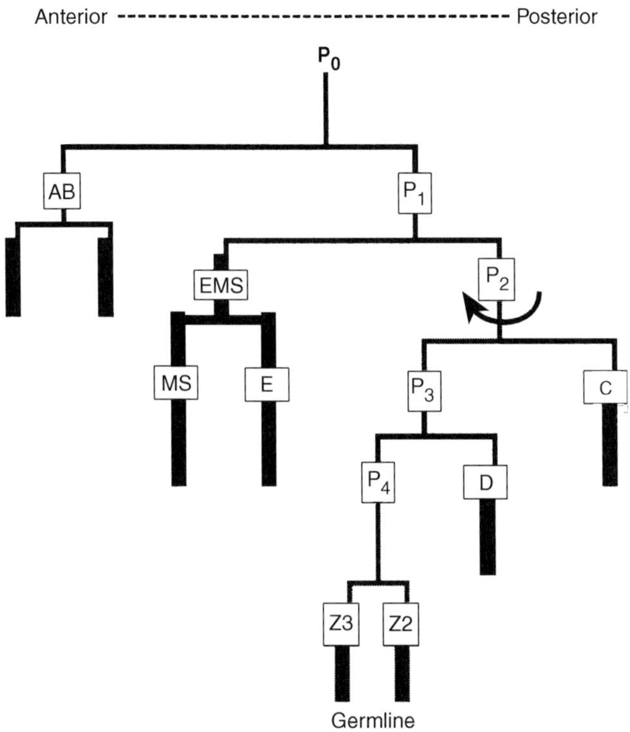

FIG. 1 The embryonic origin of the germline. This tree diagram shows the series of divisions (horizontal lines) of the zygote (P_0) into somatic (AB, EMS, C, D) and germline (P_1, P_2, P_3, P_4) blastomeres and the division of P_4 into the primordial germ cells Z2 and Z3. The onset of mRNA transcription in each lineage is indicated by thick lines. Note the reversal of polarity in P2 (arrow).

divisions occur along the anterior-posterior axis of the embryo. This axis is formed after fertilization and is thought to be determined by the site of sperm entry, which marks the posterior pole of the embryo (Goldstein and Hird, 1996). Approximately midway through the first cell cycle, after the maternal pronucleus has completed meiosis, internal cytoplasm begins to flow toward the sperm nucleus (posterior), and cortical cytoplasm flows away from it (toward the anterior) (Goldstein and Hird, 1996; Hird and White, 1993). Ultimately, this cytoplasmic reorganization leads to the posterior displacement of the mitotic spindle. As a result, the zygote divides unequally into a large anterior daughter (the somatic blastomere AB) and a smaller posterior daughter (the germline blastomere P_1). This process, with some modifications (see Fig. 1 and below), is repeated in P_1, P_2, and P_3 to yield four somatic blastomeres (AB, EMS, C, and D) and a single germline founder cell P_4. Unlike its predecessors, P_4 divides symmetrically; its two daughters are the primordial germ cells Z2 and Z3, which proliferate postembryonically to give rise to the entire germ line (Sulston *et al.*, 1983).

Genes required for the establishment of embryonic polarity have been identified in screens for maternal-effect mutations that disrupt the asymmetry of early cleavages. These screens identified six *par* (partitioning defective) genes (Kemphues and Strome, 1997). Strong loss-of-function mutations in any one of these genes cause maternal-effect lethality: embryos derived from *par*(*null*)/*par*(*null*) mothers divide symmetrically and die as amorphous masses of cells. In these mutants, all cells differentiate into somatic cell types and no recognizable germline is formed. Weaker (hypomorphic) alleles at these loci often result in maternal-effect sterility: many embryos derived from *par*(*hypo*)/*par*(*hypo*) mothers survive embryogenesis and grow up to become agametic adults (Kemphues *et al.*, 1988). These phenotypes indicate that establishment of anterior-posterior (A-P) polarity is required for germline formation and suggest that even small disturbances in A-P polarization can interfere with the production of functional germ cells.

B. P Granule Segregation

As in many animal species, germ cells in *C. elegans* contain in their cytoplasm ribonucleoprotein particles called "germ" or "P" granules. These organelles can be visualized by electron microscopy (Wolf *et al.*, 1983; Pitt *et al.*, 2000) or by staining with monoclonal antibodies that recognize different P granule epitopes (Strome and Wood, 1982). P granules are synthesized in the larval and adult germline, and are passed on to the offspring via the oocyte. During the first cleavages of the zygote, P granules segregate with the germ lineage and by the 24-cell stage, most are found in the single germline founder cell P_4 (Strome and Wood, 1982).

P granule segregation has been analyzed in live embryos by tagging these organelles with a fluorescently labeled antibody (Hird *et al.*, 1996). These studies have revealed the surprising finding that different P blastomeres utilize different mechanisms to segregate P granules. In the zygote P_0, the primary mode of segregation appears to be movement: during the period of cytoplasmic reorganization, P granules scattered throughout the cytoplasm migrate toward the posterior pole at an average rate of 4 μm/min (total embryo length along the A-P axis is 40 μm). This rate is similar to the rate of cytoplasmic flow observed at this stage, and both P granule movement and cytoplasmic flow are disrupted by microfilament inhibitors (Hird and White, 1993; Strome and Wood, 1983). These observations have suggested that cytoplasmic flow causes P granules to be swept to the posterior, where they are retained by specific attachment sites (Hird *et al.*, 1996). Other studies, however, have challenged the correlation between cytoplasmic flow and P granule segregation. Loss of either the nonmuscle myosin regulatory light chain MLC-4 or the nonmuscle myosin II heavy chain NMY-2 eliminates cytoplasmic flow but only delays P granule segregation (Shelton *et al.*, 1999). These observations suggest that although cytoplasmic flow may have an early role in P granule segregation, other mechanisms can also contribute to their asymmetry.

P granule migration through the cytoplasm is also observed in P_1; in contrast in P_2 and P_3, a different mechanism appears to be operating. In these blastomeres, P granules associate with the cytoplasmic face of the nucleus and travel with it as it is being pulled toward the side of the cell destined for the next germline blastomere (Hird *et al.*, 1996). This different mechanism may reflect the fact that in P_2 and P_3, the germline daughter is formed in the anterior, and accordingly P granules must segregate toward the anterior rather than toward the posterior. It has been suggested that this polarity reversal may be necessary for germ cells to maintain contact with cells from the intestinal lineage (Fig. 1; Schierenberg, 1987).

Finally, in all blastomeres, degradation or disassembly of the few P granules that remain on the somatic side of the cell also contributes to asymmetry (Hird *et al.*, 1996). Hence, at least three mechanisms are used to segregate P granules: movement through the cytoplasm in P_0 and P_1, association with nuclei in P_2 and P_3, and localized degradation (or disassembly) in P_0, P_1, P_2, and P_3. In P_4, which divides symmetrically into the two primordial germ cells Z2 and Z3, P granules are segregated equally to both daughters.

C. Genetics of P Granule Segregation

P granule segregation is intimately linked to the establishment of A-P polarity. Virtually all mutants that disrupt A-P polarity also affect P granule segregation. For example, all *par* mutants show defects in P granule localization; in most cases P granules are partitioned equally to all blastomeres (Kemphues and Strome, 1997). Because the PAR proteins affect several aspects of embryonic polarity, it is not known whether their role in P granule segregation is direct or indirect. An exception, however, may be made for *par-2*. Unlike other *par* genes, *par-2* is not required for the posterior segregation of P granules in P_0 but is required for the localization of P granules to the cortex in the posterior of P_1 (Boyd *et al.*, 1996). PAR-2 itself localizes to the posterior cortex of P_1, and hence is a good candidate for participating directly in the mechanism that anchors P granules there (Boyd *et al.,* 1996).

The *mes-1* locus was identified in screens for maternal-effect sterile mutants (Capowski *et al.,* 1991). In contrast to the *par* genes, *mes-1* is required specifically for the asymmetric divisions of P_2 and P_3, the two germline blastomeres with reversed anterior-posterior polarity. In *mes-1* mutants, the nuclei of these blastomeres fail to migrate to an asymmetric position, and as a result the cells divide symmetrically with P granules partitioned equally to both daughters (Strome *et al.,* 1995). *mes-1* encodes a transmembrane protein localized to the area of contact between the germline blastomeres and the adjacent gut precursor cells (L. Berkowitz and S. Strome, personal communication). This area corresponds to the region toward which the germline nuclei migrate prior to cell division, thus suggesting that *mes-1* may function directly to regulate nuclear positioning. The fact that *mes-1* mutants missegregate P granules is consistent with the hypothesis that P granules use their association with nuclei to segregate in P_2 and P_3.

In *par* and *mes-1* embryos, multiple blastomeres inherit P granules but none go on to form primordial germ cells. For example, in many *mes-1* mutants, P_4 and its sister D inherit similar amounts of P granules, which perdure in their descendents until adulthood (Strome *et al.,* 1995). [In these experiments P granules were identified by staining with the monoclonal antibodies K76 and OIC1D4, which recognize epitopes present on P granules throughout development (Strome and Wood, 1982).] Neither one of these blastomeres, however, generate primordial germ cells; instead, both form muscle, a fate normally associated only with D. These observations indicate that P granules alone (at least as defined by the K76 and OIC1D4 epitopes) are insufficient to specify germ cell fate and suggest that other factors must be segregated to (or excluded from) germline blastomeres to ensure the production of primordial germ cells.

D. Other Factors that Segregate with the Germ Lineage in Embryos

In addition to P granules, several proteins segregate with the germ lineage during early cleavages. GLD-1, MEX-1, and POS-1 are cytoplasmic proteins found both on P granules and diffusely throughout the cytoplasm of germline blastomeres (Guedes and Priess, 1997; Jones *et al.,* 1996; Tabara *et al.,* 1998). At each asymmetric cleavage, these proteins are segregated equally to both daughters but later are eliminated from the somatic blastomere and maintained only in the germline blastomere. A fourth protein, PIE-1, exhibits a different pattern of localization (Mello *et al.,* 1996; Tenenhaus *et al.,* 1998). PIE-1 is both nuclear and cytoplasmic; before each asymmetric cleavage, cytoplasmic PIE-1 becomes enriched on the germline side of the cell. During mitosis, PIE-1 also associates preferentially with the centrosome destined for the germline daughter. As a result, most PIE-1 is inherited by the germline blastomere during cell division.

mex-1, pos-1, gld-1, and *pie-1* RNAs also show a preference for the germ lineage (Guedes and Priess, 1997; Jones *et al.,* 1996; Tabara *et al.,* 1998; Tenenhaus *et al.,* 1998). In general, these maternal RNAs are present at equal levels in all blastomeres up to the four to eight-cell stage, but are maintained only in germline blastomeres in later stages. The exception is *pos-1* RNA, which shows preference for the germ lineage as early as the two-cell stage (Tabara *et al.,* 1998). These observations have suggested that mechanisms acting at the level of RNA stability and/or translation may contribute to the germline localization of these proteins. Recent data, however, have revealed that in the case of PIE-1, coding sequences are both necessary and sufficient for segregation to the germ lineage, thus suggesting that posttranslational mechanisms are also likely to be involved (K. Reese *et al.,* 2000). The mechanisms that localize the PIE-1 protein to the germ lineage are unknown but may be related to those that localize P granules, since PIE-1 and P granules co-segregate in wild-type and in mutants that disrupt asymmetric cleavages (Tenenhaus *et al.,* 1998).

Certain PAR proteins also segregate specifically with the germ lineage. PAR-1 and PAR-2 become enriched in the posterior cortex of the zygote before the first cleavage. As a result, these proteins are inherited preferentially by P_1 in the two-cell stage (Boyd *et al.,* 1996; Guo and Kemphues, 1995). Asymmetric segregation of PAR-1 and PAR-2 continues in each germline blastomere so that by the 28-cell stage these proteins are found predominantly in P_4. As described above, PAR-1 and PAR-2 are required for the proper establishment of anterior/posterior polarity in the zygote. Their continued presence in germline blastomeres suggests that they may play a similar role in these cells, but this has not yet been demonstrated conclusively (the early cleavage defects of *par* mutants make it difficult to analyze their role in later blastomeres).

III. Specification of Germ Cell Fate: Function of P Granules and Associated Factors

A. Constitutive Components of P Granules

As mentioned above, P granules alone are insufficient to specify germ cell fate. Recent evidence, however, indicates that two families of P granule components are required for normal germline development.

1. PGL Family

The founding member of this family, *pgl-1,* was identified in a screen for mutants defective in P granule staining with the monoclonal antibody K76 (Kawasaki *et al.,* 1998). *pgl-1* codes for a novel protein containing an RGG box (Kawasaki *et al.,* 1998), a motif commonly found in RNA-binding proteins, including the *Drosophila* germ granule component Vasa. PGL-1 is a constitutive component of P granules: antibodies to PGL-1 recognize P granules at all stages of development. Mutations in PGL-1 cause sterility at elevated temperatures. When grown at 25°C, *pgl-1* mutants that lack both maternal and zygotic PGL-1 have underproliferated gonads with either no gametes or poorly differentiated gametes. These phenotypes suggest that PGL-1 participates in several aspects of post-embryonic germline development, including proliferation and gametogenesis (Kawasaki *et al.,* 1998).

Using a yeast two-hybrid screen to identify proteins that interact with PGL-1, Strome and colleagues recently identified two proteins, PGL-2 and PGL-3, closely related to PGL-1 (Y. Fan, I. Kawasaki, A. Amiri, and S. Strome, personal communication). Like PGL-1, PGL-3 is a constitutive component of P granules. Elimination of PGL-3 expression by RNA-mediated interference enhances the phenotype of *pgl-1* mutants, leading to sterility even at low temperatures. These observations suggest that the PGL family of proteins interact with each other and function redundantly in P granules to regulate germline development during postembryonic stages.

2. GLH Family

GLH family members are putative RNA helicases related to the *Drosophila* germ granule component Vasa (Gruidl *et al.,* 1996). Like Vasa, GLH-1, 2, 3, and 4 contain eight helicase motifs typical of the DEAD-box family of RNA helicases (Gruidl *et al.,* 1996, and Kuznicki *et al.,* 2000). Unlike Vasa, however, the GLHs lack RGG boxes (GLH-1, 2, and 4 have several N-terminal glycine-rich repeats but these are not charged), and instead contain several CCHC zinc fingers related to the RNA-binding fingers of the retroviral nucleocapsid-binding protein of HIV. GLH-1, 2, 3, and 4 are found in P granules throughout development, although their respective distributions differ slightly in the adult (Gruidl *et al.,* 1996; Kuznicki *et al.,* 2000). Accordingly, RNA-interference experiments suggest that the GLHs have both redundant and unique functions during germline development (Kuznicki *et al.,* 2000). Elimination of all four GLHs by RNA-mediated interference results in sterility in $> 90\%$ of F1 progeny; the gonads of sterile worms contain germ cells that can form sperm but not oocytes. They also appear to lack P granules in the adult stage as evidenced by the lack of punctate staining obtained with anti-PGL-1 antibodies and with OIC1D4 and L416, 2 monoclonal antibodies against unidentified constitutive P granule epitopes. These studies suggest that the GLHs are essential structural components of P granules required for fertility.

So far, genetic analyses of the PGL and GLH families have failed to reveal any roles for these P granule components in the specification of germ cells in embryogenesis. This is in contrast to germ granule components in *Drosophila,* such as Vasa, which is required both for the formation of germ cells in embryos and for oogenesis in the adult (Schupbach and Wieschaus, 1986; Lasko and Ashburner, 1990; Nusslein-Volhard *et al.,* 1987). A likely explanation for this discrepancy is that the methods used to disrupt PGL and GLH expression (RNA-mediated interference and, in the case of PGL-1, three genetic mutants) have not been successful in eliminating the maternal load of these proteins in embryos without also causing sterility in the mother. With the development of new RNAi methods for administering dsRNA at different times in development (Timmons and Fire, 1998), it may be possible in the future to overcome these difficulties and address directly what roles P granules play during embryogenesis.

B. Proteins Transiently Associated with P Granules in Embryos

In addition to the PGL and GLH proteins, P granules in embryos also contain two other classes of predicted RNA-binding proteins. These proteins are found both diffusely throughout the cytoplasm of embryonic blastomeres and at higher concentrations on P granules, but do not seem to be present in P granules in larvae and adults. Mutations in these proteins do not eliminate P granules; instead, they typically lead to aberrant patterns of cell fates and P granule segregation in early embryos. Whether the association of these proteins with P granules is important for their function or P granule function is not yet known. These embryo-specific P granule components can be divided into two classes, depending on the type of RNA binding motif they contain.

1. CCCH Proteins: MEX-1, POS-1, and PIE-1

MEX-1, POS-1, and PIE-1 are novel proteins, each containing 2 CCCH zinc finger motifs (Guedes and Priess, 1997; Mello *et al.,* 1996; Tabara *et al.,* 1998). This type of zinc finger has been found in several proteins shown to bind RNA or to be implicated in mRNA cleavage, processing, and/or turnover (Bai and Tolias, 1996, 1998; Barabino *et al.,* 1997; Murray *et al.,*

1997; Carballo *et al.,* 1998; Rudner *et al.,* 1998). As described above, MEX-1, POS-1, and PIE-1 are maternally loaded proteins that segregate with the germ lineage. Mutations in MEX-1, POS-1, and PIE-1 all cause maternal-effect embryonic lethality and loss of primordial germ cells, but their specific phenotypes differ.

In *mex-1* mutant embryos, germline blastomeres often divide symmetrically, P granules segregate abnormally, and all blastomeres adopt aberrant fates (Mello *et al.,* 1992; Schnabel *et al.,* 1996). These phenotypes indicate that, much like the *par* genes, *mex-1* is generally required for embryonic polarity (Schnabel *et al.,* 1996). *pos-1* mutants also exhibit several abnormalities in both somatic and germ lineages, but at least some of the *pos-1* somatic defects appear to be secondary consequences of defects in the germ lineage, suggesting that *pos-1* may function more specifically in germline blastomeres (Tabara *et al.,* 1998). The earliest defect seen in *pos-1* mutants is a failure to translate APX-1, a protein synthesized in P_1 and P_2 from maternal RNA. This observation has lead to the proposal that the primary function of *pos-1* may be to promote the translation of maternal RNAs in germline blastomeres (Tabara *et al.,* 1998).

In contrast to *mex-1* and *pos-1, pie-1* mutants exhibit only mild defects in asymmetric cleavages and P granule segregation (Mello *et al.,* 1992). Instead, the most dramatic phenotype observed in *pie-1* mutants is a fate transformation: cells in the P_2 lineage adopt fates characteristic of cells in the EMS lineage (Mello *et al.,* 1992). This fate transformation depends in part on SKN-1, a maternally-inherited transcription factor present in both EMS and P_2 and required for EMS fate (Bowerman *et al.,* 1993, 1992). These observations suggest that PIE-1 normally functions in P_2 to inhibit the activity of SKN-1. Consistent with this hypothesis, PIE-1 is present in P_2 but not EMS as it segregates with the germ lineage during early cleavages (Mello *et al.,* 1996).

How does PIE-1 inhibit SKN-1 activity? Insight into this question came from the finding that mRNA transcription is generally inhibited in germline blastomeres and that this inhibition depends on PIE-1. Newly synthesized mRNAs, and a specific phosphoepitope on the carboxy-terminal domain of RNA polymerase II (RNAPII-H5), first appear in somatic nuclei in the four-cell stage but remain absent from the germ lineage until the 100-cell stage (Seydoux *et al.,* 1996; Seydoux and Dunn, 1997). By contrast, in *pie-1* mutants, newly transcribed mRNAs and RNAPII-H5 are detected in both somatic and germline blastomeres starting in the four-cell stage (Seydoux *et al.,* 1996, Seydoux and Dunn, 1997). These observations suggest that one function of PIE-1 is to keep mRNA transcription off in the germ lineage. In support of this hypothesis, PIE-1 accumulates in the nuclei of germline blastomeres (Mello *et al.,* 1996) and can inhibit transcription when introduced into HeLa cells (Batchelder *et al.,* 1999). The mechanisms used by PIE-1 to inhibit mRNA transcription are not yet known but are probably specific to RNA polymerase II [rRNA transcription is active in germline blastomeres (Seydoux and Dunn, 1997)], and have been proposed to involve targeting of a CTD-binding complex (Batchelder *et al.,* 1999). Inhibition of RNA polymerase II activity probably is an aspect of germ cell specification that has been conserved in evolution, since early germ cells in *Drosophila* also fail to transcribe mRNAs (Kobayashi *et al.,* 1988; Lamb and Laird, 1976; Van Doren *et al.,* 1998; Zalokar, 1976) and lack the RNAPII-H5 epitope (Seydoux and Dunn, 1997). However a PIE-1 homologue in *Drosophila* has not yet been described.

In addition to its nuclear localization, PIE-1 is also present in the cytoplasm of germline blastomeres, where, like MEX-1 and POS-1, it accumulates on P granules (Mello *et al.,* 1996). This localization suggests that PIE-1 may have additional functions in germline balstomeres besides inhibiting transcription. Consistent with this possibility, *pie-1* mutants fail to express at least one protein (NOS-2) that is translated in P_4 from a maternal mRNA associated with P granules (K. Subramaniam and G. Seydoux, unpublished observations; see below). An attractive hypothesis is that all three proteins, MEX-1, POS-1, and PIE-1, function in the cytoplasm of germline blastomeres to regulate the stability and/or translation of maternal RNAs associated with P granules; PIE-1 would then be unique in the family in having acquired the additional function of repressing transcription in the nucleus.

2. KH Proteins: MEX-3 and GLD-1

MEX-3 and GLD-1 are KH domain proteins that, like MEX-1, POS-1, and PIE-1, associate transiently with P granules in early embryos (Draper *et al.,* 1996; Jones *et al.,* 1996; Jones and Schedl, 1995). K homology (KH) domains are commonly found in RNA binding proteins and are thought to mediate binding to single-stranded RNA (Musco *et al.,* 1996). Missense mutations in the KH domains of MEX-3 and GLD-1 abolish their function, thus confirming that these proteins are likely to act primarily as RNA-binding proteins (Draper *et al.,* 1996; Jones *et al.,* 1996).

Both MEX-3 and GLD-1 have been implicated in translational control. MEX-3 is required to repress the translation of maternal *pal-1* RNA in the AB lineage; in the absence of MEX-3, PAL-1, a transcription factor related to *Drosophila* Caudal, is translated inappropriately in AB-derived blastomeres (Hunter and Kenyon, 1996). This ectopic expression causes cells in the AB lineage to adopt fate characteristics of the C lineage, where PAL-1 is normally expressed. *mex-3* mutants also exhibit defects in the P_3 blastomere, including abnormal P granule segregation and production of extra primordial germ cells (Draper *et al.,* 1996). These phenotypes suggest that MEX-3 may also regulate the translation of maternal RNAs in the germ lineage, although the specific RNAs involved have not yet been identified.

The function of GLD-1 is best understood in the larval/adult germline (Francis *et al.,* 1995a, 1995b), where it appears to act as a translational repressor of target RNAs during oogenesis (M.-H. Lee and T. Schedl, unpublished observations) and hermaphrodite germline sex determination (Jan *et al.,* 1999). In addition to its postembryonic function in germline development, GLD-1 is expressed in the embryonic germ lineage from the four-cell stage until the end of embryogenesis (Jones *et al.,* 1996). In germline blastomeres, GLD-1 accumulates on many particles, some of which coincide with P granules. Although the function of GLD-1 and MEX-3 in germline blastomeres is not known, the presence of these known translational regulators in P granules supports the idea that these organelles serve to regulate the expression of maternal RNAs in the early germ lineage.

C. P Granule-Associated RNAs

All P granule proteins identified so far contain RNA-binding motifs, so it is perhaps not surprising that P granules should also contain RNA. The presence of RNA in P granules was first detected by staining with acridine orange (D. Albertson, personal communication) and was later confirmed by fluorescent *in situ* hybridization (FISH) experiments using probes specific for messenger RNAs [oligo-dT probe and a probe complementary to SL1, a splice leader sequence commonly found at the 5′ end of mRNAs in *C. elegans* (Seydoux and Fire, 1994; Pitt *et al.,* 2000]. These experiments revealed that clusters of mRNAs co-localize with most if not all P granules. With one exception discussed below, the identities of these mRNAs are not known.

In *Drosophila* and *Xenopus,* several mRNAs associated with germ granules have been identified. Among them is *Drosophila nanos* and its *Xenopus* homolog *Xcat-2* (Mosquera *et al.,* 1993; Wang and Lehmann, 1991; Zhou and King, 1996; Kloc *et al.,* 1998; Kloc *et al.,* Chapter 2; Mahowald, Chapter 5). There are three *nanos*–related genes in the *C. elegans* genome and one of them, *nos-2,* was recently shown to encode an mRNA associated with P granules (Subramaniam and Seydoux, 1999). In *Drosophila, nanos* RNA is present throughout the early embryo and is enriched at the posterior pole where the germ granules reside (Bergsten and Gavis, 1999; Wang and Lehmann, 1991). Similarly, in *C. elegans nos-2* mRNA is present in all blastomeres from the one to eight-cell stage, and is enriched on P granules in germline blastomeres (Subramaniam and Seydoux, 1999). In later embryos, *nos-2* mRNA is maintained only in the germ lineage, where it is eventually translated into protein in the germline blastomere P4 and the two primordial germ cell Z2 and Z3. In somatic blastomeres, *nos-2* mRNA is not stable and is not translated (Subramaniam and Seydoux, 1999). These observations suggest that, as has been shown for *Drosophila nanos, nos-2* translation may be under negative regulation throughout the embryo, with the exception of germ cells (Gavis *et al.,* 1996; Smibert *et al.,* 1996; Dahanukar *et al.,* 1999; Dahanukar and Wharton, 1996).

Using RNA mediated interference, *nos-2* was shown to be required for the proper incorporation of primordial germ cells into the somatic gonad. In wild-type, the primordial germ cell Z2 and Z3 are joined in mid-embryogenesis by two somatic gonad precursor cells (Z1 and Z4); together these cells form a basement membrane-bound structure, known as the gonad primordium, with Z1 and Z4 at each end and Z2 and Z3 in the middle (Fig. 2) (Kimble and Hirsh, 1979). *nos-2* is required for this process: 30% of L1 larvae derived from mothers injected with *nos-2* dsRNA hatch with one primordial germ cell outside the gonad primordium (Subramaniam and Seydoux, 1999). Remarkably, *nanos* in *Drosophila* is also required for germ cells to associate with the somatic gonad (Forbes and Lehmann, 1998; Kobayashi *et al.,* 1996). The *nos-2* data demonstrate that, as in other organisms, P granules in *C. elegans* contain RNAs required for the development of the early germline.

How do P granules acquire specific mRNAs? A recent ultrastructural study suggests that P granules may capture newly synthesized mRNAs as they exit the nucleus (Pitt *et al.,* 2000). This study shows that in the distal gonad of adult hermaphrodites, P granules are perinuclear and associate preferentially with regions of the nuclear membrane rich in nuclear pores. Inside P granules, electron-dense zones can be seen opposite nuclear pores, suggesting that material coming out of the pores passes through the P granules. P granules appear to remain associated with nuclear pores even after they detach from the nucleus during late oogenesis (Pitt *et al.,* 2000). Perhaps RNAs acquired by P granules in the distal germline are stored in these organelles for use later during oogenesis and early embryogenesis when germ cells are no longer synthesizing mRNAs.

D. P Granules: Germline-Specific Translation Machines?

The presence in P granules of proteins implicated in translational regulation and of at least one mRNA (*nos-2*) translated specifically in germ cells suggests that these granules function to regulate the translation of germline mRNAs. Studies in *Drosophila* also support this hypothesis. For example, the germ granule components Vasa and Oskar are required to activate the translation of the otherwise translationally repressed *nanos* mRNA (Gavis *et al.,* 1996; Dahanukar *et al.,* 1999; Dahanukar and Wharton, 1996). *Drosophila* germ granules also contain ribosomal RNA derived from mitochondria (Kobayashi *et al.,* 1993), and recently this RNA was shown to be required for germline formation (Iida and Kobayashi, 1998). These data suggest that germ granules may provide a micro-environment that allows translationally repressed mRNAs to be translated specifically in germ cells. An important challenge for the future will be to determine how P granule components regulate translation; in particular, it will be interesting to determine what distinct functions are mediated by the constitutive and transient components of P granules. Perhaps constitutive components function directly in the process of translational activation, whereas transient components serve to select specific mRNAs for translational activation at specific times in development.

IV. Proliferation and the Decision to Initiate Meiotic Development

A. Overview of Postembryonic Gonadal Development

Postembryonic development in *C. elegans* consists of four larval stages (L1–L4) and adulthood. Figure 2 diagrams postembryonic gonadogenesis. The Z1 and Z4 somatic precursors within the gonad primordium divide by an essentially invariant pattern during larval development, generating 143 cells in the hermaphrodite and 56 in the male (Kimble and Hirsh, 1979). The Z2 and Z3 germline precursors within the gonad primordium begin proliferation after hatching, and their descendants divide by a variable pattern during larval development and adulthood. Much of the germline is syncytial with each nucleus partially enclosed by plasma membranes (Hirsh *et al.,* 1976; Hall *et al.,* 1999). By convention, the term *germ cells* refers to a nucleus, its associated cytoplasm, and surrounding membranes.

The adult hermaphrodite gonad is symmetric, consisting of two reflexed tubes (arms), one anterior and one posterior, that share a common uterus and vulva at the midbody of the animal. A spermatheca, the site of sperm storage and fertilization, connects each gonad arm to the uterus. The adult male gonad is asymmetric, with a single reflexed tube that terminates at the seminal vesicle and vas deferens. A distal-proximal axis is defined by the somatic distal tip cells (DTC) at the ends of the gonad arms and the proximal opening of the gonad to the exterior, the vulva and cloaca for the hermaphrodite and the male, respectively. The *C. elegans* hermaphrodite is considered to have a female soma, with the germline being hermaphrodite,

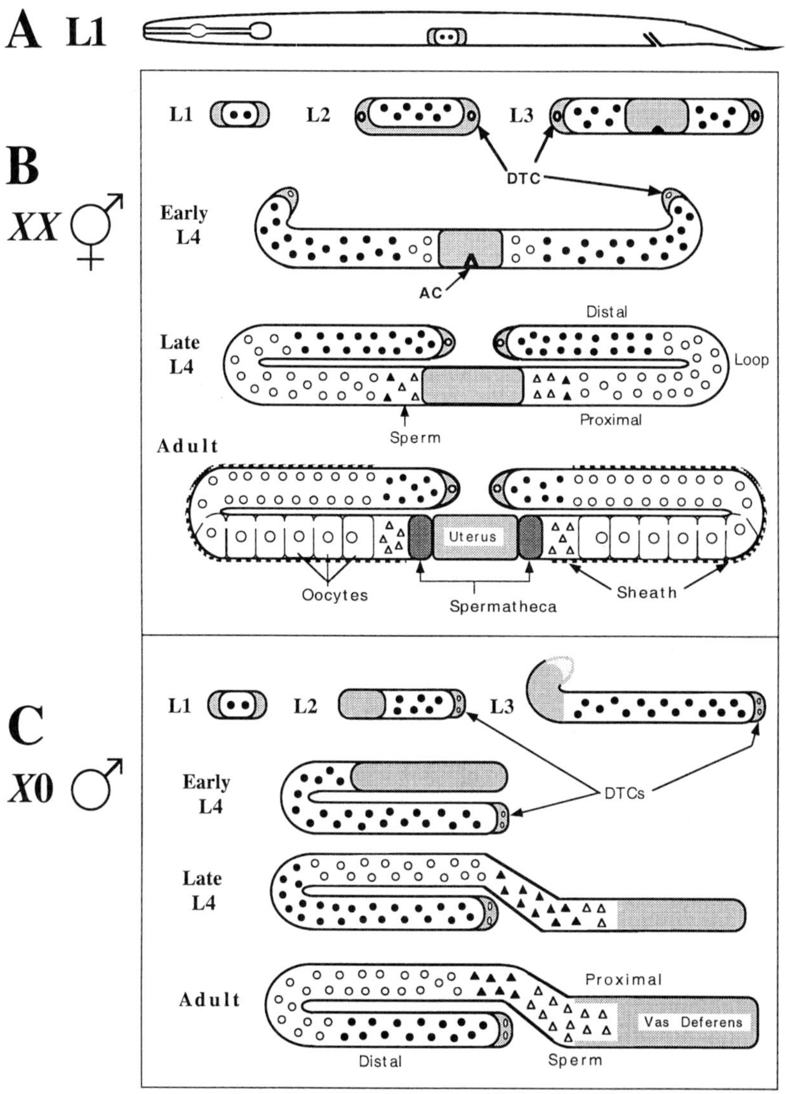

FIG. 2 Schematic diagram of postembryonic germline development in the hermaphrodite and male. (A) Newly hatched L1, equivalent in both sexes, with a mid-ventrally located gonad primordium. Gonadogenesis of the hermaphrodite (B) and the male (C), with selected stages from L1 through adult depicted. Somatic gonal cells are shaded, with the sheath cells of the hermaphrodite (stipples) only shown in the adult. DTC, distal tip cell; AC, anchor cell. Germ line: closed circles, mitotic nuclei; open circles, meiotic prophase nuclei; closed triangles, primary spermatocytes; open triangles, sperm. (Reprint, with permission, from Schedl, 1997.)

producing first sperm during the L4 stage and then switching to oogenesis, which is continuous throughout adulthood.

The germline is highly polarized tissue in its pattern of proliferation, entry, and progression through meiotic prophase and gametogenesis. Figure 3 illustrates this polarity with respect to both nuclear morphology and cell size/shape. The distal-most cells (from the tip to 20 to 25 cells proximally) are proliferative and can be considered a stem cell population (Hirsh *et al.*, 1976; Crittenden *et al.*, 1994). Germ cells >25 cells from the tip initiate meiotic development in the transition zone undergoing stages consistent with early meiotic prophase (leptotene and zygotene) where the homologous chromosomes can be observed to pair (Dernburg *et al.*, 1998). Germ cells in pachytene occupy the remainder of the distal arm. In the loop region germline nuclei make the transition from pachytene to diplotene, followed by diakinesis in the proximal arm. Overt oogenesis is usually limited to the proximal arm. More complete cellularization and growth in volume occurs as the oocyte progresses proximally.

B. Somatic Distal Tip Cell Signaling and the GLP-1/Notch Signaling Pathway

1. The DTC Is an Organizer of Germline Polarity

A major organizer of larval/adult germline polarity is a signaling pathway controlled by the somatic DTC. Ablation of the DTC(s) using a laser microbeam at any time during larval or adult life results in all proliferative germ cells undergoing meiotic development and gametogenesis (Kimble and White, 1981; see below). The DTC thus acts to promote proliferation and/or to inhibit initiation of meiotic development.

The DTC can be mispositioned by ablating certain somatic gonad cells. When this occurs, germ cells adjacent to the mispositioned DTC proliferate, while those at a distance undergo meiotic development (Kimble and White, 1981). As discussed above, the proliferative region extends 20 to 25 cell diameters from the DTC in wild-type young adult hermaphrodites. These observations indicate that the DTC is an organizer of distal–proximal polarity and that the DTC exerts a localized influence over germ cells: germ cells close to the DTC are proliferative whereas those more distant are in meiotic prophase (also, see below).

2. The GLP-1/Notch Signaling Pathway

Genetic analysis led to the identification of three genes that directly mediate the induction of germline proliferation by the DTC (*lag-2, glp-1, and*

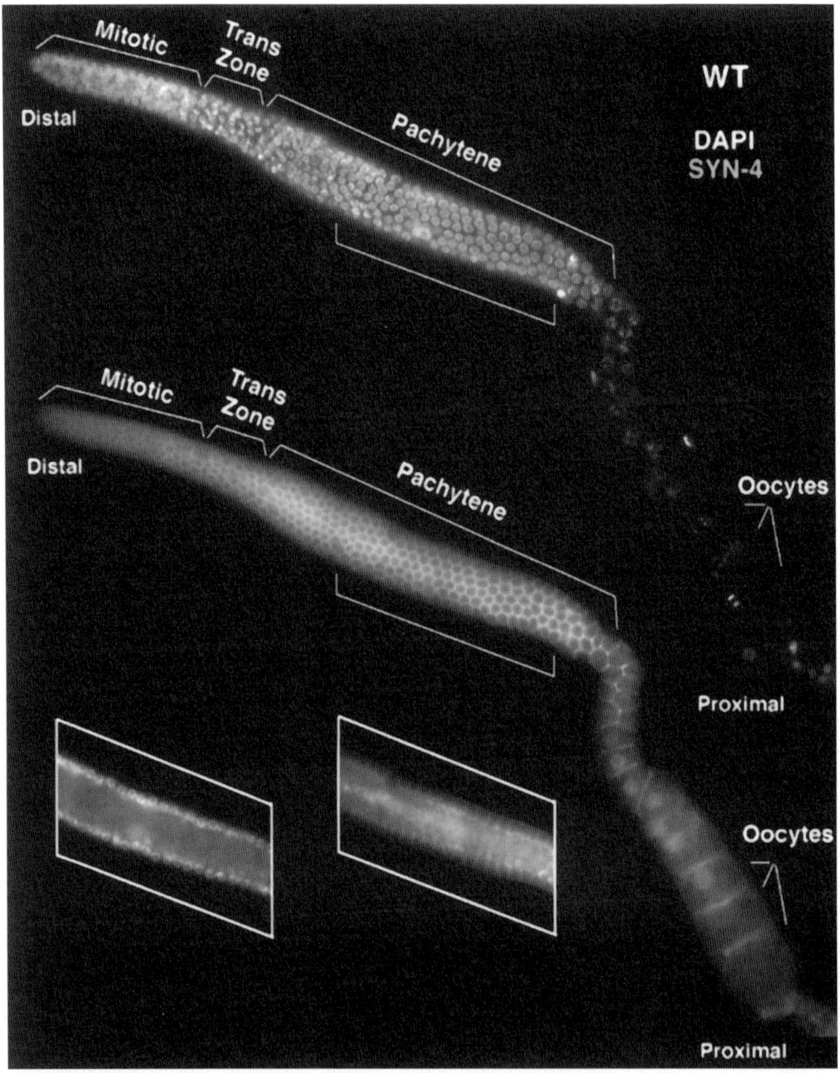

FIG. 3 Distal-proximal morphology of germline nuclei and cell membranes of an adult hermaphrodite gonad. Fluorescence micrographs of an adult hermaphrodite gonad, released by dissection and stained with DAPI, to visualize nuclear morphology (upper gonad), and anti-SYN-4 (Jantsch-Plunger and Glotzer, 1999) to visualize membranes (lower gonad). The distal portion of the gonad (surface view) contains the mitotic proliferative region, the transition (Trans) zone, and an extended pachytene region. The proximal gonad (interior view), from the loop to the spermatheca, contains growing oocytes in diplotene and diakinesis. Somatic gonad nuclei are not indicated. Insets at the bottom are interior views of the pachytene region (brackets in the above gonads) with DAPI (left) and anti-SYN-4 (right) staining to show the nuclei on the surface of the gonadal tube and the interior acellular/ anuclear core or rachis. (From M.-H. Lee and T. Schedl, unpublished.)

lag-1). These genes are *C. elegans* homologs of the Notch signaling pathway in *Drosophila* (Greenwald, 1998; Kimble *et al.,* 1998). LAG-2 is a single-pass transmembrane ligand homologous to *Drosophila* Delta and Serrate (the DSL family; Tax *et al.,* 1994; Henderson *et al.,* 1994), GLP-1 is a single-pass transmembrane receptor homologous to Notch, containing large extracellular and intracellular domains (Austin and Kimble, 1989; Yochem and Greenwald, 1989), and LAG-1 is a DNA-binding protein homologous to *Drosophila* suppressor of hairless [Su(H)] and mammalian proteins CBF-1/RBP-Jkapa/KBF2 (the CSL family; Christensen *et al.,* 1996).

The *glp-1* gene was identified in a screen for recessive mutations with a germline phenotype similar to that observed in DTC ablated animals (Austin and Kimble, 1987). In *glp-1* null (0) L1 larvae, Z2 and Z3 each divide approximately two times and then undergo meiotic development. Some partial *glp-1* loss-of-function (lf) mutants undergo additional divisions before all germ cells execute meiotic development. That GLP-1 acts as the receptor for the DTC signal is based on genetic mosaic analysis indicating that GLP-1 is required in the germline (Austin and Kimble, 1987), antibody staining showing that GLP-1 is located on membranes of proliferating germ cells but not in DTCs (Crittenden *et al.,* 1994), and its homology to Notch family receptors. Consistent with the requirement of the DTCs and the presence of mitotic germ cells throughout postembryonic life, temperature shift experiments using a temperature sensitive (ts) allele demonstrated that GLP-1 is required continuously for germ cells to remain proliferative after the L1 stage (Austin and Kimble, 1987).

The hypothesis that the GLP-1 receptor acts as a switch that regulates the decision between germline proliferation versus initiation of meiotic development (Austin and Kimble, 1987) is strongly supported by studies of a *glp-1* gain-of-function (gf) activated allele (*oz112gf,* Berry *et al.,* 1997). Germlines in *glp-1* (*oz112gf*) mutants, under the most restrictive conditions, contain only proliferating cells and are defined as tumorous because of the vast excess of germ cells and the complete absence of differentiation. The gain-of-function phenotype for the GLP-1 (*oz112*) mutant receptor is due to a combination of ligand independent signaling as well as hyperactivation (Berry *et al.,* 1997). Given the opposite phenotypes of *glp-1* null and the constitutively activated *glp-1*(*gf*), the activity state of GLP-1 signaling appears to determine whether germ cells will proliferate or undergo meiotic development.

In addition to its role in germline development, GLP-1 also functions in certain somatic cells in late embryogenesis, where it acts redundantly with a second Notch-related receptor (*lin-12;* Yochem *et al.,* 1988). *lin-12;glp-1* double mutants have a characteristic L1 larval lethal phenotype called Lag (for *l*in-12 *a*nd glp-1) (Lambie and Kimble, 1991). Screens for Lag mutants have been used to identify genes that function with *lin-12* and *glp-1* and have identified *lag-1* and *lag-2* (Lambie and Kimble, 1991). Partial loss-of-function (lf) alleles of *lag-1* and *lag-2* display the same premature onset of meiotic development phenotype characteristic of *glp-1* mutants, thus confirming that these genes function with GLP-1 during germline development. LAG-2 is homologous to Delta; consistent with its functioning as the ligand for GLP-1, *lag-2* mRNA accumulates in the DTC but not in the germline (Henderson *et al.,* 1994). LAG-1 is related to CBF1 and can bind the CBF1/Su(H) consensus DNA site RTGGGAA *in vitro* (Christensen *et al.,* 1996).

Genetic epistasis analysis with the constitutively activated *glp-1(oz112gf)* mutant supports the pathway shown in Figure 4. *glp-1(oz112gf)* is epistatic to a *lag-2* null mutant indicating that GLP-1 functions downstream of LAG-2. By contrast, a partial lf allele of *lag-1* displaying the premature initiation of meiotic development phenotype is epistatic to *glp-1(oz112gf)* indicating that LAG-1 acts downstream of GLP-1.

3. Molecular Aspects of GLP-1/Notch Receptor Signaling

Studies in mammalian systems, flies, and worms have begun to provide a molecular understanding of the mechanism by which Notch family receptors signal (reviewed in Chan and Jan, 1998). Binding of DSL ligands (LAG-2) to Notch family receptors appears to lead to proteolytic cleavage and release of the intracellular domain (NICD; see Figure 5). Following cleavage, NICD enters the nucleus, associates with CSL proteins (LAG-1), and alters the transcription of specific target genes. The presence of endogenous NICD in nuclei has not yet been observed (e.g., Crittenden *et al.,* 1994; Schroeter *et al.,* 1998), possibly due to the small amount of such products *in vivo.* However, forced expression of the NICD domains of Notch and LIN-12 in transgenic flies and worms has demonstrated that these domains are constitutively active and localize to the nucleus (Struhl *et al.,* 1993; Lieber *et al.,* 1993). Furthermore, ligand-induced cleavage and nuclear translocation has been demonstrated indirectly in the fly using sensitive genetic reporter constructs where the DNA binding domain of yeast Gal4p has been inserted into the intracellular domain of intact Notch, just C-terminal to the transmembrane domain (Struhl and Adachi, 1998: Lecourtois and Schweisguth, 1998). These constructs activate transcription from Gal4p upstream activation sequences in a Delta-dependent manner and rescue Notch null mutants.

Biochemical data suggest that Notch family receptors undergo three proteolytic cleavages, analogous to the cleavages observed in beta-amyloid precursor protein (Chan and Jan, 1999). The first cleavage is a furin-like convertase cleavage (termed S1) that occurs during transit through the trans-Golgi (Logeat *et al.,* 1998). This cleavage leads to the formation

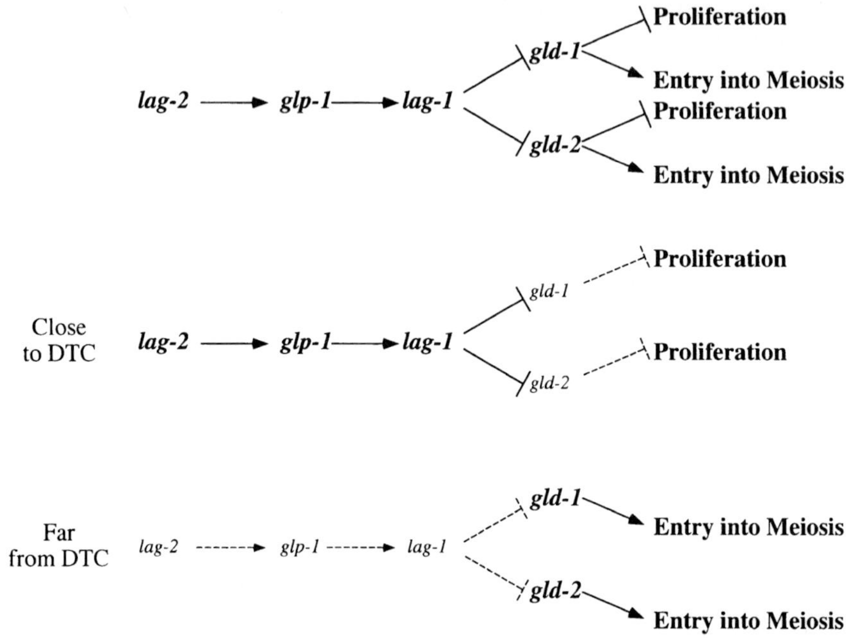

FIG. 4 Genetic pathway for the decision to proliferate or initiate meiotic development. The pathway is drawn showing wild-type gene functions that promote (arrow) or inhibit (T-bar) subsequent genes in the pathway. LAG-2, GLP-1, and LAG-1 act to promote proliferation and/or inhibit meiotic development. The GLD-1 and GLD-2 pathways act redundantly to promote meiotic development and/or inhibit proliferation. When germ cells are in the vicinity of the DTC, LAG-2 signaling from the DTC activates the germline GLP-1 receptor, forming a nuclear LAG-1/GLP-1(NICD) complex. The LAG-1/GLP-1(NICD) complex then, directly or indirectly, inactivates both the GLD-1 and GLD-2 pathways. When germ cells move farther away from the DTC, GLP-1 signaling is reduced to the point where the GLD-1 and GLD-2 pathways are active, thus resulting in the initiation of meiotic development. (See text for details.) Current genetic data suggest that the pathway for initiation of meiotic development is the same in male and female germ cells, as identical results are obtained in either sex. Germ cells in males, however, appear to begin meiotic development about a half a larval stage earlier than hermaphrodites (pachytene detected in mid L3 compared to early L4). The somatic gonad of the hermaphrodite and male differ significantly. To date, far fewer studies and mutant screens have been performed with males compared to hermaphrodites. It is possible that some aspects of the decision to proliferate or initiate meiotic development will show male versus hermaphrodite or male versus female germ-cell-specific control mechanisms.

of an intramolecular heterodimer (Figure 5) composed of an N-terminal fragment containing the majority of the extracellular domain (EGF-like repeats and the LNG repeats, Fig. 5) associated with a C-terminal fragment containing a small region of the extracellular domain, the transmembrane domain and the intracellular domain (Blaumueller *et al.,* 1997). The second

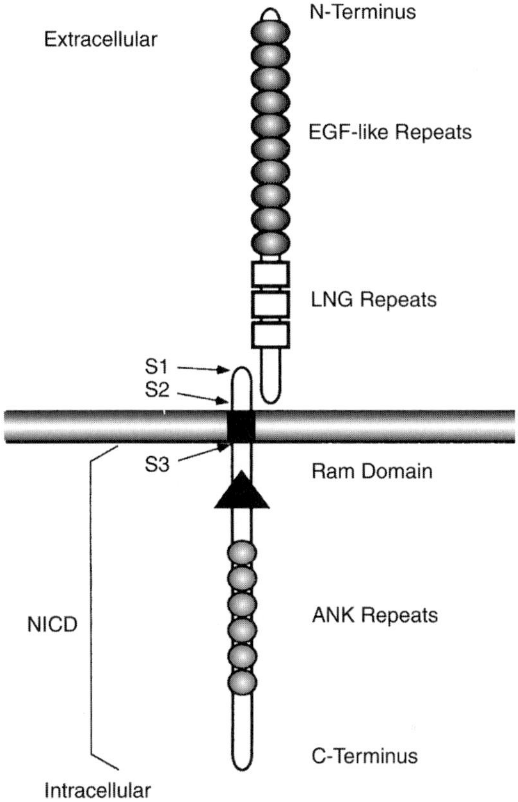

FIG. 5 Schematic diagram of the GLP-1, a member of the Notch receptor family, its domains and proposed processing sites. The mature receptor is thought to be an intramolecular heterodimer composed of an N-terminal fragment containing the majority of the extracellular domain, and a C-terminal fragment containing a small region of the extracellular domain, the transmembrane domain, and all of the intracellular domain (see text for citations). The heterodimer is thought to be generated by a furin-like convertase cleavage (S1) (although the nature of the linkage that allows the N- and C-terminal fragments to remain associated is unclear). It is proposed that ligand binding induces a cleavage (S2) in the extracellular domain of the C-terminal fragment that then leads to a third cleavage (S3), within or adjacent to the transmembrane domain, which releases GLP-1(NICD) for entry into the nucleus and association with LAG-1 (CSL). The GLP-1/NOTCH domains include the epidermal growth factor-(EGF)-like repeats (shaded ovals), some of which have been implicated in ligand binding for *Drosophila* Notch (Rebay *et al.,* 1991); LNG cysteine-rich repeats (open boxes); transmembrane domain (black box); Ram domain (back triangle), and ankyrin (ANK) or *cdc 10*/SWI6 repeats (shaded circles). The Ram and ANK domains have both been implicated in interacting with LAG-1 (Roehl *et al.,* 1996).

cleavage (S2) occurs in response to ligand binding and cuts in the extracellular juxtamembrane region of the C-terminal subunit of the heterodimer (Mumm *et al.,* 2000). S2 cleavage appears necessary for a third cleavage (S3), which occurs within or just C-terminal to the transmembrane domain, and serves to liberate NICD for entry into the nucleus (Jarriault *et al.,* 1995; Kopan *et al.,* 1996; Schroeter *et al.,* 1998). Presenilin proteins, which are mutated in familial early onset Alzheimer disease, either directly perform the S3 cleavage or are essential cofactors (De Strooper *et al.,* 1999; Ray *et al.,* 1999a,b; Struhl and Greenwald, 1999; Ye *et al.,* 1999). Indeed, the first evidence that presenilins are involved in the Notch pathway came from analysis of the *sel-12* gene, which was identified as a recessive suppressor of a weak activating mutation in *lin-12* (Levitan and Greenwald, 1995). *sel-12* acts as a general positive effector of worm Notch signaling as *sel-12*(*lf*) mutations enhance weak *lf* alleles of both *glp-1* and *lin-12.* Studies in mammalian culture cells suggest that presenilins are aspartyl proteases (Wolfe *et al.,* 1999) and thus might directly carry out the S3 cleavage at the plasma membrane (Ray *et al.,* 1999a).

Missense mutations in *lin-12, glp-1,* and Notch that lead to ligand-independent activation have begun to suggest residues that regulate receptor activation in response to ligand binding. These missense mutations map in the vicinity of the S1 and S2 cleavage sites in a domain that extends from the third LNG repeat to the transmembrane domain (Greenwald and Seydoux, 1990; Lieber *et al.,* 1993, Berry *et al.,* 1997; J. Hubbard, personal communication). One possibility is that some of these missense mutations interfere with the formation of a stable heterodimer and result in the loss of the N-terminal extracellular fragment. The remaining C-terminal fragment would be analogous to transgene constructs that are constitutively active (Greenwald, 1994) and are thought to undergo constitutive S2 cleavage (Mumm *et al.,* 2000). Alternatively, some of these mutants could form an intact heterodimer at the plasma membrane that leads to inappropriate S2 or S3 cleavage. Biochemical experiments will be required to distinguish between these possibilities. The N-terminal and C-terminal fragments of the GLP-1 heterodimer appear to have been detected in worm extracts (Crittenden *et al.,* 1994), but ligand-dependent cleavage fragments have not yet been reported.

4. Identification of Additional Genes that May Function in GLP-1 Signaling to Promote Germline Proliferation

Enhancer/suppressor genetics has been used to identify additional players in the GLP signaling pathway. Three types of screens have been conducted so far. The first two screens were designed to identify negative regulators of GLP-1 signaling (or possibly, positive regulators of entry into meiotic prophase). A first screen searched for recessive mutations that suppress the sterility/embryonic lethality of ts, loss-of-function mutations in *glp-1* (*s*uppressors *o*f *g*lp-1 or *sog;* Maine and Kimble, 1993). A second screen searched for recessive mutations that enhance a weak gf allele *glp-1* (*t*umorous *e*nhancers of *g*lp-1(gf) or *teg;* L. Berry, D. Hansen, and T. Schedl, unpublished). The Sog screen identified 6 *sog* genes (Maine and Kimble, 1993); mutations in the *sog* genes have no obvious phenotype in a *glp-1*(+) background. In contrast, the Teg screen identified three loci that display sterility in a *glp-1*(+) background. The germline defects of the *teg* mutants is not related to the Glp-1 phenotype, suggesting that they have additional functions in germline development besides their role in GLP-1 signalling. Two of the *teg* loci (*teg-2* and *teg-4*) appear to be general negative regulators of worm Notch signaling as mutations also enhance a weak gf allele of *lin-12.*

A third screen was designed to identify genes that act as positive effectors of GLP-1 signaling (or possibly, negative regulators of entry into meiotic prophase). This screen searched for recessive mutations that enhance a weak *glp-1* lf allele (*e*nhancers *o*f *g*lp-1 or *ego*). As expected, weak alleles of *lag-1* were obtained in the Ego screen (Qiao *et al.,* 1995). In addition, mutations in six other genes were identified (*glp-4, ego-1, 2, 3, 4, and 5*). Characterization of sterile mutant phenotypes for the six genes, including null alleles of *ego-1* (also see section VI.C), indicate that they not only function to promote germline proliferation but also have important roles in meiotic prophase and gametogenesis. Experiments to address whether the Ego genes act specifically with GLP-1 signaling in the germline or also function with LIN-12 have not been reported. Further understanding of the Ego, Sog, and Teg genes await molecular characterization.

Another class of genes required for germ cell proliferation has been identified. FBF-1 and FBF-2 were originally described as proteins that bind to the 3′ untranslated region of *fem-3,* a gene required for the male germ cell sex fate (Zhang *et al.,* 1997). FBF-1 and FBF-2 are related to the *Drosophila* translational regulator Pumilio and have been proposed to repress *fem-3* translation and thus allow the female fate to be specified and oogenesis to occur (Zhang *et al.,* 1997). These proteins have been shown to also be required for germline proliferation: the *fbf-1(0) fbf-2(0)* double mutant has a premature entry into meiotic prophase phenotype (S. Crittenden and J. Kimble, personal communication).

C. Initiation of Meiotic Development

What genes function downstream of the DTC-GLP-1 signaling pathway to control meiotic development? Genetic studies described below indicate that two redundant pathways, one containing GLD-1 and the other containing GLD-2, function to direct meiotic development and/or inhibit germ cell proliferation. The DTC-GLP-1 signaling pathway thus promotes proliferation by inhibiting the GLD-1 and GLD-2 pathways (see Figure 4).

gld-1 has multiple functions in germline development (Francis *et al.,* 1995a, b; also see above). Here we focus on its role in promoting meiosis. *gld-1* activity is not essential for initiation of meiotic development, as null mutant germ cells enter meiotic prophase normally. However, a nonessential (redundant) function was suggested by the observation that mutations in *gld-1* can increase the number of rounds of mitosis undergone by *glp-1(0)* germ cells before entering meiosis. Specifically, *glp-1(0)* germ cells divide twice before entering meiotic prophase whereas *gld-1(0); glp-1(0)* double mutant germ cells divide four times before entering meiotic prophase (Austin and Kimble, 1987; Francis *et al.,* 1995b). These observations led to the idea that *gld-1* has a function to promote initiation of meiotic development and/or inhibit proliferation. Since this function is not essential in a *glp-1*(+) background, *gld-1* was proposed to be redundant with another gene(s) (Francis *et al.,* 1995b).

Kadyk and Kimble (1998) have provided further support for this hypothesis by showing that *gld-2* is a gene that functions redundantly with *gld-1* to promote initiation of meiotic development. As is observed for *gld-1,* mutations in *gld-2(lf)* do not generally affect the ability of *glp-1* (+) germ cells to enter meiosis but cause *glp-1(0)* germ cells to undergo two additional rounds of division before they enter meiotic prophase significantly. The *gld-2(lf) gld-1(0)* double mutant hermaphrodites and males have a tumorous germline (Kadyk and Kimble, 1998). The tumorous phenotype appears to be due to a failure of germ cells to enter meiotic prophase as is observed for *glp-1(oz112gf)* (Berry *et al.,* 1997). Hence, GLD-1 and GLD-2 have overlapping functions to promote meiotic development, with the activity of each gene sufficient for the process. [Note that molecular markers for the very early stages of meiotic prophase do not currently exist; therefore, it remains a possibility that *glp-1(gf)* and/or *gld-2(lf) gld-1(0)* mutant germ cells transiently initiate meiotic development before becoming tumorous.]

The similarity of the *gld-2(lf) gld-1(0)* double mutant phenotype and the constitutively active *glp-1* gf mutant phenotype led to the proposal that the DTC-GLP-1 signaling pathway promotes germline proliferation by inhibiting GLD-1 and GLD-2 (Kadyk and Kimble, 1998). This model is supported by the observation that the *gld-2(lf) gld-1(0)* tumorous phenotype is epistatic to the premature entry into meiotic prophase phenotype of *glp-1(0)* and *lag-1(lf)*. GLD-1 and GLD-2 thus act downstream of and are negatively regulated by LAG-1 (see Figure 4). These results also demonstrate that *glp-1* activity is not essential for germ cell proliferation when *gld-1* and *gld-2* activities are absent.

GLD-1 and GLD-2 are likely to be members of redundant pathways rather than have redundant molecular activities. First, while *gld-1* and *gld-2* function redundantly in the control of entry into meiotic prophase, their other activities in germline development are non-overlapping (Francis *et al.*, 1995a, b; Kadyk and Kimble 1998). Second, GLD-1 and GLD-2 seem to be distinct molecularly: GLD-1 is a KH domain containing RNA-binding protein (Jones and Schedl, 1995) whereas GLD-2 is a novel protein (L. Kadyk and J. Kimble, personal communication). The RNA targets that GLD-1 regulates for initiation of meiotic development are not known.

Additional genes that function in the GLD-1 pathway for initiation of meiotic development have been sought in a screen for recessive mutations that show a *sy*nthetic *t*umorous (*syt*) phenotype with *gld-2(lf)*. This screen identified new alleles of *gld-1,* as predicted given the model shown in Figure 4, as well as mutations that define at least three new *syt* loci (D. Hansen, T. Dang, D. Schneider, L. Berry and T. Schedl, unpublished). These loci are candidates for genes that function with *gld-1* downstream of the GLP-1/Notch signaling pathway.

Three major issues to be addressed in future studies are as follows:

1. How is the presumed transcriptional control by GLP-1 (NICD)-LAG-1 connected to the regulation of the GLD-1 and GLD-2 pathways?
2. What are the processes controlled by the GLD-1 and GLD-2 pathways?
3. Where and how do the *sog, teg, ego* and *fbf-1 & fbf-2* genes act in the decision to proliferate or initiate meiotic development?

D. Temporal and Spatial Control of Proliferation and Meiotic Development

The development of the germline during postembryonic stages can be divided into three phases. Here we discuss how genes involved in regulating proliferation and the onset of meiosis function during these different phases and consider what other levels of regulation might be involved.

1. First Phase: Onset of Proliferation in the L1 Stage

The first phase begins after hatching when Z2 and Z3 start proliferating. Several observations suggest that this initial proliferation is independent of GLP-1 signaling: Z2 and Z3 each undergo approximately two rounds of division in the *glp-1(0)* mutant; temperature shift experiments with a *glp-1(ts)* allele indicate that *glp-1* function is not required until after the L1 stage (Austin and Kimble, 1987); and expression studies suggest that there is no *glp-1* RNA or protein in Z2 and Z3 at hatching (Evans *et al.,* 1994; Seydoux and Fire, 1994; Crittenden *et al.,* 1994). Instead, a major control may be the nutritional status of the newly hatched larvae. Z2 and Z3, like other blast cells, do not begin mitotically cycling until the L1 larvae has begun feeding; hatching worms in nutrient-free media results in a block in both somatic and germ blast cell proliferation. Two of the three *C. elegans nanos* homologs, *nos-1* and *nos-2,* appear to be components of a mechanism that blocks Z2 and Z3 proliferation in the absence of food (Subramaniam and Seydoux, 1999). Loss of *nos-1* and *nos-2* activity results in germ cell proliferation in about 50% of starved L1 larvae; the somatic blast cells examined remain arrested. The incomplete penetrance of the inappropriate proliferation phenotype suggests there may be additional redundant activities. *nos-3,* however, does not appear to be involved as *nos-1*(−)*nos-2*(−) animals are as affected as *nos-1*(−) *nos-2*(−) *nos-3*(−) animals (Subramaniam and Seydoux, 1999). Expression studies suggest that *nos-1* and *-2* are functioning in the germline. Genes involved in sensing the nutritional status of the L1 larvae and transducing that information to germline genes like *nos-1* and *-2* are not known.

Early germ cell proliferation appears to require the somatic gonad. When Z1 and Z4 (or their daughters) are ablated, germ cells divide once or twice but then fail to divide further or differentiate as gametes (Kimble and White, 1981; Francis *et al.,* 1995b). This effect may be trophic in nature.

2. Second Phase: Continuous Proliferation in the L1/L2 Stages

The second phase is in the late L1 stage and throughout the L2 stage, when all germ cells are proliferative. In the hermaphrodite, germ cells are centrally located within the bean-shaped gonad; somatic gonad cells primarily occupy the ventral surface, with the DTCs (Z1.aa and Z4.pp) or their precursors (Z1.a and Z4.p) residing at the anterior and posterior ends, respectively (Kimble and Hirsh, 1979). During this period the DTCs and their precursors express LAG-2 and presumably signal the germ cells via the GLP-1 receptor to inhibit the meiosis-promoting activity of the GLD-1 and GLD-2 pathways.

There is a distinct possibility that the DTC/precursors are not the only cells in the somatic gonad that can activate GLP-1 in the L2 stage. This hypothesis is supported by the following observations. First, two other somatic gonad cells express LAG-2 during the L2 stage (Wilkinson *et al.,* 1994). Using LAG-2 and the LIN-12 receptor, Z1.ppp and Z4.aaa signal each other to determine which will become the anchor cell (AC) and which will become a ventral uterine precursor cell (VU) (Greenwald *et al.,* 1983; Wilkinson *et al.,* 1994). Z1.ppp and Z4.aaa are in contact with germ cells until the late L2 stage, when the gonad primordium forms. Second, ablations of the DTC precursors in the L1 stage cause germ cells to enter meiosis only in the L3 stage, with pachytene nuclei first detected in early L4 (Kimble and White, 1981; Francis *et al.,* 1995b). By contrast, *glp-1(0)* mutations cause all germ cells to enter meiosis as early as the L2 stage. The temporal delay in the ablation experiments suggest that other cells can continue to signal the germ cells after ablation of the DTC/precursors; one possibility is that these cells are the AC/VU precursors Z1.ppp and Z4.aaa. When the gonad reorganizes in late L2, the AC/VU cells are no longer in contact with germ cells, signaling ceases, and a Glp-1-like phenotype results. Further ablation experiments will be required to determine whether the AC/VU cells, and/or possibly other somatic gonadal cells, can signal the germ cells in the L2 stage.

3. Third Phase: Balancing Proliferation and Meiotic Development

The third phase begins in the L3 stage when the germ cells are in two gonad arms and the proximal germ cells furthest from the DTCs that lead the arms cease proliferation and begin meiotic development. The transition between the bean-shaped L2 gonad and the two-armed L3 gonad occurs in late L2, when 10 somatic cells move to the center of the gonad while the DTCs remain at the ends and begin to migrate anteriorly (Z1.aa) and posteriorly (Z4.pp) (Kimble and Hirsh, 1979). These movements displace germ cells from the central region into the two forming gonad arms. From this point on each arm appears to behave autonomously.

From L3 through adulthood, it is thought that germ cells either proliferate when they are under the influence of the DTC or initiate meiotic development when they reach a certain distance and are no longer under its influence (Kimble and White, 1981; Austin and Kimble, 1987; Yochem and Greenwald, 1989; Crittenden *et al.,* 1994; Kadyk and Kimble, 1998). Close to the DTC, LAG-2 dependent GLP-1 signaling promotes proliferation by negatively regulating the GLD-1 and GLD-2 pathways (Figure 5). When germ cells move farther from the DTC, GLP-1 signaling falls below a critical threshold such that negative regulation of the GLD-1 and GLD-2 pathways is no longer sufficient to inhibit meiotic development. This distance-dependent mechanism generates a population of proliferating "stem" cells from which gametes can be continuously produced.

Genetic experiments where the GLP-1 pathway is inappropriately activated support the view that localized activation of GLP-1 signaling by the DTC leads to localized proliferation. Secreted forms of LAG-2, expressed from *lag-2* promotor constructs, can result in ectopic proliferation and the formation of tumorous germlines that are *glp-1* dependent (Fitzgerald and

Greenwald, 1995; Henderson *et al.*, 1997). This result implies that restriction of LAG-2 ligand to the DTC, by its transmembrane domain, limits the region of proliferation. Similarly, the tumorous germline phenotype of the constitutively active *glp-1(oz112gf)* is likely the result of ligand-independent activation of the mutant GLP-1 receptor in germ cells throughout the gonad arm, most of which are not under the influence of the DTC.

While the model for spatial control of proliferation/meiosis by localized DTC-GLP-1 signaling is likely to be correct in broad outline, there remain certain aspects of the proliferation/meiosis decision that are still poorly understood. In particular, one important question remaining is how the size of the proliferative zone is determined. The basic model described above predicts that the size of the proliferative zone should correspond to the region of the gonad covered by the DTC. Several studies, however, have shown that the proliferative zone extends beyond the area apparently covered by the DTC. The DTC has cytoplasmic processes that stretch 8 (+/-4, range 2 to 20) cell diameters from the distal tip in young adult hermaphrodites compared to the 20 to 25 cell diameters for the proliferative zone (Hall *et al.,* 1999; Fitzgerald and Greenwald, 1995; Crittenden *et al.,* 1994). The DTC processes are even smaller in the L4 (three to five cell diameters from the distal tip; Hall *et al.,* 1999). Hence, GLP-1 signaling in the germline appears to be distinct from other Notch signaling events in that interacting cells may not be in direct physical contact.

We consider three general models to explain how the DTC controls proliferation at a distance. In the first, LAG-2 mediated GLP-1 signaling acts throughout the mitotic region to directly promote proliferation. There are a number possible ways that this could occur. The described distribution of DTC process length is a static picture; *in vivo* the processes may be dynamic, growing and shrinking, as well as leaving fragments behind (Hall *et al.,* 1999). In this scenario, the extent of distal germline that is actually covered by the processes and fragments, and thus exposed to LAG-2, would be essentially equivalent to the size of the proliferative region. An alternative possibility is that DSL ligands are cleaved to create a diffusible form (Crittenden *et al.,* 1994; Qi *et al.,* 1999); LAG-2 dependent signaling could then exceed the extent of DTC coverage. However, to prevent ectopic proliferation, as was observed in the secreted LAG-2 transgenic experiments (Fitzgerald and Greenwald, 1995; Henderson *et al.,* 1997), the amount of LAG-2 and the distance it can diffuse would need to be carefully controlled to achieve the sharp border between the proliferative and early meiotic prophase regions. Hall *et al.* (1999) have noted that in hermaphrodites, the somatic distal sheath cells are well positioned to function in that role since they are excluded from the proliferative zone and cover germ cells only in the meiotic region. However, such a mechanism could not function in the male, as the male gonad lacks sheath cells.

In this model, the concentration of GLP-1 in the membrane of germ cells potentially could influence the size of the proliferation zone by affecting the ability of germ cells to respond to low concentrations of ligand. It is interesting in this regard that expression studies in wild-type hermaphrodites and males indicate that GLP-1 accumulation on plasma membranes is strongest in the proliferative region (Crittenden *et al.,* 1994). In mutants with tumorous germlines or ectopic proliferation [*glp-1*(*oz112gf*), *gld-1*(*0*), *gld-2, gld-1,* and *lin-12*(*0*); Berry *et al.,* 1997; Kadyk and Kimble, 1998; Crittenden *et al.,* 1994], GLP-1 staining also coincides with proliferative zones. These observations suggest the existence of a feedback loop between GLP-1 signaling or proliferation and GLP-1 accumulation (Berry *et al.,* 1997; Kadyk and Kimble, 1998; Kodoyianni *et al.,* 1992). Studies with LIN-12 have demonstrated the existence of a positive transcriptional autoregulatory loop in the VU precursor cell (Wilkinson *et al.,* 1994). This loop has been proposed to be mediated by the CSL transcription factor LAG-1, since there are multiple CSL consensus binding sites in the 5′ region of *lin-12* (Christensen *et al.,* 1996). A similar mechanism may also be operating for *glp-1,* since the *glp-1* promoter region also contains CSL consensus-binding sites. Alternatively, the regulatory loop could function at the translational or posttranslational levels. Such a feedback loop could help define the end of the proliferation zone if it caused a reduction in GLP-1 accumulation in cells exposed to relatively low levels of LAG-2.

In the second model, LAG-2 mediated GLP-1 signaling is only in the 8 (+/-4) cell diameters covered by the DTC processes but then generates an intracellular signal, cleaved GLP-1(NICD). The proliferative zone would then be defined by diffusion of GLP-1(NICD) through the germline syncytium and/or by rates of turnover of nuclear GLP-1(NICD)-LAG-1 complexes as germ cells move proximally; when germ cells reach 20 to 25 cell diameters from the tip, the level/activity of nuclear GLP-1(NICD)-LAG-1 complexes is no longer sufficient to repress GLD-1 and GLD-2. Stability of GLP-1(NICD)-LAG-2 complexes is likely to be the major factor, as diffusion appears to be highly restricted or controlled in the germline; in contrast to the *Drosophila* syncytial blastoderm embryo, where all nuclei progress through the cell cycle synchronously, the distal proliferative germ cells are asynchronous with only a scattered minority of cells in mitosis at any one time. It is possible that a combination of the first and second models are at work.

A third model is based on *in situ* hybridization studies (Jones *et al.,* 1996) that suggest that germ cells in the distal proliferative region (0 to 25 cell diameters) actually fall into two spatially distinct populations. For three oogenesis-specific mRNAs, *mex-3, cey-2,* and cyclin B (1.7kb), the hybridization signal was not detected in the distal half of the proliferative region (within ~10 cell diameters from the tip), although a moderate to strong signal was observed in the proximal half of the proliferative region (Jones *et al.,* 1996). One interpretation of these results is that germ cells in the proximal half of the proliferative region have initiated the female gametogenesis program even though they have not yet entered meiotic prophase. This possibility suggests that the proliferative region may contain two types of mitotically cycling germ cells: distal germ cells that are proliferating directly in response to GLP-1 signaling, and more proximal germ cells that are undergoing amplifying mitotic cell cycles independent of GLP-1. The number of amplifying cell cycles may be autonomously controlled (e.g., the four cystocyte divisions in *Drosophila* oogenesis) or may be spatially controlled by a diffusible signal, possibly even from the DTC. Further analysis of the distribution of molecular markers in the distal zone will be required to test this model.

V. Silencing in the Germline

In the early embryo, under the influence of the material factor PIE-1, transcription by RNA polymerase II is blocked in germ cells. As described in IVB, this initial period of transcriptional quiescence protects early germ cells from transcription factors (e.g., SKN-1) that promote somatic fates. This period, however, is short-lived: by about the 100-cell stage, soon after the division of P_4 into Z2 and Z3, PIE-1 disappears and the RNA polymerase II phosphoepitope H5 first appears in the germ lineage, thus suggesting that transcription begins at this time (Seydoux and Dunn, 1997). At least one gene (the *nanos* homolog *nos-1*) has been shown to be transcribed in Z2 and Z3 by mid-embryogenesis (Subramaniam and Seydoux, 1999), and many genes are known to be activated in the germline during larval development (e.g. *gld-1;* Jones *et al.,* 1996). In face of this transcriptional activity, how do germ cells continue to preserve their totipotency and unique germline character? Are there germline-specific mechanisms that continue to keep gene expression in check? In this section, we review recent evidence that supports the view that germ cells continue to utilize unique "silencing" mechanisms to regulate gene expression throughout their development.

A. Silencing of Transgenes in the Germline

Standard protocols for transformation in *C. elegans* give rise to large extrachromosomal arrays composed of many tandem copies of the injected transgene and reporter DNAs (Mello and Fire, 1995; Mello *et al.,* 1991). These highly repetitive "simple" arrays are expressed in somatic tissues and have been used successfully for GFP tagging and transgenic rescue of soma-expressed genes. In contrast, these arrays are expressed very poorly, if at all, in the germline. A most striking example of this silencing comes from studies of the *let-858* gene (Kelly *et al.,* 1997). This "house-keeping" gene is expressed in all cells and is required for viability and fertility. *let-858* transgenes in simple arrays, however, are expressed only in somatic cells and rescue the lethality but not the sterility of *let-858* mutants.

In an effort to overcome this germline-specific silencing, Fire and colleagues modified the standard transformation protocol by adding fragmented *C. elegans* genomic DNA to the injection mix to "dilute out" and reduce the copy number of the transgene and marker in the array. This approach was successful: *let-858* transgenes in these nonrepetitive "complex" arrays are expressed efficiently in the germline and can fully rescue the lethality and sterility of *let-858* mutants (Kelly *et al.,* 1997). Similar results were obtained with other ubiquitously expressed transgenes and with transgenes expressed only in the germline. These studies suggest that transgene silencing is, at least in part, repeat-dependent and that germ cells contain mechanisms that can recognize and silence repetitive sequences. Transgenes in complex arrays, however, often become silenced in the germline after a few generations, raising the possibility that other, repeat-independent mechanisms may also be operating (W. Kelly and A. Fire, personal communication).

mes-2, mes-3, mes-4, and *mes-6* are currently the best candidates for genes that participate directly in transgene silencing in the germline. These loci were identified originally in screens for mutants that cause maternal-effect sterility ("grandchildless" phenotype; Capowski *et al.,* 1991). Homozygous *mes/mes* hermaphrodites derived from heterozygous *mes*/+ mothers are viable and fertile, but their progeny grow up to be sterile. The sterility is due to the progressive degeneration of the germline: germ cells undergo only limited proliferation during postembryonic development before degenerating (Garvin *et al.,* 1998; Paulsen *et al.,* 1995). The penetrance and expressivity of this phenotype is sensitive to chromosome dosage: XO (male) progeny from *mes/mes* mothers are less affected than their XX (hermaphrodite) siblings (Garvin *et al.,* 1998). Garvin *et al.* reasoned that this difference is consistent with a role for the MES proteins in repressing/silencing gene expression: assuming that at least some of the genes silenced by the MES proteins are on the X, animals with only one X are expected to be less affected by the inappropriate gene expression than those with two. [It is not known currently whether the germline undergoes dosage compensation in the worm (Meyer, 1988, 1997)].

Direct evidence that the *mes* genes can influence gene expression came from Kelly and Fire, who discovered that introduction of a silenced *let-858:GFP* transgene into a *mes* mutant background lead to its activation

(Kelly and Fire, 1998). In *mes-4* mutants, activation of the transgene is observed even in the fertile *mes-4/mes-4* progeny of *mes-4/+* mothers. This observation indicates that desilencing can precede germ cell death by a full generation and therefore is unlikely to be a mere byproduct of degeneration. These results implicate the MES genes in repeat-dependent silencing and suggest that this silencing is essential for germ cell viability.

The molecular identity of the *mes* genes supports a role for these factors in transcriptional silencing. *mes-2* encodes a protein similar to the *Drosophila* Polycomb group protein, Enhancer of Zeste (Ez) (Holdeman *et al.,* 1998), and MES-6 is homologous to Extra Sex Combs (Esc), another Polycomb group protein (Korf *et al.,* 1998). MES-4 contains motifs found in Polycomb group proteins but is not a direct homolog of any one protein in the group (Y. Fang and S. Strome, unpublished results), and MES-3 is a novel protein (Paulsen *et al.,* 1995). Polycomb group proteins from *Drosophila* are best known for their role in maintaining transcriptional repression of homeotic genes (Pirrotta, 1997). They function as multimeric protein complexes that associate with chromatin at distinct loci to turn off gene expression (Franke *et al.,* 1992; Rastelli *et al.,* 1993; Carrington and Jones, 1996; Platero *et al.,* 1996). The MES proteins probably also exist in a complex, since they depend on each other for nuclear localization: MES-2 requires MES-6 to accumulate in nuclei, MES-6 requires MES-2 to accumulate in nuclei, and both require MES-3 (Holdeman *et al.,* 1998; Korf *et al.,* 1998). However, unlike the *Drosophila* Polycomb proteins, which function in somatic cells, the MES proteins are expressed preferentially in germ cells in larvae and adults and appear to have essential functions only in the germline. The genes that are the natural targets of MES silencing remain to be identified.

B. Inhibition of Transposition in the Germline

The *C. elegans* genome contains multiple copies of several families of transposons, the most active being Tc1 (Plasterk and van Luenen, 1997). In most strains, including the standard "wild-type" laboratory strain Bristol N2, Tc1 transposition occurs at high frequency in somatic tissues but is not detected in the germline (Emmons and Yesner, 1984). In contrast, in some rare natural isolates, such as the Bergerac strain, transposition is detected at high levels in both soma and germline and is the main cause of spontaneous mutations (Eide and Anderson, 1985; Moerman and Waterston, 1984). Inhibition of transposition in the germline is under genetic control. Several mutator loci (*mut*) were identified and mapped throughout the genome of the Bergerac strain (Mori *et al.,* 1988); many of these appear to be mobile themselves, a finding that suggests that they might be active transposons. Additional *mut* loci were isolated following EMS mutagenesis of Bergerac (e.g. *mut-2;* Collins *et al.,* 1987) and Bristol N2 (e.g. *mut-7;* Ketting *et al.,* 1999). Unlike the other mutators, *mut-2* and *mut-7* are not mobile, which has facilitated progress toward their molecular identification.

Molecular analysis of *mut-7* mutants supports the idea that transposition is under negative regulation in the germline. The three alleles of *mut-7* that were sequenced all appear to cause loss of function: two are premature stop codons and one is a deletion of the entire locus (Ketting *et al.,* 1999). These observations suggest that the wild-type function of *mut-7* is to inhibit transposition in the germline. How *mut-7* accomplishes this task, however, remains mysterious. MUT-7 is a novel protein that contains three small domains similar to the three catalytic domains of RNAseD and Werner Syndrome Helicase (Ketting *et al.,* 1999). RNAseD is a 3′ to 5′ RNA exonuclease and Werner Syndrome Helicase is a 5′ to 3′ exonuclease that can act on both DNA and RNA. (MUT-7 is not the worm ortholog of the Werner Syndrome Helicase as it lacks the helicase domain). Whether MUT-7 has exonuclease activity remains to be determined, but the sequence similarity is consistent with the possibility that *mut-7* interferes with transposition by acting on the transposon's DNA or RNA. Alternatively, *mut-7*'s effects on transposition may be more indirect, perhaps involving global changes in chromatin structure (see below). Further studies into the mechanisms of transposon activation in *mut-7* mutants and the biochemical activity of MUT-7 will be needed to distinguish between these possibilities. It will also be important to determine how MUT-7 inhibitory activity is restricted to the germline, since transposition is active in somatic cells.

C. RNA-Mediated Interference

Recently, double stranded RNA (dsRNA) has been shown to be a potent sequence-specific inhibitor of gene function in *C. elegans* (Fire *et al.,* 1998) and several other organisms (Hunter, 1999). In *C. elegans,* dsRNA introduced by microinjection or feeding can cause gene-specific, loss-of-function phenotypes with high penetrance. This technique, called RNA-mediated interference (RNAi), is now commonly used as a first test of function for genes (or putative genes predicted from the genomic/cDNA sequence) for which mutant alleles are not yet available.

The mechanisms underlying RNA-mediated interference are still poorly understood, but several interesting properties of RNAi have been noted (see Montgomery and Fire, 1998). First, the dsRNA is specific for the matching gene, thus indicating that RNAi requires recognition of homologous sequences. Second, RNAi likely acts posttranscriptionally. RNAi does not induce changes in the DNA sequence of the targeted gene, and dsRNA molecules corresponding to introns or promoter sequences are not effective.

Furthermore, for at least some polycistronic messages, single cistrons (mature mRNAs) can be targeted individually (Montgomery *et al.,* 1998; Korf *et al.,* 1998), while for a slowly processed polycistronic message, a single cistron can also target the precursor molecule (Bosher *et al.,* 1999). These observations suggest that RNAi can act on both the mature mRNA and the precursor RNA. Indeed, following injection of dsRNA, RNA from the targeted gene is eliminated from the cytoplasm and is reduced, although not eliminated, in the nucleus (Montgomery *et al.,* 1998). A third property of RNAi is that only a few molecules of dsRNA are needed to silence even a highly transcribed target (Fire *et al.,* 1998). This observation suggests that the input RNA is amplified or is acting catalytically. Finally, dsRNA microinjected in the intestine can exert its effect throughout the injected animal and its progeny, thus raising the possibility that there are mechanisms that transport the silencing agent (Fire *et al.,* 1998).

Genes required for RNAi have begun to be identified by testing already existing mutants for resistance to dsRNA and by selecting for new resistant, or *R*NA-interference *de*ficient (*rde*), mutants (Tabara *et al.,* 1999). One set of mutants, including *rde-1,* has no apparent phenotype besides resistance to RNAi. Molecular analysis of *rde-1* reveals that it belongs to a large gene family conserved in yeast, plants, and animals (Tabara *et al.,* 1999). The biochemical activities of members of this family are not known, but one member has functions potentially related to RNAi: *Drosophila sting* has been implicated in the silencing of repetitive sequences (Schmidt *et al.,* 1999). Conservation of *rde-1* across phyla suggests that this class of proteins, and perhaps RNAi itself, serves an important physiological role that has been maintained throughout evolution (Tabara *et al.,* 1999).

A second set of mutants resistant to RNAi display pleiotropic phenotypes in the germline, thus indicating that these genes have essential functions in this tissue. This set of mutants include certain *rde* and *mut* loci (discussed in the next section) and *ego-1. ego-1* was originally identified as an enhancer of a ts mutation in *glp-1,* a gene required for postembryonic proliferation of germ cells (see section IV.B) (Qiao *et al.,* 1995). *ego-1* null mutants show multiple defects during germline development, including premature entry into meiosis, a delayed and incomplete switch from spermatogenesis to oogenesis, abnormal meiotic prophase, and abnormal oogenesis (Smardon *et al.,* 2000). *ego-1* mutants are also partially resistant to RNAi: three out of five dsRNAs tested failed to cause a phenotype in *ego-1* mutants. EGO-1 is related to tomato RNA-dependent RNA polymerase and to *N. crassa* QDE-1 (Smardon *et al.,* 2000). *qde-1* mutants are defective in quelling, a posttranscriptional gene-silencing (PTGS) mechanism related to RNAi (Cogoni and Macino, 1997). Like RNAi, PTGS is thought to involve an RNA trigger that may be amplified. RNA-dependent RNA polymerase can transcribe RNA from RNA templates, and hence is a good candidate for the amplification agent (Schiebel *et al.,* 1998; Wassenegger and Pelissier, 1998). Whether EGO-1 plays this role in *C. elegans* remains to be tested. Nevertheless, the homology of EGO-1 to QDE-1 supports the idea that RNAi in *C. elegans* is related mechanistically to the posttranscriptional gene-silencing mechanisms that have been observed in other organisms. The observation that *ego-1* mutants are resistant to only a subset of dsRNAs might be explained by the fact that the worm genome contains several predicted genes closely related in sequences to *ego-1.*

An important question remaining is how EGO-1's role in RNAi relates to its function during germline development. One possibility is that the two functions are independent. Alternatively, EGO-1's function in RNAi may be directly linked to its function in germ cell development. Perhaps the mechanisms of RNAi are used by germ cells to regulate gene expression or to mediate homology recognition during chromosome pairing in prophase of meiosis I. The lack of germline phenotypes in the RNAi-deficient *rde-1* mutant would seem to argue against such a possibility. However, because there are multiple *rde-1*-like genes in the worm genome, a function for *rde-1* in germline development could have been masked by redundancy. Furthermore, if *rde-1* acts very early in an RNAi pathway for exogenous dsRNA, much or all of the RNAi mechanism for endogenous germline RNAs may remain intact in the *rde-1* mutant.

The presence of gene families for both *rde-1* and *ego-1* in the worm complicate the interpretation of their gene function. Gene families also present a problem with interspecies comparisons. For example, it is not clear which of the related genes in the fly is orthologous to *rde-1.* A more complete understanding of the function of *rde-1* and *ego-1* awaits the characterization of mutant phenotypes for the other family members.

D. Cross Talk between "Silencing" Mechanisms

The first evidence for cross talk between RNAi and other silencing mechanisms came from the discovery that *mut-7* mutants are resistant to RNAi and are deficient in transgene silencing (Tabara *et al.,* 1999) (see Table I). *mut-7* is not the only mutator loci that is resistant to RNAi: among 27 other EMS-induced mutators tested, 19 were found to be resistant to RNAi (Ketting *et al.,* 1999). Furthermore, in their screen for RNAi-resistant mutants, Tabara and colleagues identified two mutants, *rde-2* and *rde-3,* which were subsequently found also to activate transposition in the germline. *rde-2* was also shown to be deficient in transgene silencing (Tabara *et al.,* 1999). The discovery that *mut-7* and *rde-2* exhibit defects in all three silencing phenomena—transgene silencing, inhibition of transposition and RNAi—raises the possibility that these phenomena share a common mech-

TABLE I

Genes Implicated in Germline Silencing

Gene	Similarity	Resistance to RNAi germ	Resistance to RNAi soma	Germline transposition	Transgene desilencing	Chromosome non-disjunction	Sterility
		Phenotypes					
WT		–	–	–	–	–	–
mut-2	ND	+	+	+	ND	+	Sterile at 25°C
mut-7	RNAseD-like	+	Partial	+	+	+	Sterile at 25°C
rde-2	ND	+	Partial	+	+	+	Weakly sterile
rde-3	ND	+	+	+	ND	+	ND
ego-1	RNA-dependent RNA polymerase-like	Partial	+	ND	ND	ND	Zygotic sterile
rde-1	Piwi/sting/eIF2C-like	+	+	–	–	–	–
rde-4	ND	+	+	–	ND	–	ND
mes-2	Enhancer of zeste-like	–*		ND	+	–*	Maternal-effect sterile
mes-3	Novel	–*		ND	+	–*	Maternal-effect sterile
mes-4	Polycomb-like motifs	–*		ND	+	–*	Maternal-effect sterile
mes-6	Extra-sex comb-like	–*		ND	+	–*	Maternal-effect sterile

ND: not determined.

* These phenotypes were scored in fertile mes/mes hermaphrodites derived from mes/+ mothers.

(Data are from Tabara *et al.*, 1999; Ketting *et al.*, 1999; Kelly and Fire, 1998; Korf *et al.*, 1998; Holdeman *et al.*, 1998; Smardon *et al.*, 1999; S. Strome, personal communication; H. Tabara and C.C. Mello, personal communication.)

anistic basis. One attractive possibility is that all three forms of silencing use dsRNA as the triggering agent. In the cases of transposon and transgene silencing, dsRNA might derive from the occasional read-through in the antisense direction of one or a few copies of the transposon or transgene. These parralels have suggested that one natural function of RNAi might be to suppress transposition (Ketting *et al.,* 1999; Tabara *et al.,* 1998).

Not all RNAi-resistant mutants, however, are defective in transposon silencing; some mutator loci do not confer resistance to RNAi, and some mutator loci that are defective in transgene silencing are only partially resistant to RNAi in the soma (Table I). This incomplete overlap raises the possibility that mutants such as *mut-7* and *rde-2* have more general roles in the germline and interfere with RNAi only indirectly. As discussed in Tabara *et al.* (1999), one possibility is that these loci function in chromatin-based processes that suppress unwanted transcription from repetitive sequences. Such a function is consistent with the ability of these loci to silence multi-copy transgenes and transposons. In principle, disruption of these loci could cause an accumulation of nonspecific dsRNAs that, in turn, might overwhelm the mechanism that recognizes dsRNA as a gene-specific trigger for RNAi.

Consistent with a more general role for *mut-7* and *rde-2,* these mutants exhibit germline phenotypes not seen in other RNAi-resistant mutant (Table I), including a "high incidence of males" or Him phenotype (Ketting *et al.,* 1999; Tabara *et al.,* 1998). This phenotype reflects an increased frequency in X chromosome non-disjunction during meiosis and could potentially be caused by alterations in chromosome structure. Additionally, the observation that a *mut-7* null mutant is only partially resistant to RNAi in the soma argues against MUT-7 having an essential role in all aspects of the RNAi mechanism.

In summary, germ cells use specialized mechanisms to silence gene expression and transposition. These mechanisms have partially overlapping genetic requirements with each other and with RNAi, suggesting that they may share a common molecular basis. How this overlap is realized at the molecular level and how these mechanisms contribute to the health and development of germ cells promises to be an intense area of research in years to come.

VI. Concluding Remarks

Our understanding of the processes that guide germ cell development is progressing rapidly, but much remains to be uncovered. In this section, we highlight some of the emerging themes and remaining issues.

Germ cells use mechanisms both unique to the germline and shared by somatic cells. The use of P granules as potential regulators of mRNA stability and/or translation appears to be unique to the germline. Evidence accumulated so far indicates that these organelles and the mRNAs they contain are required for several aspects of germline development, from incorporation into the somatic gonad in embryos to oogenesis in the adult. Their exact function, however, remains unclear. In contrast to P granules, GLP-1/Notch receptor signaling is an example of a regulatory mechanism used by both germ cells and somatic cells. The downstream output, however, is singular to the germline, promoting germ cell proliferation over meiotic development. How this output is implemented at the molecular level remains to be determined.

Throughout development, germ cells use several mechanisms to silence unwanted gene expression. In the early embryo, production of mRNAs is inhibited in germline blastomeres by the maternal factor PIE-1. This inhibition is thought to insulate the nascent germline from transcriptional programs that promote somatic development. Once primordial germ cells are formed and mRNA transcription begins in the germ lineage, Polycomb-like genes act globally to repress certain genes/sequences, including repetitive sequences. Other genes, such as *rde-2* and *mut-7,* also contribute to the silencing of repetitive sequences and, by doing so, appear to protect the germline genome from the spread of transposable elements. In addition to silencing certain sequences, germ cells also activate the transcription of a number of genes. How this transcriptional program is established is unknown. DNA chip microarray hybridization is currently being used to catalog genes that are preferentially expressed in the postembryonic germline and during oogenesis and spermatogenesis (V. Reinke, H. Smith, J. Nance, S. Ward, and S. Kim, personal communication). Reverse genetics will help elucidate the function of these genes in germline development and should eventually identify coregulated gene sets that are involved in similar germline processes.

Finally, regulation of mRNA expression at the posttranscriptional level appears to play a particularly important role in the germline. As mentioned above, germ cells contain unique RNA-rich particles in their cytoplasm. These P granules may serve both as storage sites for newly transcribed mRNAs and/or as translational activators for specific RNAs at specific times in development. Another posttranscriptional mechanism, RNA-mediated interference, may act as a second line of defense against mobilization of transposable elements and, as suggested by *ego-1,* may also serve to regulate the expression of endogenous germline RNAs. Further evidence for the importance of posttranscriptional mechanisms in germ cells has come from the increasing number of RNA binding proteins that have been implicated in germline development (e.g., GLD-1, NOS-2, FBF-1 and 2). Indeed, recent microarray hybridization experiments have shown that 59% of genes predicted to encode RNA-binding proteins are expressed preferentially in the germline (V. Reinke, H. Smith, J. Nance, S. Ward, and S. Kim, personal communication). This result supports the view that much of the temporal and spatial regulation of gene expression in the germline takes place at the posttranscriptional level. Future characterization of the posttranscriptionally regulated genome (posttranscriptome), and the RNA binding proteins that regulate it, will be crucial to further our understanding of germ cell development.

Acknowledgments

We thank our many colleagues in the field for sending preprints and reprints. In particular we are grateful to Susan Strome, Karen Bennett, Jim Priess, Judith Kimble, Sam Ward, Stuart Kim, Raffi Kopan, and Eleanor Maine for sharing unpublished data. We also thank Min-Ho Lee and Dave Hansen for assistance with the figures and Dave Hansen for comments on the manuscript. Research on the embryonic origin of the germline in the Seydoux lab is supported by the Packard Foundation, the Searle Scholars Program/The Chicago Community Trust, and the National Institutes of Health. Research in the Schedl lab is supported by the National Institutes of Health.

References

Anderson, P., and Kimble, J. (1997). mRNA and Translation. *In "C. Elegans* II" (D. L. Riddle, T. Blumenthal, B. J. Meyer, and J. R. Priess, Eds.), pp. 185–208. Cold Spring Harbor Laboratory Press, Cold Spring Harbor, MA.

Austin, J., and Kimble, J. (1987). *glp-1* is required in the germ line for regulation of the decision between mitosis and meiosis in *C. elegans. Cell* **51,** 589–599.

Austin, J., and Kimble, J. (1989). Transcript analysis of *glp-1* and *lin-12,* homologous genes required for cell interactions during development of *Caenorhabditis elegans. Cell* **58,** 565–571.

Bai, C., and Tolias, P. P. (1996). Cleavage of RNA hairpins mediated by a developmentally regulated CCCH zinc finger protein. *Mol. Cell. Biol.* **16,** 6661–6667.

Bai, C., and Tolias, P. P. (1998). *Drosophila* clipper/CPSF 30K is a post-transcriptionally regulated nuclear protein that binds RNA containing GC clusters. *Nuc. Acids Res.* **26,** 1597–1604.

Barabino, S. M., Hubner, W., Jenny, A., Minvielle-Sebastia, L., and Keller, W. (1997). The 30-kD subunit of mammalian cleavage and polyadenylation specificity factor and its yeast homolog are RNA-binding zinc finger proteins. *Genes Dev.* **11,** 1703–1716.

Batchelder, C., Dunn, M. A., Choy, B., Suh, Y., Cassie, C., Shim, E. Y., Shin, T. H., Mello, C., Seydoux, G., and Blackwell, T. K. (1999). Transcriptional repression by the *Caenorhabditis elegans* germ-line protein PIE-1. *Genes Dev.* **13,** 202–212.

Bergsten, S. E., and Gavis, E. R. (1999). Role for mRNA localization in translational activation but not spatial restriction of nanos RNA. *Development* **126,** 659–669.

Berry, L., Westlund, B., and Schedl, T. (1997). Germline tumor formation caused by activation of *glp-1,* a member of the *Notch* family of receptors. *Development* **124,** 925–936.

Blaumueller, C. M., Qi, H. L., Zagouras, P., and Artavanis-Tsakonas, S. (1997). Intracellular cleavage of Notch leads to a heterodimeric receptor on the plasma membrane. *Cell* **90,** 281–291.

Bosher, J. M., Dufourcq, P., Sookhareea, S., and Labouesse, M. (1999). RNA interference can target pre-mRNA. Consequences for gene expression in a *Caenorhabditis elegans* operon. *Genetics* **153,** 1245–1256.

Bowerman, B., Draper, B. W., Mello, C. C., and Priess, J. R. (1993). The maternal gene skn-1 encodes a protein that is distributed unequally in early *C. elegans* embryos. *Cell* **74,** 443–452.

Bowerman, B., Eaton, B. A., and Priess, J. R. (1992). *skn-1,* a maternally expressed gene required to specify the fate of ventral blastomeres in the early *C. elegans* embryo. *Cell* **68,** 1061–1075.

Boyd, L., Guo, S., Levitan, D., Stinchcomb, D. T., and Kemphues, K. J. (1996). PAR-2 is asymmetrically distributed and promotes association of P granules and PAR-1 with the cortex in *C. elegans* embryos. *Development* **122,** 3075–3084.

Capowski, E., Martin, P., Gavin, C., and Strome, S. (1991). Identification of genetic loci whose products are required for normal germline development in the nematode *Caenorhabditis elegans. Genetics* **129,** 1061–1072.

Carballo, E., Lai, W. S., and Blackshear, P. J. (1998). Feedback inhibition of macrophage tumor necrosis factor-alpha production by tristetraprolin. *Science* **281,** 1001–1005.

Carrington, E. A., and Jones, R. S. (1996). The *Drosophila* Enhancer of zeste gene encodes a chromosomal protein: Examination of wild-type and mutant protein distribution. *Development* **122,** 4073–4083.

C. elegans Sequencing Consortium (1998). Genome sequence of the nematode *C. elegans:* A platform for investigating biology. *Science* **282,** 2012–2018.

Chan, Y. M., and Jan, Y. N. (1998). Roles for proteolysis and trafficking in Notch maturation and signal transduction. *Cell* **94,** 423–426.

Chan, Y. M., and Jan, Y. N. (1999). Presenilins, processing of beta-amyloid precursor protein, and Notch signaling. *Neuron* **23,** 201–204.

Christensen, S., Kodoyianni, V., Bosenberg, M., Friedman, L., and Kimble, J. (1996). *lag-1,* a gene required for *lin-12* and *glp-1* signaling in *C. elegans,* is homologous to human CBF1 and *Drosophila SU(H). Development* **122,** 1373–1383.

Cogoni, C., and Macino, G. (1997). Isolation of quelling-defective (qde) mutants impaired in posttranscriptional transgene-induced gene silencing in Neurospora crassa. *Proc. Natl. Acad. Sci. USA* **94,** 10233–10238.

Collins, J., Saari, B., and Anderson, P. (1987). Activation of a transposable element in the germ line but not the soma of *Caenorhabditis elegans. Nature* **328,** 726–728.

Crittenden, S. L., Troemel, E. R., Evans, T. C., and Kimble, J. (1994). GLP-1 is localized to the mitotic region of the *C. elegans* germ line. *Development* **120,** 2901–2911.

Dahanukar, A., Walker, J. A., and Wharton, R. P. (1999). Smaug, a novel RNA-binding protein that operates a translational switch in *Drosophila. Mol. Cell.* **4,** 209–218.

Dahanukar, A., and Wharton, R. P. (1996). The Nanos gradient in *Drosophila* embryos is generated by translational regulation. *Genes Dev.* **10,** 2610–2620.

Dernburg, A. F., McDonald, K., Moulder, G., Barstead, R., Dresser, M., and Villeneuve, A. M. (1998). Meiotic recombination in *C. elegans* initiates by a conserved mechanism and is dispensable for homologous chromosome synapsis. *Cell* **94,** 387–398.

De Strooper, B., Annaert, W., Cupers, P., Saftig, P., Craessaerts, K., Mumm, J. S., Schroeter, E. H., Schrijvers, V., Wolfe, M. S., Ray, W. J., Goate, A., and Kopan, R. (1999). A presenilin-1-dependent gamma-secretase-like protease mediates release of Notch intracellular domain. *Nature* **398,** 518–522.

Draper, B. W., Mello, C. C., Bowerman, B., Hardin, J., and Priess, J. R. (1996). MEX-3 is a KH domain protein that regulates blastomere identity in early *C. elegans* embryos. *Cell* **87,** 205–216.

Eide, D., and Anderson, P. (1985). Transposition of Tc1 in the nematode *Caenorhabditis elegans. Proc. Natl. Acad. Sci. USA* **82,** 1756–1760.

Ellis, R. E. (1998). Sex and death in the *C. elegans* germline. *In* "Cell Lineage and Fate Determination." S. A. Moody, ed. pp. 119–138. Academic Press, San Diego.

Emmons, S. W., and Yesner, L. (1984). High-frequency excision of transposable element Tc 1 in the nematode *Caenorhabditis elegans* is limited to somatic cells. *Cell* **36,** 599–605.

Evans, T. C., Crittenden, S. L., Kodoyianni, V., and Kimble, J. (1994). Translational control of maternal *glp-1* mRNA establishes an asymmetry in the *C. elegans* embryo. *Cell* **77,** 183–194.

Fire, A., Xu, S., Montgomery, M. K., Kostas, S. A., Driver, S. E., and Mello, C. C. (1998). Potent and specific genetic interference by double-stranded RNA in *Caenorhabditis elegans. Nature* **391,** 806–811.

Fitzgerald, K., and Greenwald, I. (1995). Interchangeability of *Caenorhabditis elegans* DSL proteins and intrinsic signalling activity of their extracellular domains *in vivo. Development* **121,** 4275–4282.

Forbes, A., and Lehmann, R. (1998). Nonos and Pumilio have critical roles in the development and function of *Drosophila* germline stem cells. *Development* **125,** 679–690.

Francis, R., Barton, M. K., Kimble, J., and Schedl, T. (1995a). *gld-1,* a tumor suppressor gene required for oocyte development in *Caenorhabditis elegans. Genetics* **139,** 579–606.

Francis, R., Maine, E., and Schedl, T. (1995b). Analysis of the multiple roles of *gld-1* in germline development: Interactions with the sex determination cascade and the *glp-1* signaling pathway. *Genetics* **139,** 607–630.

Franke, A., DeCamillis, M., Zink, D., Cheng, N., Brock, H. W., and Paro, R. (1992). Polycomb and polyhomeotic are constituents of a multimeric protein complex in chromatin of *Drosophila melanogaster. EMBO J.* **11,** 2941–2950.

Garvin, C., Holdeman, R., and Strome, S. (1998). The phenotype of *mes-2, mes-3, mes-4* and *mes-6,* maternal-effect genes required for survival of the germline in *Caenorhabditis elegans,* is sensitive to chromosome dosage. *Genetics* **148,** 167–185.

Gavis, E. R., Lunsford, L., Bergsten, S. E., and Lehmann, R. (1996). A conserved 90 nucleotide element mediates translational repression of nanos RNA. *Development* **122,** 2791–2800.

Goldstein, B., and Hird, S. N. (1996). Specification of the anteroposterior axis in *Caenorhabditis elegans. Development* **122,** 1467–1474.

Greenwald, I. (1994). Structure/function studies of *lin-12/Notch* proteins. *Curr. Opin. Genet. Dev.* **4,** 556–562.

Greenwald, I. (1998). LIN-12/Notch signaling: Lessons from worms and files (review). *Genes Dev.* **12,** 1751–1762.

Greenwald, I., and Seydoux, G. (1990). Analysis of gain-of-function mutations of the *lin-12* gene of *Caenorhabditis elegans. Nature* **346,** 197–199.

Greenwald, I. S., Sternberg, P. W., and Horvitz, H. R. (1983). The *lin-12* locus specifies cell fates in *Caenorhabditis elegans. Cell* **34,** 435–444.

Gruidl, M. E., Smith, P. A., Kuznicki, K. A., McCrone, J. S., Kirchner, J., Roussell, D. L., Strome, S., and Bennett, K. L. (1996). Multiple potential germ-line helicases are components of the germ-line-specific P granules of *Caenorhabditis elegans. Proc. Natl. Acad. Sci. USA* **93,** 13837–13842.

Guedes, S., and Priess, J. R. (1997). The *C. elegans* MEX-1 protein is present in germline blastomeres and is a P granule component. *Development* **124,** 731–739.

Guo, S., and Kemphues, K. J. (1995). *par-1,* a gene required for establishing polarity in *C. elegans* embryos, encodes a putative Ser/Thr kinase that is asymmetrically distributed. *Cell* **81,** 611–620.

Hall, D. H., Winfrey, V. P., Blaeuer, G., Hoffman, L. H., Furuta, T., Rose, K. L., Hobert, O., and Greenstein, D. (1999). Ultrastructural features of the adult hermaphrodite gonad of *Caenorhabditis elegans:* Relations between the germ line and soma. *Dev. Biol.* **212,** 101–123.

Hansen, D., and Pilgrim, D. (1999). Sex and the single worm: Sex determination in the nematode *C. elegans. Mech. Dev.* **83,** 3–15.

Henderson, S. T., Gao, D., Christenson, S., and Kimble, J. (1997). Functional domains of *C. elegans* LAG-2, a putative signaling ligand for LIN-12/GLP-1 receptors. *Mol. Biol. Cell.* **8,** 1751–1762.

Henderson, S. T., Gao, D., Lambie, E. J., and Kimble, J. (1994). *lag-2* may encode a signaling ligand for the GLP-1 and LIN-12 receptors of *C. elegans. Development* **120,** 2913–2924.

Hird, S. N., Paulsen, J. E., and Strome, S. (1996). Segregation of germ granules in living *Caenorhabditis elegans* embryos: Cell-type-specific mechanisms for cytoplasmic localization. *Development* **122,** 1303–1312.

Hird, S. N., and White, J. G. (1993). Cortical and cytoplasmic flow polarity in early embryonic cells of *Caenorhabditis elegans. J. Cell Biol.* **121,** 1343–1355.

Hirsh, D., Oppenheim, D., and Klass, M. (1976). Development of the reproductive system of *Caenorhabditis elegans. Dev. Biol.* **49,** 200–219.

Holdeman, R., Nehrt, S., and Strome, S. (1998). MES-2, a maternal protein essential for viability of the germline in *Caenorhabditis elegans,* is homologous to a *Drosophila* Polycomb group protein. *Development* **125,** 2457–2467.

Hubbard, E. J., and Greenstein, D. (2000). The *C. elegans* gonad: A test tube for cell and developmental biology. *Dev. Dyn.* **218,** 2–22.

Hunter, C. P. (1999). Genetics: A touch of elegance with RNAi. *Curr. Biol* **9,** R440–R442.

Hunter, C. P., and Kenyon, C. (1996): Spatial and temporal controls target pal-1 blastomere-specification activity to a single blastomere lineage in *C. elegans* embryos. *Cell* **87,** 217–226.

Iida, T., and Kobayashi, S. (1998). Essential role of mitochondrially encoded large rRNA for germ-line formation in *Drosophila* embryos. *Proc. Natl. Acad. Sci. USA* **95,** 11274–11278.

Jan, E., Motzny, C. K., Graves, L. E., and Goodwin, E. B. (1999). The STAR Protein, GLD-1, is a translational regulator of sexual identity in *Caenorhabditis elegans. EMBO J.* **18,** 258–269.

Jantsch-Plunger, V., and Glotzer, M. (1999). Depletion of syntaxins in the early *Caenorhabditis elegans* embryo reveals a role for membrane fusion events in cytokinesis. *Curr. Biol.* **9,** 738–745.

Jarriault, S., Brou, C., Logeat, F., Schroeter, E. H., Kopan, T., and Israel, A. (1995). Signalling downstream of activated mammalian Notch. *Nature* **377,** 355–358.

Jones, A. R., Francis, R., and Schedl, T. (1996). GLD-1, a cytoplasmic protein essential for oocyte differentiation, shows stage- and sex-specific expression during *Caenorhabditis elegans* germline development. *Dev. Biol.* **180,** 165–183.

Jones, A. R., and Schedl, T. (1995). Mutations in *gld-1,* a female germ cell-specific tumor suppressor gene in *Caenorhabditis, elegans* affect a conserved domain also found in Src-associated protein Sam68. *Genes Dev.* **9,** 1491–1504.

Kadyk, L. C., and Kimble, J. (1998). Genetic regulation of entry into meiosis in *C. elegans. Development* **125,** 1803–1813.

Kawasaki, I., Shim, Y. H., Kirchner, J., Kaminker, J., Wood, W. B., and Strome, S. (1998). PGL-1, a predicted RNA-binding component of germ granules, is essential for fertility in *C. elegans. Cell* **94,** 635–645.

Kelly, W. G., and Fire, A. (1998). Chromatin silencing and the maintenance of a functional germline in *Caenorhabditis elegans. Development* **125,** 2451–2456.

Kelly, W. G., Xu, S., Montgomery, M. K., and Fire, A. (1997). Distinct requirements for somatic and germline expression of a generally expressed *Caenorhabditis elegans* gene. *Genetics* **146,** 227–238.

Kemphues, K. J., Priess, J. R., Morton, D. G., and Cheng, N. S. (1988). Identification of genes required for cytoplasmic localization in early C. elegans embryos. *Cell* **52,** 311–320.

Kemphues, K. J., and Strome, S. (1997). Fertilization and establishment of polarity in the embryo. *In "C. elegans* II," Volume II (D. L. Riddle, T. Riddle, T. Blumenthal, B. J. Meyer, and J. R. Priess, Eds.) Cold Spring Harbor Laboratory Press, Cold Spring Harbor, Mass.

Ketting, R. F., Haverkamp, T. H., van Luenen, H. G., and Plasterk, R. H. (1999). Mut-7 of *C. elegans,* required for transposon silencing and RNA interference, is a homolog of Werner syndrome helicase and RNaseD. *Cell* **99,** 133–141.

Kimble, J., and Hirsh, D. (1979). The postembryonic cell lineages of the hermaphrodite and male gonads in *Caenorhabditis elegans. Dev. Biol.* **70,** 396–417.

Kimble, J., Henderson, S., and Crittenden, S. (1998). Notch/Lin-12 signaling-transduction by regulated protein slicing. *Trends Biochem. Sci.* **23,** 353–357.

Kimble, J. E., and White, J. G. (1981). On the control of germ cell development in *Caenorhabditis elegans. Dev. Biol.* **81,** 208–219.

Kloc, M., Larabell, C., Chan, P.-Y. A., and Etkin, L. D. (1998). Contribution of METRO pathway localized molecules to the organization of the germ cell lineage. *Mech. Dev.* **75,** 81–93.

Kobayashi, S., Amikura, R., and Okada, M. (1993). Presence of mitochondrial large ribosomal RNA outside mitochondria in germ plasm of *Drosophila melanogaster. Science* **260,** 1521–1524.

Kobayashi, S., Mizuno, H., and Okada, M. (1988). Accumulation and spatial distribution of poly-A^+ RNA in oocytes and early embryos of *Drosophila melanogaster. Dev. Growth Differen.* **30,** 251–260.

Kobayashi, S., Yamada, M., Asaoka, M., and Kitamura, T. (1996). Essential role of the posterior morphogen nanos for germline development in *Drosophila. Nature* **380,** 708–711.

Kodoyianni, V., Maine, E. M., and Kimble, J. (1992). Molecular basis of loss-of-function mutations in the *glp-1* gene of *Caenorhabditis elegans. Mol. Biol. Cell* **3,** 1199–1213.

Kopan, R., Schroeter, E., Weintraub, H., and Nye, J. (1996). Signal transduction by activated mNotch: Importance of proteolytic processing and its regulation by the extracellular domain. *Proc. Natl. Acad. Sci. USA* **93,** 1683–1688.

Korf, I., Fan, Y., and Strome, S. (1998). The Polycomb group in *Caenorhabditis elegans* and maternal control of germline development. *Development* **125,** 2469–2478.

Kuwabara, P. E. (1999). Development genetics in *Caenorhabditis elegans* sex determination. *Curr. Top. Dev. Biol.* **41,** 99–132.

Kuznicki, K. A., Smith, P. A., Leung-Chiu, W. M., Estevez, A. O., Scott, H. C., Bennett, K. L. (2000). Combinatorial RNA interference indicates GLH-4 can compensate for GLH-1; these two P granule components are critical for fertility in *C. elegans. Development* **127,** 2907–2916.

Lamb, M. M., and Laird, C. D. (1976). Increase in nuclear poly(A)-containing RNA at syncytial blastoderm in *Drosophila meelanogaster* embryos. *Dev. Biol.* **54,** 31–42.

Lambie, E. J., and Kimble, J. (1991). Two homologous regulatory genes, *lin-12* and *glp-1,* have overlapping functions. *Development* **112,** 231–240.

Lasko, P. F., and Ashburner, M. (1990). Posterior localization of vasa protein correlates with, but is not sufficient for, pole cell development. *Genes Dev.* **4,** 905–921.

Lecourtois, M., and Schweisguth, F. (1998). Indirect evidence for Delta-dependent intracellular processing of Notch in *Drosophila* embryos. *Curr. Biol.* **8,** 771–774.

Levitan, D., and Greenwald, I. (1995). Facilitation of *lin-12*-mediated signalling by *sel-12,* a *Caenorhabditis elegans S182* Alzheimer's disease gene. *Nature* **377,** 351–354.

L'Hernault, S., Singson, A. (2000). Developmental genetics of spermatogenesis in the nematode *Caenorhabditis elegans.* Serano Symposia USA. Ira Goldberg, ed. Springer, New York, in press.

Lieber, T., Kidd, S., Alcamo, E., Corbin, V., and Young, M. W. (1993). Antineurogenic phenotypes induced by truncated Notch proteins indicate a role in signal transduction and may point to a novel function for Notch in nuclei. *Genes Dev.* **7,** 1949–1965.

Logeat, F., Bessia, C., Brou, C., Debail, O., Jarriault, S., Seidah, N. G., and Israel, A. (1998). The Notch1 receptor is cleaved constitutively by a furin-like convertase. *Proc. Natl. Acad. Sci. USA* **95,** 8108–8112.

Maine, E. M., and Kimble, J. (1993). Suppressors of *glp-1,* a gene required for cell communication during development in *Caenorhabditis elegans,* define a set of interacting genes. *Genetics* **135,** 1011–1022.

Mello, C. C., Draper, B. W., Krause, M., Weintraub, H., and Priess, J. R. (1992). The *pie-1* and *mex-1* genes and maternal control of blastomere identity in early *C. elegans* embryos. *Cell* **70,** 163–176.

Mello, C. C., and Fire, A. (1995). DNA transformation. *Meth. Cell Biol.* **48,** 451–482.

Mello, C. C., Kramer, J. M., Stinchcomb, D., and Ambros, V. (1991). Efficient gene transfer in *C. elegans:* Extrachromosomal maintenance and integration of transforming sequences. *EMBO J.* **10,** 3959–3970.

Mello, C. C., Schubert, C., Draper, B., Zhang, W., Lobel, R., and Priess, J. R. (1996). The PIE-1 protein and germline specification in *C. elegans* embryos [letter]. *Nature* **382,** 710–712.

Meyer, B. J. (1988). Primary events in *C. elegans* sex determination and dosage compensation. *Trends Genet.* **4,** 337–342.

Meyer, B. J. (1997). Sex Determination and X Chromosome Dosage Compensation, *C. elegans* II (D. L. Riddle, T. Blumenthal, B. J. Meyer and J. R. Priess, Eds.) Cold Spring Harbor Laboratory Press, Cold Spring Harbor, MA.

Moerman, D. G., and Waterston, R. H. (1984). Spontaneous unstable *unc-22* IV mutations in *C. elegans var. Bergerac. Genetics* **108,** 859–877.

Montgomery, M. K., and Fire, A. (1998). Double-stranded RNA as a mediator in sequence-specific genetic silencing and co-suppression [see comments]. *Trends Genet.* **14,** 255–258.

Montgomery, M. K., Xu, S., and Fire, A. (1998). RNA as a target of double-stranded RNA-mediated genetic interference in *Caenorhabditis elegans. Proc. Natl. Acad. Sci. USA* **95,** 15502–15507.

Mori, I., Moerman, D. G., and Waterston, R. H. (1988). Analysis of a mutator activity necessary for germline transposition and excision of Tc1 transposable elements in *Caenorhabditis elegans. Genetics* **120,** 397–407.

Mosquera, L., Forristall, C., Zhou, Y., and King, M. L. (1993). A mRNA localized to the vegetal cortex of *Xenopus* oocytes encodes a protein with a nanos-like zinc finger domain. *Development* **117,** 377–386.

Mumm, J. S., Schroeter, E. H., Saxena, M. T., Griesemer, A., Tian, X., Pan, D. J., Ray, W. J., and Kopan, R. (2000). A ligand induced extracellular cleavage regulates gamma-secretase like proteolytic activation of Notch 1. *Mol. Cell* **5,** 197–206.

Murray, M. V., Turnage, M. A., Williamson, K. J., Steinhauer, W. R., and Searles, L. L. (1997). The *Drosophila* suppressor of sable protein binds to RNA and associates with a subset of polytene chromosome bands. *Mol. Cell. Biol.* **17,** 2291–2300.

Musco, G., Stier, G., Josph, C., Morelli, M. A., and Pastore, A. (1996). Three-dimensional structure and stability of the KH domain: molecular insights onto fragile X syndrome. *Cell* **85,** 239–245.

Nusslein-Volhard, C., Frohnhofer, H. G., and Lehmann, R. (1987). Determination of anteroposterior polarity in *Drosophila. Science* **238,** 1675–1681.

Paulsen, J. E., Capowski, E. E., and Strome, S. (1995). Phenotypic and molecular analysis of *mes-3,* a maternal-effect gene required for proliferation and viability of the germ line in *C. elegans. Genetics* **141,** 1383–1398.

Pirrotta, V. (1997). PcG complexes and chromatin silencing. *Curr. Opin. Genet. Dev.* **7,** 249–258.

Pitt, J. N., Schisa, J. A., Priess, J. R. (2000). P granules in the germ cells of *C. elegans* adults are associated with clusters of nuclear pores and contain RNA. *Dev. Biol.* **219,** 315–333.

Platero, J. S., Sharp, E. J., Alder, P. N., and Eissenberg, J. C. (1996). *In vivo* assay for protein-protein interactions using *Drosophila* chromosomes. *Chromosoma* **104,** 393–404.

Plasterk, R. H. A., van Luenen H. G. A. M. (1997). Transposons. In "*C. elegans* II" (D. L. Riddle, T. Blumenthal, B. J. Meyer, and J. R. Priess, Eds.), pp. 97–116. Cold Spring Harbor Laboratory Press, Cold Spring Harbor, MA.

Qi, H. L., Rand, M. D., Wu, X. H., Sestan, N., Wang, W. Y., Rakic, P., Xu, T., and Artavanis-Tsakonas, S. (1999). Processing of the Notch ligand delta by the metalloprotease kuzbanian. *Science* **283,** 91–94.

Qiao, L., Lissemore, J. L., Shu, P., Smardon, A., Gelber, M. B., and Maine, E. M. (1995). Enhancers of glp-1, a gene required for cell-signaling in *Caenorhabditis elegans,* define a set of genes required for germline development. *Genetics* **141,** 551–569.

Rastelli, L., Chan, C. S., and Pirrotta, V. (1993). Related chromosome binding sites for zeste, suppressor of zeste and Polycomb group proteins in *Drosophila* and their dependence on Enhancer of zeste function. *EMBO J.* **12,** 1513–1522.

Ray, W. J., Yao, M., Mumm, J., Schroeter, E. H., Saftig, P., Wolfe, M., Selkoe, D. J., Kopan, R., Goate, A. M. (1999a). Cell surface presenilin-1 participates in the gamma-secretase-like proteolysis of Notch. *J. Biol. Chem.* **17,** 36801–36807.

Ray, W. J., Yao, M., Nowotny, P., Mumm, J., Zhang, W., Wu, J. Y., Kopan, R., and Goate, A. M. (1999b). Evidence for a physical interaction between presenilin and Notch. *Proc. Natl. Acad. Sci. USA* **96,** 3263–3268.

Rebay, I., Fleming, R., Fehon, R., Cherbas, P., and Artavanis-Tsakonas, S. (1991). Specific EGF repeats of *Notch* mediate interactions with *Delta* and *Serrate:* Implications for *Notch* as a multifunctional receptor. *Cell* **67,** 687–699.

Reese, K. J., Dunn, M. A., Waddle, J. A., and Seydoux, G. (2000). Asymmetric segregation of PiE-1 in C. elegans is mediated by two complementary mechanisms that act through separate PiE-1 protein domains. *Mol. Cell* **6,** 1–20.

Roehl, H., Besenburg, M., Blelloch, R., and Kimble, J. (1996). Roles of the RAM and ANK domains in signaling by the *C. elegans* GLP-1 receptor. *EMBO J.* **15,** 7002–7012.

Rudner, D. Z., Breger, K. S., and Rio, D. C. (1998). Molecular genetic analysis of the heterodimeric splicing factor U2AF: The RS domain on either the large or small *Drosophila* subunit is dispensable *in vivo. Genes Dev.* **12,** 1010–1021.

Schedl, T. (1997). Developmental genetics of the germline in "*C. elegans* II" D. L. Riddle, T. Blumenthal, B. J. Meyer and J. R. Priess (eds). Cold Spring Harbor Lab. Press, II, 241–269. Cold Spring Harbor, NY.

Schiebel, W., Pelissier, T., Riedel, L., Thalmeir, S., Schiebel, R., Kempe, D., Lottspeich, F., Sanger, H. L., and Wassenegger, M. (1998). Isolation of an RNA-directed RNA polymerase-specific cDNA clone from tomato. *Plant Cell* **10,** 2087–2101.

Schierenberg, E. (1987). Reversal of cellular polarity and early cell-cell interaction in the embryos of *Caenorhabditis elegans. Dev. Biol.* **122,** 452–463.

Schmidt, A., Palumbo, G., Bozzetti, M. P., Tritto, P., Pimpinelli, S., and Schafer, U. (1999). Genetic and molecular characterization of sting, a gene involved in crystal formation and meiotic drive in the male germ line of *Drosophila melanogaster. Genetics* **151,** 749–760.

Schnabel, R., Weigner, C., Hutter, H., Feichtinger, R., and Schnabel, H. (1996). *mex-1* and the general partitioning of cell fate in the early *C. elegans* embryo. *Mech. Dev.* **54,** 133–147.

Schroeter, E. H., Kisslinger, J. A., and Kopan, R. (1998). Notch-1 signalling requires ligand-induced proteolytic release of intracellular domain. *Nature* **393,** 382–386.

Schupbach, T., Wieschaus, E. (1986). Maternal-effect mutations altering the anterio-posterior pattern of the *Drosophila* embryo. *Wilhem Roux's Arch. Dev. Biol.* **195,** 302–317.

Seydoux, G., and Dunn, M. A. (1997). Transcriptionally repressed germ cells lack a subpopulation of phosphorylated RNA polymerase II in early embryos of *Caenorhabditis elegans* and *Drosophila melanogaster. Development* **124,** 2191–2201.

Seydoux, G., and Fire, A. (1994). Soma-germline asymmetry in the distributions of embryonic RNAs in *Caenorhabditis elegans. Development* **120,** 2823–2834.

Seydoux, G., Mello, C. C., Pettitt, J., Wood, W. B., Priess, J. R., and Fire, A. (1996). Repression of gene expression in the embryonic germ lineage of *C. elegans. Nature* **382,** 713–716.

Shelton, C. A., Carter, J. C., Ellis, G. C., and Bowerman, B. (1999). The nonmuscle myosin regulatory light chain gene mlc-4 is required for cytokinesis, anterior-posterior polarity, and body morphology during Caenorhabditis elegans embryogenesis *J. Cell. Biol.* **146,** 439–451.

Smardon, A., Spoerke, J. M., Stacey, S. C., Klein, M. E., Mackin, N., Maine, E. M. (2000). EGO-1 is related to RNA-directed RNA polymerase and functions in germline development and RNA interference in *C. elegans. Current Biol.* **10,** 169–178.

Smibert, C. A., Wilson, J. E., Kerr, K., and Macdonald, P. M. (1996). smaug protein represses translation of unlocalized nanos mRNA in the *Drosophila* embryo. *Genes Dev.* **10,** 2600–2609.

Strome, S., Martin, P., Schierenberg, E., and Paulsen, J. (1995). Transformation of the germ line into muscle in *mes-1* mutant embryos of *C. elegans. Development* **121,** 2961–2972.

Strome, S., and Wood, W. B. (1982). Immunofluorescence visualization of germ-line-specific cytoplasmic granules in embryos, larvae, and adults of *Caenorhabditis elegans. Proc. Natl. Acad. Sci. USA* **79,** 1558–1562.

Strome, S., and Wood, W. B. (1983). Generation of asymmetry and segregation of germ-line granules in early *C. elegans* embryos. *Cell* **35,** 15–25.

Struhl, G., and Adachi, A. (1998). Nuclear access and action of Notch *in vivo. Cell* **93,** 649–660.

Struhl, G., Fitzgerald, K., and Greenwald, I. (1993). Intrinsic activity of the LIN-12 and Notch intracellular domains *in vivo. Cell* **74,** 331–345.

Struhl, G., and Greenwald, I. (1999). Presenilin is required for activity and nuclear access of Notch in *Drosophila. Nature* **398,** 522–525.

Subramaniam, K., and Seydoux, G. (1999). *nos-1* and *nos-2,* two genes related to *Drosophila* nanos, regulate primordial germ cell development and survival in *Caenorhabditis elegans. Development* **126,** 4861–4871.

Sulston, J. E., Schierenberg, E., White, J. G., and Thomson, J. N. (1983). The embryonic cell lineage of the nematode *Caenorhabditis elegans. Dev. Biol.* **100,** 64–119.

Tabara, H., Hill, R. J., Mello, C. C., Priess, J. R., and Kohara, Y. (1998). *pos-1* encodes a cytoplasmic zinc-finger protein essential for germline specification in *C. elegans. Development* **126,** 1–11.

Tabara, H., Sarkissian, M., Kelly, W. G., Fleenor, J., Grishok, A., Timmons, L., Fire, A., and Mello, C. C. (1999). The *rde-1* gene, RNA interference, and transposon silencing in *C. elegans. Cell* **99,** 123–132.

Tax, F. E., Yeargers, J. J., and Thomas, J. H. (1994). Sequence of *C. elegans lag-2* reveals a cell-signaling domain shared with *Delta* and *Serrate* of *Drosophila. Nature* **368,** 150–154.

Tenenhaus, C., Schubert, C., and Seydoux, G. (1998). Genetic requirements for PIE-1 localization and inihibition of gene expression in the embryonic germ lineage of *Caenorhabditis elegans. Dev. Biol.* **200,** 212–224.

Timmons, L., and Fire, A. (1998). Specific interference by ingested dsRNA [letter]. *Nature* **395,** 854.

Van Doren, M., Williamson, A. L., and Lehmann, R. (1998). Regulation of zygotic gene expression in *Drosophila* primordial germ cells. *Curr. Biol.* **8,** 243–246.

Wang, C., and Lehmann, R. (1991). Nanos is the localized posterior determinant in *Drosophila* [published erratum appears in *Cell,* 1992, Mar. 20;**68**(6):1177]. *Cell* **66,** 637–647.

Wassenegger, M., and Pelissier, T. (1998). A model for RNA-mediated gene silencing in higher plants. *Plant Mol. Biol.* **37,** 349–362.

Wilkinson, H. A., Fitzgerald, K., and Greenwald, I. (1994). Reciprocal changes in expression of the receptor lin-12 and its ligand lag-2 prior to commitment in a *C. elegans* cell fate decision. *Cell* **79,** 1187–1196.

Wolfe, M. S., Xia, W., Moore, C. L., Leatherwood, D. D., Ostaszewski, B., Rahmati, T., Donkor, I. O., and Selkoe, D. J. (1999). Peptidomimetic probes and molecular modeling suggest that Alzheimer's gamma-secretase is an intramembrane-cleaving aspartyl protease. *Biochemistry* **38,** 4720–4727.

Wolf, N., Priess, J., and Hirsh, D. (1983). Segregation of germline granules in early embryos of *Caenorhabditis elegans:* An electron microscopic analysis. *J. Embryol Exp. Morphol.* **73,** 297–306.

Ye, Y. H., Lukinova, N., and Fortini, M. E. (1999). Neurogenic phenotypes and altered Notch processing in *Drosophila* Presenilin mutants. *Nature* **398,** 525–529.

Yochem, J., and Greenwald, I. (1989). *glp-1* and *lin-12,* genes implicated in distinct cell-cell interactions in *Caenorhabditis elegans,* encode similar transmembrane proteins. *Cell* **58,** 553–563.

Yochem, J., Weston, K., and Greenwald, I. (1988). The *Caenorhabditis elegans lin-12* gene encodes a transmembrane protein with overall similarity to *Drosophila Notch. Nature* **335,** 547–550.

Zalokar, M. (1976). Autoradiographic study of protein and RNA formation during early development of *Drosophila* eggs. *Dev. Biol.* **49,** 425–437.

Zhang, B., Gallegos, M., Puoti, A., Durkin, E., Fields, S., Kimble, J., and Wickens, M. P. (1997). A conserved RNA-binding protein that regulates sexual fates in the *C. elegans* hermaphrodite germ line. *Nature* **390,** 477–484.

Zhou, Y., and King, M. L. (1996). Localization of Xcat-2 RNA, a putative germ plasm component, to the mitochondrial cloud in *Xenopus* stage I oocytes. *Development* **122,** 2947–2953.

Assembly of the *Drosophila* Germ Plasm

Anthony P. Mahowald
Department of Molecular Genetics and Cell Biology, The University of Chicago, Chicago, Illinois 60637

The *Drosophila melanogaster* germ plasm has become the paradigm for understanding both the assembly of a specific cytoplasmic localization during oogenesis and its function. The posterior ooplasm is necessary and sufficient for the induction of germ cells. For its assembly, localization of *gurken* mRNA and its translation at the posterior pole of early oogenic stages is essential for establishing the posterior pole of the oocyte. Subsequently, *oskar* mRNA becomes localized to the posterior pole where its translation leads to the assembly of a functional germ plasm. Many gene products are required for producing the posterior polar plasm, but only *oskar, tudor, valois, germcell-less* and some noncoding RNAs are required for germ cell formation. A key feature of germ cell formation is the precocious segregation of germ cells, which isolates the primordial germ cells from mRNA turnover, new transcription, and continued cell division. *nanos* is critical for maintaining the transcription quiescent state and it is required to prevent transcription of *Sex-lethal* in pole cells. In spite of the large body of information about the formation and function of the *Drosophila* germ plasm, we still do not know what specifically is required to cause the pole cells to be germ cells. A series of unanswered problems is discussed in this chapter.

KEY WORDS: Germ plasm, Pole cells, Polar granules, *oskar, tudor, vasa, Drosophila.*

I. Introduction

A. Historical Background

Evolution from single-celled primitive organisms to multicellular forms requires the segregation of a specialized cell lineage, in which meiosis and gametogenesis occur, from the remaining somatic lineages responsible for

0074-7696/01 $35.00

growth and survival. Within the plant kingdom a variety of mechanisms exist for deriving germ cells. In the simple colonial alga *Volvox,* sex cells are set aside very early in development (Kirk, 1998) whereas in "higher" plants the floral meristem forms only after extensive growth and depends upon a variety of environmental signals, such as light cycles, for its development (Levy and Dean, 1998). In the animal kingdom, a similar range of mechanisms is used for the segregation of the germline for production of gametes, ranging from totipotent neoblasts in planaria to precociously segregated cells in flies, worms, frogs, and fish (Nieuwkoop and Sutasurya, 1979, 1981). A strong argument, however, can be made that the precocious segregation of the germ line, as found in flies, worms, frogs, and fish, is probably the ancestral mechanism. The analysis of the mechanisms responsible in *Drosophila* for segregating the primordial germ cell population has become the paradigm for understanding the mechanisms for germ cell origin in animal development. In this chapter, I will review and update our current knowledge of the *Drosophila* germ plasm (Rongo and Lehmann, 1996), both in its development and function, and then discuss remaining key, unanswered questions.

In spite of *Drosophila melanogaster's* prominence in the early part of this century as *the* organism for understanding basic genetic mechanisms in animals, its small size, rapid development, and impermeable eggshell have limited the progress of developmental studies of the embryo during the first three quarters of this century. Precocious segregation of the germ cell lineage was clear in both nematodes and a variety of insects already at the turn of the century, and this segregation was associated with specialized cytoplasms containing organelles postulated to be "germ cell determinants" (Hegner, 1914). Subsequent studies extended these observations to *Drosophila* (Sonnenblick, 1950; and Poulson, 1950). Initially it was proposed that pole cells might also provide portions of the midgut (Poulson and Waterhouse, 1960), but subsequent clonal analyses (Underwood *et al.,* 1980; Technau and Campos-Ortega, 1986) have failed to identify any developmental fate for pole cells other than the germ line in *Drosophila.*

B. Experiments Demonstrating Function of Polar Plasm

Geigy (1931) was the first to demonstrate experimentally that the posterior polar plasm of the egg, which is incorporated into the pole cells, was necessary for the production of pole cells. UV-irradiation of the posterior cytoplasm of newly fertilized eggs resulted in sterile flies. Geigy further showed that the somatic components of the gonads formed even when germ cells were absent, thus establishing the independent origin of the germ line and somatic portions of the gonad. Okada *et al.* (1974) later showed that the transplantation of non-UV-irradiated polar plasm into irradiated embryos restored the ability to produce germ cells. That the posterior polar plasm was *sufficient* for germ cell formation was demonstrated by transplanting posterior polar plasm to ectopic sites in the embryo and showing that the pole cells formed in these sites could function as germ cells if incorporated into the gonad. Two ectopic sites were used: the anterior pole (Illmensee and Mahowald, 1974) and the midventral site where mesodermal precursors would form after gastrulation (Illmensee and Mahowald, 1976). Because embryos injected anteriorly with posterior polar plasm developed with severe abnormalities, it was impossible to determine whether presumptive pole cells might become germ cells. To test this, anterior cells were transplanted to the pole cell region of genetically marked host embryos and these embryos were tested as adults for the presence of germ cells derived from the donor embryos. Ventrally induced pole cells were tested in the same manner by transplanting presumptive cells to host embryos. The small number of fertile flies (Table I) is readily explained. First, frequently polar plasm transplantations were unsuccessful, as demonstrated by the absence of polar granules and pole cells when examined by electron microscopy (Table I). Second, because the induced pole cells could not be seen due to the damage caused by the initial cytoplasm injection, most of the cells transplanted to the posterior pole of host blastoderms were somatic cells. Finally, the transplanted cells had to compete with host pole cells in most of the experiments. Injections of anterior or ventral cells, in control experiments, never produced germline cells.

Because injections into the midventral region did not seem to disrupt development as much as anterior injections, it seemed possible that these ventral pole cells might directly contribute to the germ line. Since the ectopic pole cells were genetically identical to the host's posterior pole cells, it would be impossible to distinguish the origin of the germ cells in these embryos. To test whether ventral pole cells might contribute to the germ line, wild-type pole cells were transplanted into the presumptive mesoderm at the ventral furrow stage (Illmensee and Mahowald, 1976). These cells produced functional gametes (Table I). This experiment also demonstrated that ventrally induced pole cells probably could have reached the embryonic gonad by this novel route.

C. Developmental Changes in Polar Granule Structure

Distinctive organelles in the posterior polar plasm of *Drosophila* embryos, called polar granules, are readily seen histologically following iron hematoxylin staining. Counce (1963) analyzed a variety of *Drosophila* species and demonstrated that the quantity, amount, and distribution of polar granules

TABLE I.

Summary of Germ Plasm Transplantation Studies

Source of transplanted polar plasm	Site of transplantation	Number examined by electron microscopy	Number with PG - no PCs	Number with PC	Number tested for functional pole cells	Number with functional pole cells
Fertilized embryos[a]	Anterior	10	1	5	53	4
Fertilized embryos[b]	Mid-ventral	7	0	2	46	3
Unfertilized eggs[c]	Anterior	9	2	2	69	5
Stage 14 oocytes[c]	Anterior	8	2	3	42	3
Stage 13 oocytes[c]	Anterior	5	1	4	50	5
Stage 12 oocytes[c]	Anterior	7	5	0	104	0
Stage 10 & 11[c]	Anterior	5	3	0	145	0
D. immigrans[d]	Anterior	10	0	10	71	6
[Pole cells]*[b]	Midventral				120	4
Posterior embryos	UV-irradiated fertilized eggs[e]				582	21
Posterior *gs(1)N26*[f]	Anterior	25		4		

[a] Illmensee and Mahowald, 1974.
[b] Illmensee and Mahowald, 1976.
[c] Illmensee *et al.*, 1976.
[d] Mahowald *et al.*, 1976.
[e] Okaka *et al.*, 1974.
[f] Niki, 1984.
* Posterior pole cells were transplanted to the midventral mesoderm during gastrulation.

varied considerably between different species and at different stages of development. Mahowald (1962; 1968) extended these studies to the ultrastructural level and established a number of distinctive features of polar granules in different species, some of which continue to have significance. The most extensive ultrastructural studies have been carried out on *D. melanogaster.* Polar granules first appear at the posterior tip of the growing oocyte at Stage 9 during mid-vitellogenesis (stages according to King, 1970; Spradling, 1993). They lack membranes and are composed of an interwoven meshwork of fibrils. Initially, polar granules appear free in the posterior polar plasm, but as the oocyte nears the end of oogenesis they become localized to the surface of mitochondria. Granules also aggregate into long changes of granules, with one end attached to mitochondria (Fig. 1A). In contrast to this characteristic association with mitochondria, polar granules in *Drosophila hydei* are never found associated with mitochondria, either in the oocyte or embryo (Fig. 1B) (Mahowald, 1968). In lower dipterans (e.g., *Miastor* or *Smittia* [Mahowald, 1975; Zissler and Sander, 1973]) polar granule material forms a continuous meshwork of fibrous material, and in these species polar granules are also not associated with mitochondria.

Following fertilization (or in unfertilized but activated eggs [Mahowald *et al.*, 1983; Schwalm, 1974]), polar granules fragment into individual organelles, approximately 0.25–0.5 μM in diameter, and clusters of ribosomes, frequently as helices of ribosomes, become associated with their periphery (Fig. 1A). A similar fragmentation and association with ribosomes also occurs in *Miastor,* where the continuous fibrous meshwork disperses into small organelles resembling polar granules in *Drosophila* (Mahowald, 1975). The interpretation that polysomes are attached to polar granules (Mahowald, 1968) is supported by the observation that ribosomes remained associated with the periphery of polar granules following fractionation through sucrose gradients (Mahowald, 1977).

Following the eighth nuclear division in *D. melangaster,* three to four nuclei reach the posterior polar plasm and large cell-like bulges protrude from the posterior tip of the embryo. Following two additional divisions, these posterior bulges are cleaved from the remaining syncytial cytoplasm to form pole cells. Subsequently, pole cell divisions occur out of synchrony with the rest of the embryo and after one or two more divisions they cease mitosis until they reach the embryonic gonad. In embryos in which DNA synthesis has been inhibited by aphidicolin, centrosomes continue to multiply and organize cytasters. These centrosomes migrate to the surface of the embryo, including into the posterior polar plasm, where cell-sized protrusions separate into nucleci-less cells. Centrosomes at the surface of other regions do not segregate into cells (Raff and Glover, 1989). This nuclear

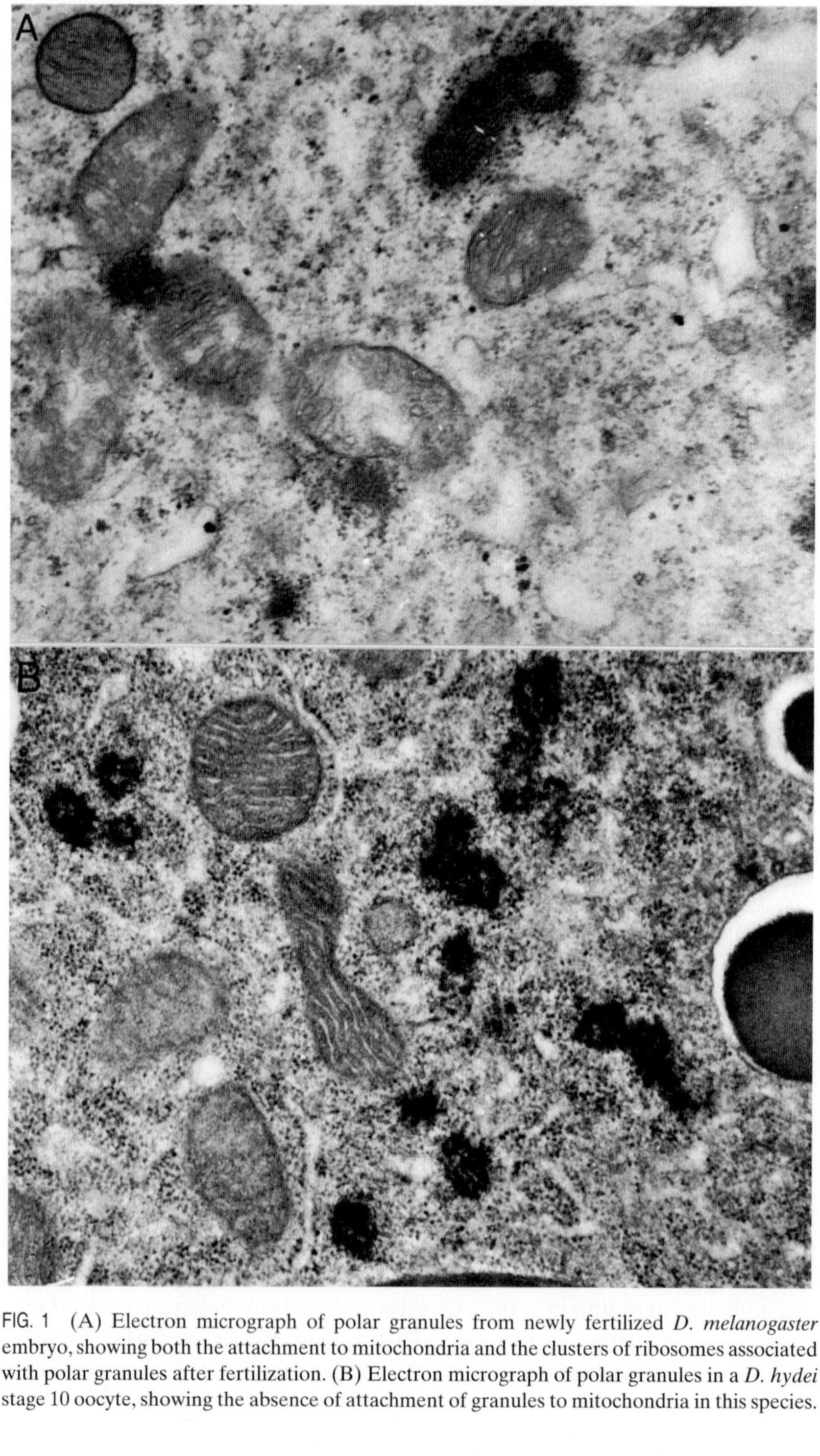

FIG. 1 (A) Electron micrograph of polar granules from newly fertilized *D. melanogaster* embryo, showing both the attachment to mitochondria and the clusters of ribosomes associated with polar granules after fertilization. (B) Electron micrograph of polar granules in a *D. hydei* stage 10 oocyte, showing the absence of attachment of granules to mitochondria in this species.

independent formation of "pole cells" is further evidence for the maternal origin of the information required to segregate pole cells precociously.

Polar granules aggregate around the centrosome during the pole cell cell divisions (Counce, 1963; Mahowald, 1962, 1968). After the completion of mitosis, polar granules fuse into large organelles, often more than 1 mM in diameter in many species. In *Drosophila immigrans* the granules fuse into one large organelle, forming a plaque one granule thick and 2 to 3 μM in size. Moreover, the fibrous substructure of the granule in *D. immigrans* becomes rod-shaped and each rod is regularly spaced from adjacent rods (Mahowald, 1968). A similar rearrangement of polar granule substructure occurs in *Coelopa* (Schwalm, 1974). Although polar granules stained positively for RNA prior to the formation of pole cells, after the granules reaggregate, they no longer show histochemical levels of RNA and polysomes are no longer associated with the surface of the organelles (Mahowald, 1971b).

During gastrulation pole cells move into the posterior midgut invagination and then, following germband elongation, they leave the midgut and migrate into the dorsal mesoderm (Figure 2, see color insert). During germband shortening, pole cells aggregate together in association with the somatic cells of the gonad to form the embryonic gonad (cf. Rongo *et al.*, 1997, for summary). During the migration of pole cells, polar granules gradually disappear and a new, peri-nuclear organelle called "nuage" forms. Because of the distinctive rod substructure of *D. immigrans* polar granules, transition stages between the polar granule and nuage can be seen (Mahowald, 1971a). In addition, Vasa antibodies react with both polar granules and nuage (Hay *et al.*, 1988), thus indicating that these organelles have some components in common.

D. Polar Granule Features Uncovered by Transplantation

Transplantation of polar plasm to the anterior pole has uncovered a number of additional features of the *Drosophila* germ plasm. In homospecific transplants, polar plasm from unfertilized eggs, mature ovarian eggs, and even late-stage oocytes (during eggshell formation or Stage 13) are able to induce functional pole cells (Illmensee *et al.*, 1976) (Table I). Polar plasm from earlier stages, however, lacks this function. Ooplasm form ovarian eggs also inhibits cell formation in the region adjacent to the injection, whereas polar plasm from fertilized embryos lacks this property (Illmensee *et al.*, 1976). The basis for this difference is unknown. A number of intriguing results were also obtained following the heterospecific transplantation of *D. immigrans* polar plasm into *D. melanogaster* embryos (Mahowald *et al.*, 1976). First, these heterospecific polar plasms not only induced pole cells, but also consistently produced many more ectopic pole cells than homospecific transplants (Table I). Although the basis for this difference is unknown, it is probably due to the more plentiful polar plasm of *D. immigrans.* These nuclear-cytoplasmic hybrid cells were fully functional as germ cells when transplanted to host embryos. Second, the reorganization of polar granules in these chimeric pole cells followed the temporal sequence characteristic of the donor cytoplasm rather than the host embryo. Hence, the reorganization of the polar granules into the rod substructure occurred only after gastrulation began, rather than during blastoderm formation. Since *D. immigrans,* development is slower than *D. melanogaster,* the later reorganization of the transplanted polar granules corresponded to the developmental time expected for *D. immigrans,* thus indicating that the reorganization of the polar granules occurs autonomously. Third, the specific-specific structural characteristics of the nuclear body corresponded to the transplanted polar plasm, rather than the nucleus. This indicates that the nuclear body is probably directly related to the cytoplasmic polar granule. Hay *et al.* (1988) showed that Vasa is located in the nuclear body by immunoelectron microscopy, a finding that suggests that this component of the polar granule moves into the nucleus (cf. section II for composition of polar granules). Fourth, the structure of the polar granule in oocytes descended from chimeric pole cells correspond to the nuclear-determined structure rather than the cytoplasm (Mahowald *et al.,* 1976). This provides convincing evidence that the polar granule is not an independently inherited organelle in spite of its morphological continuity in the life cycle (Mahowald, 1971c).

II. Molecular Features of the *Drosophila* Germ Plasm

A. Direct Biochemical Approaches

In spite of many efforts to fractionate *Drosophila* oocytes or early embryos for polar granules (Mahowald *et al.,* 1979), it has been impossible to obtain fractions substantially enriched for these organelles. Kobayashi *et al.* (1993), using the *in vivo* assay of restoring pole cell formation in UV-irradiated embryos, demonstrated that the mitochondrial large ribosomal RNA (*mlrRNA*) was able to restore the ability to form pole cells. Moreover, they showed that injection of a ribozyme that specifically cleaved *mlrRNA* significantly interfered with pole cell formation (Iida and Kobayashi, 1998). By immunoelectron microscopy, they localized both *mlrRNA* and *msrRNA* (mitochondrial small ribosomal RNA) to thc edge of polar granules prior to pole cell formation and found that this RNA disappeared at the pole cell stage (Kobayashi *et al.,* 1994; Kashikawa *et al.,* 1999). Although these experiments strongly support the conclusion that these RNAs are important for pole cell formation, there are two caveats. First, the pole cells formed in these UV-irradiated embryos fail to form functional germ cells. Second, polar granules are not associated with mitochondria in all *Drosophila* species (cf. Figure 1B), and it is unknown whether these mitochondrially derived RNAs are associated with polar granules in these species.

Although direct fractionation of early embryos has been unsuccessful for purifying polar granules, relatively pure fractions can be obtained from enriched populations of pole cells (Waring *et al.*, 1978). One protein, approximately 95,000 Daltons in mass with a pI of 7.5 to 8.0, copurified with these fractions, suggesting that it is a constituent of polar granules. At the time these experiments were done, techniques were not available to obtain amino acid sequences from such small quantities of protein, although this might be possible today. None of the proteins identified genetically with polar granules (see next section) appear be the source of this 95 K protein.

Another constituent of polar granules was identified by differential display of RNAs from either wild-type embryos or *tudor* embryos, which lack pole cells (Nakamura *et al.*, 1996). *Polar granule component,* or *Pgc,* codes for abundantly expressed, germline-specific, small RNAs that appear to lack open reading frames for protein synthesis. These RNAs are found throughout the nurse cell and oocyte cytoplasm in early stages of oogenesis, but they gradually become concentrated at the posterior pole of the embryo. *Pcg* RNAs outside the pole cells are rapidly degraded, whereas pole cell *Pgc* RNA is found in pole cells until the cells exit the midgut. By EM immuno-staining *Pgc* RNA is found distributed throughout the polar granule (as opposed to *mlrRNA* and *msrRNA* which are localized to one side of the polar granule only), and it is still present in granules at the pole cell stage, a time when the polar granular mitochondrial RNAs of polar granules have disappeared. Antisense RNA to *Pgc,* produced under control of the *hsp70* promoter during oogenesis, did not affect oogenesis or pole cell formation (Nakamura *et al.*, 1996). However, levels of *nos, gcl,* and *vasa* RNAs were reduced in pole cells and fewer pole cells reached the embryonic gonad. Fifty-three percent of embryos had adult gonads without germ cells. Hence, *Pgc* is probably not required for pole cell formation but may be required for polar granule stability and formation of the germ line.

B. Genetic Identification of Germ Plasm Components

Mutations eliminating the formation of pole cells are commonly called "grandchildless" because embryos derived from homozygous mothers lack pole cells. The original *grandchildless* (*gs*) mutation was identified in *D. subosbscura* (Spurway, 1948) and caused a degeneration of the ooplasmic polar plasm during oogenesis (Mahowald *et al.,* 1979). Saturation screens in *D. melanogaster for grandchildless* mutations have only been completed for the X-chromosome (Gans *et al.,* 1975; L. Engstrom and A. P. Mahowald, unpublished results), but no true *grandchildless* loci have been found, although some pleiotrophic mutations with a *gs* phenotype have been characterized (Thierry-Mieg, 1982; Niki, 1984).

A major advance occurred when saturation screens for female sterile mutations uncovered a large class of posterior group maternal-effect-lethal mutations, all of which result in the absence of the embryonic abdominal segments (Table II; St. Johnston, 1993). In addition, all genes in this class, except for *pumilio* and *nanos,* produced embryos without pole cells. Detailed genetic and molecular analyses have established that Pumilio and Nanos function together to prevent the translation of maternal *hunchback* mRNA, and that these genes are not needed for the formation of germ cells. Recent data, however, have established that both are essential in the germ line for normal germ cell function (Forbes and Lehmann, 1998) (I will return to this later). An elegant experiment by Ephrussi and Lehmann (1992) provided a basis for separating the remaining "abdomen/grandchildless" genes into two classes, based on whether they are needed only for germ cell formation or for localizing germ cell-forming capacity to the posterior polar plasm. Utilizing the known property of the 3′UTR (untranslated region) of the *bicoid* gene to localize RNAs to the anterior of the oocyte (Macdonald and Struhl, 1988), Ephrussi and Lehmann showed that the localization of *oskar* mRNA to the anterior pole was sufficient to induce the formation of a functional germ plasm. Ectopic Oskar protein caused the formation of a typical germ plasm at the anterior tip of the embryo, including the polar granules characteristic of the posterior germ plasm. In addition, they were able to separate the large set of genes necessary for making germ plasm into those needed in the posterior only, which are assumed to be necessary for localizing *oskar* mRNA to the posterior, and those necessary and sufficient for making functional anterior germ plasm (Figure 3). The key role of *oskar* is further supported by the observation that dosage of *oskar* gene determines both the amount of polar plasm and the number of pole cells (Smith *et al.,* 1992; Ephrussi and Lehmann, 1992).

III. Genetic and Molecular Data on the Assembly and Function of the Germ Plasm

A. *gurken* Is Required for Establishing the Posterior Pole of the Oocyte

The assembly of the germ plasm is intimately connected with processes required for forming the anterior-posterior axis of the oocyte and for localiz-

TABLE II
Genes Required for Formation of the Posterior Germ Plasm

Function	Gene	Molecular nature
Genes required for multiple processes in oogenesis, including establishing the posterior pole of the oocyte	*gurken*	TGFα-like ligand
	orb	CPEB-RNA binding protein
	spindle genes	Double-strand DNA repair
	maelstrom	Novel protein
	spire	Unknown
	cappuccino	Formin homology
	staufen	RNA-binding domain
	mago nashi	Novel; highly conserved
Genes required specifically for producing the posterior germ plasm	*oskar*	Novel
	vasa	RNA helicase
	tudor	Novel, "tudor" domain
	valois	Not yet cloned
	germcell-less	Nuclear membrane protein
Genes required for abdominal development only	*nanos*	CCHC-zinc binding domain; RNA binding
	pumilio	Binds Nanos response elements in RNA

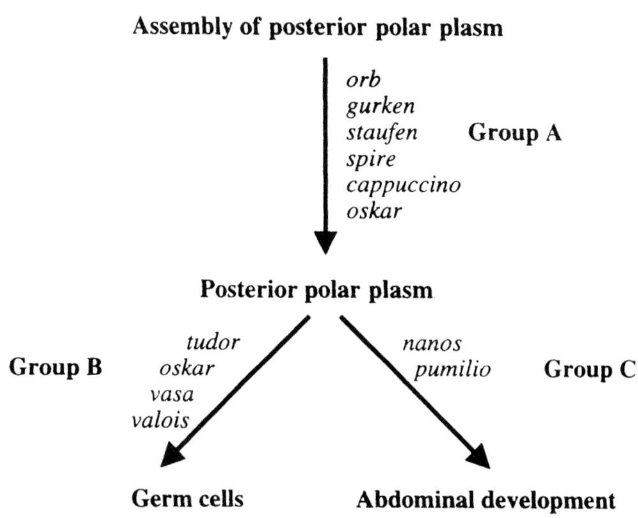

FIG. 3 Genes required in the germ plasm can be divided into three groups: those needed only for assembling germ plasm at the posterior pole (group A), those required for producing pole cells (group B), and those required for abdominal development (group C). (cf. Ephrussi and Lehmann, 1992).

ing RNAs to specific locations during oogenesis. Nearly all the RNA in the oocyte is synthesized in the 15 sister nurse cells and transported into the oocyte via the ring canals, which are intercellular bridges remaining following incomplete cytokinesis (*cf.* Spradling, 1993 for review of oogenesis). As soon as 16-cell clusters form in the germarium RNAs accumulate in one cell, which is the future oocyte (Fig. 4A). Some of this early transport is essential for determining one cell to be the oocyte. In the absence of either *BicD* or *egalitarian,* none of the 16-cells becomes the oocyte (Ran *et al.,* 1994; Mach and Lehmann, 1997). Some RNAs accumulate in the oocyte for later functions, such as *orb* (Chang *et al.,* 1999), *osk* (Ephrussi *et al.,* 1991), and *yolkless* (Schonbaum *et al.,* 2000). The initial A-P polarity of the oocyte requires the localization of *gurken* RNA and protein to the posterior pole (Fig. 4A, B) (González-Reyes *et al.,* 1995). Gurken protein, in turn, induces the adjacent follicle cells to become posterior, which is essential for later formation of a posterior polar plasm. *mago nashi* is required in the oocyte to respond to this signal from posteriorly determined follicle cells (Newmark *et al.,* 1997; Micklem *et al.,* 1997). In *mago* mutants, posterior follicle cells are determined to be posterior by the *grk* signal, but the oocyte does not form a posterior polar plasm. Mago is a highly conserved

---

FIG. 4 Steps in the assembly of the germ plasm at the posterior pole of the oocyte and embryo. Anterior is to the left, and dorsal is up. (A) and (B) are magnified by four relative to (C) to (F). The illustration is patterned after Rongo and Lehmann, 1996. (A) The germarium and stage 2 egg chamber, with two or more germline stem cells at the anterior tip, followed by dividing cystocytes leading to the production of 16-cell cysts. As soon as the cysts form, one cell becomes distinguishable by the presence of synaptonemal complexes (Carpenter, 1975) and the accumulation of specific mRNAs, such as *grk, BicD, egl, orb,* and *yl.* Two follicle cell stem cells (hatched) are present in the region with meiotic cysts, and these cells produce the follicular epithelium that surrounds the egg chambers. The localization of Orb protein (and *grk* mRNA and protein) at the posterior rim of the oocyte begins already at stage 2. (B) Stage 6 oocyte, in which the localization of Gurken between the oocyte nucleus and adjacent follicle cells is essential for establishing the posterior character of the adjoining follicle cells; these cells, in turn, signal back to the oocyte for the establishment of the posterior pole. (C) Stage 10 oocyte in which Gurken is located adjacent to the anteriorly localized nucleus, and Staufen, Oskar, Tudor, and Vasa proteins accumulate at the posterior pole. At this time, Orb is no longer associated with the polar plasm. (D) Stage 14 oocyte, in which Tudor, Vasa, and Oskar proteins and *nanos, tudor, nos, gcl, Pgc,* and *CycB* mRNAs become associated with the posterior polar plasm. (E) Early fertilized embryo, when *nanos* RNA becomes translated to produce a gradient of Nanos protein. *mlrRNA* and *msrRNA* are associated with polar granules at this time. (F) Pole cell stage, when Tudor, Oskar, Vasa, and Gcl proteins continue to be associated with polar granules. The 95 kDa protein is associated with these granules and probably is present earlier. Due to programmed lost of RNAs outside pole cells, *Pgc, hsp83,* and *CycB* mRNAs become restricted to pole cells.

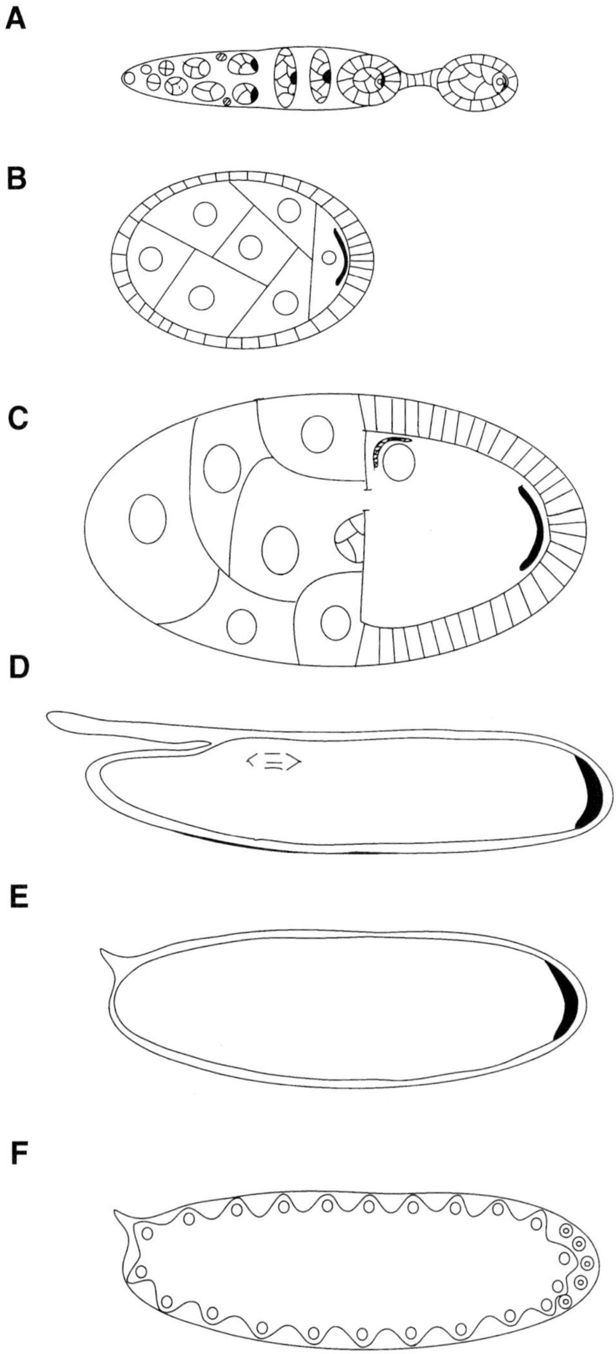
A
B
C
D
E
F

protein in worms and mammals, but its molecular function is not yet known. Both *spire* and *cappuccino* are required to localize Mago to the posterior pole, but *vasa,* which is assembled after Oskar accumulates at the pole, is not needed for Mago's localization (Newmark *et al.,* 1997). In *vasa* mutants (Breitwieser *et al.,* 1996) and mutants affecting double-stranded DNA repair (Ghabrial *et al.,* 1998), or *maelstrom* mutants (Clegg *et al.,* 2000), *grk* mRNA is either not localized or not translated, which results in the failure to establish the posterior pole of the oocyte. In *encore* oocytes, although *grk* RNA is localized properly, very little Grk protein appears in the ovary (Hawkins *et al.,* 1997), thus indicating that *encore* is also involved in posttranscriptional control of *grk* function. In all of these situations, Grk is clearly the critical inductive molecule required to establish the posterior pole of the oocyte. Following the induction of this posteriorness in the oocyte, the oocyte nucleus moves to the anterior dorsal edge of the oocyte, and *grk* mRNA and protein accumulates in this location for induction of the anterior-dorsal region of the oocyte (Fig. 4C).

B. Primary Role of Oskar in Assembly of Polar Granules

The critical role of *oskar* in forming the posterior polar plasm has already been illustrated by the induction of ectopic pole cells following localization of *oskar* mRNA to the anterior pole (Ephrussi and Lehmann, 1992), the formation of supernumerary pole cells following overexpression of *osk,* and by the absence of pole cells when *osk* is underexpressed (Smith *et al.,* 1992). Since abdominal development occurs without pole cell formation, apparently lower levels of Oskar protein are required for abdomen formation than for pole cells. *osk* mRNA is first detected in the germarium, where it is initially distributed throughout the ooplasm. Oskar protein also appears in these early stages (Chang *et al.,* 1999). At the beginning of vitellogenesis, *osk* mRNA becomes concentrated at the anterior rim of the oocyte, adjacent to the nurse cell chamber. Staufen protein also accumulates in this anterior rim, and both Staufen and *osk* mRNA move to the posterior tip of the growing oocyte, reaching the pole at stage 9 (Figure 4C, D). In the absence of Staufen, Cappuccino, and Spire, *osk* mRNA does not accumulate at the posterior pole. The initial anterior localization of these RNAs may be important, since in a variety of mutants *osk* RNA accumulates at the anterior pole rather than moving to the posterior pole. A microtubule-based transport system appears to be primarily responsible for this transport (Gillespie and Berg, 1995), although a trapping mechanism to localize *osk* mRNA at the posterior pole is also active (Glotzer *et al.,* 1997). These observations correspond to the experimental results described earlier in which functional pole cells are formed at the anterior tip of oocytes expressing *osk-bcd-3′ UTR,* even in *stau, capu,* or *spi* mutants (Ephrussi and Lehmann, 1992).

Orb protein is also critical for the localization of *osk* mRNA to the posterior pole. Orb accumulates at the posterior edge of the oocyte in previtellogenic stages prior to the arrival of *osk* mRNA (Fig. 4A, B), a finding that suggests that Orb might be necessary for the initial anchoring of *osk* mRNA to the posterior pole. Orb also appears to be necessary for *osk* mRNA translation (Chang *et al.,* 1999). *orb* codes for a member of the CPEB (cytoplasmic polyadenylation element binding) class of proteins that bind to 3′UTRs of oocyte mRNAs and inhibit polyadenylation. In the absence of *orb, osk* RNAs have only short poly (A) tails, thus suggesting that Orb might play a similar function in controlling translation of *orb* RNA, although translation of *osk* RNA apparently does not require lengthening of the poly (A) tail (Lie and Macdonald, 1999b). Oskar protein is a constituent of polar granules (Breitwieser *et al.,* 1996).

Ectopic appearance of Oskar protein leads to assembly of the polar plasm at these sites, which results in abnormal development due to the accompanying mislocalization of *nanos* mRNA (Ephrussi and Lehmann, 1992). Hence, translation of *osk* mRNA must be repressed prior to its transport to the posterior pole, and then activated at the posterior pole to produce a functional posterior polar plasm. Oskar protein is essential both for maintaining the *osk* mRNA localization and for the subsequent elaboration of polar plasm. In *osk* mutants that produce a trunkated protein, *osk* mRNA reaches the posterior pole but disperses without producing a functional polar plasm (Kim-Ha *et al.,* 1991; Ephrussi *et al.,* 1991). *Osk* mutants, which produce full-length proteins, however, result in stable localization of *osk* mRNA at the posterior, but without formation of a functional polar plasm. Oskar protein thus appears to be the linchpin for construction of the posterior polar plasm. Two types of Oskar protein are formed: p50, which is essential for both abdominal and germ cell development, and p68, which appears to enhance posterior localization of *osk* mRNA but which is insufficient by itself for either abdominal or germ cell formation (Markussen *et al.,* 1995).

Although we lack a full understanding of the mechanism for repression of *osk* mRNA translation prior to reaching the posterior pole and activation of its translation at the posterior pole, some important players have been identified. Kim-Ha *et al.* (1995) isolated a protein, called Bruno, which bound to the 3′-UTR of *osk* mRNA and prevented its translation. Bruno interacts with Vasa, but this interaction does not depend upon the 3′-UTR of *osk* mRNA (Rongo *et al.,* 1995). Bruno is coded for by the *aret* locus (Webster *et al.,* 1997). Since *aret* mutants show many phenotypes not related to mistranslation of Oskar protein, Bruno must be involved in the control of other RNAs during oogenesis. Staufen is also required not only for the transport of *osk* MRNA to the posterior pole but also for its translation (Breitwieser *et al.,* 1996). Apontic was discovered in a two-hybrid screen for proteins interacting with Bruno (Lie and Macdonald, 1999a), and its absence leads to ectopic translation of *oskar.* Repression of *osk* mRNA translation prior to reaching the posterior pole also depends upon a repression domain in the 5′-UTR of *osk* mRNA (Gunkel *et al.,* 1998). Two proteins (p50 and p68) bind to this domain; the p50 is constitutive, whereas p68 is ovary-specific and interacts with Bruno and both the 5′- and 3′-UTR of *osk* mRNA. In *Bic-C* mutants *osk* translation occurs prematurely. Bic-C protein has RNA binding activity, but whether it binds to *osk* mRNA is unknown (Saffman *et al.,* 1998). Finally, *aubergine* is also required for effective translation of *osk* mRNA (Wilson *et al.,* 1996). Both the repression and derepression mechanisms must have unique posterior pole properties, since ectopic *osk* mRNA is efficiently translated when inappropriately localized to the anterior pole (Ephrussi and Lehmann, 1991) and this translation does not require *aubergine* (Wilson *et al.,* 1996). Additional factors are probably required for this critical step in development.

In addition to *oskar,* a second, key component of the polar plasm is Vasa. *vasa* mRNA and protein are found constitutively in the germline throughout the lifetime of *Drosophila.* During oogenesis, Vasa is initially localized with the nuage associated with the nuclear envelope of the nurse cell nuclei (Hay *et al.,* 1988). This localization occurs independently of genes required for polar granule assembly, namely *cappuccino, spire, oskar,* and *staufen* (Liang *et al.,* 1994). Subsequent localization of Vasa to the posterior polar plasm requires *staufen, spire,* and *cappuccino* (Lasko and Ashburner, 1990; Hay *et al.,* 1990). Vasa also directly interacts with Oskar (Breitwieser *et al.,* 1996) and this interaction is independent of either the ATPase or helicase activity of Vasa (Liang *et al.,* 1994). Vasa is also required for *grk* mRNA translation at the posterior pole at a time when Oskar is not yet present (Tomancak *et al.,* 1998), so this function in RNA translation is at least partially independent of Oskar.

pipsqueak was initially identified as a posterior-group gene with both abdominal and grandchildless phenotypes (Siegel *et al.,* 1993). *oskar* mRNA and protein accumulate at the posterior tip, but Vasa levels are severely reduced. Pipsqueak is a nuclear protein, accumulates in all nuclei of the ovary (Horowitz and Berg, 1996), and contains a BTB domain known to be required for protein-protein interaction. The pole cell-less phenotype can be readily understood because of its effect on *vasa* translation.

A third component of the polar granule is Tudor, which is required for assembly of the polar plasm and is necessary for both abdominal and germ cell development (Figure 4D) (Boswell and Mahowald, 1985). Because weak alleles have diminished amounts of polar granule material, and Tudor itself is found in polar granules (Bardsley *et al.,* 1993), Tudor may be a prerequisite for assembling granules. Consistent with this interpretation is the requirement for *tudor* to produce anterior pole cells (Ephrussi and Lehmann, 1992). The Tudor protein is apparently cleaved into a number of polypeptides, some of which are found in somatic nuclei, some in polar granules, and others within mitochondria. Mutations affecting pole cells primarily do not affect the nuclear localization, whereas mutations eliminating pole cells also eliminate mitochondrial Tudor (Bardsley *et al.,* 1993). Since polar granules associate with mitochondria and apparently require RNAs encoded by the mitochondria for function, Tudor may play an essential role in the mitochondria–polar granule function. Further work on this intriguing component of the germ plasm is clearly needed. These experiments suggest that Oskar, Vasa, and Tudor are essential components of the polar granule and are critical for the assembly of a functional germ plasm.

D. Function of Germ Plasm Components in Germ Cell Function

Two critical events are dependent upon the posterior polar plasm: the precocious segregation of pole cells (Figure 4E, F) and their determination to become primordial germ cells. Different genes are required for each of these steps.

1. Precocious Segregation of Pole Cells

Which components of the polar plasm are critical for segregation of pole cells is unclear. The amount of *oskar* apparently determines the number of pole cells (Smith *et al.,* 1992), but there is no evidence that Oskar acts directly in the embryo to accomplish this result. Both Vasa and Tudor are known to be required for ectopic Oskar to produce a functional polar plasm (Ephrussi and Lehmann, 1992), but no evidence associates either Vasa or Tudor directly with the precocious pole cell formation. The only components required specifically for the segregation of pole cells are the mitochondrial RNAs, *mlrRNA* and presumably *msrRNA,* which can rescue pole cell formation in UV-irradiated embryos (Okada *et al.,* 1974) and whose partial elimination with a ribozyme reduced the number of pole cells formed (Iida and Kobayashi, 1998). Since segregation requires only the centrosome (Raff and Glover, 1989), new zygotic gene activity clearly is unnecessary. What other components of the maternally derived polar plasm are needed is unknown.

In addition to the known molecules associated with polar granules, a number of other processes are required for producing pole cells. As mentioned earlier, centrosomes with the associated cytasters are required for pole cell formation, even in the absence of nuclei. Inhibition of endocytosis by *shibiri*^{ts} (Swanson and Poodry, 1980) prevents pole cell formation. Temporal coordination is also important. Delaying migration of nuclei to the posterior pole of wild-type embryos results in a failure of pole cell formation (Okada, 1982). Some *grandchildless* mutations lack pole cells apparently because of a delay in nuclear migration to the posterior pole (Niki, 1984); polar plasm from these mutants can successfully induce functional pole cells at the anterior pole.

The segregation of the polar plasm accomplishes four significant results. First, specific organelles and proteins are segregated into pole cells. Segregation of polar granules has been known since the turn of the century (Hegner, 1914; Mahowald, 1962). Nanos protein, which is produced from posteriorly localized mRNA, is selectively incorporated into pole cells (Wang *et al.*, 1994) and is essential for pole cell function (see below). Second, specific maternal RNAs are sequestered in pole cells and escape the midblastula RNA degradation that characterizes most maternal RNAs (Bashirullah *et al.*, 1999). These RNAs include *gcl* (Jongens *et al.*, 1992), *Pgc* (Nakamura *et al.*, 1996), *nanos* (Gavis and Lehmann, 1992), *hsp83* (Ding *et al.*, 1993), *Dmnk* (Oishi *et al.*, 1998), and *cyc B* (Whitfield *et al.*, 1989), which are degraded in the nonpole cell portions of the embryo. A sequence on the *hsp83*-3′-UTR specifically provides protection from degradation in the pole cells (Bashirullah *et al.*, 1999). Whether similar sequences are found in other RNAs protected from degradation is unknown, but this is a reasonable possibility. Hsp90 (coded for by *hsp83*) is present throughout embryogenesis in all cells at constant levels, even though the mRNA has been degraded, and it is required for viability (Yue *et al.*, 1999). What the function of *hsp83* mRNA is in pole cells is unknown.

cycB mRNA is not translated in pole cells until after the cells are incorporated into the embryonic gonad. Repression of translation may be a critical part of the program to arrest cell division of pole cells until the formation of the embryonic gonad. *Dmnk* codes for a maternal nuclear kinase, which is initially found in both blastema and pole cell nuclei, but then the protein disappears from somatic cells, although it continues to be present in the pole cell nuclei in a large number of "speckles." Because of the number of speckles, they probably do not correspond to pole cell-specific nuclear bodies. Because of *Dmnk* cytogenetic location and possible molecular genetic changes, it may correspond to the *valois* locus (Oishi *et al.*, 1998), mutants of which cause a posterior/grandchildless phenotype. Maternally inherited *caudal* mRNA and protein are also segregated into the pole cells (Mzodlik and Gehring, 1987). Caudal is a key homeodomain protein responsible for determining the most posterior segment (Moreno and Morata, 1999). It is also expressed in the germline during larval life, but its germline function is unknown (Mzodlik and Gehring, 1987).

Third, phosphorylation of the serine-rich C-terminal domain (CTD) of RNA polymerase II is diminished in pole cells relative to the remainder of the embryo (Seydoux and Dunn, 1997). The nonphosphorated CTD of RNA polymerase II is inactive, resulting in an inhibition of new transcription even in the presence of the strong transcriptional activator from yeast, *Gal4* (Van Doren *et al.*, 1998). *nos* is clearly required for the transcriptional quiescence of pole cells (Kobayashi *et al.*, 1996). In the absence of *nos*, pole cells activate transgenes and fail to migrate successfully into lateral mesoderm to form the embryonic gonad (Asaok *et al.*, 1998; Forbes and Lehmann, 1998). The release from transcription repression in pole cells lacking *nos* is selective. Hence, *ftz, eve,* and *Sxl* are transcribed in pole cells, but *tailless, twist,* and *hunchback* are not (Deshplande *et al.*, 1999). Repression of *Sxl* function may be especially important because in the absence of both *nos* and *Sxl*, pole cells partially regain their ability to migrate to the mesoderm and form embryonic gonads, although the germ cells that form are not fertile.

In the early embryo, Nanos functions together with Pumilio to prevent the translation of maternally provided *hb* mRNA (St. Johnston, 1993). It is also intriguing that supernumerary pole cells form in *nos* embryos (Smith *et al.*, 1992), but the mechanism for this is unknown. *Pumilio* and *nanos* are also required in the germline for the maintenance of germline stem cells, although the phenotypes are slightly different, suggesting that they may function separately. Nanos may interact with Orb in the germarium, since both *orb* and *nos* show similar phenotypes (Forbes and Lehmann, 1998). New *nanos* mRNA synthesis is not detected until the first instar larval stage, so the segregation of maternally derived Nanos is essential.

Fourth, cell division in pole cells slows and becomes asynchronous relative to the synchronously dividing blastema nuclei, and then nuclear division stops the embryonic gonad forms. The cells cycle is arrested at the G_2–M transition with high levels of mitotic cyclins A and B (Su *et al.*, 1998). Pole cells lacking *nanos* continue to divide and lose both the mitotic cyclins and the segregated *cycB* mRNA (Deshplande *et al.*, 1999). The mechanism for this Nanos function is unknown.

2. Determination of Pole Cells to Become Germ Cells

Ectopic formation of pole cells by mislocalizing Oskar demonstrated that among the posterior-group genes only *tudor, vasa,* and *valois* function downstream of *oskar* in establishing the germline. In addition to these functions, *germcell-less* has been shown to be required for successful germ cell formation, and *nanos* and *pumilio* are required for establishing the embryonic gonad and for maintaining the germline during larval and adult stages. In addition, *ovo* and *otu* are required for stable formation of germ cells in the ovary, but not the testis (reviewed in Pauli and Mahowald, 1990). Both Tudor and Vasa are components of polar granules, but how they function is not known. Vasa codes for an RNA helicase and is essential during oogenesis for the translation of *grk* and *osk* mRNAs (see above). It is possible that Vasa is also important in pole cells for the activation of the maternal program necessary for pole cell formation or for control of protein synthesis in pole cells. At this time *valois* has not yet been cloned, so its mechanism of action is unknown.

germcell-less clearly has a major role in pole cell formation and function (Jongens *et al.,* 1992). *gcl* is not needed for viability, either maternally or zygotically (Robertson *et al.,* 1999). However, embryos from *gcl* mothers form reduced numbers of pole cell buds and fewer pole cells, and fewer pole cells reach the gonad. *gcl* codes for a nuclear protein that associates with the nucleoplasmic side of the nuclear envelope, especially near nuclear pores (Jongens *et al.,* 1994). Ectopic expression of *gcl* at the anterior appeared to initiate arrest of the nuclear cell cycle and the formation of spherical nuclei resembling those found in pole cells; however, these nuclei did not become segregated into individual cells (Jongens *et al.,* 1994). Hence, in spite of the apparent importance of *gcl* in pole cell segregation and subsequent migration to the gonad, it is clearly not absolutely required. It is intriguing that expression of a *gcl* antisense gene during oogenesis produced a stronger phenotype than the deleted gene (Robertson *et al.,* 1999), a result that suggests that the antisense may have affected a redundant function not yet identified.

Loss-of-function mutations in both *ovo* (Oliver *et al.,* 1987) and *ovarian tumor* (Geyer *et al.,* 1993) cause the loss of germ cells prior to ovarian differentiation in late larval and early pupal stages. Both genes are required throughout oogenesis. *ovo* has a variety of functions, initiated through different promoters. A soma-specific promoter is required for setal formation in the hypoderm (Payre *et al.,* 1999). In the germ line, a constitutive promoter is required for normal differentiation of the germ line, whereas a late ovarian promoter adds a 325 amino acid N-terminal extension to the Ovo protein; this latter isoform acts negatively, counteracting the constitutive Ovo protein (Andrews *et al.,* 1999). The dominant female sterile *ovo* alleles achieve their negative effect by introducing a new methionine codon in the N-terminal region, thus transforming the constitutive Ovo protein into a negative factor (Mével-Ninio *et al.,* 1996). Ovo itself may be critical for pole cell formation, since in the absence of inhibitory Ovo, a grandchildless phenotype ensues (Andrews *et al.,* 1999). It is unknown at what stage the germline is lost in these flies.

IV. Unanswered Questions

In spite of the extensive work accomplished on deciphering controls of early development in *Drosophila,* we still lack definitive information on the mechanism responsible for pole cell segregation. Key components are localized to the posterior polar plasm in an event clearly initiated by *grk* localization, followed by transport of *oskar* mRNA to the posterior pole and its translation to produce a functional germ plasm. What the critical molecules are for precocious pole cell formation in the embryo remains a mystery. The event requires centrosomal asters, endocytosis, and presumably microfilaments to complete cytokinesis. It is also time-dependent, but this could be due to nuclei becoming exposed to cytoplasmic signals and no longer being capable of responding to the polar plasm. It is unknown whether posterior polar plasm could induce pole cells if transplanted to the blastema surface when nuclei have already reached the surface and have become transcriptionally active.

Polar granules appear to be essential components of the posterior germ plasm, but what their full composition is and why they are essential is not yet clear. At least four RNAs are found in the organelles and four proteins, but none of these components appear to be sufficient for pole cell formation. Germcell-less protein appeared to be critical, but in complete absence of the protein pole cells still form (Robertson *et al.,* 1999). Since antisense *gcl* RNA appears to have a stronger effect than the deletion of *gcl,* it is possible that a redundant gene remains to be identified. The polar granule-associated, mitochondrially derived rRNAs are even more puzzling. It is unknown whether these RNAs are always associated with the germ plasm, especially since mitochondria are not associated with polar granules in some *Drosophila* species or in lower dipterans. It is intriguing that *mlrRNA* has been found in germinal granules of *Xenopus* (Kobayashi *et al.,* 1998), although this RNA appears only after the first few cleavages of the embryo. Because both large and small subunits are found in the *Drosophila* polar granule, Iida and Kobayashi (1998) have suggested that mitochondrial ribosomes form at the surface of polar granules. This is unlikely, since mitochondrial ribosomes can be seen as electron-dense 50 Å granules in mitochondria and such granules are not detected in polar granules (Mahowald, 1971b). It is also unlikely that the mitochondrial ribosomes could function outside mitochondria. An additional noncoding RNA (*Pgc*) is also present in polar granules, but its function remains unknown. The 95 kDa protein associated with polar granule fractions has yet to be identified with any of the known genes required for germ cell formation.

Tudor, among the known molecules associated with the polar plasm, may have the most critical role. The presence of Tudor-derived polypeptide in mitochondria suggests a potential role in generating the extra-mitochondrial ribosomal RNAs of polar granules. The discovery that both *nanos* and *pumilio* are also needed for producing functional germ cells suggests that there may be additional genes required for both germlline stability and pole cell formation. The recent discovery that the absence of the dominant-negative form of Ovo leads to a loss of the germ line further supports this assumption. The identification of this class of genes may require experiments utilizing differential displays, such as identified *Pgc,* or screening of genomic DNA chips to identify additional genes required in germ cells. Now that the *Drosophila* genome sequence is available, these experiments should be accomplished in the near future.

Acknowledgments

I am deeply grateful to those who sent manuscripts to me prior to publication, to my many collaborators over the years, and to the NIH for their support. Portions of this review were prepared while I was a resident fellow at the Rockefeller Foundation's Bellagio Center. The electron micrographs were obtained through the University of Chicago Cancer Research Center EM facility.

References

Andrews, J., Garcia-Estefania, D., Delon, I., Lü, J., Mével-Ninio, M., Spierer, A., Payre, F., Pauli, D., and Oliver, B. (1999). Antagonistic isoforms of *Drosophila* OVO transcription factors in oogenesis, embryonic patterning and as maternal products. *Development* **127,** 881–892.

Asaoka, M., Sano, H., Obara, Y., and Kobayashi, S. (1998). Maternal Nanos regulates zygotic gene expression in germline progenitors of *Drosophila melanogaster. Mech. Dev.* **78,** 153–158.

Bardsley, A., McDonald, K., and Boswell, R. E. (1993). Distribution of Tudor protein in the *Drosophila* embryo suggests separation of functions based on site of localization. *Development* **119,** 207–219.

Bashirullah, A., Halsell, S. R., Cooperstock, R. L., Kloc, M., Karaiskakis, A., Fisher, W. W., Fu, W., Hamilton, J. K., Etkin, L. D., and Litshitz, H. D. (1999). Joint action of two RNA degradation pathways controls the timing of maternal transcript elimination at the midblastula transition in *Drosophila melanogaster. EMBO J.* **18,** 2610–2620.

Boswell, R. E., and Mahowald, A. P. (1985). *tudor,* a gene required for assembly of the germ plasm in *Drosophila melanogaster. Cell* **43,** 97–104.

Breitwieser, W., Markussen, F.-H., Horstmann, H., and Ephrussi, A. (1996). Oskar protein interaction with Vasa represents an essential step in polar granules assembly. *Genes Dev.* **10,** 2179–2188.

Carpenter, A. T. C. (1975). Electron microscopy of meiosis in *Drosophila melanogaster* females. I. Structure, arrangement and temporal change of the synaptonemal complex in wild-type. *Chromosoma* **51,** 157–182.

Chang, J. S., Tan, L., and Schedl, P. (1999). The *Drosophila* CPEB homolog, Orb, is required for oskar protein expression in oocytes. *Dev. Biol.* **215,** 91–106.

Clegg, N. J., Findley, S., Mahowald, A. P., Ruohola-Baker, H. (2000). *Maelstrom* is required to position the MTOC in stage 2–6 *Drosophila* oocytes. (Submitted for publication.)

Counce, S. J. (1963). Developmental morphology of polar granules in *Drosophila* including observations on pole cell behavior and distribution during embryogenesis. *J. Morph.* **112,** 129–145.

Deshplande, G., Calhoun, G., Yanowitz, J. L., and Schedl, P. D. (1999). Novel functions of *nanos* in downregulating mitosis and transcription during the development of the *Drosophila* germline. *Cell* **99,** 271–281.

Ding, D., Parkhurst, S. M., Halsell, S. R., and Lipshitz, H. D. (1993). Dynamic Hsp83 RNA localization during *Drosophila* oogenesis and embryogenesis. *Mol. Cell Biol.* **13,** 3773–3781.

Ephrussi, A., Dickenson, L., and Lehmann, R. (1991). *oskar* organizes the germ plasm and directs localization of the posterior determinant *nanos. Cell* **66,** 37–50.

Ephrussi, A., and Lehmann, R. (1992). Induction of germ cell formation by *oskar. Nature* **358,** 387–392.

Forbes, A., and Lehmann, R. (1998). Nanos and Pumilio have critical roles in the development and function of *Drosophila* germline stem cells. *Development* **125,** 679–690.

Gans, M., Audit, C., and Mason, M. (1975). Isolation and characterization of sex linked female sterile mutants in *Drosophila melanogaster. Genetics* **81,** 683–704.

Gavis, E. R., and Lehmann, R. (1992). Localization of *nanos* RNA controls embryonic polarity. *Cell* **71,** 301–313.

Geigy, R. (1931). Action de l'ultra-violet sûr le pôle germinale dans l'oeuf de *Drosophila melanogaster. Rev. Suisse Zool.* **38,** 187–288.

Geyer, P. K., Patton, J. S., Rodesch, C., and Nagoshi, R. N. (1993). Genetic and molecular characterization of *P* element-induced mutations reveals that the *Drosophila ovarian tumor* gene has maternal activity and a variable null phenotype. *Genetics* **133,** 265–278.

Ghabrial, A., Ray, R. P., and Schüpbach, T. (1998). *Okra* and *spindle-B* encode components of the *RAD52* DNA repair pathway and affect meiosis and patterning in *Drosophila* oogenesis. *Genes Dev.* **12,** 2711–2723.

Gillespie, D. E., and Berg, C. A. (1995). *homeless* is required for RNA localization in *Drosophila* oogenesis and encodes a new member of the DE-H family of RNA-dependent ATPases. *Genes Dev.* **9,** 2495–2508.

Glotzer, J. B., Saffrich, R., Glotzer, M., and Ephrussi, A. (1997). Cytoplasmic flow localizes injected *oskar* RNA in *Drosophila* oocytes. *Curr. Biol.* **7,** 326–337.

González-Reyes, A., Elliott, H., and St Johnston, D. (1995). Polarization of both major axes in *Drosophila* by *gurken-torpedo* signalling. *Nature* **375,** 654–658.

Gunkel, N., Yano, T., Markussen, F.-H., Olsen, L. C., and Ephrussi, A. (1998). Localization-dependent translation requires a functional interaction between the 5′ and 3′ ends of *oskar* mRNA. *Genes Dev.* **12,** 1652–1664.

Hawkins, N. C., Van Buskirk, C., Grossniklaus, U., and Schüpbach, T. (1997). Posttranscriptional regulation of *gurken* by *encore* is required for axis determination in *Drosophila. Development* **124,** 4801–4810.

Hay, B., Jan, L. Y., and Jan, Y. N. (1988). Identification of a component of *Drosophila* polar granules. *Development* **103,** 635–640.

Hay, B., Jan, L. Y., and Jan, Y. N. (1990). Localization of vasa, a component of *Drosophila* polar granules, in maternal effect mutants that alter embryonic anterioposterior polarity. *Development* **109,** 425–433.

Hegner, R. W. (1914). "The Germ-Cell Cycle in Animals." Macmillan, New York.

Horowitz, H., and Berg, C. A. (1996). The *Drosophila pipsqueak* gene encodes a nuclear BTB-domain-containing protein required early in oogenesis. *Development* **122,** 1859–1871.

Iida, T., and Kobayashi, S. (1998). Essential role of mitochondrially encoded large rRNA for germ line formation in *Drosophila* embryos. *Proc. Natl. Acad. Sci. USA* **95,** 11274–11278.

Illmensee, K., and Mahowald, A. P. (1974). Transplantation of posterior polar plasm in *Drosophila.* Induction of germ cells at the anterior pole of the egg. *Proc. Natl. Acad. Sci. USA* **71,** 1016–1020.

Illmensee, K., and Mahowald, A. P. (1976). The autonomous function of germ plasm in a somatic region of the *Drosophila* egg. *Exp. Cell Res.* **97,** 127–140.

Illmensee, K., Mahowald, A. P., and Loomis, M. R. (1976). The ontogeny of germ plasm during oogenesis in *Drosophila. Dev. Biol.* **49,** 40–65.

Jongens, T. A., Ackerman, L. D., Swedlow, J. R., Jan, L. Y., and Jan, Y. N. (1994). *germ cell-less* encodes a cell type-specific nuclear pore-associated protein and functions early in the germ-cell specification pathway of *Drosophila. Genes Dev.* **8,** 2123–2136.

Jongens, T. A., Hay, B., Jan, L. Y., and Jan, Y. N. (1992). The *germ cell-less* gene product: A posteriorly localized component necessary for germ cell development in *Drosophila. Cell* **70,** 569–584.

Kashikawa, M., Amikura, R., Nakamura, A., and Kobayashi, S. (1999). Mitochondrial small ribosomal RNA is present on polar granules in early cleavage embryos of *Drosophila melanogaster. Dev. Growth Differ.* **41,** 495–502.

Kim-Ha, J., Smith, J. L., and Macdonald, P. M. (1991). *oskar* mRNA is localized to the posterior pole of the *Drosophila* oocyte. *Cell* **66,** 23–35.

Kim-Ha, J., Kerr, K., and Macdonald, P. M. (1995). Translational regulation of *oskar* mRNA by Bruno, an ovarian RNA-binding protein, is essential. *Cell* **81,** 403–412.

King, R. C. (1970). "Oogenesis in *Drosophila melanogaster.*" Academic Press, New York.

Kirk, D. L. (1998). "Volvox: *The Molecular Genetic Origins of Muticellularity and Cellular Differentiation.*" Cambridge University Press, Cambridge.

Kobayashi, S., Amikura, R., and Mukai, M. (1998). Localization of mitochondrial large ribosomal RNA in germ plasm of *Xenopus* embryos. *Curr. Biol.* **8,** 1117–1120.

Kobayashi, S., Amikura, R., and Okada, M. (1993). Presence of mitochondrial large ribosomal RNA outside mitochondria in germ plasm of *Drosophila melanogaster. Science* **260,** 1521–1524.

Kobayashi, S., Amikura, R., and Okada, M. (1994). Localization of mitochondrial large rRNA in germinal granules and the consequent segegation of germ line. *Int. J. Dev. Biol.* **38,** 193–199.

Kobayashi, S., Yamada, M., Asaoka, M., and Kitamura, T. (1996). Essential role of the posterior morphogen *nanos* for germline development in *Drosophila. Nature* **380,** 708–711.

Lasko, P. F., and Ashburner, M. (1990). Posterior localization of vasa protein correlates with, but is not sufficient for, pole cell development. *Genes Dev.* **4,** 905–921.

Levy, Y. Y., and Dean, C. (1998). Control of flowering time. *Curr. Opin. Plant Biol.* **1,** 49–54.

Liang, L., Diehl-Jones, W., and Lasko, P. (1994). Localization of *vasa* protein to the *Drosophila* pole plasm is independent of its RNA-binding and helicase activities. *Development* **120,** 1201–1211.

Lie, Y. S., and Macdonald, P. M. (1999a). Apontic binds the translational repressor Bruno and is implicated in regulation of *oskar* mRNA translation. *Development* **126,** 1129–1138.

Lie, Y. S., and Macdonald, P. M. (1999b). Translational regulation of *oskar* mRNA occurs independent of the cap and poly(A) tail in *Drosophila* ovarian extracts. *Development* **126,** 4989–4996.

Macdonald, P. M., and Struhl, G. (1988). *cis*-acting sequences responsible for anterior localization of *bicoid* mRNA in *Drosophila* embryos. *Nature* **336,** 595–598.

Mach, J. M., and Lehmann, R. (1997). An Egalitarian-BicaudalD complex is essential for oocyte specification and axis determination in *Drosophila. Genes Dev.* **11,** 423–435.

Mahowald, A. P. (1962). Fine structure of pole cells and polar granules in *Drosophila melanogaster. J. Exp. Zool.* **151,** 201–215.

Mahowald, A. P. (1968). Polar granules in *Drosophila:* II. Ultrastructural changes during early embryogenesis. *J. Exp. Zool.* **167,** 237–262.

Mahowald, A. P. (1971a). Polar granules in *Drosophila:* III. The continuity of polar granules during the life cycle of *Drosophila. J. Exp. Zool.* **176,** 329–344.

Mahowald, A. P. (1971b). Polar granules of *Drosophila.* IV. Cytochemical studies showing loss of RNA from polar granules during early stages of embryogenesis. *J. Exp. Zool.* **176,** 345–352.

Mahowald, A. P. (1971c). Origin and continuity of polar granules. *In* "Origin and Continuity of Cell Organelles" (J. Reinert and H. Ursprung, Eds.), pp. 159–169. Springer-Verlag, New York.

Mahowald, A. P. (1975). Ultrastructural changes in the germ plasm during the life cycle of *Miastor* (*Cedidomyidae, Diptera*). *Wilhelm Roux' Arch. Dev. Biol.* **176,** 233–240.

Mahowald, A. P. (1977). The germ plasm of *Drosophila:* A model system for the study of embryonic determination. *Amer. Zool.* **17,** 345–352.

Mahowald, A. P., Allis, C. D., Karrer, K. M., Underwood, E. M. and Waring, G. L. (1979). The germ plasm and pole cells of *Drosophila. In* "Determinants of Spatial Organization" (I. Konisberg and S. Subtelny, Eds.), pp. 127–146. Academic Press, New York.

Mahowald, A. P., Caulton, J. H., and Gehring, W. J. (1979). Ultrastructural studies of oocytes and embryos derived from female flies carrying the *grandchildless* mutation in *Drosophila subobscura. Dev. Biol.* **69,** 118–132.

Mahowald, A. P., Goralski, T. J., and Caulton, J. H. (1983). *In vitro* activation of *Drosophila* eggs. *Dev. Biol.* **98,** 437–445.

Mahowald, A. P., Illmensee, K., and Turner, F. R. (1976). Interspecific transplantation of polar plasm between *Drosophila* species. *J. Cell Biol.* **70,** 358–373.

Markussen, F. -H., Michon, A. -M., Breitwieser, W., and Ephrussi, A. (1995). Translational control of *oskar* generates Short OSK, the isoform that induces pole plasm assembly. *Development* **121,** 3723–3732.

Mével-Ninio, M., Fouilloux, E., Guénal, I., and Vincent, A. (1996). The three dominant female-sterile mutations of the *Drosophila ovo* gene are point mutations that create new translation-initiator AUG codons. *Development* **122,** 4131–4138.

Micklam, D. R., Dasgupta, R., Elliott, H., Gergely, F., Davidson, C., Brand, A., González-Reyes, A., and St. Johnston, D. (1997). The *mago nashi* gene is required for the polarisation of the oocyte and the formation of perpendicular axes in *Drosophila. Curr. Biol.* **7,** 468–478.

Moore, L. A., Broihier, H. T., Van Doren, M., Lunsford, L. B., and Lehmann, R. (1998). Identification of genes controlling germ cell migration and embryonic gonad formation in *Drosophila. Development* **125,** 667–678.

Morena, E., and Morata, G. (1999). *caudal* is the Hox gene that specifies the most posterior *Drosophile segment. Nature* **400,** 873–877.

Mzodlik, M., and Gehring, W. J. (1987). Expression of the *caudal* gene in the germ line of Drosophila: Formation of an RNA and protein gradient during early embryogenesis. *Cell* **48,** 465–478.

Nakamura, A., Amikura, R., Mukai, M., Kobayashi, S., and Lasko, P. F. (1996). Required for a noncoding RNA in *Drosophila* polar granules for germ cell establishment. *Science* **274,** 2075–2079.

Newmark, P. A., Gong, S. E. L., and Boswell, R. E. (1997). *mago* nashi mediates the posterior follicle cell-to-oocyte signal to organize axis formation in *Drosophila. Development* **124,** 3197–3207.

Niki, Y. (1984). Developmental analysis of the grandchildless (*gs*(*1*)*N26*) mutation in *Drosophila melanogaster:* Abnormal cleavage patterns and defects in pole cell formation. *Dev. Biol.* **103,** 182–189.

Nieuwkoop, P. D., and Sutasurya, L. A. (1979). "Primordial Germ Cells in the Chordates. Embryogenesis and Phylogenesis." Cambridge: Cambridge University Press.

Nieuwkoop, P. D., and Sutasurya, L. A. (1981). "Primordial Germ Cells in the Invertebrates. From Epigenesis to Preformation." Cambridge University Press, Cambridge.

Oishi, I., Sugiyama, S., Otani, H., Yamamura, H., Nishida, Y., and Minami, Y. (1998). A novel *Drosophila* nuclear protein serine/threonine kinase expressed in the germline during its establishment. *Mech. Dev.* **71,** 49–63.

Okada, M. (1982). Loss of the ability to form pole cells in *Drosophila* embryos with artificially delayed nuclear arrival at the posterior pole. *In* "Embryonic Development," pp. 363–372. Liss Inc., New York.

Okada, M., Kleinman, A., and Schneiderman, H. A. (1974). Restoration of fertility in sterilized Drosophila eggs by transplantation of polar cytoplasm. *Dev. Biol.* **37,** 43–54.

Oliver, B., Perrimon, N., and Mahowald, A. P. (1987). "The *ovo* locus is required for sex specific germ line maintenance in *Drosophila. Genes. Dev.* **1,** 913–923.

Pauli, D., and Mahowald, A. P. (1990). Germ-line sex determination in *Drosophila melanogaster. Trends Gen.* **6,** 259–264.

Payre, F., Vincent, A., and Carreno, S. 1999. *ovo/svb* integrates Wingless and DER pathways to control epidermis differentiation. *Nature* **400,** 271–275.

Poulson, D. (1950). Histogenesis, organogenesis and differentiation in the embryo of *Drosophila melanogaster* Meigen. *In* "Biology of *Drosophila*" (M. Demerec, Ed.), pp. 168–274. Wiley, New York.

Poulson, D. F., and Waterhouse, D. F. (1960). Experimental studies on pole cells and midgut differentiation in Diptera. *Aust. J. Biol. Sci.* **13,** 540–567.

Raff, J. W., and Glover, D. M. (1989). Centrosomes, and not nuclei, initiate pole cell formation in Drosophila embryos. *Cell* **57,** 611–619.

Ran, B., Boop, R., and Suter, B. (1994). Null alleles reveal novel requirements for *BicD* during *Drosophila* oogenesis and zygotic development. *Development* **120,** 1233–1242.

Robertson, S. E., Dockendorff, T. C., Leatherman, J. L., Faulkner, D. L., and Jongens, T. A. (1999). *Germ cell-less* is required only during the establishment of the germ cell lineage of *Drosophila* and has activities that are dependent and independent of its localization to the nuclear envelope. *Dev. Biol.* **215,** 288–297.

Rongo, C., Broihier, H. T., Moore, L., Van Doren, M., Forbes, A., and Lehmann, R. (1997). Germ plasm assembly and germ cell migration in *Drosophila. Cold Spring Harbor Symp.* **62,** 1–11.

Rongo, C., Gavis, E. R., and Lehmann, R. (1995). Localization of *oskar* RNA regulates *oskar* translation and requires Oskar Protein. *Development* **121,** 2737–2746.

Rongo, C., and Lehmann, R. (1996). Regulated synthesis, transport and assembly of the *Drosophila* germ plasm. *Trends Genet.* **12,** 102–109.

Saffman, E. E., Styhler, S., Rother, K., Li, W., Richard, S., and Lasko, P. (1998). Premature translation of oskar in oocytes lacking the RNA-binding protein Bicaudal-C. *Mol. Cell Biol.* **18,** 4855–4862.

Schonbaum, C. P., Perrino, J. J., and Mahowald, A. P. (2000). Regulation of the vitellogenin receptor during *Drosophila melanogaster* oogenesis. *Mol. Biol. Cell* **11,** 511–521.

Schwalm, F. E. (1974). Autonomous structural changes in polar granules of unfertilized eggs of *Coelopa frigida* (Diptera). *Wilhelm Roux' Arch. Dev. Biol.* **175,** 129–133.

Seydoux, G., and Dunn, M. A. (1997). Transcriptionally repressed germ cells lack a subpopulation of phosphorylated RNA polymerase II in early embryos of *Caenorhabditis elegans* and *Drosophila melanogaster. Development* **124,** 2191–2201.

Siegel, V., Jongens, T. A., Jan, L. Y., and Jan, Y. N. (1993). *pipsqueak,* an early acting member of the posterior group of genes, affects *vasa* level and germ cell-somatic interaction in the developing egg chamber. *Development* **119,** 1187–1202.

Smith, J. L., Wilson, J. E., and Macdonald, P. M. (1992). Overexpression of *oskar* directs ectopic activation of *nanos* and presumptive pole cell formation in *Drosophila* embryos. *Cell* **70,** 849–859.

Sonnenblick, B. (1950). The early embryology of *Drosophila melanogaster. In* "Biology of *Drosophila*" (M. Demerec, Ed.), pp. 62–167. Wiley, New York.

Spradling, A. C. (1993). Developmental genetics of oogenesis. *In* "The Development of *Drosophila melanogaster*" (M. Bate and A. Martinez Arias, Eds.), Vol. I, pp. 1–70. Cold Spring Harbor Press, Cold Spring Harbor.

Spurway, H. (1948). Genetics of *Drosophila subobscura.* VIII. Studies on the mutant *grandchildless. J. Genet.* **51,** 375–405.

St. Johnston, D. (1993). Pole plasm and the posterior group genes. *In* "The Development of *Drosophila melanogaster*" (M. Bate and A. Martinez Arias, Eds.), Vol. I, pp. 325–364. Cold Spring Harbor Press, Cold Spring Harbor.

Su, T. T., Campbell, S. D., and O'Farrell, P. H. (1998). The cell cycle program in germ cells of the *Drosophila* embryo. *Dev. Biol.* **196,** 160–170.

Swanson, M. M., and Poodry, C. A. (1980). Pole cell formation in *Drosophila melanogaster. Dev. Biol.* **75,** 419–430.

Technau, G., and Campos-Ortega, J. A. (1986). Lineage analysis of transplanted individual cells in embryos of *Drosophila melanogaster.* III. Commitment and proliferative capabilities of pole cells and midgut progenitors. *Wilhelm Roux' Arch. Dev. Biol.* **195,** 489–498.

Thierry-Mieg, D. (1982). *Paralog,* a control mutant in *Drosophila melanogaster. Genetics* **100,** 209–237.

Tomancak, P., Guichet, A., Zavorszky, P., and Ephrussi, A. (1998). Oocyte polarity depends on regulation of *gurken* by Vasa. *Development* **125,** 1723–1732.

Underwood, E. M., Caulton, J. H., Allis, C. D., and Mahowald, A. P. (1980). Developmental fate of pole cells in *Drosophila melanogaster. Dev. Biol.* **77,** 303–314.

Van Doren, M., Willaimson, A. L., and Lehmann, R. (1998). Regulation of zygotic gene expression in *Drosophila* primordial germ cells. *Curr. Biol.* **8,** 243–246.

Wang, C., Dickinson, L. K., and Lehmann, R. (1994). The genetics of localization in *Drosophila. Dev. Dyn.* **199,** 103–115.

Waring, G. M., Allis, C. D., and Mahowald, A. P. (1978). Isolation of polar granules and the identification of a polar granule-specific protein. *Dev. Biol.* **66,** 197–206.

Webster, P. J., Liang, L., Berg, C. A., Lasko, P., and Macdonald, P. M. (1997). Translational repressor Bruno plays multiple roles in development and is widely conserved. *Genes Dev.* **11,** 2510–2521.

Whitfield, W. G. F., Gonzalez, C., Sanchez-Herrero, E., and Glover, D. M. (1989). Transcripts of one of two *Drosophila* cyclin genes become localized in pole cells during embryogenesis. *Nature* **338,** 337–340.

Wilson, J. E., Connell, J. E., and Macdonald, P. M. (1996). *aubergine* enhances *oskar* translation in the *Drosophila* ovary. *Development* **122,** 1631–1639.

Yue, L., Karr, T. L., Nathan, D. F., Swift, H., Srinivasan, S., and Lindquist, S. (1999). Genetic analysis of viable Hsp90 alleles reveals a critical role in Drosophila spermatogenesis. *Genetics* **151,** 1065–1079.

Zissler, D., and Sander, K. (1973). The cytoplasmic architecture of the egg cell of *Smittia* spec. (Diptera, Chironomidae). I. Anterior and posterior regions. *Wilhelm Roux' Arch. Dev. Biol.* **172,** 175–186.

Early Events in the Mammalian Germ Line

Robert Anderson,* Janet Heasman,* and Christopher Wylie†
*Department of Genetics, Cell Biology and Development, and †Department of Pediatrics, University of Minnesota School of Medicine, Minneapolis, Minnesota 55455

Germ cells represent the genetic and cellular link between generations, as well as the transmitters of inherited diseases. Despite their central importance, not much is known about the molecular mechanisms whereby a germ cell lineage becomes set aside during development, or how the germ cells, once formed, migrate to the gonads and combine with somatic cells to make a gonad. This article provides a brief review of current knowledge on these issues, with particular focus on the mammalian germ line.

KEY WORDS: Germ cell, Mouse, Genital ridge, Primordial, Stem cell.

I. Introduction

The female and male gametes are possibly the most different-looking cells in the body. Despite this they share a common embryonic precursor: the primordial germ cell (PGC). PGCs are sexually indifferent cells that arise early in development near the posterior margins of the embryo in the mouse. Several days later, the somatic gonads appear as two genital ridges in a different region of the embryo, on the dorsal abdominal wall. Germ line development does not proceed outside the somatic gonads, since this special environment is required for germ cell proliferation and sexual differentiation. PGCs colonize the genital ridges by actively migrating from their place of origin to this site. Upon reaching their destination, PGCs cease their migration and coalesce with each other and with somatic cells to form the primary sex cords of the nascent gonads.

Despite their central importance to biology, little is known about the genetic mechanisms underlying the origin and behavior of the early germ line. PGCs are a very small population of migratory cells and have proven difficult to purify in sufficient numbers for molecular analysis. However, careful studies of mouse mutants (spontaneous and induced) and the development of *in vitro* culturing techniques have revealed much about PGCs in the 1990s. This chapter will focus on progress that has been made in understanding the mechanisms of germ cell determination, migration, proliferation, and survival.

II. Mammalian Germ Cells

A. The Early Germ Line

During gastrulation the epiblast, or primary ectoderm, forms all three embryonic germ layers: ectoderm, endoderm, and mesoderm (extraembryonic ectoderm and extraembryonic endoderm are both formed prior to gastrulation; Hogan *et al.,* 1994). During gastrulation, epiblast cells stream through the primitive streak and emerge on the ventral side of the embryo. The cells that emerge from the primitive streak acquire a new identity: they are now mesoderm or endoderm.

PGCs are derived from some of the first cells of the embryo to traverse and emerge from the primitive streak. Lineage analysis has shown that PGCs are derived from the proximal epiblast (primary ectoderm that is immediately adjacent to the trophectoderm and which forms part of the placenta), an area of the egg cylinder that also differentiates into extraembryonic mesoderm (allantois, yolk sac blood islands, chorion, and a portion of the amnion) (Lawson and Hage, 1994). Further, lineage analysis at E6 and E6.5 has demonstrated that descendants of individual proximal epiblast cells can contribute to both PGCs and extraembryonic mesoderm, proof that these cells arise from previously uncommitted cells. Since extraembryonic mesoderm contributes mainly to the placenta and the fetal membranes, PGCs (and potentially the blood islands) are among the only descendants of the proximal epiblast that actually contribute to adult tissues.

The first definitive PGCs appear at about E7.2 in the posterior margins of the embryo near the base of the allantois (Lawson and Hage, 1994). The initial population of PGCs is small—about 45 cells per embryo, which can be identified by their high level of alkaline phosphatase (AP) expression (upregulation of AP activity is thought to coincide with germ cell formation). Commitment to the germ line is accompanied by a dramatic decrease in the mitotic rate. The mean doubling time of epiblast cells is less than 7 hours, whereas the time for PGCs is about 16 hours (Lawson and Hage, 1994). Nevertheless, the population of PGCs undergoes tremendous expansion in the days following their formation. A day after allocation at E8.5, the population of PGCs is about 145, at E10.5 the population increases to about 1,000, and at E13.5 each embryo has about 26,000 postmigratory germ cells (Tam and Snow, 1981). After E13.5, the developmental fate of germ cells is dependent on the sex of the embryo: in the male, germ cells enter mitotic arrest; in the female, germ cells begin to enter meiosis.

B. Germ Cell Specification

1. Invertebrate Germ Cell Specification

In the fruitfly *Drosophila melanogaster* and the nematode *Caenorhabditis elegans,* germ line cells are specified by the assymetric distribution of cytoplasmic determinants, known as germ plasm (Mahowald, Chapter 5; Seydoux and Schedl, Chapter 4). Germ plasm is an electron-dense material that contains a very high concentration of mitochondria, ribosomes, protein, and RNAs (Mahowald and Henner, 1971). Transfer of *Drosophila* germ plasm to ectopic locations in the embryo—the head for instance—is sufficient to induce there the formation of functional germ cells (Illmensee and Mahowald, 1974). This property is mimicked by the germ plasm component *oskar,* which expressed ectopically can induce germ cell formation in its own right (Ephrussi and Lehmann, 1992). A large number of genes, both upstream and downstream of *oskar,* are now known to be essential for germ cell formation.

How germ plasm instructs a cell to become or maintain a germ line fate is only beginning to be understood. Some evidence suggests that germ plasm preserves or protects the germ line from the expression of "somaticizing" genes that would promote differentiation (Hunter and Kenyon, 1996). In support of this hypothesis, many of the germ plasm components identified to date act as inhibitors of transcription or translation (Mello *et al.,* 1994; Seydoux *et al.,* 1996). Further, in *C. elegans,* loss of function mutations in certain germ plasm genes can cause germ cells to adopt somatic fates.

2. Germ Cell Specification in Mammals

There is good evidence that some vertebrates, including the amphibian *Xenopus laevis,* utilize germ plasm to determine germ line cells, thus indicating that this mechanism is highly conserved throughout evolution (Heasman *et al.,* 1984; Kloc *et al.,* Chapter 2). However, in birds and mammals, a large body of evidence suggests that germ line specification is not initiated by maternally encoded determinants as it is in invertebrates and *Xenopus.*

The first piece of evidence comes from analysis of pre-implantation mouse embryos. During the first few cell divisions of the early embryo, all blastomeres are equivalent in potency and are capable of forming all cell types (including functional germ cells). This observation is not consistent with a model in which germ cell determinants are unequally partitioned (as in *Xenopus* and *Drosophila*), since all early blastomeres are functionally equivalent. The second piece of evidence comes from tissue recombination experiments of E6.5 embryos, one day prior to germ cell formation. Normally only cells of the proximal epiblast of early gastrula embryos contribute to the germ line (Lawson *et al.*, 1999). However, if cells of the distal epiblast are transplanted to a proximal position in the egg cylinder, these cells can contribute to the germ line (Gardner and Rossant, 1979; Snow, 1981). This experiment provides compelling evidence that many if not all cells of the epiblast are capable of forming germ cells, and that epiblast cells rely on proper spatial cue(s) to adopt a germ line fate.

Recently, a candidate for such a spatial cue in the proximal epiblast was identified. Bone morphogenetic protein-4 (Bmp4) is an intercellular signaling protein of the transforming growth factor-β (TGFβ) superfamily. Embryos lacking Bmp4 suffer a number of developmental defects, and most die during gastrulation. However, it was noted that in *Bmp4*$^{-/-}$ embryos that survive gastrulation, the allantois and germ line were always absent. In addition, *Bmp4*$^{+/-}$ embryos were found to have a 50% smaller founding population of PGCs. These experiments demonstrated an essential, dose-dependent role for Bmp4 in specification of the germ line (Lawson *et al.,* 1999).

In a series of elegant experiments using chimeric embryos composed of wild-type and *Bmp4*$^{-/-}$ cells, it was shown that the extraembryonic ectoderm provides an essential source of Bmp4 protein for germ cell/allantois induction (Lawson *et al.,* 1999). The proximal epiblast (the normal source of PGC precursors) is immediately adjacent to the extraembryonic ectoderm, whereas the remainder of the epiblast is farther away. By virtue of its proximity to the extraembryonic ectoderm, the proximal epiblast is exposed to the highest concentration of secreted Bmp4; a high concentration of this signal appears to be crucial for induction of germ cell precursors. It is not yet known whether Bmp4 acts directly on the epiblast or whether it modulates the expression of additional signal(s) that do so.

Since Bmp4 is required for the formation of both PGCs and the allantois, it plays a critical role in the allocation of the common PGC/allantois precursor cells. What defines the developmental fate of the PGC/allantois precursor is unclear. Several genes are essential for formation of the allantois but not for PGCs; these genes would be expected to act downstream of *Bmp4* (Lawson *et al.,* 1999). It is not yet known whether there is an analogous group of genes downstream of *Bmp4* required uniquely for the germ line.

C. Germ Cell Migration

1. Early Entry of PGCs into the Endodermal Epithelium

The exact location of PGC formation has been difficult to establish, because AP, the first marker for PGCs, is also expressed by several other tissues (Chiquoine, 1954). Based AP expression, it has been hypothesized that PGCs arise in the yolk sac endoderm, the extraembryonic mesoderm, or in the posterior primitive streak (Chiquoine, 1954; Copp *et al.*, 1986; Ginsburg *et al.*, 1990; Snow, 1981). Since PGCs share a direct lineage with the extraembryonic mesoderm (Lawson and Hage, 1994), it is very likely that PGCs arise either from the primitive streak (the original source of all extraembryonic mesoderm) or within the extraembryonic mesoderm itself.

Shortly after formation, PGCs are found in the endoderm epithelium, ventral to both the base of the allantois and the posterior primitive streak. How PGCs translocate into this endoderm has been difficult to establish on the basis of traditional histology and grafting experiments. At least three relatively simple hypotheses have been put forward to explain this phenomenon: (1) PGCs are passively incorporated by the endodermal epithelium, (2) PGCs are pushed into the endodermal epithelium by movements of gastrulation, or (3) PGCs migrate actively from their site of origin into the endodermal epithelium.

Recent studies have shed light on the place of origin of PGCs and the mechanism by which they enter the endodermal epithelium (Anderson *et al.*, 1999a). A transgenic mouse strain was generated that expresses green fluorescent protein (GFP) only in pluripotent cells, including PGCs. A great advantage of using GFP as a marker is that it can be detected in living embryos, thus allowing the observation of cell movements *in vivo*. Time-lapse confocal laser microscopy was used to show that PGCs move directly from the posterior primitive streak into the adjacent endodermal epithelium (Fig. 1, see color insert). High-power analysis of living cells revealed that virtually all early PGCs were highly polarized and extended very long filopodia—features typical of migratory cells (Anderson *et al.*, 1999a). This is in accord with electron microscopic studies, which concluded that early PGCs more closely resemble primitive streak mesoderm than endoderm. Taken together, these observations indicate that PGCs (1) arise in the posterior primitive streak and (2) migrate actively into the endodermal epithelium of the future hindgut.

2. Migration within the Hindgut

After PGCs have entered the endodermal epithelium, they remain within this tissue for about 48 hours. This endoderm is definitive, that is, derived from the primitive streak (Hogan *et al.,* 1994); at about E8, it begins to invaginate at the posterior pole of the embryo. This morphogenetic movement sweeps the PGCs into the interior and forms the hindgut diverticulum, the first step in organogenesis of the distal gastrointestinal tract.

Between E8.5 and E9.5, PGCs are confined to the hindgut endoderm. For reasons that are unclear, PGCs retain their highly motile phenotype, and continue to migrate within the confines of this organ. This behavior is retained *in vitro,* as PGCs explanted from E8.5 or E9.5 embryos and cultured on feeder layers are highly motile and invasive (Donovan *et al.,* 1986; Stott and Wylie, 1986). In *Drosophila,* the timing of PGC migration out of the endoderm of the gut is dictated by the cells of the gut, not by the PGCs (Jaglarz and Howard, 1995). As in *Drosophila,* mouse PGCs may be trapped within the endodermal epithelium until the suitable developmental stage has been reached. The hindgut endoderm undergoes substantial growth from E8 to E9.5, and intra-endodermal PGC migration may be necessary to spread the pool of PGCs along the anterior-posterior and axis of the hindgut.

3. Migration to the Urogenital Ridges

Beginning at about E9.5, PGCs begin migrating out of the hindgut endoderm into the splanchnic mesoderm that surrounds the gut. What determines the timing of this exit is unknown. "Pioneer PGCs," those that leave the hindgut at E9.5, need only migrate a short distance (about 30 microns) out of the gut to reach the adjacent urogenital ridge mesoderm (Gomperts *et al.,* 1994). Between E9.5 and E10.5, the hindgut descends ventrally into the abdominal cavity and remains attached to the dorsal body wall by a thin sheet of cells, the mesentery. The formation of the mesentery coincides with the exit of most PGCs from the hindgut and lengthens their route to the urogenital ridges substantially (>100 microns). Nevertheless, by E10.5 more than 80% of PGCs reach the urogenital ridges (Snow, 1981).

The urogenital ridges are patches of mesoderm adjacent to the coelomic cavity lateral to the root of the mesentery. Upon reaching the urogenital ridges at E10.5, PGCs appear to accumulate on the basement membrane immediately below the coelomic mesothelium (Garcia-Castro *et al.,* 1997). Shortly thereafter, PGCs begin aggregating with each other and with a population of somatic cells of mesothelial origin. By E12.5, these aggregates show definitive signs of sexual differentiation: male gonads contain large, cord-like structures; female gonads contain many small, irregularly shaped clusters. These are the forerunners of the seminiferous tubules and the ovarian follicles, respectively. A schema of PGC migration and primary sex cord formation is shown in Figure 2 (see color insert).

III. Control of PGC Behavior

A. Proliferation

Between formation and gonad assembly, PGCs are a highly migratory, scattered population of cells. Despite this, the behavior of PGCs is tightly controlled. For instance, two embryos of the same developmental stage can be expected to have populations of PGCs that vary by less than 10% (Snow, 1981). Further, in embryos where the total number of PGCs is reduced (but not obliterated) by toxin administration, the resulting population of PGCs will increase to nearly normal levels at later stages (Snow, 1981). This observation suggests that there are mechanisms to sense and control the total number of PGCs in an embryo; if a deficiency is detected, this is overcome by accelerated proliferation.

When PGCs are explanted onto tissue culture plastic or purified ECM molecules, they are unable to spread or proliferate and rapidly undergo apoptosis. However, PGCs were found to spread and survive on somatic feeder cell monolayers (Donovan *et al.,* 1986). This was a major technical advance, since for the first time this *in vitro* culturing system made it possible to manipulate the environment of living PGCs directly.

1. Positive Regulators of PGC Proliferation

A number of mutant mouse strains (derived by random mutagenesis or spontaneous mutations) have been identified with defects in coat color, hematopoeisis, and fertility. These mice have mutations in the genes encoding either the receptor tyrosine kinase kit, or its ligand, steel factor (Donovan *et al.,* 1998). Kit is a transmembrane receptor expressed by PGCs. Steel factor, expressed by somatic cells, is produced in membrane-bound and soluble forms by alternative splicing and proteolytic cleavage. Culturing experiments of PGCs explanted *in vitro* demonstrated that steel factor acts as a survival signal for PGCs (Godin *et al.,* 1991; Matsui *et al.,* 1991; Dolci *et al.,* 1993). In addition, it was shown that steel factor is most potent when presented in a membrane-bound form (Donovan *et al.,* 1998). These results demonstrated that somatic cells of the early embryo provide a critical factor for migrating PGCs.

The success of the *in vitro* work in identifying the role of steel factor in survival of PGCs led to a search for other molecules involved in PGC survival and proliferation. One such molecule, leukemia inhibitory factor (LIF), was shown to be a survival factor for PGCs (Dolci *et al.,* 1993). LIF is a member of a group of secreted cytokines that signal through a receptor complex composed of a common subunit gp130 in combination with a ligand-specific subunit. Further, LIF (and other members of the LIF cytokine family), when added in combination with steel factor, acts to stimulate PGC proliferation. Current evidence suggests that a LIF-like molecule is required for PGC survival and/or proliferation *in vivo,* since the common receptor subunit gp130 is essential for PGC proliferation, although LIF itself is not. Other factors may also affect germ cell survival. For example, the cytokine IL-4 has been shown to increase PGC survival in culture, but its role in vivo is unknown (Cooke *et al.,* 1996).

The fibroblast growth factor (FGF) family may also play a role in stimulation of PGC proliferation. When FGF (in addition to steel factor and LIF) is added to PGC cultures *in vitro,* PGCs divide rapidly and transform into immortal cell lines (called embryonic germ cells or EG cells; Resnick *et al.,* 1992). EG cells share a wide variety of antigenic and morphological characteristics with embryonic stem cells (ES cells) and embryonal carcinoma cells (EC cells); these cells will be discussed later in this chapter (Wylie, 1999). The role of FGFs and FGF receptors in the early germ line has not been evaluated by genetic analysis.

2. Negative Regulators of PGC Proliferation

Transforming growth factor-β1 (TGFβ1) and the related protein activin βA have the opposite effect on PGCs as LIF, steel factor, and FGF when added to cultures: they inhibit proliferation of PGCs (Godin and Wylie, 1991; Richards *et al.,* 1999). Biochemical evidence suggests that PGCs express receptors for both TGFβs and activins, thus indicating that that these molecules can act directly on PGCs. This *in vitro* work supports a model in which a TGFβ-like molecule acts to balance the mitogenic signals expressed in the embryo.

Figure 3 shows a schema of known cytokine actions on PGCs during their migration.

B. Guidance of PGC Migration

PGCs begin migrating from the hindgut endoderm to the urogenital ridges at E9.5. By E10.5 about 80% and by E11.5 more than 90% have reached the urogenital ridges (Snow, 1981). The efficiency of this process makes it likely that precise guidance mechanisms have evolved.

1. PGC Migration in *Drosophila*

Genetic screens in *Drosophila* have identified over 10 genes required for migration of PGCs from the gut to the gonads. Many of these genes are

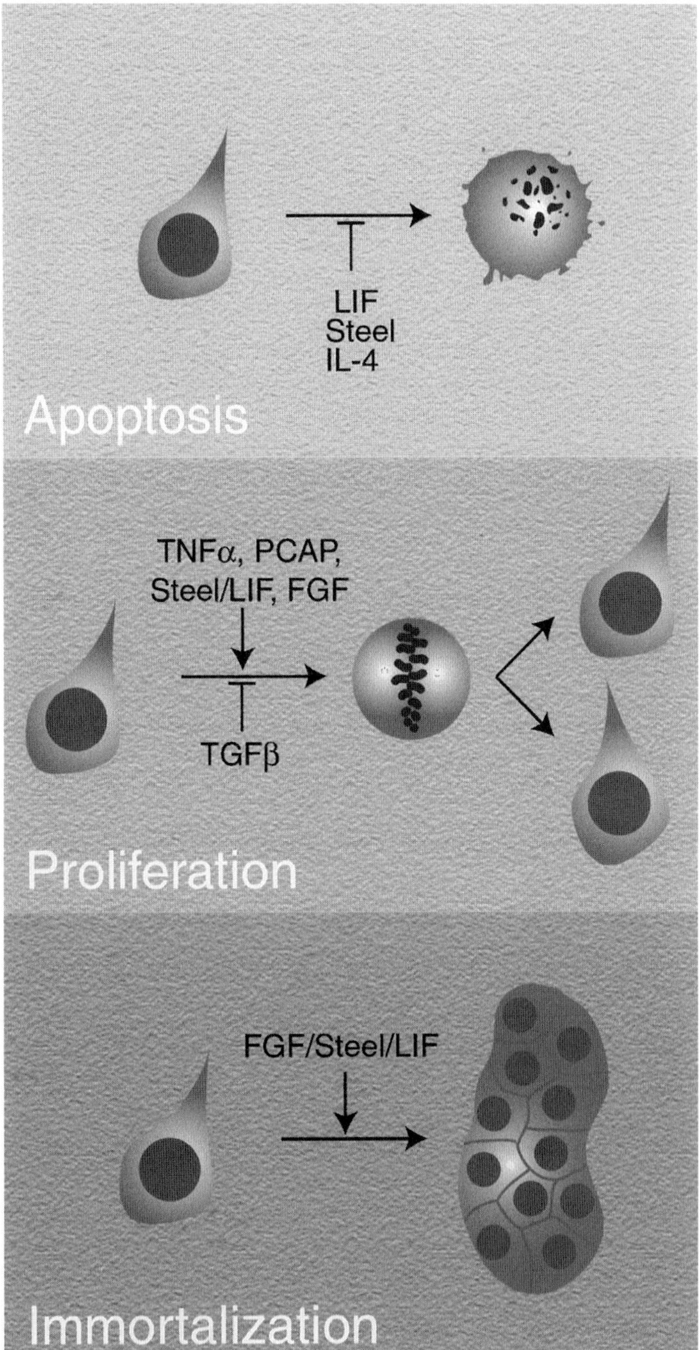

FIG. 3 A schema showing known effects of cytokines on mouse PGCs during migration.

transcription factors required for proper differentiation of the gonadal mesoderm (Wylie, 1999).

In addition, two signaling mechanisms have been identified that are required for efficient colonization of the gonads by PGCs in Drosophila. *wunen* is a gene expressed by the gut endoderm that generates a repulsive signal for PGCs, whereas *columbus* is a mesodermal gene that generates an attractive signal for PGCs (Van Doren *et al.,* 1998; Zhang *et al.,* 1997). Both genes are essential for efficient migration of PGCs into the gonads. The repulsive and attractive characteristics of these genes were elegantly demonstrated by ectopic expression experiments. When *colombus* was misexpressed in the epidermis or nervous system, PGCs were attracted to these structures; when *wunen* was misexpressed in the gonad, PGCs would not colonize it (Van Doren *et al.,* 1998; Zhang *et al.,* 1997). These experiments provide compelling evidence that guidance of PGC migration in *Drosophila* is controlled by somatic tissues.

The molecular mechanisms of *columbus* and *wunen* activity are not yet clear, although both encode proteins involved in lipid metabolism. *columbus* encodes HMG-CoA reductase, an enzyme best known for its role in cholesterol synthesis in mammals (Van Doren *et al.,* 1998). Insects do not synthesize cholesterol, so the required role of HMG-CoA reductase in *Drosophila* PGC migration acts in some other metabolic pathway. It will be very interesting to establish whether *columbus* and/or *wunen* homologs play a role in mammalian PGC migration guidance.

2. Guidance of PGC Migration in the Mouse

Much less is known about the guidance of germ cells in mice than is known in *Drosophila.* Evidence suggests that as in *Drosophila,* mouse PGCs are attracted to the urogenital ridges by a somatic signal (Godin *et al.,* 1990). This hypothesis is based on the observation via *in vitro* chemotropic assays that PGCs are attracted to genital ridge-conditioned media. It is not known whether the mammalian gut, like that of *Drosophila,* expresses a repulsive signal for PGCs.

In melanocyte precursor migration, the domain of steel factor expression determines the eventual distribution of melanocytes. Steel factor expression creates a permissive zone where melanocyte precursors are able to migrate and survive; melanocyte precursors outside this zone cannot migrate and undergo apoptosis (Wehrle-Haller *et al.,* 1996). Some evidence suggests that this mechanism also regulates the distribution of PGCs. As discussed above, steel factor is essential for PGC surivival *in vivo* and *in vitro.* In addition, there is some evidence that steel factor is required for PGC migration, since a reduced number of PGCs remain in the gut in severe *kit* alleles (Buehr *et al.,* 1993). It is an attractive hypothesis, therefore, that if a PGC reaches an area of the embryo where steel factor is not expressed, it will stop moving and undergo apoptosis, and this would be the fate of all PGCs in steel factor-deficient embryos.

Experiments *in vitro* suggest that steel factor is not a chemotropic agent for migrating PGCs (Godin *et al.,* 1991). Therefore, secretion of steel factor is probably not the mechanism by which the urogenital ridges attract PGCs. Two pieces of evidence suggest that this is accomplished by expression of a member of the TGFβ family of signaling molecules. First, in chemotropic assays, TGFβ1 is sufficient to attract PGCs *in vitro* (Godin and Wylie, 1991). Second, the chemotropic properties of genital ridge-conditioned media are abrogated by preincubation with polyclonal antibodies against TGFβ1. It is not yet known whether TGFβ family members, or which, play a role in attracting PGCs *in vivo,* but activin is a particularly good candidate, since it is highly expressed in the early gonad (Feijen *et al.,* 1994). However, there are no reported defects in PGC guidance in embryos lacking TGFβ1 or both activin βA and activin βB (Matzuk *et al.,* 1995), so the role of TGFβs in guidance of mouse PGCs is unproven.

IV. Interaction of PGCs with the Environment

A. Cell-Cell interactions

Since PGCs migrate rapidly to the urogenital ridges, it is assumed that PGCs do not tightly adhere to the cells of the somatic tissues through which they migrate. However, the compaction of PGCs during early gonad assembly probably requires high levels of intercellular adhesion. Therefore, the behavioral transition of PGCs from active migration to aggregation in the urogenital ridges is thought to be accompanied by an increase in cell-cell interactions. This transition also occurs *in vitro,* since mixed clones of E10.5 PGCs stop moving and aggregate together after 48 hours in culture (Gomperts *et al.,* 1994).

There is also clear evidence that PGCs interact with each other during migration by means of long filopodial networks. In fact, about 35% of migratory PGCs participate in networking at any given time *in vivo* (Gomperts *et al.,* 1994). The function(s) of these filopodial networks are unknown.

The basis of PGC-PGC adhesion is not yet understood at the molecular level. It is known that PGCs express several cell-cell adhesion molecules, including members of the cadherin family (Bendel-Stenzel *et al.,* 1999) and Ep-CAM (Anderson *et al.,* 1999c). It has recently been shown via antibody blocking experiments that E-cadherin is required for PGC coalescence in the genital ridges (Bendel-Stenzel *et al.,* 1999).

B. Cell-Matrix Interactions

Cell migration in embryogenesis is often dependent on cellular interactions with the extracellular matrix (ECM). The ECM is composed of secreted fibrillar and globular proteins that provide a scaffolding for the adhesion of both migratory and static cells.

Early qualitative studies indicated that PGCs were able to bind to several molecules, including fibronectin, laminin, and collagen IV (De Felici and Dolci, 1989; ffrench-Constant *et al.,* 1981). Subsequently, a quantitative study demonstrated that PGCs adhere to these molecules and that this adhesion profile changed according to the developmental stage of the PGC (Garcia-Castro *et al.,* 1997). These adhesion studies also provided the first indirect evidence that PGCs interact with the ECM through integrin-mediated and heparin sulfate proteoglycan-mediated mechanisms.

In an attempt to identify the molecular basis of PGC-ECM interactions, a wide survey of ECM receptor expression in the early germ line was performed by using flow cytometry, immunofluorescence, and RT-PCR (Anderson *et al.,* 1999b). PGCs were shown to express several integrin subunits as well as the heparin sulfate proteoglycan dystroglycan. Integrins are heterodimeric receptors capable of binding to a wide variety of ligands, whereas dystroglycan is a transmembrane laminin receptor (Hynes, 1987; Ibraghimov-Beskrovnaya *et al.,* 1992).

Through the analysis of null embryos (integrin subunits $\alpha3$, $\alpha6$, αV) or null◇wild-type chimeras (dystroglycan), several PGC receptors were shown not to be required for germ cell migration in the mouse (Anderson *et al.,* 1999b; R. Anderson and C. Wylie, unpublished data). However, the integrin subunit $\beta1$ was shown to be required for efficient colonization of the gonads by PGCs by chimeric analysis of integrin $\beta1^{-/-}$◇wild-type chimeras (Anderson *et al.,* 1999b). In support of this finding, no integrin $1^{-/-}$ cells were ever transmitted to offspring in chimeric animals. This PGC phenotype is very similar to that seen with integrin $\beta1^{-/-}$ hematopoetic precursor cells, which are unable to migrate from the yolk sac into the fetal liver (Hirsch *et al.,* 1996). The observation that PGCs and hematopoetic cells share a common migratory mechanism may be a consequence of the shared embryological origin of these cells.

It was noted that in the absence of $\beta1$ integrin, a small population of PGCs was still able to reach the urogenital ridges and enter the gonads. This observation suggests that PGC migration is not completely dependent on $\beta1$ integrin, although the process is less efficient in their absence. Therefore, additional receptors (for instance, integrin $\alpha V\beta 3$) may also be able to support PGC migration.

V. Stem Cells

Cell transplantation experiments in the mouse have shown that pluripotency is lost as cells differentiate. This was first demonstrated by grafting embryonic tissues (or cell lines derived from these tissues) into the organs of newborn mice. Cell lines derived from the inner cell mass of blastocysts (embryonic stem cells/ES cells), the epiblast (embryonic carcinoma cells/ EC cells), and PGCs (embryonic germ cells/EG cells) were all capable of differentiating into tissues of all three germ layers. No other tissues have so far been shown to retain this high degree of pluripotency. It has been subsequently shown that ES, EC, and EG cells are capable of forming any cell type and are more properly described as "totipotent."

Our understanding of pluripotency is currently in its infancy. Recent cloning experiments in sheep and mice have demonstrated that somatic nuclei are pluripotent and can form fertile animals if transplanted into the cytoplasm of eunucleated eggs (Campbell *et al.,* 1996; Wakayama *et al.,* 1988). These experiments demonstrate that it is the cytoplasm, not the nucleus, that is responsible for determining whether or not a cell is pluripotent.

The transcription factor Oct-4 is expressed by all pluripotent cells, including the entire morula, the inner cell mass of the blastocyst, the epiblast, and finally the germ line (Yeom *et al.,* 1996). Oct-4 expression thus correlates perfectly with the distribution of pluripotent cells during early development. Oct-4 is also expressed by the three types of pluripotent cell lines: ES cells, EC cells, and EG cells.

In the absence of Oct-4, all cells of the preimplantation embryo differentiate into the first somatic cell lineage, trophectoderm (Nichols *et al.,* 1998). This result demonstrated that Oct-4 expression is essential for maintenance of the pluripotent state in the early embryo. It will be interesting to determine whether Oct-4 expression is also necessary for maintenance of the germ line during embryonic, fetal, and adult stages. In addition, further analysis of the downstream targets of Oct-4 action should help better determine the molecular basis of the pluripotential state.

VI. Concluding Remarks

The biological significance of the germ line cannot be overstated. However, because of technical limitations, it has been very difficult to examine thoroughly the molecular basis of PGC formation and behavior. Advances in mouse genetics (i.e., knockouts, transgenesis) and *in vitro* culturing techniques have greatly accelerated the rate of discovery in this field. In the near future, new technologies, including PGC-specific gene targeting, PGC cDNA arrays, and *in vivo* examination of PGC behavior, will provide additional insight into the early germ line.

Acknowledgments

Work reviewed here and carried out in the authors' lab has been sponsored by NIH (RO1-HD33440), The Harrison Fund, and the Wellcome Trust. R. A. has also been sponsored by National Life and Health Insurance Medical Research Fund.

References

Anderson, R., Copeland, T., Schöler, H., Heasman, J., and Wylie, C. (1999a). The onset of germ cell migration in the mouse embryo. *Mech. Dev.* **91,** 61–68.

Anderson, R., Fässler, R. F., Georges-Labouesse, E., Hynes, R. O., Bader, B. L., Kreidberg, J. A., Schaible, K., Heasman, J., and Wylie, C. (1999b). Mouse primordial germ cells lacking β1 integrin enter the germ line but fail to migrate normally to the gonads. *Development* **126,** 1655–1664.

Anderson, R., Schaible, K., Heasman, J., and Wylie, C. (1999c). Expression of the homophilic adhesion molecule Ep-CAM in the mammalian germ line. *J. Reprod. Fertil.* **116,** 379–384.

Bendel-Stenzel, M., Gomperts, M., Anderson, R., Heasman, J., and Wylie C. (1999). The role of cadherins during primordial germ cell migration and early gonad formation in the mouse. *Mech. Dev.* **91,** 143–152.

Buehr, M., McLaren, A., Bartley, A., and Darling, S. (1993). Proliferation and migration of primordial germ cells in We/We mouse embryos. *Dev. Dynam.* **198,** 182–189.

Campbell, K. H., McWhir, J., Ritchie, W. A., and Wilmut, I. (1996). Sheep cloned by nuclear transfer from a cultured cell line. *Nature* **380,** 64–66.

Chiquoine, A. (1954). The identification, origin and migration of the primordial germ cells of the mouse embryo. *Anat. Rec.* **118,** 135–146.

Cooke, J., Heasman, J., and Wylie, C. (1996). The role of interleukin 4 in the regulation of mouse primordial germ cell numbers. *Dev. Biol.* **174,** 14–22.

Copp, A. J., Roberts, H. M., and Polani, P. E. (1986). Chimaerism of primordial germ cells in the early postimplantation mouse embryo following microsurgical grafting of posterior primitive streak cells *in vitro. J. Embryol. Exp. Morphol.* **95,** 95–115.

De Felici, M., and Dolci, S. (1989). In vitro adhesion of mouse fetal germ cells to extracellular matrix components. *Cell Differen. Dev.* **26,** 87–96.

Dolci, S., Pesce, M., and De Felici, M. (1993). Combined action of stem cell factor, leukemia inhibitory factor and cAMP on in vitro proliferation of mouse primordial germ cells. *Mol. Reprod. Dev.* **35,** 134–139.

Donovan, P. J., de Miguel, M., Cheng, L., and Resnick, J. L. (1998). Primordial germ cells, stem cells, and testicular cancer. *Apmis* **106,** 134–141.

Donovan, P., Stott, D., Cairns, L., Heasman, J., and Wylie, C. (1986). Migratory and postmigratory mouse primordial germ cells behave differently in culture. *Cell* **44,** 831–838.

Ephrussi, A., and Lehmann, R. (1992). Induction of germ cell formation by oskar. *Nature* **358,** 387–392.

Feijen, A., Goumans, M. J., and van den Eijnden-van Raaij, A. J. (1994). Expression of activin subunits, activin receptors and follistatin in postimplantation mouse embryos suggests specific developmental functions for different activins. *Development* **120,** 3621–3637.

ffrench-Constant, C., Hollingsworth, A., Heasman, J., and Wylie, C. (1981). Response to fibronectin of mouse primordial germ cells before, during, and after migration. *Development* **113,** 1365–1373.

Garcia-Castro, M., Anderson, R., Heasman, J., and Wylie, C. (1997). Interactions between germ cells and extracellular matrix glycoproteins during migration and gonad assembly in the mouse embryo. *J. Cell Biol.* **138,** 471–480.

Gardner, R. L., and Rossant, J. (1979). Investigation of the fate of 4–5 day post-coitum mouse inner cell mass cells by blastocyst injection. *J. Embryol. Exp. Morphol.* **52,** 141–152.

Ginsburg, M., Snow, M., and McLaren, A. (1990). Primordial germ cells in the mouse embryo during gastrulation. *Development* **110,** 521–528.

Godin, I., Deed, R., Cooke, J., Zsebo, K., Dexter, M., and Wylie, C. (1991). The effects of the *Steel* gene product on mouse primordial germ cells in culture. *Nature* **352,** 807–809.

Godin, I., and Wylie, C. C. (1991). TGFβ1 inhibits proliferation and has a chemotropic effect on mouse primordial germ cells in culture. *Development* **113,** 1451–1457.

Godin, I., Wylie, C., and Heasman, J. (1990). Genital ridges exert longrange effects on mouse primordial germ cell numbers and direction of migration in culture. *Development* **108,** 357–363.

Gomperts, M., Garcia-Castro, M., Wylie, C., and Heasman, J. (1994). Interactions between primordial germ cells play a role in their migration in mouse embryos. *Development* **120,** 135–141.

Heasman, J., Quarmby, J., and Wylie, C. C. (1984). The mitochondrial cloud of *Xenopus* oocytes: The source of germinal granule material. *Dev. Biol.* **105,** 458–469.

Hirsch, E., Iglesias, A., Potocnik, A., Hartmann, U., and Fassler, R. (1996). Impaired migration but not differentiation of haematopoietic stem cells in the absence of betal integrins. *Nature* **380,** 171–175.

Hogan, B., Beddington, R., Constantini, F., and Lacy, E. (1994). "Manipulating the Mouse Embryo." Cold Spring Harbor Press, Cold Spring Harbor, MA.

Hunter, C. P., and Kenyon, C. (1996). Spatial and temporal controls target pal-1 blastomere-specification activity to a single blastomere lineage in *C. elegans* embryos. *Cell* **87,** 217–226.

Hynes, R. (1987). Integrins: A family of cell surface receptors. *Cell* **48,** 459–550.

Ibraghimov-Beskrovnaya, O., Ervasti, J., Leveille, C., Slaughter, C., Sernett, S., and Campbell, K. (1992). Primary structure of dystrophin-associated glycoproteins linking dystrophin to the extracellular matrix. *Nature* **355,** 696–702.

Illmensee, K., and Mahowald, A. P. (1974). Transplantation of posterior pole plasm in *Drosophila:* Induction of germ cells at the anterior pole of the egg. *Proc. Natl. Acad. Sci. USA* **71,** 1016–1020.

Jaglarz, M., and Howard, K. (1995). The active migration of *Drosophila* primordial germ cells. *Development* **121,** 3495–3503.

Lawson, K. A., Dunn, N. R., Roelen, B. A., Zeinstra, L. M., Davis, A. M., Wright, C. V., Korving, J. P., and Hogan, B. L. (1999). BMP4 is required for the generation of primordial germ cells in the mouse embryo. *Genes Dev.* **13,** 424–436.

Lawson, K. A., and Hage, W. J. (1994). Clonal analysis of the origin of primordial germ cells in the mouse. *CIBA Found. Symp.* **182,** 68–91.

Mahowald, A. P., and Henner, S. (1971). Polar granules of *Drosophila.* 3. The continuity of polar granules during the life cycle of *Drosophila. J. Exp. Zool.* **176,** 345–352.

Matsui, Y., Toksoz, D., Nishikawa, S., Williams, D., Zsebo, K., and Hogan, B. L. (1991). Effect of Steel factor and leukaemia inhibitory factor on murine primordial germ cells in culture. *Nature* **353,** 750–752.

Matzuk, M. M., Kumar, T. R., Vassalli, A., Bickenbach, J. R., Roop, D. R., Jaenisch, R., and Bradley, A. (1995). Functional analysis of activins during mammalian development. *Nature* **374,** 354–356.

Mello, C. C., Draper, B. W., and Priess, J. R. (1994). The maternal genes apx-1 and glp-1 and establishment of dorsal-ventral polarity in the early *C. elegans* embryo. *Cell* **77,** 95–106.

Nichols, J., Zevnik, B., Anastassiadis, K., Niwa, H., Klewe-Nebenius, D., Chambers, I., Scholer, H., and Smith, A. (1998). Formation of pluripotent stem cells in the mammalian embryo depends on the POU transcription factor Oct4. *Cell* **95,** 379–391.

Resnick, J. L., Bixler, L. S., Cheng, L., and Donovan, P. J. (1992). Long-term proliferation of mouse primordial germ cells in culture. *Nature* **359,** 550–551.

Richards, A. J., Enders, G. C., and Resnick, J. L. (1999). Activin and TGFbeta limit murine primordial germ cell proliferation. *Dev. Biol.* **207,** 470–475.

Seydoux, G., Mello, C. C., Pettitt, J., Wood, W. B., Priess, J. R., and Fire, A. (1996). Repression of gene expression in the embryonic germ lineage of *C. elegans. Nature* **382,** 713–716.

Snow, M. H. (1981). Autonomous development of parts isolated from primitive-streak-stage mouse embryos. Is development clonal? *J. Embryol. Exp. Morphol.* **65 Suppl.,** 269–287.

Stott, D., and Wylie, C. (1986). Invasive behaviour of mouse primordial germ cells *in vitro. J. Cell Sci.* **86,** 133–144.

Tam, P., and Snow, M. (1981). Proliferation and migration of primordial germ cells during compensatory growth in mouse embryos *J. Embryol. Exp. Morphol.* **64,** 133–147.

Van Doren, M., Broihier, H. T., Moore, L. A., and Lehmann, R. (1998). HMG-CoA reductase guides migrating primordial germ cells *Nature* **396,** 466–469.

Wakayama, T., Perry, A., Zucotti, M., Johnson, K., and Yanagimachi, R. (1988). Full-term development of mice from enucleated oocytes injected with cumulus cell nuclei. *Nature* **394,** 369–374.

Wehrle-Haller, B., Morrison-Graham, K., and Weston, J. A. (1996). Ectopic c-kit expression affects the fate of melanocyte precursors in Patch mutant embryos. *Dev. Biol.* **177,** 463–474.

Wylie, C. (1999). Germ cells. *Cell* **96,** 165–174.

Yeom, Y., Fuhrmann, G., Ovitt, C., Brehm, A., Ohbo, K., Gross, M., Hübner, K., and Schöler, H. (1996). Germline regulatory element of Oct-4 specific for the totipotent cycle of embryonal cells. *Development* **122,** 881–894.

Zhang, N., Zhang, J., Purcell, K., Cheng, Y., and Howard, K. (1997). The *Drosophila* protein Wunen repels migrating germ cells. *Nature* **385,** 64–67.

Part II

Patterning of the Embryo

The Initial Phase of Embryonic Patterning in Mammals

R. L. Gardner
University of Oxford, Oxford, United Kingdom

Although specification of the antero-posterior axis is a critical intial step in development of the fetus, it is not known either how, or at what stage in development, this process begins. Such information is vital for understanding not only normal development in mammals but also monozygotic twinning, which, at least in man, is associated with a significantly increased incidence of birth defects. According to recent studies in the mouse, specification of the fetal anteroposterior axis begins well before gastrulation, and probably even before the conceptus implants. Moreover, evidence is accruing that the origin of relevant asymmetries depends on information that is already present in the zygote before it embarks on cleavage. Hence, early development in mammals does not differ as markedly from that in other animals as has generally been assumed. Consequently, at present, the possibility of adverse effects of techniques used to assist human reproduction cannot be disregarded.

KEY WORDS: Patterning, Polarity, Antero-posterior axis, Animal-vegetal axis, Gastrulation, Blastocyst, Anterior visceral endoderm, Epiblast.

"The biology of twinning events has implications for understanding the cellular mechanisms of embryogenesis. When differences between twin and singleton embryogenesis can be documented and understood, and their observable consequences can be assigned to specific differences, we should be able to undertake productive new approaches to human developmental biology, and particularly to the biology of anomalous development" (Boklage, 1987).

I. Introduction

During the 1990s interest in pattern-formation in mammals burgeoned to such an extent that it is no longer possible to encompass all areas of significant advance in a single review. For this reason, the scope of the present article is restricted to dealing with one primary aspect, namely the establishment of the antero-posterior (A-P) axis of the fetus. In the mouse, this axis is first clearly discernible morphologically with the appearance of the primitive streak (PS) at the onset of gastrulation. An earlier claim that the prospective posterior of the epiblast can already be distinguished from the anterior approximately one day before the PS forms (Bonnevie, 1950) has not been substantiated. However, a very early observation suggesting that the anterior end of the A-P axis can be discerned before gastrulation in the rabbit (Van Beneden, 1883) has recently been confirmed (Viebahn *et al.,* 1995).

Differentiation of its A-P axis is a crucial initial step in development of the fetus, since it marks the start of the laying down of the basic plan of the body. The aim of this chapter is to review studies that bear on how and when the orientation and polarity of the fetal A-P axis are specified. The point of making a distinction between these two parameters is simply to emphasize that there is at present no compelling reason to suppose that they should necessarily be specified simultaneously. Most of the work that will be discussed relates to the mouse, since very few relevant studies have been undertaken in other eutherian mammals.

Apart from enlarging our basic understanding of mammalian development, knowledge about formation of the fetal A-P axis is relevant to the question of the etiology of monozygotic (MZ) twinning and the rarer higher-order axial duplications within individual conceptuses that can result in the birth of triplets, quadruplets, or even quintuplets (MacArthur, 1938; Bulmer, 1970). It has become clear as a result of the more widespread use of ultrasound to image conceptuses that MZ twinning is initiated in human pregnancy much more frequently than it is maintained (Jauniaux *et al.,* 1986; Boklage, 1990; Hall, 1996). In addition to posing obstetric problems, it is a condition that carries an increased risk of mortality, congenital malformation, and other problems (Schinzel *et al.,* 1979; Little and Bryan, 1988; Hall, 1996; Bryan, 1998), regardless of whether only one twin rather than both survives to term (Szymonowicz *et al.,* 1986). For example, an overrepresentation of congenital heart defects has been found in MZ versus dizygotic (DZ) twins, even when cases of persistent ductus arteriosus and conjoined twins are discounted (Burn and Corney, 1984). The consequences are, of course, yet more dire when axial duplication is incomplete and thereby leads to the development conjoined or Siamese twins (Cunniff *et al.,* 1988). There is, moreover, evidence from various studies to suggest that MZ twinning may be increasing in frequency (Kyvik *et al.,* 1995), although whether this reflects a change in its rate of initiation or in the survival to term of both twins is not clear. It would be particularly intriguing if the former were the case given that, unlike DZ twinning, MZ twinning does not have an obvious hereditary basis (Bulmer, 1970; Stern, 1973).

Understanding how early spatial patterning that is instrumental in specifying the A-P axis is established is also of interest in the context of assisted conception in man. Here, relevant techniques include inducing ovulation of additional eggs, which are thereby caused to mature precociously, and, in the case of *in vitro* fertilization (IVF) and its variants such as intracytoplasmic sperm injection (ICSI), obtaining fertilization and growth of the initial stages of development under conditions that unquestionably differ from, and are inferior to, those obtaining *in vivo* (Fishel and Symonds, 1993). Given the scale on which these techniques are now employed and their inevitable expansion in the future, it is becoming increasingly important to ascertain whether their use carries an enhanced risk of disturbing fetal development. In considering whether it might, it is relevant to note that experience in various vertebrates other than mammals, and in numerous invertebrates, has revealed that perturbing the organization of the egg before or shortly after fertilization can cause malformation of the embryo (Witschi, 1952; Black and Gerhart, 1986). The obvious implication is that patterning information that is crucial for normal development is already localized within the fertilized egg before it begins to divide (see Davidson, 1986, and other contributions to this volume). Organisms in which a role for egg organization in embryonic patterning has clearly been established typically develop externally and may thus have evolved protective mechanisms to ensure that critical early developmental processes are insulated from environmental change. A concern would be that because they normally develop in a much more tightly regulated *in vivo* environment, the early stages of mammalian development may be more susceptible to perturbation under the suboptimal conditions obtaining *in vitro*. The prevailing view is, however, that mammals are an exception to the general rule that developmentally significant information is already present in the egg, and processes involved in the initiation of fetal development are held not to begin until after the conceptus has implanted in the uterus. Among the various reasons why this view has gained wide currency are the following:

1. The initial period of development up to and even beyond implantation is concerned with the differentiation of wholly extraembryonic tissues that mediate attachment of the future fetus to the mother and its nutrition, and which are all discarded at birth. The small residual pool of cells that will give rise to the fetus, as well as to additional extraembryonic tissue, only starts to grow once implantation has occurred, and does not show any overt signs of differentiation before the onset of gastrulation. Given such a long

interval between the start of cleavage and fetal development, the persistence from the zygotic stage of information for specifying the latter has been deemed most unlikely (Tarkowski and Wroblewska, 1967).

2. It has generally been assumed that the conceptus retains a radially symmetrical organization about its embryonic-abembryonic axis (see Fig. 1) until the PS forms at the onset of gastrulation, thereby marking the posterior end of the future fetus.

3. More persuasively, cleavage-stage mammalian conceptuses show an impressive ability to regulate their development following gain, loss, or rearrangement of cells. The existence of patterning information in the egg is assumed to be incompatible with such regulative ability, principally because axial duplication has never been recorded following production of unitary giant conceptuses via morula aggregation (Gardner, 1996a). Whether the findings that have emerged from these types of experiments really constitute conclusive evidence against a patterning role for the egg in mammals is an issue that will be taken up later (see section VII).

4. Classical studies on the sensitivity of different stages of mammalian development to teratogenic insult identified gastrulation and subsequent organogenesis as critical periods. The consequences of such insult to the regulative preimplantation stages were held to be all-or-none, with development either failing altogether or continuing entirely normally thereafter (Kimmel *et al.,* 1993; Rutledge, 1997). This was certainly the prevailing wisdom at the time when the pioneering work of Robert Edwards and the

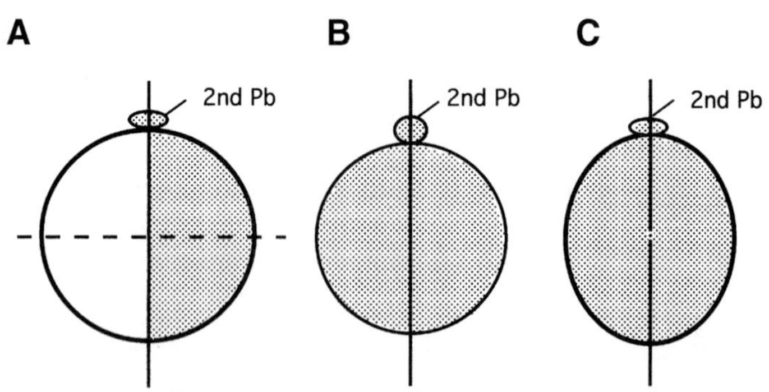

FIG. 1 Diagrams showing the relationship of the embryonic-abembryonic (Em.Ab) axis (dashed line) and the bilateral axis of the early mouse blastocyst (continuous line) to the animal-vegetal (A-V) axis of the zygote (continuous line), as deduced from the location of the second polar body (Pb). (A) Early blastocyst in lateral view showing both Em.Ab and bilateral axis. (B) Zygote with A-V axis extending from beneath the 2nd Pb to the diametrically opposite point on its surface. (C) Embryonic polar view of early blastocyst to show that its bilateral axis is parallel, and its Em.Ab axis orthogonal, to the A-V axis of the zygote.

late Patrick Steptoe on human *in vitro* fertilization (IVF) was in its early stages (Saxen and Rapola, 1969; Austin, 1973).

5. Finally, the very precocious activation of the zygotic genome that occurs in conjunction with the destruction of most maternal mRNA in mammals (Schultz, 1986; Braude *et al.,* 1988; Ram and Schultz, 1993; De Sousa *et al.,* 1998) has engendered the view that the influence of the maternal genome must be short-lived.

In the following pages, evidence will be discussed that challenges the foregoing grounds for supposing that egg organization can be discounted from having any role in patterning of the mammalian embryo. Points 2 and 4 will be shown to be patently wrong, and points 1 and 3 rather less robust than they may appear at first sight. Regarding point 5 it is important to note, as discussed elsewhere (Gardner, 1996a), that the transition from maternal to zygotic control in the mouse is not completed until after the conceptus has implanted *in utero.* However, before discussing work that relates more directly to the issue of specification of the A-P axis of the fetus, brief consideration will be given to several disparate, and largely neglected, findings that show that fetal organization can be affected by the conditions to which the oocyte, zygote, or early cleavage stages are exposed.

II. Sensitivity of Fetal Development to Conditions to which the Early Conceptus Is Exposed

A. Induced Ovulation, IVF, and Gamete Aging

The first category of findings relate to increases in the incidence of partial or complete axial duplication. Thus, a higher than normal rate of monozygotic twinning has been linked with assisted reproduction, including induction of ovulation (Wenstrom *et al.,* 1993), even where this is not accompanied by IVF (Derom *et al.,* 1987). The increase in IVF has been attributed to hardening of the zona pellucida of conceptuses produced *in vitro,* with consequent elevated risk of externalization and eventual subdivision of part of the inner cell mass (ICM) and trophectoderm (Edwards, 1985; Leroy, 1985). That induced ovulation per se can lead to hardening of the zona has been clearly demonstrated in certain strains of mice (Krzanowska, 1972). How herniation through a restricted break in the continuity of the zona might lead to twinning is illustrated schematically in Figure 2. In one IVF program, all cases of MZ twinning were reported to have occurred following replacement of conceptuses whose zona were either thin naturally or had been breached mechanically to assist either fertilization or hatching

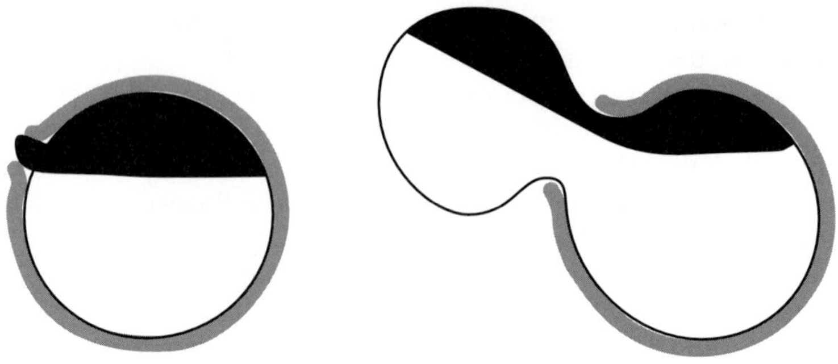

FIG. 2 Scheme of possible mode of twinning via herniation of part of the ICM as well as the trophectoderm through a restricted breach in the zona pellucida. Shearing of the externalized part of mouse blastocysts has been observed following their transfer to recipient uteri (author's unpublished observations). However, it is questionable whether this would occur during normal pregnancy in rodents because the blastocyst is very closely invested by the uterine luminal epithelium well before it begins to implant (Reinius, 1967; Enders and Schlafke, 1969). The condition of the uterine luminal epithelium prior to implantation in the human is not known (Edwards *et al.,* 1985).

(Alikani *et al.,* 1994). It is noteworthy in this context, however, that pre-hatching blastocysts containing two separate ICMs have been recorded in the mouse (Chida, 1990). Such blastocysts were not only found more commonly following fertilization *in vitro* than *in vivo,* but all that attached following hatching *in vitro* continued to exhibit two distinct ICMs during subsequent outgrowth. Given the small size of the mouse blastocyst and the great difficulty in keeping a transplanted second ICM separate from the host one (Papaioannou and Gardner, 1979), the occurrence of this phenomenon is most intriguing. Whatever its explanation, it clearly shows that twinning can occur in blastocysts independently of their partial herniation through the zona during hatching. Nonetheless, in a study in which mouse blastocysts with two separate ICMs were produced experimentally, orientation of the resulting egg cylinders following implantation was clearly perturbed to such an extent that it became most unlikely both twins could ever survive to term (Ozdzenski *et al.,* 1997).

Overripenning of eggs through delaying fertilization, which can cause partial axial duplication in lower vertebrates (Witschi, 1952), has been found to produce similar effects, albeit less commonly, in mammals. Cases of partial duplication of the CNS and somites were found among midgestation rat fetuses in which ovulation had been delayed by daily injection of sodium pentobarbital (Butcher *et al.,* 1969). Delaying ovulation in the rabbit resulted in six sets of monozygotic twins at the blastocyst stage (a 1.5% rate of twinning), the four better-developed of which clearly consisted of pairs of similarly developed blastocysts within a common zona (Bomsel-Helmreich and Papiernik-Berkhauer, 1976). Given the ease with which cleavage stage conceptuses aggregate to form fully integrated blastocysts in mammals, this is a most surprising finding. Bomsel-Helmreich and Papiernik-Berkhauer (1976) and others (e.g. Harlap *et al.,* 1985) have also amassed indirect evidence in support of an association between oocyte aging and monozygotic twinning in man.

The relative rarity of defects produced by aged gametes in mammals may be due to a rather limited window of time during which the aging process is compatible with extensive further development. An obvious way in which mammals differ from lower vertebrates is in the dependence of the conceptus on maintaining synchrony with the mother for its development to progress very far. In view of this, it is perhaps not surprising that most conceptuses resulting from a union of gametes of which one or both are aged arrest during, if not before, implantation (Blandau 1954; Braden, 1959; Marston and Chang, 1964; Fugo and Butcher, 1966; Butcher and Fugo, 1967; Thompson and Zamboni, 1975).

B. Chemical and Other Perturbations

Of particular interest are findings relating to the effects of several alkylating agents, particularly ethylene oxide and ethyl methane sulfonate, which are well-established conventional mutagens in terms of their effects on germ cells. Specific exposure of early zygotic stages to both these and certain other agents has been found to induce high incidences of fetal anomalies and death in the mouse. The pattern of mortality and morbidity is distinct from that resulting from gene mutation, is not explicable in terms of numerical or structural chromosomal changes, and appears to be due to a direct effect on the zygote rather than an indirect one via modification of the maternal environment (Katoh *et al.,* 1989; Rutledge *et al.,* 1992; Polifka *et al.,* 1996). Furthermore, it has been found by zygote reconstitution that the cytoplasm as well as the pronuclei have to be exposed to such alkylating agents for their effects to be manifest (Generoso *et al.,* 1990). The anomalies produced by these agents are held to resemble those characterizing the large class of sporadic defects of unknown etiology that are enriched among human stillbirths. That cases of very gross disturbance in patterning or axiation were not included among the abnormalities recorded may reflect the fact that scoring was invariably done only shortly before term. It would be particularly interesting to examine the conceptuses earlier in order to find out what anomalies occur among those that were represented as resorptions at such an advanced stage of pregnancy.

Zygotes derived from mouse oocytes that were exposed briefly to a vitrification solution containing dimethyl sulfoxide, acetamide, 1,2-propanediol, and polyethylene glycol also showed a significant rise in the incidence of malformed fetuses on day 15 postcoitum (dpc) (Kola *et al.*, 1988). Here again, the pattern does not accord with single gene mutations, and neither structural nor numerical chromosomal abnormalities nor alterations in the maternal environment can account for it.

Anomalies in axial patterning have been observed on day 9 of gestation following very brief exposure of eight-cell or two-cell mouse conceptuses to 300 mM LiCl (Rogers and Varmuza, 1996). Although these were said to occur in the absence of any alteration in morphogenesis to the blastocyst stage, no data on overall or differential cell counts were provided to support this contention. Such a high extracellular concentration of LiCl, which is necessary to produce a modest internal concentration of Li^+ in the relatively impermeable embryos of amphibia, presumably resulted in a much higher intracellular concentration in mouse blastomeres. Similar defects could also be obtained by exposing mouse cleavage stages to a lower concentration of LiCl for a much longer period, namely 90 mM for five hours. To account for the variable but generally limited axial defects observed in postgastrula embryos, Rogers and Varmuza (1996) proposed that through interfering with the inositol signalling pathway, Li^+ caused epigenetic modification of the genome via protein kinase C-mediated induction of topoisomerase activity. However, were this the case, one might have expected the defects to be more extensive following exposure of the two-cell than eight-cell stage to LiCl, which was the opposite of what was observed. What is more significant, the "inositol depletion" hypothesis for the action of Li^+ has been challenged by the finding that the effect of this cation in various systems is not mimicked by a far more potent inhibitor of inositol monophosphatase (Klein and Melton, 1996). Studies undertaken both *in vivo* and *in vitro* have suggested an alternative primary mode of action of Li^+ on morphogenesis in different organisms, namely as a noncompetitive inhibitor of glycogen synthase kinase-3, an enzyme that antagonizes wnt signalling by promoting the degradation of β-catenin (Klein and Melton, 1996; Stambolic *et al.*, 1996; Hedgepath *et al.*, 1997). This attractive alternative explanation for the morphogenetic effects of Li^+ does not, however, explain why simultaneous presentation of an intermediate in the polyphosphoinoside cycle such as myo-inositol, or an analog of diacylglycerol, can negate the teratogenic effects of Li^+ (Busa and Gimlich, 1989).

That allocation of cells in the blastocyst can be affected by the conditions to which the very early conceptus is exposed is evident from altering the Ca^{2+} oscillations accompanying completion of meiosis and first cleavage in the mouse (Bos-Mikich *et al.*, 1997) and from maternal diabetes in the BB/E rat (Lea *et al.*, 1996). The significant point is that the changes in ICM versus trophectoderm cell number that occur in such circumstances do not accord with the expectations of simply altering the "inside-outside" cell ratios during cleavage (Tarkowski and Wroblewska, 1967). Although the postimplantation consequences of such changes in blastocyst cell number and distribution have not yet been explored, increased rates of malformation have been found in both naturally and experimentally diabetic rats and mice that depend, at least in part, on the early maternal enviornment (Pampfer *et al.,* 1990; Moley *et al.,* 1991; Otani *et al.,* 1991; Lea *et al.,* 1996). Finally, exposure of cleavage stages to a raised concentration of ammonium ions during *in vitro* culture has been found to cause exencephaly (Lane and Gardner, 1994). However, in this particular case, malformation may be a relatively nonspecific consequence of retarded postimplantation development.

Although the effects of both alkylating agents on the zygote and Li^+ on cleavage stages have been attributed to epigenetic modification of the genome, in neither case has a persuasive mechanism been advanced to explain how this might occur. Bearing in mind that exposure of the cytoplasm is necessary for the effects of the alkylating agents (Generoso *et al.,* 1990), the basis of all such malformations is still far from clear, and the notion that the action of these and other chemicals is mediated via the genome remains entirely conjectural. What, nevertheless, the foregoing findings demonstrate unequivocally is that development of the fetus is susceptible to perturbations occurring very early in development, or even prior to fertilization. This not only renders untenable the view of early development encapsulated in point 4 in section I, but it also raises doubt about whether cleavage blastomeres really are developmentally naive and strictly equivalent until at least the eight-cell stage.

III. Polarity and the Establishment of Axes of the Conceptus

A. Polarity of the Oocyte and Zygote

Discussion of this topic in mammals is complicated by their pattern of early development. Since the latter is, as noted earlier, concerned initially with the formation of extraembryonic tissues, it is necessary to consider patterning of the entire early conceptus as well as the small part of it from which the fetus will eventually be formed. Given the diversity in morphology of the immediate postblastocyst stages, our understanding of early development in mammals is very narrowly based. Hence, the overwhelming majority of studies have been undertaken in the mouse, with only modest attention being paid to other species, particularly the human and some larger animals of economic importance.

As discussed in detail elsewhere (Gardner 1996a), the mammalian egg is a polarized cell, though the extent to which its polarity is of functional significance seems to vary among species. Until recently, eccentricity of the germinal vesicle was the only documented sign of polarity of the oocyte before it embarked on final maturation. However, nuage-like material has now been found to be localized cortically in rat oocytes (Young *et al.*, 1999), particularly before germinal vesicle breakdown. Thereafter, this material becomes less condensed and thus comparatively inconspicuous. A difficulty in investigating its significance is that it can be visualized only in fixed and stained material. The fact that mouse oocytes can be fertilized and develop normally following removal of up to half, but not three-quarters, of their cytoplasm (Wakayama and Yanagimachi, 1998) is more likely to depend on the quantity rather than the quality of the cytoplasmic constituents that are removed.

In the mouse, once the oocyte begins to mature an area of approximately one-fifth of its surface centered on the animal pole becomes relatively microvillus-free (Johnson *et al.*, 1975) and is evidently not conducive to sperm attachment (Talansky *et al.*, 1991). In the human, neither regionalization of the oocyte surface nor obvious restriction in the site of sperm attachment has been observed (Santella *et al.*, 1992). Once the fertilizing sperm has induced the completion of meiosis, polarity is only evident morphologically in zygotes of either species through localization of the polar bodies (Pbs). While the first Pb seldom remains intact for long, at least in the mouse, the second Pb survives regularly to the early blastocyst stage (Lewis and Wright, 1935; Gardner, 1997). However, in both mouse and human zygotes with both Pbs intact, these bodies are not infrequently well apart (Fig. 3) rather than immediately adjacent to each other (Zamboni, 1970; Payne *et al.*, 1997). This implies that the animal-vegetal (A-V) axis may shift in orientation between the secondary oocyte and zygote stage, though how this is acccomplished is far from clear. The genome of the second Pb normally undergoes only partial rather than complete replication of its DNA (Howlett and Bolton, 1985), but has nonetheless been found by nuclear transplantation to be competent to support normal development to term (Wakayama *et al.*, 1997). Although the second Pb can affect the autonomous cortical activity of anucleate fragments of mouse zygotes (Waksmundzka *et al.*, 1984), there is no evidence to suggest that it plays any role in early development normally.

B. Cleavage and Blastulation

First cleavage is said to be meridional in the mouse (Howlett and Bolton, 1985), as in other eutherian mammals. However, departures from this orien-

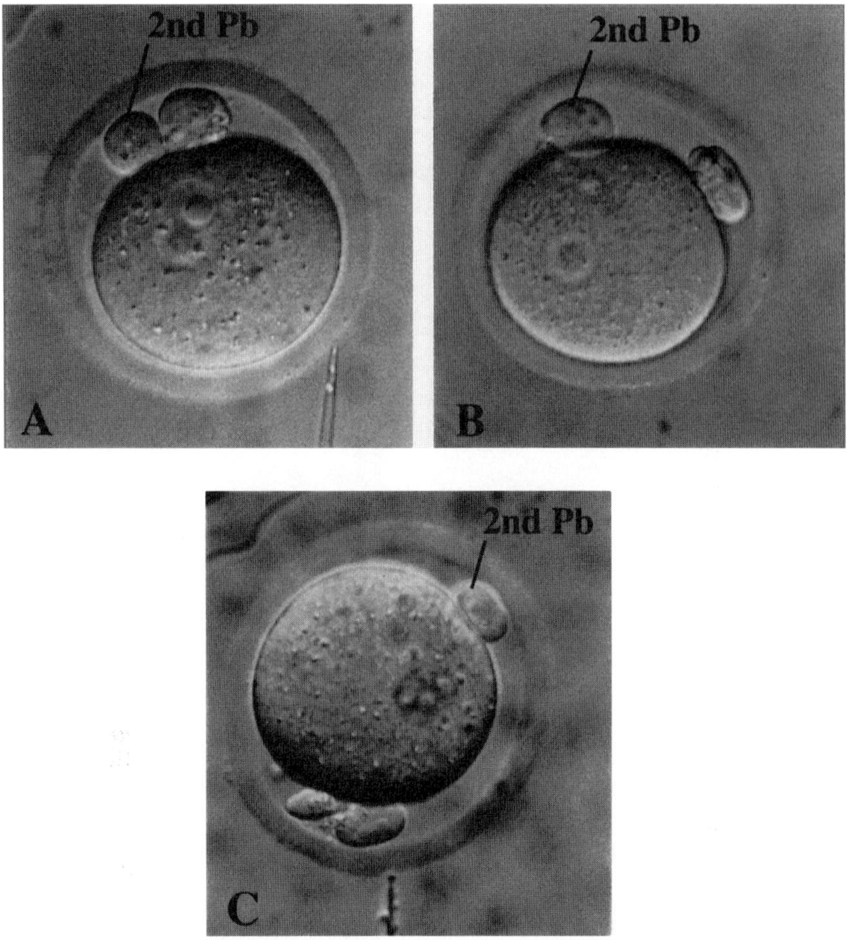

FIG. 3 If still intact after fertilization, the first Pb usually lies adjacent to second, as in (A). That this is not always the case implies that the A-V axis may undergo moderate (B) or extensive (C) reorientation during the period between completion of the two meiotic divisions. The first polar body typically differs from the second in lacking a nuclear membrane and in embarking on cytokinesis, which, as shown in (C), may go to completion (Longo, 1987).

tation are not uncommon and, when relatively modest, tend to be masked by displacement of the second Pb into the interblastomeric groove. This is evident from examination of early two-cell conceptuses that have been divested of the zona pellicuda either before or shortly after first cleavage (Fig. 4). Nevertheless, cases where the plane of cytokinesis is essentially equatorial rather than meridional (Evsikov *et al.,* 1994) are sufficiently unusual that it has yet to be established whether this is compatible with

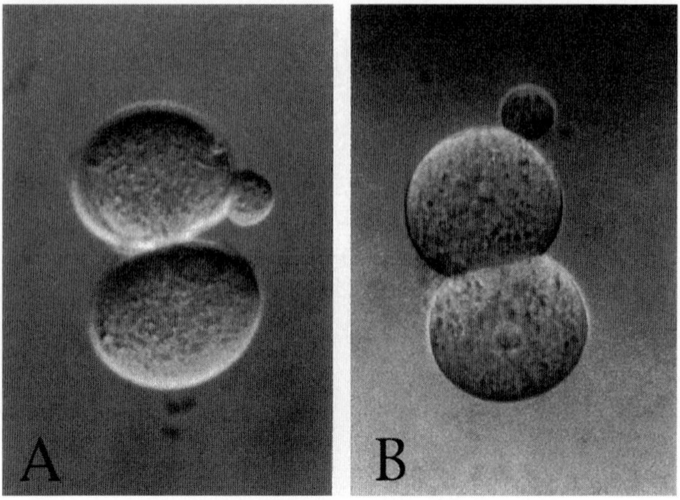

FIG. 4 Mouse conceptuses that were divested of the zona pellucida shortly after first cleavage, showing modest (A) versus very marked (B) departure of cytokinesis from a meridional plane.

normal development. Although the form of the early four-cell stage is rather variable, a more or less regular tetrahedral arrangement of blastomeres occurs most commonly, both in the mouse and in other mammals. How this arises has only been examined closely in the rabbit where second cleavage is evidently meridional for both blastomeres, but the second one to divide undergoes rotation through almost 90° either before or during cytokinesis (Gulyas, 1975). Whether this is also true for the mouse has yet to be investigated by critical time-lapse analysis. What has been demonstrated in the mouse is that in the absence of the zona pellucida the number of contacts between blastomeres is typically reduced at the early four-cell stage, and this may affect postimplantation development adversely (Graham and Deussen, 1978; Suzuki *et al.,* 1995).

Although all external blastomeres are polarized from the late eight-cell stage onward (Johnson and Ziomek, 1981; Reeve, 1981), the conceptus as a whole does not exhibit obvious intrinsic polarity until blastocyst formation begins with the localized accumulation of extracellular fluid between some of its inner and outer cells. This is generally regarded as the first stage at which the conceptus departs from spherical symmetry. Because the blastocoele forms eccentrically, attachment between inner and outer cells only persists opposite it, resulting in the generation of an unambiguous axis of polarity that is known as the embryonic-abembryonic (Em.Ab) axis. The embryonic pole of this axis consists of the consolidated inner cell mass (ICM) and the immediately overlying outer cells, the polar trophectoderm, and the abembryonic pole of cells of the remaining, mural trophectoderm that encloses the blastocoele (Fig. 1). Although the entire outer cell layer has differentiated as a trophectodermal epithelium by the stage when the conceptus begins to cavitate, there is no evidence to suggest that its polar and mural regions are distinct at this juncture. Mural cells, whose fate is to differentiate as polytene giant trophoblasts, can clearly substitute for polar cells, which, through retaining a diploid proliferative status, normally serve as the antecedents of the entire population of diverse trophoblast cells that are produced following implantation.

Growth of the blastocyst is accompanied by a progressive decline in cell proliferation in the mural relative to the polar trophectoderm. Nevertheless, polar cell number increases very little compared with mural because there is a net flow of trophectoderm cells from the former to the latter region (Copp, 1979; Cruz and Pedersen, 1985). Hence, surplus polar cells must adopt a mural fate as, indeed, do all polar cells when denied contact with viable ICM tissue (Gardner and Beddington, 1988; Nichols *et al.,* 1998). The default pathway of trophectoderm differentiation is evidently to form giant cells via polyteny (Brower, 1987; Varmuza *et al.,* 1988; Keighren and West, 1993). Nevertheless, as shown by blastocyst reconstitution experiments, mural trophectoderm cells do not become committed to endoreduplicating their genome until blastocyst growth is relatively advanced (Gardner *et al.,* 1973; Papaioannou, 1982). By this stage mural cells clearly differ from polar in various surface properties (Carollo and Weitlauf, 1981; Chavez *et al.,* 1984; Lehtonen and Reima, 1986; Yamagata and Yamazaki, 1991; Paria *et al.,* 1995), which presumably reflect their role in mediating attachment of the blastocyst to the uterine epithelium during implantation (Mehrotra, 1984). However, whether synthesis and secretion of a zona lysin is also a specific property of mouse mural trophectoderm (Perona and Wassarman, 1986) is more contentious (Yamazaki and Kato, 1989). The situation in the human is clearly different from that in the mouse inasmuch that both attachment and loss of mitotic potential first occur in the polar rather than the mural region (Boyd and Hamilton, 1970).

The Em.Ab axis of the mouse blastocyst is clearly of morphogenetic significance in terms of localizing the future placenta and in ensuring that it is oriented correctly with respect to the uterine blood supply during implantation (Kirby *et al.,* 1967; Smith, 1980). Although this axis also coincides with the dorso-ventral axis of the future fetus, whether the ICM is already polarized with respect to it is unknown, but seems unlikely. By the late blastocyst stage, however, the abembryonic part of the ICM adjacent to the blastocoele has differentiated into primitive endoderm and its embryonic part, though remaining unspecialized morphologically, has become restricted in fate as the epiblast (Gardner and Rossant, 1979).

C. The Em.Ab Axis

At present, how and when the Em.Ab axis is specified are questions that remain open despite various attempts to address them. The earliest step in the process of blastocoele formation to be detected so far is a colcemid-inhibitable focal clustering of refractile vesicles that are abundant within both internal and external cells of the late morula (Wiley and Eglitis, 1980). This clustering is normally followed by the appearance of one or more refractile furrows or clefts between outer and inner cells. Intracellular vesicles are already present in the zygote and increase in abundance until the four-cell stage, whereafter they appear to become larger but less numerous, consistent with their amalgamation by fusion (Calarco and Brown, 1969). The involvement of secondary lysosomes in the process of forming an extracellular cavity has been inferred from the inhibitory effect of chloroquine on this process. However, there is no consensus as to whether initial cavitation starts in a few external cells (Aziz and Alexandre, 1991), or involves the participation of both external and internal cells (Wiley and Eglitis, 1980).

The timing of blastulation has been shown not to depend on total cell number or on the number of cleavage divisions completed since fertilization (Smith and McLaren, 1977; Surani *et al.,* 1980; Eglitis and Wiley, 1981; O'Brian *et al.,* 1984). The number of DNA replications has also been discounted (Alexandre, 1979; Dean and Rossant, 1984). However, studies on asynchronous combinations of blastomeres indicate that timing of the appearance of the blastocoele can be affected by cell interaction (Prather and First, 1986). Finally, the notion that nucleocytoplasmic ratio is critical has been questioned in favor of a clock that operates according to biological rather than chronological time (Alexandre, 1979; Evsikov *et al.,* 1990). That genetic factors are nevertheless involved is suggested by the finding that the timing of blastocoele formation is advanced in aggregation chimaeras constructed from particular genotypic combinations (Evsikov and Solomko, 1998).

Two specific models for specification of the orientation as opposed to the time of formation of the the Em.Ab axis have been advanced, both of which depend on the considerable temporal heterogeneity in the cycles of blastomeres that arises once cleavage is advanced (Barlow *et al.,* 1972; Graham and Deussen, 1978; Chisholm *et al.,* 1985). One is that competence to form a blastocoele is acquired by blastomeres only when they reach a particular cycle, so that siting of this cavity will depend on where the most advanced cells happen to be. However, while a tendency for early dividing cells to be associated with the nascent blastocoele has been reported, other factors had to be invoked to explain why the association was rather marginal (Garbutt *et al.,* 1987). According to the other model, the outer blastomeres that are relatively late dividing during the transition from the 16- to the 32-cell stage tend to become stretched as a consequence of the accumulation of internal cells. Such stretching is held to reduce the likelihood of outer blastomeres undergoing cytokinesis and thus increases the chance of their becoming binucleate (Surani and Barton, 1984). This model presupposes that binucleation is an initial step toward the terminal differentiation that mural trophectoderm cells undergo in forming primary trophoblastic giant cells. However, binucleation has only been reported in minority of morulae in two strains of mice (Soltynska *et al.,* 1985). Were this a normal feature of early development in the mouse, it is most unlikely to have been overlooked in other strains. Moreover, even in blastocysts its occurrence is sporadic and seems to be associated particularly with shifts between incubation and room temperature, thus suggesting that under suboptimal conditions cytokinesis is more readily disrupted than mitosis (author's unpublished observations). Furthermore, not only are primary trophoblastic giant cells typically mononucleate (Gardner and Johnson, 1972), but *in situ* hybridization studies provide no evidence to suggest that they embark on polyteny via a tetraploid state (Brower, 1987; Varmuza *et al.,* 1988; Keighren and West, 1993).

Hence, it is still far from clear what the critical initial step in establishment of the Em.Ab axis is. Efforts have been directed toward localizing cells that are responsible for the accumulation and secretion of blastocoelic fluid (Wiley and Eglitis, 1980; Aziz and Alexandre, 1991), on the assumption that where the blastocoele forms is dictated by the siting of such cells. Yet to be discounted, however, is the possibility that it is not the capacity to secrete fluid that is localized, but where contacts between inner and outer cells can readily be broken so as to permit fluid to accumulate extracellularly. In other words, whether the locus of the blastocoele determines that of the ICM or whether the locus of the ICM determines that of the blastocoele has yet to be resolved. Recent findings discussed below raise doubt about two assumptions relating to the Em.Ab axis. The first is whether this really is the first axis to be established following the onset of cleavage, and the second is whether its location depends on essentially stochastic processes operating during this initial phase of development.

D. Bilateral Symmetry

The first relevant findings discredit the notion that the mouse conceptus remains radially symmetrical about its Em.Ab axis until the PS forms at the onset of gastrulation. Early this century, Huber (1915) noted that rat blastocysts *in utero* appeared oval rather than circular when sectioned orthogonally to their Em.Ab axis. Smith (1980) presented more detailed histological evidence that the mouse blastocyst was bilaterally symmetrical before it implants, and also claimed the two ends of the bilateral axis could be differentiated—i.e., that the axis was polarized. Smith's observations were subsequently confirmed on living advanced blastocysts (Gardner, 1990), thereby enabling the possibility to be discounted that, rather than being intrinsic to the blastocyst, asymmetry was either an artefact of histology or imposed on the blastocyst by the uterus. Subsequently, early blastocysts were also found consistently to be oval rather than round when viewed along their Em.Ab axis in the living state, regardless of whether the zona pellucida was present or not. But at this stage, the bilateral axis did not exhibit any of the elements of polarity that characterize it later (Gardner, 1997).

In the nearly two-thirds of early blastocysts where a Pb, almost invariably the second, had survived, it was not only typically located at the junction between the polar and mural trophectoderm but aligned with one end of the axis of bilateral symmetry (Fig. 1). Such a circumscribed location of surviving Pbs was hard to reconcile with earlier claims based on time-lapse recordings that these bodies were motile (e.g., Lewis and Wright, 1935; Borghese and Cassini, 1963), and attempts to detect their movement yielded negative results (Gardner, 1997). The nature and properties of its linkage to the blastocyst were consistent with the second Pb remaining attached to the conceptus via the intercellular bridge formed during the cytokinetic phase of second meiosis. It was therefore concluded that for as long as it survives, the second Pb is anchored to the conceptus at its site of abstriction and therefore provides an enduring marker of the animal pole of the zygote (Gardner, 1997). This implies that the early blastocyst's bilateral axis is aligned with, and its Em.Ab axis is orthogonal to, the AV axis of the zygote (see Fig. 1). Hence, if specification of the Em.Ab axis of the blastocyst does depend on differences in the cleavage rates of blastomeres then, at least in undisturbed development, these would have to be rooted in regional organization of the zygote.

A further question raised by these findings is which of the two axes, the Em.Ab or that of bilateral symmetry, is established earlier in development. The first signs of appearance of the Em.Ab axis are not evident until the relatively late morula stage (Wiley and Eglitis, 1980; Aziz and Alexandre, 1991). What about the axis of bilateral symmetry? Huber (1915) noted that cleavage stage rat conceptuses tended to be somewhat flattened rather than spherical when fixed and sectioned *in vivo,* but considered that this might be imposed on them by maternal tissue rather than reflect their intrinsic shape. However, explanted mouse conceptuses not only tend to exhibit a greater and a lesser diameter as early as the four-cell stage, but to have surviving Pbs aligned with their greater diameter (author's unpublished observations). Furthermore, marking experiments that are in progress suggest that the bilateral axis and the Em.Ab axis are both already specified at the beginning of cleavage (author's unpublished observations). If confirmed, the question of how the consistent relationship between the two axes (Fig. 1) is established needs to be addressed.

As noted elsewhere (Gardner, 1999a), the axis of bilateral symmetry could depend on the orientation of first cleavage, which is typically meridional (Howlett and Bolton, 1985), rather than the AV axis of the zygote. Determining whether or not the second Pb marks the bilateral axis in blastocysts obtained from zygotes induced to undergo equatorial rather than meridional first cleavage (Zernicka-Goetz, 1998) would seem to offer a way of distinguishing between these two possibilities. Of course, in the absence of any discernible departure from radial symmetry of the zygote about its A-V axis, it is impossible to say whether the meridional plane of first cleavage is fixed or variable with respect to any underlying cytoplasmic organization that might be of developmental significance later on. In this regard, it should be borne in mind that the orientation of first cleavage was held to be variable with respect to an axis of bilateral symmetry that was defined principally by histochemical means in rodent zygotes (Jones-Seaton, 1950; Dalcq, 1957).

Rather than being intrinsic to the egg or zygote, the necessary asymmetry for establishing the orientation of either axis of the blastocyst with respect to the meridion of the zygote could be imposed from without, for example, by the site of sperm penetration. Although it is claimed that the fertilizing sperm can attach to the surface of the oocyte anywhere except, in the case of the mouse, to the microvillus-free region around the animal pole (Talansky *et al.*, 1991), the case that it is otherwise random has not been substantiated. Moreover, whether orientation of the meridional first cleavage plane bears any consistent relationship to the site of attachment of the fertilizing sperm also remains unexplored. Available indications from using sperm to which a fluorochrome has been coupled covalently (Gabel *et al.*, 1979), or where entry of the tail into the vitellus had been prevented (Bennett, 1982), are that the site of sperm attachment does not coincide with the plane of first cleavage. The possibility that this plane, or that of the future bilateral axis of the conceptus, is nevertheless fixed in relation to the site of sperm entry obviously cannot be discounted thereby.

IV. Significance of Bilateral Symmetry of the Early Conceptus

A. Relationship between Bilateral Axis of Conceptus and Fetus

Regardless of its mode of origin, the Em.Ab axis of the blastocyst has obvious relevance developmentally since, as noted earlier, both the trophectoderm and ICM undergo regional differentiation with respect to it. The morphogenetic significance of the early origin of bilateral symmetry appears at first sight to be rather less easy to explain. Smith (1980, 1985) argued that the set of three axes established by the late blastocyst stage—anterior-posterior (A-P), dorso-ventral and, by extrapolation, left-right—are conserved through gastrulation and thereby serve to specify the corresponding axes of the fetus. The case for conservation of the conceptus's set of axes was based on the identification in histologically sectioned material of compatible or consistent asymmetries at successive stages. Although it seems very likely that these asymmetries are conserved because they affect the structure of the late blastocyst and subsequent stages so profoundly, direct evidence that this is indeed the case is still lacking. What is more contentious, however, is the question of the relationship of these axes to those of the nascent fetus. This was reexamined in a study in which the posterior end of the A-P axis was marked indelibly in seventh dpc conceptuses after they had been divested of uterine tissue so that its location in relation to the site of the PS could be determined in specimens with an optimal plane of sectioning. The relationship between the two A-P axes was found to be nonrandom and fully consistent with their sharing a common orientation. However, contrary to the findings of Smith (1985), the axes were as often of the opposite as the same polarity (Gardner *et al.*, 1992). A curious further finding was that the posterior of the conceptus tended to be displaced somewhat to the left of the projected A-P axis of the fetus regardless of whether it was by the PS or opposite it. Although this skewing was modest; with a mean of only 4% or 3.6°, it was significant for all conceptuses combined, as also for the set of approximately half of them in which the polarity of the two A-P axis was opposed (see Fig. 5). While it suggests that one axis may be curved relative to the other, it is far from clear what the morphogenetic consequence of such curvature might be, particularly since it would be of opposite handedness where the two axes were antiparallel as opposed to parallel (Gardner *et al.,* 1992).

B. Bilaterality and Trophectoderm Growth

The possible morphogenetic significance of the bilateral symmetry of the early blastocyst in the shorter term has emerged from a clonal analysis of growth of the polar trophectoderm. The findings suggested that the flow of surplus polar cells into the mural trophectoderm that occurs during growth of the blastocyst (Copp, 1979; Cruz and Pedersen, 1985) was restricted to only part rather than all of the junction between the two regions (Gardner, 1996b). This has been confirmed in further experiments in which readily endocytosed fluorescent microspheres (Fleming and George, 1987)

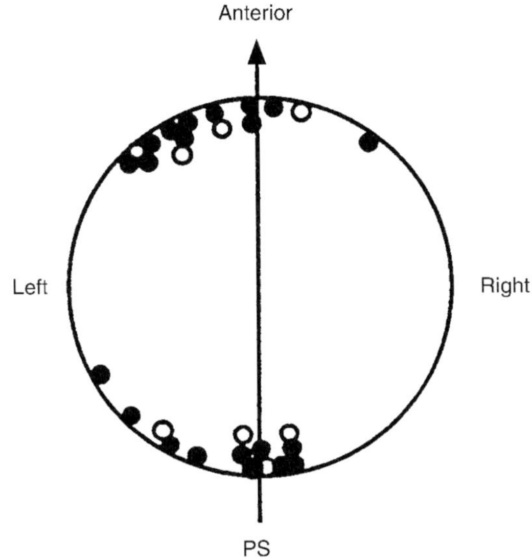

FIG. 5 Relationship between A-P axis of conceptus and fetus early in gastrulation. Each small circle represents a single conceptus. In some the location of the conceptus's posterior relative to the fetal A-P axis was based on a single estimate (filled circles) and in others on the mean of two estimates (open circles). Note the leftward bias, regardless of whether the two axes are parallel or antiparallel. Reproduced from Gardner *et al.* (1992) with permission of Wiley-Liss, Inc., a subsidiary of John Wiley and Sons, Inc.

were used to label the entire polar trophectoderm selectively. Spread of polar cells into the mural trophectoderm typically took the form of a single coherent patch, which varied in extent both circumferentially and in how far it extended toward the abembryonic pole (Gardner, 2000a). Elsewhere, the label was either wholly confined to the polar trophectoderm or had spread at most to the immediately adjacent mural cells. It is intriguing that in the exceptional cases where there were two foci of label in the mural trophectoderm rather than just one, these were diametrically opposite each other (Gardner, 2000a). This finding suggested that the orientation of the polar to mural flow of trophectoderm cells might be dictated by the bilateral axis of the early blastocyst. The results of additional experiments that were designed to investigate this possibility have provided results that are consistent with this being the case (Gardner, 2000b). Hence, the bilateral axis of the conceptus appears to be conserved between the early and late blastocyst stage. Accordingly, the direction of flow of polar cells may serve to confer the polarity that the bilateral axis exhibits in the advanced blastocyst, as was postulated in an earlier review (Gardner, 1998). An essential feature of this proposal is that the longer trophectoderm cells have resided in the mural region, the less deformable they become, possibly because of the basal lamina components that are deposited on their blastocoelic surface (Van Blerkom *et al.,* 1979; Wartiovaara *et al.,* 1979; Carnegie, 1991). In consequence, the increase in hydrostatic pressure in the blastocoele caused by activity of the "sodium pump" (Manejwala and Schultz, 1989; Fleming, 1992; Jones *et al.,* 1997) will stretch disproportionately those trophectoderm cells that have entered the mural region more recently. The effect of these new mural cells being together in a coherent patch will be tilting away from the abembryonic pole of the blastocyst of the ICM/PT complex on the side on which they lie (Gardner, 1998), thus establishing what Smith (1980, 1985) defines as the anterior end of the bilateral axis.

One further point to emerge from the study of the direction of polar to mural cell movement is that it was as often directed away from as toward the Pb in the minority of cases where the latter persisted throughout post-labelling culture (Gardner, 2000b). Hence, while the polar to mural flow of cells appears to depend on the A-V axis of the zygote (or the plane of first cleavage) for its orientation, its polarity is evidently not specified until later. As yet, there are no clues as to how or when this occurs. As far as the question of how information for orienting axes might be perpetuated, the possibility that it resides in the extracellular matrix cannot be ignored (Jost, 1992).

In summary, bilateral symmetry may be established very early in development, probably well before the blastocyst stage. It evidently has consequences for morphogenesis, at least in directing the flow of cells from polar to mural trophectoderm during growth of the blastocyst. It may thus explain the tilting of the ICM/PT complex in the late blastocyst which almost certainly accounts for the corresponding tilt of the proximal relative to the distal region in early post-implantation conceptuses. Furthermore, the orientation of the fetal A-P axis accords with the axis defined by this tilt, implying the existence of a link between the two. However, there is at present no evidence to suggest that morphogenesis of the trophectoderm plays a direct role in specifying the orientation of the fetal A-P axis. As will be apparent from studies discussed in the next section that entail working backwards from early gastrulation, it is the extraembryonic endoderm rather than the trophectoderm that has been implicated in this axiation process.

V. Gene Expression and Molecular Asymmetries before Gastrulation

A. The Transition from Primitive to Visceral Endoderm

As discussed below, there is now compelling evidence that cells derived from the primitive endoderm play a vital role in patterning the early embryo in the mouse. However, since there is some uncertainty about the developmental status of these cells during the period of development with which this review is concerned, this issue needs to be considered briefly at this juncture. Part of the primitive endoderm that continues to be associated with the ICM encases the epiblast until the beginning of gastrulation and is gradually replaced thereafter by the definitive embryonic endoderm that originates from the epiblast (Lawson *et al.,* 1986, 1987, 1991). Hence this derivative of the primitive endoderm makes no enduring cellular contribution to the fetus (Gardner and Rossant, 1979). Before gastrulation, most if not all its cells retain the potential to form parietal endoderm (PE) when cloned by blastocyst injection (Cockroft and Gardner, 1987). Furthermore, they do not display unequivocal early markers of visceral endodermal (VE) differentiation such as apical localization of villin (Maunoury *et al.,* 1988; Ezzell *et al.,* 1989), synthesis of AFP (Dziadek and Adamson, 1978), or the presence of binding sites for *Dolichos biflorus* agglutinin (Sato and Muramatsu, 1985). Whether these cells have undergone any differentiation beyond the primitive endoderm stage is not possible to say in the absence of any molecular markers that are specific to the latter tissue. Hence, prior to gastrulation, it would seem most appropriate to refer to the primitive endoderm-derived cells that continue to invest the egg-cylinder, instead of migrating over the inner surface of the mural trophectoderm, as nascent visceral endoderm (nVE).

The primitive endoderm or nVE is indispensible for development of the definitive embryo or fetus. This is evident from the fact that isolated early epiblast typically degenerates when it is deprived of its investing endoderm (Hogan and Tilly, 1977; Gardner, 1985), unless it is provided with appropriate growth factors or feeder cells—i.e., conditions that are employed for deriving ES cells from it (Brook and Gardner, 1997). That this is also true for the epiblast in the intact conceptus is evident from the effects of disrupting certain genes that are expressed in the primitive endoderm or nVE (see below). The nVE or VE has been implicated in controlling the differentiation as well as survival of the epiblast, largely from studies on embryoid bodies obtained from embryonal carcinoma (EC) or embryonic stem (ES) cells. Recent tissue recombination experiments have shown that this tissue is also effective in inducing freshly isolated epiblast to engage in hematopoiesis *in vitro* (Belaoussoff *et al.,* 1998).

B. Molecular Studies Relating to Specification of the A-P Axis of the Fetus

1. Genes Expressed in the Extraembryonic Endoderm

Differential expression of a growing list of genes encoding diverse proteins has been documented for this cell lineage. Although expression of some of these genes occurs throughout both derivatives of the primitive endoderm, that of others is spatially restricted to either the parietal or the visceral layer. Among the genes that are expressed very early in the primitive endodermal lineage are those that appear to be involved in its differentiation. They include genes encoding members of the GATA class of transcription factors. These are a family of zinc finger-containing proteins that recognize a consensus DNA sequence known as the GATA motif, which is a vital regulatory element in the promoters and enhancers of a variety of genes. Members of this family have been implicated in endoderm differentiation in a diverse range of organisms extending from *Caenorhabditis elegans* through to vertebrates. Those expressed very early in the primitive endodermal lineage include GATA-6 and GATA-4. By using LacZ as a reporter gene in mutant heterozygotes, GATA-6 expression has first been detected in part of the ICM and mural trophectoderm of 3.5 dpc blastocysts, and in cells of the primitive endoderm of 4.5 dpc implanting blastocysts. Following implantation, enzyme activity was restricted to the PE at the start of gastrulation but subsequently appeared in the nascent embryo proper, both in the developing heart and its precursor tissues, as well as in the gut (Koutsourakis *et al.,* 1999). Conceptuses that are homozygous for disruption of the gene evidently cease developing before gastrulation. Although both layers of the extraembryonic endoderm form, the part of the visceral layer overlying the epiblast is deficient and the embryonic region is markedly reduced in size, and by 7.0 dpc the epiblast exhibits enhanced apopotosis (Morrisey *et al.,* 1998; Koutsourakis *et al.,* 1999). The pattern of cell death within the epiblast resembled that found in an earlier study in conceptuses that were homozygous for disruption of HNF-4 (Chen *et al.,* 1994), whose expression can be transactivated by GATA-6 in NIH-3T3 cells (Morrisey *et al.,* 1998). Embryoid bodies formed by GATA-6 −/− ES cells fail to undergo VE differentiation *in vitro,* as judged by both morphological and molecular criteria (Morrisey *et al.,* 1998). Although −/− ES cells injected into +/+ blastocysts were able to contribute to both the cardiomyocyte population and gut, the indispensibility for development of expression of the gene in extraembryonic tissues was confirmed by the failure to obtain any postimplantation conceptuses in reciprocal blastocyst injection experiments (Koutsourakis *et al.,* 1999).

In normal development, a low level of GATA-4 expression was detected at 5.5 dpc exclusively within the PE and VE (Morrisey *et al.,* 1998). It was reduced but not abrogated altogether in GATA-6 null conceptuses (Morrisey *et al.,* 1998). ES cells that are homozygous for disruption of GATA-4 show failure of VE type differentiation *in vitro,* according to both morphological and molecular indices (Soudais *et al.,* 1995). However, since the conditions of culture employed were not conducive for the differentation of parietal endoderm even in wild type embryoid bodies, it is not clear at what stage extraembryonic endoderm differentiation is blocked *in vitro* in the absence of GATA-4. Investigation of the effects of homozygosity for disruption of GATA-4 on development *in vivo* has yielded rather variable results; only about one-third of −/− conceptuses arrest prior to gastrulation. These failed to show expression of AFP. The demise of the remaining −/− conceptuses that progressed to early somite stages was presumably due to failure of proper heart development (Molkentin *et al.,* 1997). That it is extraembryonic endodermal rather than mesodermal expression of the gene which is crucial for morphogenesis of the heart, was demonstrated by injection of −/− ES cells into wild type eight-cell stage host conceptuses that were positive for the ROSA26 marker transgene (Narita *et al.,* 1997). Hence, although extraembryonic expression of GATA-6, presumably in the primitive endoderm, is indispensible for initial differentiation of the VE and its consequent support of the epiblast, the requirement for GATA-4 activity in this context seems less clear.

Very early postimplantation lethality is seen in conceptuses that are homozygous for disruption of a third gene, *Evx-1,* whose transcription was found initially to be restricted to the nVE at 5.0 dpc. Somewhat later, but before the onset of gastrulation, transcripts seem to be confined to the future posterior region of the embryo, both in the nVE and, in a more restricted domain, in the epiblast (Dush and Martin, 1992). However, since in −/− conceptuses the viability of the trophectoderm, which clearly does not depend on the extraembryonic endoderm for its survival (see Gardner, 1998), is also compromised (Spyropoulos and Capecchi, 1994), the gene must also be expressed in the trophectoderm lineage.

There are a two further genes expressed principally or exclusively throughout the nVE or early VE whose disruption has been found to perturb subsequent development of the epiblast. Both seem to be required for realization of the differentiated state of the VE rather than its initial establishment. These are an uncharacterized gene identified through its disruption by the *Hβ58* transgene and the gene encoding the HNF-4 transcription factor. Conceptuses that are homozygous for the *Hβ58* transgene insertion show gross perturbation of the development of the epiblast during gastrulation without obvious impairment of the subsequent differentiation of the visceral endoderm itself (Radice *et al.,* 1991; Lee *et al.,* 1992). Those homozygous for disruption of the *HNF-4* gene, which is expressed early in the extraembryonic endoderm (Duncan *et al.,* 1994), failed to progress beyond early gastrulation (Chen *et al.,* 1994), evidently because the VE does not express at an appropriate level several genes that are markers of its advanced differentiation (Duncan *et al.,* 1997). The *Stat 3* gene, which encodes a protein involved in signalling and transcriptional activation in the cytokine pathway, is expressed specifically in nVE from 6 dpc, and conceptuses that are homozygous for its disruption rapidly degenerate shortly after the start of gastrulation (Takeda *et al.,* 1997). A further gene that is expressed in the embryonic nVE is that encoding the TGFβ2 growth factor. This protein is already detectable in preimplantation stages, mainly in the trophectoderm of the blastocyst. In early postimplantation stages it is confined to the nVE, where its distribution seems to be both patchy and variable until after the start of gastrulation, when it becomes more uniform (Slager *et al.,* 1991).

A further set of genes is expressed in only part of the nVE or VE during its early differentiation. Three of these, plus an entity that has so far been defined only by its specific reactivity with a polyclonal antiserum, are of particular interest in the present context insofar as their asymmetric expression in the tissue unquestionably occurs well before gastrulation begins. Transcripts of the first three of the genes, a divergent homeobox-containing gene, *Hex* (Thomas *et al.,* 1997, 1998), also known as *Prd,* were detected first in the primitive endoderm of the implanting blastocyst and subsequently in nVE cells at the distal tip of the early egg cylinder. Between the latter stage and gastrulation, the transcripts were found to be localized progressively more asymmetrically toward one side of the egg cylinder and were eventually confined to a limited focus of cells lying diametrically opposite the PS by the onset of gastrulation (Thomas *et al.,* 1997, 1998). It should be noted, however, that a second group failed to detect any signal for *Hex* expression by *in situ* hybridization before 7.5 dpc, when transcripts were detected only in the chorion and VE, and were more abundant in the former than in the latter tissue (Keng *et al.,* 1998). This group differed from Thomas *et al.,* (1998) in using a probe prepared from the 3′ untranslated region rather than the coding region of the gene so as to avoid the risk of including sequences bearing a close similarity to those of other homeobox-containing genes.

The second gene that is expressed locally in the nVE before gastrulation shares modest homology with *Cerberus* in *Xenopus* (Boummeester *et al,* 1996). It has variously been termed Cerberus-like (*Cerl*) (Belo *et al.,* 1997), mouse cerberus (*mCer*) (Biben *et al.,* 1998), or Cerberus-related (*Cerr*) (Shawlot *et al.,* 1998) by its discoverers. Although its anterior expression in the tissue is clearly evident by the onset of gastrulation, Belo *et al.* (1997) found that transcripts were already localized asymmetrically approximately one day earlier. In addition, asymmetric distribution in the nVE before gastrulation has been demonstrated for an uncharacterized antigen recognized by the VE 1 antibody (Rosenquist and Martin, 1995). Since both *Cerl* transcripts and this VE 1-reactive antigen are localized anteriorly once the PS forms, they, like *Hex,* are assumed to be anterior before gastrulation. However, in contrast to *Hex,* neither has been found to have an initially symmetrical location distally in the tissue, though Cerl transcripts appear to be more proximal at than before the onset of gastrulation (Belo *et al.*, 1997). The third gene, *Mrg1,* is expressed in the nVE with a higher level of transcripts in a ring in the proximal embryonic region within which the signal peaks at a site that is found to be located most rostrally once gastrulation begins. This localized expression persists through gastrulation, but transcripts appear later also in the presumptive cardiac primordia (Dunwoodie *et al.,* 1998).

In the case of other genes with asymmetrical expression in the VE, transcripts have not been detected unequivocally before gastrulation. The first of these genes is goosecoid (*gsc*) (Blum *et al.,* 1992), whose disruption is without discernible effect on gastrulation but causes perturbation of craniofacial and rib development (Rivera-Perez *et al.,* 1995; Yamada *et al.,* 1995). The second is *Hesx1,* whose disruption has serious but variable effects on development of the forebrain which are often disparate laterally (Dattani *et al.,* 1998). The third and fourth of these genes are *Lim1* and *HNF-3β*. Disruption of both produces severe anterior truncations, though with rather variable penetrance in the case of *HNF-3β* (Ang and Rossant, 1994; Weinstein *et al.,* 1994; Shawlot and Behringer, 1995). Expression of *HNF-3β* in the VE has been found to be necessary for subsequent elongation of the PS rather than its initial formation (Dufort *et al.,* 1998).

2. Genes Expressed in Early Epiblast

One of the genes expressed locally in epiblast before the beginning of gastrulation is *Brachyury.* Its transcripts have been detected as a ring in the most proximal region of the tissue at 5.0 dpc (Thomas and Beddington, 1996). Another is *Cripto,* whose product was originally identified as a growth factor secreted by human NTERA2 embryonal carcinoma cells (Ciccodicola *et al.,* 1989). At first, this gene was assigned the epidermal growth factor family but on subsequent discovery of the related gene, *Cryptic,* was placed in a distinct family (Shen *et al.,* 1997). *Cripto* mRNA has been detected in the preimplantation conceptus (Johnson *et al.,* 1994). Postimplantation, its transcripts are initially distributed uniformly in epiblast but later become graded with the highest concentration proximally, and are localized to the PS by onset of gastrulation (Dono *et al.,* 1993; Ding *et al.,* 1998). The effects of disrupting *Cripto* have been interpreted somewhat differently by the two groups who have targeted this gene. Hence, though Ding *et al.* (1998) regard the homozygous mutant phenotype as primarily due to failure of the A-P axis of the fetus to rotate from initial alignment with Em.Abemb axis of egg cylinder (i.e., the DV axis of the fetus), Xu *et al.* (1999) are of view that the fetal A-P axis fails to form, and place particular emphasis on altered adhesiveness and migration of mutant cells. Both groups agree, however, that abundant but disorganized neuroepithelium forms despite the absence of all except the extraembryonic component of the mesoderm.

It is noteworthy that there is unusual sequence conservation between mouse and human in part of the 3′ UTR in addition to the coding region of *Cripto* (Dono *et al.,* 1993), suggesting that its expression may be subject to posttranscriptional regulation. However, the human gene differs from its murine counterpart in lacking a putative signal peptide (Dono *et al.,* 1993).

A third gene, whose transcripts are present in a circumferentially restricted distribution in epiblast, is *Fgf8.* Expression in this case also is assumed to be posterior, since the PS is positive once it differentiates. Transcripts for *Fgf8* were also reported to be present in the overlying nVE at this stage, though details of their distribution in this tissue were not provided (Crossley and Martin, 1995). *Fgf4* transcripts occur throughout the early postimplantation epiblast but are downregulated in all areas except the future posterior by the stage when the PS has formed (Niswander and Martin, 1992). A fifth gene, *Evx-1,* a murine relative of the *Drosophila even-skipped* gene, is also expressed in putative posterior epiblast, but in conjunction with the adjacent plus more distal nVE (Dush and Martin, 1992). However, the description of the pattern of early expression of this gene is evidently only partial, since, as noted earlier, its homozygous disruption causes very early postimplantation demise of trophectoderm as well as ICM derivatives of the blastocyst.

3. Genes Expressed in both nVE and Epiblast

One such gene, *Nodal,* is expressed transiently in both the nVE and epiblast before gastrulation with, according to use of LacZ as a reporter, highest levels in the posterior quadrant distal to embryonic-extraembryonic junction. Conceptuses that are homozygous for its disruption exhibit both impaired gastrulation and later rostral truncation. The presence of wild-type cells in the fetus alleviates the former but not the latter condition, which requires such cells in the extraembryo and hence presumably the VE (Iannaccone *et al.,* 1992; Conlon *et al.,* 1994; Varlet *et al.,* 1997). Transcripts of a second gene, *Otx2,* occur throughout the epiblast prior to gastrulation (Simeone *et al.,* 1993; Ang *et al.,* 1996) and subsequently in the nVE as well (Acampora *et al.,* 1995; Rhinn *et al.,* 1998). However, using LacZ as a reporter, expression was found to be modest and mosaic or patchy in the epiblast before gastrulation and showed progressive anteriorization in the VE during gastrulation. In conceptuses that are homozygous for disruption of the gene expression of the reporter remained largely confined to the VE and, instead of becoming anteriorized, continued to be restricted to the distal third to one-half of the tissue. The main feature of mutant homozygotes is failure of development of the fore- and midbrain and the most rostral part of the hindbrain. As with *Nodal,* rostral truncations are partially rescued by the presence of wild-type extraembryonic tissues, but detailed patterning of the brain is not restored thereby (Rhinn *et al.*, 1998). Mutant heterozygotes for *Otx2* also show variably penetrant defects that are more marked on some genetic backgrounds than others (Acampora *et al.*, 1995; Matsuo *et al.*, 1995; Ang *et al.*, 1996).

4. Genes Expressed in the Extraembryonic Ectoderm

Bmp4 RNA has recently been detected in both the ICM and trophectoderm of the 3.5 dpc blastocyst, and in the polar trophectoderm and epiblast of the 4.5 dpc blastocyst (Coucouvanis and Martin, 1999). After implantation, transcripts are found in the extraembryonic ectoderm before it cavitates but are confined symmetrically to the ring of this tissue closest to the epiblast by the threshold of gastrulation (Waldrip *et al.*, 1998; Coucouvanis and Martin, 1999; Lawson *et al.*, 1999). The effects of disrupting this gene are somewhat variable. While most homozygotes arrest at the egg cylinder stage without exhibiting embryonic expression of *Brachyury,* others progress to an advanced stage of gastrulation (Winnier *et al.*, 1995). Evidence has been provided that extraembryonic ectodermal expression of *Bmp4* is also required for the establishment of primordial germ cells (Lawson *et al.*, 1999). A more uniform phenotype has been found in conceptuses that are homozygous for the targeted disruption of the BMP type I receptor (*BmprI*), which shows generalized expression in both embryonic and extraembryonic tissues following the onset of gastrulation (Mishina *et al.*, 1995). The embryonic region of mutant homozygotes is about half the normal size and, while becoming multilayered, it fails to show evidence of formation of mesoderm even one day after the time when gastrulation should have started. Indications that growth of the epiblast is compromized were also obtained in ectopic grafts of mutant conceptuses (Mishina *et al.*, 1995). The early and more consistent phenotype produced by disrupting the gene for its receptor molecule than for BMP 4 itself argues that vital early signalling occurs via additional ligands of the TGFβ family of growth factors. A second gene whose expression has been detected in the extraembryonic ectoderm prior to gastrulation is *Eomesodermin.* It is expressed throughout the early postimplantation tissue and extends to the adjacent proximal epiblast once gastrulation starts (Ciruna and Rossant, 1999).

5. Other Gene Expression Patterns in Early Postimplantation Conceptuses

A further gene whose pattern of expression before gastrulation is distinct from all others considered so far is *Pem.* The protein encoded by this X-linked gene is first evident in morulae and in a fraction of nuclei of both trophectoderm and ICM cells of blastocysts. Immediately postimplantation, it occurs in the nuclei only of cells of the two purely extraembryonic lineages, namely derivatives of the polar trophectoderm, and in both visceral and parietal endoderm. Its early expression is thus limited to the two wholly extraembryonic lineages of the blastocyst (Lin *et al.,* 1994) that, in female conceptuses, exhibit nonrandom inactivation of the paternal X-chromosome (Chapman, 1986). The protein disappears from both the parietal endoderm and embryonic, as opposed to extraembryonic, VE region as gastrulation proceeds (Lin *et al.,* 1994). It is intriguing that even in the tissues in which the gene is expressed, not all cells are positive for the protein. The only other cells in which it has been found to be expressed during embryogenesis are primordial germ cells during their migratory phase. In adults it occurs in the Sertoli cells of the testis and in mural cumulus cells in the ovary, but is present only phasically in both (Pitman *et al.,* 1998). However, apart from indications of a slight reduction in the size of litters sired by *Pem-/γ* males, no adverse effect of hemi- or homozygosity for disruption of the *Pem* gene has been discerned (Pitman *et al.,* 1998). In contrast, forcing expression of the gene in ES cells by coupling its coding region to the Pgk-1 promoter blocked the ability of such cells to undergo differentiation either as embryoid bodies *in vitro* or as subcutaneous teratomas *in vivo* (Fan *et al.,* 1999). This was true regardless of whether the full-length protein or merely the N terminal region devoid of the homeodomain was expressed. Fan *et al.,* (1999) also noted that *Pem-/γ* ES cells were somewhat compomised in their ability to differentiate into embryoid bodies compared to their wild-type counterparts.

There are a number of additional genes whose product or pattern of early expression have yet to be characterized, or which may play a general role in cell physiology rather than a more specific one in early patterning. Among these, *Eed* is interesting inasmuch that homozygous mutant conceptuses show upregulation of expression of *Evx-1* throughout the PS, which, in accordance with the view that the latter gene specifies posterior mesoderm, is accompanied by failure of anterior mesoderm migration (Schumacher *et al.,* 1996; Faust *et al.,* 1998). Whether the latter accounts for the subsequent complete lack of anterior development seems doubtful in view of the situation in *Cripto −/−* conceptuses, where the differentiation of neural folds seems to occur in the complete absence of embryonic mesoderm (Ding *et al.,* 1998; Xu *et al.,* 1999). Other genes in this general category include various t alleles (Gluecksohn-Schoenheimer, 1940; Bennett, 1975), *Fug 1,* whose product is likely to be involved in chromatin structure or RNA processing (De Gregori *et al.,* 1994), the murine homolog of the human *BRCA1* gene (Liu *et al.,* 1996), and *Raly.* The latter encodes an HnRNP whose disruption appears to account for the perimplantation lethality associated with homozygosity for the classical *Ay* mutation (Cuenot, 1908; Castle and Little, 1910; Michaud *et al.,* 1993). The *Mov34* mutation, caused by retroviral insertion in an as yet uncharacterized gene, also results in very early postimplanatation failure of development (Soriano *et al.*, 1987). Further details of these and other genes whose mutation is associated with peri-implantation lethality are discussed elswhere (Copp, 1995; Rinkenberger *et al.*, 1997; Rappolee, 1999).

C. Appraisal of Molecular Studies

A number of points emerge from the foregoing survey of genes that show differential expression in the early postimplantation conceptus. First, several genes are expressed asymmetrically in relation to the proximo-distal (= Em.Ab) axis of the future embryonic region up to one day or more before it embarks on gastrulation. In each case, whether expression is in the epiblast or nVE, it is localized either to the prospective anterior or posterior region of the embryo. Accordingly, these genes are obvious candidates for playing a role in the initial establishment of the fetal A-P axis as opposed to its subsequent regional differentiation.

Second, prominent among the genes whose products have been characterized are those encoding either growth or transcription factors. Evidence for the indispensibility of signalling by members of the TGFβ family of growth factors has already been noted from the effects of disruption of the ligands Nodal and BMP4, as well as BMPR1. In addition, although expression of activins is clearly not crucial at this early stage (Schrewe *et al.*, 1994; Matzuk *et al.*, 1995), that of their type 1 receptor, ActR1B, whose transcripts occur in both the extraembryonic ectoderm and the epiblast before gastrulation, unquestionably is (Gu *et al.*, 1998). Further testimony to the importance for gastrulation of signalling by ligands of the TGFβ family is provided by the phenotypes of conceptuses that are homozygous for disruption of genes encoding two components that act downstream of their receptors, SMAD 2 (Waldrip *et al.*, 1998; Nomura and Li, 1998) and SMAD 4 (Sirard *et al.*, 1998). Both *Smad* genes are expressed weakly throughout the conceptus before gastrulation, and absence of the former seems to result in failure of all except the most posterior mesoderm to differentiate, and also, as discussed later, failure of molecular features of the anterior visceral endoderm to materialize (Waldrip *et al.*, 1998). Conceptuses that are homozygous for disruption of *Smad 4* are already aberrant by 5.5 dpc with indistinct division between the embryonic and extraembryonic region and an abnormal looking nVE. Twenty-four hours later they lack mesoderm and the extraembryonic region also looks disorganized. There appears, furthermore, to be an overall reduction in cell proliferation in both the embryonic and extraembryonic regions that is not accompanied by any obvious increase in apoptosis. The failure of gastrulation in *Smad 4* −/− conceptuses can be corrected by ensuring that the extraembryo is wild type, but this does not prevent subsequent anterior truncation of the fetus (Sirard *et al.*, 1998). Although the overall phenotype of *Smad 4* −/− conceptuses bears some resemblance to that caused by distruption of the genes encoding BPM4 or its type I receptor, the fact that it is not identical confirms that signalling via other TGFβ-type ligands is required during early postimplantation development. As discussed earlier, there is evidence for expression of the genes for both Nodal and TGFβ2 at this early stage (Slager *et al.*, 1991; Varlet *et al.*, 1997). Expression of the gene for a further downstream component in this signal transduction pathway, FAST 1, has been noted throughout the epiblast before gastrulation, and although transcripts are regarded as likely to be present also in the nVE, they are claimed to be absent from the extraembryonic ectoderm (Weisberg *et al.*, 1998).

Regarding other growth factors, not only are at least two members of the fibroblast growth factor (FGF) family expressed before gastrulation, but conceptuses that are homozygous for targeted disruption of both *Fgf4* and *Fgf Receptor* 2 show peri-implantation death (Feldman *et al.*, 1995; Arman *et al.*, 1998). Moreover, mosaic expression of a putative dominant negative form of FGF receptor in pre-implantation conceptuses prevents affected blastomeres from completing the fifth cell cycle (Chai *et al.*, 1998). The foregoing findings, coupled with the ability of FGF 4 in conjunction with heparin and medium conditioned by embryonic fibroblasts to sustain the proliferation of trophoblast cells in a diploid state (Tanaka *et al.*, 1998), strongly implicate this growth factor in morphogenesis of the extraembryonic ectoderm. A member of the FGF family has also been implicated in PS induction in the rabbit (Hrabe de Angelis and Kirchner, 1993), and what is most interesting, exposure of isolated of anterior mouse epiblast to FGF2 *in vitro* induces it to convert from an ectodermal to a mesoderm fate (Burdsal *et al.*, 1998). *Cripto* is a further gene that has been designated a growth factor and whose product appears, at least in the mouse, to be a secreted molecule (Ciccodicola *et al.*, 1989; Dono *et al.*, 1993). *Cerl* belongs to an ancient gene family encoding proteins that have a TGFβ-like cysteine knot in common and that, at least in the case of Cerl itself, are not processed from the proprotein (Pearce *et al.*, 1999). From work in *Xenopus* in particular, Cerberus protein has been recognized as one of a growing number of diffusible BMP antagonists.

Growth factors are likely to be involved in the morphogenesis as well as the growth of tissues during early postimplantation development and, indeed, it may well be impossible to differentiate their roles in these two processes. Thus, the transformation of the epiblast from a solid ball of cells into a cup-shaped pseudostratified epithelium surrounding the proamniotic cavity, which is a key step in its differentiation, is accompanied by a rapid increase in cell number (Solter *et al.*, 1971; Snow, 1976). It has been proposed, on the basis of analysis of the corresponding events in the differentiation of embryoid bodies (EBs) derived from EC or ES cells, that this process is controlled via two signals from the adajacent endoderm cells (Coucouvanis and Martin, 1995, 1999). One is a cell death signal that induces demise of cells in the core of the epiblast and the other a viability signal that acts via the intervening basal lamina to promote survival of epithelially organized cells at the periphery of the tissue. BMP expression is held to be necessary for the differentiation of endoderm to the stage when it is competent to transmit such signals. It is further suggested that it may also provide the apoptotic signal because BMPs have been implicated in induction of apoptosis in other systems.

There are, however, a number of problems with this rather elegant scheme. First, it is based largely on studying cell lines that form EBs with a mixed parietal- and visceral-like outer endodermal layer. Second, as discussed earlier (see section 5A) *in vivo* the VE cannot be regarded as differentiated before 6.5 dpc, which is approximately one day after transformation of the epiblast is completed. Third, to explain why conceptuses that are homozygous for disruption of the *Smad 4* gene nevertheless undergo cavitation, the authors have to postulate functional redundancy of this category of signal transducer. Fourth, it is not clear that apoptosis is an essential feature of the cavitation of epiblast as opposed to EBs since, though varying between mouse strains (Manova *et al.,* 1998), it seems to be much less conspicuous in epiblast than the typically much larger EBs (Coucouvanis and Martin, 1995, 1999). Moreover, apoptosis in the epiblast of normal conceptuses developing *in vivo* does not peak until 6.0 dpc when epithelization and cavitation of the epiblast has been completed (Manova *et al.,* 1998). Finally, if cavitation of the epiblast, like that of certain other epithelia, depends on vectorial transport of fluid, its conversion into a tight epithelium might be expected to anticipate rather than accompany cavitation. That epithelialization may occur before the proamniotic cavity forms was indicated in a recent clonal analysis of epiblast growth (Gardner and Cockroft, 1998).

The third, and most significant point to emerge from the differential gene expression studies is the recognition of a special region of the nVE and early VE that, because it lies diametrically opposite the site where the PS forms, has been termed the anterior visceral endoderm (AVE). As discussed earlier, this region is the focus of expression before gastrulation of *Hex, Cerl,* and *Mrg1* as well as the antigen recognized by VE-1 and of additional genes thereafter. Where does the AVE originate and why is it considered to be important? If Hex expression can be regarded as a legitimate lineage marker for the AVE, this part of the nVE can be traced back to the central cells in the deep layer of the primitive endoderm of the fifth dpc implanting blastocyst (Thomas *et al.,* 1997, 1998). In the 5.5 dpc egg cylinder stage,

Hex-expressing cells occur either at or just to one side of the distal tip of the embryonic region (Thomas *et al.,* 1997, 1998) and are said to be more columnar than those of the remainder of nVE (R. Beddington, unpublished observations cited in Thomas *et al.,* 1998). Subsequently, transcripts are restricted to one side of the egg cylinder where they extend from the distal tip to half way up the embryonic region. Finally, by the onset of gastrulation, the side where the transcripts are expressed is clearly opposite the PS and consists of a positive patch of cells that extends from the embryonic-extraembryonic junction to about 30 microns from the distal tip of the egg cylinder (Thomas *et al.,* 1997, 1998).

The case that the shift in expression of *Hex1* in the egg cylinder is due to movement of cells rather than a change in those engaged in its transcription is based on cell-marking experiments. Hence, labelling of the distal VE in 5 dpc conceptuses was followed by the consistent displacement of the patch of positive cells to one side of the egg cylinder during subsequent culture. Reflecting the difficulty of obtaining satisfactory further development of such early postimplantation stages *in vitro,* only four reached the stage of forming a PS. However, in each of the four, the patch of label in the VE was reported to be located opposite the PS, i.e., anteriorly with respect to the A-P axis of the fetus (Thomas *et al.,* 1997, 1998). Differential growth within the VE was suggested as one possible means whereby this displacement is achieved, although there is at present no evidence to support such a possibility. What remains to be established is whether the cells in the VE actually move relative to those in the adjacent epiblast. There is clearly differential growth between these two tissues during early postimplantation development, and the slower rate of increase in cell number in the endoderm presumably accounts for the fact that its embryonic part gradually adopts a squamous form (Snell and Stevens, 1966). Such differences in growth would seem to require the existence of a region or regions where "slippage" or relative movement between the two tissues and considerable stretching of their shared basal lamina could occur. Were this to involve much of the future PS region, for example, it could result in distalward movement of the epiblast relative to its enclosing nVE with the consequent displacement or rolling of the original distal-most cells of the latter progressively toward the future anterior. Continuation of such slippage beyond the start of gastrulation could, furthermore, account for the eventual extension of the originally proximal PS along the entire length of the posterior surface of the embryo. That remodelling of basal laminae is an important factor in morphogenesis has been highlighted very recently by the discovery that *Gon-1,* a gene that is involved in controlling the shape of the gonad in *C. elegans,* encodes a secreted metalloprotease (Blelloch and Kimble, 1999).

Hence, the question of how distal nVE cells are shifted consistently toward the future anterior surface of the embryo remains a matter of conjecture. Nevertheless, this important finding has fostered the interesting hypothesis that the A-P axis of the embryo is initially aligned with the proximo-distal axis of the egg-cylinder and hence with the Em.Ab axis of the blastocyst (see Beddington and Robertson, 1998, 1999). The phenotype of conceptuses that are homozygous for the disruption of both *Smad 2* and *Cripto,* in particular, are consistent with this notion. Hence, while in *Smad 2* −/− homozygotes the usual pre-gastrulation ring of expression of *Brachyury* in the proximal epiblast and of *Bmp4* in the extraembryonic ectoderm appear, there is failure of molecular features of the AVE to materialize and of all except the most posterior mesoderm to differentiate (Waldrip *et al.,* 1998). According to the molecular analysis of *Cripto* −/− conceptuses by Ding *et al.* (1998), the AVE remains localized at the distal tip of the egg cylinder while early markers of the PS show mislocalization proximally.

Collectively, the genetic and experimental embryological studies point to the existence of discrete head and trunk organizers in mammals, the former being the AVE and the latter, Hensen's node (see Beddington and Robertson, 1998, 1999). Among other things, this explains the failure of ectopically grafted nodes to organize a secondary axis that includes the most rostral structures (Beddington, 1994).

The foregoing studies also render untenable the notion that the A-P axis of the mouse fetus is laid down strictly from posterior to anterior. That this may apply generally in mammals is suggested by confirmation of a very early finding (Van Beneden, 1883) that the anterior end of the fetal axis is differentiated morphologically even before the PS is discernible (Viebahn *et al.,* 1995). Although a case has been made for the existence of distinct head and trunk organizers throughout vertebrates, attempts to find a counterpart of the AVE in avian hypoblast have so far yielded negative results (Knoetgen *et al.,* 1999). Indeed, doubts have been expressed as to whether the hypoblast normally has any role in axis induction in the chick (Khaner, 1995).

Particularly since great emphasis is now being placed on the precocious differentiation of the anterior end of the fetal A-P axis, it is relevant to reiterate that not all molecules expressed asymmetrically in the embryonic region of the egg cylinder before gastrulation are located anteriorly. Three genes that are expressed locally in the epiblast before gastrulation, *Evx-1, Cripto,* and *fgf8,* all have their transcripts confined to the future posterior region of this tissue by the stage when the PS appears. Given the marked lack of coherence in growth of the epiblast compared to the nVE once it has acquired an epithelial organization (Gardner and Cockroft, 1998), maintenance of such localization presumably depends on continual regulation of transcription. However, there is at present no information to suggest by how long an interval local alterations in the relationship between the epiblast and nVE necessary for the morphogenetic activity of the PS anticipate its formation.

D. Future Directions

While an increasing number of genes are being implicated in early patterning of the embryo, the task of assigning them specific roles is difficult for several reasons. First, one can be sure that the list is still very far from complete. Second, data on expression patterns have various shortcomings. Most are based on use of *in situ* hybridization to detect transcripts and, particularly in late pre- and early postimplantation stages, the quality of preparations is often too poor to allow very precise delimitation, especially with regard to detecting possible axial asymmetries. Furthermore, given the scope for posttranscriptional control of expression, probing for transcripts is not necessarily a valid alternative to studying the distribution of the protein resulting from their translation. Even looking at proteins, as can be done with specific antibodies (e.g., for *Pem*) or by using LacZ as a reporter, does not necessarily give an accurate picture of expression, since they often require proteolytic cleavage or association with other gene products in order to acquire an active form. Some idea of the possible complexities of posttranslational processing can be gleaned from recent *in vitro* studies on TGFβ family members (Constam and Robertson, 1999). Antibodies that are specific to the active forms of proteins would clearly be valuable in this regard.

In considering the sort of studies that might help carry the analysis of specification of the fetal A-P axis forward at the molecular level, focusing on genes whose mutation causes either partial or complete duplication of this axis would seem to be a particularly promising strategy. *Fused* is such a gene in the mouse. Four mutant alleles have been identified at this locus. These include three spontaneous mutations, all of which exhibit dominant effects. Two, *Fused* (*Fu*) and *Knobbly* (Fu^{kb}) (Jacobs-Cohen *et al.,* 1984), are due to insertion of transposons into introns 6 and 7, respectively (Vasicek *et al.,* 1997), whereas the third allele Kinky (Fu^{ki}), remains undefined and is now thought to be extinct. The fourth and most recent mutant allele, Fu^{Tg1} is due to insertion of a transgene into the locus (Perry *et al.,* 1995; Zeng *et al.,* 1997). Only *Fu* homozygotes are viable because, in this case, transposon insertion is evidently compatible with the production of substantial mRNA of normal length. Although such homozygotes show some overgrowth and duplication of the posterior neural tube during fetal stages, much more obvious axial duplication has been described for conceptuses that are homozygous for the other three mutant alleles. This is particularly true for Fu^{ki} where, in the extreme, the entire A-P axis from the allantoic bud to anteriormost structures of the neuraxis is evidently replicated (Gluecksohn-Schoenheimer, 1949). Neither ectopic grafts of Henson's node (Beddington, 1994) nor overexpression of the chicken *wnt8* gene (Popperl *et al.*, 1997) cause duplications that include such anterior components. The implication is that the *Fused* gene, which has been renamed *Axin* (= *ax*is *in*hibition) to avoid confusion with its unrelated namesake in *Drosophila,* plays a critical role in regulating the initial establishment of the embryonic A-P axis by spatially restricting the expression of genes that are involved it is construction. To achieve this for the entire A-P axis, i.e., the head as well as the trunk organizer, this gene must be active well before gastrulation begins.

Axin mRNA is not only present in all adult tissues that have been examined and uniformly distributed in all embryonic and extraembryonic tissues of the postimplantation conceptus, but is also detectable in preimplantation conceptuses of all stages by RT-PCR (Zeng *et al.,* 1997). It has been shown by mRNA injection that murine Axin can suppress both primary and secondary axis formation in *Xenopus* in a manner that its consistent with its acting as a negative regulator of the Wnt signalling pathway. According to available evidence it lies upstream of β-catenin but downstream of *Wnt, Dsh,* and GSK-3. Axin protein evidently stimulates β-catenin degradation by forming a complex with it, together with GSK3 β and adenomatous polyposin coli (APC) (Kishida *et al.,* 1999). Given the growing evidence implicating β-catenin in axis formation in vertebrates (Heasman *et al.,* 1994; Heasman, 1997), it would be interesting to know more about the synthesis and distribution of the cytoplasmically mobilizable fraction of this protein during early mouse development. Although it is has been shown that zygotic expression of the β-catenin gene is required for gastrulation (Haegel *et al.,* 1995), such a gene disruption experiment does not enable the role of the protein in cell adhesion to be differentiated from that in transcriptional activation or suppression. Examining how expression of both *β-catenin* and other genes implicated in patterning that are expressed before gastrulation are affected in Fu^{kb} or Fu^{Tg1} homozygotes could prove very instructive in providing deeper insight into the sequence of events whereby the orientation and polarity of the fetal A-P axis are specified.

A curious and unexplained observation concerning the Fu^{kb} mutation was the repeated occurrence of two or more conceptuses within a common decidual swelling. Because such pairs or groups of conceptuses were all normal or a mixture of abnormal and normal rather than all abnormal, this phenomenon was attributed to the genotype of the mother rather than the conceptuses (Jacobs-Cohen, 1984).

VI. When Is the Fetal A-P Axis Specified?

The findings discussed in the previous section demonstrate unequivocally that relevant asymmetries are evident well before gastrulation begins. However, they leave open the question of whether these asymmetries arise when the conceptus implants in the uterus or thereafter or depend on spatial cues that are established before implantation. That the uterus has some role in the process is evident from the fact that before fetuses turn, and are thus free to rotate within their membranes, their A-P axis is typically transverse to the long axis of the horn in which they lie (Snell and Stevens, 1966; Smith, 1980). However, since normal-looking early somite-stage fetuses have been obtained *in vitro* from conceptuses explanted prior to implantation (Hsu *et al.,* 1974), the uterus is clearly not required for inducing the A-P axis. It might, nonetheless, provide cues that serve to orient this axis during normal development.

Although much attention has been devoted to investigating the mode of attachment of the blastocyst to the uterine epithelium during implantation, the possible significance of its already being bilaterally symmetrical before it implants (Smith, 1980, 1985; Gardner, 1990, 1997) has so far been neglected. Hence, potentially very relevant conclusions drawn from a careful histological analysis of implantation from this perspective (Smith, 1980) have yet to be confirmed. The first conclusion was that when blastocysts initially attach to the uterine luminal epithelium they do so in one of just two orientations. Whereas about half attach to the left wall with their abembryonic pole pointing toward the cervix and embryonic pole toward the ovary, the remainder attach to the right wall with their abembryonic pole pointing toward the ovary and embryonic pole toward the cervix (Smith, 1980). The former orientation was defined as type L and the latter as type R, and the mirror images of these two orientations were said not to occur. That differentiation was between right and left rather that medial versus lateral walls was based on the same two orientations applying to both uterine horns. A further conclusion was that regardless of whether the orientation of implantation was type L or R, blastocysts invariably have their same side, the right side, directed toward the wall to which they initially attached (Smith, 1980). This initial attachment appeared to induce a local change in the conformation of the uterine epithelium that eventually leads to rotation of the Em.Ab axis of the blastocyst from parallel to the long axis of the uterus so that its embryonic pole becomes directed mesometrially and its abembryonic pole antimesometrially. Concurrent rotation of the blastocyst about its Em.Ab axis orients its A-P axis transversely with respect to the uterine horn (Smith, 1980).

The foregoing analysis implies that preexisting asymmetries in both the blastocyst and the uterus are involved in the process of implantation. Furthermore, providing the A-P axis of the conceptus is conserved from before implantation, its orientation with respect to the uterus will clearly be fixed at this juncture. Hence the most parsimonious explanation for the fact that the fetal A-P axis is also tranverse to the uterus is that it depends on the conceptus's A-P axis for its orientation (Gardner *et al.,* 1992). This obviously leaves open the question of the source of cues whereby the polarity of the fetal A-P axis is specified. This cannot be the uterus since, as noted earlier, the fetal axis is as often directed toward its left as toward its right. Therefore, it seems likely that specification of both the orientation and the polarity of the A-P axis of the fetus depends on information that already resides within the conceptus before it implants. If this is the case, the question arises as to how the preimplantation conceptus acquires this information. In attempting to grapple with this issue on the assumption that mammalian eggs possess "no morphogenetically significant and stable heterogeneities of their cytoplasm" (Smith, 1980), Smith proposed that the appropriate positional information was supplied to the developing conceptus by the uterus. It was reasoned that by virtue of its lying in the uterus with its Em.Ab axis horizontal prior to implantation, the blastocyst would have different parts of its surface exposed to distinct surfaces of the uterine lumen, as long as it did not rotate as it moved along the lumen to its site of implantation. It was argued, furthermore, that asymmetry of the blastocyst might prevent its rotating about its Em.Ab axis even if the investing zona pellucida did so (Smith, 1980), a notion that seems rather implausible. However, conceptuses normally enter the uterus as morulae, so the process by which they are assigned to fairly uniformly spaced crypts may be completed by the early blastocyst stage. Hence, if most of the preimplantation phase of their development occurs when blastocysts are closely invested by luminal epithelium within their individual crypts, the imposition of asymmetry via the uterus is conceivable, though clearly dispensible (Hsu *et al.,* 1974).

In considering other ways in which the necessary asymmetries might originate, there is the evidence that the second Pb, which is typically aligned with the bilateral axis of the early blastocyst, provides an enduring marker of the animal pole of the zygote (Gardner, 1997). It is difficult to dismiss such a consistent relationship between the A-V axis of the zygote and the bilateral axis of the blastocyst (Fig. 1) as entirely fortuitous, particularly since the latter may determine the polarity of polar trophectoderm growth (Gardner, 2000a, b). As noted earlier (see section III. D), the mouse conceptus seems to be bilaterally rather than radially symmetrical from very early in cleavage. Furthermore, the Em.Ab axis, which is conserved through implantation and may represent the original orientation of the future fetal A-P axis, bears a consistent relationship to the bilateral axis. Such findings, in conjunction with those discussed in section II concerning the effects on fetal development of perturbing the very early conceptus, engender sufficient doubt about the lack of involvement of egg organization in specifying the A-P axis of the fetus as to warrant reexamination of the basis for such a conclusion.

VII. How Compelling Is the Case against Dependence of Early Patterning on Information in the Zygote?

A. Regulative Development

A key question is whether, as is widely assumed, the findings on the regulative capacity of the cleaving conceptus really do exclude the possibility that specification of the fetal A-P axis might depend on cues that are present in the egg or zygote. The argument most commonly deployed against this possibility is that chimeras formed by aggregating pairs or larger groups of morulae would often be expected to contain a corresponding number of disparate sets of axial cues and therefore show either full or partial duplication of the fetal A-P axis. At least in the mouse, this is clearly not the case, but in other mammals the requisite examination of chimeras at early postimplantation stages does not appear to have been undertaken. It is noteworthy, however, that according to direct as opposed to indirect evidence, monozygotic twinning seems to be rare in the mouse compared to the human (Gadda, 1961; Wallace and Williams, 1965). Bodemann (1935) described one pair of rather similar size twins with a common parietal yolk sac and ectoplacental cone that were both at the early allantoic bud stage, and whose A-P axes were parallel but of opposite polarity. Bateman (1960) reported two pairs of twins with separate chorions within a common parietal yolk sac at 9.5 dpc. In each case one twin had its chorion oriented correctly toward the mesometrium and the other, with an anti-mesometrial chorion, was unquestionably too retarded to develop further.

Only one case of similarly developed mid-gestation mouse twins has been described, and these shared a common visceral yolk sac (Runner, 1984). These twins, unlike those identified among a series of earlier postimplantation diploid parthenotes (Kaufman, 1982), were within separate amnia. That the rarity of monozygotic twinning in mice is not simply due to failure of detection is supported by recent findings in which DNA fingerprinting was used to screen a large number of offspring (McLaren *et al.,* 1995). Apart from among conceptuses induced to develop parthenogenetically (Kaufman, 1982, 1985), mono-amniotic twinning in mice has only be reported following early postimplantation treatment of pregnant dams with vinchristine (Kaufman and O'Shea, 1978). Duplication of the ICM has been induced by *in vitro* exposure of blastocysts to the related vinca alkaloid, vinblastine (Naruse *et al.,* 1983). However, given the lack of information on how twinning is induced by such microtubule-inhibiting agents, the relevance of these findings to the situation in aggregation chimeras remains uncertain. ES cells in which the *v-src* oncogene is expressed at high levels can also induce twinning in host conceptuses following blastocyst injection. However, in this case the entire egg cylinder is duplicated (Boulter *et al.,* 1991). Reproducible axial duplications within a single amnion has been obtained in conceptuses carrying a transgene designed to produce ectopic expression of the chicken *Wnt8c* gene (Popperl *et al.,* 1997). However, unlike in cases of homozygosity for certain mutations at the *Axin* (*Fused*) locus, the axis was only partially duplicated, with fusion anteriorly that was coupled with reduction of rostral tissue. Such rostral truncation was also seen in transgenic embryos in which axial duplication had not taken place.

Among the factors that may militate against axial duplication in the mouse are the cup-like shape and small size of the pregastrulation epiblast, which, even in aggregation chimaeras, is somehow restored to its normal cell number before gastrulation begins (Buehr and McLaren, 1974; Lewis and Rossant, 1982; Rands, 1986). As illustrated in Figure 6, these features

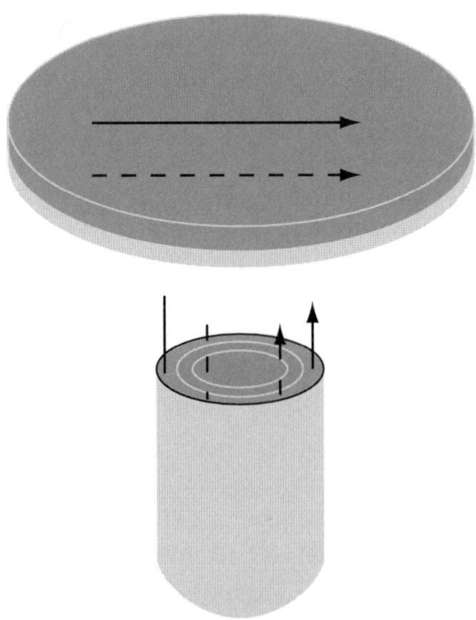

FIG. 6 Diagrams illustrating how the typically flat blastoderm of amniotes (upper) is remodeled into a small cup in the mouse (lower) in which the epiblast (dark) is surrounded by the endoderm or hypoblast layer (light). This reduces the distance from the primary embryonic axis (continuous arrow) at which a secondary one (discontinuous arrow) resulting from duplicated axial information can arise, thereby diminishing the likelihood of its forming.

of the mouse epiblast might be expected to facilitate inhibition or assimilation of supernumerary axes, evidence of which has been documented in other species, including the chick (Cooke, 1972; Khaner and Eyal-Giladi, 1989; Ziv *et al.,* 1992; Yuan *et al.,* 1995). Hence, there are grounds for supposing that duplication of axes is not an inevitable consequence of the coexistence of more than one set of axial information. It is also relevant to note that morulae tend to lie with their bilateral plane horizontal (author's unpublished observations), so the assumption that they bear a random relationship to each other during aggregation (e.g., Mintz, 1964, 1965) may not be valid.

In principle, studies on the developmental potential of individual blastomeres should be more incisive for investigating whether specification of the fetal A-P axis depends on information that is localized spatially in egg or zygote. This is because segregation of such "determinants" during cleavage would be expected to result in the production of blastomeres that altogether lack the information required to establish an axis. In practice, neither in the mouse nor in other mammals are existing studies on the development potential of isolated blastomeres informative in this respect. This is because, in the studies undertaken so far, it has not been possible to decide whether the failure of the majority of blastomeres isolated at different cleavage stages to develop normally is, as generally assumed, due to chance rather than to their cytoplasmic endowment. Without being able to identify corresponding blastomeres in different conceptuses, the only way to exclude the latter possibility decisively is by showing that all blastomeres from an individual conceptus can develop normally in isolation. This has be achieved reproducibly only at the two-cell stage (Tsunoda and McLaren, 1983). Beyond this, just one case of success has been reported, namely the development of a normal calf from each blastomere of a four-cell bovine conceptus (Johnson *et al.,* 1995). If the orientation of cleavage planes in the bovine is similar to that proposed for the rabbit (Gulyas, 1975), it is during the third rather than second cleavage that segregation of information localized along the A-V axis of the zygote is likely to begin.

Because normal development has been obtained following removal of much of the animal or vegetal polar region from mouse zygotes, it has been argued that information for specifying axes cannot be localized prior to cleavage (Zernicka-Goetz, 1998). As discussed elsewhere (Gardner 1999a), this study a based on two untested assumptions about cytoplasmic localization. One is that it occurs prior to removal of one or other pole, which was done three to five hours before first cleavage, and the other that it accords with the A-V axis of the zygote. The lack of any enduring effect of perturbing the cytoplasm of the mouse zygote by centrifugation (Mulnard and Puissant, 1984) or by rapid agitation with a probe (Evsikov *et al.,* 1994) also does not provide conclusive evidence against its playing a role in patterning (Gardner, 1999a).

There is, of course, no compelling reason to suppose that if information for specifying axes exists in the zygote it should necessarily be localized. A consistent overall polarity of stable cytoskeletal units that is heritable through cleavage could serve just as well. It is interesting in this connection that novel large-scale matrices composed of intermediate filament components, which include discrete subsets of cytokeratins, have been described in mammalian eggs and zygotes (Capco and McGaughey, 1986; Gallicano *et al.,* 1991, 1992). Not only do these persist until shortly before gastrulation, but they also undergo changes in organization that correlate with major early morphogenetic events (Schwarz *et al.,* 1995). That intracellular microinjection of the Troma 1 anticytokeratin monoclonal antibody at the two-cell stage had no discernible effect on development to the blastocyst stage (Emerson, 1988) might suggest that these particular intermediate filament complexes can be discounted from playing a role in early patterning. However, it remains to be determined to what extent they are disrupted by an antibody directed against one of their later synthesized components and, what is more important, whether bilateral symmetry is perturbed thereby.

B. Molecular Asymmetries in the Oocyte and Zygote

The notion that early patterning depends on egg organization in mammals as in other metazoa, is not novel (Dalcq, 1957) but was based originally on regional cytoplasmic differentiation in eggs and early blastomeres that was made visible largely through the use of relatively nonspecific histochemical staining procedures. Recent application of *in situ* staining with specific antibodies has revealed that eight different proteins are not only localized in the cortex of both the mouse and human egg and zygote, but retain an asymmetric distribution throughout cleavage (Antczuk and Van Blerkom, 1997, 1999). These are leptin, STAT3, Bax, BCL-x, TGFβ2, vascular endothelial growth factor (VEGF), c-KIT, and the epidermal growth factor receptor (EGFR). In oocytes, staining for both leptin and STAT3, which is a downstream component in leptin signalling via its receptor (OB-R), was also evident in follicle cells lying adjacent to their positive cortical region. Such a distribution of the proteins in follicle cells is particularly intriguing because of the failure in two studies to detect mRNA for leptin in oocytes by RT-PCR (Cioffi *et al.,* 1997; Matsuoka *et al.,* 1999). This finding argues that in mammals, as in other metazoa, at least some of the proteins found in the mature oocytes that are not derived from maternal serum (Glass, 1971) are products of the biosynthetic activity of surrounding follicle cells. Furthermore, the colocalization of leptin in the cortex of the oocyte and in the follicle raises the further interesting possibility that the follicle may also play a role in establishing asymmetries in the oocyte and thus in patterning the conceptus or early embryo.

How macromolecules might be transferred from follicle cells to the oocyte remains to be established, since continuity at their sites of mutual contact seems to be limited to gap junctions that permit only small molecules to be transferred (Gilula *et al.,* 1978). It is intriguing, however, that leptin VEGF, and TGFβ2 have been found to be enriched at the periphery of human follicle cells whence they appear to be released in vesicles (Antczuk *et al.,* 1997). Although both the leptin receptor (OB-R) and its mRNA are present in the oocyte, there are conflicting data on the distribution of the protein. According to Matsuoka *et al.* (1999), it occurs uniformly at the surface of the GV and metaphase II oocyte. However, Antczuk and Van Blerkom (1997) found it to be distributed diffusely in the cytoplasm and mainly confined to pronuclei or nuclei following fertilization. Nevertheless, evidence that this signal transduction pathway is functionally competent before fertilization in the mouse was provided by demonstrating that STAT3 acquired the capacity to bind antiphosphotyrosine antibody following brief incubation of metaphase II oocytes in the presence of exogenous leptin (Matsuoka *et al.,* 1999).

Whereas leptin has been implicated in signal transduction processes that have diverse effects on reproduction (Messinis and Milingos, 1999), it is still far from clear what role it might be playing in the context of the oocyte and preimplantation conceptus. Early development is not discernibly impaired in conceptuses that are homozygous for mutations that cause marked truncation of this ligand (Zhang *et al.,* 1994), or its receptor (Ghilardi *et al.,* 1996). In contrast, however, homozygosity for disruption of the gene encoding STAT3 causes conceptuses to degenerate shortly after implantation (Takeda *et al.,* 1997). As pointed out by Antczuk and Van Blerkom (1997), the fact that mRNA for the STAT3 was not detected before day 6 in normal conceptuses does not mean that it is not required for preimplantation development. Maternal supply of the protein might be sufficient, providing it was long-lived. Efforts to establish what leptin and the other localized proteins might be doing in early development are needed in order to find out whether they are involved specifically in patterning rather than more general aspects of growth or metabolism during early development. Given the seemingly generous maternal provision of all these proteins, this is going to require the production of maternal effect mutations. Some success has already been achieved in this realm by using transgenic antisense RNA technology to reduce the activity of maternally encoded tissue-type plasminogen activator (Richards *et al.,* 1993).

VIII. Concluding Remarks

Processes leading to the establishment of the fetal A-P axis unquestionably begin well before the start of gastrulation and almost certainly before the conceptus implants *in utero*. This is evident from both morphological and molecular asymmetries that bear a consistent relationship to this axis and that are discernible in pre- and early postimplantation stages, respectively. If asymmetry of the blastocyst is crucial, then the key issues are its origin and how it imposes polarity on ICM derivatives so as to generate molecular asymmetries in the nVE and epiblast of the pregastrulation egg cylinder. Conceptuses that have never been exposed to the uterus not only exhibit bilateral symmetry at the early blastocyst stage (Gardner, 1997), but can also engage in normal differentiation of the fetal A-P axis *in vitro* (Hsu *et al.*, 1974). Moreover, the existence of a consistent axial relationship between the blastocyst and zygote (Gardner, 1997) seems to require that specification of both the bilateral and Em.Ab axes of the conceptus depends on information that is already present in the zygote before it begins to cleave. When these findings are considered in conjunction with the evidence suggesting that the fetal A-P axis is initially aligned with the Em.Ab axis of the conceptus, the notion of axial continuity between the zygote and gastrula stage seems rather less improbable than has generally been assumed. It is notable that independently of the studies discussed here, a relationship between oocyte organization and the determination of body symmetry has been invoked to account for an association among twinning, symmetry development, and the pedigree of certain malformations in man (Boklage, 1987).

Recently, it has be argued that egg organization is indispensible for essentially all aspects of early development in mammals (Edwards and Beard, 1997). This is a perspective that, notwithstanding the points raised in section VII, may appear difficult to reconcile with the demonstrably marked regulative capacity of the preimplantation mouse conceptus. Nevertheless, the disparate findings reviewed in section II cannot be ignored and must clearly be accommodated in any scheme that seeks to explain how and when specification of the A-P axis of the fetus takes place.

In biology, the truth is often found to lie somewhere between starkly contrasting views. It is important to recognize that studies on the regulative capacity of the early conceptus have necessarily depended on experimental intervention, while many of those on axial relationships have not. What the seemingly conflicting data may therefore be indicating is that mammals resemble other highly regulative species like sea urchins in exhibiting a disparity between what happens in normal undisturbed development and following experimental intervention. If this is true, an obvious concern regarding assisted conception in our own species is just how efficient such regulative mechanisms in mammals are in responding to, and correcting, subtle but nevertheless potentially serious perturbations. In this context, it is relevant to note that in a recent study of the early embryo of the mollusc, *Ilyanassa,* regulation was more often found to be incomplete following minor as opposed to more gross experimental manipulation (Sweet, 1998).

Acknowledgments

I wish to thank Ann Yates and Tim Davies for their invaluable assistance in preparing the manuscript and acknowledge the support of the Royal Society and the Wellcome Trust.

References

Acampora, D., Mazan, S., Lallemand, Y., Avantaggiato, V., Maury, M., Simeone, A., and Brulet, P. (1995). Forebrain and midbrain regions are deleted *Otx 2* $^{-/-}$ in mutants due to a defective anterior neurectoderm specification during gastrulation. *Development* **121,** 3279–3290.

Alexandre, H. (1979). The utilization of an inhibitor of spermidine and spermine synthesis as a tool for the study of the determination of cavitation in the preimplantation mouse embryo. *J. Embryol. Exp. Morph.* **53,** 145–162.

Alikani, M., Noyes, N., Cohen, J., and Rosenwaks, Z. (1994). Monozygotic twinning in the human is associated with the zona pellucida architecture. *Hum. Reprod.* **9,** 1318–1321.

Ang, S-L., Jin, O., Rhinn, M., Daigle, N., Stevenson, L., and Rossant, J. (1996). A targeted mouse Otx2 mutation leads to severe defects in gastrulation and formation of axial mesoderm and to deletion of rostral brain. *Development* **122,** 243–252.

Ang, S-L., and Rossant, J. (1994). HNF-3 beta is essential for node and notochord formation in mouse development. *Cell* **78,** 561–574.

Antczuk, M., and Van Blerkom, J. (1997). Oocyte influences on early development: the regulatory proteins leptin and STAT3 are polarized in mouse and human oocytes and differentially distributed within cells of the preimplantation stage embryo. *Mol. Hum. Reprod.* **12,** 1067–1086.

Antczuk, M., and Van Blerkom, J. (1999). Temporal and spatial aspects of fragmentation in early human embryos—possible effects on developmental competence and association with differential elimination of regulatory proteins from polarized domains. *Hum. Reprod.* **14,** 429–447.

Antczuk, M., Van Blerkom, J., and Clark, A. (1997). A novel mechanism of VEGF, leptin and TGFβ2 sequestration in subpopulations of human ovarian follicle cells. *Hum. Reprod.* **12,** 2226–2234.

Arman, E., Haffner-Krausz, R., Chen, Y., Heath, J. K., and Lonai, P. (1998). Targeted disruption of fibroblast growth factor (FGF) receptor 2 suggests a role for FGF signaling in pregastrulation mammalian development. *Proc. Natl. Acad. Sci. USA* **95,** 5082–5087.

Austin, C. R. (1973). Embryo transfer and sensitivity to teratogenesis. *Nature* **244,** 333–334.

Aziz, M., and Alexandre, H. (1991). The origin of the nascent blastocoele in preimplantation mouse embryos: ultrastructural cytochemistry and the effects of chloroquine. *Roux's Arch. Dev. Biol.* **200,** 77–85.

Barlow, P., Owen, D. A. J., and Graham, C. F. (1972). DNA synthesis in the preimplantation embryo. *J. Embryol. Exp. Morph.* **27,** 431–445.

Bateman, A. J. (1960). Dichorial one-egg twins in the mouse. *Nature* **187,** 339–340.

Beddington, R. S. (1994). Induction of a second neural axis by the mouse node. *Development* **120,** 613–620.

Beddington, R. S. P., and Robertson, E. J. (1998). Anterior patterning in the mouse. *Trends Genet.* **14,** 277–284.

Beddington, R. S. P., and Robertson, E. J. (1999). Axis development and early asymmetry in mammals. *Cell* **96,** 195–209.

Belaoussoff, M., Farrington, S. M., and Baron, M. H. (1998). Hematopoietic induction and respecification of A-P identity by visceral endoderm signalling in the mouse embryo. *Development* **125,** 5009–5018.

Belo, J. A., Bouwmeester T., Leyns, L., Kertesz, N., Gallo, M., Follettie, M., and De Robertis, E. M. (1997). Cerberus-like is a secreted factor with neuralizing activity expressed in the anterior primitive endoderm of the mouse gastrula. *Mech. Dev.* **68,** 45–57.

Bennett, D. (1975). The T locus of the mouse. *Cell* **6,** 441–454.

Bennett, J. (1982). Sperm entry point is related to early division of mouse blastomeres. *J. Cell Biol.* **95,** 163 (Abstr.).

Biben, I. C., Stanley, E., Fabri, L., Kotecha, K., Rhinn, M., Drinkwater, K., Wang, C., Nash, A., Hilton, D., Ang, S-L., Mohun, T., and Harvey R. P. (1998). A murine cerberus (*mCer-1*) expressed in anterior primitive endoderm and embryonic mesendoderm during gastrulation. *Dev. Biol.* **194,** 135–151.

Black, S. D., and Gerhart, J. C. (1986). High-frequency twinning of *Xenopus laevis* embryos from eggs centrifuged before first cleavage. *Dev. Biol.* **116,** 228–240.

Blandau, R. J. (1954). The effects on development when egg and sperm are aged before fertilization. *Ann. N. Y. Acad. Sci.* **57,** 526–532.

Blelloch, R., and Kimble, J. (1999). Control of organ shape by a secreted metalloprotease in the nematode *Caenorhabditis elegans. Nature* **399,** 586–590.

Blum, M., Gaunt, S. J., Cho, K. W. Y., Steinbeisser, H., Blumberg, B., Bittner, D., and De Robertis, E. M. (1992). Gastrulation in the mouse: The role of the homeobox gene *goosecoid. Cell* **69,** 1097–1106.

Bodemann, E. (1935). A case of uniovular twins in the mouse. *Anat. Rec.* **62,** 291–294.

Boklage, C. E. (1987). The organization of the oocyte and embryogenesis in twinning and fusion malformations. *Acta Genet. Med. Gemellol.* **36,** 421–431.

Boklage, C. E. (1990). Survival probability of human conceptions from fertilization to term. *Int. J. Fertil.* **35,** 75–94.

Bomsel-Helmreich, O., and Papiernik-Berkhauer, E. (1976). Delayed ovulation and monozygotic twinning. *Acta Genet. Med. Gemellol.* **25,** 73–76.

Bonnevie, K. (1950). New facts on mesoderm formation and proamnion derivatives in the normal mouse embryo. *J. Morph.* **86,** 495–546.

Borghese, E., and Cassini, A. (1963). Cleavage of the mouse egg. *In* "Cinemicrography in Cell Biology," (G. G. Rose, Ed.), pp. 263–277. Academic Press, New York.

Bos-Mikich, A., Whittingham, D. G., and Jones, K. T. (1997). Meiotic and mitotic Ca^{++} oscillations affect cell composition in resulting blastocysts. *Dev. Biol.* **182,** 172–179.

Boulter, C. A., Aguzzi, A., Williams, R. L., Wagner, E. F., Evans, M. J., and Beddington, R. (1991). Expression of v-*src* induces aberrant development and twinning in chimaeric mice. *Development* **111,** 357–366.

Bouwmeester, T., Kim, S-H, Sasai, Y., Lu, B., and De Robertis, E. M. (1996). Cerberus is a head-inducing secreted factor expressed in the anterior endoderm of Spemann's organizer. *Nature* **382,** 595–601.

Boyd, J. D., and Hamilton, W. J. (1970). "The Human Placenta." Heffer and Son, Cambridge.

Braden, A. W. H. (1959). Are nongenetic defects of the gametes important in the etiology of prenatal mortality? *Fertil. Steril.* **10,** 285–298.

Braude, P., Bolton, V., and Moore, S. (1988). Human gene expression first occurs between the four- and eight-cell stage of preimplantation development. *Nature* **332,** 459–461.

Brook, F. A., and Gardner, R. L. (1997). The origin and efficient derivation of embryonic stem cells in the mouse. *Proc. Nat. Acad. Sci. USA* **94,** 5709–5712.

Brower, D. (1987). Chromosome organization in polyploid mouse trophoblast nuclei. *Chromosoma* **95,** 76–80.

Bryan, E. M. (1998). A spare or an individual? Cloning and the implications of monozygotic twinning. *Hum. Reprod. Update* **4,** 812–815.

Buehr, M., and McLaren, A. (1974). Size regulation in chimaeric mouse embryos. *J. Embryol. Exp. Morph.* **31,** 229–234.

Bulmer, M. G. (1970). "The Biology of Twinning in Man." Clarendon Press, Oxford.

Burdsal, C. A., Flannery, M. L., and Pedersen, R. A. (1998). FGF-2 alters the fate of mouse epiblast from ectoderm to mesoderm in vitro. *Dev. Biol.* **198,** 231–244.

Burn, J., and Corney, G. (1984). Congenital heart defects and twinning. *Acta Genet. Med. Gemellol.* **33,** 61–70.

Busa, W. B., and Gimlich, R. L. (1989). Lithium-induced teratogenesis in frog embryos prevented by a polyphosphoinoside cycle intermediate or a diacylglycerol analogue. *Dev. Biol.* **132,** 315–324.

Butcher, R. L., Blue, J. D., and Fugo, N. W. (1969). Overripeness and the mammalian ova. III. Fetal development at midgestation and at term. *Fertil. Steril.* **20,** 222–231.

Butcher, R. L., and Fugo, N. W. (1967). Overripeness and the mammalian ova: II. Delayed ovulation and chromosome anomalies. *Fertil. Steril.* **18,** 297.

Calarco, P. G., and Brown, E. A. (1969). An ultrastructural and cytological study of preimplantation development in the mouse. *J. Exp. Zool.* **171,** 253–284.

Capco, D. G., and McGaughey, R. W. (1986). Cytoskeletal reorganization during early mammalian development: analysis using embedment-free sections. *Dev. Biol.* **115,** 446–458.

Carnegie, J. A. (1991). Immunolocalization of fibronectin and laminin within rat blastocysts cultured under serum-free conditions. *J. Reprod. Fert.* **91,** 423–434.

Carollo, J. R., and Weitlauf, H. M. (1981). Regional changes in the binding of [3H] concanavalin A to mouse blastocysts at implantation: an autoradiographic study. *J. Exp. Zool.* **218,** 247–251.

Castle, W. E., and Little, C. C. (1910). On a modified mendelian ratio among yellow mice. *Science* **32,** 868–870.

Chai, N., Patel, Y., Jacobson, K., McMahon, J., McMahon, A., and Rapolee, D. A. (1998). FgF is an essential regulator of the fifth cell division in preimplantation mouse embryos. *Dev. Biol.* **198,** 105–115.

Chapman, V. M (1986). X-chromosome regulation in oogenesis and early mammalian development. *In* "Experimental Approaches to Mammalian Embryonic Development" (J. Rossant, P. A. Pedersen, Eds.), pp. 365–435. Cambridge University Press, Cambridge.

Chavez, D. J., Enders, A. C., and Schlafke, S. (1984). Trophectoderm cell subpopulations in the periimplantation mouse blastocysts. *J. Exp. Zool.* **231,** 267–271.

Chen, W. S., Manova, K, Weinstein, D. C., Duncan, S. A., Plump, A. S., Prezioso, V. R., Bachvarova, R. F., and Darnell, J. E., Jr. (1994). Disruption of the HNF-4 gene expressed in visceral endoderm, leads to cell death in the embryonic ectoderm and impaired gastrulation of mouse embryos. *Genes Dev.* **8,** 2466–2477.

Chida, S. (1990). Monozygous double inner cell masses in mouse blastocysts following fertilization in vitro and in vivo. *J. in Vitro Fert. Embryo Transf.* **7,** 177–179.

Chisholm, J. C., Johnson, M. H., Warren, P. D., Fleming, T. P., and Pickering, S. J. (1985). Developmental variability within and between mouse expanding blastocysts and their ICMs. *J. Embryol. Exp. Morph.* **86,** 311–336.

Ciccodicola, A., Dono, R., Obici, S., Simeone, A., Zollo, M., and Persico, M. G. (1989). Molecular characterization of a gene of the "EGF family" expressed in undifferentiated human NTERA2 teractocarcinoma cells. *EMBO J.* **8,** 1987–1991.

Cioffi, J., Van Blerkom, J., Antczak, M., Shafer, A., Wittmerr, S., and Snodgrass, H. R. (1997). The expression of leptin and its receptors in preovulatory human follicles. *Mol. Hum. Reprod.* **3,** 467–472.

Ciruna, B. G., and Rossant, J. (1999). Expression of the T-box gene eomesodermin during early mouse development. *Mech. Dev.* **81,** 199–203.

Cockroft, D. L., and Gardner, R. L. (1987). Clonal analysis of the developmental potential of 6th and 7th day visceral endoderm cells in the mouse. *Development* **101,** 143–155.

Conlon, F. L., Lyons, K. M., Takaesu, N., Barth, K. S., Kispert, A., Hermann, B., and Robertson, E. J. (1994). A primary requirement for nodal in the formation and maintenance of the primitive streak in the mouse. *Development* **120,** 1919–1928.

Constam, D. B., and Robertson, E. J. (1999). Regulation of bone morphogenetic protein activity by pro domains and proprotein convertases. *J. Cell Biol.* **144,** 139–149.

Cooke, J. (1972). Properties of the primary organization field in the embryo of Xenopus laevis. II. Positional information for axial organization in embryos with two head organizers. *J. Embryol. Exp. Morph.* **28,** 27–46.

Copp, A. J. (1979). Interaction between inner cell mass and trophectoderm of the mouse blastocyst. II. The fate of the polar trophectoderm. *J. Embryol. Exp. Morph.* **51,** 109–120.

Copp, A. J. (1995). Death before birth: Clues from gene knockouts and mutations. *Trends Genet.* **11,** 87–93.

Coucouvanis, E., and Martin, G. R. (1995). Signals for death and survival: A two-step mechanism for cavitation in the vertebrate embryo. *Cell* **83,** 279–287.

Coucouvanis, E., and Martin, G. R. (1999). BMP signalling plays a role in visceral endoderm differentiation and cavitation in the early mouse embryo. *Development* **126,** 535–546.

Crossley, P. H., and Martin, G. R. (1995). The mouse *Fgf 8* gene encodes a family of polypeptides and is expressed in regions that direct outgrowth and patterning in the developing embryo. *Development* **121,** 439–451.

Cruz, Y. P., and Pedersen, R. A. (1985). Cell fate in the polar trophectoderm of mouse blastocysts as studied by microinjection of lineage tracers. *Dev. Biol.* **112,** 73–83.

Cuenot, L. (1908). Sur quelques anomalies apparentes des proportions mendeliennes. *Arch. Zool. Exp. Gen.* **9,** 7–15.

Cunniff, C., Jones, K. L., Jones, M. C., Saunders, B., Shepard, T., and Benirschke, K. (1988). Laterality defects in conjoined twins: Implications for normal asymmetry in human embryogenesis. *Am. J. Med. Genet.* **31,** 669–677.

Dalcq, A. M. (1957). "Introduction to General Embryology." Oxford University Press, London.

Dattani, M., Martinez-Barbera, J-P., Thomas, P. Q., Brickman, J. M., Gupta, R., Krauss, S., Wales, J., Hindmarsh, P. C., Beddington, R. S. P., and Robinson, I. C. (1998). Mutations in the homeobox gene Hesx1 associated with septo-optic dysplasia in human and mouse. *Nat. Genet.* **19,** 125–133.

Davidson, E. H. (1986). "Gene Activity in Early Development." 3rd ed. Academic Press: Orlando.

Dean, W. L., and Rossant, J. (1984). Effects of delaying DNA replication on blastocyst formation in the mouse. *Differentiation* **26,** 134–137.

De Gregori, J., Russ, A., von Melchner, H., Rayburn, H., Priyaranjan, P., Jenkins, N. A., Copeland, N. G., and Ruley, H. E. (1994). A murine homolog of the yeast RNA1 gene is required for postimplantation development. *Genes Dev.* **8,** 265–276.

Derom, C., Vlietinck, R., Derom, R., Van Den Berghe, H., and Thiery, M. (1987). Increased monozygotic twinning rate after ovulation induction. *Lancet i.* 1236–1238.

De Sousa, P. A., Watson, A. J., Schultz, G. A. and Bilodeau Goeseels, S. (1998). Oogenetic and zygotic gene expression directing early bovine embryogenesis: A review. *Mol. Reprod. Dev.* **51,** 112–121.

Ding, J., Yang, L., Yan, Y-T., Chen, A., Desai, N., Wynshaw-Boris, A., and Shen, M. M. (1998). *Cripto* is required for correct orientation of the anterior-posterior axis in the mouse embryo. *Nature* **395,** 702–707.

Dono, R., Scalera, L., Pacifico, F., Acampora, D., Persico, M. G., and Simeone, A. (1993). The murine cripto gene: Expression during medoderm induction and heart morphogenesis. *Development* **118,** 1157–1168.

Dufort, D., Schwartz, L., Harpal, K., and Rossant, J. (1998). The transcription factor *HNF3 β* is required in the visceral endoderm for normal primitive streak morphogenesis. *Development* **125,** 3015–3025.

Duncan, S. A., Manova, K., Chen, W. S., Hoodless, P., Weinstein, D. C., Bachvarova, R. F., and Darnell, J. E. Jr. (1994). Expression of transcription factor HNF-4 in the extraembryonic endoderm, gut, and nephrogenic tissue of the developing mouse embryo: HNF-4 is a marker for primary endoderm in the implanting blastocyst. *Proc. Nat. Acad. Sci. USA* **91,** 7598–7602.

Duncan, S. A., Nagy, A., and Chan, W. (1997). Murine gastrulation requires HNF-4 regulated gene expression in the visceral endoderm: Tetraploid rescue of HNF-4 −/− embryos. *Development* **124,** 279–287.

Dunwoodie, S. L., Rodriguez, T. A., and Beddington, R. S. P. (1998). Msg1 and Mrg1, founding members of a gene family, show distinct patterns of gene expression during mouse embryogenesis. *Mech. Dev.* **72,** 27–40.

Dush, M. K., and Martin, G. R. (1992). Analysis of mouse Evx genes: Evx-1 displays graded expression in the primitive streak. *Dev. Biol.* **151,** 273–287.

Dziadek, M., and Adamson, E. D. (1978). Localization and synthesis of alphafoetoprotein in postimplantation mouse embryos. *J. Embryol. Exp. Morph.* **43,** 289–313.

Edwards, R. G. (1985). Normal and abnormal implantation in the human uterus. *In* "Implantation of the Human Embryo: Proceedings of the Second Bourn Hall Meeting" (R. G. Edwards, J. M. Purdy, P. C. Steptoe, Eds.), pp. 303–333. Academic Press, London.

Edwards, R. G., and Beard, H. K. (1997). Oocyte polarity and cell determination in early mammalian embryos. *Mol. Hum. Reprod.* **3,** 863–905.

Edwards, R. G., Purdy, J. M., and Steptoe, P. C. Eds. (1985). "Implantation of the Human Embryo: Proceedings of the Second Bourn Hall Meeting." See Discussion on twinning pp. 408. Academic Press, London.

Eglitis, M. A., and Wiley, L. M. (1981). Tetraploidy and early development: Effects on developmental timing and embryonic metabolism. *J. Embryo. Exp. Morph.* **66,** 91–108.

Emerson, J. A. (1988). Disruption of the cytokeratin filament network in the preimplantation mouse embryo. *Development* **104,** 219–234.

Enders, A. C., and Schlafke, S. (1969). Cytological aspects of trophoblast-uterine interaction in early implantation. *Am. J. Anat.* **125,** 1–30.

Evsikov, S. V., Morozova, L. M., and Solomko, A. P. (1990). The role of the nucleocytoplasmic ratio in development regulation of the early mouse embryo. *Development* **109,** 323–328.

Evsikov, S. V., Morozova, L. M., and Solomko, A. P. (1994). Role of ooplasmic segregation in mammalian development. *Roux's Arch. Dev. Biol.* **203,** 199–204.

Evsikov, S. V., and Solomko, A. P. (1998). Change in the cavitation start time in chimeric CB6F1—BALB/c embryos. *Tsitol. Genet.* **32,** 84–87.

Ezzell, R. M., Chafel, M. M., and Matsudaira, P. T. (1989). Differential localization of villin and fimbrin during development of the mouse visceral endoderm and intestinal epithelium. *Development* **106,** 407–419.

Fan, Y., Melhem, M. F., and Chaillet, J. R. (1999). Forced expression of the homeobox-containing gene Pem blocks differentiation of embryonic stem cells. *Dev. Biol.* **210,** 481–496.

Faust, C., Lawson, K. A., Schork, N. J., Thiel, B., and Magnuson, T. (1998). The Polycomb-group gene eed is required for normal morphogenetic movements during gastrulation in the mouse embryo. *Development* **125,** 4495–4506.

Feldman, B., Poueymirou, W., Papaioannou, V. E., De Chiara, T. M., and Goldfarb, M. (1995). Requirement of FGF-4 for postimplantation mouse development. *Science* **267,** 246–249.

Fishel, S., and Symonds, M., Eds (1993). "Gamete and Embryo Micromanipulation in Human Reproduction." Edward Arnold, London.

Fleming, T. P. (1992). Trophectoderm biogenesis in the preimplantation mouse embryo. *In* "Epithelial Organization and Development," (T. P. Fleming, Ed.), pp. 111–136. Chapman & Hall, London.

Fleming, T. P., and George, M. A. (1987). Fluorescent latex microparticles: A non-invasive short-term cell lineage marker suitable for use in the mouse early embryo. *Roux's Arch. Dev. Biol.* **196,** 1–11.

Fugo, N. W., and Butcher, R. L. (1966). Overripeness and the mammalian ova: Overripeness and early embryonic development. *Fertil. Steril.* **17,** 804–814.

Gabel, C. A., Eddy, E. M., and Shapiro, B. M. (1979). After fertilization, sperm surface components remain as a patch in sea urchin and mouse embryos. *Cell* **18,** 207–215.

Gallicano, G. I., McGaughely, R. W., and Capco, D. G. (1991). Cytoskeleton of the mouse egg and embryo: Reorganization of planar elements. *Cell Motil. Cytoskel.* **18,** 143–154.

Gallicano, G. I., McGaughely, R. W., and Capco, D. G. (1992). Cytoskeletal sheets appear as universal components of mammalian eggs. *J. Exp. Zool.* **263,** 194–203.

Garbutt, C. L., Chisholm, J. C., and Johnson, M. H. (1987). The establishment of the embryonic-abembryonic axis in the mouse embryo. *Development* **100,** 125–134.

Gardner, R. L. (1985). Regeneration of endoderm from primitive ectoderm: Fact or artifact? *J. Embryol. Exp. Morph.* **88,** 303–326.

Gardner, R. L. (1990). Location and orientation of implantation. *In* "Establishing a Successful Human Pregnancy." Serono Symposia Publications, Vol. 66, (R. G. Edwards, Ed.), pp. 225–238. Raven Press, New York.

Gardner, R. L. (1996a). Can developmentally significant spatial patterning of the egg be discounted in mammals. *Hum. Reprod. Update* **2,** 3–27.

Gardner, R. L. (1996b). Clonal analysis of growth of the polar trophectoderm in the mouse. *Hum. Reprod.* **11,** 1979–1984.

Gardner, R. L. (1997). The early blastocyst is bilaterally symmetrical and its axis of symmetry is aligned with the animal-vegetal axis of the zygote in the mouse. *Development* **124,** 289–301.

Gardner, R. L. (1998). Axial relationships between egg and embryo in the mouse. *Curr. Top. Dev. Biol.* **39,** 35–71.

Gardner, R. L. (1999a). Scrambled or bisected mouse eggs and the basis of patterning in mammals. *BioEssays* **21,** 271–274.

Gardner, R. L. (2000a). Flow of cells from polar to mural trophectoderm is polarized in the mouse blastocyst. *Hum. Reprod.* **15,** 694–701.

Gardner, R. L. (2000b). Manuscript in preparation.

Gardner, R. L., and Beddington, R. S. P. (1988). Multi-lineage 'stem cells' in the mammalian embryo. *J. Cell Sci. Suppl.* **10,** 11–27.

Gardner, R. L., and Cockroft, D. L. (1998). Complete dissipation of coherent clonal growth occurs before gastrulation in mouse epiblast. *Development* **125,** 2397–2402.

Gardner, R. L., and Johnson, M. H. (1972). An investigation of inner cell mass and trophoblast tissues following their isolation from the mouse blastocyst. *J. Embryol. Exp. Morph.* **28,** 279–312.

Gardner, R. L., Meredith, M. M., and Altman, D. G. (1992). Is the anterior-posterior axis of the fetus specified before implantation in the mouse? *J. Exp. Zool.* **264,** 437–443.

Gardner, R. L., Papaioannou, V. E., and Barton, S. C. (1973). Origin of the ectoplacental cone and secondary giant cells in mouse blastocysts reconstituted from isolated trophoblast and inner cell mass. *J. Embryol. Exp. Morph.* **30,** 561–572.

Gardner, R. L., and Rossant, J. (1979). Investigation of the fate of 4.5 day postcoitum mouse inner cell mass cells by blastocyst injection. *J. Embryol. Exp. Morph.* **52,** 141–152.

Gedda, L. (1961). "Twins in History and Science." C. C. Thomas, Springfield, Illinois.

Generoso, W. M., Rutledge, J. C., and Aronson, J. (1990). Developmental anomalies: Mutational consequences of mouse zygote exposure. *In* "Biology of Mammalian Germ Cells: Banbury Report 34," pp. 311–319. Cold Spring Harbor Laboratory Press, New York.

Ghilardi, N., Ziegler, S., Wiestner, A., Stoffe, R., Heim, M. H., and Skoda, R. C. (1996). Defective STAT signaling by leptin receptor in diabetic mice. *Proc. Natl. Acad. Sci. USA* **93,** 6231–6235.

Gilula, N. B., Epstein, M. L., and Beers, W. H. (1978). Cell-to-cell communication and ovulation: A study of the cumulus cell-oocyte complex. *J. Cell Biol.* **78,** 58–75.

Glass, L. E. (1971). Transmission of maternal proteins into oocytes. *Adv. Biosci.* **6,** 29–58.

Gluecksohn-Schoenheimer, S. (1940). The effect of an early lethal (t^o) in the house mouse. *Genetics* **25,** 391–400.

Gluecksohn-Schoenheimer, S. (1949). The effects of a lethal mutation responsible for duplications and twinning in the mouse. *J. Exp. Zool.* **110,** 47–76.

Graham, C. F., and Deussen, Z. A. (1978). Features of cell lineage in preimplantation mouse development. *J. Embryol. Exp. Morph.* **48,** 53–72.

Gu, Z., Nomura, M., Simpson, B. B., Lei, H., Feijen, A., Van den Eijnden-van Raaij, J., Donahoe, P. K., and Li, E. (1998). The type I activin receptor ActRIB is required for egg cylinder organization and gastrulation in the mouse. *Genes Dev.* **12,** 844–857.

Gulyas, B. (1975). A reexamination of cleavage patterns in eutherian mammalian eggs: Rotation of blastomere pairs during second cleavage in the rabbit. *J. Exp. Zool.* **193,** 235–248.

Haegel, H., Larue, L., Ohsugi, M., Federov, L., Herrenknecht, K., and Kemler, R. (1995). Lack of beta-catenin affects mouse development at gastrulation. *Development* **121,** 3529–3537.

Hall, J. G. (1996). Twins and twinning. *Am. J. Med. Genet.* **61,** 202–204.

Harlap, S., Shahar, S., and Baras, M. (1985). Overripe ova and twinning. *Am. J. Hum. Genet.* **37,** 1206–1215.

Heasman, J. (1997). Patterning the *Xenopus* blastula. *Development* **124,** 4179–4191.

Heasman, J., Crawford, A., Goldstone, K., Garner-Hamrick, P., Gumbiner, B., McCrea, P., Kintner, C., Noro, C. Y., and Wylie, C. (1994). Overexpression of cadherins and underexpression of β-catenin inhibit dorsal mesoderm induction in early *Xenopus* embryos. *Cell* **79,** 791–803.

Hedgepath, C. M., Conrad, L. J., Zhang, J., Huang, H.-C., Lee, V. M. Y., and Klein, P. S. (1997). Activation of the Wnt signaling pathway: A molecular mechanism for lithium action. *Dev. Biol.* **185,** 82–91.

Hogan, B., and Tilly, R. (1977). In vitro culture and differentiation of normal mouse blastocysts. *Nature* **265,** 626–629.

Howlett, S. K., and Bolton, V. N. (1985). Sequence and regulation of morphological and molecular events during the first cycle of mouse embryogenesis. *J. Embryo. Exp. Morph.* **87,** 175–206.

Hrabe de Angelis, M., and Kirchner, C. (1993). Fibroblast growth factor induces primitive streak formation in rabbit pre-implantation embryos in vitro. *Anat. Embryol.* **187,** 269–273.

Hsu, Y.-C., Baskar, J., Stevens, L. C., and Rash, J. E. (1974). Development in vitro of mouse embryos from the two-cell egg stage to the early somite stage. *J. Embryol. Exp. Morph.* **31,** 235–245.

Huber, G. C. (1915). The development of the albino rat, Mus norvegicus albinus. I. From the pronuclear stage to the stage of the mesoderm anlage: end of the first to the end of the 9th day. *J. Morph.* **26,** 247–358.

Iannaccone, P. M., Zhou, X., Khokha, M., Boucher, D., and Kuehn, M. R. (1992). Insertional mutation of a gene involved in growth regulation of the early mouse embryo. *Dev. Dyn.* **194,** 198–208.

Jacobs-Cohen, R. J., Spiegelman, M., Cookingham, J. C., and Bennett, D. (1984). Knobbly, a new dominant mutation in the mouse that affects embryonic ectoderm organization. *Genet. Res.* **43,** 43–50.

Jauniaux, E., Elkazen, N., and Leroy, F. (1986). Clinical and morphological aspects of the vanishing twin phenomenon. *Obstet. Gynecol.* **68,** 577–581.

Johnson, M. H., Eager, D. D., Muggleton-Harris, A., and Grave, H. M. (1975). Mosaicism in the organization of concanavalin A receptors on the surface membrane of mouse eggs. *Nature* **257,** 321–322.

Johnson, M. H., and Ziomek, C. A. (1981). The foundation of two distinct lineages within the mouse morula. *Cell* **24,** 71–80.

Johnson, S. E., Rothstein, J. L., and Knowles, B. B. (1994). Expression of epidermal growth factor family gene members in early mouse development. *Dev. Dyn.* **201,** 216–226.

Johnson, W. H., Loskutoff, N. M., Plante, Y., and Betteridge, K. J. (1995). Production of four identical calves by separation of blastomeres from an in vitro derived four-cell embryo. *Vet. Rec.* **137,** 15–16.

Jones, D. H., Davies, T. C., and Kidder, G. M. (1997). Embryonic expression of the putative gamma subunit of the sodium pump is required for acquisition of fluid transport capacity during mouse blastocyst development. *J. Cell Biol.* **139,** 1545–1552.

Jones-Seaton, A. (1950). Etude de l'organization cytoplasmic de l'oeuf de rongeurs principalement quant a la basophile ribonucleique. *Arch. Biol.* **61,** 291–444.

Jost, H. J. (1992). Regulation of vertebrate left-right asymmetries by extracellular matrix. *Nature* **357,** 158–161.

Katoh, M., Cacheiro, N. L. A., Cornett, C. V., Cain, K. T., Rutledge, J. C., and Generoso, W. M. (1989). Fetal anomalies produced subsequent to treatment of zygotes with ethylene oxide or ethyl methanesulfonate are not likely due to the usual genetic causes. *Mutation Res.* **210,** 337–344.

Kaufman, M. H. (1982). Two examples of monoamniotic monozygotic twinning in diploid parthenogenetic mouse embryos. *J. Exp. Zool.* **224,** 277–282.

Kaufman, M. H. (1985). Experimental aspects of monozygotic twinning. *In* "Implantation of the Human Embryo: Proceedings of the Second Bourn Hall Meeting" (R. G. Edwards, J. M. Purdy, and P. C. Steptoe, Eds.), pp. 371–391. Academic Press, London.

Kaufman, M. H., and O'Shea, K. S. (1978). Induction of monozygotic twinning in the mouse. *Nature* **276,** 707–708.

Keighren, M., and West, J. D. (1993). Analysis of cell ploidy in histological sections of mouse tissued by DNA-DNA in situ hybridization with digoxigenin labelled probes. *Histochem. J.* **25,** 30–44.

Keng, V. W., Fujimori, K. E., Myint, Z., Tamamaki, N., Nojyo, Y., and Noguchi, T. (1998). Expression of Hex mRNA in early murine development. *FEBS Lett.* **426,** 183–186.

Khaner, O. (1995). The rotated hypoblast of the chicken embryo does not initiate an ectopic axis in the epiblast. *Proc. Natl. Acad. Sci. USA* **92,** 10733–10737.

Khaner, O., and Eyal-Giladi, H. (1989). The chick's marginal zone and primitive streak formation. I. Coordinative effects of induction and inhibition. *Dev. Biol.* **134,** 206–214.

Kimmel, C. A., Generoso, W. M., Thomas, R. D., and Bakshi, K. S. (1993). A new frontier in understanding the mechanism of developmental abnormalities. *Toxicol. Applied Pharmacol.* **119,** 159–165.

Kirby, D. R. S., Potts, D. M., and Wilson, I. B. (1967). On the orientation of the implanting blastocyst. *J. Embryol. Exp. Morph* **17,** 527–532.

Kishida, M., Koyama, S., Kishida, S., Matsubara, K., Nakashima, S., Higano, K., Takada, R., Takada, S., and Kikuchi, A. (1999). Axin prevents Wnt-3a-induced accumulation of beta-catenin. *Oncogene* **18,** 978–985.

Klein, P. S., and Melton, D. A. (1996). A molecular mechanism for the effect of lithium on development. *Proc. Natl. Acad. Sci. USA* **93,** 8455–8459.

Knoetgen, H., Viebahn, C., and Kessel, M. (1999). Head induction in the chick by primitive endoderm of mammalian but not avian origin. *Development* **126,** 815–825.

Kola, I., Kirby, C., Shaw, J., Davey, A., and Trounson, A. (1988). Vitrification of mouse oocytes results in aneuploid zygotes and malformed fetuses. *Teratology* **38,** 467–474.

Koutsourakis, M., Langeveld, A., Patient, R., Beddington, R., and Grossveld, F. (1999). The transcription factor GATA 6 is essential for early extraembryonic development. *Development* **126,** 723–732.

Krzanowska, H. (1972). Rapidity of removal in vitro of the cumulus oophorus and the zona pellucida in different strains of mice. *J. Reprod. Fert.* **31,** 7–14.

Kyvik, K. O., Green, A., and Beck-Nielson, H. (1995). The new Danish twin register: Establishment and analysis of twinning rates. *Int. J. Epidemiol.* **24,** 589–596.

Lane, M., and Gardner, D. K. (1994). Increase in postimplantation development of cultured mouse embryos by amino acids and induction of fetal retardation and exencephaly by ammonium ions. *J. Reprod. Fert.* **102,** 305–313.

Lawson, K. A., Meneses, J. J., and Pedersen, R. A. (1986). Cell fate and cell lineage in the endoderm of the presomite mouse embryo. *Dev. Biol.* **115,** 325–339.

Lawson, K. A., Pedersen, R. A., and Van den Geer, S. (1987). Cell fate, morphogenetic movements and population kinetics of embryonic endoderm at the time of germ layer formation in the mouse. *Development* **101,** 627–652.

Lawson, K. A., Meneses, J. J., and Pedersen, R. A. (1991). Clonal analysis of epiblast fate during germ layer formation in the mouse embryo. *Development* **113,** 891–911.

Lawson, K. A., Dunn, N. R., Roelen, B. A., Zeinstra, L. M., Davis, A. M., Wright, C. V., Korning, J. P., and Hogan, B. L. (1999). Bmp4 is required for the generation of primordial germ cells in the mouse embryo. *Genes Dev.* **13,** 424–436.

Lea, R. G., McCracken, J. E., McIntyre, S. S., Smith, W., and Baird, J. D. (1996). Disturbed development of the preimplantation embryo in the insulin-dependent diabetic BB/E rat. *Diabetes* **45,** 1463–1470.

Lee, J. J., Radice, G., Perkins, C. P., and Constantini, F. (1992). Identification and characterization of a novel, evolutionarily conserved gene disrupted by the murine Hβ58 embryonic lethal transgene insertion. *Development* **115,** 277–288.

Lehtonen, E., and Reima, I. (1986). Changes in the distribution of vinculin during preimplantation mouse development. *Differentiation* **32,** 125–134.

Leroy, F. (1985). Early embryology and placentation of human twins. *In* "Implantation of the Human Embryo: Proceedings of the Second Bourn Hall Meeting" (R. G. Edwards, J. M. Purdy, and P. C. Steptoe, Eds.), pp. 393–405. Academic Press, London.

Lewis, N. E., and Rossant, J. (1982). Mechanism of size regulation in mouse embryo aggregates. *J. Embryol. Exp. Morph.* **72,** 169–181.

Lewis, W. H., and Wright, E. S. (1935). On the early development of the mouse egg. *Cont. Embry. Carnegie Inst. Wash.* **25,** 113–143.

Lin, T.-P., Laborsky, P. A., Grabel, L. B., Kozak, C. A., Pitman, J. L., Kleeman, J., and MacLeod, C. L. (1994). The *Pem* homeobox gene is X-linked and exclusively expressed in extraembryonic tissues during early murine development. *Dev. Biol.* **166,** 170–179.

Little, J., and Bryan, E. M. (1988). Congenital anomalies. *In* "Twinning and Twins" (I. MacGillivray, D. M. Campbell, and B. Thompson, Eds.), pp. 207–240. John Wiley and Sons, Chichester.

Liu, C.-Y., Flesken-Nikitin, A., Li, S., Zeng, Y., and Lee, W.-H. (1996). Inactivation of the mouse Brca1 gene leads to failure in the morphogenesis of the egg cylinder in early postimplantation development. *Genes Dev.* **10,** 1835–1843.

Longo, F. J. (1987). "Fertilization." Chapman & Hall, London.

MacArthur, J. W. (1938). Genetics of quintuplets. I. Diagnosis of the Dionne quintuplets as a monozygotic set. *J. Hered.* **29,** 323–329.

Manejwala, F. M., and Schultz, R. M. (1989). Blastocoel expansion in the preimplantation mouse embryo: Stimulation of sodium uptake by cAMP and possible involvement of cAMP-dependent protein kinase. *Dev. Biol.* **136,** 560–563.

Manova, K., Tomihara-Newberger, C., Wang, S., Godelman, A., Kalantry, S., Witty-Blease, K., De Leon, V., Chen, W. S., Lacey, E., and Bachvarova, R. F. (1998). Apoptosis in mouse embryos: Elevated levels in pregastrulae and in the distal anterior region of gastrulae of normal and mutant mice. *Dev. Dyn.* **213,** 293–308.

Marston, J. H., and Chang, M. C. (1964). The fertilizable life of ova and their morphology following delayed insemination in mature and immature mice. *J. Exp. Zool.* **155,** 237–252.

Matsuo, I., Kuratani, S., Kimura, C., Tadaka, N., and Aizawa, S. (1995). Mouse Otx2 functions in the formation and patterning of the rostral head. *Genes Dev.* **9,** 2646–2658.

Matsuoka, T., Tahara, M., Yokoi T., Masumoto, N., Takeda, T., Yamaguchi, M., Tasaka, K., Kurachi, H., and Murota, Y. (1999). Tyrosine phosphorylation of STAT 3 by leptin through leptin receptor in mouse metaphase 2 stage oocyte. *BBRC* **256,** 480–484.

Matzuk, M. M., Kumar, T. R., Vassalli, J. R., Bickenbach, D. R., Roop, R., Jaenisch, R., and Bradley, A. (1995). Functional analysis of activins during mammalian development. *Nature* **374,** 354–356.

Maunoury, R., Robine, S., Pringault, E., Huet, C., Guenet, J. L., Gaillard, J. A., and Louvard, D. (1988). Villin expression in the visceral endoderm and in the gut anlage during early mouse embryogenesis. *EMBO J.* **7,** 3321–3329.

Mehrotra, P. K. (1984). Role of trophoblast cells in attachment of mouse blastocyst and its further development. *In* "Proceedings of the Fifth All India Symposium on Developmental Biology" (S. K. Agarwal, and S. C. Goel, Eds.), pp. 21–27. Indian Society of Developmental Biologists, Poona.

Messinis, I. E., and Milingos, S. D. (1999). Leptin in human reproduction. *Hum. Reprod. Update* **5,** 52–63.

McLaren, A., Molland, P., and Signer, E. (1995). Does monozygotic twinning occur in mice? *Genet. Res.* **66,** 195–202.

Michaud, E. J., Bultman, S. J., Stubbs, L. J., and Woychik, R. P. (1993). The embryonic lethality of homozygous lethal yellow mice (*A y*/*A y*) is associated with the disruption of a novel RNA-binding protein. *Genes Dev.* **7,** 1203–1213.

Mintz, B. (1964). Formation of genetically mosaic mouse embryos and the early development of "lethal" (t^{12}/t^{12})-normal mosaics. *J. Exp. Zool.* **157,** 273–292.

Mintz, B. (1965). Experimental genetic mosaicism in the mouse. *In* "Preimplantation Stages of Pregnancy; A Ciba Foundation Symposium" (G. E. W. Wolstenholme and M. O'Connor, Eds.), pp. 194–207. Churchill: London.

Mishina, Y., Suzuki, A., Ueno, N., and Behringer, R. R. (1995). Bmpr encodes a type I bone morphogenetic protein receptor that is essential for gastrulation during mouse embryogeneses. *Genes Dev.* **9,** 3027–3037.

Moley, K. H., Vaughn, W. K., DeCherney, A. H., and Diamond, M. P. (1991). Effect of diabetes mellitus on mouse pre-implantation development. *J. Reprod. Fertil.* **93,** 325–332.

Molkentin, J. D., Lin, Q., Duncan, S. A., and Olsen, E. N. (1997). Requirement of the transcription factor GATA 4 for heart tube formation and ventral morphogenesis. *Genes Dev.* **11,** 1061–1072.

Morrisey, E. E., Tang, Z., Sigrist, K., Lu, M. M., Jiang, F., Ip, H. S., and Parmacek, M. S. (1998). GATA 6 regulates HNF4 and is required for differentiation of visceral endoderm in the mouse embryo. *Genes Dev.* **12,** 3579–3590.

Mulnard, J. G., and Puissant, F. (1984). Development of mouse embryos after ultracentrifugation at the pronuclei stage. *Arch. Biol.* **97,** 301–315.

Narita, N., Bielinska, M., and Wilson, D. B. (1997). Wild type endoderm abrogates the ventral developmental defects associated with GATA 4 deficiency in the mouse. *Dev. Biol.* **189,** 270–274.

Naruse, I., Kano, S., and Shoji, R. (1983). Experimental induction of two inner cell masses in mouse embryos by vinblastine treatment in vitro. *Teratology* **28,** 215–218.

Nichols, J., Zevnik, B., Anastassiadis, K., Niwa, H., Klewe-Nebenius, D., Chambers, I., Scholer, H., and Smith, A. (1998). Formation of pluripotent stem cells in the mammalian embryo depends on the POU transcription factor Oct 4. *Cell* **95,** 379–391.

Niswander, L., and Martin, G. R. (1992). Fgf-4 expression during gastrulation, myogenesis, limb and tooth development in the mouse. *Development* **114,** 755–768.

Nomura, M., and Li, E. (1998). Smad 2 role in mesoderm formation, left-right patterning and craniofacial development. *Nature* **393,** 786–790.

O'Brien, M. J., Critser, E. S., and First, N. L. (1984). Developmental potential of isolated blastomeres from from early murine embryos. *Theriogenology* **22,** 601–607.

Otani, H., Tanaka, O., Tatewaki, R., Naora, H., and Yoneyama, T. (1991). Diabetic environment and genetic predisposition as causes of congenital malformations in NOD mouse embryos. *Diabetes* **40,** 1245–1250.

Ozdzenski, W., Szczesny, E., and Tarkowski, A. K. (1997). Postimplantation development of mouse blastocysts with two separate inner cell masses. *Anat. Embryol.* **195,** 467–471.

Pampfer, S., De Hertogh, R., Vanderheyden, I., Michiels, B., and Vercheval, M. (1990). Decreased inner cell mass proportion in blastocysts from diabetic rats. *Diabetes* **39,** 471–476.

Papaioannou, V. E. (1982). Lineage analysis of inner cell mass and trophectoderm using microsurgically reconstituted blastocysts. *J. Embryol. Exp. Morph.* **68,** 199–209.

Papaioannou, V. E., and Gardner, R. L. (1979). Investigation of the lethal yellow A*y*/A*y* embryo using mouse chimaeras. *J. Embryol. Exp. Morph.* **52,** 153–163.

Paria, B. C., Das, S. K., and Dey, S. K. (1995). The preimplantation mouse embryo is a target for cannabinoid ligand-receptor signalling. *Proc. Natl. Acad. Sci. USA* **92,** 9460–9464.

Payne, D., Flaherty, S. P., Barry, M. F., and Matthews, C. D. (1997). Preliminary observations on polar body extrusion and pronuclear formation in human oocytes using time-lapse video cinematography. *Hum. Reprod.* **12,** 532–541.

Pearce, J. J., Penny, G., and Rossant, J. (1999). A mouse Cerberus/Dan-related gene family. *Dev. Biol.* **209,** 98–110.

Perona, R. M., and Wassarman, P. M. (1986). Mouse blastocysts hatch in vitro using a trypsin-like proteinase associated with cells of the mural trophectoderm. *Dev. Biol.* **114,** 42–52.

Perry, W. L. 3rd, Vasicek, T. J., Lee, J. J., Rossi, J. M., Zeng, L., Zhang, T., Tilghman, S. M., and Constantini, F. (1995). Phenotype and molecular analysis of a transgenic insertion allele of the mouse fused locus. *Genetics* **141,** 321–332.

Pitman, J. L., Lin, T. P., Kleeman, J. E., Erickson, G. F., and MacLeod, C. L. (1998). Normal reproductive and macrophage function in Pem homeobox-containing gene-deficient mice. *Dev. Biol.* **202,** 196–214.

Polifka, J. E., Rutledge, J. C., Kimmel, G. L., Dellarco, V., and Generoso, W. M. (1996). Exposure to ethylene oxide during early zygotic period induces skeletal anomalies in mouse fetuses. *Teratology* **53,** 1–9.

Popperl, H., Schmidt, C., Wilson, V., Hume, C. R., Dodd, J., Krumlauf, R., and Beddington, R. S. P. (1997). Misexpression of Cwnt8C in the mouse induces an ectopic axis and causes truncation of the anterior neuectoderm. *Development* **124,** 2997–3005.

Prather, R. S., and First, N. L. (1986). Reprogramming of murine blastocoele formation. *J. Exp. Zool.* **237,** 347–350.

Radice, G., Lee, J. J., and Constantini, F. (1991). Hβ58, an insertional mutation affecting early postimplantation development of the mouse embryo. *Development* **111,** 801–811.

Ram, P. T., and Schultz, R. M. (1993). Reporter gene expression in G2 of the 1-cell mouse embryo. *Dev. Biol.* **156,** 552–556.

Rands, G. F. (1986). Size regulation in the mouse embryo. I. The development of quadruple aggregates. *J. Embryol. Exp. Morph.* **94,** 139–148.

Rappolee, D. A. (1999). It's not just baby's babble/Babel: Recent progress in understanding the language of early mammalian development: A minireview. *Mol. Reprod. Dev.* **52,** 234–240.

Reeve, W. J. D. (1981). The distribution of ingested horseradish peroxidase in the 16-cell mouse embryo. *J. Embryol. Exp. Morph.* **66,** 191–207.

Reinius, S. (1967). Ultrastructure of blastocyst attachment in the mouse. *Zeit. Zellforchung.* **77,** 257–266.

Rhinn, M., Dierich, A., Shawlot, W., Behringer, R. R., Le Meur, M., and Ang, S.-L. (1998). Sequential roles for Otx2 in visceral endoderm and neurectoderm for forebrain and midbrain induction and specification. *Development* **125,** 845–856.

Richards, W. G., Carroll, P. M., Kinloch, R. A., Wassarman, P. M., and Strickland, S. (1993). Creating maternal effect mutations in transgenic mice: Antisense inhibition of an oocyte gene product. *Dev. Biol.* **160,** 543–553.

Rinkenberger, J. L., Cross, J. C., and Werb, Z. (1997). Molecular genetics of implantation in the mouse. *Dev. Genet.* **21,** 6–20.

Rivera-Perez, J. A., Mallo, M., Gendron-Maguire, M., Gridley, T., and Behringer, R. R. (1995). goosecoid is not an essential component of the mouse gastrula organizer but is required for craniofacial and rib development. *Development* **121,** 3005–3012.

Rogers, I., and Varmuza, S. (1996). Epigenetic alterations brought about by lithium treatment disrupts mouse embryo development. *Mol. Reprod. Dev.* **45,** 163–170.

Rosenquist, T. A., and Martin, G. R. (1995). Visceral endoderm -1 (VE-1): An antigen marker that distinguishes anterior from posterior embryonic visceral endoderm in the early post-implantation mouse embryo. *Mech. Dev.* **49,** 117–121.

Runner, M. N. (1984). New evidence for monozygotic twins in the mouse: Twinning initiated in the late blastocyst can account for mirror image asymmetries. *Anat. Rec.* **209,** 399–406.

Rutledge, J. C. (1997). Development toxicity induced during early stages of mammalian embryogenesis. *Mutat. Res.* **396,** 113–127.

Rutledge, J. C., Generoso, W. M., Shourbaji, A., Cain, K. T., Gans, M., and Oliva, J. (1992). Development anomalies derived from exposure of zygotes and first cleavage embryos to mutagens. *Mutation Res.* **296,** 167–177.

Santella, L., Alikani, M., Talansky, B. E., Cohen, J., and Dale, B. (1992). Is the human oocyte plasm membrane polarized? *Human Reprod.* **7,** 999–1003.

Sato, M., and Muramatsu, T. (1985). Reactivity of five N-acetylgalactosamine-recognizing lectins with preimplantation embryos, early postimplantation embryos, and teratocarcinoma cells of the mouse. *Differentiation.* **29,** 29–38.

Saxen, L., and Rapola, J. (1969). "Congenital Defects." Holt, Rinehart and Winston, New York.

Schinzel, A. A., Smith, D. W., and Miller, J. R. (1979). Monozygotic twinning and structural defects. *J. Pediatr.* **95,** 921–930.

Schrewe, H., Gendorn-Maguire, M., Harbison, M. L., and Gridley, T. (1994). Mice homozygous for a null mutation of activin β_B are viable and fertile. *Mech. Dev.* **47,** 43–51.

Schultz, G. A. (1986). Utilization of genetic information in the preimplantation mouse embryo. *In* "Experimental Approaches to Mammalian Embryonic Development," (J. Rossant and P. A. Pedersen, Eds.), pp. 239–265. Cambridge University Press, Cambridge.

Schumacher, A., Faust, C., and Magnuson, T. (1996). Positional cloning of a global regulator of anterior-posterior patterning in mice. *Nature* **383,** 250–253.

Schwarz, S. M., Gallicano, G. I., McGaughey, R. W., and Capco, D. G. (1995). A role for intermediate filaments in the establishment of the primitive epithelia during mammalian embryogenesis. *Mech. Dev.* **53,** 305–321.

Shawlot, W., and Behringer, R. R. (1995). Requirement for *Lim 1* in head-organizer function. *Nature* **374,** 425–430.

Shawlot, W., Deng, J. M., and Behringer, R. R. (1998). Expression of the mouse cerberus-related gene, Cerr 1, suggests a role in anterior neural induction and somitogenesis. *Proc. Natl. Acad. Sci. USA* **95,** 6198–6203.

Shen, M. M., Wang, H., and Leder, P. (1997). A differential display strategy identifies Cryptic, a novel EGF-related gene expressed in the axial and lateral mesoderm during mouse gastrulation. *Development* **124,** 429–442.

Simeone, A., Acampora, D., Mallamaci, A., Stornaiuolo, A., D'Apice, M. R., Nigro, V., and Boncinelli, E. (1993). A vertebrate gene related to *orthodenticle* contains a homeodomain of the bicoid class and demarcates anterior neurectoderm in the gastrulating mouse embryo. *EMBO J.* **12,** 2735–2747.

Sirard, C., De La Pompa, J. L., Elia, A., Itie, A., Miortsos, C., Cheung, A., Hahn, S., Wakeham, A., Schwartz, L., Kern, S. E., Rossant, J., and Mak, T. W. (1998). The tumor suppressor gene Smad4/Dpc4 is required for gastrulation and later for anterior development of the mouse embryo. *Genes Dev.* **12,** 107–119.

Slager, H. G., Lawson, K. A., van den Eijmdem-vanRaaij, A. J. M., DeLaat, S. W., and Mummery, C. L. (1991). Differential localization of TGF-β2 in mouse preimplantation and postimplantation development. *Dev. Biol.* **145,** 205–218.

Smith, L. J. (1980). Embryonic axis orientation in the mouse and its correlation with blastocysts relationship to uterus: Part I. Relationships between 82 hours and 4 1/2 days. *J. Embryol. Exp. Morph.* **55,** 257–277.

Smith, L. J. (1985). Embryonic axis orientation in the mouse and its correlation with blastocysts relationship to uterus: Part II. Relationships from 4 1/2 to 9 1/2 days. *J. Embryol. Exp. Morph.* **89,** 15–35.

Smith, R., and McLaren, A. (1977). Factors affecting the time of formation of the mouse blastocoele. *J. Embryol. Exp. Morph.* **41,** 79–92.

Snell, G. D., and Stevens, L. C. (1966). Early embryology. *In* "Biology of the Laboratory Mouse," 2nd Ed. (E. L. Green, Ed.), pp. 205–254. McGraw-Hill, New York.

Snow, M. H. L. (1976). Embryo growth during the immediately postimplantation period. *In* "Embryogenesis in Mammals: Ciba Foundation Symposium 40" (new series), (K. Elliot and M. O'Connor, Eds.), pp. 53–66. Elsevier: Amsterdam.

Solter, D., Skreb, N., and Damjanov, I. (1971). Cell cycle analysis in the mouse egg-cylinder. *Exp. Cell Res.* **64,** 331–334.

Soriano, P., Gridley, T. and Jaenisch, R. (1987). Retroviruses and insertional mutagenesis in mice: Proviral integration at the Mov 34 locus leads to early embryonic death. *Genes Dev.* **1,** 366–375.

Soltynska, M. S., Balakier, H., Witkowska, A., and Karasiewicz, J. (1985). Binucleate cells in mouse morulae. *Wilhelm Roux' Arch. Dev. Biol.* **194,** 173–177.

Soudais, C., Bielinska, M., Heikinheimo, M., MacArthur, C. A., Narita, N., Saffitz, J. E., Simon, M. C., Leiden, J. M., and Wilson, D. B. (1995). Targeted mutagenesis of the transcription factor GATA 4 in mouse embryonic stem cells disrupts visceral endoderm differentiation in vitro. *Development* **121,** 3877–3888.

Spyropoulos, D. D., and Capecchi, M. R. (1994). Targeted disruption of the *even-skipped* gene, *evx-1,* causes early postimplantation lethality of the mouse conceptus. *Genes Dev.* **8,** 1949–1961.

Stambolic, V., Reul, L., and Woodgett, J. R. (1996). Lithium inhibits glycogen synthase kinase-3 activity and mimics wingless signalling in intact cells. *Curr. Biol.* **6,** 1664–1668.

Stern, C. (1973). "Principles of Human Genetics." 3rd Ed. W. H. Freeman & Company, San Francisco.

Surani, M. A. H., and Barton, S. C. (1984). Spatial distribution of blastomeres is dependent on cell division order and interactions in mouse morulae. *Dev. Biol.* **102,** 335–343.

Surani, M. A. H., Barton, S. C., and Burling, A. (1980). Differentiation of 2-cell and 8-cell mouse embryos arrested by cytoskeletal inhibitors. *Exp. Cell Res.* **125,** 275–286.

Suzuki, H., Togashi, M., Adachi, J., and Toyoda, Y. (1995). Development ability of zona-free mouse embryos is influenced by cell association at the 4-cell stage. *Biol. Reprod.* **53,** 78–83.

Sweet, H. C. (1998). Specification of first quartet micromeres in Ilyanassa involves inherited factors and position with respect to the inducing D macromere. *Development* **125,** 4033–4044.

Szymonowicz, W., Preston, H., and Yu, V. Y. H. (1986). The surviving monozygotic twin. *Arch. Dis. Child.* **61,** 454–458.

Takeda, N., Noguchi, K., Shi, W., Tanaka, T., Matsumoto, M., Yoshida, N., Kishimoto, T., and Akira, S. (1997). Targeted disruption of the mouse Stat3 gene leads to early embryonic lethality. *Proc. Natl. Acad. Sci. USA* **94,** 3801–3804.

Talansky, B. E., Malter, H. E., and Cohen, J. (1991). A preferential site for sperm-egg fusion in mammals. *Mol. Reprod. Dev.* **28,** 183–188.

Tanaka, S., Kunath, T., Hadjantonakia, A.-K., Nagy, A., and Rossant, J. (1998). Promotion of trophoblast stem cell proliferation by FGF 4. *Science* **282,** 2072–2075.

Tarkowski, A. K., and Wroblewska, J. (1967). Development of blastomeres of mouse eggs isolated at the 4- and 8-cell stage. *J. Embryol. Exp. Morph.* **18,** 155–180.

Thomas, P., and Beddington, R. (1996). Anterior primitive endoderm may be responsible for patterning the anterior neural plate in the mouse embryo. *Curr. Biol.* **6,** 1487–1496.

Thomas, P., Brickman, J., Popperl, H., Krumlauk, R., and Beddington, R. S. P. (1997). Axis duplication and anterior identity in the mouse embryo. *Cold Spring Harb. Symp. Quant. Biol.* **62,** 135–144.

Thomas, P. Q., Brown, A., and Beddington, R. S. P. (1998). *Hex:* A homeobox gene revealing peri-implantation asymmetry in the mouse embryo and an early transient marker of endothelial cell precursors. *Development* **125,** 85–94.

Thompson, R. S., and Zamboni, L. (1975). Anomalous patterns of mammalian oocyte maturation and fertilization. *Am. J. Anat.* **142,** 233–264.

Tsunoda, Y., and McLaren, A. (1983). Effect of various procedures on the viability of mouse embryos containing half the normal number of blastomeres. *J. Reprod. Fert.* **69,** 315–322.

Van Beneden, E. (1883). Recherches sur la maturation de l'oeuf et la fecundation. *Arch. Biol.* **4,** 265–640.

Van Blerkom, J., Chavez, D. J., and Bell, H. (1979). Molecular and cellular aspects of facultative delayed implantation in the mouse. *In* "Maternal Recognition of Pregnancy: Ciba Foundation Symposium 64" (new series), (J. Whelan, Ed.), pp. 141–163. Excerpta Medica, Amsterdam.

Varlet, I., Collignon, J., and Robertson, E. J. (1997). Nodal expression in the primitive endoderm is required for specification of the anterior axis during gastrulation. *Development* **124,** 1033–1044.

Varmuza, S., Prideaux, V., Kothary, R., and Rossant, J. (1988). Polytene chromosomes in mouse trophoblastic giant cells. *Development* **102,** 127–134.

Vasicek, T. J., Zeng, L., Guan, X. J., Zhang, T., Constantini, F., and Tilghman, S. M. (1997). Two dominant mutations in the mouse fused gene are the result of transposon insertion. *Genetics* **147,** 777–786.

Viebahn, C., Mayer, B., and Hrabe de Angelis, M. (1995). Signs of the principle body axes prior to primitive streak formation in the rabbit embryo. *Anat. Embryol.* **192,** 159–169.

Wakayama, T., Hayashi, Y., and Ogura, A. (1997). Participation of the female pronucleus derived from the second polar body in full embryonic development in mice. *J. Reprod. Fert.* **110,** 263–266.

Wakayama, T., and Yanagimachi, R. (1998). Fertilisability and developmental ability of mouse oocytes with reduced amounts of cytoplasm. *Zygote* **6,** 341–346.

Waksmundzka, M., Krysiak, E., Karasiewicz, J., Czoloska, R., and Tarkowski, A. K. (1984). Autonomous cortical activity in mouse eggs controlled by a cytoplasmic clock. *J. Embryol. Exp. Morph.* **79,** 77–96.

Waldrip, W. R., Bikoff, E. K., Hoodless, P. A., Wrana, J. L., and Robertson, E. J. (1998). Smad 2 signalling in extraembryonic tissues determines anterior-posterior polarity of the early mouse embryo. *Cell* **92,** 797–808.

Wallace, M. E., and Williams, D. A. (1965). Monozygotic twinning in mice. *J. Med. Genet.* **2,** 26–31.

Wartiovaara, J., Leivo, I., and Vaheri, A. (1979). Expression of cell surface-associated glycoprotein, fibronectin, in the early mouse embryo. *Dev. Biol.* **69,** 247–257.

Weinstein, D. C., Ruiz i Altaba, A., Chen, W. S., Hoodless, P., Prezioso, V. R., Jessell, T. M., and Darnell, J. J. (1994). The winged-helix transcription factor HNF-3 beta is required for notochord development in the mouse embryo. *Cell* **78,** 575–588.

Weisberg, E., Winnier, G. E., Chen, X., Farnsworth, C. L., Hogan, B. L., and Whitman, M. (1998). A mouse homologue of FAST1 transduces TGFβ superfamily signals and is expressed during early embryogenesis. *Mech. Dev.* **79,** 17–27.

Wenstrom, K. D., Syrop, C. H., Hammitt, D. G., and Van Voorhis, B. J. (1993). Increased risk of monchorionic twinning associated with assisted reproduction. *Fertil. Steril.* **60,** 510–514.

Wiley, L. M., and Eglitis, M. A. (1980). Effects of colcemid on cavitation during mouse blastocoele formation. *Exp. Cell Res.* **127,** 89–101.

Winnier, G., Blessing, M., Laborsky, P. A., and Hogan, B. L. (1995). Bone morphogenetic protein-4 is required for mesoderm formation and patterning in the mouse. *Genes Dev.* **9,** 2105–2116.

Witschi, E, (1952). Overripeness of the egg as a cause of twinning and teratogenesis: A review. *Cancer Res.* **12,** 763–786.

Xu, C., Liguori, G., Persico, M. G., and Adamson, E. D. (1999). Abrogation of the Cripto gene in mouse leads to failure of postgastrulation morphogenesis and lack of differentiation of cardiomyocytes. *Development* **126,** 483–494.

Yamada, G., Mansouri, A., Torres, M., Stuart, E. T., Blum, M., Schultz, M., De Robertis, E. M., and Gruss, P. (1995). Targeted mutation of the murine goosecoid gene results in craniofacial defects and neonatal death. *Development* **121,** 2917–2922.

Yamagata, T., and Yamazaki, K. (1991). Implanting mouse embryo stain with LNF-1 bearing fluorescent probe at their mural trophectoderm side. *Biochem. Biophys. Res. Commun.* **181,** 1004–1009.

Yamazaki, K., and Kato, Y. (1989). Sites of zona shedding by mouse embryo other than mural trophectoderm. *J. Exp. Zool.* **249,** 347–34.

Young, J. K., Allworth, A. E., and Baker, J. H. (1999). Evidence for polar cytoplasm/nuage in rat oocytes. *Anat. Embrtol.* **200,** 43–48.

Yuan, S., Darnell, D. K., and Schoenwolf, G. C. (1995). Identification of inducing, responding, and suppressing regions in an experimental model of notochord formation in avian embryos. *Dev. Biol.* **172,** 567–584.

Zamboni, L. (1970). Ultrastructure of mammalian oocytes and ova. *Biol. Reprod. Suppl.* **2,** 44–63.

Zeng, L., Fagotto, F., Zhang, T., Hsu, W., Vasicek, T. J., Perry, W., Lee, J. J., Tilghman, S. M., Gumbiner, B. M., and Constantini, F. (1997). The mouse fused locus encodes axin, an inhibitor of the Wnt signalling pathway that regulates embryonic axis formation. *Cell* **90,** 181–192.

Zernicka-Goetz, M. (1998). Fertile offspring derived from mammalian eggs lacking either animal or vegetal poles. *Development* **125,** 4803–4808.

Zhang, Y., Proenca, R., Maffei, M., Barone, M., Leopold, L., and Friedman, J. M. (1994). Positional cloning of the mouse obese gene and its human homologue. *Nature* **372,** 425–432.

Ziv, T., Shimoni, Y., and Mitrani, E. (1992). Activin can generate ectopic axial structures in chick blastoderm explants. *Development* **115,** 689–694.

Cadherins and Catenins, Wnts and SOXs: Embryonic Patterning in *Xenopus*

Allison L. St. Amand and Michael W. Klymkowsky
Molecular, Cellular, and Developmental Biology, University of Colorado, Boulder, Colorado 80309

Wnt signaling plays a critical role in a wide range of developmental and oncogenic processes. Altered gene regulation by the canonical Wnt signaling pathway involves the cytoplasmic stabilization of β-catenin, a protein critical to the assembly of cadherin-based cell-cell adherence junctions. In addition to binding to cadherins, β-catenin also interacts with transcription factors of the TCF-subfamily of HMG box proteins and regulates their activity. The *Xenopus* embryo has proven to be a particularly powerful experimental system in which to study the role of Wnt signaling components in development and differentiation. We review this literature, focusing on the role of Wnt signaling and interacting components in establishing patterns within the early embryo.

KEY WORDS: Wnts, SOXs, TCFs, Catenins, Cadherins, *Xenopus,* Dorsal determination, Neural patterning.

I. Introduction

The mechanisms that enabled eukaryotic cells to go beyond simply sticking together and become integrated multicellular organisms are key to understanding both metazoan development and its evolutionary origins. Adhesion and juxtacrine signaling are clearly central to this process. While it is likely that a number of distinct adhesion/signaling systems were involved and have co-evolved, we will focus our discussion on a major and well-studied adhesion system, that based on cadherin-type proteins. Cadherins form the structural foundation of the two major adherence junctions found between vertebrate cells, i.e. adherens junctions and desmosomes (*macula adhaerens*). The appearance of cadherin-based adherence junctions appears to have occurred quite early in the evolution of the metazoans (Ax, 1996).

Cellular adhesion via cadherins depends upon interactions with cytoplasmic cadherin-associated proteins, catenins. The various catenins play a dual role. They mediate cytoskeletal attachment and modulate gene expression through their interactions with transcription factors. The cadherin-catenin system has been implicated in the regulation of the cell cycle (Shtutman *et al.,* 1999; Tetsu and McCormick, 1999) and tumor progression (Birchmeier, 1995; Hirohashi, 1998; Christofori and Semb, 1999). The dual function of the cadherin-catenin system places it at the nexus through which cell adhesion is transformed into differential cellular behavior (see Klymkowsky and Parr, 1995).

We will further focus this review on the early development of the anuran vertebrate *Xenopus laevis.* Because of its experimental accessibility, *Xenopus* has emerged as a powerful system in which to study the molecular and cellular events of early development and to test ideas about pattern formation and morphogenesis. In order to manage the growing deluge of often incomplete information, we will limit our discussion to the early embryo, i.e. the role of the cadherin-catenin system in the establishment of the dorsal-ventral axis and the patterning of the neuroectoderm (see Elinson, 1997). We are well aware that multiple signaling/gene regulatory systems combine to bring about these developmental events. We suspect that their interactions lead to nonlinear and nonintuitive behaviors. Nevertheless, we hope that this review of one aspect of the process will be useful.

II. The Cadherin Superfamily and Cell

Cadherins, and the cadherin-related (although perhaps misnamed) protocadherins, have been implicated in a number of cell-cell adhesion processes (Shapiro *et al.,* 1995). Cadherin-superfamily members are single pass integral membrane proteins. Their N-terminal extracellular domains are characterized by the presence of multiple cadherin repeats (Nagar *et al.,* 1996) that mediate homotypic, Ca^{2+} dependent adhesion, i.e. cadherins of the same type adhere to one another but not to cadherins of different types. Through their C-terminal cytoplasmic domains and its associated proteins, cadherins interact with actin and intermediate filaments (IFs).

Cadherin-mediated homotypic adhesion, and its regulation, plays a critical role in morphogenesis (Takeichi, 1995); in particular the processes of epithelial to mesenchymal and mesenchymal to epithelial transitions. An analogous cytoskeleton anchorage function, at sites of cell-extracellular matrix interactions (focal contacts and hemidesmosomes), is performed by the structurally distinct integrins. It is clear that there are interactions between these two types of adherence junctions (see Monier and Duband, 1997; Novak *et al.*, 1998); however, we will ignore them for the sake of simplicity.

A. The Types of Cadherins

Cadherins are typically classified as either classical (or type I) or desmosomal (Fig. 1). Both have structurally similar extracellular domains composed of 5 cadherin repeats. The classical and desmosomal cadherins differ most dramatically in the nature of their intracellular domains. Classical (type I) cadherins are characterized by a well-conserved C-terminal cytoplasmic domain that interacts with specific accessory polypeptides and mediates that attachment of actin filaments at cell-cell adhesion sites known as adherens junctions. A number of classical cadherins have been identified, e.g. epidermal (E-), placental (P-), neural (N-), muscle (M-), vascular-endothelial (VE-), brain (B-), retinal (R-) and liver (L-) types, among others (see Tanihara *et al.*, 1994; Gallin 1998).

As usual, there are exceptions to this rule, sometimes referred to as type II cadherins. The cytoplasmic domain of liver/intestinal (LI-) cadherin (Berndorff *et al.*, 1994) is distinct from that of the other classical cadherins, does not appear to interact with the same set of catenins, and does not appear to be required for LI-cadherin mediated cell-cell adhesion (Kreft *et al.*, 1997). Truncated (T-) cadherins (Ranscht and Dours, 1991) lack both a transmembrane and cytoplasmic domain and are anchored to the plasma membrane through a glycosylphosphoinositol linkage (Vestal and Ranscht, 1992; Sacristan *et al.*, 1993). T-cadherins are expressed by smooth muscle cells, bind to serum lipoprotein (Kuzmenko *et al.*, 1998; Tkachuk *et al.*, 1998), and have been implicated in the control of neurite path finding (Fredette *et al.*, 1996). We will largely ignore these nonconventional cadherins, since it they have yet to be characterized in *Xenopus*.

1. Desmosomal Cadherins

A third class of cadherins, the desmosomal cadherins, are associated with a second type of cell-cell adherence junctions, desmosomes (Burdett, 1998; Kowalczyk *et al.*, 1999). Desmosomes anchor IFs, most conspicuously in epithelia and in cardiac muscle cells. Two types of desmosomal cadherins, i.e. the desmogleins and the desmocollins, have been identified. Desmocollins appear more closely related to the classical cadherins (Marsden *et al.*, 1997). The desmocollins come in two alternative sliced variants, the a- and b-forms. The cytoplasmic protein plakoglobin (γ-catenin—see below) binds

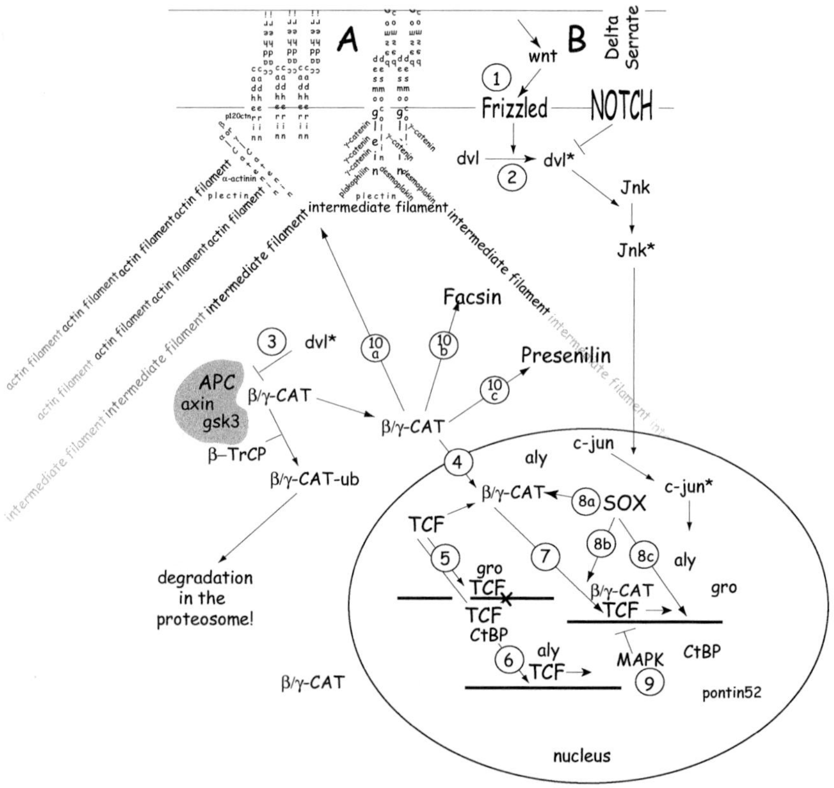

FIG. 1 The relationship between gene expression and cellular adhesion. (A) Cell-cell adhesion is mediated by homotypic interactions between cadherins. At adherens junctons, the cadherin's cytoplasmic tail domain interacts directly with two types of cadherins, p120[ctn] binds to a proximal site, while β-catenin or plakoglobin (γ-catenin) interacts in a mutually exclusive manner with a more distal site. Both β-catenin and plakoglobin also bind to α-catenin, which in turn interacts with actin filaments; interactions with dimeric proteins such as α-actinin and the plectin cross-link actin filaments and further stabilize the adherence junction. An analogous system is used to anchor IFs to desmosomal cadherins. Binding of plakoglobin to desmosomal cadherins (desmocollins and desmogleins) blocks plakoglobin's ability to interact with α-catenin. The p120[ctn]-like plakophilins bind to the desmosomal cadherin tail domains and to IFs. Desmoplakin dimers also bind to desmosomal cadherin tail domain and to IFs. Plectin is also localized to desmosomes and may strengthen EF anchorage. (B) The canonical Wnt pathway acts through the binding of secreted Wnt proteins with Frizzled-type membrane receptors (1). In this pathway, the binding of a Wnt to its receptor activates the cytoplasmic protein Dishevelled (dvl) (2). Notch, another integral membrane-signaling molecule, can inhibit Wnt signaling by binding to and sequestering dvl. Dvl acts in a number of ways within the responding cell (and can be activated independently of Wnt signaling). Activated dvl can activate Jnk, with in turn modulates the activities of downstream factors, including the c-jun transcription factor. Within the classical Wnt pathway, activated dvl (3) disrupts the activity of the APC/axin/GSK3 complex, which normally functions to target $\beta\gamma$ catenins ($\beta\gamma$-CAT) for ubiquitination

to the longer a-form, but not to the shorter b-form (Mathur *et al.,* 1994; Troyanovsky *et al.,* 1994). Desmocollins have relatively short cytoplasmic tail domains, whereas the tail domains of the desmogleins are much larger. A number of desmosomal cadherin genes have been found with distinct patterns of expression (Legan *et al.,* 1994; Nuber *et al.,* 1995).

2. Protocadherins

Protocadherins form a related subfamily of cadherin-repeat membrane proteins (Sano *et al.,* 1993; Suzuki, 1996). They typically have a larger extracellular domain composed of six, seven, or more cadherins repeats. For example, the *Drosophila* and human FAT proteins have 34 extracellular cadherin repeats (and five EGF-repeats) (Dunne *et al.,* 1995; Bisoffi and Betschart, 1996; Ponassi *et al.,* 1999), while the product of the *Drosophila Daschous* gene has 27 cadherin repeats (Clark *et al.,* 1995). The intracellular domains of the protocadherins are structurally distinct from those of either the classical or desmosomal cadherins (Suzuki, 1996). In *Xenopus,* specific protocadherins have been found associated with gastrulation (Kim *et al.,* 1998) and neural plate formation (Bradley *et al.,* 1998). Moreover, a large and complex family of protocadherins (Wu and Maniatis, 1999) appears to be involved with synaptic specificity and may generate intracellular signals through their interactions with the proto-oncogene product Fyn (Kohmura *et al.,* 1998). Little is as yet known, however, about the details of protocadherin-based interactions.

---

by the β-TrCP complex. Ubiquitinated $\beta\gamma$-catenins are then degraded in the proteosome. Activation of the classical Wnt pathway leads to the cytoplasmic accumulation of $\beta\gamma$-catenins (4), which can then enter the nucleus in an importin/karyopherin-independent manner. Within the nucleus, $\beta\gamma$-catenin can interact with a number of different factors. In particular, $\beta\gamma$-catenins can bind to the an N-terminal "catenin-binding domain" present in most (but not all) members of the TCF-family of M4G-box transcription factors. TCFs can also interact with a number of other regulatory factors. Among these are the co-repressors (5) groucho (gro), C-terminal binding protein (CtBP) and the co-activator ALY (6). In the absence of bound catenin, therefore, TCFs can act as either activators or repressors of target genes. The binding of catenins to TCF (7) enhances their ability to activate target genes or repress the repressive effects of TCFS. SOX proteins (8), a subset of the HMG box family of transcription factors, can regulate their own target genes (often in combination with other transcription factors 8c), and can modulate TCF activity in two ways. First, they can, at least in theory, bind to the same DNA-binding sites as those recognized by TCFs (8b); second, some SOXs can compete with TCFs for binding to catenins (8a). The binding of the catenin/TCF complex to DNA has itself been shown to be modulated by phosphorylation of TCFs by MAP kinases (9). Finally, cytoplasmic proteins, like cadherin (10a), fascin (10b), and presenilin (10c), bind to catenins and modulate their stability and their ability to enter the nucleus.

3. Flamingos

The most recently described type of cadherin-related protein is the *Drosophila* Flamingo protein, and a related protein in the mouse (Usui *et al.,* 1999). Flamingo differs from the other cadherins in a number of ways. Its extracellular domain contains nine cadherin repeats (along with other sequences), its transmembrane domain is composed of seven-transmembrane elements, and its cytoplasmic domain does not appear to interact with the conventional catenins (Usui *et al.,* 1999). Flamingo has been suggested to interact with G-proteins, and has been shown to mediate homotypic adhesion (Usui *et al.,* 1999).

B. Cytoplasmic Interactions and Classical Cadherin-Based Adhesion

Cadherins interact with one another in two distinct modes. A *cis* interaction occurs within the plane of the plasma membrane and is independent of cytoplasmic interactions (Chitaev and Troyanovsky, 1998; Pertz *et al.,* 1999; Takeda *et al.,* 1999). *Trans* associations, i.e. between cells, are stabilized by cytoplasmic interactions with catenins. Using immunoprecipitation, Ozawa *et al.* (1989) identified three polypeptides associated with uvomorulin (E-cadherin): α-, β-, and γ-catenin (plakoglobin). Binding of catenins is required to stabilize the formation of a two-dimensional "zippered patch" that interacts with the actin cytoskeleton (Fig. 1) (Yap *et al.,* 1997; Vleminckx and Kemler, 1999). It is not clear, however, exactly how catenin binding leads to the stabilization of cadherin adhesion. A likely hypothesis, given that many adherens junction-associated proteins are themselves dimeric, is that cross-linking of actin filaments, attached to the cadherin-catenin complex, plays a major role.

β-catenin and plakoglobin are quite similar proteins (Fig. 2A) (Gelderloos *et al.,* 1997) and bind in a mutually exclusive manner to a conserved and phosphorylated region at the C-terminus of the classical cadherin's cytoplasmic domain (Stappert and Kemler, 1994) (Fig. 1). Their primary function, in terms of cell-cell adhesion, is to link the cadherin tail domain to α-catenin, a vinculin-like protein, that in turn binds to actin filaments. In the absence of endogenous β-catenin/plakoglobin, a cadherin-α-catenin chimeric polypeptide can support adherens junction formation and cell-cell adhesion (Nagafuchi *et al.,* 1994). This leads to the hypothesis that β-catenin and plakoglobin function as the targets of pathways that regulate cell-cell adhesion. For example, IQGAP, which is regulated by the small

GTPases Cdc42 and Rac (Pettitt *et al.*, 1996), binds to β-catenin and blocks its interactions with α-catenin (Fukata *et al.*, 1999; see below).

In addition to α-catenin, other actin-associated proteins, e.g. vinculin (Hazan *et al.*, 1997; Weiss *et al.*, 1998) and α-actinin (Knudsen *et al.*, 1995) are localized to adherens junctions. Protein phosphotases (Kypta *et al.*, 1996; Brady *et al.*, 1998; Costa *et al.*, 1998), the epidermal growth factor-receptor (Hoschuetzky *et al.*, 1994), the hepatocyte growth factor/scatter factor receptor c-met (Hiscox and Jiang, 1999), the zonulae adherens-associated protein ZO-1, and the "plakin"-type protein plectin (Ruhrberg and Watt, 1997; Wiche, 1998) have all been localized to adherens junctions A number of these proteins are dimeric and their cross-linking functions could be involved in stabilizing cadherin interactions. For example, plectin contains a dimerization domain and its actin-binding domain has been implicated in the regulation of actin filament dynamics (Andra *et al.*, 1998; Wiche, 1998).

It seems likely that the formation of adherens junctions is regulated by a number of different intracellular signaling pathways (Williams, 1997) and that adhesion in turn modulates intracellular signaling (Sap, 1997). For example, the "simple" act of adhesion should facilitate juxtacrine signaling, such as that mediated by Notch/Notch ligand interactions (Fig. 1) (Lendahl, 1998; Artavanis-Tsakonas *et al.*, 1999). Finally, formation of tight and gap junctions also seems to be regulated by cadherin junction formation, although both positive and negative effects have been reported (see Wang and Rose, 1997; Jongen *et al.*, 1991).

Other proteins are associated with cadherins at the adherens junction. Of these, $p120^{ctn}$ (Reynolds *et al.* 1992, 1994) has been studied most extensively. $p120^{ctn}$ is structurally (although distantly) related to β-catenin and plakoglobin (Fig. 2A), and is a member of the "armadillo-repeat" family of cell adhesion proteins (see Reynolds and Daniel, 1997; Hatzfeld, 1999). $p120^{ctn}$ binds directly to a membrane-proximal site in the cadherin tail domain (Yap *et al.*, 1998; Ohkubo and Ozawa, 1999), but does not interact with α-catenin (Daniel and Reynolds, 1995). $p120^{ctn}$'s function at the adherens junction is not clear. Aono *et al.* (1999) report that $p120^{ctn}$ acts to antagonize cadherin-based adhesion in colon carcinoma cells, whereas Yap *et al.* (1998) suggested that $p120^{ctn}$ binding is involved in adhesive strengthening. $p120^{ctn}$ was originally identified as a p60-v-src/tyrosine kinase substrate (Reynolds *et al.*, 1992), and expression of p60-v-src leads to a change in cadherin-based adhesion from strong to weak in a β-catenin independent manner (Hamaguchi *et al.*, 1993; Takeda *et al.*, 1995). A rather large number of splice variants of $p120^{ctn}$ have been identified (see Reynolds and Daniel, 1997; Keirsebilck *et al.*, 1998), although their physiological significance remains unclear.

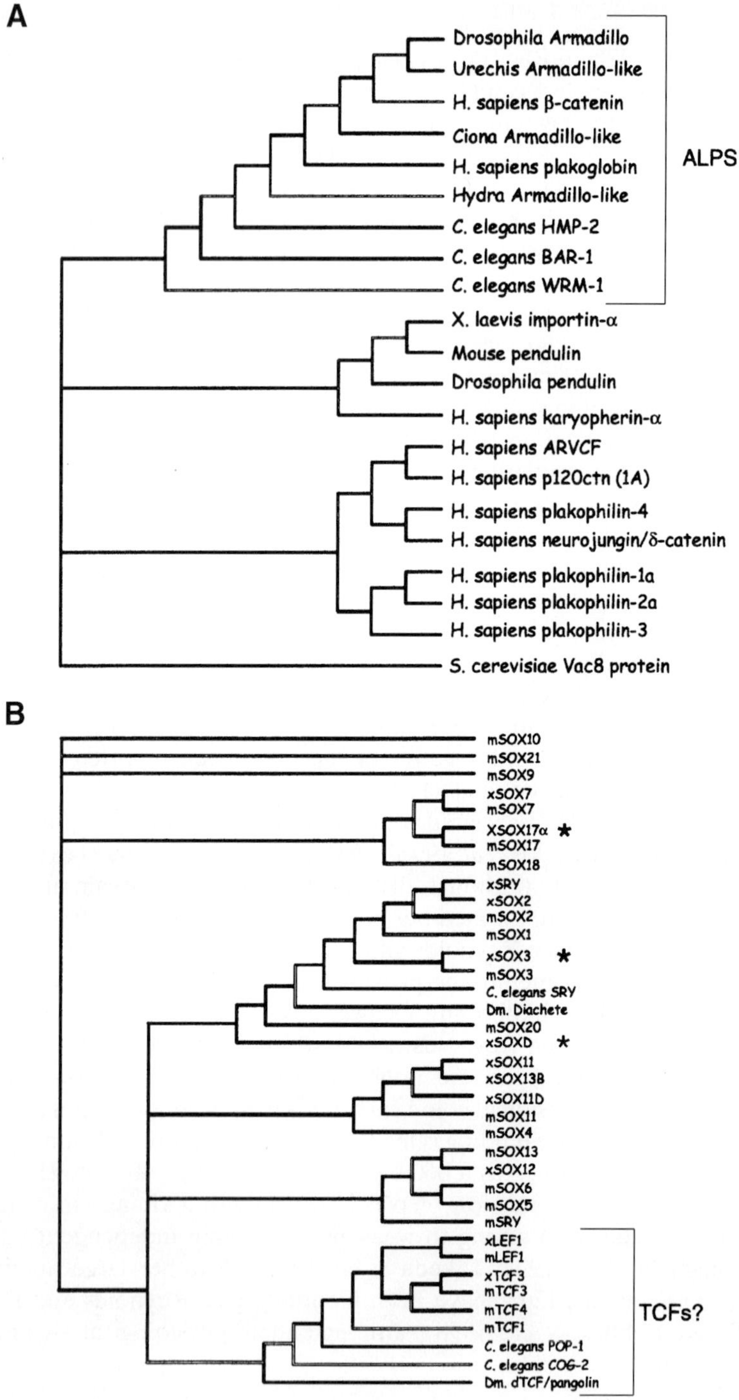
A
Drosophila Armadillo
Urechis Armadillo-like
H. sapiens β-catenin
Ciona Armadillo-like
H. sapiens plakoglobin
Hydra Armadillo-like
C. elegans HMP-2
C. elegans BAR-1
C. elegans WRM-1
ALPS
X. laevis importin-α
Mouse pendulin
Drosophila pendulin
H. sapiens karyopherin-α
H. sapiens ARVCF
H. sapiens p120ctn (1A)
H. sapiens plakophilin-4
H. sapiens neurojungin/δ-catenin
H. sapiens plakophilin-1a
H. sapiens plakophilin-2a
H. sapiens plakophilin-3
S. cerevisiae Vac8 protein
B
mSOX10
mSOX21
mSOX9
xSOX7
mSOX7
XSOX17α *
mSOX17
mSOX18
xSRY
xSOX2
mSOX2
mSOX1
xSOX3 *
mSOX3
C. elegans SRY
Dm. Diachete
mSOX20
xSOXD *
xSOX11
xSOX13B
xSOX11D
mSOX11
mSOX4
mSOX13
xSOX12
mSOX6
mSOX5
mSRY
xLEF1
mLEF1
xTCF3
mTCF3
mTCF4
mTCF1
C. elegans POP-1
C. elegans COG-2
Dm. dTCF/pangolin
TCFs?

C. Desmosomal Organization

With the exception of plakoglobin (Cowin *et al.,* 1986), a different set of catenins is involved in the anchorage of IFs to cadherins. The assembly of a desmosome requires both types of desmosomal cadherins, together with plakoglobin (Marcozzi *et al.,* 1998), and appears to depend upon the presence of adherens junctions (Lewis *et al.,* 1997). Plakoglobin binds to the tail domains of the desmosomal cadherins, a single plakoglobin molecular binds to the a-type desmocollin tail domain, and up to six plakoglobins can bind to each desmoglein tail domain (Kowalczyk *et al.,* 1996; Witcher *et al.,* 1996). Binding to desmosomal cadherins blocks plakoglobin's ability to interact with α-catenin, which presumably explains why actin filaments are not anchored to desmosomes. Plakoglobin binds to the plakin-family member desmoplakin (Kowalczyk *et al.,* 1997), which in turn interacts with

---

FIG. 2 Armadillo and TCF/SOX family relationships: We have carried out a rather naive (and by no means complete) parsimony-based sequence similarity analysis of a number of (A) armadillo-repeat containing and (B) HMG-box containing proteins. The full coding sequence of each polypeptide was obtained from GenBank and aligned by using the ClustalW program; the resulting alignment was then processed using the PAUP 4.0 tree building program. (A) Analysis of the armadillo-family of proteins: In the case of the armadillo-repeat family of proteins, we included as many invertebrate armadillo-like proteins as we could identify; these included *Drosophila melanogaster* armadillo, and armadillo-like proteins isolated from the cnidarian *Hydra magnipapillata,* the spoonworm *Urechis caupo,* the sea squirt *Ciona savignyj,* and the three nematode *Caenorhabditis elegans* armadillo-like proteins. These were compared with *H. sapiens* β-catenin and plakoglobin. In addition, we included nuclear localization sequence receptor proteins (NLS receptors) from *X laevis* - α-importin, mouse and *Drosophila* pendulins, and human karyopherin-(α). We also included all of the human p120^{ctn}-like proteins currently identified; these included the p120^{ctn} -splice variant 1A, ARVCF, plakophilins 1 through 4 and δ-catenin/neurojungin. As an outgroup, we used the armadillo-repeat containing protein Vac8 from the yeast *S. cerevisiae.* This analysis grouped the armadillo-like proteins ("ALPS"), the NLS-receptors, and the p120^{ctn}-like proteins in distinct subfamilies (see also Hatzfeld, 1999). (B) Analysis of TCF/SOX proteins. For this analysis, we compared mouse and *Xenopus* SOX/SRY and TCF family proteins, together with SOX/TCF proteins identified in *C. elegans* (SRY, COG-2, and POP-1), and *Drosophila* (Diachete and dTCF/pangolin). As an outgroup, we arbitrarily chose the mouse SOX protein mSOX10. Again, there was a clear grouping of the proteins into a number of different classes. Among the vertebrate TCFS, TCF1 was clearly the most divergent, although the invertebrate TCFS, POP-1, and DTCF were clearly members of this group. Surprisingly, the COG-2 protein, identified as a SOX protein, was found within the TCF group. Among the SOXS, distinct subgroups were apparent, although surprisingly XSry and Mouse SRY proteins were found in different groups. Catenin-binding SOXs (marked with an "*") were found in a number of SOX subgroups. XSOXD (marked by a grey *) can block TCF activation of the TOPFLASH reporter. Clearly, a more rigorous analysis is required to draw more robust conclusions.

IFs (Stappenbeck and Green, 1992). Desmoplakin also appears to be able to bind directly to the desmosomal cadherin tail domains (Smith and Fuchs, 1998).

Mutational studies indicate that plakoglobin is required to maintain the distinction between adherens junction and desmosomes (Bierkamp *et al.*, 1996; Ruiz *et al.*, 1996). In the absence of plakoglobin, desmosomes are aberrant and there is mixing of desmosomal and adherens junction components, particularly in the heart (Ruiz *et al.*, 1996; Bierkamp *et al.*, 1999). Plakoglobin has been specifically implicated in the development of adhesive strength in endothelial cells, required to resist shear stress (Edelmann *et al.*, 1999). In the absence of plakoglobin, β-catenin can associate with desmosomal components (Bierkamp *et al.*, 1999), but clearly cannot completely replace plakoglobin's functions, particularly in mechanically stressed tissues (Bierkamp *et al.*, 1996; Ruiz *et al.*, 1996).

A single pro-ortholog of the vertebrate β-catenin/plakoglobin proteins, Armadillo, has been identified in *Drosophila melanogaster* (Fig. 2A). Since *Drosophila* does not appear to have cytoplasmic IFs (Bartnik and Weber, 1989; Klymkowsky and Evans, 1995) it is likely that it has only a single generic type of cadherin-based adherence junction. The septate junctions found in *Drosophila* and other arthropods are structurally distinct from adherens junctions (see Fawcett, 1981) and may provide some of the functions provided by desmosomes (see Tepass and Hartenstein, 1994).

From a phylogenic perspective, however, the situation is complex, since the nematode *Caenorhabditis elegans,* which appears to be more closely related to arthropods than to vertebrates (Aguinaldo *et al.*, 1997), has an extensive family of cytoplasmic IF proteins (Bartnik *et al.*, 1986; Dodemont *et al.*, 1990; Dodemont *et al.*, 1994). Cadherin-based junctions in *C. elegans* have only recently been described (see Pettitt *et al.*, 1996; Costa *et al.*, 1998). Three armadillo-like proteins, i.e. HMP-2 (Costa *et al.*, 1998), BAR-1 (Eisenmann *et al.*, 1998), and WRM-1 (Rocheleau *et al.*, 1997) have been identified in *C. elegans.* All three are quite divergent from *Drosophila* armadillo but clearly more closely related to the armadillo-like proteins (ALPs) than to other armadillo-repeat proteins (Fig. 2A).

D. Other Desmosomal Catenins and Plakins

$p120^{ctn}$-like proteins (Fig. 2A) are also found at the desmosome. These include the plakophilins 1-, 2-, and 3 (Hatzfeld *et al.*, 1994; Heid *et al.*, 1994; Mertens *et al.*, 1996; Schmidt *et al.*, 1999), NPRAP/neurojungin/δ-catenin (Paffenholz and Franke, 1997; Zhou *et al.*, 1997; Paffenholz *et al.*, 1999), and plakophilin-4 (Hatzfeld and Nachtsheim, 1996; Hatzfeld, 1999). The "armadillo-related" protein, ARVCF, associated with velo-cardio-facial syndrome, is expressed ubiquitously (Sirotkin *et al.,* 1997) and is closely related to p120^{ctn} (Fig. 2A), but whether it is specifically associated with adherens junctions or desmosomes has not been reported. That plakophilin-1, at least, plays a critical role in desmosome organization is indicated by the aberrations in desmosomal organization seen in people carrying null mutations in the gene. They display a skin-blistering disease similar to that seen following disruption of the epidermal IF-desmosomal network (McGrath *et al.,* 1997, 1999).

Two members of the plakin family of polypeptides (see Ruhrberg and Watt, 1997; Klymkowsky, 1999a) have been found to be widely associated with desmosomes: desmoplakin and plectin (HD-1). Both bind to IFs through structurally similar C-terminal domains (Stappenbeck and Green, 1992; Wiche *et al.,* 1993). Desmoplakin binds directly to the tail domains of desmosomal cadherins (Smith and Fuchs, 1998) and to plakoglobin (Kowalczyk *et al.,* 1997). Mouse embryos homozygous for a desmoplakin null mutation die early in development (embryonic day 6.5) (Gallicano *et al.,* 1998); this early embryonic lethality is surprising, since the formation of polarized epithelial is not affected by the absence of IFs (Emerson, 1988; Baribault and Oshima, 1991), and embryos do not appear to require an intact IF system during this period of development (Baribault *et al.,* 1993, 1994; Magin *et al.,* 1998).

Plectin is found associated with desmosomes and hemidesmosomes, and associated with desmoplakin (Eger *et al.,* 1997). Mutations in plectin produce a less severe, but nevertheless fatal neonatal phenotype. Mice homozygous for a null mutation in plectin display defects in epidermal integrity and muscle structure. While desmosome and hemidesmosomes appear structurally normal, the numbers of hemidesmosomes are reduced (Andra *et al.,* 1997). Humans carrying mutations in plectin develop a form of epidermolysis associated with muscular dystrophy (Chavanas *et al.,* 1996; McLean *et al.,* 1996; Pulkkinen *et al.,* 1996; Smith *et al.,* 1996; Gache *et al.,* 1996; Mellerio *et al.,* 1997). In addition to stabilizing IF-membrane interactions, plakins have been found to mediate cytoskeletal interactions (see Klymkowsky, 1999a); for example plectin is known to modulate actin filament dynamics (Andra *et al.,* 1998) whereas dystonin/BPAG regulates microtubule stability (Yang *et al.,* 1999).

E. Junctional complexity

Adherens junctions can assume a variety of forms, from discs to belts. Desmosomes are typically discs up to 0.5 μm in diameter. In different tissues, adherence junctions are composed of different classical (Marrs and Nelson, 1996; Yap *et al.,* 1997) and desmosomal cadherin isoforms (Koch *et al.*, 1992). Cadherin-based junctions have also been described that combine features of both adherens junctions and desmosomes. *Complexus adherents* are junctions of variable size and shape, which have plakoglobin and desmoplakin associated with their cytoplasmic face (Schmelz and Franke, 1993); desmosomal cadherins are not present and their place appears to be taken by cadherin-5/vascular-endothelial (VE-) cadherin. Adherence junctions between vascular endothelial cells contain VE-cadherin, and exclude N-cadherin, which is also expressed by these cells (Navarro *et al.*, 1998). VE-cadherin junctions mature over time (Lampugnani *et al.*, 1995); when initially formed they contain β-catenin but little plakoglobin. With time, however, plakoglobin comes to be associated with these junctions. If the cells are induced to migrate, e.g. in response to wounding, plakoglobin disappears from the junctions before β-catenin does, thus suggesting a hierarchy in the assembly and diassembly of these structures.

Different types of junctions are associated with different accessory proteins. For example, adhesion junctions in nonneural cells are associated with the actin-binding protein 1-afadin (Mandai *et al.*, 1997) through interactions with the protein ponsin (Mandai *et al.*, 1999). Afadin in turns recruits the immunoglobulin-like cell adhesion molecule nectin/PRR (poliovirus receptor-related protein) (Takahashi *et al.*, 1999). It seems likely that the nature of the nectin-like proteins expressed in a cell will in turn modulate the behavior of the adherens junction to which they are associated. At this point, however, studies on the role of afadin, nectin, and ponsin in *Xenopus* have not yet appeared.

II. Wnts, Catenins, and the Regulation of Gene Expression

The idea that catenins are involved in the specification of gene expression in general, and the determination of embryonic patterning in particular, first emerged from studies of the *Drosophila* segment polarity gene *armadillo*. Mutations in *armadillo* produce a phenotype similar to that seen for mutations in the *wingless* gene, i.e. loss of segment polarity in the early embryo (Wieschaus and Riggleman, 1987). Armadillo appears to be the pro-ortholog (Holland, 1999) of vertebrate plakoglobins (Peifer and Wieschaus, 1990) and β-catenins (Peifer *et al.*, 1992), and we will refer these three proteins as armadillo-like proteins or ALPs (Fig. 2A). The *C. elegans* proteins HMP-2, BAR-1, and WRM-1 are divergent members of the ALP family and have been implicated in gene regulation, although perhaps via different mechanisms (see below). It remains to be shown whether specific ALPs are associated with specific types of adherence junctions in *C. elegans*, although given the presence of IFs it seems likely that there are both desmosomal- and adherens-like junctions. The ALPs are distinct, both structurally and functionally, from the p120^{ctn}/plakophilin group of armadillo-repeat containing proteins (Fig. 2A).

The similarities between the ALPs extend beyond primary sequence. Like plakoglobin and β-catenin, Armadillo forms a complex with *Drosophila* cadherins, α-catenin, and APCs (Peifer, 1993; Cox *et al.,* 1996; Pai *et al.,* 1996; Hayashi *et al.,* 1997; McCartney *et al.,* 1999). Both β-catenin and plakoglobin rescue the adhesion defects associated with mutations in *armadillo* (White *et al.,* 1998). There are, however, also real differences between ALPs. Plakoglobin is ineffective at rescuing the signaling phenotype of *armadillo* mutations, and β-catenin rescues only weakly (White *et al.,* 1998). Moreover, both plakoglobin and β-catenin mimic the behavior of dominant-negative *armadillo* mutations (White *et al.,* 1998), suggesting that β-catenin and Armadillo are not complete functional homologs and should not be considered as such in experimental studies. The C-terminal tail domains of β-catenin, plakoglobin, and Armadillo have been identified as transcriptional activation domains (Orsulic and Peifer, 1996; Hsu *et al.,* 1998b; Simcha *et al.,* 1998); however a BLAST search using the C-terminal tail domain of Armadillo reveals no significant homology between the proteins in this region (Gelderloos *et al.,* 1997). We would suggest that referring to Armadillo as *Drosophila* β-catenin, or using *Drosophila* armadillo in vertebrate systems, is likely to produce a misleading impression and results.

A. Catenins and Wnt Signaling

Wingless encodes a secreted protein that is a member of the Wnt (w*ingless*/*i*nt-*1*) family of proteins (Nusse and Varmus, 1992). Four *Drosophila* Wnts have been identified (Dierick and Bejsovec, 1999), while over 16 Wnts have been described in vertebrates (Parr and McMahon, 1994; Wodarz and Nusse, 1998; Loureiro, 1999). In *Drosophila* the intracellular processing, transport, and secretion of Wnts is regulated by the product of the *porcupine* gene (see Dierick and Bejsovec, 1999); the function of porcupine is unclear, but related proteins have been found in *C. elegans* (MOM-1) (Rocheleau *et al.,* 1997) and in the human (OATL1-accession number AAA74510). Wnt activity is regulated by glycosylation and interactions with extracellular components, i.e. sulfated glycosaminoglycans (Binari *et al.,* 1997; Hacker *et al.,* 1997; Haerry *et al.,* 1997). In *Drosophila* it is clear that expression of engrailed/hedgehog in neighboring cells can modulate the transcellular transport, and therefore the effective signaling domain, of Wnts (Bejsovec and Wieschaus, 1995; Dierick and Bejsovec, 1999; Sanson *et al.,* 1999).

Once secreted, Wnts bind to surface receptors of the Frizzled protein family. In *Drosophila* two Frizzled-type genes, Frizzled and Frizzled-2, have been identified; both are required for normal transduction of a Wnt signal (Bhat, 1998; Kennerdell and Carthew, 1998). In addition, heparan sulfate proteoglycans appear to act as coreceptors (Lin and Perrimon, 1999; Tsuda *et al.,* 1999) In vertebrates over 10 Frizzleds have been identified (Koike *et al.,* 1999). Whether the vertebrate Wnt receptor consists of multiple Frizzled and accessory polypeptides remains to be determined. It is clear, however, that different Wnts interact with different Frizzleds differently. For example, the binding of some Wnts to some Frizzleds activates the "canonical" Wnt pathway (see Perrimon, 1994; Fig. 1), whereas other Wnts act through a distinct pathway that does not appear to involve β-catenin (Kengaku *et al.,* 1998) and may involve G-proteins and protein kinase C (Slusarski *et al.,* 1997; Sheldahl *et al.,* 1999).

A number of extracellular Wnt-inhibitors have been identified. Some of these antagonists are structurally related to the extracellular domain of the frizzleds (Finch *et al.,* 1997; Leyns *et al.,* 1997; Mayr *et al.,* 1997; Wang *et al.,* 1997a; Xu *et al.,* 1998), while others are novel polypeptides (Salic *et al.,* 1997; Hsieh *et al.,* 1999). Some, like *Xenopus* Cerberus and mammalian DAN, bind simultaneously to, and inhibit, Wnts and other signaling molecules (e.g. transforming growth factors; TGFs) (Stanley *et al.,* 1998; Pearce *et al.,* 1999; Piccolo *et al.,* 1999). Wnt antagonists can block certain Wnts and not others (Wang *et al.,* 1997b), and in some cases the antagonists appear to act through interactions with Frizzleds, as well as Wnts (see Bafico *et al.,* 1999). It is clear, therefore, that there is a complex interplay among Wnt secretion, diffusion, and transport, anti-Wnt activities, and cellular Wnt responses.

1. Wnt Signaling and the Regulation of ALP Stability

In the typical "resting" cell, the level of soluble β-catenin and plakoglobin is kept quite low through the action of the adenomatous polyposis coli (APC) proteins. Two distinct APCs have been identified in human, mouse, *Xenopus,* and *Drosophila* (see McCartney *et al.,* 1999 and references therein). APCs are large proteins that have been found to interact with microtubules (Munemitsu *et al.,* 1994; Smith *et al.,* 1994) and the plasma membrane (Conway *et al.,* 1997; McCartney *et al.,* 1999). APCs can be found in both the cytoplasm and the nucleus (Neufeld and White, 1997; Klymkowsky *et al.,* 1999). APCs bind to ALPs not bound to cadherins (Rubinfeld *et al.,* 1993; Hulsken *et al.,* 1994; Rubinfeld *et al.,* 1995) and act to destabilize them (Munemitsu *et al.,* 1995) (Fig. 1). The ALP-APC complex includes the constitutively active protein kinase, glycogen synthase kinase 3 (GSK3) (Rubinfeld *et al.,* 1996), and the accessory protein Axin (Axil/

Conductin) (Behrens *et al.,* 1998; Ikeda *et al.,* 1998; Kishida *et al.,* 1998; Sakanaka *et al.,* 1998; Yamamoto *et al.,* 1998). Together with the E3-ubiquitin ligase polypeptides β-TrCP (Slimb) (Jiang and Struhl, 1998; Hart *et al.,* 1999; Latres *et al.,* 1999; Liu *et al.,* 1999a), this complex targets catenins for ubiquitination and subsequent proteolytic degradation by the proteosome (Aberle *et al.,* 1997) (Fig. 1). Mutations in APC are found in the majority of cases of familial and spontaneous colon cancers (Morin *et al.,* 1997; Sparks *et al.,* 1998), and lead to the stabilization of β-catenin, which in turn mimics that activation of the Wnt pathway.

In the canonical Wnt signaling pathway (Fig. 1), the binding of Wnts to Frizzleds leads to the activation of the downstream protein Dishevelled (Dvl). Activation has been associated with the phosphorylation of Dvl by casein kinases (Willert *et al.,* 1997; Peters *et al.,* 1999). Activated Dvl binds to Axin (Li *et al.,* 1999a; Smalley *et al.,* 1999), inhibits Axin's phosphorylation by GSK3, which destabilizes Axin (Yamamoto *et al.,* 1999b). Dvl binding to Axin induces the disassembly of the ALP-APC complex, stabilizing cytoplasmic ALP (Willert *et al.,* 1999). To complicate matters, activated Dvl interacts with, and activates jun-N-terminal kinase (JNK), which regulates the c-jun transcription factor (Boutros *et al.,* 1998; Li *et al.,* 1999b). Dvl has also been reported to interact with the cytoplasmic domain of the *Drosophila* Notch protein (Axelrod *et al.,* 1996), an interaction that inhibits Dvl's Wnt pathway activity (Fig. 1). Finally, there are multiple isoforms of Dvl in mammals (Sussman *et al.,* 1994; Klingensmith *et al.,* 1996; Tsang *et al.,* 1996), and perhaps *Xenopus* as well; whether they are equivalent in their activities remains to be determined.

Phosphorylation of β-catenin and plakoglobin is involved in their turnover, and mutations that abolish their phosphorylation lead to their stabilization (Yost *et al.,* 1996; Rubenstein *et al.,* 1997). In colon cancers lacking mutations in APC, ~50% have mutations in β-catenin (Morin *et al.,* 1997; Sparks *et al.,* 1998). β-catenin mutations are found in a substantial percentage of melanoma cell lines (Rubinfeld *et al.,* 1997), hepatocellular carcinoma lesions and cell lines (de La Coste *et al.,* 1998; Miyoshi *et al.,* 1998), pilomatricomas (Chan *et al.,* 1999), Wilm's tumor (Koesters *et al.,* 1999) and a number of other tumor types. With few exceptions, these are point mutations, which when engineered into wild type β-catenin dramatically stabilize the cytoplasmic form of the protein (see Polakis, 1999). Orford *et al.* (1999) report that overexpression of β-catenin in epithelial cells is sufficient to suppress anoikis (epithelial, substrate-dependent apoptosis) and leads to growth in soft agar and continued proliferation following γ-ray irradiation.

A search for similar "oncogenic" mutations in plakoglobin has been largely negative. Simcha *et al.* (1996) report that plakoglobin acts as a tumor suppressor, and loss of heterozygosity at the plakoglobin locus has been associated with breast and ovarian cancer (Aberle *et al.,* 1995). Reduced levels of plakoglobin transcription, consistent with a role as a tumor suppressor, have been found in cervical carcinoma (Denk *et al.,* 1997), and reduced plakoglobin expression has been linked to unfavorable prognosis in patients with non-small cell lung carcinoma (Pantel *et al.,* 1998). We have found that soluble plakoglobin is significantly more stable than β-catenin when expressed in human 293T cells, and that a mutation in human plakoglobin, analogous to oncogenic mutations in β-catenin, does not significantly increase plakoglobin's stability under these conditions (Williams *et al.,* 2000). However, Caca *et al.* (1999) have reported a mutation in plakoglobin analogous to the oncogenic mutations in β-catenin, associated with a gastric cancer. It is possible that differences in the rate of plakoglobin turnover in different cell types lead to different behaviors, i.e. tumor suppressor-like activity in some cell types, and proto-oncogene-like behavior in others.

2. ALPs and the Nucleus: A Modest Hypothesis

When the cytoplasmic levels of β-catenin or plakoglobin rise, these proteins enter the nucleus. Two models have been proposed to explain their nuclear localization. The first was based on the binding of ALPs to TCF-type transcription factors. ALPs bind to the N-terminal domain of the TCFs through their central "armadillo" repeat domain; TCFs, in turn, are transported to the nucleus via interactions with importins (Prieve *et al.,* 1996, 1998). ALPs interact with other nuclear factors, as well, e.g. certain SOX proteins (Zorn *et al.,* 1999a; see below), the TATA-box-binding protein binding protein Pontin52 (Bauer *et al.,* 1998; Hecht *et al.,* 1999), and the *Drosophila* zinc-finger protein teashirt (Gallet *et al.,* 1998). The other armadillo-family members (AFPs), e.g. p120^{ctn} and plakophilin (Fig. 2A), have not been reported to interact with any of these proteins (see Daniel and Reynolds, 1995).

ALPs are structurally similar to the karyopherin/importin family of nuclear localization sequence binding proteins (Malik *et al.,* 1997) (Fig. 2A). β-catenin can enter and leave nuclei in a karyopherin/importin- and RAN-independent manner (Fagotto *et al.,* 1998; Yokoya *et al.,* 1999). The crystal structure of the armadillo repeat region of β-catenin (Huber *et al.,* 1997) and importin-β (Cingolani *et al.,* 1999) indicates that both form a superhelix of helical domains characterized by a positively charged groove. It is through this grove that ALPs bind to TCFs, cadherins, and APC. Deletion analysis of ALPs indicates that regions of the polypeptide that are not expected to bind to TCFs, and which are inactive in Wnt signaling, still enter nuclei (Orsulic and Peifer, 1996; Rubenstein *et al.,* 1997; Prieve and Waterman, 1999). We would suggest that ALPs were adapted to a cell adhesion function from a previous role as nuclear transport factors (Malik *et al.,* 1997). This would allow them to integrate the regulation of both gene expression (cell behavior) and cell adhesion (morphogenesis). In this light the use of ALPs in both adhesion and the regulation of gene expression would not be accidental.

A number of cell adhesion proteins have been found to localize to the nucleus. For example plakophilin-1, the band 6 protein of desmosomal preparations, is expressed in a wide range of cells that do not make desmosomes and is localized to nuclei (Schmidt *et al.,* 1997). Subsequently, plakophilins-2 (Mertens *et al.,* 1996) and −3 (Bonn *et al.,* 1999; Schmidt *et al.,* 1999) have been found to localize to nuclei. Our own analysis of the nuclear localization of plakophilin-1 indicates that the prime determinant of nuclear localization resides in the N-terminal, non-armadillo-repeat head domain (Klymkowsky, 1999b). This same domain has been found to bind directly to keratin-type IFs (Hatzfeld *et al.,* 1994; Klymkowsky, 1999b). However, the armadillo-repeat domain is also able to enter nuclei, although it may be inhibited from doing so by interactions with cadherins and other cytoplasmic components (Klymkowsky, 1999b). Regulation of the nuclear import of β-catenin and plakoglobin has been observed in the early *Xenopus* embryo (Schneider *et al.,* 1996; Merriam *et al.,* 1997), which may also reflect the regulation of ALP/cadherin interactions In contrast, $p120^{ctn}$ has not been reported to localize to nuclei, even though it has been found to bind to Kaiso, a BTB/POZ domain-zinc finger transcription factor (Daniel and Reynolds, 1999). The absence of $p120^{ctn}$ from the nucleus appears to be due to the presence of nuclear export sequences (van Hengel *et al.,* 1999). This leads to the intriguing possibility that $p120^{ctn}$ acts not as a co-activator for Kaiso- and Kaiso-like transcription factors, but rather as an inhibitory, cytoplasmic anchor. Whether other $p120^{ctn}$/plakophilin-like proteins interact with transcription factors remains unresolved.

3. ALPs as Transcriptional Regulatory Factors

In vertebrates, there are four known members of the T-cell enhancer factor (TCF) family of transcription factors, i.e. TCF1, LEF1, TCF3, and TCF4 (Fig. 2B). All contain a highly conserved high mobility group (HMG) box that mediates sequence specific binding to DNA via interactions with the minor groove (van de Wetering and Clevers, 1992). A conserved region at the N-terminal region of the TCFs has been implicated in binding to ALPs (Huber *et al.,* 1996; Molenaar *et al.,* 1996; Behrens *et al.,* 1996). It is interesting that this region is missing from the sequences of mouse TCF1 currently present in GenBank (see van de Wetering *et al.,* 1996; Roose *et al.,* 1999). In the absence of a ALP-binding domain, TCF1 would act as an unregulable (at least by ALPs) transcription factor.

TCFs differ in their ability to interact with accessory factors. LEF1 binds to a ubiquitously expressed co-activator polypeptide, ALY (Bruhn *et al.*, 1997) and this binding may explain LEF-1's ability to act as a context-dependent, ALP-independent transcription activator (Giese and Grosschedl, 1993; Hsu *et al.*, 1998; Klymkowsky *et al.*, 1999). Whether other TCFs interact with ALY or ALY-like proteins remains unknown. XTCF3, the *Xenopus* homolog of mammalian TCF3, binds directly to Groucho-type corepressor polypeptides (Roose *et al.*, 1998), whereas LEF1 does not (Gradl *et al.*, 1999). TCF3 and TCF4 are more similar to each other than to LEF-1 (or TCF1) (Fig. 2B), suggesting that TCF4 may also interact with Groucho-type co-repressors. The Groucho-binding domain of XTCF3 is distinct from its β-catenin binding domain (Roose *et al.*, 1998) and it is possible that both ALPs and Grouchos can bind simultaneously to the same TCF. A distinct binding site has been identified for the C-terminal binding, co-repressor protein CtBP in dTCF (Waltzer and Bienz, 1998) and XTCF3 (Brannon *et al.*, 1999). It has yet to be reported whether other TCFs interact with CtBP. Finally, *Drosophila* dTCF binds to the teashirt protein through its C-terminal domain (Gallet *et al.*, 1998, 1999). Teashirt is a zinc-finger containing transcription factor involved in a number of differentiation events (see Mathies *et al.*, 1994; Pan and Rubin, 1998). Recently, teashirt-like proteins have been identified in the mouse (AAF64095).

Catenin-regulation of TCF activity is commonly measured via artificial reporter plasmids. A widely used plasmid of this type is TOPFLASH (Korinek *et al.*, 1997) or the optimized TOPFLASH (OT) reporter (generated by B. Vogelman *et al.;* see (Klymkowsky *et al.*, 1999; Zorn *et al.*, 1999a), which consists of three TCF DNA binding sites upstream of a minimal *c-fos* promoter and the fire fly luciferase coding region. Different TCFs differ in their ability to regulate OT (Klymkowsky *et al.*, 1999). For example, while mammalian LEF1 activates OT in the absence of exogenous β-catenin, perhaps due to interactions with ALY, XTCF3 and human TCF4 do not. Coexpression β-catenin and LEF1 or TCF4 produces a synergistic enhancement of OT activation, while XTCF3 repressed β-catenin's ability to activate OT in these studies (Klymkowsky *et al.*, 1999). These results are different from those reported by Molenaar *et al.* (1996), a difference that may be due to differences in reporter and cells used in the assay. For example, Prieve and Waterman (1999) found that β-catenin and LEF-1 were able to activate a TCF-responsive reporter in the Jurkat transformed T-cell line (which expresses both TCF1 and LEF1—see Klymkowsky *et al.*, 1999) but not in normal T-lymphocytes, even though both β-catenin and LEF1 were localized to nuclei in these cells. This observation, together with the ability of different TCFs to associate with a variety of coregulatory factors (see above), suggests that the constellation of co-activator and corepressor factors expressed in a cell can dramatically influence β-catenin/TCF activity.

B. Signaling Pathway "Cross-Talk"

Yet another complication in how a cell responds to a Wnt signal is the cross-talk between signaling pathways (Fig. 1). It is well known that GSK3 activity can be regulated by a number of other signaling systems (Woodgett, 1994; Yost *et al.,* 1997). For example, binding of Hepatocyte growth factor/scatter factor (HGF/SF) to its receptor c-met (see To and Tsao, 1998). inhibits GSK-3β activity in mammary epithelial cells (Papkoff and Aikawa, 1998). Fibroblast (FGF) and epidermal (EGF) growth factors may also modulate the Wnt pathway through the MAP kinase pathway activation of Rsk, which can inhibit GSK3 activity (Saito *et al.,* 1994; Papkoff and Aikawa, 1998). A similar interaction involving the LIT-1 kinase and the ALP WRM-1 has been described in *C. elegans* (Meneghini *et al.,* 1999; Rocheleau *et al.,* 1999). Other protein regulators of GSK3 activity are the *Xenopus* GSK3-binding protein GBP and the mammalian protein FRAT1 (Yost *et al.,* 1998). However, GSK3 typically phosphorylates proteins that have been prephosphorylated by other kinases; the substrates of GSK3 phosphorylation in a particular cell are therefore determined, at least in part, by the activities of other kinases and phosphotases, which in turn are regulated by other signaling pathways. For example, GSK3 phosphorylation has been found to regulate nuclear export of NF-ATc (Beals *et al.,* 1997), suggesting that the generic inhibition of GSK3 activity is likely to have pleiotrophic effects beyond its effects on ALP stability.

Staal *et al.* (1999) report that inhibition of GSK3 does not activate TCF-responsive reporters in T-cells, but does so in fibroblasts—an observation that suggests that kinases other than GSK3 may also regulate β-catenin stability. Ishitani *et al.* (1999) report that activation of the TAK1-NLK-MAPK-related pathway antagonizes Wnt signaling by phosphorylating TCFs, thereby inhibiting the ability of ALP-TCF complexes (but not the TCFs themselves) to bind to DNA (see Fig. 1).

Another potential regulatory interaction involves the sequestration of catenins by cadherins. Cadherin overexpression stabilizes β-catenin (see Kurth *et al.,* 1999) and inhibits β-catenin signaling (Heasman *et al.,* 1994; Karnovsky and Klymkowsky, 1995; Fagotto *et al.,* 1996; see below). Overexpression of E-cadherin has been found to regulate the expression of metalloproteinase in prostate cancer cells (Luo *et al.,* 1999); the matrilysin gene is known to be regulated by catenins and TCFs (Crawford *et al.,* 1999).

β-catenin and a number of AFPs bind to the endoplasmic reticulum protein presenilin-1 (Zhou *et al.,* 1997; Murayama *et al.,* 1998; Yu *et al.,* 1998;

Levesque *et al.,* 1999; Wei *et al.,* 1999; Usui *et al.,* 1999.). This interaction has been reported to stabilize (Zhang *et al.,* 1998b) or enhance (Kang *et al.,* 1999) the proteolytic degradation of β-catenin, while mutations in presenilin destabilize (Nishimura *et al.,* 1999) or stabilize β-catenin (Kang *et al.,* 1999). The apparent discrepancy between these studies remains unexplained but presumably reflects differences in cellular context. Whether β-catenin bound to presenilin can still interact with TCFs is not yet clear, nor is it obvious what role presenilins, which are ubiquitously expressed, play in the regulation of Wnt signaling or AFP function during development.

Presenilin has been implicated in the regulation of the Notch juxtacrine signaling system (Latinkic *et al.,* 1997; Chan and Jan, 2000; Joubin and Stern, 1999). Assuming that interactions among ALPs, TCFs, cadherins, presenilins, and various regulatory cofactors are themselves regulated, the possibilities for multiple cross-interactions are rather mind-numbing (Fig. 1). In the end, only rigorous numerical modeling will lead to a rigorous understanding of the regulatory complexity of this (and other) gene regulatory pathways.

C. SOXs as Wnt Regulators

SOXs are HMG-box transcription factors related to sex-determining factor on the Y-chromosome (SRY) protein (Pevny and Lovell-Badge, 1997). More than 20 SOX proteins have been identified in mammals (Fig. 2B). These proteins are expressed in a wide variety of tissues during development and have been implicated in a number of cell fate decisions, including sex determination (Wegner, 1999). They have been reported to bind to concensus sequence (CAA^{A}/TG^{A}/T^{A}/T which includes the reported DNA binding motif recognized by the TCFs (CAAAG^{A}/T^{A}/T) (van de Wetering and Clevers, 1992; Brannon *et al.,* 1997). Like the TCFs, SOXs bind to DNA and bend it. While some SOXs function as classical transcription factors, most appear to act as context-dependent transcriptional regulators and interact with other DNA-binding proteins to activate (Yuan *et al.,* 1995; Ambrosetti *et al.,* 1997; Kuhlbrodt *et al.,* 1998) or repress (Botquin *et al.,* 1998) target genes (see also Kamachi *et al.,* 1995, 1999).

SOXs could modulate the Wnt signaling pathway in one obvious way, i.e. by competing with TCFs for DNA binding sites in the promoter regions of target genes (Fig. 1). However, we have found an unexpected mechanism by which certain SOX proteins can block β-catenin/TCF-mediated gene activation, namely by competing with TCFs for binding to β-catenin (Zorn *et al.,* 1999a) (Fig. 1). Given the structural similarities between β-catenin and the importins/karyopherins (see above), it seems an intriguing possibility that ALPs could modulate the activity of a number of nuclear proteins, and vice versa, through interactions with nuclear localization sequences.

IV. Catenins and Dorsal-Ventral Axis Determination

That Wnt signaling might be involved in dorsal-ventral axis specification in vertebrates was first suggested by the observation that ventral injection of Wnt RNA leads to axis duplication, i.e. ectopic dorsal axis formation (McMahon and Moon, 1989; Smith and Harland, 1991; Sokol *et al.,* 1991; Chakrabarti *et al.,* 1992). Much data have subsequently been assembled that point to the regulation of β-catenin availability as a key to embryonic patterning in *Xenopus, zebrafish,* sea urchin, and mouse (Heasman *et al.,* 1994; Schneider *et al.,* 1996; Larabell *et al.,* 1997; Rowning *et al.,* 1997; Yost *et al.,* 1998; Miller *et al.,* 1999; Haegel *et al.,* 1995; Wikramanayake *et al.,* 1998; Liu *et al.,* 1999a). There appear to be significant differences, however, among these species in the mechanisms used to establish β-catenin asymmetry. In the mouse secreted Wnt3 and appropriate Frizzleds appear to be required (Liu *et al.,* 1999b) whereas cytoplasmic reorganization, rather than a secreted Wnt, appears to play this role in *Xenopus* (Moon and Kimelman, 1998). Nevertheless, the underlying similarity among species is striking. In the next part, we will review the process of dorsal-ventral axis specification in *Xenopus* from the perspective of asymmetries in β-catenin. A thorough review of early amphibian development can be found in Elinson (1997).

A. The Polarized *Xenopus* Egg

The *Xenopus* oocyte is a highly asymmetric cell (see Gerhart, 1980; Dent and Klymkowsky, 1989; Klymkowsky and Karnovsky, 1994; Gard and Klymkowsky, 1998; Moon and Kimelman, 1998). Oocyte maturation activates this asymmetry in preparation for fertilization; anchored RNAs are released, a number of RNAs are polyadenylated, and their translation efficiency increases (Simon *et al.,* 1992; Kloc *et al.,* Chapter 2). Since zygotic transcription does not begin in earnest until the midblastula transition (MBT) (Newport and Kirschner, 1982), shortly before gastrulation begins, it is the asymmetric localization and translation of specific mRNAs and proteins that underlies the patterning the early embryo.

The most obvious asymmetries within the egg are the animal-vegetal differences in cortical pigmentation and yolk platelets (Fig. 3, see color insert) (Gerhart, 1980; Danilchik and Gerhart, 1987). Yolk asymmetry arises during oogenesis due to the asymmetric centripetal transport of yolk platelets (Danilchik and Gerhart, 1987) that reflects an underlying asymmetry in cytoskeletal organization (Roeder and Gard, 1994; Gard, 1995; Gard and Klymkowsky, 1998). Messenger RNA localization occurs through a cytoskeletally driven process associated with the dispersion of the mitochondrial mass that forms early in oogenesis (Klymkowsky and Karnovsky, 1994; Kloc *et al.,* 1996, 1998). A number of different RNAs have been found localized to the animal and vegetal hemispheres of the oocyte (Fig. 3) (Rebagliati *et al.,* 1985). Of the animally localized RNAs, there are examples that encode a ubiquitin-like fusion protein (An1) (Weeks and Melton, 1987), a mitochondrial ATPase subunit (An3) (Weeks and Melton, 1987), a ATP-dependent RNA helicase-like polypeptide (An2) (Gururajan *et al.,* 1991), a Wnt, Xwnt8b (Cui *et al.,* 1995), and XSOX3 (Penzel *et al.,* 1997). In the vegetal hemisphere, RNAs encoding an activin-like polypeptide (Vg1) (Weeks and Melton, 1987), a T-box transcription factor (VegT a.k.a. Antipodean/Xombi/Brat)(Stennard *et al.,* 1996; Zhang and King, 1996; Horb and Thomsen, 1997; Zhang *et al.,* 1998a), a Wnt, Xwnt-11 (Ku and Melton, 1993), a putative RNA-binding protein (XCat-2) (Zhou and King, 1996) and a novel germ plasm-associated polypeptide XPat (Hudson and Woodland, 1998) have been identified. With RT-PCR, Hudson *et al.* (1996) identified three transcripts encoding β-TrCP, one of which was localized to the animal hemisphere of the oocyte and two were localized to the vegetal hemisphere. That many more asymmetries in RNA localization will be identified seems almost certain.

1. The Gray Crescent and Cortical Rotation

In many amphibian embryos, a "gray crescent" appears soon after fertilization and marks the future dorsal side of the embryo (Gerhart, 1980) (Fig. 3B). Typically, the first cleavage furrow bisects this region, producing the initial left-right division of the embryo (Klein, 1987). The motor that drives gray crescent formation is the assembly of a microtubule array, localized to the region between the thin egg cortex and underlying cytoplasm (Elinson and Rowning, 1988; Houliston and Elinson, 1991a,b; Schroeder and Gard, 1992). The directionality of this microtubule array is biased by the position of the sperm aster, but even in the absence of sperm, an activated egg will form an oriented cortical microtubule system through a positive feed-back loop in which cortical movement leads to the reorientation of microtubules. The disassembly of the oocyte keratin filament network, which occurs during oocyte maturation (Klymkowsky *et al.,* 1991), may be required to allow this rotation to occur (Klymkowsky, 1995; Gard and Klymkowsky, 1998).

At the boundary between the cortex and the deeper cytoplasm, a transiently stable array of microtubules assembles (see Cha and Gard, 1999) and associates with kinesin (Houliston and Elinson, 1991; Houliston *et al.,* 1994) to drive the cortex ~30° with respect to the underlying cytoplasm, after which the microtubule array disassembles. Some organelles appear to be driven as much as 60° (Rowning *et al.,* 1997). Since the fertilized egg is not a rigid structure but undergoes a series of contractions and relaxations during the first cleavage cycle (see Elinson, 1983), cortical rotation leads to a patterned swirling of the cytoplasm in both the vegetal and animal hemispheres, and produces a distinctive pro-dorsal/ventral asymmetry (Danilchik and Denegre, 1991). A number of cytoplasmic components are concentrated to the "pro-dorsal" side of the early embryo, including mitochondria (Marinos, 1986; Yost *et al.,* 1995) and components of the Wnt signaling pathway, including Dvl (Miller *et al.,* 1999) and β-catenin (Larabell *et al.,* 1997; Rowning *et al.,* 1997). That cortical rotation is necessary and sufficient to establish the dorsal-ventral axis in *Xenopus* has been illustrated by a number of elegant studies (see Chung and Malacinski, 1980; Scharf and Gerhart, 1980; Gerhart *et al.,* 1981; Vincent *et al.,* 1986). This implies that it is the altered juxtaposition of maternal components that establishes the dorsal-ventral axis. The underlying molecular mechanisms, however, remain rather mysterious.

The cytoplasmic patterns established by cortical rotation during the first cell cycle are "fixed" spatially by embryonic cleavage, but their effects on gene expression are not apparent until the onset of zygotic transcription at the MBT (Newport and Kirschner, 1982). Prior to the MBT, however, a dorsal-ventral asymmetry in gap junction conduction, which can be modulated by β-catenin, appears (Olson and Moon, 1992; Guger and Gumbiner, 1995; Krufka *et al.,* 1998). The mechanism by which β-catenin influences gap junction conduction is unknown (see above).

At MBT a number of genes begin to be expressed in specific embryonic regions. The homeobox-containing genes *Siamois* and *Twin* are expressed in the dorsal-vegetal region of the embryo known as the Nieuwkoop center (Lemaire *et al.,* 1995; Laurent *et al.,* 1997), whereas other homeobox genes—e.g. XVent-1 and -2, Xvex (Shapira *et al.,* 1999), Vox, and Mixer (Henry and Melton, 1998)—are expressed in the vegetal region. The regulation of the dorsalizing genes *Siamois* and *Twin* is dependent upon the catenin-signaling pathway, whereas the expression of the ventralizing genes, e.g. *Mixer* and *XVent,* depends upon the action of the BMP pathway (see Graff *et al.,* 1994; Hemmati-Brivanlou and Thomsen, 1995; Schmidt *et al.,* 1995). FGFs are also present and appear to be required as a competence factor, allowing cells to respond to other signals (see Slack *et al.,* 1996; Isaacs, 1997). Inhibiting the BMP pathway or activating the catenin pathway on the ventral side of the embryo induces a second dorsal axis.

It is known that BMPs are made in the early embryo and that cortical rotation is required for the expression of certain ventral endodermal markers, as well dorsal ones (Henry *et al.*, 1996). Both BMP and WNT/catenin pathways are involved in the induction of anterior endomesoderm (Zorn *et al.*, 1999b). The responses of cells to BMPs and Wnt signaling are modulated by the products of maternal genes. For example, the vegetally localized T-box containing protein VegT appears to determine where, along the animal-vegetal axis, ectoderm, mesoderm, and endoderm form (Stennard *et al.*, 1996; Horb and Thomsen, 1997; Zhang *et al.*, 1998a).

B. Wnts and Dorsal-Ventral Axis Specification in *Xenopus*

Mouse embryos homozygous for a deletion in *Wnt3* fail to form the primitive streak, the dorsal-most structure (Liu *et al.*, 1999b). Two observations, however, argue that secreted Wnts *are not* involved in dorsal axis specification in *Xenopus*. First, expression of a dominant-negative mutant form of XWnt8 does not block dorsal specification, although it does disrupt the differentiation of somites and other tissues (Hoppler *et al.*, 1996). Second, expression of a dominant-negative form of the Wnt pathway component Dvl, which blocks axis duplication in response to injection of XWnt8 or XDvl RNAs but not β-catenin RNA, did not block endogenous dorsal axis formation (although it does disrupt normal embryonic patterning) (Sokol, 1996). In contrast, injection of a dominant negative form of GSK3 induces axis duplication, whereas expression of the wild-type protein ventralizes the embryo (Dominguez *et al.*, 1995; He *et al.*, 1995; Pierce and Kimelman, 1995). Together these results suggest that it is not a response to secreted Wnts, but rather the regulation of GSK3 activity that is involved in dorsal axis formation. This view is supported by the observation that the GSK3-binding protein GBP/FRAT, which specifically blocks axin-mediated GSK3-phosphorylation of β-catenin (Li *et al.*, 1999a; Thomas *et al.*, 1999), is required for dorsal axis formation (Yost *et al.*, 1998). In this light, the observation of dishevelled asymmetry in the early embryo (Miller *et al.*, 1999) may simply be a reflection of the global redistribution of cytoplasmic components (see above) rather than a functional role of dorsal determination.

1. The role of Catenins in Pro-dorsal Specification

The first suggestive evidence that catenins were involved in the determination of the dorsal-ventral axis in *Xenopus* was the observation by McCrea *et al.* (1993) that injection of an anti-β-catenin antibody induced an ectopic dorsal axis. In retrospect, this result can be understood in terms of the anti-β-catenin antibody stabilizing (rather than inhibiting) β-catenin. This model is supported by the observation that overexpression of cadherins that sequester β-catenin or the anti-sense oligonucleotide depletion of maternal β-catenin mRNAs lead to suppression of dorsal axis formation (Heasman *et al.,* 1994), while injection of β-catenin (Funayama *et al.,* 1995) or plakoglobin (Karnovsky and Klymkowsky, 1995) mRNAs produced an axis duplication phenotype.

Both β-catenin and plakoglobin are maternal mRNAs and proteins in *Xenopus* (DeMarais and Moon, 1992; Fouquest *et al.,* 1992; Fagotto and Gumbiner, 1994). However, antisense depletion of maternal plakoglobin mRNA does not significantly affect axis formation, although it does produce an transient effect on embryonic morphology (Kofron *et al.,* 1997). The difference between the plakoglobin and β-catenin RNA depletion phenotypes is due in large part to the fact that there is 10 to 20 times more β-catenin mRNA (Fig. 4) in the early embryo and the fact that the plakoglobin that is present is in an insoluble form (C. Dufton, J. Gelderloos, and M. W. Klymkowsky, unpublished observations) that is incapable of taking

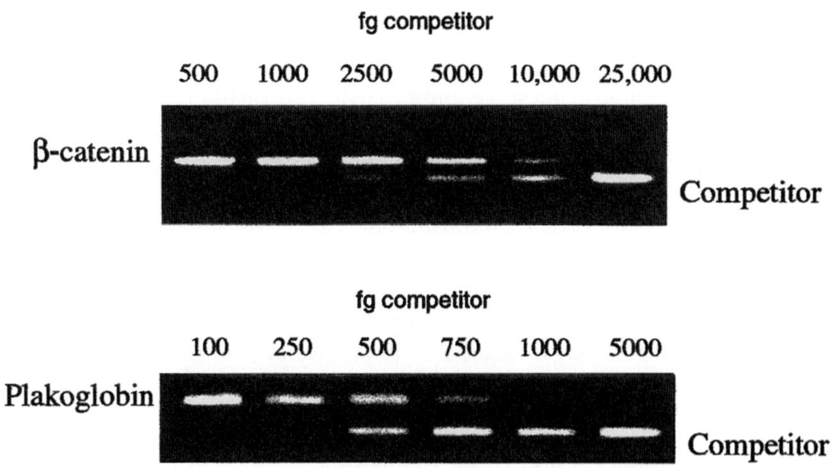

FIG. 4 β-catenin and plakoglobin mRNA levels in the early embryo: Total RNA from stage 7/8 *Xenopus* embryos was reverse transcribed and then split into two. Individual PCR reactions were carried out with constant amounts of cDNA and specific oligonucleotides and known amounts of either the plakoglobin or β-catenin competitor plasmid DNA. Reactions were subject to 40 cycles of PCR. Competitor (lower) and endogenous (upper) PCR products were then resolved on a 2.5% agarose gel. Because competitor plasmid DNAs were double stranded, where the intensities of the two bands were equal indicates where the amount of the specific cDNA was equal to twice the amount of added plasmid DNA. The results of this analysis indicated that there is between 10 to 20 times as much β-catenin as plakoglobin RNA in the early embryo (taken from unpublished work of C. Dufton, J. Gelderloos, and M. W. Klymkowsky).

part in signaling interactions and unable to associate with exogenous cadherins (Kurth *et al.,* 1999).

Original studies of β-catenin localization (Fagotto and Gumbiner, 1994) found that the protein was present ubiquitously throughout the early embryo, localized to nuclear and peri-nuclear regions, with no apparent dorsal-ventral asymmetry. The perinuclear localization of β-catenin could well be due to interactions with cadherins and presenilins in the endoplasmic reticulum (see Kurth *et al.,* 1999). Subsequent whole-mount immunocytochemical studies revealed that β-catenin is preferentially localized to the nuclei of dorsal blastomeres and that this nuclear localization does not occur until the MBT (Schneider *et al.,* 1996). Similarly, exogenous plakoglobin (Merriam *et al.,* 1997) and β-catenin (B. Gumbiner, personal communication) are cytoplasmic prior to MBT, after which they enter nuclei. This implies that the proteins are actively anchored in the cytoplasm during early development, although whether they are able to interact with TCFs in this cytoplasmic complex remains unclear.

It is worth noting that early studies of Armadillo localization in the *Drosophila* embryo failed to report its nuclear localization (see Riggleman *et al.,* 1990), and changes in fixation conditions have been found to produce profound changes in the intracellular distribution of a range of proteins (see Carl and Klymkowsky, 1999). Our own analyses indicate that easily detectable levels of nuclear β-catenin are found in essentially all regions of the late blastula/gastrula stage embryo (Merriam *et al.,* 1997; unpublished observations). In the unpertubed embryo relatively small differences in the dorsal-ventral levels β-catenin are critical for dorsal specification (Larabell *et al.,* 1997; Rowning *et al.,* 1997). It therefore seems likely that other dorsal-ventral asymmetries, established in parallel during cortical rotation, cooperate with the asymmetry in β-catenin to activate dorsalizing genes. In the experimental situation, which typically involves the injection of RNAs, it is common to deliver super-physiological levels of a protein; in that case, it is clear that β-catenin or plakoglobin are sufficient to induce dorsalizing genes.

2. Catenin Asymmetry and Gene Regulation

While four distinct TCFs have been identified in mammals, three have been identified in *X. laevis,* XTCF3 (Molenaar *et al.,* 1996), XLEF1 (Molenaar *et al.,* 1998) and XTCF4 (Konig *et al.,* 2000). Four different XTCF3 cDNAs were originally described, and these appear to fall into two classes that presumably reflect alleles of the two alloalleles in the tetraploid species *X. laevis.* XTCF3 is a maternal mRNA, whereas XLEF1 expression begins only after the MBT and XTCF4 expression begins after nemulation (Molenaar *et al.,* 1996; Molenaar *et al.,* 1998; Konig *et al.,* 2000). In the absence of evidence for other maternal TCFs, XTCF3 appears to be the major regulator of catenin-regulated target genes in the early embryo.

Early in the analysis of TCF activity in *Xenopus,* two conflicting observations were made (discussed in Klymkowsky, 1997). Behrens *et al.* (1996) found that injection of mammalian LEF1 RNA lead to axis duplication, and that mouse LEF1 and β-catenin synergized with one another to induce axis duplication. Moreover, even after removal of its catenin-binding domain, mouse LEF1 was still able to induce axis duplication, albeit less efficiently that the wild-type polypeptide. In contrast Molenaar *et al.* (1996) found that injection of XTCF3 RNA *did not* induce axis duplication. A mutated form of XTCF3, missing its catenin-binding domain (ΔN-XTCF3) ventralized the embryo by *suppressing* dorsal axis formation.

Subsequent analysis of the *Siamois* promoter, a target of catenin-signaling in the early embryo, indicated that the TCF-binding sites were required for the ventral suppression of *Siamois* expression, rather than its dorsal expression (Brannon *et al.,* 1997). Fan *et al.* (1998) found that ΔN-XTCF3, which is probably more accurately described as a non-catenin regulable form of XTCF3, rather than a dominant negative, inhibited *Siamois* promoter activity. Our own analysis of TCF activity indicates that XTCF3 differs from LEF1 and TCF4 in that it suppressed rather than enhanced β-catenin's ability to activate the OT reporter in cultured human 293T cells (Klymkowsky *et al.,* 1999), while injection of XTCF3 RNA suppresses dorsal axis formation in the embryo (Zorn *et al.,* 1999a). XTCF3 has been found to associate two types of transcriptional co-repressors, i.e. Grouchos (Roose *et al.,* 1998) and CtBP (Brannon *et al.,* 1999), whereas LEF-1 does not appear to interact with Grouchos (Gradl *et al.,* 1999). Whether LEF-1 interacts with CtBP has not yet been reported. These interactions could explain, at least in part, the repressive behavior of XTCF3 (see Bienz, 1998).

C. SOXs as Regulators of Catenin Signaling in the Embryo

The discovery that certain SOXs interact with β-catenin (Zorn *et al.,* 1999a) raises another intriguing means of regulating the patterning of the early embryo. A number of SOXs are present as maternal RNAs: XSOX3 (Koyano *et al.,* 1997; Penzel *et al.,* 1997), XSOX11 (XLS13A and XLS13B) (Hiraoka *et al.,* 1997), XSOX7 (Shiozawa *et al.,* 1996) XSOX4 (Miyata *et al.,* 1996), and XSOX-B1 (Sakai *et al.,* 1997). Other SOXs are are expressed soon after the beginning of zygotic transcription, e.g. XSOX17α and β (Hudson *et al.,* 1997), XSOXD (Mizuseki *et al.,* 1998), and XSOX2 (Mizuseki *et al.,* 1998). Using RT-PCR, we amplified the coding region of XSOX3 from *Xenopus* egg RNA (Vourgourakis *et al.,* 1999). *In situ* hybridization indicates that XSOX3 RNA is localized to the animal hemisphere throughout early embryogenesis, as described previously by Penzel *et al.* (1997). It is not known whether the other maternal SOXs bind to β-catenin but XSOX17α and β, XSOX3 (Zorn *et al.,* 1999a) and XSOXD (Vourgourakis *et al.,* 1999) inhibit β-catenin's ability to activate the OT reporter. XSOX3 (Penzel *et al.,* 1997), XSOX2 (Mizuseki *et al.,* 1998), and XSOXD (Mizuseki *et al.,* 1998) are all expressed in the animal hemisphere of the late blastula stage embryo, whereas XSOX17α and β (Hudson *et al.,* 1997) are expressed in the ventral endoderm (Fig. 3C). Given that nuclear β-catenin levels are reported to be higher on the entire dorsal side of the embryo, it is likely that regional expression of SOXs, modifies a particular cell's response to elevated β-catenin levels. For example, the restricted region of *Siamois* expression in the late blastula (Lemaire *et al.,* 1995; Carnac *et al.,* 1996), could be determined, in part, by the presence of XSOX3 and XSOX17α/β in the embryonic ectoderm and endoderm, respectively (Fig. 3). Similarly, in later stages of development, the presence of SOXs could modulate a cell's response to a "canonical" Wnt signal by competing with TCFs for binding to β-catenin or DNA (Fig. 1) (see below).

D. Cadherins and Catenin Patterning

At least three distinct maternal cadherins have been identified in *Xenopus* XB/U-, EP- and XmN-cadherins (Muller *et al.,* 1994; Tashiro *et al.,* 1996). Of these XB/U- and EP-cadherins have been shown to mediate homotypic adhesion (Munchberg *et al.,* 1997). During oocyte maturation, vesicular secretion is blocked and new membrane proteins are not delivered to the plasma membrane (Leaf *et al.,* 1990). The egg membrane differs in a number of ways from that of the oocyte, particularly in terms of its impermeability (see Dent and Klymkowsky, 1989). As cleavage begins, cleavage furrows are generated by the fusion of membrane vesicles. Cells that inherit part of the egg membrane have a distinct polarity (Muller and Hausen, 1995), and tight junctions are formed between blastomeres at their apical surface (Merzdorf *et al.,* 1998). Cadherins are delivered to the cleavage membranes between blastomeres and are required for cell-cell adhesion. Expression of the isolated cytoplasmic domain of N-cadherin leads to the loss of blastomere adhesion and subsequent defects in morphogenesis (Kintner, 1992), presumably by titrating β-catenin, which is required for robust cell-cell adhesion (see Kurth *et al.,* 1999) and the regulation of gene expression (see Fagotto *et al.,* 1996). Similar results have been found with the cytoplasmic domain of EP- (Dufour *et al.,* 1994) and E-cadherins (Levine *et al.,* 1994) or the extracellular domain of XB/U-cadherin (Kuhl *et al.,* 1996). In addition to cadherins, there appears to be a glycolipid mediated adhesion system between blastomeres (Turner *et al.,* 1992), although this system has not been studied extensively.

Aside from cleavage, there is little cell movement in the early embryo until the MBT (Newport and Kirschner, 1982). The initiation of cell motility leads to the morphogenetic movements of gastrulation and neurulation (see Keller, 1985; Keller and Danilchik, 1988). Initiation of cell movement implies the regulation of adhesion. At present little is known about how this regulation is achieved or coordinated. On possibility is the system of small GTPases, i.e. the Rho, Rac, and Cdc42 polypeptides (see Braga *et al.,* 1997; Drechsel *et al.,* 1997; Nobes and Hall, 1999). The GTPase-activating protein IQGAP binds to β-catenin, and inhibits its interaction with α-catenin (Fukata *et al.,* 1999). This interaction is inhibited by the binding of Cdc42/Rac-GTP to IQGAP, suggesting that activation of Cdc42/Rac stabilizes cell-cell adhesion, whereas accumulation of the GDP-forms of the these proteins would be expected to destabilize cell adhesion.

V. Neural Plate and Crest Formation

At the end of the blastula stage of development, the embryo has been divided into a number of distinct regions (Heasman, 1997) (Fig. 3). Gastrulation and neurulation transform the spherical blastula into the basic vertebrate body plan, which is dominated by the dorsal neural tube. Gastrulation begins on the dorsal side of the embryo with the formation of the blastopore, the dorsal lip (Keller, 1985). As gastrulation proceeds, dorsal ectodermal cells begin to move toward the anterior-dorsal midline of the embryo and undergo a process of thickening or columnarization to form the neural plate (Keller *et al.,* 1992a,b; Poznanski *et al.,* 1997; Elinson, 1997; Davidson and Keller, 1999). At the boundary between the neural plate and the surrounding ectoderm, ridges of cells, the neural folds, rise (Fig. 5). The neural plate then begins to extend along its rostral-caudal axis, narrows laterally, and curls up so that the neural folds meet along the dorsal midline. As the sheets of the epidermal ectoderm fuse above the neural tube, the neural tube detaches from the future epidermis and the cells located at the most dorsal aspect of the neural folds generate a lineage of their own, the neural crest (Anderson, 1997). Neural crest cells delaminate from the neural tube/embryonic epidermis and migrate throughout the embryo to form substantial portions of the skull and face, portions of the heart, the peripheral nervous system, and the pigmented cells of the epidermis (Sadaghiani and Thiebaud, 1987; Mayor *et al.,* 1999). This process of neural specification

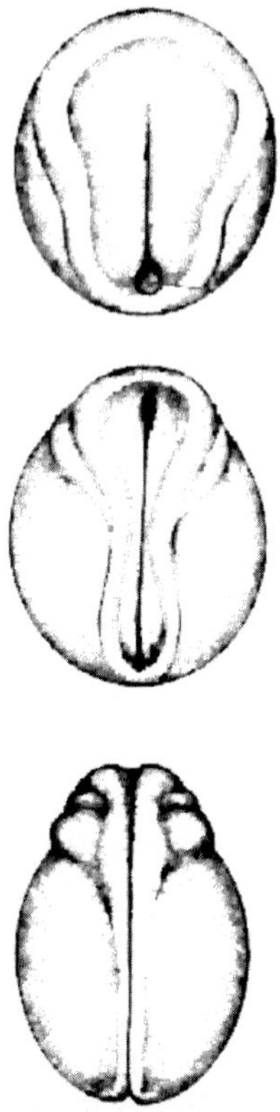

FIG. 5 A schematic of the formation of the neural folds of *Xenopus*. Shown in schematic form is the process of neural tube closure (viewed from the dorsal aspect of the embryo). Modified from Ellinson (1997).

and differentiation in *Xenopus* requires the spatial and temporal orchestration of numerous signaling molecules, transcription factors, cell adhesion, and cytoskeletal systems.

During gastrulation/neurulation, the neural plate/neural tube is patterned along two axes: the lateral-medial (later; dorsal-ventral) and rostral-caudal. Neural determination is thought to be initially anterior in character and that planar and vertical signals later impose regional differentiation (Ruiz i Altaba, 1992). Along its lateral-medial (dorsal-ventral) axis, the neural plate forms the neural crest, sensory neurons, roof plate, inter neurons, motor neurons, and floor plate cells (Mayor *et al.,* 1999). Ventral neural fates, i.e. floor plate, motor neurons, and ventral interneurons along the spinal cord, appears to be imposed by sonic hedgehog (Shh), expressed in the notochord (the most dorsal mesodermal tissue) and the floorplate (Brewster *et al.,* 1998). Dorsal structures, to some extent, seem to be maintained by BMPs and BMP antagonists (Liem *et al.,* 1995; Brewster *et al.,* 1998). Along the rostral-caudal axis the neural tube is divided into forebrain, midbrain, hindbrain, and spinal cord, as well as cranial and trunk neural crest (Mayor *et al.,* 1999). This patterning is controlled by caudalizing factors such as FGFs, retinoids, and Wnts (Cox and Hemmati-Brivanlou, 1995; Saint-Jeannet *et al.,* 1997; Sasai and De Robertis, 1997; Holowacz and Sokol, 1999).

Planar signals, primarily BMP-antagonists, initially originate from the Spemann organizer and influence the adjacent dorsal ectoderm (Hemmati-Brivanlou and Melton 1994, Harland and Gerhart, 1997). These signals increase the bias of the dorsal ectoderm to respond to vertical signals emanating from the underlying mesoderm and possibly planar signals from the adjacent ectoderm. The competence of dorsal ectoderm to respond to neuralizing signals is highest at early gastrula, and by the midgastrula stage, neural determination is thought to have occurred (Chang and Hemmati-Brivanlou, 1998).

A. Prepatterning the Ectoderm

1. BMP Activity Gradient

Gradients of BMP signaling are the foundation of neural/ectodermal differentiation. BMP2/4 are expressed in the future ectoderm (embyronic epidermis/neuroectoderm) in the late blastula (Dale and Jones, 1999). High levels of BMPs are sufficient to repress neural and to induce epidermal differentiation. On its own, as in the standard "animal cap" assay, the ectoderm will differentiate into atypical epidermis (Hemmati-Brivanlou and Melton, 1997). A decrease in BMP signaling, such as occurs when cells are dissociated from one another, leads to neuronal differentiation.

BMP4 signals are transduced through the phosphorylation of SMAD proteins by activated BMP receptor (serine/threonine kinase) proteins (Kawabata *et al.,* 1998). Phosphorylation leads to the translocation of the SMADs to the nucleus where they modulate gene expression (Nakayama *et al.,* 1998; Christian and Nakayama, 1999). In *Xenopus,* SMAD1 inhibits the expression of neural specific genes and induces the expression the epidermal genes (Wilson *et al.,* 1997). A number of BMP inhibitors, e.g. noggin, chordin, follistatin, Cerberus, and gremlin, are secreted by the cells of the organizer, bind to BMPs, and act as dorsalizing factors (Harland

and Gerhart, 1997; Hsu *et al.,* 1998a; Piccolo *et al.,* 1999). *Xenopus* GDF, a recently discovered noggin antagonist, is a member of the TGFβ family and acts to induce epidermis and inhibits neural tissue in dissociated cells (Chang and Hemmati-Brivanlou, 1999) GDF6 binds directly to noggin, forms heterodimers with BMP2, and perhaps synergizes with BMPs to pattern the ectoderm.

By the early gastrula stage, a dorsoventral gradient of BMP antagonists, secreted from cells of the organizer, has been established (Wilson *et al.,* 1997; Mayor *et al.,* 1999). The highest levels of the BMP inhibitors are found in the dorsal ectoderm and will eventually become the ventral and lateral neural tube; the neural folds, characterized by intermediate levels of BMP activity, give rise to the neural crest and dorsal neural tube; while the epidermis is associated with high levels of BMP activity (Fig. 6) (Groves and Bronner-Fraser, 1999; Mayor *et al.,* 1999). While BMPs set the stage for epidermal/neuroectodermal patterning, they are not the only players. The neural crest domain is formed between the neural plate, with high concentration of noggin, and the presumptive epidermis, expressing high levels of BMPs. Increasing the size of the neural plate domain, by over-expressing proneural genes such as XASH3 or X-NGNR-1, results in a diminished neural crest domain (Turner and Weintraub, 1994; Ma *et al.,* 1996). If induction of the neural crest was dependent only on BMP inhibitors released by the neural plate (e.g. planar signals) it would be expected that an increase in the neural plate region would result in an increase in the neural crest region not a decrease. In addition, the paraxial mesoderm, which lies beneath the neural folds, can induce neural crest in the absence of neural plate (Mayor *et al.,* 1995; Mancilla and Mayor, 1996; LaBonne and Bronner-Fraser, 1998). Hence, underlying the epidermal/neuroectodermal BMP gradient are various other signaling systems. For simplicity, we will restrict our discussion to the roles of Wnts.

2. Wnt Signaling Role in Patterning of the Neurula

Many Wnts are expressed in the neural tube along the rostral-caudal and medial-lateral axes (Figs. 6 and 7, see color insert) (Wolda and Moon, 1992). Xwnt1 and Xwnt3a have similar patterns of expression in neural tissue (Wolda *et al.,* 1993). Xwnt3a transcripts have been detected at the lateral edges of the neural plate around stage 13 (Bang *et al.,* 1999). During neurulation Xwnt3a is expressed exclusively along the dorsal midline of the developing neural tube (Wolda *et al.,* 1993). The expression of Xwnt1 overlaps that of Xwnt3a. Xwnt1 expression begins later in neurula and is uniquely detected along the mid and hindbrain boundary (Wolda *et al.,* 1993; Chang and Hemmati-Brivanlou, 1998). Expression of the maternal gene Xwnt7b is detected in the dorsal neural tube, including the roof plate, and the epidermis after neural tube closure (Chang and Hemmati-Brivanlou, 1998). *In situ* analysis of Xwnt4 in embryos show that it is expressed along dorsal midline of the midbrain, hindbrain, and floorplate of the neural tube (McGrew *et al.*, 1992). Xwnt8 is expressed in the ventral lateral mesoderm during gastrulation (Christian *et al.*, 1991), which appears to be involved with the induction of the neuroectodermal patterning genes Pax3 and Msx1 (see below) (Bang *et al.*, 1999). Xwnt3a, Xwnt1, and Xwnt4 expression is dependent upon neural induction, suggesting that they participate in the patterning of the nervous system during development. Xwnt8 and Xwnt7b, which are expressed prior to neural induction, may have roles in both induction and patterning.

3. Rostral-Caudal Patterning

BMP antagonists secreted from Spemann's organizer transform dorsal ectoderm into anterior neural tissue (McGrew *et al.*, 1995). Neural tissue then becomes caudalized by signals coming from the late organizer and the underlying mesoderm. For example, FGFs and retinoids transform the rostral to caudal neural tissue at the neural plate stage (Cox and Hemmati-Brivanlou, 1995; Sasai and De Robertis, 1997; Holowacz and Sokol, 1999). Among the caudalizing factors are members of the Wnt family. Much of the evidence that Wnt signaling is involved in anterior-posterior patterning comes from studies with Xwnt3a (Saint-Jeannet *et al.*, 1997). Overexpression of Xwnt3a in both embryos and neuralized animal caps assays induces Krox20 and Xlhbox6, posterior neural markers, and represses anterior neural genes such as Oxt2 (Fig. 7A). Hence, Xwnt3a appears to be able to pattern neural tissue by repressing the expression of anterior neural genes and increasing the expression of posterior neural genes. Conversely, expression of a "dominant negative" Wnt (dnWnt) reduced the expression of posterior neural genes and increased the expression of anterior neural genes (McGrew *et al.*, 1997). The ability of Xwnt3a to repress the expression of anterior neural genes in noggin-treated animal caps is dependent on a functional FGF pathway (McGrew *et al.*, 1997). As mentioned above, FGF is a strong caudalizing factor that is necessary for the posteriorization of the neural tube. FGF and Wnt signaling appear to cooperate in caudalizing neural tissue. More recently, McGrew and coworkers (1999) found that the *Xenopus engrailed-2* promoter contains three TCF/LEF consensus-binding sites that are required for its transcriptional activation. Engrailed-2 is a midbrain marker of the dorsal neural tube. A combination of Xwnt3a (or β-catenin) and noggin can induce an *engrailed-2* promoter luciferase reporter construct in animal cap assays, thus supporting a role of Wnt signal in the development of the neural tube.

4. Dorsal-Ventral/Medial-Lateral Patterning

Noggin induces expression of the forebrain marker Oxt2 and the panneural marker Nrp1, but represses the expression of the neural crest marker AP2 in both animal cap assays and intact embryo experiments (Saint-Jeannet *et al.,* 1997). Ectopic coexpression of XWnt3a with noggin results in a strong inhibition of Nrp1 and Oxt2 expression while at the same time inducing the most dorsal neural structure, the neural crest, as determined by Slug, Krox20, and AP2 expression (Saint-Jeannet *et al.,* 1997) (Fig. 7B). The ability of XWnt3a to induce neural crest markers was inhibited by the overexpression of GSK3, arguing that Xwnt3a acts through the canonical Wnt pathway (Saint-Jeannet *et al.,* 1997).

In addition to activating neural crest genes, Xwnt3a can induce dorsal and suppress ventral neural genes (Saint-Jeannet *et al.,* 1997). This was demonstrated in a neuralized animal cap assay where two dorsally restricted genes Pax3 (Bang *et al.,* 1997) and Xwnt1 (Wolda *et al.,* 1993) were strongly activated by injected Xwnt3a RNA. Two ventrally restricted genes, Shh (Stolow and Shi, 1995) and FKH1 (Dirksen and Jamrich, 1995), were not induced. The induction of Krox20, a hindbrain/neural crest marker, and Pax3 implies that Xwnt3a has both posteriorizing and dorsalizing activities. However, Xwnt3a does not change the anteroposterior pattern of Krox20 expression but rather its dorsal expression (Saint-Jeannet *et al.,* 1997). Overexpression of Xwnt3a in whole embryos led to increases in the Slug and Krox20 expression in the neural crest but the hindbrain expression of Krox20 was unaffected.

Xwnt7b appears to act as a dorsalizing rather than a caudalizing factor. In noggin-treated animal caps Xwnt7b induces neural crest markers but produces only a slight caudalization of neural tissues (Chang and Hemmati-Brivanlou, 1998). In neuralized animal cap assays midbrain marker En2 is induced only at highest levels of Xwnt7b overexpression while the spinal cord marker, X1hbox6, is never induced; nor does Xwnt7b overexpression repress Otx2. Xwnt7b does have the ability to induce the expression of Krox20; however, it is not clear whether this induction is specific to the neural crest region, as is the case with Xwnt3a. In cell dissociation assays Xwnt7b induces Slug and Twist expression (Chang and Hemmati-Brivanlou, 1998). However, overexpression of Xwnt7b in whole embryos does not result in expansion of neural crest domain into the neural plate, thus suggesting that other factors, perhaps by proneural factors or SOXs (see below), suppress neural crest formation in the neural plate region. Therefore, while Xwnt3a has both general dorsalizing and caudalizing activities in the neuroectoderm, Xwnt7b appears to act mainly to maintain dorsal and particularly neural crest gene expression. This suggests that Wnt signals play a critical role in specification and maintenance of the most dorsal neural tissue, the neural crest.

B. The Neural Crest

The formation of the neural crest is a classic example of a epithelial to mesenchymal transition (EMT). EMT involves the remodeling of adherence junctions and the acquisition of cellular motility (Savagner *et al.,* 1997). In *X. laevis,* neural crest formation is dependent upon the Zinc-finger transcription factors Slug and Snail (Carl *et al.,* 1999; Mayor *et al.,* 1999). Expression of Slug, and neural crest induction, appears to be regulated by intermediate levels of BMPs at the lateral edges of the neural plate (Baker and Bronner-Fraser, 1997; LaBonne and Bronner-Fraser, 1998; Mayor *et al.,* 1999). However, neural folds isolated at stage 12, when they begin to express Slug, are unable to maintain Slug expression autonomously unless they are co-cultured with dorsal mesoderm (Baker and Bronner-Fraser, 1997). LaBonne and Bronner-Fraser (1998) demonstrated that chordin can induce low levels of Slug expression in a concentration-dependent manner in animal caps, but co-culturing chordin and mesodermal tissue was required for high levels of Slug expression. This effect can also been seen when explants overexpressing chordin or noggin were co-cultured with explants overexpressing XeFGF or various Wnts including Wnt3a, 7b, and XWnt8 (Saint-Jeannet *et al.,* 1997; Chang and Hemmati-Brivanlou, 1998; LaBonne and Bronner-Fraser, 1998).

Xwnt8's ability to induce neural crest markers remains somewhat controversial. Saint-Jeannet *et al.* (1997) and Chang and Hemmati-Brivanlou (1998) found that Xwnt8 had little to no ability to induce neural crest markers in neuralized animal caps. In contrast, LaBonne and Bronner-Fraser (1998) found that ectopic expression of Xwnt8 in both intact embryos and chordin-treated animal caps resulted in an increase in Slug expression and the formation of neural crest cells. Xwnt8 is expressed in the ventral lateral mesoderm during early gastrula and could be potentially involved the initial induction of the presumptive neural crest and Slug domains (Christian *et al.,* 1991; Mayor *et al.,* 1995).

Pax3 is a member of the paired type homeobox gene family and is specifically expressed in the dorsal-posterior neural tube (Bang *et al.,* 1997). Its expression begins early (stage 11) in a broad mediolateral domain of the neural plate and then is restricted quickly (stage 11.5) to two posterior lateral domains, specifying a domain where the dorsal hindbrain, spinal cord, and neural crest will arise (Bang *et al.,* 1997). The late organizer and the posterior paraxial mesoderm are capable of inducing Pax3 (Bang *et al.,* 1997). By using both a Wnt overexpression and Wnt dominant negative approach it was found that Pax3 expression requires Wnt-dependent inductive signals emanating from the posterior paraxial mesoderm (Bang *et al.,* 1999). Overexpression of Xwnt8 in embryos leads to an enlargment of the Pax3, while ectopic expression of a dominant negative Wnt blocks the expression of Pax3. Furthermore, injection of dominant negative Wnt blocks the expression of Msx1 (Suzuki *et al.,* 1997; Feledy *et al.,* 1999), another early marker of the lateral neural plate, and the development of Rohon-Beard cells that arise from the neural crest domain (Bang *et al.,* 1999)

Hence, the specification of the posterior lateral neural plate, a domain that includes the presumptive neural crest, is to some extent patterned by Xwnt8 signaling. The secreted inhibitors of the Wnt signaling, FrzB (Wang *et al.,* 1997a) and sizzled (Salic *et al.,* 1997), which pattern the mesoderm, could also play a role in restricting the activity of Wnt signaling in the neural plate (see below). A potential mediator of the Xwnt8 signal from paraxial mesoderm, *Xenopus* brain factor 2 (Xbf2), is expressed in the dorsolateral mesoderm, where it helps specify this territory by downregulating BMP4 and its downstream genes (Gomez-Skarmeta *et al.,* 1999). Overexpression of Xbf2 in embryos causes partial axis duplication, while in animal caps Xbf2 induces neural tissue. Xbf2 also seems to participate in the control of the neural crest migration.

The effects of Wnt signaling on the neural crest formation are direct. It has been shown that β-catenin overexpression dramatically increases Slug expression in a cell-autonomous fashion. FGF signaling may be indirect in its ability to induce Slug and the neural crest (Mayor *et al.,* 1997). FGF has been shown to be able to induce the expression of Xwnt8 in mesoderm induction assays (Christian *et al.,* 1991, 1992) and overexpression of a dominant negative Wnt blocks the ability of FGF and noggin to induce neural crest markers (LaBonne and Bronner-Fraser, 1998). Therefore, it maybe that the ability for FGF to induce Slug expression in the neural crest by previous studies was a consequence of its ability to induce Xwnt8.

As mentioned above, ectopic expression of Xwnt3a, in neuralized animal caps results in the increase expression of neural crest markers through GSK3 (Saint-Jeannet *et al.,* 1997). However, Xwnt3a transcripts are not expressed until after neural induction, and Slug expression therefore cannot be involved in the initial induction of the neural crest. Clues to Xwnt3a's role in neural crest differentiation have come from other model systems. In the zebrafish activation of the Wnt pathway in neural crest cells induced cartilage and pigment cells, while blocking Wnt signaling favored the differentiation of glia and neurons (Dorsky *et al.,* 1998). Wnt3a null mice have severe defects in the differentiation of the paraxial mesoderm (Galceran *et al.,* 1999). Double null mutants in both TCF1 and LEF1, downstream effectors of the Wnt signal, display similar phenotypes (Galceran *et al.,* 1999).

C. Modulating Wnt Signaling

As in the case in the early embryo, a number of factors are likely to modulate the responses of secreted Wnts in the neural plate. First, there are the secreted inhibitors of the Wnt signaling. The Wnt/Nodal/BMP inhibitor Cerberus is expressed in the endoderm and Spemann organizer, and its expression leads to head induction (Piccolo *et al.,* 1999). It has not been reported whether "dominant-negative" forms of Wnts, used in some studies, alter Cerberus activity.

Frizzled proteins appear to be the membrane-anchored receptors for the Wnt growth factors (Dierick and Bejsovec, 1999). Clearly the pattern of frizzled expression within neural plate and neural crest cells will determine how these cells respond to specific Wnt signals. However, little about the specific patterns of Frizzled expression in *Xenopus* has been published (Itoh *et al.,* 1998; Shi *et al.,* 1998). Likewise, the pattern of Wnt antagonist expression is clearly important. For example, Sizzled transcripts can be detected in the ventral marginal zone as a narrow domain during gastrulation (Salic *et al.,* 1997), whereas XFrzb is expressed in the Spemann organizer (Wang *et al.,* 1997a) while FrzA is expressed in the somitic mesoderm, developing heart, and later in neural crest derivatives (Xu *et al.,* 1998). Both sizzled and Frzb are effective antagonists of Xwnt-8, which has an expression pattern that overlaps that of Frzb and sizzled (Salic *et al.,* 1997; Wang *et al.,* 1997a,b). As discussed above Frzb and Sizzled may play a role in defining the Pax3 and Msx1 domain by localizing Xwnt8 activity to the posterior lateral mesoderm (Bang *et al.,* 1999).

1. TCFs

A second regulator of Wnt signaling is the pattern of TCF expression. Both of the TCFs characterized in *Xenopus* are expressed in the neural tube and neural crest (Molenaar *et al.,* 1998). At later stages both genes are expressed in the central nervous system, eyes, otic vesicles, head mesenchyme, neural crest and derivatives, branchial arches, developing heart, tailbud, and limb buds. TCF-binding sites have been identified in the En2 promoter (McGrew *et al.,* 1999). However, XLEF1 and XTCF3 can have quite different effects on specific promoters (Vourgourakis *et al.,* 1999) and their relative binding affinities for specific DNA sites or their various co-activators/co-repressor remains to be determined. If *X. laevis* also has homologs of the TCF1 and TCF4 genes, which seems likely given the isolation of a TCF4 homolog from zebrafish (Dorsky *et al.,* 1999), the story could be more complicated yet.

2. SOXs

A number of SOX genes are expressed in the neural plate/neural crest. XSOX2, XSOX3, and XSOXD are expressed in the neural plate. XSOX3 (Zorn *et al.,* 1999a) and XSOXD (Vourgourakis *et al.,* 1999) have been shown to inhibit the ability of β-catenin to activate the OT reporter.

XSOXD expression begins late in the blastulae, becomes restricted to the prospective ectoderm, and later to the dorsal ectoderm by the midgastrula stage (Mizuseki *et al.*, 1998). Ectopic expression of XSOXD leads to ectopic formation of neural tissue and neuronal markers (e.g. Xngnr1), whereas the expression of a "dominant-negative" form of the protein inhibits neural differentiation (Mizuseki *et al.*, 1998). At later stages XSOXD is restricted to only the lateral edges of the neural plate and the neural crest domain. This exclusion from the medial neural plate may be caused by the expression of the proneural genes. It has been demonstrated that both XASH3 and Xngnr1 can activate the expression of Delta which acts via the Notch pathway to suppress neural markers (Turner and Weintraub, 1994; Ma *et al.*, 1996; Coffman *et al.*, 1993; Chitnis *et al.*, 1995; Artavanis-Tsakonas *et al.*, 1999). Like other neuroectodermal genes, XSOXD expression is enhanced by chordin and is downregulated by BMP4. Hence, XSOXD is a necessary mediator of the induction of anterior neural tissues (Mizuseki *et al.*, 1998). Ectopic expression of XSOXD upregulates Slug expression while suppressing keratin formation, a process characteristic of EMT (Savagner *et al.*, 1997).

XSOX2 has a panneural expression pattern in the neural plate, neural crest, and later the central nervous system (Mizuseki *et al.*, 1998). Like XSOXD, XSOX2 is induced by chordin and suppressed by BMPs (Mizuseki *et al.*, 1998). Unlike XSOXD, ectopic XSOX2 expression has little to no neuralizing activity; however, it does have the ability to make the ectoderm responsive to other signals such as FGF. Alone FGF does not neuralize animal caps, but together with XSOX2 can induce the expression of NCAM, XSlug, Otx2, Krox20, and Xlhbox6, thus arguing that XSOX2 acts to make the ectoderm competent to respond to extracellular neuralizing signals.

XSOX3 is expressed as both a maternal transcript (see above) and in the animal hemisphere of the late blastula stage embryo (Penzel *et al.*, 1997). Later expression is detected in the prospective central nervous system and lens epithelium during neurula. XSOX3 can bind to β-catenin, and inhibits its ability to activate the OT reporter *in vitro*. However, function of XSOX3 during neurulation in *Xenopus* has yet to be determined.

In the mouse a number of SOXs have been found to influence neural crest development. SOX10 expression is high in the neural crest at the time of their emergence and continues to be expressed in cells that will become part of the peripheral nervous system (Kuhlbrodt *et al.*, 1998b). Mutations in SOX10 lead to Hirshsprung's disease (Kuhlbrodt *et al.*, 1998c). Mouse SOX11, which is expressed in the central nervous system and tissue involved in epithelial mesenchymal transitions, such as branchial arches, ganglia, trunk, and limbs, is expressed strongly in mesenchymal cells (Wegner, 1999). Mouse SOX9 is important for cartilage development (Bi *et al.*, 1999)

development, both the quantitative and qualitative details of their interactions remain to be described and refined.

Acknowledgments

We thank Kirk Harris for help with the SOX/TCF phylogeny (and we absolve him of all its inadequecies, since they are our "fault" alone), Yanni Vourgourakis and Eric Jensen for comments on the manuscript and sharing their unpublished observations, and Chris Dufton and Julie Gelderloos for their work of the analysis of plakoglobin in the early embryo. Our work is supported by grants from the NIH (GM54001) and the March of Dimes Birth Defects Foundation, Inc. A. St. A. is supported through an NIH training grant and the March of Dimes.

References

Aberle, H., Bauer, A., Stappert, J., Kispert, A., and Kemler, R. (1997). β-catenin is a target for the ubiquitin-proteasome pathway. *EMBO J.* **16,** 3797–3804.

Aberle, H., *et al.* (1995). The human plakoglobin gene localizes on chromosome 17q21 and is subjected to loss of heterozygosity in breast and ovarian cancers. *Proc. Natl. Acad. Sci. USA* **92,** 6384–6388.

Aguinaldo, A. M., Turbeville, J. M., Linford, L. S., Rivera, M. C., Garey, J. R., Raff, R. A., and Lake, J. A. (1997). Evidence for a clade of nematodes, arthropods and other moulting animals. *Nature* **387,** 489–493.

Akitaya, T., and Bronner-Fraser, M. (1992). Expression of cell adhesion molecules during initiation and cessation of neural crest cell migration. *Dev. Dyn.* **194,** 12–20.

Ambrosetti, D. C., Basilico, C., and Dailey, L. (1997). Synergistic activation of the fibroblast growth factor 4 enhancer by Sox2 and Oct-3 depends on protein-protein interactions facilitated by a specific spatial arrangement of factor binding sites. *Mol. Cell Biol.* **17,** 6321–6329.

Anderson, D. J. (1997). Cellular and molecular biology of neural crest cell lineage determination. *Trends Genet.* **13,** 276–280.

Andra, K., Lassmann, H., Bittner, R., Shorny, S., Fassler, R., Propst, F., and Wiche, G. (1997). Targeted inactivation of plectin reveals essential function in maintaining the integrity of skin, muscle, and heart cytoarchitecture. *Genes Dev.* **11,** 3143–3156.

Andra, K., Nikolic, B., Stocher, M., Drenckhahn, D., and Wiche, G. (1998). Not just scaffolding: Plectin regulates actin dynamics in cultured cells. *Genes Dev.* **12,** 3442–3451.

Aono, S., Nakagawa, S., Reynolds, A. B., and Takeichi, M. (1999). p120(ctn) acts as an inhibitory regulator of cadherin function in colon carcinoma cells. *J. Cell Biol.* **145,** 551–562.

Artavanis-Tsakonas, S., Rand, M. D., and Lake, R. J. (1999). Notch signaling: Cell fate control and signal integration in development. *Science* **284,** 770–776.

Ax, P. (1996). "Multicellular Animals: A New Approach to the Phylogenetic Order in Nature." Berlin, Springer.

Axelrod, J. D., Matsuno, K., Artavanis, T. S., and Perrimon, N. (1996). Interaction between Wingless and Notch signaling pathways mediated by dishevelled. *Science* **271,** 1826–1832.

Bafico, A., Gazit, A., Pramila, T., Finch, P. W., Yaniv, A., and Aaronson, S. A. (1999). Interaction of frizzled related protein (FRP) with Wnt ligands and the frizzled receptor suggests alternative mechanisms for FRP inhibition of Wnt signaling. *J. Biol. Chem.* **274,** 16180–16187.

Baker, C. V., and Bronner-Fraser, M. (1997). The origins of the neural crest. Part I: embryonic induction. *Mech. Dev.* **69,** 3–11.

Bang, A. G., Papalopulu, N., Goulding, M. D., and Kintner, C. (1999). Expression of Pax-3 in the lateral neural plate is dependent on a Wnt-mediated signal from posterior non-axial mesoderm. *Dev. Biol.* **212,** 366–380.

Bang, A. G., Papalopulu, N., Kintner, C., and Goulding, M. D. (1997). Expression of Pax-3 is initiated in the early neural plate by posteriorizing signals produced by the organizer and by posterior non-axial mesoderm. *Development* **124,** 2075–2085.

Baribault, H., and Oshima, R. G. (1991). Polarized and functional epithelia can form after the targeted inactivation of both mouse keratin 8 alleles. *J. Cell Biol.* **115,** 1675–1684.

Baribault, H., Penner, J., Iozzo, R. V., and Wilson, H. M. (1994). Colorectal hyperplasia and inflammation in keratin 8-deficient FVB/N mice. *Genes Dev.* **8,** 2964–2973.

Baribault, H., Price, J., Miyai, K., and Oshima, R. G. (1993). Mid-gestational lethality in mice lacking keratin 8. *Genes Dev.* **7,** 1191–1202.

Bartnik, E., Osborn, M., and Weber, K. (1986). Intermediate filaments in muscle and epithelial cells of nematodes. *J. Cell Biol.* **102,** 2033–2041.

Bartnik, E., and Weber, K. (1989). Widespread occurrence of intermediate filaments in invertebrates: Common principles and diversion. *Eur. J. Cell Bio.* **50,** 17–33.

Bauer, A., Huber, O., and Kemler, R. (1998). Pontin52, an interaction partner of β-catenin, binds to the TATA box binding protein. *Proc. Natl. Acad. Sci. USA* **95,** 14787–14792.

Beals, C. R., Sheridan, C. M., Turck, C. W., Gardner, P., and Crabtree, G. R. (1997). Nuclear export of NF-ATc enhanced by glycogen synthase kinase-3. *Science* **275,** 1930–1934.

Behrens, J., Jerchow, B. A., Wrutele, M., Grimm, J., Asbrand, C., Wirtz, R., Kohl, M., Wedlich, D., and Birchmeier, W. (1998). Functional interaction of an axin homolog, conductin, with β-catenin, APC, and GSK3β. *Science* **280,** 596–599.

Behrens, J., von Kries, J. P., Kuhl, M., Bruhn, L., Wedlish, D., Grosschedl, R., and Birchmeier, W. (1996). Functional interaction of β-catenin with the transcription factor LEF-1. *Nature* **382,** 638–642.

Bejsovec, A., and Wieschaus, E. (1995). Signaling activities of the *Drosophila* wingless gene are separately mutable and appear to be transduced at the cell surface. *Genetics* **139,** 309–320.

Berndorff, D., Gessner, R., Kreft, B., Schnoy, N., Lajous, P. A., Loch, N., Reutter, W., Hortsch, M., and Tauber, R. (1994). Liver-intestine cadherin: Molecular cloning and characterization of a novel $Ca^{(2+)}$-dependent cell adhesion molecule expressed in liver and intestine. *J. Cell Biol.* **125,** 1353–1369.

Bhat, K. M. (1998). frizzled and frizzled 2 play a partially redundant role in wingless signaling and have similar requirements to wingless in neurogenesis. *Cell* **95,** 1027–1036.

Bi, W., Deng, J. M., Zhang, Z., Behringer, R. R., and de Crombrugghe, B. (1999). Sox9 is required for cartilage formation. *Nat. Genet.* **22,** 85–89.

Bienz, M. (1998). TCF: Transcriptional activator or repressor? *Curr. Opin. Cell Biol.* **10,** 366–372.

Bierkamp, C., Mclaughlin, K. J., Schwarz, H., Huber, O., and Kemler, R. (1996). Embryonic heart and skin defects in mice lacking plakoglobin. *Dev. Biol.* **180,** 780–785.

Bierkamp, C., Schwarz, H., Huber, O., and Kemler, R. (1999). Desmosomal localization of β-catenin in the skin of plakoglobin null-mutant mice. *Development* **126,** 371–381.

Binari, R. C., Staveley, B. E., Johnson, W. A., Godavarti, R., Sasisekharan, R., and Manoukian, A. S. (1997). Genetic evidence that heparin-like glycosaminoglycans are involved in wingless signaling. *Development* **124,** 2623–2632.

Birchmeier, W. (1995). E-cadherin as a tumor (invasion) suppressor gene. *Bioessays* **17,** 97–99.

Bisoffi, M., and Betschart, B. (1996). Ascaris suum: Molecular cloning of an intermediate filament. *Trop. Med. Int. Health* **1,** 640–645.

Bonn, S., van Hagel, J., Nollet, F., Kools, P., and van Roy, F. (1999). Plakophilin-3, a novel armadillo-like protein present in nuclei and desmosomes of epithelial cells. *J. Cell Sci.* 2265–2276.

Botquin, V., Hess, H., Fuhrmann, G., Anastassiadis, C., Gross, M. K., Vriend, G., and Scholer, H. R. (1998). New POU dimer configuration mediates antagonistic control of an osteopontin preimplantation enhancer by Oct-4 and Sox-2. *Genes Dev.* **12,** 2073–2090.

Boutros, M., Paricio, N., Strutt, D. I., and Mlodzik, M. (1998). Dishevelled activates JNK and discriminates between JNK pathways in planar polarity and wingless signaling. *Cell* **94,** 109–118.

Bradley, R. S., Espeseth, A., and Kintner, C. (1998). NF-protocadherin, a novel member of the cadherin superfamily, is required for *Xenopus* ectodermal differentiation. *Curr. Biol.* **8,** 325–334.

Brady-Kalnay, S. M., *et al.* (1998). Dynamic interaction of PTPμ with multiple cadherins in vivo. *J. Cell Biol.* **141,** 287–296.

Braga, V. M., Machesky, L. M., Hall, A., and Hotchin, N. A. (1997). The small GTPases Rho and Rac are required for the establishment of cadherin-dependent cell-cell contacts. *J. Cell Biol.* **137,** 1421–1431.

Brannon, M., Brown, J. D., Bates, R., Kimelman, D., and Moon, R. T. (1999). XCtBP is a XTcf-3 co-repressor with roles throughout *Xenopus* development. *Development* **126,** 3159–3170.

Brannon, M., Gomperts, M., Sumoy, L., Moon, R. T., and Kimelman, D. (1997). A β-catenin/XTcf-3 complex binds to the siamois promoter to regulate dorsal axis specificiation in Xenopus. *Genes Dev.* **11,** 2359–2370.

Brewster, R., Lee, J., and Ruiz, i Altaba, A. (1998). Gli/Zic factors pattern the neural plate by defining domains of cell differentiation. *Nature* **393,** 579–583.

Broders, F., and Thiery, J. P. (1995). Contribution of cadherins to directional cell migration and histogenesis in Xenopus embryos. *Cell Adhes. Commun.* **3,** 419–440.

Bruhn, L., Munnerlyn, A., and Grosschedl, R. (1997). ALY, a context-dependent coactivator of LEF-1 and AML-1, is required for TCRα enhancer function. *Genes Dev.* **11,** 640–653.

Burdett, I. D. (1998). Aspects of the structure and assembly of desmosomes. *Micron* **29,** 309–328.

Caca, K., Kolligs, F. T., Ji, X., Hayes, M., Qian, J., Yahanda, A., Rimm, D. L., Costa, J., and Fearon, E. R. (1999). β- and γ-catenin mutations, but not E-cadherin inactivation, underlie T-cell factor/lymphoid enhancer factor transcriptional deregulation in gastric and pancreatic cancer. *Cell Growth Differ.* **10,** 369–376.

Carl, T., and Klymkowsky, M. W. (1999). Whole-mount visualization of endogenous and exogenous proteins in *Xenopus* and other organisms. *In* "A Comparative Methods approach to the study of oocytes and embryos," (J. Richter, Ed.) pp. 291–315. Oxford University Press, New York.

Carl, T. F., Dufton, C., Hanken, J., and Klymkowsky, M. W. (1999). Inhibition of neural crest migration in Xenopus using antisense slug RNA. *Dev. Biol.* **213,** 101–115.

Carnac, G., Kodjabachian, L., Gurdon, J. B., and Lemaire, P. (1996). The homeobox gene Siamois is a target of the Wnt dorsalization pathway and triggers organizer activity in the absence of mesoderm. *Development* **122,** 3055–3065.

Cha, B. J., and Gard, D. L. (1999). XMAP230 is required for the organization of cortical microtubules and patterning of the dorsoventral axis in fertilized *Xenopus* eggs. *Dev. Biol.* **205,** 275–286.

Chakrabarti, A., Matthews, G., Colman, A., and Dale, L. (1992). Secretory and inductive properties of *Drosophila* wingless protein in *Xenopus* oocytes and embryos. *Development* **115,** 355–369.

Chan, E. F., Gat, U., McNiff, J. M., and Fuchs, E. (1999). A common human skin tumour is caused by activating mutations in β-catenin. *Nat. Genet.* **21,** 410–413.

Chan, Y. M., and Jan, Y. N. (2000). Presenilins, processing of beta-amyloid precursor protein, and notch signaling. *Neuron.* **23,** 201–204.

Chang, C., and Hemmati-Brivanlou, A. (1998). Cell fate determination in embryonic ectoderm. *J. Neurobiol.* **36,** 128–151.

Chang, C., and Hemmati-Brivanlou, A. (1998). Neural crest induction by Xwnt7B in Xenopus. *Dev. Biol.* **194,** 129–134.

Chang, C., and Hemmati-Brivanlou, A. (1999). *Xenopus* GDF6, a new antagonist of noggin and a partner of BMPs. *Development* **126,** 3347–3357.

Chavanas, S., Pulkkinen, L., Gache, Y., Smith, F. J., McLean, W. H., Uitto, J., Ortonne, J. P., and Meneguzzi, G. (1996). A homozygous nonsense mutation in the PLEC1 gene in patients with epidermolysis bullosa simplex with muscular dystrophy. *J. Clin. Invest.* **98,** 2196–2200.

Chitaev, N. A., and Troyanovsky, S. M. (1998). Adhesive but not lateral E-cadherin complexes require calcium and catenins for their formation. *J. Cell Biol.* **142,** 837–846.

Chitnis, A., Henrique, D., Lewis, J., Ish-Horowicz, D., and Kintner, C. (1995). Primary neurogenesis in *Xenopus* embryos regulated by a homologue of the *Drosophila* neurogenic gene Delta. *Nature* **375,** 761–766.

Christian, J. L., McMahon, J. A., McMahon, A. P., and Moon, R. T. (1991). Xwnt-8, a Xenopus Wnt-1/int-1-related gene responsive to mesoderm-inducing growth factors, may play a role in ventral mesodermal patterning during embryogenesis. *Development* **111,** 1045–1055.

Christian, J. L., and Nakayama, T. (1999). Can't get no SMADisfaction: Smad proteins as positive and negative regulators of TGF-β family signals. *Bioessays* **21,** 382–390.

Christian, J. L., Olson, D. J., and Moon, R. T. (1992). Xwnt-8 modifies the character of mesoderm induced by bFGF in isolated *Xenopus* ectoderm. *EMBO J.* **11,** 33–41.

Christofori, G., and Semb, H. (1999). The role of the cell-adhesion molecule E-cadherin as a tumour-suppressor gene. *Trends Biochem. Sci.* **24,** 73–76.

Chung, H. M., and Malacinski, G. M. (1980). Establishment of the dorsal/ventral polarity of the amphibian embryo: Use of ultraviolet irradiation and egg rotation as probes. *Dev. Biol.* **80,** 120–133.

Cingolani, G., Petosa, C., Weis, K., and Muller, C. W. (1999). Structure of importin-beta bound to the IBB domain of importin-alpha. *Nature* **399,** 221–229.

Clark, H. F., Brentrup, D., Schneitz, K., Bieber, A., Goodman, C., and Noll, M. (1995). Dachsous encodes a member of the cadherin superfamily that controls imaginal disc morphogenesis in *Drosophila. Genes Dev.* **9,** 1530–1542.

Coffman, C. R., Skoglund, P., Harris, W. A., and Kintner, C. R. (1993). Expression of an extracellular deletion of Xotch diverts cell fate in *Xenopus* embryos. *Cell* **73,** 659–671.

Conway, S. J., Henderson, D. J., and Copp, A. J. (1997). Pax3 is required for cardiac neural crest migration in the mouse: Evidence from the splotch (Sp2H) mutant. *Development* **124,** 505–514.

Costa, M., Raich, W., Agbunag, C., Leung, B., Hardin, J., and Priess, J. R. (1998). A putative catenin-cadherin system mediates morphogenesis of the *Caenorhabditis elegans* embryo. *J. Cell Biol.* **141,** 297–308.

Cowin, P., Kapprell, H. P., Franke, W. W., Tamkun, J., and Hynes, R. O. (1986). Plakoglobin: A protein common to different kinds of intercellular adhering junctions. *Cell* **46,** 1063–1073.

Cox, R. T., Kirkpatrick, C., and Peifer, M. (1996). Armadillo is required for adherens junction assembly, cell polarity, and morphogenesis during *Drosophila* embryogenesis. *J. Cell Biol.* **134,** 133–148.

Cox, W. G., and Hemmati-Brivanlou, A. (1995). Caudalization of neural fate by tissue recombination and bFGF. *Development* **121,** 4349–4358.

Crawford, H. C., Fingleton, B. M., Rudolph, O. L., Goss, K. J., Rubinfeld, B., Polakis, P., and Matrisian, L. M. (1999). The metalloproteinase matrilysin is a target of β-catenin transactivation in intestinal tumors. *Oncogene* **18:** 2883–2891.

Cui, Y., Brown, J. D., Moon, R. T., and Christian, J. L. (1995). Xwnt-8b: A maternally expressed *Xenopus* Wnt gene with a potential role in establishing the dorsoventral axis. *Development* **121:** 2177–2186.

Dale, L., and Jones, C. M. (1999). BMP signalling in early *Xenopus* development. *Bioessays* **21,** 751–760.

Daniel, J. M., and Reynolds, A. B. (1995). The tyrosine kinase substrate p120cas binds directly to E-cadherin but not to the adenomatous polyposis coli protein or alpha-catenin. *Mol. Cell Biol.* **15,** 4819–4824.

Daniel, J. M., and Reynolds, A. B. (1999). The catenin p120(ctn) interacts with Kaiso, a novel BTB/POZ domain zinc finger transcription factor. *Mol. Cell Biol.* **19,** 3614–3623.

Danilchik, M. V., and Denegre, J. M. (1991). Deep cytoplasmic rearrangements during early development in *Xenopus laevis. Development* **111,** 845–856.

Danilchik, M. V., and Gerhart, J. C. (1987). Differentiation of the animal-vegetal axis in *Xenopus laevis* oocytes. I. Polarized intracellular translocation of platelets establishes the yolk gradient. *Dev. Biol.* **122,** 101–112.

Davidson, L. A., and Keller, R. E. (1999). Neural tube closure in *Xenopus laevis* involves medial migration, directed protrusive activity, cell intercalation and convergent extension. *Development* **126,** 4547–4556.

de La Coste, A., *et al.* (1998). Somatic mutations of the β-catenin gene are frequent in mouse and human hepatocellular carcinomas. *Proc. Natl. Acad. Sci. USA* **95,** 8847–8851.

DeMarais, A. A., and Moon, R. T. (1992). The armadillo homologs β-catenin and plakoglobin are differentially expressed during early development of *Xenopus laevis. Dev. Biol.* **153:** 337–346.

Denk, C., Hulsken, J., and Schwarz, E. (1997). Reduced gene expression of E-cadherin and associated catenins in human cervical carcinoma cell lines. *Cancer Lett.* **120,** 185–193.

Dent, J. A., and Klymkowsky, M. W. (1989). Whole-mount analyses of cytoskeletal reorganization and function during oogenesis and early embryogenesis. *In* "The Cell Biology of Fertilization." (H. Schatten and G. Schatten, Eds.), pp. 63–103. *Monographs in Cell Biology.* Academic Press, San Diego.

Dierick, H., and Bejsovec, A. (1999). Cellular mechanisms of wingless/Wnt signal transduction. *Curr. Top. Dev. Biol.* **43,** 153–190.

Dirksen, M. L., and Jamrich, M. (1995). Differential expression of fork head genes during early *Xenopus* and zebrafish development. *Dev. Genet.* **17,** 107–116.

Dodemont, H., Riemer, D., Ledger, N., and Weber, K. (1994). Eight genes and alternative RNA processing pathways generate an unexpectedly large diversity of cytoplasmic intermediate filament proteins in the nematode *Caenorhabditis elegans. EMBO J.* **13,** 2625–2638.

Dodemont, H., Riemer, D., and Weber, K. (1990). Structure of an invertebrate gene encoding cytoplasmic intermediate filament (IF) proteins: Implications for the origin and the diversification of IF proteins. *EMBO J.* **9,** 4083–4094.

Dominguez, I., Itoh, K., and Sokol, S. Y. (1995). Role of glycogen synthase kinase 3β as a negative regulator of dorsoventral axis formation in *Xenopus* embryos. *Proc. Natl. Acad. Sci. USA* **92,** 8498–8502.

Dorsky, R. I., Moon, R. T., and Raible, D. W. (1998). Control of neural crest cell fate by the Wnt signalling pathway. *Nature* **396,** 370–373.

Dorsky, R. I., Snyder, A., Cretekos, C. J., Grunwald, D. J., Geisler, R., Haffter, P., Moon, R. T., and Raible, D. W. (1999). Maternal and embryonic expression of zebrafish lef1. *Mech. Dev.* **86,** 147–150.

Drechsel, D. N., Hyman, A. A., Hall, A., and Glotzer, M. (1997). A requirement for Rho and Cdc42 during cytokinesis in *Xenopus* embryos. *Curr. Biol.* **7,** 12–23.

Dufour, S., Beauvais, J. A., Delouv, E. A., and Thiery, J. P. (1999). Differential function of N-cadherin and cadherin-7 in the control of embryonic cell motility. *J. Cell Biol.* **146,** 501–516.

Dufour, S., Saint, J. J., Broders, F., Wedlich, D., and Thiery, J. P. (1994). Differential perturbations in the morphogenesis of anterior structures induced by overexpression of truncated XB- and N-cadherins in *Xenopus* embryos. *J. Cell Biol.* **127,** 521–535.

Dunne, J., *et al.* (1995). Molecular cloning and tissue expression of FAT, the human homologue of the *Drosophila* fat gene that is located on chromosome 4q34-q35 and encodes a putative adhesion molecule. *Genomics* **30,** 207–223.

Edelmann, L., *et al.* (1999). A common molecular basis for rearrangement disorders on chromosome 22q11. *Hum. Mol. Genet.* **8,** 1157–1167.

Eger, A., Stockinger, A., Wiche, G., and Foisner, R. (1997). Polarisation-dependent association of plectin with desmoplakin and the lateral submembrane skeleton in MDCK cells. *J. Cell Sci.* **110,** 1307–1316.

Eisenmann, D. M., Maloof, J. N., Simske, J. S., Kenyon, C., and Kim, S. K. (1998). The β-catenin homolog BAR-1 and LET-60 Ras coordinately regulate the Hox gene lin-39 during *Caenorhabditis elegans* vulval development. *Development* **125,** 3667–3680.

Elinson, R. P. (1983). Cytoplasmic phases in the first cell cycle of the activated frog egg. *Dev. Biol.* **100,** 440–451.

Elinson, R. P. (1997). Amphibians. *In* "Embryology: Constructing the Organism." (S. F. Gilbert and A. M. Raunio, Eds.) Sunderland, MA., Sinauer and Associates, Inc. 409–436.

Elinson, R. P., and Rowning, B. (1988). A transient array of parallel microtubules in frog eggs: potential tracks for a cytoplasmic rotation that specifies the dorso-ventral axis. *Dev. Biol.* **128,** 185–197.

Emerson, J. A. (1988). Disruption of the cytokeratin filament network in the preimplantation mouse embryo. *Development* **104,** 219–234.

Espeseth, A., Johnson, E., and Kintner, C. (1995). *Xenopus* F-cadherin, a novel member of the cadherin family of cell adhesion molecules, is expressed at boundaries in the neural tube. *Mol. Cell Neurosci.* **6,** 199–211.

Fagotto, F., Funayama, N., Gluck, U., and Gumbiner, B. M. (1996). Binding to cadherins antagonizes the signaling activity of β-catenin during axis formation in *Xenopus. J. Cell Biol.* **132,** 1105–1114.

Fagotto, F., Gluck, U., and Gumbiner, B. M. (1998). Nuclear localization signal-independent and importin/karyopherin-independent nuclear import of β-catenin. *Curr. Biol.* **8,** 181–190.

Fagotto, F., and Gumbiner, B. M. (1994). β-catenin localization during *Xenopus* embryogenesis: Accumulation at tissue and somite boundaries. *Development* **120,** 3667–3679.

Fan, M. J., Gruning, W., Walz, G., and Sokol, S. Y. (1998). Wnt signaling and transcriptional control of siamois in *Xenopus* embryos. *Proc. Natl. Acad. Sci. USA* **95,** 5626–5631.

Fawcett, D. W. (1981). "The Cell." W. B. Saunders Co., Philadephia.

Feledy, J. A., Beanan, M. J., Sandoval, J. J., Goodrich, J. S., Lim, J. H., Matsuo, T. M., Sato, S. M., and Sargent, T. D. (1999). Inhibitory patterning of the anterior neural plate in *Xenopus* by homeodomain factors Dlx3 and Msx1. *Dev. Biol.* **212,** 455–464.

Finch, P. W., *et al.* (1997). Purification and molecular cloning of a secreted, Frizzled-related antagonist of Wnt action. *Proc. Natl. Acad. Sci. USA* **94,** 6770–6775.

Fouquet, B., Zimbelmann, R., and Franke, W. W. (1992). Identification of plakoglobin in oocytes and early embryos of *Xenopus laevis:* Maternal expression of a gene encoding a junctional plaque protein. *Differentiation* **51,** 187–194.

Fredette, B. J., Miller, J., and Ranscht, B. (1996). Inhibition of motor axon growth by T-cadherin substrata. *Development* **122,** 3163–3171.

Fujimori, T., Miyatani, S., and Takeichi, M. (1990). Ectopic expression of N-cadherin perturbs histogenesis in *Xenopus* embryos. *Development* **110,** 97–104.

Fukata, M., *et al.* (1999). Cdc42 and Rac1 regulate the interaction of IQGAP1 with β-catenin. *J. Biol. Chem.* **274,** 26044–26050.

Funayama, N., Fagotto, F., McCrea, P., and Gumbiner, B. M. (1995). Embryonic axis induction by the armadillo repeat domain of β-catenin: Evidence for intracellular signaling. *J. Cell Biol.* **128,** 959–968.

Gache, Y., Chavanas, S., Lacour, J. P., Wiche, G., Owaribe, K., Meneguzzi, G., and Ortonne, J. P. (1996). Defective espression of plectin/HD1 in epidermolysis bullosa simplex with muscular dystophy. *J. Clin. Invest.* **97,** 2289–2298.

Galceran, J., Farinas, I., Depew, M. J., Clevers, H., and Grosschedl, R. (1999). Wnt3a-/–like phenotype and limb deficiency in Lef1(-/-)Tcf1(-/-) mice. *Genes Dev.* **13,** 709–717.

Gallet, A., Angelats, C., Erkner, A., Charroux, B., Fasano, L., and Kerridge, S. (1999). The C-terminal domain of armadillo binds to hypophosphorylated teashirt to modulate wingless signalling in *Drosophila. EMBO J.* **18,** 2208–2217.

Gallet, A., Erkner, A., Charroux, B., Fasano, L., and Kerridge, S. (1998). Trunk-specific modulation of wingless signalling in *Drosophila* by teashirt binding to armadillo. *Curr. Biol.* **8,** 893–902.

Gallicano, G. I., Kouklis, P., Bauer, C., Yin, M., Vasioukhin, V., Degenstein, L., and Fuchs, E. (1998). Desmoplakin is required early in development for assembly of desmosomes and cytoskeletal linkage. *J. Cell Biol.* **143,** 2009–2022.

Gallin, W. J. (1998). Evolution of the "classical" cadherin family of cell adhesion molecules in vertebrates. *Mol. Biol. Evol.* **15,** 1099–1107.

Gard, D. L. (1995). Axis formation during amphibian oogenesis: Reevaluating the role of the cytoskeleton. *Curr. Top. Dev. Biol.* **30,** 215–252.

Gard, D. L., and Klymkowsky, M. W. (1998). Intermediate filament organization during oogenesis and early development in the clawed frog, *Xenopus laevis. In* "Intermediate Filaments." Plenum, New York. 35–69.

Gelderloos, J., Witcher, L., Cowin, P., and Klymkowsky, M. W. (1997). Plakoglobin: The other ARM of vertebrates. *In* "Cytoskeletal-Membrane Interactions and Signal Transduction" (P. Cowin and M. W. Klymkowsky, Eds.) Landes Bioscience, Austin, Texas. 13–30.

Gerhart, J., Ubbels, G., Black, S., Hara, K., and Kirschner, M. W. (1981). A reinvestigation of the role of the grey crescent in axis formation in *Xenopus laevis. Nature* **292,** 511–516.

Gerhart, J. C. (1980). Mechanisms regulating pattern formation in the amphibian egg and early embryo. "Biological Regulation and Development." Plenum, New York.

Giese, K., and Grosschedl, R. (1993). LEF-1 contains an activation domain that stimulates transcription only in a specific context of factor-binding sites. *EMBO J.* **12,** 4667–4676.

Gomez-Skarmeta, J., de la Calle, I., Mustienes, E., Modolell, J., and Mayor, R. (1999). *Xenopus* brain factor-2 controls mesoderm, forebrain and neural crest development. *Mech. Dev.* **80,** 15–27.

Gradl, D., Kuhl, M., and Wedlich, D. (1999). The Wnt/Wg signal transducer β-catenin controls fibronectin expression. *Mol. Cell Biol.* **19,** 5576–5587.

Graff, J. M., Thies, R. S., Song, J. J., Celeste, A. J., and Melton, D. A. (1994). Studies with a *Xenopus* BMP receptor suggest that ventral mesoderm-inducing signals override dorsal signals in vivo. *Cell* **79,** 169–179.

Groves, A. K., and Bronner-Fraser, M. (1999). Neural crest diversification. *Curr. Top. Dev. Biol.* **43,** 221–258.

Guger, K. A., and Gumbiner, B. M. (1995). β-Catenin has Wnt-like activity and mimics the Nieuwkoop signaling center in *Xenopus* dorsal-ventral patterning. *Dev. Biol.* **172,** 115–125.

Gururajan, R., Perry-O'Keefe, H., Melton, D. A., and Weeks, D. L. (1991). The *Xenopus* localized messenger RNA An3 may encode an ATP-dependent RNA helicase. *Nature* **349,** 717–719.

Hacker, U., Lin, X., and Perrimon, N. (1997). The *Drosophila* sugarless gene modulates Wingless signaling and encodes an enzyme involved in polysaccharide biosynthesis. *Development* **124,** 3565–3573.

Hadeball, B., Borchers, A., and Wedlich, D. (1998). *Xenopus* cadherin-11 (Xcadherin-11) expression requires the Wg/Wnt signal. *Mech. Dev.* **72,** 101–113.

Haegel, H., Larue, L., Ohsugi, M., Fedorov, L., Herrenknecht, K., and Kemler, R. (1995). Lack of beta-catenin affects mouse development at gastrulation. *Development* **121,** 3529–3537.

Haerry, T. E., Heslip, T. R., Marsh, J. L., and O'Connor, M. B. (1997). Defects in glucuronate biosynthesis disrupt Wingless signaling in *Drosophila. Development* **124,** 3055–3064.

Hamaguchi, M., Matsuyoshi, N., Ohnishi, Y., Gotoh, B., Takeichi, M., and Nagai, Y. (1993). p60v-src causes tyrosine phosphorylation and inactivation of the N-cadherin-catenin cell adhesion system. *EMBO J.* **12,** 307–314.

Harland, R., and Gerhart, J. (1997). Formation and function of Spemann's organizer. *Annu. Rev. Cell Dev. Biol.* **13,** 611–667.

Hart, M., *et al.* (1999). The F-box protein β-TrCP associates with phosphorylated β-catenin and regulates its activity in the cell. *Curr. Biol.* **9,** 207–210.

Hatzfeld, M. (1999). The armadillo family of structural proteins. *Int. Rev. Cytol.* **186,** 179–224.

Hatzfeld, M., Kristjansson, G. I., Plessmann, U., and Weber, K. (1994). Band 6 protein, a major constituent of desmosomes from stratified epithelia, is a novel member of the armadillo multigene family. *J. Cell Sci.* **117,** 2259–2270.

Hatzfeld, M., and Nachtsheim, C. (1996). Cloning and characterization of a new armadillo family member, p0071, associated with the junctional plaque: Evidence for a subfamily of closely related proteins. *J. Cell Sci.* **109,** 2767–2778.

Hayashi, S., Rubinfeld, B., Souza, B., Polakis, P., Wieschaus, E., and Levine, A. J. (1997). A *Drosophila* homolog of the tumor suppressor gene adenomatous polyposis coli down-regulates β-catenin but its zygotic expression is not essential for the regulation of Armadillo. *Proc. Natl. Acad. Sci. USA* **94,** 242–247.

Hazan, R. B., Kang, L., Roe, S., Borgen, P. I., and Rimm, D. L. (1997). Vinculin is associated with the E-cadherin adhesion complex. *J. Biol. Chem.* **272,** 32448–32453.

He, X., Saint, J. J., Woodgett, J. R., Varmus, H. E., and Dawid, I. B. (1995). Glycogen synthase kinase-3 and dorsoventral patterning in *Xenopus* embryos. *Nature* **374,** 617–622.

Healy, C., Uwanogho, D., and Sharpe, P. T. (1999). Regulation and role of Sox9 in cartilage formation. *Dev. Dyn.* **215,** 69–78.

Heasman, J. (1997). Patterning the *Xenopus* blastula. *Development* **124,** 4179–4191.

Heasman, J., Crawford, A., Goldstone, K., Garner-Hamrick, P., Gumbiner, B., McCrea, P., Kintner, P., Noro, C. Y., and Wylie, C. (1994). Overexpression of cadherins and underexpression of β-catenin inhibit dorsal mesoderm induction in early *Xenopus* embryos. *Cell* **79,** 791–803.

Hecht, A., Litterst, C. M., Huber, O., and Kemler, R. (1999). Functional characterization of multiple transactivating elements in β-catenin, some of which interact with the TATA-binding protein in vitro. *J. Biol. Chem.* **274,** 18017–18025.

Heid, H. W., *et al.* (1994). Cell type-specific desmosomal plaque proteins of the plakoglobin family: Plakophilin 1 (band 6 protein). *Differentiation* **58,** 113–131.

Hemmati-Brivanlou, A., and Melton, D. (1997). Vertebrate embryonic cells will become nerve cells unless told otherwise. *Cell* **88,** 13–17.

Hemmati-Brivanlou, A., and Melton, D. A. (1994). Inhibition of activin receptor signaling promotes neuralization in *Xenopus. Cell* **77,** 273–281.

Hemmati-Brivanlou, A., and Thomsen, G. H. (1995). Ventral mesodermal patterning in *Xenopus* embryos: Expression patterns and activities of BMP-2 and BMP-4. *Dev. Genet.* **17,** 78–89.

Henry, G. L., Brivanlou, I. H., Kessler, D. S., Hemmati-Brivanlou, A., and Melton, D. A. (1996). TGF-β signals and a pattern in *Xenopus laevis* endodermal development. *Development* **122,** 1007–1015.

Henry, G. L., and Melton, D. A. (1998). Mixer, a homeobox gene required for endoderm development. *Science* **281,** 91–96.

Hiraoka, Y., Komatsu, N., Sakai, Y., Ogawa, M., Shiozawa, M., and Aiso, S. (1997). XLS13A and XLS13B: SRY-related genes of *Xenopus laevis. Gene* **197,** 65–71.

Hirohashi, S. (1998). Inactivation of the E-cadherin-mediated cell adhesion system in human cancers. *Am. J. Pathol.* **153,** 333-339.

Hiscox, S., and Jilang, W. G. (1999). Association of the HGF/SF receptor, c-met, with the cell-surface adhesion molecule, E-cadherin, and catenins in human tumor cells. *Biochem. Biophy. Res. Commun.* **261,** 406–411.

Holland, P. W. H. (1999). The effect of gene duplication on homology. *In* "Homology" (B. K. Hall, Ed.) J. Wiley and Sons, New York. 226–235.

Holowacz, T., and Sokol, S. (1999). FGF is required for posterior neural patterning but not for neural induction *Dev. Biol.* **205,** 296–308.

Hoppler, S., Brown, J. D., and Moon, R. T. (1996). Expression of a dominant-negative Wnt blocks induction of MyoD in *Xenopus* embryos. *Genes Dev.* **10,** 2805–2817.

Horb, M. E., and Thomsen, G. H. (1997). A vegetally localized T-box transcription factor in *Xenopus* eggs specifies mesoderm and endoderm and is essential for embryonic mesoderm formation. *Development* **124,** 1689–1698.

Hoschuetzky, H., Aberle, H., and Kemler, R. (1994). β-catenin mediates the interaction of the cadherin-catenin complex with epidermal growth factor receptor. *J. Cell Biol.* **127,** 1375–1380.

Houliston, E., and Elinson, R. P. (1991a). Evidence for the involvement of microtubules, ER, and kinesin in the cortical rotation of fertilized frog eggs. *J. Cell Biol.* **114,** 1017–1028.

Houliston, E., and Elinson, R. P. (1991b). Patterns of microtubule polymerization relating to cortical rotation in Xenopus laevis eggs. *Development* **112,** 107–117.

Houliston, E., Le Guellec, R., Kress, M., Philippe, M., and Le Guellec, K. (1994). The kinesin-related protein Eg5 associates with both interphase and spindle microtubules during *Xenopus* early development. *Dev. Biol.* **164,** 147–159.

Hsieh, J. C., Kodjabachian, L., Rebbert, M. L., Rattner, A., Smallwood, P. M., Samos, C. H., Nusse, R., Dawid, I. B., and Nathans, J. (1999). A new secreted protein that binds to Wnt proteins and inhibits their activities. *Nature* **398,** 431–436.

Hsu, D. R., Economides, A. N., Wang, X., Eimon, P. M., and Harland, R. M. (1998a). The *Xenopus* dorsalizing factor Gremlin identifies a novel family of secreted proteins that antagonize BMP activities. *Mol. Cell* **1,** 673–683.

Hsu, S. C., Galceran, J., and Grosschedl, R. (1998b). Modulation of transcriptional regulation by LEF-1 in response to wnt-1 signaling and association with β-catenin. *Mol. Cell Biol.* **18,** 4807–4818.

Huber, A. H., Nelson, W. J., and Weis, W. I. (1997). Three-dimensional structure of the armadillo repeat region of β-catenin. *Cell* **90,** 871–882.

Huber, O., Korn, R., McLaughlin, J., Ohsugi, M., Herrmann, B. G., and Kemler, R. (1996). Nuclear localization of β-catenin by interaction with transcription factor LEF-1. *Mech. Dev.* **59,** 3–10.

Hudson, C., Clements, D., Friday, R. V., Stott, D., and Woodland, H. R. (1997). Xsox17α and -β mediate endoderm formation in *Xenopus. Cell* **91,** 397–405.

Hudson, C., and Woodland, H. R. (1998). Xpat, a gene expressed specifically in germ plasm and primordial germ cells of *Xenopus laevis. Mech. Dev.* **73,** 159–168.

Hudson, J. W., Alarcon, V. B., and Elinson, R. P. (1996). Identification of new localized RNAs in the *Xenopus* oocyte by differential display PCR. *Dev. Genet.* **19,** 190–198.

Hulsken, J., Birchmeier, W., and Behrens, J. (1994). E-cadherin and APC compete for the interaction with β-catenin and the cytoskeleton. *J. Cell Biol.* **127,** 2061–2069.

Ikeda, S., Kishida, S., Yamamoto, H., Murai, H., Koyama, S., and Kikuchi, A. (1998). Axin, a negative regulator of the wnt signaling pathway, forms a complex with GSK-3β and β-catenin and promotes GSK-3β-dependent phosphorylation of β-catenin. *EMBO J.* **17,** 1371–1384.

Isaacs, H. V. (1997). New perspectives on the role of the fibroblast growth factor family in amphibian development. *Cell Mol. Life Sci.* **53,** 350–361.

Ishitani, T., *et al.* (1999). The TAK1-NLK-MAPK-related pathway antagonizes signalling between β-catenin and transcription factor TCF. *Nature* **399,** 798–802.

Itoh, K., Jacob, J., and Sokol, S. (1998). A role for *Xenopus* Frizzled 8 in dorsal development. *Mech. Dev.* **74,** 145–157.

Jiang, J., and Struhl, G. (1998). Regulation of the Hedgehog and Wingless signalling pathways by the F- box/WD40-repeat protein Slimb. *Nature* **391,** 493–496.

Jongen, W. M., Fitzgerald, D. J., Asamoto, M., Piccoli, C., Slaga, T. J., Gros, D., Takeichi, M., and Yamasaki, H. (1991). Regulation of connexin 43-mediated gap junctional intercellular communication by Ca^{2+} in mouse epidermal cells is controlled by E-cadherin. *J. Cell Biol.* **114,** 545–555.

Joubin, K., and Stern, C. D. (1999). Molecular interactions continuously define the organizer during the cell movements of gastrulation. *Cell* **98,** 559–571.

Kamachi, Y., Cheah, K., and Kondoh, H. (1999). Mechanism of regulatory target selection by the SOX high-mobility-group domain proteins as revealed by comparison of SOX1/2/3 and SOX9. *Mol. Cell Biol.* **19,** 107–120.

Kamachi, Y., Sockanathan, S., Liu, Q., Breitman, M., Lovell-Badge, R., and Kondoh, H. (1995). Involvement of SOX proteins in lens-specific activation of crystallin genes. *EMBO J.* **14,** 3510–3519.

Kang, D. E., Soriano, S., Frosch, M. P., Collins, T., Naruse, S., Sisodia, S. S., Leibowitz, G., Levine, F., and Koo, E. H. (1999). Presenilin 1 facilitates the constitutive turnover of β-catenin: Differential activity of Alzheimer's disease-linked PS1 mutants in the β-catenin-signaling pathway. *J. Neurosci.* **19,** 4229–4237.

Karnovsky, A., and Klymkowsky, M. W. (1995). Anterior axis duplication in *Xenopus* induced by the over-expression of the cadherin-binding protein plakoglobin. *Proc. Natl. Acad. Sci. USA* **92,** 4522–4526.

Kawabata, M., Imamura, T., and Miyazono, K. (1998). Signal transduction by bone morphogenetic proteins. *Cytokine Growth Factor Rev.* **9,** 49–61.

Keirsebilck, A., Bonne, S., Staes, K., van Hagel, J., Nollet, F., Reynolds, A., and van Roy, F. (1998). Molecular cloning of the human p120ctn catenin gene (CTNND1): Expression of multiple alternatively spliced isoforms. *Genomics* **50,** 129–146.

Keller, R. (1985). The cellular basis of amphibian gastrulation. *In* "The Cellular Basis of Morphogenesis" (L. W. Browder, Ed.) Plenum, New York. 241–327.

Keller, R., and Danilchik, M. (1988). Regional expression, pattern and timing of convergence and extension during gastrulation of *Xenopus laevis. Development* **103,** 193–209.

Keller, R., Shih, J., and Sater, A. (1992a). The cellular basis of the convergence and extension of the *Xenopus* neural plate. *Dev. Dyn.* **193,** 199–217.

Keller, R., Shih, J., Sater, A. K., and Moreno, C. (1992b). Planar induction of convergence and extension of the neural plate by the organizer of *Xenopus. Dev. Dyn.* **193,** 218–234.

Kengaku, M., *et al.* (1998). Distinct WNT pathways regulating AER formation and dorsoventral polarity in the chick limb bud. *Science* **280,** 1274–1277.

Kennerdell, J. R., and Carthew, R. W. (1998). Use of dsRNA-mediated genetic interference to demonstrate that frizzled and frizzled 2 act in the wingless pathway. *Cell* **95,** 1017–1026.

Kim, S. H., Yamamoto, A., Bouwmeester, T., Agius, E., and DeRobertis, E. M. (1998). The role of paraxial protocadherin in selective adhesion and cell movements of the mesoderm during *Xenopus* gastrulation. *Development* **125,** 4681–4690.

Kintner, C. (1992a). Cadherins and the morphogenesis of epithelial tissues in *Xenopus* embryos. *Cold Spring Harb. Symp. Quant. Biol.* **57,** 335–344.

Kintner, C. (1992b). Regulation of embryonic cell adhesion by the cadherin cytoplasmic domain. *Cell* **69,** 225–236.

Kishida, S., Yamamoto, H., Ikeda, S., Kishida, M., Sakamoto, I., Koyama, S., and Kikuchi, A. (1998). Axin, a negative regulator of the wnt signaling pathway, directly interacts with adenomatous polyposis coli and regulates the stabilization of β-catenin. *J. Biol. Chem.* **273,** 10823–10826.

Klein, S. L. (1987). The first cleavage furrow demarcates the dorsal-ventral axis in *Xenopus* embryos. *Dev. Biol.* **120,** 299–304.

Klingensmith, J., Yang, Y., Axelrod, J. D., Beier, D. R., Perrimon, N., and Sussman, D. J. (1996). Conservation of dishevelled structure and function between flies and mice: Isolation and characterization of Dvl2. *Mech. Dev.* **58,** 15–26.

Kloc, M., Larabell, C., Chan, A. P., and Etkin, L. D. (1998). Contribution of METRO pathway localized molecules to the organization of the germ cell lineage. *Mech. Dev.* **75,** 81–93.

Kloc, M., Larabell, C., and Etkin, L. D. (1996). Elaboration of the messenger transport organizer pathway for localization of RNA to the vegetal cortex of *Xenopus* oocytes. *Dev. Biol.* **180,** 119–130.

Klymkowsky, M. W. (1995). Intermediate filament organization, reorganization and function in the clawed frog *Xenopus. Curr. Top. Dev. Biol.* **31,** 455–486.

Klymkowsky, M. W. (1997). Minireviews, minidogmas and mythinformation. *BioEssays* **20,** 537–539.

Klymkowsky, M. W. (1999a). Weaving a tangled web . . . the interconnected cytoskeleton. *Nat. Cell Biol.* **1,** 121–123.

Klymkowsky, M. W. (1999b). Plakophilin, armadillo repeats and nuclear localization. *Micro. Res. Tech.* **45,** 43–54.

Klymkowsky, M. W., and Evans, R. M. (1995). Intermediate filaments: A medical overview. *Principals of Medical Biology.* JAI Press. 147–188.

Klymkowsky, M. W., and Karnovsky, A. (1994). Morphogenesis and the cytoskeleton: studies of the *Xenopus* embryo *Dev. Biol.* **165,** 372–384.

Klymkowsky, M. W., Maynell, L. A., and Nislow, C. (1991). Cytokeratin phosphorylation, cytokeratin filament severing and the solubilization of the maternal mRNA Vg1. *J. Cell. Biol.* **114,** 787–797.

Klymkowsky, M. W., and Parr, B. (1995). A glimpse into the body language of cells: The intimate connection between cell adhesion and gene expression. *Cell* **83,** 5–8.

Klymkowsky, M. W., Williams, B. O., Barish, G. D., Varmus, H. E., and Vourgourakis, Y. E. (1999). Membrane-anchored plakoglobins have multiple mechanisms of action in Wnt signaling. *Mol. Biol. Cell* **10,** 3151–3169.

Knudsen, K. A., Soler, A. P., Johnson, K. R., and Wheelock, M. J. (1995). Interaction of α-actinin with the cadherin/catenin cell-cell adhesion complex via α-catenin. *J. Cell Biol.* **130,** 67–77.

Koch, P. J., Goldschmidt, M. D., Zimbelmann, R., Troyanovsky, R., and Franke, W. W. (1992). Complexity and expression patterns of the desmosomal cadherins. *Proc. Natl. Acad. Sci. USA* **89,** 353–357.

Koesters, R., *et al.* (1999). Mutational activation of the β-catenin proto-oncogene is a common event in the development of Wilms' tumors. *Cancer Res.* **59,** 3880–3882.

Kofron, M., Spagnuolo, A., Klymkowsky, M., Wylie, C., and Heasman, J. (1997). The roles of maternal α-catenin and plakoglobin in the early *Xenopus* embryo. *Development* **124,** 1553–1560.

Kohmura, N., *et al.* (1998). Diversity revealed by a novel family of cadherins expressed in neurons at a synaptic complex. *Neuron* **20,** 1137–1151.

Koike, J., Takagi, A., Miwa, T., Hirai, M., Terada, M., and Katoh, M. (1999). Molecular cloning of frizzled-10, a novel member of the frizzled gene family. *Biochem. Biophys. Res. Commun.* **262,** 39–43.

Konig, A., Gradl, D., Kuhl, M., and Wedlich, D. (2000). The HMG-box transcription factor XTcf-4 demarcates the forebrain–midbrain boundary. *Mech Dev.* **93,** 211–214.

Korinek, V., Barker, N., Morin, P. J., van Wichen, D., de Weger, R., Kinzler, K. W., Vogelstein, B., and Clevers, H. (1997). Constitutive transcriptional activation by a β-catenin-Tcf complex in APC-/- colon carcinoma. *Science* **275,** 1784–1787.

Kowalczyk, A. P., Borgwardt, J. E., and Green, K. J. (1996). Analysis of desmosomal cadherin-adhesive function and stoichiometry of desmosomal cadherin-plakoglobin complexes. *J. Invest. Dermatol.* **107,** 293–300.

Kowalczyk, A. P., Bornslaeger, E. A., Borgwardt, J. E., Palka, H. L., Dhaliwal, A. S., Corcoran, C. M., Denning, M. F., and Green, K. J. (1997). The amino-terminal domain of desmoplakin binds to plakoglobin and clusters desmosomal cadherin-plakoglobin complexes. *J. Cell Biol.* **139,** 773–784.

Kowalczyk, A. P., Bornslaeger, E. A., Norvell, S. M., Palka, H. L., and Green, K. J. (1999). Desmosomes: Intercellular adhesive junctions specialized for attachment of intermediate filaments. *Int. Rev. Cytol.* **185,** 237–302.

Koyano, S., Ito, M., Takamatsu, N., Takiguchi, S., and Shiba, T. (1997). The *Xenopus* Sox3 gene expressed in oocytes of early stages. *Gene* **188,** 101–107.

Kreft, B., Berndorff, D., Bottinger, A., Finnemann, S., Wedlich, D., Hortsch, M., Tauber, R., and Gessner, R. (1997). LI-cadherin-mediated cell-cell adhesion does not require cytoplasmic interactions. *J. Cell Biol.* **136,** 1109–1121.

Krufka, A., Johnson, R. G., Wylie, C. C., and Heasman, J. (1998). Evidence that dorsal-ventral differences in gap junctional communication in the early *Xenopus* embryo are generated by β-catenin independent of cell adhesion effects. *Dev. Biol.* **200,** 92–102.

Ku, M., and Melton, D. A. (1993). Xwnt-11: A maternally expressed *Xenopus* wnt gene. *Development* **119,** 1161–1173.

Kuhl, M., Finnemann, S., Binder, O., and Wedlich, D. (1996). Dominant negative expression of a cytoplasmically deleted mutant of XB/U-cadherin disturbs mesoderm migration during gastrulation in *Xenopus* laevis. *Mech. Dev.* **54,** 71–82.

Kuhlbrodt, K., Herbarth, B., Sock, E., Enderich, J., Hermans-Borgmeyer, I., and Wegner, M. (1998a). Cooperative function of POU proteins and SOX proteins in glial cells. *J. Biol. Chem.* **273,** 16050–16057.

Kuhlbrodt, K., Herbarth, B., Sock, E., Hermans-Borgmeyer, I., and Wegner, M. (1998b). Sox10, a novel transcriptional modulator in glial cells. *J. Neurosci.* **18,** 237–250.

Kuhlbrodt, K., Schmidt, C., Sock, E., Pingault, V., Bondurand, N., Goossens, M., and Wegner, M. (1998c). Functional analysis of Sox10 mutations found in human Waardenburg-Hirschsprung patients. *J. Biol. Chem.* **273,** 23033–23038.

Kurth, T., Fesenko, I. V., Schneider, S., Munchberg, F. E., Joos, T. O., Spieker, T. P., and Hausen, P. (1999). Immunocytochemical studies of the interactions of cadherins and catenins in the early *Xenopus* embryo. *Dev. Dyn.* **215,** 155–169.

Kuzmenko, Y. S., Stambolsky, D., Kern, F., Bochkov, V. N., Tkachuk, V. A., and Resink, T. J. (1998). Characteristics of smooth muscle cell lipoprotein binding proteins (p105/p130) as T-cadherin and regulation by positive and negative growth regulators. *Biochem. Biophys. Res. Commun.* **246,** 489–494.

Kypta, R. M., Su, H., and Reichardt, L. F. (1996). Association between a transmembrane protein tyrosine phosphotase and the cadherin-catenin complex. *J. Cell Biol.* **134,** 1519–1529.

LaBonne, C., and Bronner-Fraser, M. (1998). Neural crest induction in *Xenopus:* Evidence for a two-signal model. *Development* **125,** 2403–2414.

Lampugnani, M. G., Corada, M., Caveda, L., Breviario, F., Ayalon, O., Geiger, B., and Dejana, E. (1995). The molecular organization of endothelial cell to cell junctions: Differential association of plakoglobin, β-catenin, and α-catenin with vascular endothelial cadherin (VE-cadherin). *J. Cell Biol.* **129,** 203–217.

Larabell, C. A., Torres, M., Rowning, B. A., Yost, C., Miller, J. R., Wu, M., Kimelman, D., and Moon, R. T. (1997). Establishment of the dorso-ventral axis in *Xenopus* embryos is presaged by early asymmetries in β-catenin that are modulated by the Wnt signaling pathway. *J. Cell Biol.* **136,** 1123–1136.

Latinkic, B. V., Umbhauer, M., Neal, K. A., Lerchner, W., Smith, J. C., and Cunliffe, V. (1997). The *Xenopus* Brachyury promoter is activated by FGF and low concentrations of activin and suppressed by high concentrations of activin and by paired-type homeodomain proteins. *Genes Dev.* **11,** 3265–3276.

Latres, E., Chiaur, D. S., and Pagano, M. (1999). The human F box protein β-Trcp associates with the Cull/Skp1 complex and regulates the stability of β-catenin. *Oncogene* **18,** 849–854.

Laurent, M. N., Blitz, I. L., Hashimoto, C., Rothbacher, U., and Cho, K. W. (1997). The *Xenopus* homeobox gene twin mediates Wnt induction of goosecoid in establishment of Spemann's organizer. *Development* **124,** 4905–4916.

Leaf, D. S., Roberts, S. J., Gerhart, J. C., and Moore, H. P. (1990). The secretory pathway is blocked between the trans-Golgi and the plasma membrane during meiotic maturation in *Xenopus* oocytes. *Dev. Biol.* **141,** 1–12.

Legan, P. K., Yue, K. K., Chidgey, M. A., Holton, J. L., Wilkinson, R. W., and Garrod, D. R. (1994). The bovine desmocollin family: A new gene and expression patterns reflecting epithelial cell proliferation and differentiation. *J. Cell Biol.* **126,** 507–518.

Lemaire, P., Garrett, N., and Gurdon, J. B. (1995). Expression cloning of Siamois, a *Xenopus* homeobox gene expressed in dorsal-vegetal cells of blastulae and able to induce a complete secondary axis. *Cell* **81,** 85–94.

Lendahl, U. (1998). A growing family of Notch ligands. *Bioessays* **20,** 103–107.

Levesque, G., *et al.* (1999). Presenilins interact with armadillo proteins including neural-specific plakophilin-related protein and β-catenin. *J. Neurochem.* **72,** 999–1008.

Levi, G., Ginsberg, D., Girault, J. M., Sabanay, I., Thiery, J. P., and Geiger, B. (1991a). EP-cadherin in muscles and epithelia of *Xenopus laevis* embryos. *Development* **113,** 1335–1344.

Levi, G., Gumbiner, B., and Thiery, J. P. (1991b). The distribution of E-cadherin during *Xenopus laevis* development. *Development* **111,** 159–169.

Levine, E., Lee, C. H., Kintner, C., and Gumbiner, B. M. (1994). Selective disruption of E-cadherin function in early *Xenopus* embryos by a dominant negative mutant. *Development* **120,** 901–909.

Lewis, J. E., Wahl, J. III., Sass, K. M., Jensen, P. J., Johnson, K. R., and Wheelock, M. J. (1997). Cross-talk between adherens junctions and desmosomes depends on plakoglobin. *J. Cell Biol.* **136,** 919–934.

Leyns, L., Bouwmeester, T., Kim, S. H., Piccolo, S., and De Robertis, E. M. (1997). Frzb-1 is a secreted antagonist of Wnt signaling expressed in the Spemann organizer. *Cell* **88,** 747–756.

Li, L., Yuan, H., Weaver, C. D., Mao, J., Farr, I. G., Sussman, D. J., Jonkers, J., Kimelman, D., and Wu, D. (1999a). Axin and Frat1 interact with Dvl and GSK, bridging Dvl to GSK in Wnt-mediated regulation of LEF-1. *EMBO J.* **18,** 4233–4240.

Li, L., Yuan, H., Xie, W., Mao, J., Caruso, A. M., McMahon, A., Sussman, D. J., and Wu, D. (1999b). Dishevelled proteins lead to two signaling pathways. Regulation of lef-1 and c-jun n-terminal kinase in mammalian cells. *J. Biol. Chem.* **274,** 129–134.

Liem, K. J., Tremml, G., Roelink, H., and Jessell, T. M. (1995). Dorsal differentiation of neural plate cells induced by BMP-mediated signals from epidermal ectoderm. *Cell* **82,** 969–979.

Lin, X., and Perrimon, N. (1999). Dally cooperates with *Drosophila* Frizzled 2 to transduce Wingless signalling. *Nature* **400,** 281–284.

Liu, C., Kato, Y., Zhang, Z., Do, V. M., Yankner, B. A., and He, X. (1999a). β-Trcp couples β-catenin phosphorylation-degradation and regulates *Xenopus* axis formation. *Proc. Natl. Acad. Sci. USA* **96,** 6273–6278.

Liu, J. P., and Jessell, T. M. (1998). A role for rhoB in the delamination of neural crest cells from the dorsal neural tube. *Development* **125,** 5055–5067.

Liu, P., Wakamiya, M., Shea, M. J., Albrecht, U., Behringer, R. R., and Bradley, A. (1999b). Requirement for Wnt3 in vertebrate axis formation. *Nat. Genet.* **22,** 361–365.

Loureiro, J. J. (1999). The Wnts. *Curr. Biol.* **9,** R4.

Luo, J., Lubaroff, D. M., and Hendrix, M. J. (1999). Suppression of prostate cancer invasive potential and matrix metalloproteinase activity by E-cadherin transfection. *Cancer Res.* **59,** 3552–3556.

Ma, Q., Kintner, C., and Anderson, D. J. (1996). Identification of neurogenin, a vertebrate neuronal determination gene. *Cell* **87,** 43–52.

Magin, T. M., Schroder, R., Leitgeb, S., Wanninger, F., Zatloukal, K., Grund, C., and Melton, D. W. (1998). Lessons from keratin 18 knockout mice: formation of novel keratin filaments, secondary loss of keratin 7 and accumulation of liver-specific keratin 8-positive aggregates. *J. Cell Biol.* **140,** 1441–1451.

Malik, H. S., Eickbush, T. H., and Goldfarb, D. S. (1997). Evolutionary specialization of the nuclear targeting apparatus. *Proc. Natl. Acad. Sci. USA* **94,** 13738–13742.

Mancilla, A., and Mayor, R. (1996). Neural crest formation in *Xenopus laevis:* Mechanisms of Xslug induction. *Dev. Biol.* **177,** 580–589.

Mandai, K., *et al.* (1997). Afadin: A novel actin filament-binding protein with one PDZ domain localized at cadherin-based cell-to-cell adherens junction. *J. Cell Biol.* **139,** 517–528.

Mandai, K., Nakanishi, H., Satoh, A., Takahashi, K., Satoh, K., Nishioka, H., Mizoguchi, A., and Takai, Y. (1999). Ponsin/SH3P12: An 1-afadin- and vinculin-binding protein localized at cell-cell and cell-matrix adherens junctions. *J. Cell Biol.* **144,** 1001–1017.

Marcozzi, C., Burdett, I. D., Buxton, R. S., and Magee, A. I. (1998). Coexpression of both types of desmosomal cadherin and plakoglobin confers strong intercellular adhesion. *J. Cell Sci.* **111,** 495–509.

Marinos, E. (1986). Observations on the mitochondrial distribution in normal, rotated and cold-treated 2-cell stage embryos of *Xenopus laevis. Cell Diff.* **18,** 163–171.

Marrs, J. A., and Nelson, W. J. (1996). Cadherin cell adhesion molecules in differentiation and embryogenesis. *Int. Rev. Cytol.* **165,** 159–205.

Marsden, M. D., Collins, J. E., Greenwood, M. D., Adams, M. J., Fleming, T. P., Magee, A. I., and Buxton, R. S. (1997). Cloning and transcriptional analysis of the promoter of the human type 2 desmocollin gene (DSC2). *Gene* **186,** 237–247.

Mathies, L. D., Kerridge, S., and Scott, M. P. (1994). Role of the teashirt gene in *Drosophila* midgut morphogenesis: Secreted proteins mediate the action of homeotic genes. *Development* **120,** 2799–2809.

Mathur, M., Goodwin, L., and Cowin, P. (1994). Interactions of the cytoplasmic domain of the desmosomal cadherin Dsg1 with plakoglobin. *J. Biol. Chem.* **269,** 14075–14080.

Mayor, R., Guerrero, N., and Martinez, C. (1997). Role of FGF and noggin in neural crest induction. *Dev. Biol.* **189,** 1–12.

Mayor, R., Morgan, R., and Sargent, M. G. (1995). Induction of the prospective neural crest of Xenopus *Development* **121,** 767–777.

Mayor, R., Young, R., and Vargas, A. (1999). Development of neural crest in *Xenopus. Curr. Top. Dev. Biol.* **43,** 85–113.

Mayr, T., Deutsch, U., Kuhl, M., Drexler, H. C., Lottspeich, F., Deutzmann, R., Wedlich, D., and Risau, W. (1997). Fritz: A secreted frizzled-related protein that inhibits Wnt activity. *Mech. Dev.* **63,** 109–125.

McCartney, B. M., Dierick, H. A., Kirkpatrick, C., Moline, M. M., Baas, A., Peifer, M., and Bejsovec, A. (1999). *Drosophila* APC2 is a cytoskeletally-associated protein that regulates wingless signaling in the embryonic epidermis. *J. Cell Biol.* **146,** 1303–1318.

McCrea, P. D., Brieher, W. M., and Gumbiner, B. M. (1993). Induction of a secondary body axis in *Xenopus* by antibodies to β-catenin. *J. Cell Biol.* **123,** 477–484.

McGrath, J. A., *et al.* (1999). Skin fragility and hypohidrotic ectodermal dysplasia resulting from ablation of plakophilin 1. *Br. J. Dermatol.* **140,** 297–307.

McGrath, J. A., McMillan, J. R., Shemanko, C. S., Runswick, S. K., Leigh, I. M., Lane, E. B., Garrod, D. R., and Eady, R. A. (1997). Mutations in the plakophilin 1 gene result in ectodermal dysplasia/skin fragility syndrome. *Nat. Genet.* **17,** 240–244.

McGrew, L. L., Hoppler, S., and Moon, R. T. (1997). Wnt and FGF pathways cooperatively pattern anteroposterior neural ectoderm in *Xenopus. Mech. Dev.* **69,** 105–141.

McGrew, L. L., Lai, C. J., and Moon, R. T. (1995). Specification of the anteroposterior neural axis through synergistic interaction of the Wnt signaling cascade with noggin and follistatin. *Dev. Biol.* **172,** 337–342.

McGrew, L. L., Otte, A. P., and Moon, R. T. (1992). Analysis of Xwnt-4 in embryos of Xenopus laevis: A Wnt family member expressed in the brain and floor plate. *Development* **115,** 463–473.

McGrew, L. L., Takemaru, K., Bates, R., and Moon, R. T. (1999). Direct regulation of the *Xenopus* engrailed-2 promoter by the Wnt signaling pathway, and a molecular screen for Wnt-responsive genes, confirm a role for Wnt signaling during neural patterning in *Xenopus. Mech. Dev.* **87,** 21–32.

McLean, W. H., *et al.* (1996). Loss of plectin causes epidermolysis bullosa with muscular dystrophy: cDNA cloning and genomic organization. *Genes Dev.* **10,** 1724–1735.

McMahon, A. P., and Moon, R. T. (1989). Ectopic expression of the proto-oncogene int-1 in *Xenopus* embryos leads to duplication of the embryonic axis. *Cell* **58,** 1075–1084.

Mellerio, J. E., *et al.* (1997). Recessive epidermolysis bullosa simplex associated with plectin mutations: infantile respiratory complications in two unrelated cases. *Br. J. Dermatol.* **137,** 898–906.

Meneghini, M. D., Ishitani, T., Carter, J. C., Hisamoto, N., Ninomiya, T. J., Thorpe, C. J., Hamill, D. R., Matsumoto, K., and Bowerman, B. (1999). MAP kinase and Wnt pathways converge to downregulate an HMG-domain repressor in Caenorhabditis elegans. *Nature* **399,** 793–797.

Merriam, J. M., Rubenstein, A. B., and Klymkowsky, M. W. (1997). Cytoplasmically anchored plakoglobin induces a WNT-like phenotype in *Xenopus. Dev. Biol.* **185,** 67–81.

Mertens, C., Kuhn, C., and Franke, W. W. (1996). Plakophilins 2a and 2b: Constitutive proteins of dual location in the karyoplasm and the desmosomal plaque. *J. Cell Biol.* **135,** 1009–1025.

Merzdorf, C. S., Chen, Y. H., and Goodenough, D. A. (1998). Formation of functional tight junctions in *Xenopus* embryos. *Dev. Biol.* **195,** 187–203.

Miller, J. R., Rowning, B. A., Larabell, C. A., Yang, S. J., Bates, R. L., and Moon, R. T. (1999). Establishment of the dorsal-ventral axis in *Xenopus* embryos coincides with the dorsal enrichment of dishevelled that is dependent on cortical rotation. *J. Cell Biol.* **146,** 427–438.

Miyata, S., Miyashita, K., and Hosoyama, Y. (1996). SRY-related genes in *Xenopus* oocytes. *Biochim. Biophys. Acta* **1308,** 23–27.

Miyoshi, Y., Iwao, K., Nagasawa, Y., Aihara, T., Sasaki, Y., Imaoka, S., Murata, M., Shimano, T., and Nakamura, Y. (1998). Activation of the β-catenin gene in primary hepatocellular carcinomas by somatic alterations involving exon 3. *Cancer Res.* **58,** 2524–2527.

Mizuseki, K., Kishi, M., Matsui, M., Nakanishi, S., and Sasai, Y. (1998). *Xenopus* Zic-related-1 and Sox-2, two factors induced by chordin, have distinct activities in the initiation of neural induction. *Development* **125,** 579–587.

Mizuseki, K., Kishi, M., Shiota, K., Nakanishi, S., and Sasai, Y. (1998). SoxD: An essential mediator of induction of anterior neural tissues in *Xenopus* embryos. *Neuron* **21,** 77–85.

Molenaar, M., Roose, J., Peterson, J., Venanzi, S., Clevers, H., and Destree, O. (1998). Differential expression of the HMG box transcription factors XTcf-3 and XLef-1 during early *Xenopus* development. *Mech. Dev.* **75,** 163–166.

Molenaar, M., van de Wetering, M., Oosterwegel, M., Peterson-Maduro, J., Godsave, S., Korinek, V., Roose, J., Destrée, O., and Clevers, H. (1996). XTcf-3 transcription factor mediates β-catenin-induced axis formation in *Xenopus* embryos. *Cell* **86,** 391–399.

Monier, G. F., and Duband, J. L. (1997). Cross talk between adhesion molecules: Control of N-cadherin activity by intracellular signals elicited by β1 and β3 integrins in migrating neural crest cells. *J. Cell Biol.* **137,** 1663–1681.

Moon, R. T., and Kimelman, D. (1998). From cortical rotation to organizer gene expression: Toward a molecular explanation of axis specification in *Xenopus. Bioessays* **20,** 536–545.

Morin, P. J., Sparks, A. B., Korinek, V., Barker, N., Clevers, H., Vogelstein, B., and Kinsler, W. (1997). Activation of β-catenin-Tcf signaling in colon cancer by mutations in β-catenin or APC. *Science* **275,** 1787–1790.

Muller, H. A., and Hausen, P. (1995). Epithelial cell polarity in early *Xenopus* development. *Dev. Dyn.* **202,** 405–420.

Muller, H. A., Kuhl, M., Finnemann, S., Schneider, S., van der Poel, S. Z., Hausen, P., and Wedlich, D. (1994). *Xenopus* cadherins: The maternal pool comprises distinguishable members of the family. *Mech. Dev.* **47,** 213–232.

Munchberg, F. E., Spieker, T. P., Joos, T. O., and Hausen, P. (1997). A paired oocyte adhesion assay reveals the homophilic binding properties of the *Xenopus* maternal cadherins, XB/U- and EP-cadherin. *Mech. Dev.* **64,** 87–94.

Munemitsu, S., Albert, I., Souza, B., Rubinfeld, B., and Polakis, P. (1995). Regulation of intracellular β-catenin levels by the adenomatous polyposis coli (APC) tumor-suppressor protein. *Proc. Natl. Acad. Sci. USA* **92,** 3046–3050.

Munemitsu, S., Souza, B., Muller, O., Albert, I., Rubinfeld, B., and Polakis, P. (1994). The APC gene product associates with microtubules *in vivo* and promotes their assembly *in vitro. Cancer Res.* **54,** 3676–3681.

Murayama, M., *et al.* (1998). Direct association of presenilin-1 with β-catenin. *FEBS Lett.* **433,** 73–77.

Nagafuchi, A., Ishihara, S., and Tsukita, S. (1994). The roles of catenins in the cadherin-mediated cell adhesion: Functional analysis of E-cadherin-α catenin fusion molecules. *J. Cell Biol.* **127,** 235–245.

Nagar, B., Overduin, M., Ikura, M., and Rini, J. M. (1996). Structural basis of calcium-induced E-cadherin rigidification and dimerization. *Nature* **380,** 360–364.

Nakagawa, S., and Takeichi, M. (1998). Neural crest emigration from the neural tube depends on regulated cadherin expression. *Development* **125,** 2963–2971.

Nakayama, T., Snyder, M. A., Grewal, S. S., Tsuneizumi, K., Tabata, T., and Christian, J. L. (1998). *Xenopus* Smad8 acts downstream of BMP-4 to modulate its activity during vertebrate embryonic patterning. *Development* **125,** 857–867.

Navarro, P., Ruco, L., and Dejana, E. (1998). Differential localization of VE- and N-cadherins in human endothelial cells: VE-cadherin competes with N-cadherin for junctional localization. *J. Cell Biol.* **140,** 1475–1484.

Neufeld, K. L., and White, R. L. (1997). Nuclear and cytoplasmic localization of the adenomatous polyposis coli protein. *Proc. Natl. Acad. Sci. USA* **94,** 3034–3039.

Newport, J., and Kirschner, M. (1982). A major developmental transition in early *Xenopus laevis* embryos: 1. Characterization and timing of cellular changes at the midblastula stage. *Cell* **30,** 675–686.

Nishimura, M., *et al.* (1999). Presenilin mutations associated with Alzheimer disease cause defective intracellular trafficking of β-catenin, a component of the presenilin protein complex. *Nat. Med.* **5,** 164–169.

Nobes, C. D., and Hall, A. (1999). Rho GTPases control polarity, protrusion, and adhesion during cell movement. *J. Cell Biol.* **144,** 1235–1244.

Novak, A., Hsu, S. C., Leung, H. C., Radeva, G., Papkoff, J., Montesano, R., Roskelley, C., Grosschedl, R., and Dedhar, S. (1998). Cell adhesion and the integrin-linked kinase regulate the LEF-1 and β-catenin signaling pathways. *Proc. Natl. Acad. Sci. USA* **95,** 4374–4379.

Nuber, U. A., Schafer, S., Schmidt, A., Koch, P. J., and Franke, W. W. (1995). The widespread human desmocollin Dsc2 and tissue-specific patterns of synthesis of various desmocollin subtypes. *Eur. J. Cell Biol.* **66,** 69–74.

Nusse, R., and Varmus, H. (1992). Wnt genes. *Cell* **69,** 1073–1087.

Ohkubo, T., and Ozawa, M. (1999). p120(ctn) binds to the membrane-proximal region of the E-cadherin cytoplasmic domain and is involved in modulation of adhesion activity. *J. Biol. Chem.* **274,** 21409–21415.

Olson, D. J., and Moon, R. T. (1992). Distinct effects of ectopic expression of Wnt-1, activin B, and bFGF on gap junctional permeability in 32-cell *Xenopus* embryos. *Dev. Biol.* **151,** 204–212.

Orford, K., Orford, C. C., and Byers, S. W. (1999). Exogenous expression of β-catenin regulates contact inhibition, anchorage-independent growth, anoikis, and radiation-induced cell cycle arrest. *J. Cell. Biol.* **146,** 855–868.

Orsulic, S., and Peifer, M. (1996). An *in vivo* structure-function study of armadillo, the β-catenin homolog, reveals both separate and overlapping regions of the protein required for cell adhesion and for wingless signaling. *J. Cell Biol.* **134,** 1283–1300.

Ozawa, M., Baribault, H., and Kemler, R. (1989). The cytoplasmic domain of the cell adhesion molecule uvomorulin associates with three independent proteins structurally related in different species. *EMBO J.* **8,** 1711–1717.

Paffenholz, R., and Franke, W. W. (1997). Identification and localization of a neurally expressed member of the plakoglobin/ armadillo multigene family. *Differentiation* **61,** 293–304.

Paffenholz, R., Kuhn, C., Grund, C., Stehr, S., and Franke, W. W. (1999). The arm-repeat protein NPRAP (neurojungin) is a constituent of the plaques of the outer limiting zone in the retina, defining a novel type of adhering junction. *Exp. Cell Res.* **250,** 452–464.

Pai, L. M., Kirkpatrick, C., Blanton, J., Oda, H., Takeichi, M., and Peifer, M. (1996). Drosophila α-catenin and E-cadherin bind to distinct regions of *Drosophila* Armadillo. *J. Biol. Chem.* **271,** 32411–32420.

Pan, D., and Rubin, G. M. (1998). Targeted expression of teashirt induces ectopic eyes in Drosophila. *Proc. Natl. Acad. Sci. USA* **95,** 15508–15512.

Pantel, K., *et al.* (1998). Reduced expression of plakoglobin indicates an unfavorable prognosis in subsets of patients with non-small-cell lung cancer. *J. Clin. Oncol.* **16,** 1407–1413.

Papkoff, J., and Aikawa, M. (1998). WNT-1 and HGF regulate GSK3 beta activity and β-catenin signaling in mammary epithelial cells. *Biochem. Biophys. Res. Commun.* **247,** 851–858.

Parr, B. A., and McMahon, A. P. (1994). Wnt genes and vertebrate development. *Curr. Opin. Genet. Dev.* **4,** 523–528.

Pearce, J. J., Penny, G., and Rossant, J. (1999). A mouse cerberus/Dan-related gene family. *Dev. Biol.* **209,** 98–110.

Peifer, M. (1993). The product of the *Drosophila* segment polarity gene armadillo is part of a multi-protein complex resembling the vertebrate adherens junction. *J. Cell Sci.* **105,** 993–1000.

Peifer, M., McCrea, P. D., Green, K. J., Wieschaus, E., and Gumbiner, B. M. (1992). The vertebrate adhesive junction proteins β-catenin and plakoglobin and the *Drosophila* segment polarity gene armadillo form a multigene family with similar properties. *J. Cell Biol.* **118,** 681–691.

Peifer, M., and Wieschaus, E. (1990). The segment polarity gene armadillo encodes a functionally modular protein that is the *Drosophila* homolog of human plakoglobin. *Cell* **63,** 1167–1176.

Penzel, R., Oschwald, R., Chen, Y., Tacke, L., and Grunz, H. (1997). Characterization and early embryonic expression of a neural specific transcription factor xSOX3 in Xenopus laevis. *Int. J. Dev. Biol.* **41,** 667–677.

Perrimon, N. (1994). the genetic basis of patterned baldness in *Drosophila. Cell* **76,** 781–784.

Pertz, O., Bozic, D., Koch, A. W., Fauser, C., Brancaccio, A., and Engel, J. (1999). A new crystal structure, Ca^{2+} dependence and mutational analysis reveal molecular details of E-cadherin homoassociation. *EMBO J.* **18,** 1738–1747.

Peters, J. M., McKay, R. M., McKay, J. P., and Graff, J. M. (1999). Casein kinase I transduces Wnt signals. *Nature* **401,** 345–350.

Pettitt, J., Wood, W. B., and Plasterk, R. H. (1996). cdh-3, a gene encoding a member of the cadherin superfamily, functions in epithelial cell morphogenesis in Caenorhabditis elegans. *Development* **122,** 4149–4157.

Pevny, L. H., and Lovell-Badge, R. (1997). Sox genes find their feet. *Curr. Opin. Genet. Dev.* **7,** 338–344.

Piccolo, S., Agius, E., Leyns, L., Bhattacharyya, S., Grunz, H., Bouwmeester, T., and De Robertis, E. M. (1999). The head inducer Cerberus is a multifunctional antagonist of Nodal, BMP and Wnt signals. *Nature* **397,** 707–710.

Pierce, S. B., and Kimelman, D. (1995). Regulation of Spemann organizer formation by the intracellular kinase Xgsk3. *Development* **121,** 755–765.

Polakis, P. (1999). The oncogenic activation of β-catenin. *Curr. Opin. Genet. Dev.* **9,** 15–21.

Ponassi, M., Jacques, T. S., Ciani, L., and ffrench-Constant, C. (1999). Expression of the rat homologue of the *Drosophila* fat tumour suppressor gene. *Mech. Dev.* **80,** 207–212.

Poznanski, A., Minsuk, S., Stathopoulos, D., and Keller, R. (1997). Epithelial cell wedging and neural trough formation are induced planarly in *Xenopus,* without persistent vertical interactions with mesoderm. *Dev. Biol.* **189,** 256–269.

Prieve, M. G., Guttridge, K. L., Munguia, J., and Waterman, M. L. (1998). Differential importin-α recognition and nuclear transport by nuclear localization signals within the high-mobility-group DNA binding domains of lymphoid enhancer factor 1 and T-cell factor 1. *Mol. Cell Biol.* **18,** 4819–4832.

Prieve, M. G., Guttridge, K. L., Munguia, J. E., and Waterman, M. L. (1996). The nuclear localization signal of lymphoid enhancer factor-1 is recognized by two differentially expressed Srp1-nuclear localization sequence receptor proteins. *J. Biol. Chem.* **271,** 7654–7658.

Prieve, M. G., and Waterman, M. L. (1999). Nuclear localization and formation of β-catenin-lymphoid enhancer factor 1 complexes are not sufficient for activation of gene expression. *Mol. Cell Biol.* **19,** 4503–4515.

Pulkkinen, L., Smith, F. J., Shimizu, H., Murata, S., Yaoita, H., Hachisuka, H., Nishikawa, T., McLean, W. H., and Uitto, J. (1996). Homozygous deletion mutations in the plectin gene (PLEC1) in patients with epidermolysis bullosa simplex associated with late-onset muscular dystrophy. *Hum. Mol. Genet.* **5,** 1539–1546.

Ranscht, B., and Dours, Z. M. (1991). T-cadherin, a novel cadherin cell adhesion molecule in the nervous system lacks the conserved cytoplasmic region. *Neuron* **7,** 391–402.

Rebagliati, M. R., Weeks, D. L., Harvey, R. P., and Melton, D. A. (1985). Identification and cloning of localized maternal RNAs from *Xenopus* eggs. *Cell* **42,** 769–777.

Reynolds, A. B., Daniel, J., McCrea, P. D., Wheelock, M. J., Wu, J., and Zhang, Z. (1994). Identification of a new catenin: The tyrosine kinase substrate p120cas associates with E-cadherin complexes. *Mol. Cell Biol.* **14,** 8333–8342.

Reynolds, A. B., and Daniel, J. M. (1997). p120cnt: A src-substrate during catenin. *In* "Cytoskeletal-Membrane Interactions and Signal Transduction" (P. Cowin and M. W. Klymkowsky, Eds.) R.G. Landes, Ltd., Austin, Texas. 31–48.

Reynolds, A. B., Herbert, L., Cleveland, J. L., Berg, S. T., and Gaut, J. R. (1992). p120, a novel substrate of protein tyrosine kinase receptors and of p60v-src, is related to cadherin-binding factors β-catenin, plakoglobin and armadillo. *Oncogene* **7,** 2439–2445.

Riggleman, B., Schedl, P., and Wieschaus, E. (1990). Spatial expression of the *Drosophila* segment polarity gene armadillo is posttranscriptionally regulated by wingless. *Cell* **63,** 549–560.

Rocheleau, C. E., Downs, W. D., Lin, R., Wittmann, C., Bei, Y., Cha, Y. H., Ali, M., Priess, J. R., and Mello, C. C. (1997). Wnt signaling and an APC-related gene specify endoderm in early C. elegans embryos. *Cell* **90,** 707–716.

Rocheleau, C. E., Yasuda, J., Shin, T. H., Lin, R., Sawa, H., Okano, H., Priess, J. R., Davis, R. J., and Mello, C. C. (1999). WRM-1 activates the LIT-1 protein kinase to transduce anterior/posterior polarity signals in C. elegans. *Cell* **97,** 717–726.

Roeder, A. D., and Gard, D. L. (1994). Confocal microscopy of F-actin distribution in *Xenopus* oocytes. *Zygote* **2,** 111–124.

Roose, J., *et al.* (1999). Synergy between tumor suppressor APC and the β-catenin-tcf4 target tcf1. *Science* **285,** 1923–1926.

Roose, J., Molenaar, M., Peterson, J., Hurenkamp, J., Brantjes, H., Moerer, P., van de Wetering, M., Destree, O., and Clevers, H. (1998). The *Xenopus* Wnt effector XTcf-3 interacts with Groucho-related transcriptional repressors. *Nature* **395,** 608–612.

Rowning, B. A., Wells, J., Wu, M., Gerhart, J. C., Moon, R. T., and Larabell C. A. (1997). Microtubule-mediated transport of organelles and localization of β-catenin to the future dorsal side of *Xenopus* eggs. *Proc. Natl. Acad. Sci. USA* **94,** 1224–1229.

Rubenstein, A., Merriam, J., and Klymkowsky, M. W. (1997). Localizing the adhesive and signaling functions of plakoglobin. *Dev. Genet.* **20,** 91–102.

Rubinfeld, B., Albert, I., Porfiri, E., Fiol, C., Munemitsu, S., and Polakis, P. (1996). Binding of GSK3β to the APC-β-catenin complex and regulation of complex assembly. *Science* **272,** 1023–1026.

Rubinfeld, B., Robbins, P., El-Gamil M., Albert, I., Porfiri, E., and Polakis, P. (1997). Stabilization of β-catenin by genetic defects in melanoma cell lines. *Science* **275,** 1790–1792.

Rubinfeld, B., Souza, B., Albert, I., Muller, O., Chamberlain, S. H., Masiarz, F. R., Munemitsu, S., and Polakis, P. (1993). Association of the APC gene product with β-catenin. *Science* **262,** 1731–1734.

Rubinfeld, B., Souza, B., Albert, I., Munemitsu, S., and Polakis, P. (1995). The APC protein and E-cadherin form similar but independent complexes with α-catenin, β-catenin, and plakoglobin. *J. Biol. Chem.* **270,** 5549–5555.

Ruhrberg, C., and Watt, F. M. (1997). The plakin family: Versatile organizers of cytoskeletal architecture. *Curr. Opin. Genet. Dev.* **7,** 392–397.

Ruiz, P., *et al.,* (1996). Targeted mutation of plakoglobin in mice reveals essential functions of desmosomes in the embryonic heart. *J. Cell Biol.* **135,** 215–225.

Ruiz i Altaba, A. (1992). Planar and vertical signals in the induction and patterning of the *Xenopus* nervous system. *Development* **116,** 67–80.

Sacristan, M. P., Vestal, D. J., Dours-Zimmermann, M. T., and Ranscht, B. (1993). T-cadherin 2: Molecular characterization, function in cell adhesion, and coexpression with T-cadherin and N-cadherin. *J. Neurosci. Res.* **34,** 664-680.

Sadaghiani, B., and Thiebaud, C. H. (1987). Neural crest development in the *Xenopus laevis* embryo, studied by interspecific transplantation and scanning electron microscopy. *Dev. Biol.* **124,** 91–110.

Saint-Jeannet, J. P., He, X., Varmus, H. E., and Dawid, I. B. (1997). Regulation of dorsal fate in the neuraxis by Wnt-1 and Wnt-3a. *Proc. Natl. Acad. Sci. USA* **94,** 13713–13718.

Saito, Y., Vandenheede, J. R., and Cohen, P. (1994). The mechanism by which epidermal growth factor inhibits glycogen synthase kinase 3 in A431 cells. *Biochem. J.* **303,** 27–31.

Sakai, Y., Hiraoka, Y., Konishi, M., Ogawa, M., and Aiso, S. (1997). Isolation and characterization of *Xenopus laevis* xSox-B1 cDNA. *Arch Biochem. Biophys.* **346,** 1–6.

Sakanaka, C., Weiss, J. B., and Williams, L. T. (1998). Bridging of β-catenin and glycogen synthase kinase-3β by axin and inhibition of β-catenin-mediated transcription. *Proc. Natl. Acad. Sci. USA* **95,** 3020–3023.

Salic, A. N., Kroll, K. L., Evans, L. M., and Kirschner, M. W. (1997). Sizzled: A secreted Xwnt8 antagonist expressed in the ventral marginal zone of *Xenopus* embryos. *Development* **124,** 4739–4748.

Sano, K., Tanihara, H., Heimark, R. L., Obata, S., Davidson, M., St. John, T., Taketani, S., and Suzuki, S. (1993). Protocadherins: A large family of cadherin-related molecules in central nervous system. *EMBO J.* **12,** 2249–2256.

Sanson, B., Alexandre, C., Fascetti, N., and Vincent, J. P. (1999). Engrailed and hedgehog make the range of Wingless asymmetric in *Drosophila* embryos. *Cell* **98,** 207–216.

Sap, J. (1997). Interactions between protein tyrosine phosphotases and cell adhesion molecules. *In* "Cytoskeletal-Membrane Interactions and Signal Transduction." Landes Bioscience, Austin, Tx. 127–142.

Sasai, Y., and De Robertis, E. M. (1997). Ectodermal patterning in vertebrate embryos. *Dev. Biol.* **182,** 5–20.

Savagner, P., Yamada, K. M., and Thiery, J. P. (1997). The zinc-finger protein slug causes desmosome dissociation, an initial and necessary step for growth factor-induced epithelial-mesenchymal transition. *J. Cell Biol.* **137,** 1403–1419.

Scharf, S. R., and Gerhart, J. C. (1980). Determination of the dorsal-ventral axis in eggs of Xenopus laevis: Complete rescue of uv-impaired eggs by oblique orientation before first cleavage. *Dev. Biol.* **79,** 181–198.

Schmelz, M., and Franke, W. W. (1993). Complexus adhaerentes, a new group of desmoplakin-containing junctions in endothelial cells: The syndesmos connecting retothelial cells of lymph nodes. *Eur. J. Cell Biol.* **61,** 274–289.

Schmidt, A., Langbein, L., Pratzel, S., Rode, M., Rackwitz, H. R., and Franke, W. W. (1999). Plakophilin 3—A novel cell-type-specific desmosomal plaque protein. *Differentiation* **64,** 291–306.

Schmidt, A., Langbein, L., Rode, M., Pratzel, S., Zimbelmann, R., and Franke, W. W. (1997). Plakophilins 1a and 1b: Widespread nuclear proteins recruited in specific epithelial cells as desmosomal plaque components. *Cell Tissue Res.* **290,** 481–499.

Schmidt, J. E., Suzuki, A., Ueno, N., and Kimelman, D. (1995). Localized BMP-4 mediates dorsal/ventral patterning in the early *Xenopus* embryo. *Dev. Biol.* **169,** 37–50.

Schneider, S., Steinbeisser, H., Warga, R. M., and Hausen, P. (1996). β-catenin translocation into nuclei demarcates the dorsalizing centers in frog and fish embryos. *Mech. Dev.* **57,** 191–198.

Schroeder, M. M., and Gard, D. L. (1992). Organization and regulation of cortical microtubules during the first cell cycle of *Xenopus* eggs. *Development* **114,** 699–709.

Sela-Donenfeld, D., and Kalcheim, C. (1999). Regulation of the onset of neural crest migration by coordinated activity of BMP4 and Noggin in the dorsal neural tube. *Development* **126,** 4749–4762.

Shapira, E., Marom, K., Yelin, R., Levy, A., and Fainsod, A. (1999). A role for the homeobox gene Xvex-1 as part of the BMP-4 ventral signaling pathway. *Mech. Dev.* **86,** 99–111.

Shapiro, L., Kwong, P. D., Fannon, A. M., Colman, D. R., and Hendrickson, W. A. (1995). Considerations on the folding topology and evolutionary origin of cadherin domains. *Proc. Natl. Acad. Sci. USA* **92,** 6793–6797.

Sheldahl, L. C., Park, M., Malbon, C. C., and Moon, R. T. (1999). Protein kinase C is differentially stimulated by wnt and frizzled homologs in aG-protein-dependent manner. *Curr. Biol.* **9,** 695–698.

Shi, D. L., Goisset, C., and Boucaut, J. C. (1998). Expression of Xfz3, a *Xenopus* frizzled family member, is restricted to the early nervous system. *Mech. Dev.* **70,** 35–47.

Shiozawa, M., Hiraoka, Y., Komatsu, N., Ogawa, M., Sakai, Y., and Aiso, S. (1996). Cloning and characterization of *Xenopus laevis* xSox7 cDNA. *Biochim. Biophys. Acta* **1309,** 73–76.

Shtutman, M., Zhurinsky, J., Simcha, I., Albanese, C., D'Amico, M., Pestell, R., and Ben Ze'ev, A. (1999). The cyclin D1 gene is a target of the β-catenin/LEF-1 pathway. *Proc. Natl. Acad. Sci. USA* **96,** 5522–5527.

Simcha, I., Geiger, B., Yehuda-Levenberg, S., Salomon, D. and Ben-Ze'ev, A. (1996). Suppression of tumorigenicity by plakoglobin: An augmenting effect of N-cadherin. *J. Cell Biol.* **133,** 199–209.

Simcha, I., Shtutman, M., Salomon, D., Zhurinsky, J., Sadot, E., Geiger, B., and Ben Ze'ev, A. (1998). Differential nuclear translocation and transactivation potential of β-catenin and plakoglobin. *J. Cell Biol.* **141,** 1433–1448.

Simon, R., Tassan, J. P., and Richter, J. D. (1992). Translational control by poly(A) elongation during *Xenopus* development: differential repression and enhancement by a novel cytoplasmic polyadenylation element. *Genes Dev.* **6,** 2580–2591.

Simonneau, L., Broders, F., and Thiery, J. P. (1992). N-cadherin transcripts in *Xenopus laevis* from early tailbud to tadpole. *Dev. Dyn.* **194,** 247–260.

Sirotkin, H., *et al.* (1997). Identification of a new human catenin gene family member (ARVCF) from the region deleted in velo-cardio-facial syndrome. *Genomics* **41,** 75–83.

Slack, J. M., Isaacs, H. V., Song, J., Durbin, L., and Pownall, M. E. (1996). The role of fibroblast growth factors in early *Xenopus* development. *Biochem. Soc. Symp.* **62,** 1–12.

Slusarski, D. C., Corces, V. G., and Moon, R. T. (1997). Interaction of Wnt and a Frizzled homologue triggers G-protein-linked phosphatidylinositol signalling. *Nature* **390,** 410–413.

Smalley, M. J., *et al.* (1999). Interaction of axin and dv1-2 proteins regulates dv1-2-stimulated TCF- dependent transcription. *EMBO J.* **18,** 2823–2835.

Smith, E. A., and Fuchs, E. (1998). Defining the interactions between intermediate filaments and desmosomes. *J. Cell Biol.* **141,** 1229–1241.

Smith, F. J., *et al.* (1996). Plectin deficiency results in muscular dystrophy with epidermolysis bullosa. *Nat. Genet.* **13,** 450–457.

Smith, K. J., Levy, D. B., Maupin, P., Pollard, T. D., Vogelstein, B., and Kinzler, K. W. (1994). Wild-type but not mutant APC associates with the microtubule cytoskeleton. *Cancer Res.* **54,** 3672–3675.

Smith, W. C., and Harland, R. M. (1991). Injected Xwnt-8 RNA acts early in *Xenopus* embryos to promote formation of a vegetal dorsalizing center. *Cell* **67,** 753–765.

Sokol, S., Christian, J. L., Moon, R. T., and Melton, D. A. (1991). Injected Wnt RNA induces a complete body axis in *Xenopus* embryos. *Cell* **67,** 741–752.

Sokol, S. Y. (1996). Analysis of Dishevelled signalling pathways during *Xenopus* development. *Curr. Biol.* **6,** 1456–1467.

Sparks, A. B., Morin, P. J., Vogelstein, B., and Kinzler, K. W. (1998). Mutational analysis of the APC/β-catenin/Tcf pathway in colorectal cancer *Cancer Res.* **58,** 1130–1134.

Staal, F. J., Burgering, B. M., van de Wetering, M., and Clevers, H. C. (1999). Tcf-1-mediated transcription in T lymphocytes: Differential role for glycogen synthase kinase-3 in fibroblasts and T cells. *Int. Immunol.* **11,** 317–323.

Stanley, E., *et al.* (1998). DAN is a secreted glycoprotein related to *Xenopus cerberus. Mech. Dev.* **77,** 173–184.

Stappenbeck, T. S., and Green, K. J. (1992). The desmoplakin carboxyl terminus coaligns with and specifically disrupts intermediate filament networks when expressed in cultured cells. *J. Cell Biol.* **116,** 1197–1209.

Stappert, J., and Kemler, R. (1994). A short core region of E-cadherin is essential for catenin binding and is highly phosphorylated. *Cell Adhes. Commun.* **2,** 319–327.

Stennard, F., Carnac, G., and Gurdon, J. B. (1996). The *Xenopus* T-box gene, Antipodean, encodes a vegetally localised maternal mRNA and can trigger mesoderm formation. *Development* **122,** 4179–4188.

Stolow, M. A., and Shi, Y. B. (1995). *Xenopus* sonic hedgehog as a potential morphogen during embryogenesis and thyroid hormone-dependent metamorphosis. *Nucl. Acids Res.* **23,** 2555–2562.

Sussman, D. J., Klingensmith, J., Salinas, P., Adams, P. S., Nusse, R., and Perrimon, N. (1994). Isolation and characterization of a mouse homolog of the *Drosophila* segment polarity gene dishevelled. *Dev. Biol.* **166,** 73–86.

Suzuki, A., Ueno, N., and Hemmati-Brivanlou, A. (1997). *Xenopus* msx1 mediates epidermal induction and neural inhibition by BMP4. *Development* **124,** 3037–3044.

Suzuki, S. T. (1996). Protocadherins and diversity of the cadherin superfamily. *J. Cell Sci.* **109,** 2609–2611.

Takahashi, K., *et al.* (1999). Nectin/PRR: An immunoglobulin-like cell adhesion molecule recruited to cadherin-based adherens junctions through interaction with Afadin, a PDZ domain-containing protein. *J. Cell Biol.* **145,** 539–549.

Takeda, H., Nagafuchi, A., Yonemura, S., Tsukita, S., Behrens, J., Birchmeier, W., and Tsukita, S. (1995). V-src kinase shifts the cadherin-based cell adhesion from the strong to the weak state and β catenin is not required for the shift. *J. Cell Biol.* 1839–1847.

Takeda, H., Shimoyama, Y., Nagafuchi, A., and Hirohashi, S. (1999). E-cadherin functions as a cis-dimer at the cell-cell adhesive interface in vivo. *Nat. Struct. Biol.* **6,** 310–312.

Takeichi, M. (1995). Morphogenetic roles of classic cadherins. *Curr. Opin. Cell Biol.* **7,** 619–627.

Tanihara, H., Sano, K., Heimark, R. L., St. John, T., and Suzuki, S. (1994). Cloning of five human cadherins clarifies characteristic features of cadherin extracellular domain and provides further evidence for two structurally different types of cadherin. *Cell Adhes. Commun.* **2,** 15–26.

Tashiro, K., Tooi, O., Nakamura, H., Koga, C., Ito, Y., Hikasa, H., and Shiokawa, K. (1996). Cloning and expression studies of cDNA for a novel *Xenopus* cadherin (XmN-cadherin), expressed maternally and later neural-specifically in embryogenesis. *Mech. Dev.* **54,** 161–171.

Tepass, U., and Hartenstein, V. (1994). The development of cellular junctions in the *Drosophila* embryo. *Dev. Biol.* **161,** 563–596.

Tetsu, O., and McCormick, F. (1999). β-catenin regulates expression of cyclin D1 in colon carcinoma cells. *Nature* **398,** 422–426.

Thomas, G. M., Frame, S., Goedert, M., Nathke, I., Polakis, P., and Cohen, P. (1999). A GSK3-binding peptide from FRAT1 selectively inhibits the GSK3- catalysed phosphorylation of Axin and β-catenin. *FEBS Lett.* **458,** 247–251.

Tkachuk, V. A., Bochkov, V. N., Philippova, M. P., Stambolsky, D. V., Kuzmenko, E. S., Sidorova, M. V., Molokoedov, A. S., Spirov, V. G., and Resink, T. J. (1998). Identification of an atypical lipoprotein-binding protein from human aortic smooth muscle as T-cadherin. *FEBS Lett.* **421,** 208–212.

To, C. T., and Isao, M. S. (1998). The roles of hepatocyte growth factor/scatter factor and met receptor in human cancers. *Oncol. Rep.* **5,** 1013–1024.

Troyanovsky, S. M., Troyanovsky, R. B., Eshkind, L. G., Leube, R. E., and Franke, W. W. (1994). Identification of amino acid sequence motifs in desmocollin, a desmosomal glycoprotein, that are required for plakoglobin binding and plaque formation. *Proc. Natl. Acad. Sci. USA* **91,** 10790–10794.

Tsang, M., Lijam, N., Yang, Y., Beier, D. R., Wynshaw, B. A., and Sussman, D. J. (1996). Isolation and characterization of mouse dishevelled-3. *Dev. Dyn.* **207,** 253–262.

Tsuda, M., *et al.* (1999). The cell-surface proteoglycan Dally regulates Wingless signalling in *Drosophila. Nature* **400,** 276–280.

Turner, A. P., Brown, D., Heasman, J., Cook, G. M., Evans, J., Vickers, L., and Wylie, C. C. (1992). Involvement of a neutral glycolipid in differential cell adhesion in the Xenopus blastula. *EMBO J.* **11,** 3845–3855.

Turner, D. L., and Weintraub, H. (1994). Expression of achaete-scute homolog 3 in *Xenopus* embryos converts ectodermal cells to a neural fate. *Genes Dev.* **8,** 1434–1447.

Usui, T., Shima, Y., Shimada, Y., Hirano, S., Burgess, R. W., Schwarz, T. L., Takeichi, M., and Uemura, T. (1999). Flamingo, a seven-pass transmembrane cadherin, regulates planar cell polarity under the control of frizzled. *Cell* **98,** 585–595.

Vallin, J., Girault, J. M., Thiery, J. P., and Broders, F. (1998). *Xenopus* cadherin-11 is expressed in different populations of migrating neural crest cells. *Mech. Dev.* **75,** 183–186.

van de Wetering, M., Castrop, J., Korinek, V., and Clevers, H. (1996). Extensive alternative splicing and dual promoter usage generate Tcf-1 protein isoforms with differential transcription control properties. *Mol. Cell Biol.* **16,** 745–752.

van de Wetering, M., and Clevers, H. (1992). Sequence-specific interaction of the HMG box proteins TCF-1 and SRY occurs within the minor groove of a Watson-Crick double helix. *EMBO J.* **11,** 3039–3044.

van Hengel, J., Vanhoenacker, P., Staes, K., and van Roy, F. (1999). Nuclear localization of the p120(ctn) armadillo-like catenin is counteracted by a nuclear export signal and by E-cadherin expression. *Proc. Natl. Acad. Sci. USA* **96,** 7980–7985.

Vestal, D. J., and Ranscht, B. (1992). Glycosyl phosphatidylinositol-anchored T-cadherin mediates calcium-dependent, homophilic cell adhesion. *J. Cell Biol.* **119,** 451–461.

Vincent, J.-P., Oster, G. F., and Gerhart, J. C. (1986). Kinematics of gray crescent formation in *Xenopus* eggs: The displacement of subcortical cytoplasm relative to the egg surface. *Dev. Biol.* **113,** 484–500.

Vleminckx, K., and Kemler, R. (1999). Cadherins and tissue formation: Integrating adhesion and signaling. *Bioessays* **21,** 211–220.

Vourgourakis, Y. E., Parson, C., Jensen, E., Zorn, A., and Klymkowsky, M. W. (1999). Soxs, catenins and dorsal-ventral patterning in *Xenopus. Mol. Biol. Cell* 10 Supplement 7a.

Waltzer, L., and Bienz, M. (1998). *Drosophila* CBP represses the transcription factor TCF to antagonize Wingless signalling. *Nature* **395,** 521–525.

Wang, S., Krinks, M., Lin, K., Luyten, F. P., and Moos, M. J. (1997a). Frzb, a secreted protein expressed in the Spemann organizer, binds and inhibits Wnt-8. *Cell* **88,** 757–766.

Wang, S., Krinks, M., and Moos, M. J. (1997b). Frzb-1, an antagonist of Wnt-1 and Wnt-8, does not block signaling by Wnts -3A, -5A, or -11. *Biochem. Biophys. Res. Commun.* **236,** 502–504.

Wang, Y., and Rose, B. (1997). An inhibition of gap-junctional communication by cadherins. *J. Cell Sci.* **110,** 301–309.

Weeks, D. L., and Melton, D. A. (1987). A maternal mRNA localized to the animal pole of *Xenopus* eggs encodes a subunit of mitochondrial ATPase. *Proc. Natl. Acad. Sci. USA* **84,** 2798–2802.

Weeks, D. L., and Melton, D. A. (1987). A maternal mRNA localized to the vegetal hemisphere in *Xenopus* eggs codes for a growth factor related to TGF-β. *Cell* **51,** 861–867.

Wegner, M. (1999). From head to toes: The multiple facets of Sox proteins. *Nucl. Acids Res.* **27,** 1409–1420.

Wei, X., Makori, N., Peterson, P. E., Hummler, H., and Hendrickx, A. G. (1999). Pathogenesis of retinoic acid-induced ear malformations in primate model. *Teratology* **60,** 83–92.

Weiss, E. E., Kroemker, M., Rudiger, A. H., Jockusch, B. M., and Rudiger, M. (1998). Vinculin is part of the cadherin-catenin junctional complex: Complex formation between α-catenin and vinculin. *J. Cell Biol.* **141,** 755–764.

White, P., Aberle, H., and Vincent, J. P. (1998). Signaling and adhesion activities of mammalian β-catenin and plakoglobin in *Drosophila. J. Cell Biol.* **140,** 183–195.

Wiche, G. (1998). Role of plectin in cytoskeleton organization and dynamics. *J. Cell Sci.* **111,** 2477–2486.

Wiche, G., Gromov, D., Donovan, A., Castanon, M. J., and Fuchs, E. (1993). Expression of plectin mutant cDNA in cultured cells indicates a role of COOH-terminal domain in intermediate filament association. *J. Cell Biol.* **121,** 607–619.

Wieschaus, E., and Riggleman, R. (1987). Autonomous requirements for the segment polarity gene armadillo during *Drosophila* embryogenesis. *Cell* **49,** 177–184.

Wikramanayake, A. H., Huang, L., and Klein, W. H. (1998). β-catenin is essential for patterning the maternally specified animal-vegetal axis in the sea urchin embryo. *Proc. Natl. Acad. Sci. USA* **95,** 9343–9348.

Willert, K., Brink, M., Wodarz, A., Varmus, H., and Nusse, R. (1997). Casein kinase 2 associates with and phosphorylates dishevelled. *EMBO J.* **16,** 3089–3096.

Willert, K., Shibamoto, S., and Nusse, R. (1999). Wnt-induced dephosphorylation of axin releases β-catenin from the axin complex. *Genes Dev.* **13,** 1768–1773.

Williams, C. L. (1997). Regulation of cadherin activity by G-proteins and the actin cytoskeleton. In "Cytoskeletal-Membrane Interactions and Signal Transduction." (P. Cowin and M. W. Klymkowsky, Eds.) Landes Bioscience, Austin, Texas. 111–126.

Williams, B. O., Barish, G. D., Klymkowsky, M. W., and Varmus, H. E. (2000). A comparative evaluation of β-catenin and plakoglobin signaling activity. *Oncogene* (in press).

Wilson, P. A., Lagna, G., Suzuki, A., and Hemmati-Brivanlou, A. (1997). Concentration-dependent patterning of the *Xenopus* ectoderm by BMP4 and its signal transducer Smad1. *Development* **124,** 3177–3184.

Witcher, L. L., Collins, R., Puttagunta, S., Mechanic, S. E., Munson, M., Gumbiner, B., and Cowin, P. (1996). Desmosomal cadherin binding domains of plakoglobin. *J. Biol. Chem.* **271,** 10904–10909.

Wodarz, A., and Nusse, R. (1998). Mechanisms of Wnt signaling in development. *Annu. Rev. Cell Dev. Biol.* **14,** 59–88.

Wolda, S. L., Moody, C. J., and Moon, R. T. (1993). Overlapping expression of Xwnt-3A and Xwnt-1 in neural tissue of *Xenopus laevis* embryos. *Dev. Biol.* **155,** 46–57.

Wolda, S. L., and Moon, R. T. (1992). Cloning and developmental expression in *Xenopus laevis* of seven additional members of the Wnt family. *Oncogene* **7,** 1941–1947.

Woodgett, J. R. (1994). Regulation and functions of the glycogen synthase kinase-3 subfamily. *Semin. Cancer Biol.* **5,** 269–275.

Wu, Q., and Maniatis, T. (1999). A striking organization of a large family of human neural cadherin-like cell adhesion genes. *Cell* **97,** 779–790.

Xu, Q., D'Amore, P. A., and Sokol, S. Y. (1998). Functional and biochemical interactions of Wnts with FrzA, a secreted Wnt antagonist. *Development* **125,** 4767–4776.

Yamamoto, H., Kishida, S., Kishida, M., Ikeda, S., Takada, S., and Kikuchi, A. (1999). Phosphorylation of axin, a Wnt signal negative regulator, by glycogen synthase kinase-3β regulates its stability. *J. Biol. Chem.* **274,** 10681–10684.

Yamamoto, H., Kishida, S., Uochi, T., Ikeda, S., Koyama, S., Asashima, M., and Kikuchi, A. (1998). Axil, a member of the Axin family, interacts with both glycogen synthase kinase 3β and β-catenin and inhibits axis formation of *Xenopus* embryos. *Mol. Cell Biol.* **18,** 2867–2875.

Yang, Y., Bauer, C., Strasser, G., Wollman, R., Julien, J.-P., and Fuchs, E. (1999). Integrators of the cytoskeleton which stabilize microtubules. *Cell* **98,** 229–238.

Yap, A. S., Brieher, W. M., and Gumbiner, B. M. (1997). Molecular and functional analysis of cadherin-based adherens junctions. *Ann. Rev. Cell Dev. Biol.* **13,** 119–146.

Yap, A. S., Niessen, C. M., and Gumbiner, B. M. (1998). The juxtamembrane region of the cadherin cytoplasmic tail supports lateral clustering, adhesive strengthening, and interaction with p120ctn. *J. Cell Biol.* **141,** 779–789.

Yokoya, F., Imamoto, N., Tachibana, T., and Yoneda, Y. (1999). β-catenin can be transported into the nucleus in a Ran-unassisted manner. *Mol. Biol. Cell* **10,** 1119–1131.

Yost, C., Farr, G. 3rd, Pierce, S. B., Ferkey, D. M., Chen, M. M., and Kimelman, D. (1998). GBP, an inhibitor of GSK-3, is implicated in *Xenopus* development and oncogenesis. *Cell* **93,** 1031–1041.

Yost, C., Pierce, S. B., Torres, M., Moon, R. T., and Kimelman, D. (1997). Glycogen synthase kinase-3: A component of multiple signaling pathways. *In* "Cytoskeletal-Membrane Interactions and Signal Transduction." (P. Cowin and M. W. Klymkowsky, Eds.) R. G. Landes, Ltd., Austin, Texas. 73–86.

Yost, C., Torres, M., Miller, J. R., Huang, E., Kimelman, D., and Moon, R. T. (1996). The axis-inducing activity, stability, and subcellular distribution of β-catenin is regulated in *Xenopus* embryos by glycogen synthase kinase 3. *Genes Dev.* **10,** 1443–1454.

Yost, H. J., Phillips, C. R., Boore, J. L., Bertman, J., Whalon, B., and Danilchik, M. V. (1995). Relocation of mitochondria to the prospective dorsal marginal zone during *Xenopus* embryogenesis. *Dev. Biol.* **170,** 83–90.

Yu, G., *et al.* (1998). The presenilin 1 protein is a component of a high molecular weight intracellular complex that contains β-catenin. *J. Biol. Chem.* **273,** 16470–16475.

Yuan, H., Corbi, N., Basilico, C., and Dailey, L. (1995). Developmental-specific activity of the FGF-4 enhancer requires the synergistic action of Sox2 and Oct-3. *Genes Dev.* **9,** 2635–2645.

Zhang, J., Houston, D. W., King, M. L., Payne, C., Wylie, C., and Heasman, J. (1998a). The role of maternal VegT in establishing the primary germ layers in *Xenopus* embryos. *Cell* **94,** 515–524.

Zhang, J., and King, M. L. (1996). *Xenopus* VegT RNA is localized to the vegetal cortex during oogenesis and encodes a novel T-box transcription factor involved in mesodermal patterning. *Development* **122,** 4119–4129.

Zhang, Z., *et al.* (1998b). Destabilization of β-catenin by mutations in presenilin-1 potentiates neuronal apoptosis. *Nature* **395,** 698–702.

Zhou, J., Liyanage, U., Medina, M., Ho, C., Simmons, A. D., Lovett, M., and Kosik, K. S. (1997). Presenilin 1 interaction in the brain with a novel member of the Armadillo family. *Neuroreport* **8,** 1489–1494.

Zhou, Y., and King, M. L. (1996). Localization of Xcat-2 RNA, a putative germ plasm component, to the mitochondrial cloud in *Xenopus* stage I oocytes. *Development* **122,** 2947–2953.

Zorn, A. M., Barish, G. D., Williams, B. O., Lavender, P., Klymkowsky, M. W., and Varmus, H. E. (1999a). Regulation of Wnt signaling by Sox proteins: XSox17 α/β and XSox3 physically interact with β-catenin. *Mol. Cell* **4,** 487–498.

Zorn, A. M., Butler, K., and Gurdon, J. B. (1999b). Anterior endomesoderm specification in *Xenopus* by Wnt/β-catenin and TGF-β signalling pathways. *Dev. Biol.* **209,** 282–297.

Establishment of Left-Right Asymmetry

H. Joseph Yost
Huntsman Cancer Institute, Center for Children, University of Utah,
Salt Lake City, Utah 84112

The vertebrate body plan has bilateral symmetry and left-right asymmetries that are highly conserved. The molecular pathways for left-right development are beginning to be elucidated. Several distinct mechanisms to initiate the vertebrate left-right axis have been proposed. These mechanisms appear to converge on highly conserved expression patterns of genes in the transforming growth factor-β (TGFβ) family of cell-cell signaling factors, nodal and *lefty-2,* and subsequently the expression of the transcription regulator Pitx2, in left lateral plate mesoderm. It is possible that downstream signaling pathways diverge in distinct classes of vertebrates.

KEY WORDS: Left-right asymmetry, Heart, Brain, Intestine, Midline, Lefty, Vg1, BMP, TGFβ, Nodal, Pitx2.

I. Introduction

One of the most striking features of the body plans of most animals is symmetry. In vertebrates, most of the external body is bilaterally symmetric across a plane that runs through the dorsal-ventral axis. In contrast, many internal organs, including the heart, gut and parts of the brain, have highly conserved left-right asymmetries. A challenging and outstanding fundamental problem in developmental biology is to understand how left-right asymmetries and bilateral symmetries are consistently and concurrently generated (Yost, 1995).

Consider a hypothetical vertebrate body plan, made of several layers or sheets of cells, with both bilateral symmetry and left-right asymmetry (Figure 1). This is analogous to an amniote embryo, in which the dorsal-ventral axis is established by the relationships of the sheets of cells making up the three germ layers. A midline is established along the anterior-posterior

0074-7696/01 $35.00

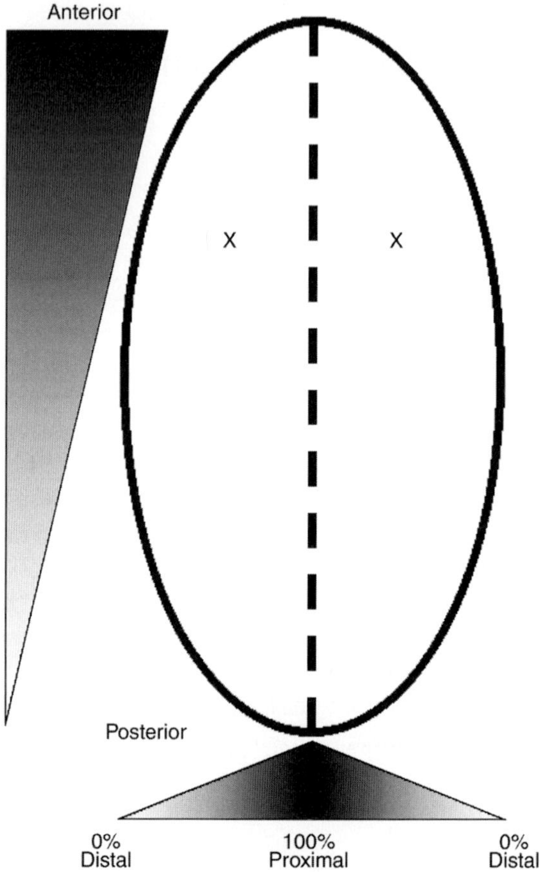

FIG. 1 A coordinate diagram of bilateral symmetry and left-right asymmetry. Schematic diagram, with anterior at the top, posterior at the bottom, and the left and right sides divided by a midline (dotted line) along the anterior-posterior axis. The "X" marks a position that is 70% along the length of the A-P axis and 50% along the proximal (midline) to distal axis.

axis by the process of gastrulation. Proximal-distal axes are determined in relationship to the midline. Any given position in the embryo could be assigned coordinates according to the dorsal-ventral, anterior-posterior, and proximal-distal axes. However, although this information might be sufficient to establish anterior-posterior (A-P) and proximal-distal identities (P-D), it is not sufficient to establish left-right identities. In the given example, for any given set of axial coordinates (70% A-P; 50% P-D), there are two positions (marked by "X" in Figure 1), one on the left and one on the right. Sufficient information is available to an external viewer, a vertebrate embryologist, to predict which side of the embryo will be the right side and which will be the left side. But how does a cell that finds itself at position 70% A-P; 50% P-D decide whether it is on the left side or the right side, and whether it should execute a bilaterally symmetric fate or a left-right asymmetric fate? Clearly, to make a functional three-dimensional organism with both bilateral symmetry and left-right asymmetry there must be developmental mechanisms that coordinate the pathways that specify left-right identity with pathways that specify anterior-posterior and proximal-distal identities (Danos and Yost, 1995; Yost, 1995).

Since the discovery of asymmetrically expressed genes in chick embryos (Levin *et al.,* 1995), over 35 genes have been implicated in vertebrate left-right development (Figure 2, see color insert; Ramsdell and Yost, 1998; Wood, 1997). As discussed below, most of the evidence for the roles of these genes is based on asymmetric expression patterns, mutation analyses, and/or ectopic expression studies. These studies make evident three important principles. First, the developmental specification of left-right identity occurs early in development, concurrent with the specification of identities along the anterior-posterior and proximal-distal axes, and well before the morphogenesis of asymmetric organs. Second, cells that will contribute only to bilaterally symmetric structures are important players in the establishment of the left-right axis, and can express asymmetric molecular markers. An example of this is lateral plate mesoderm, which expresses asymmetric markers but develops mostly bilaterally symmetric structures. The striking implication in the overview discussed above (Figure 1) is that cells can have duel roles and duel identities, both in establishing left-right identity and in contributing to bilateral symmetry. Third, many of the earliest genes in the left-right pathway are cell-cell signaling factors, consistent with the idea that the development of the left-right axis involves inductive or antagonistic interactions among groups of cells throughout the embryo. It is not until later in the left-right pathway that asymmetrically expressed transcription factors arise, perhaps to maintain left-right identities. Regardless of this enormous progress and the availability of molecular markers that appear before morphological left-right development, we still do not understand the fundamental process by which evolutionarily conserved left-right asymmetry is generated from apparent bilateral symmetry. However, there are several emerging models that appear quite promising.

II. Left-Right Organogenesis

More is known about the left-right development of the heart, briefly reviewed below, than about asymmetric development of the other organ primorida. Studies of gut left-right development and molecular markers for gut asymmetry are just emerging (Branford *et al.,* 2000; Essner *et al.,* 2000; Logan *et al.,* 1998; Milewski *et al.,* 1998). To date, studies of brain left-right development are predominantly histological descriptions. All vertebrates that have been examined have left-right asymmetry in the habenula nucleus of the midbrain, but the function of this nucleus is unknown (Engbretson *et al.,* 1981; Gugliemotti and Fiorino, 1998; Gurusinghe and Ehrlich, 1985; Morgan, 1991; Yost, 1998b). The discovery of three asymmetrically expressed genes in the zebrafish midbrain, *cyclops* (*nodal*) (Rebagliati *et al.,* 1998b; Sampath *et al.,* 1998), *lefty-1,* and *pitx2c* (Bisgrove *et al.,* 1999; Essner *et al.,* 2000) should facilitate genetic studies of left-right brain asymmetry.

The heart serves as a strong example for left-right organ morphogenesis. Among classes of vertebrates there are variations in the number of chambers in the adult heart and the extent to which other developmentally derived structures such as septa and coronary arteries are formed. In contrast, it is clear from descriptive embryology over the last two centuries that cardiac tube looping along the left-right axis is highly conserved in vertebrates (Burggren, 1988). The strong conservation of left-right orientation suggests that common pathways for left-right development will be found in all vertebrates, at least for the heart. In general, asymmetric organs are formed from bilaterally symmetric primorida. For the heart, apparently bilaterally symmetric mesoderm tissue forms on either side of the embryo shortly after gastrulation. The bilateral heart primordia coalesce at the midline to form a bilaterally symmetric heart tube. The heart tube loops to the right of the embryo and bends upon itself, bringing together the inflow and outflow tracts, and positioning the segments of the heart tube for remodeling (Figure 3, from Ramsdell and Yost, 2000). The cardiac tube must loop appropriately to allow subsequent derived structures (such as septation, valves, and outflow tract) to form normally in the inner curvature of the looped heart (van den Hoff *et al.,* 1999).

In *situs inversus totalis,* an extreme case of left-right developmental defects, the left-right orientations of organ primordia (at least heart and gut, this is unknown for the brain) are fully inverted but concordant. In the heart this allows mirror image alignment of the inflow and outflow tracks. Just as in the normal embryo in which all of the organs have concordant left-right orientation, if all of the left-right asymmetries in the *situs inversus totalis* embryo are concordantly inverted, there is little or no effect on health. Models of full inversion of both the heart and gut include the *inv/inv* mouse (Mochizuki *et al.,* 1998; Morgan *et al.,* 1998; Yokoyama *et al.,* 1993) and ectopic expression of the TGFβ signaling factor Vg1 in specific cell lineages on the right side of a 16-cell *Xenopus laevis* (frog) embryo (Hyatt and Yost, 1998; Ramsdell and Yost, 1999), discussed below. These

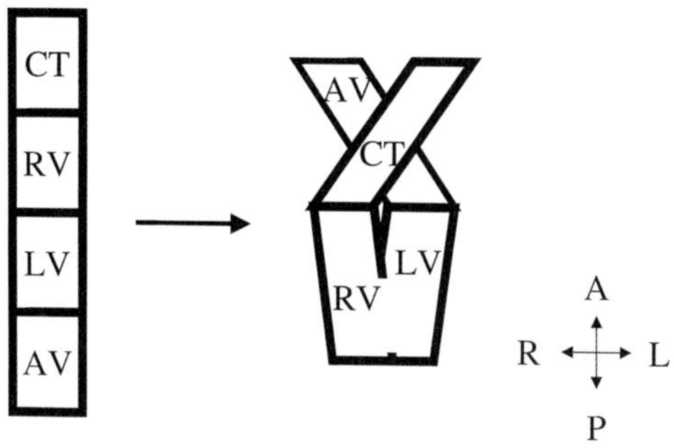

FIG. 3 Cardiac looping and subsequent inflow and outflow remodeling. The primary cardiac tube (at the left of the figure) is segmented along the anterior-posterior axis. The tube loops and bends upon itself, bringing the most anterior segments CT (conotruncus and aortic sac regions) into close juxtaposition with the most posterior segments AV (atrioventricular canal and sinoatrium regions), and placing the more anterior RV (right ventricle) segment next to, and on the right side of, the LV (left ventricle) segment. The "wedging" of the CT in between the two atria that form from the common AV results in connections of both ventricles with the CT regions. Slight alterations in left-right signaling might induce subtle changes in the extent of cardiac tube looping, such that the tube is slightly over-looped or under-looped, which could result in severe defects in this wedge region. For further description of the looping process, see Ramsdell and Yost, 2000.

full inversions presumably reflect an alteration of very early steps in the genetic pathway that establishes the left-right axis.

The more frequent cases of laterality defects are what are called "randomization" of left-right asymmetries in which underlying left-right axis information is presumably lost or scrambled. In these cases, each organ primordium (e.g., heart, gut, and brain) independently and stochastically forms aberrant left-right asymmetries. The result is discordance of left-right asymmetries among the organs. These cases lead to complex congential heart disease and a variety of other syndromes, which we are just beginning to recognize as "laterality" defects (Bowers *et al.,* 1996). Concordance occurs when the organ primorida attain left-right orientations (or underlying asymmetric gene expression patterns) in a coordinated fashion, as seen in normal embryos or in embryos with a fully inverted left-right axis (*situs inversus totalis*). Discordance occurs when the left-right orientation (or underlying asymmetric gene expression patterns) of organ primordia within an individual embryo are randomized with respect to each other. Among laterality defects, it is discordance that often leads to clinical problems.

Although the genetic pathways that regulate left-right asymmetry are beginning to be clarified, very little is known about the biomechanics of organ left-right morphogenesis. It is likely that a combination of several cellular processes drive a bilaterally symmetric cardiac tube to bend and loop or a gut tube to coil. Altered cell shapes and cytoskeletal components (Itasaki *et al.*, 1991), as well as asymmetric accumulation of extracellular components such as flectin (Tsuda *et al.*, 1998, 1996), during heart tube looping might help direct the orientation of looping.

III. Conserved Asymmetric Expression Patterns of TGFβ Family Members

At the end of gastrulation, the left lateral plate mesoderm expresses nodal, *lefty-2* and *pitx2* in all vertebrate embryos that have been examined (Figure 2). The current view of left-right organogenesis of the heart and gut is that complex and sometimes antagonistc interactions among several members of the TGFβ family expressed in the lateral plate mesoderm establish left-side and right-side identities, which are then propagated by asymmetric expression of transcription regulators. TGFβ family members implicated in left-right development include nodal, Bone Morphogenic Proteins (BMPs), Vg1-related proteins, TGFβ antagonists lefty and cerberus/caronte, and a modulator of nodal, cryptic/oep.

Nodal is a member of the TGFβ family of cell signaling factors that is expressed early in the node of gastrula-stage embryos. later, a member of the nodal family is expressed in the left lateral plate in chick (cNR1), mice (nodal), frogs (Xnr1) and zebrafish (cyclops) (Collignon *et al.*, 1996; Levin *et al.*, 1995; Lowe *et al.*, 1996; Lohr *et al.*, 1997; Rebagliati *et al.*, 1998a; Sampath *et al.*, 1998). The asymmetric expression of nodal in the lateral plate is highly conserved and is altered in a variety of experiments that perturb the initial steps in left-right development (see section V). Nodal function is modulated or facilitated by cryptic in mice, and one-eyed-pinhead (OEP) in fish, which is a member of the EGF-CFC family. EGF-CFC proteins are extracellular proteins that bind to peptide growth factors. While the expression of these genes is bilaterally symmetric in embryos, genetic evidence indicates that they are essential for the function of asymmetrically expressed nodal. Mouse *cryptic* mutants or zebrafish OEP mutants fail to express left-sided markers in the lateral plate as well as *lefty-1* in the midline (Gaio *et al.*, 1999; Yan *et al.*, 1999).

Members of the Lefty family are asymmetrically expressed in the midline (*lefty-1*) and lateral plate (*lefty-2*) in mice embryos (Meno *et al.*, 1997, 1996). Promoter analyses in mice indicate that asymmetric *lefty-2* expression in the left-lateral plate is driven by an asymmetric enhancer (ASE). In contrast, the asymmetric appearance of *lefty-1* in the neural floorplate is generated by the suppression of *lefty-2* on the right side, driven by a right-side specific silencer (RSS) element that overrides bilateral, neural-specific enhancer elements (Saijoh *et al.,* 1999). Both of the asymmetric transcription regulatory elements are downstream of *iv* and *inv* genes, as demonstrated by altered expression in the respective mutants. At least two *lefty* genes have been identified in zebrafish with symmetric expression in the midline and asymmetric expression in gut, heart, and diencephalon (Bisgrove *et al.,* 1999; Thisse and Thisse, 1999). In chick and *Xenopus,* only one *lefty* gene has been identified, but it appears to have both midline and left lateral plate expression domains (Ishimaru *et al.,* 2000; Yokouchi *et al.,* 1999).

Altered gene expression patterns in *lefty-1*$^{-/-}$ mice demonstrate a role for *lefty-1* in the midline. Mutant embryos have bilateral expression of nodal, *lefty-2,* and *Pitx2,* and thoracic left isomerism (Meno *et al.,* 1998). This bilateral expression is analogous to the effects of surgical removal of the midline in early neurula-stage *Xenopus* embryos (Lohr *et al.,* 1997), and suggests that expression of *lefty-1* in the midline serves as a barrier between signaling pathways on the left and right sides of the embryo (Meno *et al.,* 1998). Ectopic expression experiments of mouse *lefty* in chick embryos (Yoshioka *et al.,* 1998) or expression of zebrafish *lefty/antivin* in zebrafish embryos (Bisgrove *et al.,* 1999; Thisse and Thisse, 1999) indicate that there is a feedback loop between nodal and *lefty,* which are expressed in overlapping domains both in the midline and in the lateral plate. Nodal induces *lefty* expression, and *lefty* suppresses nodal expression. Coinjections of *lefty* with other members of the TGFβ family suggest that *lefty* is an antagonist of nodal (Bisgrove *et al.,* 1999) and/or activin (Thisse and Thisse, 1999) function in zebrafish. More recently, *lefty* has been shown to antagonize nodal in *Xenopus* and zebrafish (Cheng *et al.,* 2000; Thisse *et al.,* 2000). *Lefty* is likely to be a broad range antagonist of different subfamilies of TGFβ, since it interacts with the Vg1 pathway and antagonizes BMPs in context-dependent mechanisms in *Xenopus* (Branford *et al.,* 2000). Analysis of humans with laterality defects has identified two mutations in human LEFTY-A that alter a highly conserved cysteine-knot region (Kosaki *et al.,* 1999) important for protein interactions. Lefty-2 expression in the lateral plate mesoderm is altered in several laterality mutants, including *iv* (Meno *et al.,* 1996), *inv* (Meno *et al.,* 1996), *ft* (Heymer *et al.,* 1997), *nt* (Melloy *et al.,* 1998), *lefty-1* (Meno *et al.,* 1998), and *KIF3B* (Nonaka *et al.,* 1998). Hence, *lefty-1* in the midline might serve to antagonize other midline signals or signals that would otherwise cross the midline, and *lefty-2* might serve to counterbalance other TGFβ signals in the lateral plate.

Because fewer markers have been found to be expressed on the right side, significantly less is know about signaling pathways on the right side of the embryo. However, it appears that right-sided identity is not simply a default state in the absence of left-side signals. In *Xenopus,* BMP overexpression on the left side antagonizes the left-sided Vg1 pathway, whereas expression on the right side has no effect (Hyatt and Yost, 1998; Ramsdell and Yost, 1999), thus suggesting that BMPs might play a role in a right-side signal, and suppression of BMP signaling is important for asymmetric development on the left side. Using ectopic expression of dominant negative receptors and constitutively active receptors for BMP pathways, it was found that activation of a BMP pathway through the Alk2 (but not Alk3 or Alk4) receptor and Smad transcription coregulators was necessary for right-side signaling in *Xenopus*. This BMP/ALK2/Smad pathway is antagonistic of the left-sided Vg1 pathway, and normal left-side development is dependent on suppression of the BMP pathway (Ramsdell and Yost, 1999).

The results in *Xenopus* indicating that a BMP pathway is downregulated for left-side development (Ramsdell and Yost, 1999) concur with recent results with Caronte/Cerberus in chick (Rodriguez Esteban *et al.*, 1999; Yokouchi *et al.*, 1999; Zhu *et al.*, 1999). Members of the Cerberus/Dan family have been shown to antagonize members of the TGFβ and Wnt families of cell-cell signaling proteins. In chick, Caronte/Cerberus is expressed on the left side in head and lateral mesoderm. This asymmetric Caronte/Cerberus expression in the lateral mesoderm is dependent on Shh and independent of Nodal, whereas asymmetric Caronte/Cerberus expression in the head is dependent on both Shh and Nodal. The endogenous left-side expression of Caronte/Cerberus can be suppressed by FGF-8, a right-side signal near the node. Ectopic misexpression of Caronte/Cerberus on the right side induces Nodal and Pitx2 and alters left-right morphogenesis. Caronte/Cerberus can antagonize BMP. By RNA *in situ* hybridization analysis, BMP expression appears to be symmetric. Together, results in *Xenopus* and chick suggest that left-right development is dependent on both suppression of a bilaterally symmetric BMP pathway on the left side, mediated by Caronte/Cerberus in chick and the Vg1 pathway in *Xenopus,* and unhindered activation of the BMP pathway on the right side. Again, the midline appears to be crucial in the distinction between left- and right-side signaling pathways, since Caronte/Cerberus also has a role in asymmetric lefty expression in the midline (Yokouchi *et al.*, 1999).

Hence, it appears that left-right development during the neurula stages involves complex and context-dependent interactions among several members of the TGFβ superfamily, including Nodal, BMPs, Vg1 and activin, and several antagonists of TGFβ signaling, including Lefty and Caronte/Cerberus, and a facilitator of nodal function, cryptic/OEP.

IV. Asymmetric Expression of Transcription Factors PITX2, CSnR, NKX

Most of the factors described above in the early steps of the left-right pathway are cell-cell signaling factors. It is reasonable to expect these cell interactions to result in asymmetric downstream gene expression for the developmental left-right decisions to be propagated by asymmetrically expressed transcription factors. Asymmetrically expressed transcription factors have been found on both the left (e.g., Pitx2) and right (cSNR) sides.

Pitx2 is a member of the bicoid family of homeobox transcription factors. It was originally identified in humans as a gene mutated in Reiger syndrome (Semina *et al.,* 1996); the name is derived from its expression in the pituitary. Left-side expression of *pitx2* occurs in the cardiac tube during looping morphogenesis and in the stomach and intestines during asymmetric coiling. These asymmetric expression patterns are highly conserved and have been found in chick (Campione *et al.,* 1999; Logan *et al.,* 1998; Piedra *et al.,* 1998; Ryan *et al.,* 1998; St. Amand *et al.,* 1998; Yoshioka *et al.,* 1998), mouse (Meno *et al.,* 1998; Piedra *et al.,* 1998; Ryan *et al.,* 1998; Yoshioka *et al.,* 1998), frog (Campione *et al.,* 1999; Ryan *et al.,* 1998), and zebrafish (Bisgrove *et al.,* 1999; Campione *et al.,* 1999; Essner *et al.,* 2000). Because the expression pattern is late and not as transient as that of Lefty and Nodal, it is likely that Pitx2 serves as an effector molecule, controlling the transcription of genes that are involved in left-right organ morphogenesis.

Clearly Pitx2 is downstream of genes that are expressed in the node and the midline, as well as downstream of Nodal and Lefty expression in the lateral plate mesoderm. In chick, ectopic expression of Shh on the right induces *pitx2* expression (Logan *et al.,* 1998; Piedra *et al.,* 1998; Ryan *et al.,* 1998; St. Amand *et al.,* 1998), and Shh-blocking antibodies decrease normal left-sided expression of *pitx2* (Logan *et al.,* 1998). In chick and frog, ectopic expression of nodal on the right side induces Pitx2 (Campione *et al.,* 1999; Logan *et al.,* 1998; Piedra *et al.,* 1998; Ryan *et al.,* 1998). Analysis of Pitx2 expression patterns in mice mutants indicates that Pitx2 is genetically downstream of *inv* (Ryan *et al.,* 1998), *iv* (Campione *et al.,* 1999; Piedra *et al.,* 1998; Ryan *et al.,* 1998), and *lefty-1* (Meno *et al.,* 1998).

The functions of *pitx2* in left-right morphogenesis have been examined with two approaches, ectopic expression and targeted mutagenesis. There are multiple isoforms of Pitx2 in humans, mice, frog, and zebrafish, but it is not clear whether their functions are conserved in left-right development. In mice and frog, isoform Pitx2c is asymmetrically expressed in the heart and gut lateral plate mesoderm, whereas Pitx2a, Pitx2b, and Pitx2c have symmetrical patterns (Kitamura *et al.,* 1999; Schweickert *et al.,* 2000). In contrast, zebrafish Pitx2c is expressed in the left dorsal diencephalon and the intestine, but not the heart, and Pitx2a is expressed in the left heart primordium, but not asymmetrically in the brain or intestine (Essner *et al.,* 2000). Ectopic expression in the chick (Logan *et al.,* 1998; Ryan *et al.,* 1998) or frog (Campione *et al.,* 1999; Essner *et al.,* 2000) on the right side perturbs left-right organogenesis of both the heart and gut, with some distinctions made between the effects of different Pitx2 isoforms (Essner *et al.,* 2000;

Schweickert *et al.*, 2000). In chick, the increased frequency of cardiac isomerism (Logan *et al.*, 1998; Ryan *et al.*, 1998) is interpreted as the result of bilateral expression of a factor that is proximal to looping morphogenesis.

It was reasonable to predict that Pitx2 was a global regulator of left-right morphogenesis, since it is expressed in many organ primodia during left-right morphogenesis. However, targeted mutagenesis of Pitx2 resulted in a limited range of defective laterality phenotypes (Gage *et al.*, 1999; Kitamura *et al.*, 1999; Lu *et al.*, 1999). In *pitx2* mutants, the most prominent laterality defect is right isomerization of lung lobation (right pulmonary isomerism) and defects in ventral body wall closure. Other left-right organogenesis appears to be normal in *pitx2* mutants. Cardiac tubes loop in the normal rightward orientation but give rise to subsequent cardiac morphogenetic defects, including septal defects, tricuspid, and mitral valve defects and single atrium. It is possible that more subtle defects in cardiac looping could lead to misalignment of inflow and outflow segments, which would then impact septa and valve formation. Similar suble looping defects are seen in *Xenopus* embryos in which Pitx2 is overexpressed (Essner *et al.*, 2000). In addition, *pitx2* mutants have normal stomach positioning and normal orientation of body turning. Homozygous mutant phenotypes not related to laterality defects include craniofacial defects. Eye defects in the heterozygotes are consistent those seen in Rieger syndrome. Hence, there are likely to be other transcription factors expressed asymmetrically in the heart and gut that participate in left-right development. Another transcription factor, NKX3.2, is also expressed asymmetrically on the left side in chick (Schneider *et al.*, 1999), and as further discussed below (section VII); NKX3.2 is expressed on the opposite side in mouse embryos (Schneider *et al.*, 1999).

In the chick, cSnR, a zinc finger protein related to the *Drosophila* snail family of transcription factors, is expressed in lateral mesoderm on the right side at approximately the same developmental stages that *pitx2* is expressed on the left side (Isaac *et al.*, 1997). Results from ectopic placement beads soaked in activin (Boettger *et al.*, 1999; Isaac *et al.*, 1997) or Fibroblast Growth Factor (FGF8) (Boettger *et al.*, 1999) indicate that the right-sided signals activin and FGF8 can induce cSnR expression (Boettger *et al.*, 1999). Antisense cSnR oligonucleotide treatment diminishes cSnR asymmetric expression and results in isomerization of cardiac looping at high concentrations and reversed cardiac looping at lower concentrations (Patel *et al.*, 1999). High concentrations of antisense cSnR also result in ectopic expression of Pitx2 on the right side. Ectopic placement of nodal on the right side downregulates cSnR expression. These results suggest that the left and right pathways that control the expression of transcription factors Pitx2 and cSnr, respectively, are parallel and mutually exclusive.

V. The Origins of Left-Right Asymmetry

One of the most challenging questions in the study of left-right development is to discover what mechanism initiates the left-right pathway. Because of the highly conserved left-right orientations of organs in vertebrates, there must be some molecular or cellular process by which apparent bilateral symmetry in the embryo is broken to give consistently oriented left-right asymmetry. Four distinct mechanisms that might initiate left-right asymmetry are reviewed here.

A. Left-Right Coordinator Model: Mutual Antagonism between Two Sides

Full inversion of the left-right axis, resulting in left-sided markers expressed on the right side and concordant reversal of organ orientation, is relatively rare. Left-right inversion is observed in approximately 85% of mutant *inv/inv* mice embryos (Yokoyama *et al.*, 1993), leading to the suggestion that full inversion is due to an early step in the left-right pathway. The *inv* gene is upstream of asymmetric markers in the lateral plate, including nodal (Adachi *et al.*, 1999; Collignon *et al.*, 1996; Lowe *et al.*, 1996), *lefty-2* (Adachi *et al.*, 1999; Meno *et al.*, 1996), *pitx2* (Ryan *et al.*, 1998), and NKX3.2 (Schneider *et al.*, 1999). The protein encoded by the IV gene, inversin, appears to be a cytoplasmic protein that has ankyrin/Swi6 repeats (Mochizuki *et al.*, 1998; Morgan *et al.*, 1998). Regrettably, neither the ubiquitous early expression of the gene nor the sequence of the protein have given significant hints to the cellular function of the protein and how it might establish left-right asymmetry.

In *Xenopus,* the role of early antagonism among closely related members of the TGFβ cell-cell signaling family has been implicated in the initiation of the left-right axis, and formulated into a "Left-Right Coordinator" model in which one side of the embryo, the left side, acquires left-sided identity and inhibits the contralateral side from acquiring a left-side identity (Hyatt and Yost, 1998). In *Xenopus,* left-side identity appears to be driven by the Vg1 signaling pathway (Hyatt *et al.*, 1996; Hyatt and Yost, 1998). Expression of Vg1 on the right side gives full inversion of left-right downstream markers such as nodal, and concordant inversion of heart and gut orientation, in a phenotype similar to that of the *inv/inv* mouse mutant. Vg1 RNA is stored in the oocyte as a maternal RNA and does not appear to have a left-right asymmetric distribution. Ectopic expression results with a variety chimeric TGFβ-Vg1 proteins suggests that if there is an asymmetric expression of Vg1 protein, it is driven at the level of protein processing (Hyatt and

Yost, 1998; Ramsdell and Yost, 1999). Right-side identity is mediated by activation of a BMP/Alk2/Smad pathway, as demonstrated by the highly specific effects of embryo injections of both constitutively activated forms and dominant-negative forms of ligands, receptors, and transcription coactivators (Ramsdell and Yost, 1999). Strikingly, experiments in which distinct pathway components are coexpressed shows that the left-sided Vg1 pathway and right-sided BMP/Alk2/Smad pathway are mutually antagonistic (Ramsdell and Yost, 1999). These results suggest that cross-signaling between the two sides of the embryo early in development generates the left-right axis. As reviewed above (section III), there is also growing evidence that complex antagonisms of TGFβ signaling pathways in lateral plate are crucial for later steps in left-right development.

B. Asymmetry in the Node

The first report of a molecular left-right asymmetry was the left-side expression of Sonic hedgehog (Shh) asymmetric expression in the chick node, as assessed by *in situ* RNA hybridization (Levin *et al.,* 1995). Left-sided expression of sonic in the node appears to be dependent on signals from outside the node (Pagan-Westphal and Tabin, 1998). On the right side, a member of the TGFβ family, activin βB, is expressed early (Stern *et al.,* 1995). Subsequently, the activin receptor (cAct-RIIA) and FGF8 are induced (Boettger *et al.,* 1999; Levin *et al.,* 1995). Experiments with ectopically placed beads soaked in either activin or FGF8 indicate that activin can induce cAct-RIIA and FGF8, and that these factors can suppress Shh expression on the same side, thus suggesting that the left-side and right-side signaling pathways are mutually exclusive (Boettger *et al.,* 1999; Levin *et al.,* 1995). Correspondingly, the right-side signal FGF8 can induce cSnR and suppress nodal and Pitx2 (Boettger *et al.,* 1999). In addition, ActRIIB mutant mice have randomized heart orientation, pulmonary isomerism, and defects in vertebral development, thus suggesting an early role both in anterior-posterior and left-right patterning (Oh and Li, 1997).

In *Xenopus* embryo injection experiments, the effects of ectopic expression of components of the Vg1 or BMP-signaling pathways are strongest when targeted to the lateral cells of the early embryo (Hyatt *et al.,* 1996; Hyatt and Yost, 1998; Ramsdell and Yost, 1999). Comparable injections in cells that are in or near the organizer region (node) at the dorsal midline of the embryo have more limited effects on left-right development (Hyatt *et al.,* 1996). This suggests that an early signaling center for left-right development, the Left-Right Coordinator, functions outside the node of the embryo. Whether the early Vg1 and BMP/Alk2/Smad pathways have effects on asymmetry in the node cannot be assessed because the molecular asymmetries found in chick node have not been detected in frog, fish, or mice nodes. However, a similar conclusion that left-right information is passed from non-node (perhaps lateral) cells to node cells is derived from the results of transplantation experiments in the chick embryo, in which nodes were inverted from their normal orientation (Pagan-Westphal and Tabin, 1998). In embryos in which the node was inverted early, the node acquired asymmetry (as assessed by molecular markers) with respect to the rest of the embryo, thus suggesting that signals outside the node are responsible for establishing subsequent asymmetries within the node.

C. Gap Junctions

An intriguing model for the role of gap junction communication in the initiation of left-right asymmetry has been put forward by Levin and Mercola (Levin and Mercola, 1998, 1999). Early in embryogenesis, cells are electrically coupled through gap junctions formed by connexins. There is a gradient of coupling, with more gap junction communication dorsally than ventrally in the frog blastula. Defects in left-right development can be induced by blocking dorsal communication (with dominant negative connexins) or opening ventral gap junction communication (with connexin overexpression). In addition, pharmacological agents that altered junction communication indicated that the critical period was before the end of gastrulation (Levin and Mercola, 1998). Perturbation of gap junction communication in chick embryos, either with pharmacological agents or with antisense-mediated inhibition of connexin 43, results in bilaterally symmetric expression of Shh in the node and bilateral expression of nodal in lateral plate mesoderm. Since connexin43 is not expressed in the node or primitive streak, these results concur with the idea that information outside the node is important for conveying asymmetric information to the node, and that an early step in the pathway that precedes asymmetric gene expression in the node is dependent on junctional communication in cells outside the node (Levin and Mercola, 1999).

D. Monocilia in the Node

In the above three models the initial break in symmetry for left-right axis formation occurs outside the node. Left-right information is then conveyed to the node and transferred to paraxial and lateral tissues. In contrast, Nonaka *et al.* (1998) proposed that the symmetry-breaking event actually occurs in the node by an intriguing and novel mechanism.

In chick and mice, cells in the node and throughout the embryonic midline (notochord) have a single monocilium that projects from the surface of each cell into extracellular space. The monocilium has a 9+0 microtubule structure that is distinct from the standard 9+2 configuration found in cilia, which led to some uncertainty as to whether monocilia are capable of motility (Bellomo *et al.,* 1996; Sulik *et al.,* 1994). In a series of elegant experiments, Hirokawa *et al.* demonstrated that monocilia in the nodes of neurula-stage mouse embryos are motile and, what is more important, produce an overall leftward flow of extracellular fluid that is capable of moving small latex beads (Nonaka *et al.,* 1998). The "nodal flow" is proposed to generate asymmetric distribution of an unidentified molecule, which initiates the left-right pathway.

Several genes that encode microtubule-associated motor proteins have been implicated in both monocilia function and left-right development. The kinesin complex, KIF3, is a plus-end-directed microtubule motor. Targeted mutagenesis of genes that encode subunits of this complex, either KIF3B (Nonaka *et al.,* 1998) or KIF3A (Nonaka *et al.,* 1998), results in defective monocilia formation, indicating that the KIF3 complex is important for growth of monocilia and intracellular positioning of organelles. The absence of monocilia results in defective nodal flow. These mutant mice embryos have altered expression patterns of downstream left-right markers Nodal and Lefty, and have randomized heart orientation, thus demonstrating a perturbation in left-right development.

Left-Right Dynein (LRD), encoded by the IV (inversus viscerum) gene, is expressed in the mouse node (Supp *et al.,* 1997). Dyneins are minus-end directed microtubule motors, implicated in organelle translocation in the cytoplasm and, as axonemal dyneins, in cilia and flagella movement. By sequence, LRD is more closely related to axonemal dyneins, thus suggesting that LRD expression in the node might be involved in monocilia movement. In embryos with either a spontaneous mutation (lrd^{iv}) (Okada *et al.,* 1999) or targeted deletion of the ATP-binding (adenosine triphosphate) domain ($lrd^{\Delta P1}$) (Supp *et al.,* 1999), nodal monocilia cilia are present but immotile, nodal flow is greatly diminished, and left-right asymmetry is randomized. In a related finding, targeted mutation of the winged helix factor hepatocyte nuclear factor/forkhead homolog HFH-4 results in loss of node monocilia, loss of LRD expression when assayed later in the embryonic lung, and left-right developmental defects (Chen *et al.,* 1998), thus suggesting that HFH-4 regulates LRD expression. Hence, mutation of any of three distinct motor protein genes, or possibly loss of LRD expression by mutation of HFH-4, results in altered monocilia function, altered nodal flow, and perturbed left-right development.

What is the relationship between nodal flow and the generation of left-right asymmetry? The four mutants discussed above evince a strong correlation between nodal flow and left-right development. It is proposed that nodal flow is the causative agent of left-right axis initiation, at least in mouse (Vogan and Tabin, 1999). Although this appears to be the most parsimonious explanation, other models are equally plausible and several observations will be challenging to explain with current models (Wagner and Yost, 2000). In addition, there are four unresolved issues regarding the causative link of nodal flow with the initiation of the left-right axis. First, if nodal flow were the first step in a left-right pathway, one would expect mutants that have disrupted nodal flow to have similar alterations in downstream gene expression patterns. However, the limited analysis that has been reported to date suggests that this is not the case. The lateral plate expression patterns of Lefty-2 and Pitx2 are bilateral in kif3A mutants (Marszalek *et al.,* 1999; Takeda *et al.,* 1999) and randomized (left, right, bilateral, or absent) in lrd^{iv} mutants (Campione *et al.,* 1999; Meno *et al.,* 1996; Piedra *et al.,* 1998; Ryan *et al.,* 1998; Supp *et al.,* 1997). Second, nodal flow does not become active until late in gastrulation in the mouse (Okada *et al.,* 1999). In the chick, with the greatest number of early asymmetric markers available, it is clear that left-right asymmetries are well established during early gastrulation (Levin, 1998). This difference in timing suggests that nodal flow might serve to convey, magnify, or maintain preexisting asymmetry, not to initiate asymmetry. Third, with the exception of lrd, the mutants discussed above have complex pleiotropic developmental defects that could be causative of left-right development. For example, KIF3A-deficient embryos have neural tube/midline defects in Shh and Brachyury expression (Takeda *et al.,* 1999). Midline defects that do not alter node formation can induce left-right developmental defects in frogs, zebrafish, and mice (Yost, 1998a) (see section VI). Hence, it is possible that pleiotropic defects in the midline are the proximal cause of left-right perturbations in the kinesin, LRD, and HFH-4 mutants, and that defects in the node, including defects in nodal flow, lead directly to midline defects and only indirectly to left-right development defects. Roles for the midline in left-right development are described below. Fourth, the *inv* mutant, which inverts the left-right axis upstream of lateral plate markers, has normal cilia movement and overall leftward nodal flow (Okada *et al.,* 1999). This suggests that either nodal flow is upstream of the inv-dependent step, or that nodal flow is not used to initiate the left-right axis.

VI. Multiple Roles for the Midline

The embryonic midline consists of the notochord, hypochord, neural tube, and midline cells anterior to the notochord, and divides the embryo into left and right halves. Although it is likely that the pathway for left-right development is initiated before the completion of midline development, it is clear from work in several vertebrate organisms that normal left-right asymmetries in the heart and gut are dependent on development of a normal embryonic midline (Ramsdell and Yost, 2000; Yost, 1998a). The midline is upstream of asymmetric gene expression of nodal, *lefty,* and *pitx2* in the lateral plate. Several mechanisms by which the midline regulates asymmetric gene expression in lateral tissue have been proposed, but none of the models are capable of explaining the diverse laterality defects that are seen in model organisms in which midline defects are induced either by tissue extirpation or by genetic manipulation. Therefore, it is likely that there are multiple roles for the midline in left-right development, and that distinct steps within midline development control diverse aspects of left-right development.

The role of the midline was first elucidated by experiments in *Xenopus,* in which midline and anterior-posterior development were perturbed with a corresponding loss of normal left-right heart development (Danos and Yost, 1995). Removal of the midline during open neural plate stages, but not after neural tube closure, results in bilaterally symmetric expression of lateral plate markers (Lohr *et al.,* 1998, 1997) and randomization of heart orientation (Danos and Yost, 1996). The conclusion that the midline is involved in left-right development, derived from embryological manipulations, is in agreement with results from analysis of midline mutants in zebrafish (Chen *et al.,* 1997; Danos and Yost, 1996) and mutations of midline genes in mice (King *et al.,* 1998; Melloy *et al.,* 1998; Meno *et al.,* 1998).

A growing number of midline genes in mice and zebrafish have been implicated in left-right development (Yost, 1998a). It is likely that a large number of genes are necessary for both midline development and for left-right development. The most downstream component of this shared midline/laterality pathway is probably Lefty-1, which is expressed in the left prospective floorplate in mice embryos (Meno *et al.,* 1997, 1996; Oulad-Abdelghani *et al.,* 1998), and in the midline in zebrafish (Bisgrove *et al.,* 1999; Thisse and Thisse, 1999) and *Xenopus* (Branford *et al.,* 2000; Cheng *et al.,* 2000). Targeted mutagenesis of *lefty-1* results in bilateral expression of lateral plate markers and thoracic left isomerism (Meno *et al.,* 1998). Lefty is proposed to serve as a molecular barrier between the left and right sides. In the absence of this barrier, signals can cross the midline to induce bilaterally symmetric expression of genes that are normally only expressed on one side. As an example of a gene upstream of Lefty-1 in the midline, spontaneous no turning mutant lacks *lefty-1* midline expression and has bilateral expression of *nodal* and *lefty-2* in the lateral plate, as well as randomized cardiac looping (Melloy *et al.,* 1998). Expression patterns in the midline suggest that no turning is defective in the maintenance of notochord development. Disrupted notochord then leads to loss of *lefty-1* expression in the midline and loss of the midline barrier. Similarly, mutants in the T gene (Brachyury) have midline defects and randomized organ orientation but normal expression of lateral plate markers (King *et al.*, 1998). It is likely that the developmental timing of the midline defect and the spatial extent of midline deficiency are correlated with the severity of defects in left-right gene expression and perturbation of organ left-right morphogenesis. In retrospective analysis of clinical records, there is a strong correlation between midline defects and laterality defects in humans (Goldstein *et al.*, 1998), thus suggesting that at least some of the links between midline development and left-right development are highly conserved.

VII. Speculation: Evolution of the Left-Right Axis

A. Left-Right Phylotypic Stages?

The diverse models and ideas about the embryological event that initiates the left-right axis, as outlined in section V, indicate the wealth of information

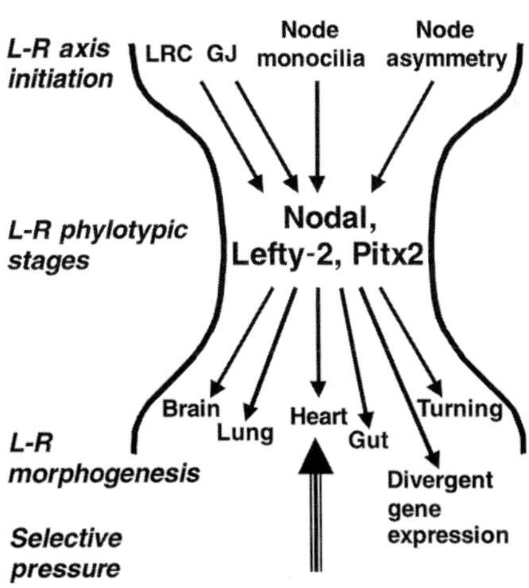

FIG. 4 Left-right phylotypic stages: At what developmental periods are the vertebrate left-right pathways conserved? Indicated at the top, diverse early mechanisms might initiate the left-right axis. In the "left-right phylotypic stages," all vertebrates go through a highly conserved phase of asymmetric nodal, *lefty* and *pitx2* expression in left lateral plate mesoderm. After these stages, is not clear whether divergent mechanisms and pathways diverge. Strong selective pressure (indicated by upward arrow) on only one of the many downstream pathways might be sufficient to maintain the highly conserved asymmetric expression patterns in the left-right phylotypic stages. Adapted and updated from Yost, 1999.

we still need to gain regarding the transition from symmetry to left-right asymmetry. A common and feasible operating assumption is that once we gain more molecular markers, new mutants, and more detailed cell biology and embryology, the various models will converge to a unified theme of how the left-right axis is initiated. This perspective assumes that the initial steps in a developmental pathway will be shared by all vertebrates, through common evolutionary descent, and that distinctions among the classes of vertebrates arose through fine-tuning and alterations in later steps in the developmental pathway. Similarly, a paleontological argument has been made for the early formation of left-right asymmetry (Jefferies *et al.,* 1996). A contrary proposal was recently put forward for consideration that suggests that the most highly conserved steps in the vertebrate left-right pathway will not be found at the initiation of the left-right pathway, but at a midpoint in the pathway called the "left-right phylotypic stages" (Yost, 1999). These stages are indicated by the highly conserved left-right asymmetric gene expression of *nodal, lefty,* and *pitx2* in the lateral plate mesoderm. This suggests that earlier steps in left-right development are capable of diverging in different classes of vertebrates as long as the earlier steps result in the stereotypic asymmetric gene expression in the lateral plate. In addition, developmental steps subsequent to the "left-right phylotypic stages" are capable of diverging as long as at least one end-point morphology provides feedback for selection pressure to maintain the asymmetric lateral plate expression.

B. Divergent Shh and FGF8 Signals Preceding Conserved LPM Expression Patterns?

One of the first molecular asymmetries described was the left-sided expression of sonic hedgehog (Shh) in the node of chick embryos (Levin *et al.,* 1995), which has been shown to be upstream of nodal and *lefty* expression in left lateral plate mesoderm. On the right side, asymmetric FGF8 expression is thought to inhibit nodal and lefty expression (Boettger *et al.,* 1999). The question is whether the early roles of Shh and FGF8 in chick left-right development are conserved in other vertebrates.

Left-right asymmetric expression of Shh has not been seen in mice, zebrafish, or *Xenopus.* In contrast, amphioxus hedgehog gene, AmphiHh, is expressed in the left side of the pharyngeal endoderm (Shimeld, 1999). Experimental manipulations, including ectopic expression of Shh and use of Shh-blocking antibodies, indicate that Shh plays an important early role in the left-right pathway in chick. Similarly, ectopic expression of Shh in zebrafish induces altered expression of BMP4 in the heart and altered left-right orientation of the heart and gut. Ectopic expression of hedgehog-related proteins on the right side in *Xenopus* induces ectopic nodal expression and alters left-right organ orientation (Sampath *et al.,* 1997). Together, these ectopic expression results indicate that manipulation of a vertebrate hedgehog signal can alter an early step in the left-right pathway. Initially, $Shh^{(-/-)}$ mouse embryos were reported to have extensive midline defects and cyclopia, but laterality defects were not reported (Chiang *et al.,* 1996). As more molecular left-right markers became available, $Shh^{(-/-)}$ embryos were found to be defective in expression of the midline *lefty-1* gene (Tsukui *et al.,* 1999), but the effects of left-right development are controversial. Some groups describe randomized body turning, pulmonary left isomerism, and altered heart looping or position (Meyers and Martin, 1999; Tsukui *et al.,* 1999), whereas others report midline defects with normal cardiac looping (Izraeli *et al.,* 1999). It is possible that alterations in left-right development in $Shh^{(-/-)}$ mice do not reflect an early role for Shh in the node but a later role for Shh in the maintenance of the midline.

Asymmetric expression of FGF8 on the right side of the chick node appears to be an early step in the right-side pathway in chick (Boettger *et al.,* 1999). FGF8 is induced by activin, induces the right side transcription regulator cSnR, and inhibits expression of the left-side pathway, as indicated by suppression of nodal and Pitx2 expression (Boettger *et al.,* 1999). These results suggest that FGF8 serves as a "right determinant" in chick. Concurring with a role for FGF8 in left-right development, FGF8 mouse mutants display altered heart orientation. However, approximately half the mutants displayed loss of left-side markers (*nodal, lefty-2,* or *pitx2*) and approximately half the mutants had right pulmonary isomerism. In contrast to the ability of ectopic FGF8 to suppress nodal expression in chick embryos (Boettger *et al.,* 1999), implantation of FGF8 beads in the right lateral plate mesoderm (LPM) induced ectopic nodal expression in mouse embryos (Meyers and Martin, 1999), thus suggesting that FGF8 serves as a "left determinant" in mice.

C. Divergent Signals after Conserved LPM Expression Patterns?

It is quite striking that the tinman-related transcription factor Nkx3.2 is expressed asymmetrically in both chick and mouse embryos, but on opposite sides (Schneider *et al.,* 1999). Results from ectopic expression experiments in chick indicate that NKX3.2 can be induced by left-side signals (Shh, nodal and *lefty-2,* and inhibition of FGF signaling) and repressed by right-side signals (FGF8). In addition, retinoic acid (RA) can induce expression of NKX3.2 when ectopically placed in the right lateral plate mesoderm. RA has been shown to alter midline development and left-right development in mice (Tsukui *et al.,* 1999). Together, these results indicate that NKX3.2 is part of the left-side pathway and is regulated at the same level as Pitx2 in the chick, downstream of nodal and *lefty* expression. In contrast, NKX3.2 is expressed asymmetrically in the right lateral plate mesoderm in mouse embryos. NKX3.2 expression is inverted (expressed in left LPM) in the inversion of embryo turning (*inv*) mutant, in which *nodal* and *lefty-2* expression patterns are also inverted (expressed in right LPM). These results suggest that NKX3.2 is part of the right-side pathway in mouse embryos, on the side opposite of nodal, *lefty-2,* and *pitx2.*

Acknowledgments

I thank M. L. Condic and members of my laboratory for discussions. H. J. Y. is an Established Investigator of the American Heart Association.

References

Adachi, H., Saijoh, Y., Mochida, K., Ohishi, S., Hashiguchi, H., Hirao, A., and Hamada, H. (1999). Determination of left/right asymmetric expression of nodal by a left side-specific enhancer with sequence similarity to a lefty-2 enhancer. *Genes Dev.* **13,** 1589–1600.

Bellomo, D., Lander, A., Harragan, I., and Brown, N. A. (1996). Cell proliferation in mammalian gastrulation: The ventral node and notochord are relatively quiescent. *Dev. Dyn.* **205,** 471–485.

Bisgrove, B. W., Essner, J. J., and Yost, H. J. (1999). Regulation of midline development by antagonism of lefty and nodal signaling. *Development* **126,** 3253–3262.

Boettger, T., Wittler, L., and Kessel, M. (1999). FGF8 functions in the specification of the right body side of the chick. *Curr. Biol.* **9,** 277–280.

Bowers, P. N., Brueckner, M., and Yost, H. J. (1996). The genetics of left-right development and heterotaxia. *Semin. Perinatol.* **20,** 577–588.

Branford, W. W., Essner, J. J., and Yost, H. J. (2000). Regulation of gut and heart left-right asymmetry by context dependent interactions between Xenopus Lefty and BMP4 signaling. *Dev. Biol.* **223,** 291–306.

Burggren, W. W. (1988). Cardiac design in lower vertebrates: What can phylogeny reveal about ontogeny? *Experientia* **44,** 919–930.

Campione, M., Steinbeisser, H., Schweickert, A., Deissler, K., van Bebber, F., Lowe, L. A., Nowotschin, S., Viebahn, C., Haffter, P., Kuehn, M. R., and Blum, M. (1999). The homeobox gene Pitx2: Mediator of asymmetric left-right signaling in vertebrate heart and gut looping. *Development* **126,** 1225–1234.

Chen, J., Knowles, H. J., Hebert, J. L., and Hackett, B. P. (1998). Mutation of the mouse hepatocyte nuclear factor/forkhead homologue 4 gene results in an absence of cilia and random left-right asymmetry. *J. Clin. Invest.* **102,** 1077–1082.

Chen, J. N., van Eeden, F. J., Warren, K. S., Chin, A., Nusslein-Volhard, C., Haffter, P., and Fishman, M. C. (1997). Left-right pattern of cardiac BMP4 may drive asymmetry of the heart in zebrafish. *Development* **124,** 4373–4382.

Cheng, A. M., Thisse, B., Thisse, C., and Wright, C. V. (2000). The lefty-related factor Xatv acts as a feedback inhibitor of Nodal signaling in mesoderm induction and L-R axis development in Xenopus. *Development* **127,** 1049–1061.

Chiang, C., Litingtung, Y., Lee, E., Young, K. E., Corden, J. L., Westphal, H., and Beachy, P. A. (1996). Cyclopia and defective axial patterning in mice lacking Sonic hedgehog gene function. *Nature* **383,** 407–413.

Collignon, J., Varlet, I., and Robertson, E. J. (1996). Relationship between asymmetric nodal expression and the direction of embryonic turning. *Nature* **381,** 155–158.

Danos, M. C., and Yost, H. J. (1995). Linkage of cardiac left-right asymmetry and dorsal-anterior development in *Xenopus. Development* **121,** 1467–1474.

Danos, M. C., and Yost, H. J. (1996). Role of notochord in specification of cardiac left-right orientation in zebrafish and Xenopus. *Dev. Biol.* **177,** 96–103.

Engbretson, G. A., Reiner, A., and Brecha, N. (1981). Habenular asymmetry and the central connections of the parietal eye of the lizard. *J. Comp. Neurol.* **198,** 155–165.

Essner, J. J., Branford, W. W., Zhang, J., and Yost, H. J. (2000). Mesendoderm and left-right brain, heart and gut development are differentially regulated by pitx2 isoforms. *Development* **127,** 1081–1093.

Gage, P. J., Suh, H., and Camper, S. A. (1999). Dosage requirement of Pitx2 for development of multiple organs. *Development* **126,** 4643–4651.

Gaio, U., Schweickert, A., Fischer, A., Garratt, A. N., Muller, T., Ozcelik, C., Lankes, W., Strehle, M., Britsch, S., Blum, M., and Birchmeier, C. (1999). A role of the cryptic gene in the correct establishment of the left-right axis. *Curr. Biol.* **9,** 1339–1342.

Goldstein, A. M., Ticho, B. S., and Fishman, M. C. (1998). Patterning the heart's left-right axis: From zebrafish to man. *Dev. Genet.* **22,** 278–287.

Gugliemotti, V., and Fiorino, L. (1998). Asymmetry in the left and right habenulo-interpeduncular tracts in the frog. *Brain Res. Bull.* **45,** 105–110.

Gurusinghe, C. J., and Ehrlich, D. (1985). Sex-dependent structural asymmetry of the medial habenular nucleus of the chicken brain. *Cell Tissue Res.* **240,** 149–152.

Heymer, J., Kuehn, M., and Ruther, U. (1997). The expression pattern of nodal and lefty in the mouse mutant Ft suggests a function in the establishment of handedness. *Mech. Dev.* **66,** 5–11.

Hyatt, B. A., Lohr, J. L., and Yost, H. J. (1996). Initiation of vertebrate left-right axis formation by maternal vg1. *Nature* **384,** 62–65.

Hyatt, B. A., and Yost, H. J. (1998). The left-right coordinator: The role of Vg1 in organizing left-right axis formation. *Cell* **93,** 37–46.

Isaac, A., Sargent, M. G., and Cooke, J. (1997). Control of vertebrate left-right asymmetry by a snail-related zinc finger gene. *Science* **275,** 1301–1304.

Ishimaru, Y., Yoshioka, H., Tao, H., Thisse, B., Thisse, C., Wright, C. V. E., Hamada, H., Ohuchi, H., and Noji, S. (2000). Asymmetric expression of antivin/lefty1 in the early chick embryo. *Mech. Dev.* **90,** 115–118.

Itasaki, N., Nakamura, H., Sumida, H., and Yasuda, M. (1991). Actin bundles on the right side in the caudal part of the heart tube play a role in dextro-looping in the embryonic chick heart. *Anat. Embryol.* **183,** 29–39.

Izraeli, S., Lowe, L. A., Bertness, V. L., Good, D. J., Dorward, D. W., Kirsch, I. R., and Kuehn, M. R. (1999). The SIL gene is required for mouse embryonic axial development and left-right specification. *Nature* **399,** 691–694.

Jefferies, R. P. S., Brown, N. A., and Daley, P. E. J. (1996). The early phylogeny of chordates and echinoderms and the origin of chrodate left-right asymmetry and bilateral symmetry. *Acta Zool.* **27,** 101–122.

King, T., Beddington, R. S. P., and Brown, N. A. (1998). The role of the brachyury gene in heart development and left-right specification in the mouse. *Mech. Dev.* **79,** 29–37.

Kitamura, K., Miura, H., Miyagawa-Tomita, S., Yanazawa, M., Katoh-Fukui, Y., Suzuki, R., Ohuchi, H., Suehiro, A., Motegi, Y., Nakahara, Y., Kondo, S., and Yokoyama, M. (1999). Mouse Pitx2 deficiency leads to anomalies of the ventral body wall, heart, extra- and periocular mesoderm and right pulmonary isomerism. *Development* **126,** 5749–5758.

Kosaki, K., Bassi, M. T., Kosaki, R., Lewin, M., Belmont, J., Schauer, G., and Casey, B. (1999). Characterization and mutation analysis of human LEFTY A and LEFTY B, homologues of murine genes implicated in left-right axis development. *Am. J. Hum. Genet.* **64,** 712–721.

Levin, M. (1998). Left-right asymmetry and the chick embryo. *Semin. Cell Dev. Biol.* **9,** 67–76.

Levin, M., Johnson, R. L., Stern, C. D., Kuehn, M., and Tabin, C. (1995). A molecular pathway determining left-right asymmetry in chick embryogenesis. *Cell* **82,** 803–814.

Levin, M., and Mercola, M. (1998). Gap junctions are involved in the early generation of left-right asymmetry. *Dev. Biol.* **203,** 90–105.

Levin, M., and Mercola, M. (1999). Gap junction-mediated transfer of left-right patterning signals in the early chick blastoderm is upstream of shh asymmetry in the node. *Development* **126,** 4703–4714.

Logan, M., Pagan-Westphal, S. M., Smith, D. M., Paganessi, L., and Tabin, C. J. (1998). The transcription factor Pitx2 mediates situs-specific morphogenesis in response to left-right asymmetric signals. *Cell* **94,** 307–317.

Lohr, J. L., Danos, M. C., Groth, T. W., and Yost, H. J. (1998). Maintenance of asymmetric nodal expression in Xenopus laevis. *Dev. Genet.* **23,** 194–202.

Lohr, J. L., Danos, M. C., and Yost, H. J. (1997). Left-right asymmetry of a nodal-related gene is regulated by dorsoanterior midline structures during *Xenopus* development. *Development* **124,** 1465–1472.

Lowe, L. A., Supp, D. M., Sampath, K., Yokoyama, T., Wright, C. V. E., Potter, S. S., Overbeek, P., and Kuehn, M. R. (1996). Conserved left-right asymmetry of nodal expression and alterations in murine situs inversus. *Nature* **381,** 158–161.

Lu, M. F., Pressman, C., Dyer, R., Johnson, R. L., and Martin, J. F. (1999). Function of Rieger syndrome gene in left-right asymmetry and craniofacial development. *Nature* **401,** 276–278.

Marszalek, J. R., Ruiz-Lozano, P., Roberts, E., Chien, K. R., and Goldstein, L. S. (1999). Situs inversus and embryonic ciliary morphogenesis defects in mouse mutants lacking the KIF3A subunit of kinesin-II. *Proc. Natl. Acad. Sci. USA* **96,** 5043–5048.

Melloy, P. G., Ewart, J. L., Cohen, M. F., Desmond, M. E., Kuehn, M. R., and Lo, C. W. (1998). No turning, a mouse mutation causing left-right and axial patterning defects. *Dev. Biol.* **193,** 77–89.

Meno, C., Ito, Y., Saijoh, Y., Matsuda, Y., Tashiro, K., Kuhara, S., and Hamada, H. (1997). Two closely-related left-right asymmetrically expressed genes, lefty-1 and lefty-2: Their distinct expression domains, chromosomal linkage and direct neuralizing activity in *Xenopus* embryos. *Genes Cells* **2,** 513–524.

Meno, C., Saijoh, Y., Fujii, H., Ikeda, M., Yokoyama, T., Yokoyama, M., Toyoda, Y., and Hamada, H. (1996). Left-right asymmetric expression of the TGFβ-family member lefty in mouse embryos. *Nature* **381,** 151–155.

Meno, C., Shimono, A., Saijoh, Y., Yashiro, K., Mochida, K., Ohishi, S., Noji, S., Kondoh, H., and Hamada, H. (1998). lefty-1 is required for left-right determination as a regulator of lefty-2 and nodal. *Cell* **94,** 287–297.

Meyers, E. N., and Martin, G. R. (1999). Differences in left-right axis pathways in mouse and chick: Functions of FGF8 and SHH. *Science* **285,** 403–406.

Milewski, W. M., Duguay, S. J., Chan, S. J., and Steiner, D. F. (1998). Conservation of PDX-1 structure, function, and expression in zebrafish. *Endocrinology* **139,** 1440–1449.

Mochizuki, T., Saijoh, Y., Tsuchiya, K., Shirayoshi, Y., Takai, S., Taya, C., Yonekawa, H., Yamada, K., Nihei, H., Nakatsuji, N., Overbeek, P. A., Hamada, H., and Yokoyama, T. (1998). Cloning of inv, a gene that controls left/right asymmetry and kidney development. *Nature* **395,** 177–181.

Morgan, D., Turnpenny, L., Goodship, J., Dai, W., Majumder, K., Matthews, L., Gardner, A., Schuste, R. G., Vien, L., Harrison, W., Elder, F., Penman-Splitt, M., Overbeek, P., and Strachan, T. (1998). Inversin, a novel gene in the vertebrate left-right axis pathway, is partially deleted in the inv mouse. *Nature Genet.* **20,** 149–156.

Morgan, M. J. (1991). The asymmetrical genetic determination of laterality: Flatfish, frogs and human handedness. *Ciba Found. Symp.* **162,** 234–247.

Nonaka, S., Tanaka, Y., Okada, Y., Takeda, S., Harada, A., Kanai, Y., Kido, M., and Hirokawa, N. (1998). Ranomization of left-right asymmetry due to loss of nodal cilia generating leftward flow of extraembryonic fluid in mice lacking KIF3B motor protein. *Cell* **95,** 829–837.

Oh, S. P., and Li, E. (1997). The signaling pathway mediated by the type IIB activin receptor controls axial patterning and lateral asymmetry in the mouse. *Genes Dev.* **11,** 1812–1826.

Okada, Y., Nonaka, S., Tanaka, Y., Saijoh, Y., Hamada, H., and Hirokawa, N. (1999). Abnormal nodal flow precedes situs inversus in iv and inv mice. *Mol. Cell* **4,** 459–468.

Oulad-Abdelghani, M., Chazaud, C., Bouillet, P., Mattei, M. G., Dolle, P., and Chambon, P. (1998). Stra3/lefty, a retinoic acid-inducible novel member of the transforming growth factor-beta superfamily. *Int. J. Dev. Biol.* **42,** 23–32.

Pagan-Westphal, S. M., and Tabin, C. J. (1998). The transfer of left-right positional information during chick embryogenesis. *Cell* **93,** 25–35.

Patel, K., Isaac, A., and Cooke, J. (1999). Nodal signalling and the roles of the transcription factors SnR and Pitx2 in vertebrate left-right asymmetry. *Curr. Biol.* **9,** 609–612.

Piedra, M. E., Icardo, J. M., Albajar, M., Rodriguez-Rey, J. C., and Ros, M. A. (1998). Pitx2 participates in the late phase of the pathway controlling left-right asymmetry. *Cell* **94,** 319–324.

Ramsdell, A. F., and Yost, H. J. (1998). Molecular mechanisms of vertebrate left-right development. *Trends Genet.* **14,** 459–465.

Ramsdell, A. F., and Yost, H. J. (1999). Cardiac looping and the vertebrate left-right axis: Antagonism of left-sided Vg1 activity by a right-sided ALK2-dependent BMP pathway. *Development* **126,** 5195–5205.

Ramsdell, A. F., and Yost, H. J. (2000). Cardiac looping and the left-right axis: Integrating morphological, molecular and genetic analyses of vertebrate left-right development. *In* "Development of the Cardiovascular System" (R. J. Romanek and R. Runyon, Eds.), Vol. 1, in press. JAI Press, Greenwich, CT.

Rebagliati, M. R., Toyama, R., Fricke, C., Haffter, P., and Dawid, I. B. (1998a). Zebrafish nodal-related genes are implicated in axial patterning and establishing left-right asymmetry. *Dev. Biol.* **199,** 261–272.

Rebagliati, M. R., Toyama, R., Haffter, P., and Dawid, I. B. (1998b). Cyclops encodes a nodal-related factor involved in midline signaling. *Proc. Natl. Acad. Sci. USA* **95,** 9932–9937.

Rodriguez Esteban, C., Capdevila, J., Economides, A. N., Pascual, J., Ortiz, A., and Izpisua Belmonte, J. C. (1999). The novel Cer-like protein Caronte mediates the establishment of embryonic left-right asymmetry. *Nature* **401,** 243–251.

Ryan, A. K., Blumberg, B., Rodriguez-Esteban, C., Yonei-Tamura, S., Tamura, K., Tsukui, T., de la Pena, J., Sabbagh, W., Greenwald, J., Choe, S., Norris, D. P., Robertson, E. J., Evans, R. M., Rosenfeld, M. G., and Izpisua Belmonte, J. C. (1998). Pitx2 determines left-right asymmetry of internal organs in vertebrates. *Nature* **394,** 545–551.

Saijoh, Y., Adachi, H., Mochida, K., Ohishi, S., Hirao, A., and Hamada, H. (1999). Distinct transcriptional regulatory mechanisms underlie left-right asymmetric expression of lefty-1 and lefty-2. *Genes Dev.* **13,** 259–269.

Sampath, K., Cheng, A. M., Frisch, A., and Wright, C. V. (1997). Functional differences among *Xenopus* nodal-related genes in left-right axis determination. *Development* **124,** 3293–3302.

Sampath, K., Rubinstein, A. L., Cheng, A. M., Liang, J. O., Fekany, K., Solnica-Krezel, L., Korzh, V., Halpern, M. E., and Wright, C. V. (1998). Induction of the zebrafish ventral brain and floorplate requires cyclops/nodal signalling. *Nature* **395,** 185–189.

Schneider, A., Mijalski, T., Schlange, T., Dai, W., Overbeek, P., Arnold, H. H., and Brand, T. (1999). The homeobox gene it NKX3.2 is a target of left-right signalling and is expressed on opposite sides in chick and mouse embryos. *Curr. Biol.* **9,** 911–914.

Schweickert, A., Campione, M., Steinbeisser, H., and Blum, M. (2000). Pitx2 isoforms: Involvement of Pitx2c but not Pitx2a or Pitx2b in vertebrate left-right asymmetry. *Mech. Dev.* **90,** 41–51.

Semina, E. V., Reiter, R., Leysens, N. J., Alward, W. L., Small, K. W., Datson, N. A., Siegel-Bartelt, J., Bierke-Nelson, D., Bitoun, P., Zabel, B. U., Carey, J. C., and Murray, J. C. (1996). Cloning and characterization of a novel bicoid-related homeobox transcription factor gene, RIEG, involved in Rieger syndrome. *Nat. Genet.* **14,** 392–399.

Shimeld, S. M. (1999). The evolution of the hedgehog gene family in chordates: Insights from amphioxus hedgehog. *Dev. Genes. Evol.* **209,** 40–47.

St. Amand, T. R., Ra, J., Zhang, Y., Hu, Y., Baber, S. I., Qiu, M., and Chen, Y. (1998). Cloning and expression pattern of chicken Pitx2: A new component in the SHH signaling pathway controlling embryonic heart looping. *Biochem. Biophys. Res. Commun.* **247,** 100–105.

Stern, C. D., Yu, R. T., Kakizuka, A., Kintner, C. R., Mathews, L. S., Vale, W. W., Evans, R. M., and Umesono, K. (1995). Activin and its receptors during gastrulation and the later phases of mesoderm development in the chick embryo. *Dev. Biol.* **172,** 192–205.

Sulik, K., Dehart, D. B., Iangaki, T., Carson, J. L., Vrablic, T., Gesteland, K., and Schoenwolf, G. C. (1994). Morphogenesis of the murine node and notochordal plate. *Dev. Dyn.* **201,** 260–278.

Supp, D. M., Brueckner, M., Kuehn, M. R., Witte, D. P., Lowe, L. A., McGrath, J., Corrales, J., and Potter, S. S. (1999). Targeted deletion of the ATP binding domain of left-right dynein confirms its role in specifying development of left-right asymmetries. *Development* **126,** 5495–5504.

Supp, D. M., Witte, D. P., Potter, S. S., and Brueckner, M. (1997). Mutation of an axonemal dynein affects left-right asymmetry in inversus viscerum mice. *Nature* **389,** 963–966.

Takeda, S., Yonekawa, Y., Tanaka, Y., Okada, Y., Nonaka, S., and Hirokawa, N. (1999). Left-right asymmetry and kinesin superfamily protein KIF3A: New insights in determination of laterality and mesoderm induction by kif3A−/− mice analysis. *J. Cell Biol.* **145,** 825–836.

Thisse, B., Wright, C. V., and Thisse, C. (2000). Activin- and Nodal-related factors control antero-posterior patterning of the zebrafish embryo. *Nature* **403,** 425–428.

Thisse, C., and Thisse, B. (1999). Antivin, a novel and divergent member of the TGFβ superfamily, negatively regulates mesoderm induction. *Development* **126,** 229–240.

Tsuda, T., Majumder, K., and Linask, K. K. (1998). Differential expression of flectin in the extracellular matrix and left-right asymmetry in mouse embryonic heart during looping stages. *Dev. Genet.* **23,** 203–214.

Tsuda, T., Philp, N., Zile, M. H., and Linask, K. K. (1996). Left-right asymmetric localization of flectin in the extracellular matrix during heart looping. *Dev. Biol.* **173,** 39–50.

Tsukui, T., Capdevila, J., Tamura, K., Ruiz-Lozano, P., Rodriguez-Esteban, C., Yonei-Tamura, S., Magallon, J., Chandraratna, R. A., Chien, K., Blumberg, B., Evans, R. M., and Belmonte, J. C. (1999). Multiple left-right asymmetry defects in Shh$^{(-/-)}$ mutant mice unveil a convergence of the shh and retinoic acid pathways in the control of Lefty-1. *Proc. Natl. Acad. Sci. USA* **96,** 11376–11381.

van den Hoff, M. J., Moorman, A. F., Ruijter, J. M., Lamers, W. H., Bennington, R. W., Markwald, R. R., and Wessels, A. (1999). Myocardialization of the cardiac outflow tract. *Dev. Biol.* **212,** 477–490.

Vogan, K. J., and Tabin, C. J. (1999). A new spin on handed asymmetry. *Nature* **397,** 295, 297–298.

Wagner, M. K., and Yost, H. J. (2000). The roles of nodal cilia. *Curr. Biol.* **10,** R149–R151.

Wood, W. B. (1997). Left-right asymmetry in animal development. *Annu. Rev. Cell Dev. Biol.* **13,** 3–82.

Yan, Y. T., Gritsman, K., Ding, J., Burdine, R. D., Corrales, J. D., Price, S. M., Talbot, W. S., Schier, A. F., and Shen, M. M. (1999). Conserved requirement for EGF-CFC genes in vertebrate left-right axis formation. *Genes Dev.* **13,** 2527–2537.

Yokouchi, Y., Vogan, K. J., Pearse, R. V. 2nd, and Tabin, C. J. (1999). Antagonistic signaling by Caronte, a novel Cerberus-related gene, establishes left-right asymmetric gene expression. *Cell* **98,** 573–583.

Yokoyama, T., Copeland, N. G., Jenkins, N. A., Montgomery, C. A., Elder, F. F., and Overbeek, P. A. (1993). Reversal of left-right asymmetry: A situs inversus mutation. *Science* **260,** 679–682.

Yoshioka, H., Meno, C., Koshiba, K., Sugihara, M., Itoh, H., Ishimaru, Y., Inoue, T., Ohuchi, H., Semina, E. V., Murray, J. C., Hamada, H., and Noji, S. (1998). Pitx2, a bicoid-type homeobox gene, is involved in a lefty-signaling pathway in determination of left-right asymmetry. *Cell* **94,** 299–305.

Yost, H. J. (1995). Vertebrate left-right development. *Cell* **82,** 689–692.

Yost, H. J. (1998a). The genetics of midline and cardiac laterality defects. *Curr. Opin. Cardiol.* **13,** 185–189.

Yost, H. J. (1998b). Left-right development from embryos to brains. *Dev. Genet.* **23,** 159–163.

Yost, H. J. (1999). Diverse initiation in a conserved left-right pathway? *Curr. Opin. Genet. Dev.* **9,** 422–426.

Zhu, L., Marvin, M. J., Gardiner, A., Lassar, A. B., Mercola, M., Stern, C. D., and Levin, M. (1999). Cerberus regulates left-right asymmetry of the embryonic head and heart. *Curr. Biol.* **9,** 931–938.

Initiation and Early Patterning of the Endoderm

Debbie Clements, Maria Rex, and Hugh R. Woodland
Department of Biological Sciences, University of Warwick, Coventry, United Kingdom

We review the early stages of endoderm formation in the major animal models. In Amphibia maternal molecules are important in initiating endoderm formation. This is followed by successive signaling events that establish and then pattern the endoderm. In other organisms there are differences in endodermal development, particularly in the initial, prephylotypic stages. Later many of the same key families of transcription factors and signaling cassettes are used in all animals, but more work will be needed to establish exact evolutionary homologies.

KEY WORDS: Endoderm development, Gut development, *Xenopus, C. elegans, Drosophila,* Zebrafish, Chick, Mouse, HNF, GATA, Sox17, TGFβ, Nodal.

I. Introduction

The endoderm is the innermost cell layer of the three Metazoan "germ layers," forming the inner layer of the gut and its associated organs, in vertebrates the lungs, thyroid, thymus, liver, and pancreas. An exact definition of endoderm is difficult; Skaer has discussed this in relation to *Drosophila* and argues that eventually definitions in terms of gene expression patterns will be advisable, but it is not clear whether this will actually clear up all problems, particularly in comparing different animals (Skaer, 1993). Typically, in many eggs that divide in entirety, endoderm develops from the pole of the egg at which the yolk protein reserves accumulate, called the vegetal pole. This region is also where primordial germ cells (PGCs) form. Throughout the animal kingdom, there is usually this association between germ cells and the early endoderm; only later do PGCs migrate into the site of the gonads.

0074-7696/01 $35.00

It may be significant that the fate of the vegetal pole is eventually to perform the two most primitive Metazoan functions—feeding and reproduction. This suggests that it is the most ancient part of the embryo, which is perhaps why in many organisms the vegetal pole controls development, organizing the rest of the embryo through signalling processes. Since this observation is true of both Deuterostomes and nematodes, it is presumably primitive. E. G. Conklin remarked, when asked whether the chicken came before the egg, that a chicken is an egg's way of making another egg, thus emphasising the primacy of the reproductive function of the yolky region.

In this review we focus first on the initiation of gut formation and its early patterning in Amphibia, *C. elegans* and *Drosophila.* Patterning within the endoderm is considered in those organisms for which there is significant molecular information, particularly vertebrates, but we restrict ourselves to the first steps in this patterning, since space is too limited to consider organogenesis. Currently information on endoderm development is sporadic through the animal kingdom, which means that no remotely complete picture exists for any single organism. However, there are common features at the molecular level that are at least beginning to be discerned. It seems likely that the main differences in endoderm formation, at least within major animal groups, arise from specializations before the phylotypic stage. For example, in the zebrafish, a teleost, the most vegetal region is the syncytial yolk cell, which does not contribute to the embryo, an evolutionary change paralleled in birds. Nevertheless, in zebrafish the yolk cell retains a signaling role like the vegetal pole of amphibians, and the endoderm develops from adjacent cells in the blastoderm. Birds have more extreme specializations along these lines, although their first signaling centers have not been identified, and mammals have evolved from reptilian ancestors with this kind of development, in turn adding their own derived features. While little is known about the earliest phases of endoderm initiation in mammals, it is among mammals that most is known about later patterning, particularly because in the mouse gene inactivation by homologous recombination has been widely applied. Particularly where they apply to later events, at and after the phylotypic stage, it is reasonable to expect that these findings have wide application, at least in chordates. Finally, *Drosophila* and *C. elegans* appear to have highly derived specializations in their overall development, and are atypical of their major clade, the Ecdysozoa. However, the detailed genetic analysis possible in these organisms is revealing features common to all animals. Unfortunately, space constraints have not permitted a discussion of other invertebrates, or indeed lower chordates.

II. The Ancestral Vertebrate Scheme? Development of the Endoderm in Amphibians

A. Fate Maps and Derivation of the Endoderm

The amphibian egg is moderately yolky and the cleavage is complete, as is seen in lower chordates, such as ascidians and amphioxus. The overall pattern of development is very similar to that in these lower chordates, so it probably represents the basic chordate plan. However, it is as yet unclear how far early events are comparbable at the molecular level; information from Agnatha and primitive jawed fish, as well as ascidians and amphioxus, would be very valuable.

At the 32-cell stage of development the *Xenopus* embryo consists of four tiers of cells (Fig. 1A, see color insert), from A at the animal pole to D at the vegetal pole, each comprising two sets of four cells, one set on each side of the bilaterally symmetrical embryo (each set of four is labelled 1–4, from dorsal to ventral). At this point the fate map is statistical, unlike that of *C. elegans,* which is rigidly determined. The D-tier contributes most of the endoderm and the C-tier forms a smaller proportion (Dale and Slack, 1987; Moody, 1987). By the gastrula stage, the presumptive endoderm represents most of the vegetal yolk mass plus a layer of superficial cells on the animal side of the blastopore; the latter includes the bottle cells that trigger gastrulation by involuting over the blastopore lip (Fig. 1B; Keller, 1991). Defining the limits of the endoderm at early stages depends on gene expression patterns; considerable progress in identifying endoderm markers has been made in recent years (Table I). Judged by marker expression in the gastrula, the limits of the endoderm are sharp and at these sharp borders the endoderm abuts both mesoderm and ectoderm (Hudson *et al.,* 1997). However, a recent report suggests that there is a layer of cells between the *Xbra*-expressing mesoderm and the *Xsox17β* expressing endoderm expressing neither marker; it is not known to which germ layer these cells belong (Zorn *et al.,* 1999). An explanation of how the endoderm is specified in *Xenopus* must explain the establishment of these sharp boundaries, as well as explaining processes within the main vegetal mass of cells.

B. Specification of the Endoderm

When vegetal poles are explanted and cultured in isolation, they express endodermal markers, but not those of the other germ layers, indicating that the vegetal pole contains molecules capable of initiating endodermal development (Gamer and Wright, 1995; Henry *et al.,* 1996; Henry and

TABLE I

Markers Useful for the Analysis of Endoderm Formation in the *Xenopus* Embryo

Marker	Expression-stage	Expression-pattern	References
Xsox17α	MBT-st 35	Pan-endodermal	Hudson *et al.* (1997)
Xsox17β	MBT-st 25	Pan-endodermal	Hudson *et al.* (1997)
Mix.1	MBT-st 13	Endodermal, but transiently overlaps with the mesodermal marker Xbra.	Rosa (1989); Lemaire *et al.* (1998)
Mixer	St. 9.5-13	Endodermal	Henry and Melton (1998); Yasuo and Lemaire (1999)
Bix1-4 (Milk = Bix2)	MBT-st 13	Vegetal pole/endoderm but also in marginal zone.	Tada *et al.* (1998); Ecochard *et al.* (1998)
Endodermin	St 10-	Pan-endodermal + Organiser	Sasai *et al.* (1996)
HNF1α	St 9-	Gastrula endoderm, later in liver gut and pronephros	Bartkowski *et al.* (1993); Weber *et al.* (1996)
HNF1β	St 10	In vegetal pole at gastrula stages, midgut and hindgut at tailbud. Also in neuroectoderm from stage 13, pronephric anlage and Wolffian duct.	Demartis *et al.* (1994)
HNF3α (Xfkh2)	Expressed by st. 10.5	Vegetal pole and marginal zone at gastrula stages, foregut, brain, hypochord and floor plate at taibud	Bolce *et al.* (1993)
HNF3β	Expressed by st. 10	Dorsal marginal zone/Organiser at gastrula, pharynx, ventral midbrain and floor plate at tailbud.	Altaba *et al.* (1993)
GATA 5	Gastrula-	Pan-endodermal at gastrula, in endoderm and developing heart at tailbud	R. Patient, pers. comm.
XHex	Gastrula-	Anterior, dorsal endoderm at gastrula, later in the liver, thyroid precursor and vascular endothelium.	Newman *et al.* (1997); Jones *et al.* (1999)
Xlhbox8	Useful from st. 30-	Pancreas and duodenum.	Wright *et al.* (1989)
IFABP	St 32-	Intestine	Henry *et al.* (1996); Ishizuyaoka *et al.* (1994)

Melton, 1998; Hudson *et al.*, 1997; Jones *et al.*, 1993). The pancreas marker *Xlhbox8* is expressed in only a part of this explant, indicating that there is a prepattern in the developing endoderm (Gamer and Wright, 1995; Zorn *et al.*, 1999). In spite of the appearance of both general and regional endoderm markers, vegetal explants do not adopt gut morphology; this requires the presence of mesoderm (see below).

When vegetal cells are disaggregated and cultured they initiate endoderm marker expression, but this largely decays by mid-gastrula, indicating that cell-cell signaling is needed to maintain endodermal gene expression (Clements *et al.*, 1999; Yasuo and Lemaire, 1999). This is consistent with the results of Heasman et al. (1984), who tested the determination of individual vegetal cells by transplanting them into the blastocoel, and found that the endodermal fate was not fixed until gastrulation. Hence, while the vegetal pole of the embryo is preprogrammed to form endoderm by endogenous maternal molecules and these molecules generate at least some of the gut pattern (i.e., localized formation of pancreas/duodenum), local cell signaling is required for endodermal differentiation to ensue, and interaction with mesoderm is required for there to be gut morphology. The next sections consider the molecules responsible in more detail.

C. Initiation of Endodermal Development by Maternal Gene Products

Since vegetal poles can autonomously express endodermal markers when isolated before MBT (Gamer and Wright, 1995), there must be maternal agents that cause endodermal expression in this region. There are currently two classes of candidates for maternal molecules initiating endoderm development, transforming growth factor-β (TGFβ) family members and the T-box transcription factor VegT.

1. Signaling Molecules that Induce Endodermal Gene Expression

It has been known for some time that various members of the TGFβ family induce animal cap cells to express the full range of known embryonic endodermal markers (Gamer and Wright, 1995; Henry *et al.*, 1996; Henry and Melton, 1998; Hudson *et al.*, 1997; Jones *et al.*, 1993). A gut epitope recognised by monoclonal 4G6 was also induced by FGF (Jones *et al.*, 1993), but otherwise FGF does not significantly induce endoderm-specific markers. To date, the TGFβ family members known to induce endoderm are activin and Vg1 (Gamer and Wright, 1995; Henry *et al.*, 1996; Henry and Melton, 1998; Hudson *et al.*, 1997; Jones *et al.*, 1993), the Nodal family members Xnr1 and 2 (A. M. Zorn *et al.,* 1999, unpublished), BMP4 (unpublished) and Derrière (Sun *et al.,* 1999). Since work on mesoderm induction has shown that the vegetal pole is the source of the TGFβ-family signaling that induces the mesoderm at the equator of the embryo, molecules of this family are in the right place, at the right time, potentially to induce endoderm in an autocrine fashion, although there is a question as to whether the signal is strong enough to induce endoderm (see below).

Most of the various TGFβ family inducers that can initiate endodermal gene expression in the animal cap test system are zygotic, and currently only activin and Vg1 are known to be maternally expressed. While activin mRNAs are absent from the egg, activin-like proteins are present (Asashima *et al.,* 1991), but nothing is known of their activity or localization. The only TGFβ family member known to be present as an mRNA is Vg1, and it is located in the vegetal pole, as one would expect of an endodermal determinant (Vize and Thomsen, 1994; Weeks and Melton, 1987). This molecule presents a puzzle, since the mRNA has no biological activity when expressed in the animal cap even though the processed ligand domain will induce endoderm, as well as mesoderm (Dale *et al.,* 1993; Henry *et al.,* 1996; Thomsen and Melton, 1993). This seems to result from failure to process Vg1 precursor. Although processed Vg1 was detected by Tannahill and Melton (1989), the amounts were small and could result from degradation during extraction; others failed to find this short form (Dale *et al.,* 1993). It is possible that in development there is very tightly controlled local Vg1 processing, and this idea is supported to some degree by observations on zDVR1, a zebrafish homolog of Vg1. Like the *Xenopus* protein, zDVR1 is not processed in zebrafish, nor is *Xenopus* Vg1, but surprisingly zDVR is processed in *Xenopus* embryos into a biologically active ligand (Dohrmann *et al.,* 1996). This suggests the existence of processing control regions that are subtly different in the two species.

Specific dominant negative Vg1 ligands have been reported to disrupt the formation of anterior mesoderm and to prevent the appearance of the pancreas marker, *Xlhbox8,* suggesting that Vg1 may have a role in both endoderm and mesoderm formation in *Xenopus* (Joseph and Melton, 1998). However, the specificity of these reagents has not been tested against the full range of TGFβ family members currently known to be present in the embryo, so further work is needed to firmly implicate Vg1 in endoderm formation.

Supporting the view that TGFβ signaling is involved in endoderm formation is the observation that a dominant negative TGFβ receptor of promiscuous specificity inhibits the expression of some late gut markers (Gamer and Wright, 1995; Henry *et al.,* 1996), and diverts vegetal pole cells from an endodermal to a neural and mesodermal fate (Hemmati-Brivanlou and Melton, 1994). The problem with these experiments in the current context is that they do not distinguish a maternal role for TGFβ family members from a zygotic one, acting as reinforcing agents downstream of locally acting transcription factors. These later effects undoubtedly occur and are discussed below.

2. VegT

As far as transcription factors are concerned, the single current candidate for a maternal endoderm determinant is the T-box molecule VegT, which induces animal cap cells to express both meso- and endodermal markers (Horb and Thomsen, 1997; Lustig *et al.,* 1996; Stennard *et al.,* 1996; Zhang *et al.,* 1998a; Zhang and King, 1996). VegT is localized predominantly but not entirely in the vegetal pole. The latter may explain why low levels of endodermal markers are initially expressed in the animal cap of the blastula. It was shown that destruction of 90–95% of the maternal VegT ablates endodermal markers, when assayed at stage 32 (Zhang *et al.,* 1998a), indicating that VegT has a critical role in endoderm formation.

Does VegT act directly, or by inducing TGFβ family signaling? It seems that it does both (Clements *et al.,* 1999; Yasuo and Lemaire, 1999). That it induces *Xsox17* cell autonomously has been shown by cell disaggregation and inhibitor experiments, but this direct effect is short-lived unless reinforced by signaling. By early stage 9 blastula VegT induces *Xsox17* expression in cells disaggregated before MBT, but *Xsox17* disappears by late stage 9. This dependence on cell contact can be correlated with the fact that vegetal pole expression of *Sox17* is activated in cells disaggregated before the late blastula stage, but then immediately declines (Clements *et al.,* 1999; Yasuo and Lemaire, 1999); it becomes stable by late blastula. The induction of *Xsox17* by VegT is largely inhibited by the promiscuous TGFβ receptor inhibitor, tAR (see above), indicating that TGFβ family signaling is at least partly responsible for these reinforcement interactions (Clements *et al.,* 1999).

In contrast to endoderm induction, the ability of VegT to induce mesoderm depends wholly on TGFβ family signals (Clements *et al.,* 1999; Yasuo and Lemaire, 1999), including Xnr1, Xnr2, Xnr4, Activinβ_B, and Derrière (Clements *et al.,* 1999; Sun *et al.,* 1999). However, as far as endoderm is concerned, the VegT-generated TGFβ signal is likely to be too weak to induce endoderm on its own, because the number of molecules of the TGFβ family signals that are induced by VegT is very low (Clements *et al.,* 1999) and endoderm induction requires levels of TGFβ class signaling that are higher than needed for mesoderm. Nevertheless, these low potency signals are presumably sufficient for endoderm maintenance after initial induction by VegT.

A model for the way these reinforcing signals operate is provided by a clonal analysis of VegT-injected cells (Clements *et al.,* 1999). This is idealized in Fig. 2A. Single A-tier animal blastomeres were injected with VegT mRNA at the 32-cell stage, along with mRNA encoding nuclear-localized β-galactosidase as a lineage marker, and the expression of markers was examined at the early gastrula stage. After injection into small 32-cell blastomeres it would be predicted that all daughter cells would inherit similar levels of the injected VegT, so if induction was autonomous it would be anticipated that all β-gal-positive cells would express the endodermal marker *Xsox17.* Instead, only cells at the center of the clone do so, the outer part of the clone and beyond expressing the mesodermal marker *Xbra.* Because of cell mixing, some cells become separated from the clone and these may even express neither the endodermal nor the mesodermal marker. This is illustrated in Fig. 2A. Given the results of the disaggregation and inhibitor experiments just discussed, we interpret the results as follows. Initially all the injected cells express *Xsox17,* as well as a full range of signaling molecules (Fig. 2, panel 2). The latter only build up at high enough levels to support stable endodermal gene expression in the center of the clone (Fig. 2, panel 3). At the periphery, and beyond for a limited distance, there is sufficient for mesodermal differentiation, and cells isolated from the clone appear not to be able to build up enough even to form mesoderm. In this connection it is interesting that isolated VegT-injected cells show a marked upregulation of the cell autonomously induced signals downstream of VegT, that is Xnr4, Activin$\beta_{B,}$ and Derriére, as well as *Xsox17,* to some extent. This suggests that there is a signal in the animal cap that opposes the effects of VegT. This signal may be present to counteract the low levels of VegT present in the animal cap (Stennard *et al.,* 1996), and it may also act in limiting the borders of mesoderm and endoderm in the embryo.

The properties of VegT in establishing domains of endodermal and mesodermal differentiation have obvious relevance to the way this vegetal molecule might pattern the embryo. However, although ablation of maternal VegT indicates that it is important in both endodermal and mesodermal differentiation, there are still questions to be answered. For example, so far it is known that VegT-depleted embryos do not express endodermal markers at the tadpole stage (Zhang *et al.,* 1998a), but is this also true at the initiation stage, before the reinforcement cell interactions of the late blastulae? Another question is why VegT is much slower to induce Xnr1, Xnr2, and mesodermal markers than their normal appearance in the embryo (Clements *et al.,* 1999). This indicates either that there are inhibitors of VegT in animal caps, there are faster-acting parallel pathways in the embryo, or that there are molecules that synergize with VegT in the vegetal pole; if the last two apply, the molecules responsible are unknown. Evidence for pre-MBT inductive events, which may constitute a faster parallel, possibly

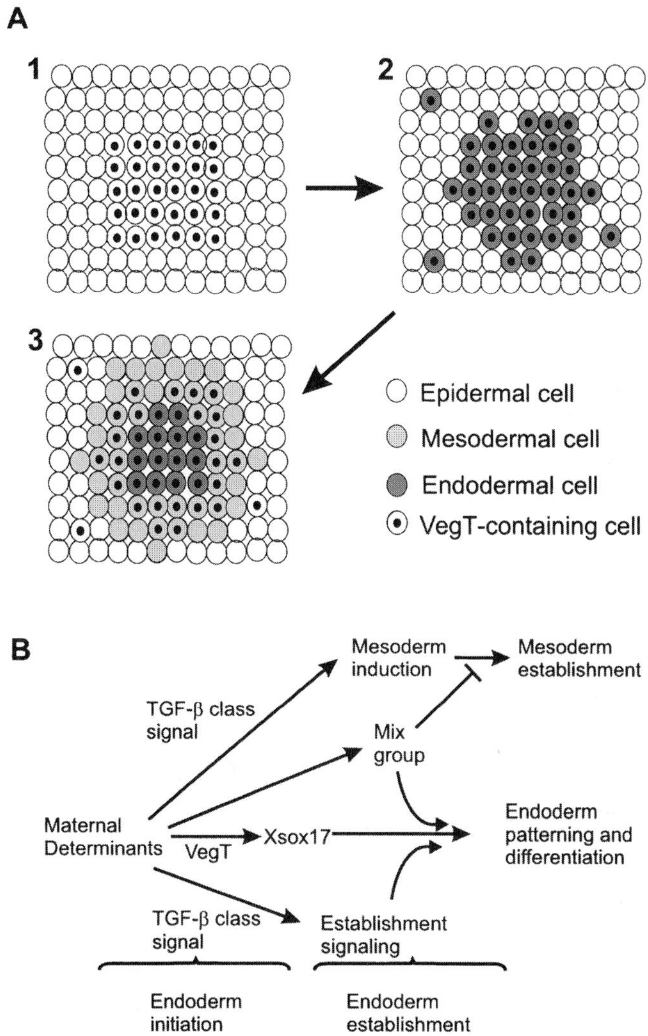

FIG. 2 (A) A model experiment to illustrate patterning by VegT. VegT mRNA was injected into an animal pole of the 32-cell stage *Xenopus* embryo. The resultant clone was marked by co-injected nuclear β-galactosidase (cells marked with dots, 1). At the gastrula stage (3) the clone expresses endodermal *Xsox17* in the center and mesodermal *Xbra* at the periphery and beyond, some scattered clonal members being negative. (2) At an earlier stage it is hypothesized from other data that Xsox17 would be expressed strictly cell-autonomously in response to VegT and mesoderm would not yet have formed (Clements *et al.,* 1999). (B) Summary of interactions involved in endoderm formation in *Xenopus.*

synergizing pathway, was obtained by several labs (Ding *et al.,* 1998; Jones and Woodland, 1987).

Finally we can ask whether the VegT-based pathway for forming endoderm is general. As we discuss below, while TGFβ class signaling is important in other vertebrates, there is as yet no maternally expressed T-box gene identified elsewhere, so it is possible that this process is an amphibian or even anuran specialization. The overall scheme of events involved in the early phase of endoderm formation in *Xenopus* are summarized in Fig. 2B.

D. Zygotic Molecules and Initiation of the Endoderm

In this section we consider the zygotic transcription factors believed to be involved in initiating the formation of endoderm as a whole in the blastula.

1. Xsox17

The first zygotic molecules to be identified with a definite role in initiation of endoderm formation were the vegetal Sry-related transcription factors *Xsox17* α and β, which for our purposes have the same activity and will be referred to collectively as *Xsox17* (Hudson *et al.,* 1997). Transcripts are present at very low levels in all regions of the stage 7.5 blastula, that is, before MBT (Clements *et al.,* 1999), but they are undetectable in the unfertilized egg (unpublished). After MBT (stage 8.5) transcript levels rise dramatically in the entire vegetal pole and in the equator, but the low stage 7 levels persist in the animal pole, only to disappear at stage 10.5, although the amount in the animal poles varies in different batches of embryos. The disappearance of *Xsox17* from the animal pole requires contact with the rest of the embryo, because animal caps taken from blastulae show persistent low level *Xsox17* expression through neurula stages. This persistence of low-level expression is also true of several other early endoderm markers, such as *endodermin,* and may relate to the variable low amounts of *VegT* mRNA found in blastula animal caps (Stennard *et al.,* 1996). The refinement of the pattern during early gastrulation means that *Xsox17* genes become very specifically expressed in the future endoderm by this time (Hudson *et al.,* 1997; Henry and Melton, 1998; Zorn *et al.,* 1999). Hence, *Xsox17*α and β mark the presumptive endoderm of the gastrula, although there are apparently a few cells between the *Xsox17* and *Xbra*-expressing cells that have not been assigned to either endoderm or mesoderm (Zorn *et al.,* 1999). *Xsox17* genes continue pan-endodermal expression until tailbud stages; *Xsox17*β ceases embryonic expression at stage 25, while *Xsox17*β continues to be widely expressed in endoderm until stage 35.

In the adult *Xsox17β* is expressed in the anterior gut (mouth, oesophagus, liver, gall bladder), as well as the testis and kidney. *Xsox17α* is expressed in testis, fat body, and skin (in these unpublished analyses the CNS was not examined). Mouse *Sox17* is also expressed in the testis (Kanai *et al.,* 1996), but it is not known if it is expressed in embryonic endoderm. mRNAs synthesized from these genes *in vitro* and microinjected into fertilized eggs induce animal caps to express both early and late endoderm markers (*endodermin, HNF1β, Xlhbox8, IFABP.*) Conversely, a repressor derivative, made by fusing the repressor domain of Engrailed to the Xsox17 DNA-binding domain (Xsox17::En^R), represses growth factor induction of endodermal markers and eliminates late markers of endoderm in the whole embryo. Early markers are partially blocked in the vegetal pole, which may indicate there are parallel pathways of endoderm formation, or that the injected Xsox17::En^R cannot fully compete with the very high levels of endogenous *Xsox17* (Hudson *et al.,* 1997). Given that *Xsox17* is expressed from the first stages of genome activation, it seems that its expression is a major step in endoderm formation.

2. Other HMG Domain Proteins

Another transcription factor that induces endoderm in *Xenopus* is murine Tcf-4 (Lee *et al.,* 1999). The related molecules Tcf1 and 3 are involved in the formation of the Spemann organizer (acting downstream of the Wnt/β-catenin pathway), the anterior part of which is in the endoderm and responsible for brain induction. Whether Tcf1 and 3 can induce endoderm has not been reported. The Tcfs, like SOX proteins, are members of the HMG box family and bind to a similar DNA motif.

3. Mix-Related Proteins

A second class of genes that induce endoderm is the large *Mix* group of paired box-like homeodomain transcription factors, which includes *Mix.1* (Lemaire *et al.,* 1998; Rosa, 1989), *Mix.2* (Vize, 1996), *Milk/Mix.3/Bix2* (Ecochard *et al.,* 1998; Mead *et al.,* 1998; Tada *et al.,* 1998), *Mixer* (Henry and Melton, 1998), and *Bix1* (*Mix.4*), *3,* and *4* (Tada *et al.,* 1998). There are detailed expression patterns published only of certain members. *Mix.1, Milk,* and *Mixer* transcripts are present in the equatorial part of the vegetal pole, and *Mix.1* and *Milk,* but not *Mixer,* initially extend into the mesodermal region (Henry and Melton, 1998; Lemaire *et al.,* 1998). Such fine resolution *in situs* have not been reported for *Bix1,* but it appears to be found throughout the vegetal pole (Tada *et al.,* 1998). *Mix.1* and *Bix1* transcripts are present from MBT, but *Mixer* transcripts only appear at the start of gastrulation. *Mix.1* and *Milk* are immediate early response genes for activin

(Ecochard *et al.,* 1998; Rosa, 1989). *Bix1* is directly induced by Xbra, so may also be a VegT target, while the *Bix4* promoter contains T-box binding sites (Tada *et al.,* 1998) and is indeed a direct VegT target *in vivo* (Casey *et al.,* 1999).

The reported results of the activities of the Mix-related proteins in animal caps are not wholly consistent in the literature, which is probably explained by the relatively weak endodermal inductions usually observed. *Mix.1* has been reported not to induce *Xsox17* or *endodermin* by some labs (Henry and Melton, 1998; Mead *et al.,* 1998), but we have found that it does indeed induce *Xsox17,* albeit weakly (for example, compared to activin; unpublished), and it also induces *endodermin* (Ecochard *et al.,* 1998; unpublished). *Milk* and *Mixer* induce these markers a little more strongly, and these, unlike *Mix.1,* also induce late endoderm markers (Henry and Melton, 1998; Mead *et al.,* 1998). While it has been reported that *Mix.1* opposes the stronger induction of endoderm by *Milk* (*Mix3*), this conclusion must be viewed with caution, since these authors failed to find endoderm induction by *Mix.1.* It has been suggested that *Mix.1* is involved in repressing mesodermal gene expression (indirectly, since it is an activator) in the endodermal domain, and hence in establishing the mesoderm/endoderm boundary (Lemaire *et al.,* 1998). Apparently *siamois* cooperates in this, and also makes *Mix.1* into a much more potent inducer. However, *in vivo* this could only apply to the dorsal side of the embryo (perhaps the *vent* genes, which also contain homeodomains, are involved ventrally). It is possible that such synergies apply to the whole *Mix* group.

Xsox17 does not induce *Mix.1* or *Mixer,* which might seem to make them upstream of *Xsox17.* However, *Mixer* cannot be required for the initiation of *Xsox17* expression because it is first expressed after *Xsox17*; Henry and Melton (1998) suggest that it is actually involved in its maintenance (although see zebrafish section). *Mix.1* and *Bix1* are expressed earlier, but not before *Xsox17,* so they may also have maintenance roles as well as being involved in defining the mesodermal/endodermal border.

E. Reinforcement of Endodermal Gene Expression

As already discussed, disaggregated vegetal pole cells initiate endoderm differentiation, as defined by *Xsox17* and *Mix.1* expression, but signaling is immediately required to maintain endodermal gene expression (Clements *et al.,* 1999; Yasuo and Lemaire, 1999). *Mix.1* and *Mixer* require signaling to maintain expression, so they could mediate the effects of the TGFβ reinforcement signals in maintaining *Xsox17* expression. Candidate signaling molecules include members of the TGFβ family, including Xnr1, Xnr2, Xnr4, activin βB, and Derrière, which are all present in the vegetal pole at this stage. A requirement for TGFβ class signaling in establishing endodermal differentiation also fits with the observation that expression of the TGFβ receptor antagonist tAR in vegetal blastomeres causes them to become mesodermal or neural (Hemmati-Brivanlou and Melton, 1994). Similarly, Osada and Wright (1999) found that a dominant negative Xnr2 ligand (which also blocks Xnr1 and Xnr4 signaling) inhibited endodermal gene expression. Hence, vegetal cells generate a signal that reinforces endodermal development. Yasuo and Lemaire (1999) hypothesize that this community effect is important in generating precise borders between the endoderm and the adjacent mesoderm.

During this phase the Mix-related homeodomain protein Mixer is expressed vegetally, and has also been proposed to play a role in maintenance of *Xsox17* expression (see above; Henry and Melton, 1998). *Xsox17* expression persists beyond stage 12, to stage 25 (Xsox17β) and 35 (Xsox17β), a period when Mix proteins are no longer expressed. The factors responsible for maintaining *Xsox17* expression at these later stages remain to be determined.

F. First Steps in Endoderm Patterning

A considerable amount is known about the early patterning of the mesoderm. So far studies on endoderm have focused on how the same processes that pattern the mesoderm are involved in endoderm. The mesoderm of the embryo is dorsoventrally patterned by the Spemann signaling center (the organizer), which acts principally by generating inhibitors of BMP, such as Noggin and Chordin (Moon and Kimelman, 1998). The part of the organizer with the most anterior fate lies in the anterior endoderm (Bouwmeester and Leyns, 1997). The Spemann signaling center is itself induced by the Nieuwkoop center, which lies in the dorsal endoderm. There is evidence that the activity of these centers is involved in endodermal patterning. The formation of the Nieuwkoop center involves a series of events. First there is a UV-sensitive cortical rotation at the one-cell stage that generates elevated β-catenin levels dorsally. This creates the Nieuwkoop center, where β-catenin acts as a cofactor with Tcf/Lef HMG box transcription factors to activate transcription factor genes such as *siamois.* These lead both to further signaling and induction of a cascade of Spemann organizer transcription factors (e.g. *Goosecoid, Lim-1*), which in turn leads to the mesoderm dorsalization and the neural induction signals of the organizer.

Regional endoderm markers are somewhat limited in *Xenopus. Xlhbox8* is a marker of the pancreas, homologous to the insulin-controlling gene *pdx-1* (*stf-1, idx-1, ipf-1*) (Peshavaria *et al.*, 1994; Wright *et al.*, 1989), which is essential for pancreatic development in mice (Ahlgren *et al.*, 1996; Offield *et al.,* 1996). In adult *Xenopus* it is expressed in the adult pancreas, duodenum, and stomach (unpublished; its orthologue is also expressed in the zebrafish pyloric caeca, Milewski *et al.,* 1998). *Xlhbox8* has been used in a number of studies of patterning of the embryonic endoderm. Gamer and Wright (1995) showed that this marker was expressed in a dorsal domain in explanted vegetal poles, and hence that region-specific endodermal determination was occurring before MBT. They also found that UV irradiation downregulated its expression, implicating cortical rotation as the source of the regionalization. This was supported by Henry *et al.* (1996), who showed that cortical rotation is required for *Xlhbox8* expression, but not the more posterior *IFABP* (*intestinal fatty acid binding protein*). IFABP mRNA is found in the duodenum and upper intestine from about stage 35 onward (Henry *et al.,* 1996; unpublished); thus, its expression overlaps that of *Xlhbox8.* However, while Henry *et al.* (1996) demonstrated that inhibition of fibroblast growth factor (FGF) signaling by the dominant negative receptor XFD reduced *Xlhbox8* expression, Gamer and Wright (1995) found that FGF itself had the same effect. *IFABP* was not affected by inhibition of FGF or TGFβ signaling, whereas *Xlhbox8* was downregulated by both. Although these results on *Xlhbox8* may appear incompatible, it is possible that both FGF activity and its absence are required at different stages of pancreas development.

As mentioned, the anterior part of the organizer could also pattern the anterior endoderm. A major activity of the organizer is BMP inhibition via a plethora of molecules. It has been reported that BMP inhibition by Noggin induces the endodermal marker *endodermin* (Sasai *et al.,* 1996), but *endodermin* is also expressed in the mesodermal organizer (Sasai *et al.,* 1996) and generally in adult frog organs (unpublished), so it is not the ideal endoderm marker. Its induction by Noggin is weak and transitory and other endodermal expression, e.g. *Xsox17,* is absent (Lemaire *et al.,* 1998; unpublished). Although Noggin was reported to induce *Xlhbox8* (Sasai *et al.,* 1996), other labs have failed to reproduce this result (Lemaire *et al.,* 1998; unpublished), so it is not clear that Noggin produces a true endodermal induction. Nevertheless, as already mentioned, there is evidence for a role of the cortical rotation/β-catenin system in liver and pancreas differentiation, and hence presumably the antero-posterior patterning of the gut depends on the Nieuwkoop and Spemann centers. It has also been argued that BMP inhibition is necessary, but not sufficient for induction of the liver marker *XHex* (Zorn *et al.,* 1999). These workers found that a number of early anterior endoderm markers can be induced by molecules of the Nieuwkoop center, although this requires the cooperation of general vegetal molecules. Thus ectopic expression of the dorsal determinants β-catenin and Siamois in ventral vegetal poles led to activation of dorsal endoderm specific markers (*XHex, cerberus, chordin, Siamois,* and *gsc*). The endogenous expression of some of these markers is dependent on cell contact (*XHex*, *cerberus*), while others are activated in disaggregated cells (*gsc*, *Xwnt8;* Lemaire and Gurdon, 1994). Overall, it has been concluded that anterior endoderm markers are activated in response to a general endoderm induction pathway (such as the Vg1 or VegT pathways outlined above), plus dorsal determinants activated by cortical rotation. In this way the expression of these markers would be limited to dorsal vegetal cells.

In summary, the asymmetrical redistribution of maternal determinants by cortical rotation is necessary for initiation of liver and pancreas development. There is a puzzle about this result. UV irradiation stops cortical rotation; as a result Nieuwkoop center molecules remain concentrated at the vegetal pole, where they induce *siamois* and other Nieuwkoop and Spemann center genes (Brannon and Kimelman, 1996; Darras *et al.*, 1997; Laurent *et al.*, 1997; Medina *et al.*, 1997). Hence, it is not clear why pancreas and liver do not form in a vegetally displaced position. Possibly there are maternal components in the equator of the egg that are also necessary to cooperate with the Nieuwkoop center molecules. Whatever these are, they are not mesodermal, because the UV inhibition of *Xlhbox8* expression is shown by isolated vegetal poles that totally lack mesoderm. Alternatively, there could be undiscovered Nieuwkoop center components or there might be inhibitors of pancreas at the extreme vegetal pole.

G. Patterning of the Gut—the Role of Mesoderm

Classical work on endodermal differentiation was principally on urodeles and relied on morphological appearance. It therefore did not detect any inherent capacity of vegetal cells to differentiate by molecular criteria alone. Okada (1957), for example, showed that anterior vegetal cells of gastrulae and neurulae of *Triturus pyrrhogaster* fated to form pharynx did not do so when cultured in an epidermal envelope, but did do so if combined also with anterior or posterior neural plate, or dorsal mesoderm. He also showed that the differentiated state of the presumptive endoderm was labile even in the neurula, because when neurula pharyngeal endoderm was combined with lateral plate mesoderm, histologically recognizable intestine, liver, pancreas and stomach formed, i.e. there was a posterior transformation. Later it was shown that presumptive mid-and hindgut also required mesoderm to develop (Okada, 1960), and the addition of lateral plate allowed intestine to develop from posterior vegetal fragments and favored its development from midvegetal fragments. In contrast, dorsal mesoderm and neural plate tended to anteriorize both. While in some of these experiments neural plate was explanted with the endoderm, the effects were attributed to "mesenchyme" that appeared within the epidermal wraps. In fact, Okada in all cases attributed the anterior development to environment provided by the mesenchyme, which he believed was not mesodermally derived. It would be interesting to know more about the status and origin of these cells with modern methods. Presumably the neural or mesodermal tissues co-explanted with the vegetal fragments produced signaling molecules that permitted or diverted the fate of the endoderm, but as yet there is no evidence as to what they were.

These experiments were in fact foreshadowed in early transplantations by Holtfreter, Mangold, and Balinsky. When endoderm was removed from one part of the presumptive endoderm and placed in another, it was possible to entirely reverse the antero-posterior polarity of the endoderm. It was also observed that intact gut develops when pieces were removed from the presumptive endoderm that *in vivo* would themselves only have formed part of the gut (see Balinsky and Fabian, 1981; Okada, 1957).

III. Formation of the Gut

A. C. elegans

The nematode worm *C. elegans* has provided a simple genetic model for the formation of an endoderm lineage. In the early embryo, cell fate is initially determined by maternal factors inherited from the oocyte, although very soon cell interactions also become important. All of the endoderm (the 20 cells that constitute the intestine) is derived from the E founder cell, born in the third cell division (Fig. 3) (Deppe *et al.,* 1978; Sulston *et al.,* 1983). The genetic pathway for the segregation of the endodermal fate into this cell is outlined below.

1. Maternal Pathways

There is no maternal endoderm-specific determinant in *C. elegans.* Instead, two maternal pathways—each of which plays a general role in pattering the embryo—converge to produce the endoderm lineage.

a. SKN-1, PIE-1 and the Determination of EMS Fate SKN-1 is a bZIP/ homeodomain hybrid transcription factor that is expressed both maternally and zygotically (Blackwell *et al.,* 1994; Bowerman *et al.,* 1992). It is a nuclear protein, distributed unequally in early embryos (Bowerman *et al.,* 1993); at the first cell division, P1 accumulates more SKN-1 than AB (Fig. 4). Several cell-fate regulators are partitioned unequally at this stage in a cytoplasmic rearrangement event triggered by sperm entry and requiring

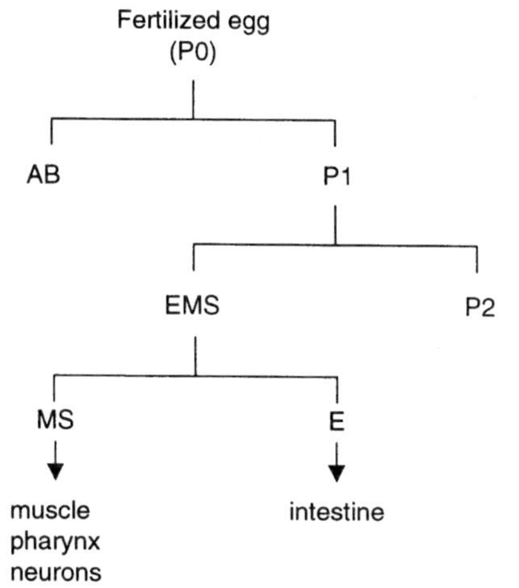

FIG. 3 Abbreviated cell lineage of *C. elegans.*

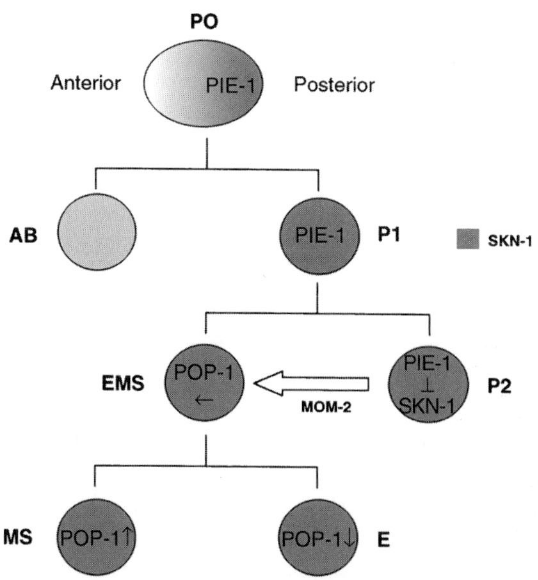

FIG. 4 Formation of the endoderm lineage in *C. elegans.* Distribution of SKN-1 proteins is shown by the stippling. In P_0 PIE-1 is posteriorly localized. POP-1 is expressed in EMS, but down-regulated in E by MOM-2 (a Wnt family signal). See text for further details.

the products of the six *par* genes and *mex-1* (Rose and Kemphues, 1998; Seydoux and Schedl, Chapter 4).

While the higher level of SKN1 is inherited equally by both daughters of P1 - P2 and EMS, SKN1 function is not required in P2, which will give rise to the germline. The Zn-finger protein PIE-1 represses SKN-1 activity in this cell (Mello *et al.,* 1992), but SKN-1 is essential for the EMS cell to develop. In *skn-1* mutants the derivatives of both E and MS—muscle, pharynx and intestine—are missing. SKN1 thus acts as a determinant of EMS fate and its function is likely to be to activate transcription of zygotic genes in both the E and the MS lineages (Bowerman *et al.,* 1993).

The *pie-1* mutant phenotype is the opposite of the *skn-1* phenotype. In *pie-1* mutants, SKN-1 is not repressed in P2; two EMS cells and hence two gut lineages develop, since P2 adopts the EMS fate. PIE-1 is found in the nucleus and, like SKN-1, it is distributed unequally at the first cell division, partitioning into P1. Unlike SKN-1, however, PIE-1 continues to segregate into the posterior blastomere at each cell division in this lineage, and consequently it becomes restricted to the germline. Its function appears to be down-regulation of zygotic transcription in the germline blastomeres; in *C. elegans* there is no zygotic transcription in the germline for a considerable period, and this suppression, which is associated with a modification of RNA polymerase II, is dependent on PIE-1 (Seydoux and Dunn, 1997; Seydoux *et al.,* 1996; Tenehaus *et al.,* 1998). *pie-1* mutants show zygotic transcription within the germline blastomeres and subsequently lack germline cells (Mello *et al.,* 1992).

b. POP-1 and Formation of the Endoderm Lineage At the 4-cell stage an interaction between P2 and EMS polarises EMS such that its daughter cells develop differently, as E (posterior, nearest to P2, producing intestine) and MS (anterior, generating muscle and pharynx) (Goldstein, 1992, 1993, 1995). EMS cells isolated before this induction divide to produce two MS cells (Goldstein, 1993, 1995).

At least seven genes—*mom1-5, lit-1* and *pop-1*—are known to be required for this polarization (Lin *et al.,* 1998; Rocheleau *et al.,* 1997; Thorpe *et al.,* 1997). Mutants in the *mom* genes and *lit-1* have variably reduced endoderm and increased amounts of mesoderm, while the converse is true for *pop-1* mutants. Of these genes, *mom-1* is related to *porcupine* (which is required for secretion of Wingless-type signals in *Drosophila;* Kadowaki *et al.,* 1996), *mom-2* is predicted to encode a Wnt/wingless family member, while *mom-5* is related to the Wnt receptor, *frizzled,* all of which are components of a Wnt signaling pathway (Fig. 5) (Rocheleau *et al.,* 1997). POP-1 is an HMG domain protein, similar to vertebrate Tcf-1 and Lef-1 and *Drosophila* Pangolin/dTcf, all of which are targets of Wnt signaling pathways (Lin *et al.,* 1995). In addition, RNA-mediated interference of

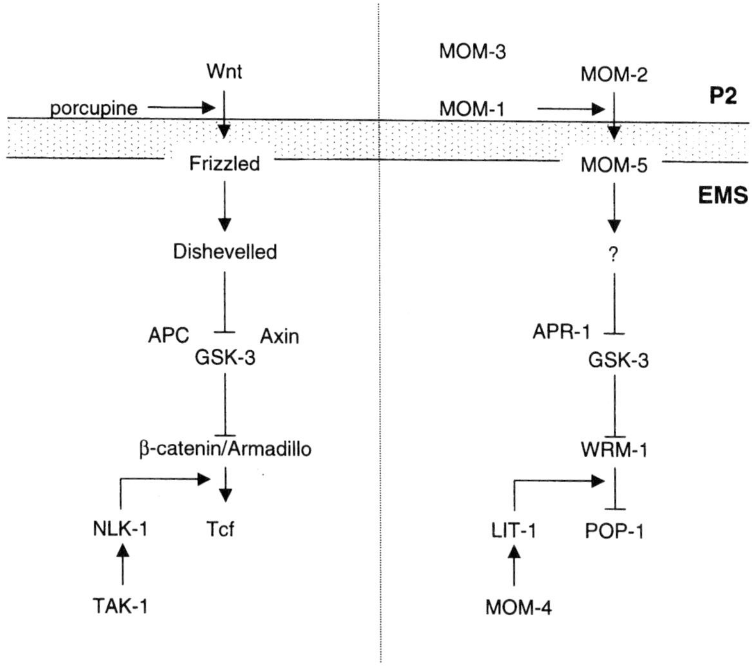

FIG. 5 Comparison of the canonical Wnt/Wingless signaling pathway of vertebrates and *Drosophila* with the MOM pathway of *C. elegans* (see text for details).

the *apr-1* (APC-related), *wrm-1* (β-catenin/Armadillo-related) and *gsk-3* transcripts also resulted in endoderm defects, although similar experiments on two *dishevelled*-related genes did not (Rocheleau *et al.,* 1997; Schlesinger *et al.,* 1999).

Figure 5 compares the canonical Wnt signaling pathway with that described in *C. elegans.* Note that, unusually, the effect of Wnt signaling in the nematode embryo is a reduction in POP-1 in the target cell, whereas normally the consequence of this pathway is activation of a Tcf/Lef homolog.

There is, however, a complication to this model: *mom-4* and *lit-1* are not part of the Wnt signalling pathway but are components of the MAPK cascade. LIT-1 is a MAPK-like protein, related to the *Drosophila* protein Nemo and vertebrate NLK (Nemo-like kinase), while MOM-4 is a MAP3K with homology to vertebrate TAK-1 (TGFβ-activated kinase, Meneghini *et al.,* 1999; Rocheleau *et al.,* 1999). It has been demonstrated that the Wnt signaling pathway can result in activation of LIT-1, and that activated LIT-1 can then phosphorylate POP-1, but the significance of this phosphorylation is not yet known. Also, an interaction between LIT-1 and WRM-1 is required for down-regulation of POP-1 activity (Rocheleau *et al.,* 1999). Hence, anterior-posterior (A-P) polarization appears to involve a Wnt signaling pathway potentiated by a MAPK pathway.

The consequence of the convergence of these pathways is the down-regulation of POP-1 activity in the E cell. This unequal distribution of POP-1 is essential to distinguish the anterior MS cell from the posterior E cell. As might be expected, the EMS cell in *pop-1* mutants gives rise to two E-type cells and two endoderm lineages, while the resulting embryos lack the MS cell derivatives, muscle and pharynx. This pathway is not specific to the polarization of EMS, but rather is a general mechanism for the generation of anterior-posterior asymmetries in other sister blastomeres born after the four-cell stage. A monoclonal antibody to a POP1 epitope revealed that POP1 is always present at higher levels in anterior cells than posterior (Lin *et al.,* 1998).

2. Zygotic Genes

The maternal pathways outlined above result in the generation of one cell at the eight-cell stage—the E cell—which will give rise to the entire endoderm. A zygotic endoderm-lineage specific transcript has been identified—*end-1*—and it encodes a member of the GATA family of transcription factors, members of which are also involved in endoderm formation in other species (Zhu *et al.,* 1998, 1997). This suggests a conserved mechanism for the establishment of the endodermal lineage in all triploblasts.

End-1 is specifically expressed in the E cell and its progeny to the 4E stage (i.e. when the E lineage consists of four cells), and its promoter has been shown to interact directly with SKN-1 (Blackwell *et al.,* 1994). Moreover, once expression of *end-1* has initiated, the cell appears to be committed to the endoderm lineage; ectopic expression of *end-1* in nonendodermal blastomeres—AB, P2 and MS—causes their progeny to form intestine (Zhu *et al.,* 1998). Hence, END-1 appears to act as an endoderm determinant. *End-1* was identified within a genomic region known as the endoderm-determining region (EDR), since deletion led to a complete loss of endoderm (Zhu *et al.,* 1997). *End-1* is able to rescue this phenotype in EDR-deficient homozygotes, but RNA-mediated inhibition of *end-1* does not result in progeny lacking intestine. This suggests that there are other genes involved in endoderm formation within the EDR, but *end-1* can compensate for lack of all of them, i.e. there is redundancy.

Recently a second GATA factor, *end-3,* and a member of the steroid hormone receptor family, *end-2,* have been identified within the EDR. Like *end-1* they can each rescue the EDR-deficient phenotype, but their roles within the embryo may be more complex. *End-3* may play a role in mesoderm formation, since it is also expressed in the pharynx, while ectopic expression of *end-2* actually decreases endoderm and increases mesodermal derivatives (Newman-Smith and Rothman, 1999).

A third GATA factor, *elt-2* has also been identified as having a role in endoderm formation (Fukushige *et al.,* 1998; Hawkins and McGhee, 1995). *Elt-2* appears to be downstream of *end-1,* and is expressed exclusively in the E lineage from the 2E stage. Ectopic expression of *elt-2* induces expression of other gut markers, and *elt-2* mutants have lethal gut defects. These GATA-related factors may directly regulate the Forkhead-domain transcription factor gene *Ce-fkh-1* (*pha-4*), which is expressed throughout the intestine (Azzaria *et al.,* 1996; Kalb *et al.,* 1998) and which has GATA sites within its promoter (Kalb *et al.,* 1998). GATA factors also seem to directly control the expression of later gut-specific genes, since the expression of *cpr-1* (a cysteine protease) appears to be induced in the gut via GATA sites in its promoter (Britton *et al.,* 1998).

3. Patterning of the Intestine

The *C. elegans* intestine is a simple tube that has a clear anterior-posterior polarity. Twenty cells are arranged in 9 "ints" or rings (Wood, 1988). Intl, which consists of four cells, is the most anterior and attaches to the pharyngeal-intestinal valve, while int2–9 consist of two cells each. There are also differences in nucleation and the length of microvilli along the A-P axis (Sulston and Horvitz, 1977). Another indication of this regionalization is expression from a deleted *ges-1* (*gut esterase*) promoter. *Ges-1* is normally expressed throughout the gut, although its expression is higher anteriorly than posteriorly. A deleted promoter lacking 50 bp is, however, expressed only in anterior cells (Egan *et al.,* 1995), and so provides a useful indication of A-P patterning.

While down-regulation of POP-1 is necessary for formation of the E-cell lineage, this pathway also plays a general role in patterning the embryo and hence in patterning the endoderm after the establishment of the E lineage. *Pop-1* is present as both a maternal and a zygotic transcript. Embryos lacking maternal *pop-1* have two endoderm lineages (since MS develops as E), but these are divided into anterior and posterior gut, a patterning event that depends upon zygotic *pop-1.* Using RNA-mediated interference to eliminate both maternal and zygotic *pop-1* causes the MS to E transition, but the deleted *ges-1* promoter is no longer expressed in the intestinal cells (Schroeder and McGhee, 1998), suggesting that they are posterior in nature. Zygotic *pop-1* thus appears to be required for anterior gut fate. In other species discussed here, interactions between the endoderm and the mesoderm are important to provide positional information and promote morphogenesis in the developing gut. In *C. elegans* however, mesoderm is not required, since patterning can occur even if mesoderm lineages are ablated (Schroeder and McGhee, 1998). A-P polarity is generated autonomously within the endoderm.

B. *Drosophila*

The *Drosophila* alimentary canal is composed of four regions: the foregut, midgut, hindgut, and Malpighian tubules. Of these, only the midgut is considered to be of endodermal origin, the remainder being derived from the ectoderm, although the status of the Malphigian tubules is debatable. The formation of the larval midgut is discussed here. Skaer (1993) has provided a detailed review of the development of the entire alimentary canal of the larva and adult fly.

1. Establishment of the Midgut Primordia—the Role of Huckebein (Hkb)

The midgut is derived from two primordia, the anterior and posterior midgut primordia (AMG and PMG) located at the poles of the blastoderm embryo (Poulson, 1950; Fig. 6, see color insert). The AMG is just anterior to the ventral furrow, the invagination of the presumptive mesoderm, while the PMG is found at the posterior pole of the embryo, within the amnioproctodeum, which also contains the (ectodermal) hindgut primordium. These primordia internalize during the morphogenetic movements of gastrulation. The cells then undergo an epithelium to mesenchyme transition and migrate toward the center of the embryo, each primordium forming two "arms" that pass on the left and right sides of the yolk. The tips of the arms from the primordia meet and fuse, and then the cells spread dorsally and ventrally over the yolk to enclose it. Finally, they form a columnar epithelium that becomes flattened.

The establishment of the two primordia depends upon the terminal gap-gene, *huckebein.* Huckebein is an Sp1/egr-like Zn-finger protein that has both glutamine-rich (potentially activating) and alanine-rich (potentially repressing) domains (Bronner *et al.,* 1994). It is expressed during the syncytial blastoderm stage at the terminal regions of the embryo, immediately adjacent to the ends of the ventral furrow, i.e. in the region of the AMG and PMG. *hkb* expression is induced indirectly by the Torso signaling cascade, which functions via Ras, Raf, and MAPK, although there is also some bicoid-dependent expression anteriorly (Bronner and Jackle, 1991). Spatially regulated activation of the Torso pathway at the termini of the embryo requires the maternal *torso-like* gene product (Casanova *et al.,*

1995; Furriols *et al.,* 1998), which is set in the vitelline membrane and generates the ligand Trunk in the perivitelline space.

In *hkb* mutant embryos the ectodermal and mesodermal precursors of the embryo expand at the expense of the endoderm, the ventral furrow extending further anteriorly and posteriorly than in wild-type embryos. Both MG primordia are absent. The aminoproctodeal invagination looks normal but does not contain the endodermal anlage (the endodermal markers *prospero* and the *PS11* enhancer trap are absent). The ectodermal hindgut forms but is a closed tube, while the foregut is missing, since Hkb activity is required for invagination of the stomodeum (Bronner *et al.,* 1994).

Huckebein interacts with Snail and Twist to restrict the ventral furrow and to separate the endoderm from the mesoderm (Reuter and Leptin, 1994). *twist* and *snail* are expressed ventrally, induced by high levels of Dorsal protein (Ip *et al.,* 1992; Thisse *et al.,* 1991; Thisse and Thisse, 1992), and are required for formation of the ventral furrow and the entire mesoderm (Grau *et al.,* 1984; Simpson, 1983). However, their domains of expression are more extensive than the region of the furrow (both *snail* and *twist* extend further anteriorly, *twist* extends further posteriorly). Hence, it would seem that some factor is required to repress the function of these proteins outside the domain of the ventral furrow in the regions that give rise to the endoderm. In *hkb* null mutant embryos the ventral furrow extends to encompass the entire *snail* and *twist*-expressing domain, and so Hkb is a good candidate for this factor. Hkb also down-regulates mesodermal targets of Snail and Twist, and represses *snail* transcription at the posterior pole. Anteriorly, Bicoid seems to prevent Hkb from repressing *snail,* since in *bcd* mutant embryos, *snail* expression is restricted anteriorly as well as posteriorly (Reuter and Leptin, 1994).

2. Downstream of Hkb

The gene *serpent* (*srp*)—first identified because mutants fail to undergo germ band retraction (Jurgens *et al.,* 1984)—is a GATA factor also called ABF (Abel *et al.,* 1993; Rehorn *et al.,* 1996). Since *srp* is required for development of the midgut, hematopoesis, fat body formation, and differentiation of the yolk cell and amnioserosa (Rehorn *et al.,* 1996; Reuter, 1994), it may be functionally analogous to the combined activities of several vertebrate GATA family members. *srp* is expressed in both the AMG and the PMG (Abel *et al.,* 1993), but this expression is not seen in *hkb* mutant embryos, although other aspects of *srp* expression are unaffected (in the amnioserosa primordium, yolk nuclei, and haemocyte primordia) (Reuter, 1994). Hence, at least with respect to midgut formation, *srp* is downstream of *hkb.* Soon after invagination of the AMG and PMG, *srp* expression is down-regulated in this region.

Srp function is required for proper differentiation and morphogenesis of the midgut (Reuter, 1994). In *srp* mutants the AMG and PMG invaginate, but they do not subsequently undergo the epithelial-to-mesenchymal transition that normally precedes migration. The PMG does form a cavity lined by a columnar epithelium similar to the hindgut epithelium, but expression of the posterior midgut marker *caudal* is not observed, nor is expression of two midgut-specific enhancer traps. The AMG also fails to form an anterior midgut, leaving the foregut as a blind-ended tube. Reuter (1994) interprets the phenotype of *srp* as a transformation of midgut to hindgut, and hence *srp* may function as an endodermal determinant, much as *end-1* is proposed to do in the *C. elegans* embryo.

3. Migration of the Midgut Primordia—the Role of the Visceral Muscle (VM)

How do the two midgut primordia, migrating from opposite ends of the embryo, find each other in order to fuse and form the midgut proper? The mesenchymal endodermal cells are in contact with the visceral muscle, which could potentially provide directional cues, or at least promote migration.

In *twist snail* double mutants the ventral furrow fails to form and all mesoderm is absent (Grau *et al.,* 1984; Leptin, 1991; Simpson, 1983). In these embryos the AMG and PMG still undergo the transition to mesenchyme, but then fail to migrate and subsequently do not form an epithelium (Reuter *et al.,* 1993; Tepass and Hartenstein, 1994). *tinman* mutants fail to differentiate visceral muscle specifically (Evans *et al.,* 1995), and once again the endoderm does not migrate and the midgut epithelium does not form. Since *tinman* is not expressed in the AMG or PMG and is a transcription factor and therefore functions cell autonomously, any effect on the endoderm in *tinman* mutants must be a consequence of the lack of visceral muscle (Reuter *et al.,* 1993; Tepass and Hartenstein, 1994).

These data imply that the visceral muscle plays two important roles in the formation of the midgut, promoting both migration of the primordia and differentiation of the epithelium. Since AMG cells can migrate all the way to the posterior end of the embryo (in embryos from *tor* mothers, which lack the PMG) it seems unlikely that the VM provides directional cues. Reuter *et al.* (1993) favor a model in which the MG primordia spread over the available surface of the VM until meeting fortuitously in the center of the embryo. The yolk membrane may then provide a further substratum for migration.

4. Specification of PMECs—the Role of Neurogenic and Proneural Genes

Once the midgut primordia have undergone the epithelial-mesenchymal transition, three cell types can be identified within the migrating cell population: PMECs (principal midgut epithelial cells) that will give rise to the midgut epithelium, ICPs (interstitial cell precursors), and AMPs (adult midgut precursors) (Tepass and Hartenstein, 1994). The specification of these three cell types does not require signals from the mesoderm. In *twist*$^-$snail$^-$ double mutants no mesoderm is formed, but PMECs, ICPs, and AMPs still develop. However, in these mutants the midgut epithelium subsequently fails to form because the PMECs do not undergo a mesenchyme-epithelium transition (Tepass and Hartenstein, 1995). In the ectoderm, the neurogenic and proneural genes have roles in the segregation of epidermal and neural precursors. However, many mutants of these genes have been noted to have disturbed midgut formation. In *Delta* mutants, and *Notch* mutants lacking both maternal and zygotic transcripts, all the MG primordial cells express *asense,* which is specific to ICPs and AMPs, but none express the epithelial marker *labial* (see below). No epithelium forms, suggesting that ICPs and AMPs develop in excess, while PMECs (a proportion of which should express *labial*) are absent. In embryos lacking *neuralized,* or zygotic *Notch* only, some PMECs do form but fewer than normal. In mutants of the proneural genes, missing the whole of the AS-C complex, plus *daughterless,* no ICPs and reduced numbers of AMPs form, but a normal epithelium is generated by the PMECs (Tepass and Hartenstein, 1995).

Double mutants of neurogenic and proneural genes are phenotypically similar to the proneural gene mutant, and so the proneural genes are epistatic to the neurogenic (Tepass and Hartenstein, 1995). The genes *lethal of scute* (*l'sc*) and *scute* (*sc*) from the proneural AS-C are activated in the entire midgut equivalence group (*achaete* is not expressed in the midgut), then *l'sc* and *sc* are up-regulated in the segregating ICPs and AMPs, and down-regulated in the PMECs. Genes of the neurogenic E(SPL)-C are expressed in the entire midgut and continue to be expressed in the PMECs. Hence, neurogenic genes promote formation of PMECs and epithelia, while proneural genes specify mesenchymal ICPs and AMPs (Tepass and Hartenstein, 1995).

5. Patterning of the Midgut Epithelium

The AMG and PMG do not possess any intrinsic A-P patterning information—they are derived from the termini of the embryo, which lack input from Hox genes (which normally provide positional information along the A-P axis)—yet once formed, the midgut epithelium becomes regionally differentiated. The most closely studied example of regionalization in the midgut is the induction of localized expression of the homeodomain protein *labial* in the endodermal layer of the central midgut (reviewed in Bienz, 1996, and outlined in Fig. 7). The expression of *labial* leads to specification

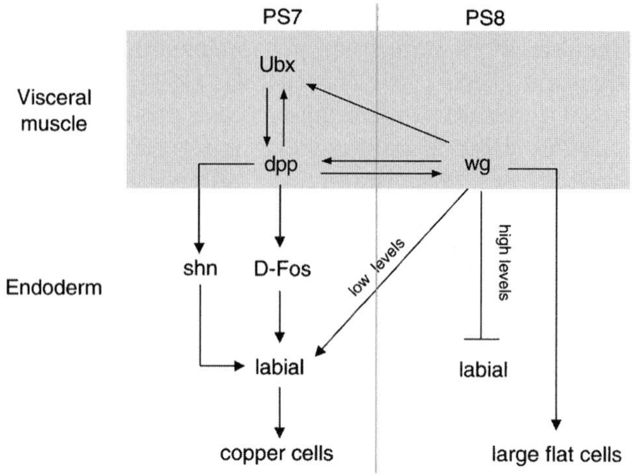

FIG. 7 Signaling in the developing gut of *Drosophila.* See text for details.

of the copper cells (Hoppler and Bienz, 1994) and formation of the second midgut constriction.

Shortly after fusion of the gut primordia, *Labial* expression in the endoderm is induced by Ubx expressed in the adjacent visceral muscle (VM). *Ubx* is expressed in a narrow parasegmental domain (ps 7), where it directly induces expression of the TGFβ family member *dpp* (Sun *et al.,* 1995). Dpp then induces *wg* in non-*Ubx*-expressing VM cells immediately posterior to this. The *labial* promoter contains both Dpp and Wg responsive elements (Tremml and Bienz, 1992), but also requires Labial activity, i.e. it is autoregulated. Low levels of Labial protein found in the midgut primordia may be sufficient to fulfil this requirement and allow its later expression in the midgut proper.

Labial is induced by low levels of Wg but repressed by high levels (Hoppler and Bienz, 1995). Consequently the *labial* domain in the endoderm is graded and is delimited by a low and a high threshold of Wg expression. Adjacent to the VM domain of high Wg expression, large flat cells, rather than copper cells, differentiate. The AP-1 related gene *D-Fos* is also expressed in the endodermal epithelium of the midgut, at low levels in the whole endoderm, but at high levels in a domain overlapping, but slightly broader than, the *labial*-expressing domain (Riese *et al.,* 1997). Like *labial, D-Fos* is induced by Dpp but, unlike *labial,* it is not induced by Wg. A dominant negative mutant of *D-Fos* reduces the number of *labial*-expressing cells in the midgut and consequently reduces the number of copper cells (Riese *et al.,* 1997).

Although both are induced by Dpp, D-Fos does not seem to mediate the induction of *labial* by Dpp. The Zn-finger protein encoded by *shnurri* (*shn*) is required for the Dpp induction of *labial,* since *labial* is not induced in *shn* mutants, even in the presence of ectopic Dpp (Arora *et al.,* 1995; Grieder *et al.,* 1995). *D-Fos,* however, shows an expanded domain of expression in these embryos, hence its induction by Dpp is independent of Shn (Riese *et al.,* 1997).

6. Regulation of *Ubx* in the Visceral Muscle

While Ubx regulates *labial* expression in the endoderm, it also regulates its own expression in the VM, since it is itself induced by the Dpp and Wg signals that it generates. Analysis of the *Ubx* midgut enhancer has resulted in the identification of the response elements that mediate these inductions.

DTcf (encoded by *pangolin*) is an HMG protein and a target of the Wg signaling pathway, just as Tcf1-3 are in *Xenopus* (Brunner *et al.,* 1997). It has been demonstrated that DTcf binds to the *Ubx* promoter, and hence *Ubx* is induced by Wg. In *dTcf/pangolin* mutants, *Ubx* is down-regulated and the second midgut constriction is missing, a phenotype similar to that of *wg* mutants in the midgut. The *labial* promoter also binds Tcf, but *labial* is actually down-regulated by high levels of Wg.

A partner element to the Tcf binding site in the *Ubx* midgut enhancer is a Dpp-response element (actually a cAMP response element, or CRE). Mutation of this site abolishes Dpp but not Wg induction of *Ubx,* but it is not yet known what binds to this site.

7. Other Regionally Localized Genes

Antp and *abd-A* are expressed in localized domains within the VM surrounding the midgut epithelium, and mutations in these genes affect midgut development (Tremml and Bienz, 1989). In *Antp* mutants the third midgut constriction is missing, while *abd-A* mutants lack both the first and second constrictions. The maintenance of *Antp* expression in the VM seems to require the TGFβ molecule Gbb (glass bottomed boat, also called TGFβ 60A; Wharton *et al.,* 1999). In *gbb* mutants, as in *Antp* mutants, the first midgut constriction fails, although its formation is initiated. *gbb* is expressed in the developing midgut epithelium and the VM and is required in both. In mutants, in addition to loss of *Antp* expression in the VM, the endodermal expression domain of the P1 enhancer trap is reduced (Sun et al., 1995).

The winged-helix transcription factor encoded by *forkhead* (*fkh*) is expressed in all gut primordia (Jurgens and Weigel, 1988) and is required for gut formation, since the entire gut is absent in *fkh* mutants. Vertebrate homologues of Forkhead—including HNF3α, β and γ—are also involved in endodermal development (see below). The role of Fkh in the fore- and hindgut has been most closely investigated. In these primordia it induces *wg*, *hh* (hedgehog) and *dpp* expression; Wg and Hh stimulate morphogenesis at the ends of the gut segment, while Dpp suppresses morphogenesis in the center.

The proventriculus, an organ that regulates food passage from the foregut to the midgut, forms at the foregut/anterior midgut boundary. The gene *defective proventriculus* (*dve*) is expressed in the proventriculus endoderm in response to Wg signals from the foregut/midgut boundary (Fuss and Hoch, 1998). It is also expressed in the central midgut in response to Dpp and Wg signals (like *labial*) and at the midgut/hindgut boundary. The Dve protein contains two homeodomains, although a short transcript lacking both domains is also expressed. *dve* is required for the formation of the proventricular epithelium, which is derived from both ectoderm and endoderm (Campos-Ortega and Hartenstein, 1997), since in mutants the proventriculus has a collapsed outer wall, blocking food passage.

IV. Other Organisms

A. Zebrafish

The zebrafish, like other teleosts, has a highly derived mode of cleavage in which only the animal pole divides, forming a blastodisc that subsequently forms the embryo. The remainder constitutes a single large cell, the yolk cell. Some of the neighboring blastodisc cells fuse with this to create the yolk syncytial layer (YSL), which adopts a signaling function analagous to the Nieuwkoop center in *Xenopus* (Mizuno *et al.*, 1999, 1996; Rodaway *et al.*, 1999); this is also true in the goldfish (Mizuno *et al.*, 1997). Meanwhile the cleaving blastoderm spreads gradually over the yolk cell by epiboly, eventually covering it. While this happens gastrulation takes place on one side of the spreading blastodisc.

1. Derivation of the Endoderm

In the zebrafish *Danio rerio* the endoderm, like the mesoderm, is derived from a layer of cells near the blastoderm margin (Kimmel *et al.*, 1990). The endoderm and mesoderm both involute into an internal layer—the hypoblast—and it seems that the early involuting cells will become endoderm, while the later involuting cells form the mesoderm. The endoderm has only recently been fate mapped (Warga and Nusslein-Volhard, 1999; reproduced in Fig. 1D) by labeling single cells in the early blastula margin, when the embryo consists of 1000 to 4000 cells. It was found that each cell could give rise to mesoderm only, endoderm only, or both mesoderm and endoderm, consistent with a common origin for the two germ layers. However, endoderm was found to be derived mainly from the dorsal side of the embryo, near (within 4 cell diameters of) the blastoderm margin, while mesoderm arose on the ventral side, and dorsally further from the margin. Within the endoderm there was also some regionalization, since anterior organs tended to arise more dorsally.

Mesodermal and endodermal cells cannot be distinguished before the onset of gastrulation, either histologically or by using molecular markers (Warga and Kimmel, 1990). However, by 75% epiboly some of the epiblast cells flatten and form a sparse layer between the embryo and the yolk cell, and these cells will give rise to the endoderm. More superficial, rounded hypoblast cells will contribute to mesoderm (Warga and Nusslein-Volhard, 1999).

2. Induction of Early Endoderm Markers

GATA and Forkhead-related transcription factors have been shown to be involved in endoderm formation in a diverse range of species, such as *Drosophila* and *C. elegans,* as already discussed. Zebrafish homologs of mammalian GATA5 (Rodaway *et al.,* 1999), HNF3α, and HNF3γ (Odenthal and Nusslein-Volhard, 1998) have been shown to be expressed in the endoderm.

gata5 is expressed from 30% epiboly in cells around the blastoderm margin (Rodaway *et al.,* 1999). Some of these cells are among the first to involute. By 80% epiboly, expression is seen in flattened cells close to the yolk cell surface, i.e. cells identified as endodermal by Warga and Nusslein-Volhard (1999), and then in a layer of cells between the embryo and the yolk cell. *gata5* is also expressed in the lateral mesoderm and heart. Rodaway et al. (1999) compared the expression of *gata5* with that of *no tail* (*ntl*), a homolog of *Brachyury* and a marker of mesoderm (Schulte-Merker *et al.,* 1992). They found that at 30% epiboly *ntl* and *gata5* were coexpressed to a depth of two to three blastomeres from the blastoderm margin (although *gata5,* unlike *ntl,* was also expressed in the YSL under the blastoderm margin). By 50% epiboly, *gata5* was expressed in a subset of *ntl*-expressing cells, and these were the cells closest to the margin. These data support a common origin for mesoderm and endoderm. *gata5* is encoded by the *faust* locus (Reiter *et al.,* 1999); *faust* mutants have defects in both heart and gut formation.

Forkhead2 (Fkd2), a HNF3γ homologue, is the only one of the nine Forkhead-domain proteins cloned in zebrafish that becomes restricted to the endoderm (Odenthal and Nusslein-Volhard, 1998), although other members of the family (including *axial,* a HNF3β homologue; Strahle *et al.,* 1993) are also expressed there. Using an anti-fkd2 antibody, Warga and Nusslein-Volhard (1999) investigated the early expression of this protein. At the onset of epiboly Fkd2 is found in presumptive dorsal cells (fated to become endoderm, according to their fate map) and yolk cell nuclei. At 60% epiboly it is found in scattered hypoblast cells, which the authors identify as endodermal precursors, at 90% epiboly it is in the prechordal plate (which gives rise to endoderm in the zebrafish), notochord, the floor plate of the nervous system, and the lateral hypoblast. Later it is found in a monolayer of cells above the yolk cell, and finally expression becomes restricted to endodermal cells in the gut.

Dorsally and laterally *gata5* and *fkd2* are coexpressed in cells around the blastoderm margin (Rodaway *et al.,* 1999)—these are presumably the cells that will give rise to the endoderm. Some ventral cells only express *gata5,* and these may give rise to lateral mesoderm and heart. Yolk cell transplantation experiments indicated that *gata5* is induced by a signal from the yolk cell, and this signal is blocked by a wide specificity dominant negative TGFβ receptor (Rodaway *et al.,* 1999). It has also been demonstrated that ectopic expression of a constitutively active activin receptor kinase (TARAM-A) can cause cells not normally fated to become endoderm to behave like endodermal cells and express endodermal markers (Peyrieras *et al.,* 1998). Hence the endogenous endoderm-inducing signal from the yolk cell is likely to be a member of the TGFβ family.

Feldman *et al.* (1998) showed that double mutants of the two zebrafish nodal-related genes *cyclops* (*cyc, znr1*) and *squint* (*sqt, znr2*) completely lacked endoderm and mesoderm, and these embryos also lack *gata5* expression (Rodaway *et al.,* 1999). *cyc*$^-$ mutants, however, have normal *gata5* expression, while down-regulation of *gata5* is seen in 25% of the progeny from *sqt*$^-$ crosses. Since *squint* is found in the YSL from MBT (Feldman *et al.,* 1998) but *cyclops* is not, *squint* seems to be the better candidate for a yolk cell-derived signal that initiates endoderm formation. Both *squint* and *cyclops,* however, are found subsequently in the blastoderm margin, and this expression may be required to maintain *gata5* expression. Warga and Nusslein-Volhard (1999) investigated the expression of fkd2 in *cyc*$^-$, but not *sqt*$^-$, embryos. They found that, although the endodermal precursors in the prechordal plate appeared normal, the plate itself was thinner than usual and the number of Fkd2-positive cells was reduced by half.

The phenotype of the *squint cyclops* double mutant (lack of mesoderm and endoderm) is similar to the phenotype of *one-eyed pinhead* (*oep*) mutants (Schier *et al.,* 1997). *oep* encodes an EGF-CFC protein homologous to mouse *cripto,* which is expressed maternally and zygotically (Zhang *et al.,* 1998b). In *oep* mutants overexpression of TARAM-A, but not of *squint* or *cyclops,* can rescue the lack of endoderm (Gritsman *et al.,* 1999; Peyrieras *et al.,* 1998) and hence oep appears to act as an extracellular cofactor for the nodal-related signaling molecules.

Alexander and Stainier (1999) have cloned zebrafish homologs of both *Xsox17* and *Mixer,* genes known to be required for endoderm formation in *Xenopus* (Hudson *et al.,* 1997; Henry and Melton, 1998). Both are expressed in the marginal cells, but the expression domain of *Mixer* is more extensive than that of *Sox17.* Overexpression of *Mixer* induces ectopic *Sox17,* implying that *Mixer* is upstream of *Sox17,* although a *Mixer* mutant is not yet available to confirm this. *Mixer* appears to be induced via a TGFβ signal, since it is induced by TARAM-A and eliminated by the nodal inhibitor antivin; *cyc sqt* double mutants lack both *Mixer* and *Sox17* expression, although both are expressed in the single mutants. As expected, since Oep is needed for Cyc and Sqt action, *oep* mutants show reduced or absent *Sox17* and *Mixer* expression, but *Sox17* expression can be rescued by overexpression of *Mixer* in these mutants, hence *Mixer* is downstream of *oep.* Finally, mutants of the uncloned *casanova* have no *Sox17* expression, and expression cannot be rescued by either *Mixer* or TARAM-A, thus suggesting that *cas* is downstream of *Mixer.*

B. Chick

1. Initiation of Endoderm Development and Formation of the Gut

Cleavage in avian embryos creates a blastodisc. Most of the cells of the blastodisc remain superficial, forming the epiblast, while some cells migrate into the subgerminal cavity to form the primary hypoblast. The space between these two layers is the blastocoel. The primary hypoblast of the chick embryo gives rise to the extraembryonic yolk sac and does not contribute any cells to the developing embryo. In this respect it resembles the primitive endoderm of the mouse, and the anterior hypoblast corresponds to the anterior visceral endoderm (AVE). Physical disruption of the hypoblast causes some of the same abnormalities associated with physical or genetic disruption of the murine visceral endoderm (VE).

The mesoderm and gut endoderm arise from an epiblast-derived structure, the primitive streak (Stern and Canning, 1990). At the rostral end of the primitive streak is a bulbous mass of cells, called Hensen's node. This region acts as an organizer and can self-differentiate into a number of embryonic tissues and induce a second axis when grafted into a host embryo. Early Hensen's node can contribute to gut endoderm, as well as to notochord and floorplate of the neural tube (Selleck and Stern, 1991).

Little is known of the molecules involved in the early phase of endodermal initiation in birds. It is possible that TGFβ factors are involved, and *cVg1,* a homolog of *Vg1,* is expressed zygotically in the posterior part of the blastodisc, before streak formation (Seleiro *et al.,* 1996; Shah *et al.,* 1997). There are the same problems with the processing of *cVg1* as with *Xenopus* Vg1, but the processed ligand domain has similar biological properties to the *Xenopus* protein and will induce a secondary primitive streak in chick blastoderms (Joubin and Stern, 1999; Shah *et al.,* 1997). It therefore has the potential to be involved in the initiation of endoderm formation.

The first cells to migrate through the primitive streak are anterior definitive endoderm. The foregut is derived from these cells by a ventral infolding which forms the anterior intestinal portal (AIP); this then lengthens posteriorly. A second wave of endodermal invaginations creates the caudal intestinal portal (CIP), which extends anteriorly to form the hindgut. As the tubes form, splanchnic mesoderm is recruited to surround the endoderm; the tubes then meet and fuse at the yolk stalk. Luminal gut differentiation creates three distinct regions, the foregut, midgut and hindgut (Romanoff, 1960). Ultimately the foregut gives rise to the oesophagus and crop, the stomach (including gizzard and proventriculus), and the accessory organs (liver and pancreas); the midgut forms the small intestine and the hindgut forms the ceca, large intestine, allantois, and cloaca. As in other vertebrates the endoderm also forms the respiratory system, which opens into the pharynx, and epithelial outpockets of the pharynx give rise to the tonsils, thyroid, thymus, and parathyroid glands.

2. Patterning and Gene Expression in the Early Gut

The adaptability of the avian embryo to grafting experiments during organogenesis has enabled work to be conducted on inductive signaling between the endoderm and mesoderm. These processes are involved in organizing the A-P axis of the gut and the correct integration of mesoderm and endoderm.

a. The Role of Sonic Hedgehog. *Sonic hedgehog* (*shh*) is expressed in the endoderm of the AIP and CIP at the earliest stages of gut development (Narita *et al.,* 1998). Juxtaposed to *shh* expression in the caudal endoderm is undifferentiated mesoderm, which becomes the visceral mesoderm of the hindgut. Since ectopic expression of Shh elsewhere in the mesoderm induces BMP4, Shh may serve as a signal from endoderm to recruit visceral mesoderm by inducing expression of BMP4, which then initiates growth or specification of the visceral mesoderm (Roberts *et al.,* 1995).

Not all gut mesoderm responds to Shh in the same way, suggesting that there is a prepattern in the future gut mesoderm prior to the expression of Shh in the endoderm. Hence, while misexpression of Shh upregulates its receptor Patched and causes cell proliferation in all gut mesoderm, BMP4 cannot be induced by Shh in the foregut region that will form stomach (Roberts *et al.,* 1998). Mesoderm is thicker in the stomach than elsewhere in the gut and ectopic expression of BMP4 reduces it, suggesting that BMP4 limits the amount of gut mesoderm in nonstomach regions. Further evidence for a mesodermal prepattern is the observation that Shh can only induce *Hoxd-13* in the mesoderm of the hindgut (Roberts *et al.,* 1998).

Inhibition of Shh expression is necessary for the pancreas to form; Shh is expressed everywhere in the endoderm except the pancreatic buds, which form in the midgut specifically where Shh is not expressed (Apelqvist *et al.,* 1997). The suppression of *shh* expression depends on a signal from the notochord, which acts permissively on a restricted field in this part of the endoderm (Kim *et al.,* 1997). Both FGF2 and activinβ_B can mimic this, and may be the signal *in vivo.* Additional evidence that Shh is acting in the embryo is that antibodies against Shh and a specific Shh inhibitor, cyclopamine, also induce pancreas in the pancreatic field (Hebrok *et al.,* 1998; Kim and Melton, 1998).

HNF3β is expressed in the early gut endoderm (Ang *et al.,* 1993) and may be involved in the induction of *shh,* since elsewhere HNF3β has been proposed to regulate the production of Shh in the floor plate and notochord (Epstein *et al.,* 1999; but see also Muller *et al.,* 1999).

b. Homeodomain Proteins The Hox genes are also likely to be causally involved in patterning the gut, although the evidence for a causative role relies on the mouse (see below). The posterior *Abd-B*-like *Hox* genes have expression patterns in visceral mesoderm that correspond to morphological boundaries in the mesoderm of the gut. In addition, like BMP4, they seem be downstream of Shh, since ectopic expression of Shh induces these *Hox* genes, just as it induces ectopic BMP4 elsewhere in the mesoderm (see above) (Roberts *et al.,* 1995). The *Abd-B* group is also expressed in the endoderm of the hindgut, but *Hoxa-10, 11* and *13* are coexpressed in the entire midgut and allantois, rather than in a nested pattern as is found in mesoderm. By day four *Hoxa-11* and *12* expression becomes restricted to the allantois, and *Hoxa-13* expression is lost from the large intestine and caecum on day five, but persists in the cloaca (Yokouchi *et al.,* 1995). The relative roles of the posterior Hox genes in the endoderm and visceral mesoderm are unclear.

Once the mesodermal type is specified, signals from the mesoderm pattern the morphological differentiation of the endoderm. *Hoxd-13* is restricted in its mesodermal expression to the most posterior region of the hindgut, which forms the cloaca, and it seems to be required to establish this identity. When expressed in more anterior midgut mesoderm, *Hoxd-13* transforms the endoderm to the cloacal type, in terms of both morphology and mucin production. Therefore, the restricted expression of the Hox gene in the mesoderm must bring about signaling that leads to regionally specific differentiation of gut endoderm (Roberts *et al.,* 1998). The other AbdB class Hoxa genes are also expressed in very restricted domains of the visceral mesoderm. *Hoxa-9* is expressed in the posterior small intestine and ceca, *Hoxa-10* in the ceca, *Hoxa-11* in the ceca, large intestine, and cloaca, and *Hoxa-13* in the cloaca. If these molecules behave like *Hoxd-13* they may determine the identity of a wide range of mid- and hindgut endodermal cell types (Yokouchi *et al.,* 1995).

The behavior of *Hoxd-13* provides a molecular parallel with grafting experiments on Amphibia, like those of Okada, that implicate the mesoderm in patterning the endoderm (see above). However, the nature of the signals to the endoderm is unknown. Comparable grafting experiments to those in Amphibia have also been performed in birds. For example, when foregut mesoderm is combined with midgut endoderm, the mesoderm respecifies the endoderm to have foregut endoderm morphology (Yasugi, 1993). There are also regional differences in competence to respond signals from the mesoderm that determine the outcome of these interactions. Hence, proventricular (glandular stomach) and gizzard (muscular stomach) mesenchymes exert different influences on endodermal epithelia differentiation. While the former induces gland formation and expression of embryonic chick pepsinogen, the latter inhibits these outcomes. In contrast, intestinal endoderm cannot respond to proventricular mesenchyme (Yasugi *et al.,* 1991).

The effects of the mesoderm have also been investigated using the homeobox gene *CdxA,* a homolog of the *caudal* gene in *Drosophila,* as a marker. Caudal is also a regulator of *AbdB* in the chick and *CdxA* is expressed solely in intestinal epithelium from early phases of digestive tube formation (Frumkin *et al.,* 1991). Grafting showed that the rostral limit of *CdxA* expression, at the boundary between stomach and duodenum, is determined by interactions with the mesoderm (Ishii *et al.,* 1997). *cSox2* provides a more anterior marker, being expressed in gut epithelium from the pharynx to the stomach, abutting the *CdxA* domain. Its caudal limit is coincident with that of the region competent for proventricular differentiation (Ishii *et al.,* 1998). As for *CdxA, cSox2* expression is regulated by the mesoderm; gizzard mesenchyme induces *cSox2* expression and suppresses *CdxA* expression in intestinal epithelium, while small intestine mesenchyme exerts influence on stomach epithelium to induce *CdxA* and suppress *cSox2* expression. Such interactions may thus be involved in establishing the boundary between gizzard and small intestine. Later the level of transcripts of *cSox2* decrease in epithelia invaginating into surrounding mesenchyme to form glandular or tubular structures, e.g. lung and thyroid primordia and the glandular epithelium of the proventriculus. *cSox2* expression may therefore be involved in fine patterning of endoderm, but direct evidence is lacking (Ishii *et al.,* 1998).

c. The GATA Family of Transcription Factors *Gata4, 5,* and *6* are expressed in overlapping zones in the heart and endoderm (Laverriere *et al.,* 1994). *Gata6* may play a role in maintaining cells in a proliferative state, while *Gata4* and *5* (which has spliced isoforms that change the number of zinc fingers; MacNeill *et al.,* 1997) may be more important in executing the pathways of differentiation. In the epithelium of the chick intestine *Gata6* is expressed strongly in proliferating cells in the crypts of the villi, whereas *Gata4* and *5* are most strongly expressed in the differentiated epithelial cells at the distal tips (Gao *et al.,* 1998). This pattern is similar to that seen in mouse and probably reflects contrasting roles in proliferation and differentiation of the epithelium.

d. Other Molecules There are other molecules expressed in early endoderm for which no functional analysis has yet been performed in chick. *cHex* is a marker with a highly dynamic expression pattern in the chick embryo. At the time of laying it is expressed in Koller's sickle and the forming hypoblast. Subsequently it is expressed in the hypoblast cells that have been displaced anterior to the primitive streak, followed by expression in endoderm overlapping the cardiogenic region. Later *cHex* is expressed in pharyngeal endoderm next to the forming myocardium, in the endocardium, and in the liver and thyroid primordia. It is also expressed in blood islands and vascular endothelial cells (Yatskievych *et al.,* 1999). In mouse and *Xenopus* Hex has been shown to be required for the induction of head by the AVE/anterior endoderm and formation of vascular tissue (Newman *et al.,* 1997; Thomas *et al.,* 1998). Its expression in chick is consistent with a similar role.

Pax1 and *9* are expressed in both the pharyngeal endoderm and the paraxial mesoderm. Grafting shows that the expression in the endoderm is intrinsically programmed, whereas in the mesoderm it depends on interaction with midline structures (Muller *et al.,* 1996). In the mouse *Pax1* has been shown to have an essential role in the thymus (Dahl *et al.,* 1997).

3. The Avian Endoderm as a Signaling Center

As already discussed, in amphibians the presumptive endoderm, i.e. the vegetal cells, generate TGFβ signals that induce the mesoderm to form and are possibly important in initiating endoderm formation *in vivo* (in parallel to VegT); they certainly act in the subsequent reinforcement phase. It is not known how such processes occur in the chick embryo. By analogy with zebrafish, it is conceivable that signals arise in the noncellular part of the vegetal region, but since partial cleavage evolved independently in birds and teleosts, this is speculative. However, as in other species, the head organizer is located in the general region of the anterior mesendoderm, but the anterior definitive endoderm does not induce nervous system, rather this ability is found in the young head process and prechordal plate (Knoetgen *et al.,* 1999). This was shown in cross-species transplantation experiments; AVE from rabbit embryos, but not chick hypoblast, induces the formation of neuroectoderm, as shown by the expression of GANF, when transplanted into the area pellucida or area opaca of chick embryos. GANF is a member of the Anf (anterior neural folds family), equivalent to Hesx1 in mouse. What direct role Spemann organizer molecules have in patterning the chick endoderm is unclear at present.

C. Mouse

1. First Steps in Endoderm Formation

In mammals there are two types of endoderm; primitive, extraembryonic endoderm is formed very early in development and is eventually replaced by definitive endoderm, which arises during gastrulation and, as in other vertebrates, forms the gut, liver, and other endodermal derivatives. The primitive endoderm consists of two extraembryonic membranes, the parietal and visceral endoderm (Gardner, 1983). The parietal endoderm (PE) arises in close association with the trophectoderm and secretes extracellular matrix proteins, while the visceral endoderm (VE) develops in close association with the inner cell mass and synthesizes proteins that facilitate uptake, digestion, and secretion of nutrients, as well as coordinating blood cell differentiation and blood vessel formation in the adjoining mesoderm (Bielinska *et al.,* 1999). It also has important signaling functions analogous to the signaling centers of amphibians. At the egg cylinder stage the layer of visceral endoderm covers the epiblast (which gives rise to the entire fetus) and extraembryonic ectoderm, and plays a role in the formation of the proamniotic cavity. BMP signaling, including BMP4 produced in the ectoderm and BMP2 from the endoderm, promotes differentiation of the visceral endoderm and acts on ectodermal core cells to cause cell death and columnar epithelial differentiation, i.e. cavitation (Coucouvanis and Martin, 1999). Less is known about the parietal endoderm, but GATA6 activity is needed for it to form (Koutsourakis *et al.,* 1999), and there is evidence for an involvement of parathyroid hormone related peptide (PTHrP) signaling from the trophectoderm (Karperien *et al.,* 1996; Verheijen *et al.,* 1999).

2. Formation of the Extraembryonic Endoderm

The differentiation of F9 embryonal carcinoma (EC) cells *in vitro* has provided a model for parietal versus visceral endoderm differentiation (Hogan *et al.,* 1983). Both GATA 4 and 6 are expressed in primitive endoderm (Arceci *et al.,* 1993; Koutsourakis *et al.,* 1999). When mouse F9 EC cells are induced to differentiate into either visceral or parietal endoderm by aggregation and treatment with retinoic acid (RA), GATA4 and GATA6 are induced (Arceci *et al.,* 1993; Bielinska and Wilson, 1997).

Mouse embryonic stem (ES) cells differentiate into embryoid bodies, covered with a layer of VE, when cultured in suspension. With RA present they form visceral endoderm, but they form parietal endoderm when cultured with dibutyryl cAMP in addition, and in both cases GATA6 is induced (Bielinska *et al.,* 1999). In embryoid bodies differentiated from GATA 4-/- ES cells by aggregation alone, no visceral endoderm forms (Soudais *et al.,* 1995). However, in the presence of RA, with or without dibutyryl cAMP, both sorts of primitive endoderm do form, presumably because GATA6 is induced in this case. This suggests that there is functional redundancy between GATA4 and GATA6 in forming primitive endoderm (Bielinska *et al.,* 1999).

GATA 4-/- mouse embryos do contain visceral endoderm, but this endoderm expresses elevated levels of GATA6, perhaps compensating for the loss of GATA4. The mutant mice die between E7 and 10 because of a loss of lateral-to-ventral folding resulting in foregut and heart tube defects and partial development outside of the yolk sac (Kuo *et al.,* 1997). In chimeras between wild-type and GATA4-/- ES cells, GATA4+/+ cells contributed only to the visceral endoderm and regions of the fore and hindgut (Narita *et al.,* 1997b). Ventral morphogenesis thus requires expression of GATA4 in the endoderm rather than the mesoderm. GATA6-/- mice also show a role for this protein in the primitive endoderm (see previous discussion). The GATA genes have been shown to regulate HNF transcription (see later).

Indian hedgehog (*Ihh*) is expressed in primitive endoderm in the mouse embryo and *Ihh* is induced when F9 EC cells are induced to differentiate as primitive endoderm (Becker *et al.,* 1997). Additionally, F9 cells transfected with a construct encoding *Ihh* show characteristics of primitive endoderm in the absence of an inducing agent, suggesting that induction of *Ihh* may promote differentiation of visceral and parietal endoderm. Various other genes are expressed in visceral endoderm and some are essential for it to form, for example *Evx1* (Spyropoulos and Capecchi, 1994); these have recently been reviewed (Bielinska *et al.,* 1999).

3. The Visceral Endoderm as a Signaling Center

The anterior visceral endoderm (AVE) of the mouse and the anterior *Xenopus* endoderm are both believed to be important for inducing head development. The transcription factors Lim1 and Otx2 are both expressed in the AVE and both have been shown to affect head development, presumably by regulating inductive interactions (Acampora *et al.,* 1995; Belo *et al.,* 1997; Shawlot and Behringer, 1995). Head development is also dependent on the expression of the TGFβ factor *nodal* in the primitive endoderm (Varlet *et al.,* 1997).

Evidence that the AVE functions as an early organizer is that prior to the formation of the primitive streak these cells express genes that are responsible, directly or indirectly, for signals that specify the fate of underlying embryonic tissues (Beddington and Robertson, 1998). Some of these same genes are subsequently expressed in underlying epiblast derivatives, such as axial mesendoderm and definitive gut endoderm, where they control their later differentiation. The AVE produces signals involved in formation of the major embryonic axes and are needed for heart formation and yolk sac positioning. The homeobox gene *Hex* is an early marker of endodermal A-P asymmetry, since it is first expressed in a patch of visceral endoderm at the distal tip of the egg cylinder and these cells move to an anterior position while still expressing *Hex.* Later in development it is expressed in definitive endoderm (Thomas *et al.,* 1998).

The winged helix transcription factor HNF3β, homologous to Forkhead in *Drosophila,* is expressed in the visceral endoderm of the E6 embryo and later in the node and its derivatives (prechordal plate, notochord, and definitive endoderm). Homozygous inactivation of *HNF3*β genes results in embryos whose defects derive from both expression in the visceral endoderm and later in the embryo proper (Dufort *et al.,* 1998; Weinstein *et al.,* 1994). HNF3β-dependent signaling from the visceral endoderm is necessary for elongation of the primitive streak, whereas its later autonomous effects from expression in the embryo proper are associated with lack of the node and notochord, leading to defects in dorsal-ventral (D-V) patterning of the neural tube and somites. Definitive endoderm forms, but fore- and midgut are lacking, again an effect autonomous to the embryo itself. Therefore, HNF3β has distinct functions in VE, where it is needed for signaling involved in primitive streak morphogenesis, and in the epiblast, where it helps to specify the node, notochord, and gut.

It was previously explained that early stages in endoderm formation in amphibians depend on TGFβ family members, particularly the Nodal subfamily, Derrière, and possibly activin and Vg-1. These signals are also responsible for mesoderm induction. There is evidence that the TGFβ family is also involved in formation of, and signaling from, extraembryonic endoderm in the mouse. For example, Alk2 is a type I TGFβ receptor that binds both activins and BMPs, but probably transduces just BMP signals. Homozygous inactivation of the *alk2* gene, which is expressed in the VE, leads to abnormal morphology of the visceral endoderm and embryonic primitive streak defects of mesoderm (Gu *et al.*, 1999). Chimera experiments indicate that it is visceral endodermal expression of Alk2 that is responsible for the streak abnormalities. Receptor effects must be cell autonomous, so TGFβ signaling, probably BMP, must be needed both to organize the visceral endoderm and to generate its capacity to signal to the embryonic tissue. Unfortunately these conclusions do not fit with any single BMP knockout yet performed, or with knockout of *alk3,* which encodes a BMP receptor (Mishina *et al.,* 1995). This may be because of redundancy or because the true range of ligands for Alk2 may not be known.

Mutants of *alk4* are more severe than *alk2,* lacking a streak altogether (although some mesoderm forms) and having defects in the visceral endoderm, but not the parietal endoderm (Gu *et al.,* 1998). Alk4 is an activin receptor acting through Smads 2 and 3, but it is proposed to act as the Nodal receptor, because *nodal*-deficient mutant mice have a similar embryonic phenotype to *alk4*$^{-/-}$ mice, resulting from disrupted *nodal* expression in the visceral endoderm (Conlon *et al.,* 1994; Varlet *et al.,* 1997). Abnormalities in the visceral endoderm itself were not reported. Hence, the signal from the visceral endoderm that leads to primitive streak formation involves Nodal as a mandatory component, but more obscure TGFβ family signaling is involved in forming the visceral endoderm.

4. Formation of the Definitive Endoderm

Gastrulation involves the formation of the mesoderm and the definitive endoderm from the epiblast; both endodermal and mesodermal precursors migrate through the primitive streak. At the early primitive streak stage there is a subpopulation of cells in the axial endoderm overlying the anterior of the streak that later contributes to the gut endoderm. By the mid-streak stage the anterior end of the streak reaches the distal tip of the embryo, where the node forms (this is equivalent to Hensen's node in the chick). The head process arises as cranial extensions of the node and contributes to definitive endoderm by displacing primary endoderm cells in the midline (Lawson *et al.,* 1991). In the chick, cells in the early Hensen's node can contribute to gut endoderm, as well as notochord and floorplate of the neural tube (Selleck and Stern, 1991) and this may also be true in the mouse. Therefore there may be two phases of definitive endoderm formation:

(1) from anterior of the streak before node formation, and (2) from the head process.

It is not clear what relation the three kinds of mammalian endoderm bear to those in a totally cleaving amniote, like *Xenopus*. It is possible that the definitive endoderm, which gastrulates like mesoderm, may be homologous to the *Xenopus* suprablastoporal endoderm. When it is known how both form in molecular terms the situation may become clearer.

a. Initiation of the Definitive Endoderm Most of what is known about the early development of the definitive endoderm comes from the study of genes or groups of genes that are expressed in it, rather than from experimental embryology. Currently accounts of these genes are rather disconnected, so they will be discussed separately. There is a large amount known about the control of transcription of genes encoding differentiated products, mainly in the adult endoderm, as well as signaling genes involved in tissue maintenance and cancer; these subjects are too large to be considered here.

i. HNF Transcription Factors The diverse groups of HNF transcription factors, originally identified in mature liver, have important roles in various parts of the endoderm throughout the life, as well as in other organs. Each group contains several members, making redundancy a serious problem in establishing their roles in a particular situation, as does the fact that they are expressed elsewhere in the embryo. HNF1α and β contain homeodomains; HNF3α, β, and γ are winged helix proteins similar to *Drosophila* Forkhead (which plays a critical role in gut development in the fly); HNF4α, β, and γ are ligand-binding orphan nuclear receptors; while HNF6 is a homeodomain protein related to *Drosophila* Cut.

HNF3β is a key endodermal gene expressed at the onset of gastrulation (see above) in the progenitors of the definitive endoderm, and transcription of *HNF3α* is also seen in midline endoderm. Expression of both persists in definitive endoderm, as well as in the notochord and in the ventral region of the neural tube. Later, HNF3α, β, and γ are expressed in various endoderm-derived tissues, such as the liver (Altaba, 1996; Altaba *et al.,* 1993; Ang and Rossant, 1994; Monaghan *et al.,* 1993; Sasaki and Hogan, 1993), and they are involved in regulating the differentiated products in the gut. HNF3α and β may thus be involved in both the initiation and maintenance of the definitive endodermal lineage (Ang *et al.,* 1993). HNF3γ is not expressed until the time of hindgut differentiation, probably under the control of HNF1α and β (Hiemisch *et al.,* 1997), and *HNF3γ^{-}/$^{-}$* mice are normal, except for somewhat reduced levels of transcription of several gut enzyme genes (Kaestner *et al.,* 1998). In this discussion we obviously emphasize functional redundancy between members of paralogous groups, but redundancy might extend to unrelated genes. For example, HNF3β and Goosecoid seem to overlap in the control of Shh (Filosa *et al.*, 1997).

Gene knockouts of HNF1α die postnatally, with kidney and liver defects (Pontoglio *et al.*, 1996), and they also have defects in pancreatic β-cells. This fits with the fact that humans carrying HNF1α mutations suffer from early onset diabetes (MODY3; Yamagata *et al.*, 1996). HNF4α, which is believed to be downstream of HNF1α (Duncan *et al.*, 1998), is responsible for MODY1 (Yamagata *et al.*, 1996) and is important in controling many endodermal genes directly (Duncan *et al.*, 1994). Early effects of HNF1α deficiency might have been expected; their absence may be because of overlap with HNF1β, for which mouse gene inactivations are not available, but human mutations in HNF1β cause MODY5 (Horikawa *et al.*, 1997; Yamagata *et al.*, 1996). It is likely that only double knockouts will be truly informative as regards early roles of the *HNF1* genes.

HNF4α knockout disrupts signaling from the visceral endoderm that controls development of the embryo proper and leads to failures of gene expression within the visceral endoderm (Duncan *et al.*, 1997). HNF1α and β activate HNF4α in the pancreas (Duncan *et al.*, 1998), so if they are generally essential for HNF4 activation we should expect that double knockouts of HNF1α and β will be similar to HNF4α knockouts, i.e. early lethals. HNF1α is itself regulated negatively by HNF3α and positively by HNF1β. HNF3β acts as a strong activator and HNF3α is a weak activator, inhibiting HNF3β by competing for binding sites with its more vigorous sister (Duncan *et al.*, 1998). Since the expression of HNF3α itself depends on HNF3β (Duncan *et al.*, 1998), the HNF group form a complex network of control genes. The key role of HNF3β is consistent with the effects on the visceral endoderm of its knockout, already mentioned in the visceral endoderm signaling section (Dufort *et al.*, 1998; Kaestner *et al.*, 1998; Weinstein *et al.*, 1994). The HNF3β knockout mice fail to develop any fore- and midgut, but they do form hindgut, so HNF3β is essential for foregut and midgut development, not for the hindgut. This may be because of redundancy with HNF1α. Inactivation of HNF3α has later effects in differentiated tissues, fitting its later expression (Kaestner *et al.*, 1999). An interesting evolutionary parallel is provided by the control of relative HNF3α and HNF3β levels by insulin in the pancreas (Duncan *et al.*, 1998). In the *C. elegans* dauer larva DAF2, an insulin receptor homolog, controls the HNF3 homolog, DAF16. (Lin *et al.*, 1997; Ogg *et al.*, 1997).

ii. GATA Transcription Factors GATA4, 5, and 6 are expressed in overlapping patterns in the heart and in endodermal derivatives. We have already discussed the possible role of these genes in forming primitive endoderm. The disposition of GATA4, 5, and 6 expression in the mature intestine suggests that these transcription factors have different activities. In chickens and humans GATA6 is expressed in proliferating cells, whereas

GATA4 is associated with differentiation; GATA5 is intermediate (Gao *et al.,* 1998). A similar kind of relationship is seen in the erythroid lineage, where GATA2 is associated with proliferation (Briegel *et al.,* 1993) and GATA1 is associated with differentiation (Briegel *et al.,* 1996; Pevny *et al.,* 1995).

During foregut invagination and heart tube formation, GATA4 is expressed in cardiogenic splanchnic mesoderm and associated endoderm (Kuo *et al.,* 1997). The defects in *GATA 4-/-* mutant mice, which die between E7 and 10, are principally the result of disrupted morphogenetic movements, rather than the loss of particular cell types in heart or gut. Presumably this represents overlapping function among GATA4, 5, and 6. In chimeras between wild-type and GATA 4-/- ES cells, only GATA 4+/+ cells contributed to the visceral endoderm and regions of the fore and hindgut (Narita *et al.,* 1997a). Hence, morphogenesis of the gut requires expression of GATA4 in the endoderm rather than the mesoderm.

GATA6-/- mice die extremely early in development (see earlier). They exhibit down-regulated expression of GATA4 and the absence of HNF4 and HNF3β gene expression (Morrisey *et al.,* 1998). HNF4-deficient embryos also show widespread programmed cell death within the embryonic ectoderm. Since forced expression of GATA6 activates the HNF4 promoter in non-endodermal cells, GATA6 probably lies upstream of HNF4 development in both visceral and later endoderm (Morrisey *et al.,* 1998). GATA4-deficient embryos do not show the absence of HNF4 seen in GATA6-/- embryos, suggesting that HNF4 is differently regulated by GATA4 and 6.

iii. Homeodomain-Containing Transcription Factors The *Hox* genes are well known as regulators of antero-posterior patterning in the nervous system, but they also pattern the gut, indeed it is possible that this was their original function. Discerning their function precisely is made difficult by the fact that they are expressed in different but related fashions in endoderm and splanchnic mesoderm, and they may have both germ layer-autonomous and nonautonomous roles (see also the chick section).

As might be expected from their roles in axial tissues, the lower number *Hox* genes have more anterior roles and the higher, more posterior. *Hox* group 3 paralogs have important roles in the pharynx, regulating the development of the thymus, thyroid, and parathyroid glands, which are branchial pouch endodermal derivatives, primarily through regulating cell movement (Manley and Capecchi, 1998). Although their roles overlap, *Hoxa3* has a nonredundant role in the thyroid and parathyroids. More posteriorly *Hoxa5* has an essential role in lung mesenchyme (Aubin *et al.,* 1997) and *Hox6–9* show expression patterns in the mid-gut corresponding to major morphological divisions (Sekimoto *et al.,* 1998). Most posteriorly, *Hoxd* genes have a particularly important role in gut sphincters. Deletion of *Hoxd-13* caused loss of the anal sphincter (Kondo *et al.,* 1996) and there were more serious effects on the rectum in *Hoxa-13/Hoxd-13* double inactivations (Warot *et al.*, 1997). When *Hoxd4–13* are simultaneously inactivated the ileocaecal sphincter is also lost and there are abnormalities in the stomach (Zakany and Douboule, 1999).

The effects of retinoic acid on expanding the anterior expression of posterior (5′) Hox gene expression, together with truncating anterior axial structures, are well known. Retinoic acid has similar effects on anterior gut morphology in various species (Hinman and Degnan, 1998; Mulder *et al.*, 1998; Suzuki *et al.*, 1999). This may be directly because of altered Hox gene expression, indeed a retinoic acid response element has been shown to be essential for expression of *Hoxb-1* in the anterior gut (Huang *et al.*, 1998). In *Xenopus,* rather general effects on gut morphology have been found with late retinoic acid treatment, perhaps suggesting later roles for Hox genes (Zeynali and Dixon, 1998).

It has recently been proposed, through studies of amphioxus, that there was an early duplication of the proto-Hox cluster to produce a "Parahox cluster" (Brooke *et al.*, 1998). In mouse the homologous genes are *mGsh* (expressed in Rathke's pouch, the gut component of the pituitary), *pdx-1/idx-1/ipf-1* (also called *Xlhbox8* in *Xenopus*; expressed in the pancreas and duodenum), and *Cdx1-4* (expressed in the hindgut). A *pdx-1* knockout indicates that it is essential for correct pancreatic/duodenal development (Jonsson *et al.*, 1994; Offield *et al.*, 1996). The *caudal* genes are responsible for transcription of a variety of mid- and hindgut terminal differentiation products (Drummond *et al.*, 1996; Tamai *et al.*, 1999; Lambert *et al.*, 1996; Laser *et al.*, 1996; Troelsen *et al.*, 1997) and inactivation of the *cdx-1* gene causes abnormalities in the mid- and hindgut (Tamai *et al.*, 1999). It was proposed that the Parahox duplication had a major role in allowing the increased complexity of the gut in Triploblastica. This is very speculative, and a preliminary report suggests that the genes already existed in diploblastic Cnidaria (Finnerty and Martindale, 1999).

A variety of families of homeodomain-containing genes also have roles in the gut. As in the chick, *Pax1/9* are expressed in the pharynx and have an essential role in the thymus (Dahl *et al.*, 1997). This is an ancient situation since ascidian Pax1/9 is expressed in the gill epithelia (Ogasawara *et al.*, 1999).

Nkx2 genes, which have major roles in heart development and are homologs of *Drosophila tinman,* are also expressed in the anterior developing gut, in the stomach liver, tongue, and anterior larynx (Kasahara *et al.*, 1998; Reecy *et al.*, 1997). *Hex* is a homeobox gene expressed in the liver, oesophagus, pharynx, and thyroid (Keng *et al.*, 1998; Thomas *et al.*, 1998; Yatskievych *et al.*, 1999). *Barx* is expressed in the stomach endoderm and mesenchyme, as well as the pharyngeal and mouth mesoderm (Tissierseta *et al.,* 1995). Doubtless other such genes will be discovered to have roles in gut development.

iv. HMG Domain Transcription Factors Although *Sox17* has been cloned from the mouse (Kanai *et al.,* 1996), its embryonic expression pattern is yet to be described. *Sox 2* is expressed throughout the gut endoderm and *Sox 3* transcripts are detected in the posterior region of the foregut (Wood and Episkopou, 1999).

5. Liver Development

The liver forms by an inductive interaction with the developing heart region. However, before this, heart development itself depends on a signal from the anterior region of endoderm that will later form the liver (Mohun and Sparrow, 1997). This may involve activin/TGFβ or BMP 2/4, which have different capacities to induce heart, depending on the target tissue (Andree *et al.,* 1998; Ladd *et al.,* 1998; Schultheiss *et al.,* 1997; Yatskievych *et al.,* 1997). Liver development then occurs in several steps (Zaret, 1998):

1. There is interaction of cardiac mesoderm with foregut endoderm, a process in which FGF is implicated and serum albumin and α-fetoprotein gene expression is induced (Gualdi *et al.,* 1996; Jung *et al.,* 1999). The cardiac mesoderm expresses FGFs 1, 2, and 8. Of these, FGF1 and 2, but not FGF8, can induce liver gene expression in isolated foregut endoderm, and the hepatogenic response is restricted to endoderm tissue coexpressing FGF receptors 1 and 4. Subsequently FGF8 contributes to the morphogenetic outgrowth of the hepatic endoderm (Jung *et al.,* 1999).
2. There is an inductive interaction with the mesenchyme of the septum transversum that leads to cell proliferation and formation of the liver bud (Zaret, 1998). Contributors to this signaling are indicated by gene knockout phenotypes in which the hepatic endoderm differentiates (as judged by expression of marker genes such as serum albumin) and the liver bud appears, but then there is little subsequent cell proliferation. Candidate genes include those encoding hepatocyte growth factor (HGF), *c-met, Hlx,* and *c-jun.* In summary, the divergent homeodomain protein Hlx, expressed in the septum mesenchyme around the liver bud, is essential to generate the signal (Hentsch *et al.,* 1996), which includes HGF (Schmidt *et al.,* 1995). HGF is expressed in the septum, although it is not known whether it is a direct target of Hlx, and generally stimulates cell division and tissue organization. Its receptor, c-met, is expressed on hepatocytes (Zarnegar *et al.,* 1994) and it is proposed to stimulate cell division through the transcription factor AP-1, of which c-jun is a subunit. C-jun is essential for hepatogenesis (Hilberg *et al.,* 1993).

3. The next phase of liver morphogenesis involves cell interactions, growth, and organ organization; when this phase is disrupted by gene mutation, cell death results (see review by Zaret, 1998).

a. Transcriptional Cascades in Liver Formation

i. The Role of HNFs One of the first responses of endoderm to hepatic induction is an increase in the level of expression of HNF3 α,β, and γ genes (Ang *et al.*, 1993; Monghan *et al.*, 1993). HNF3β binds to its own promoter, and can maintain its own synthesis as well as inducing expression of genes that are downstream in the regulatory cascade (Rausa *et al.*, 1997). However, Rausa *et al.* propose that maintenance of HNF3β expression in embryonic hepatocytes is due to a collaboration between HNF6 and C/EBPβ, which is expressed early in hepatocyte specification and later by C/EBPα, whose expression starts at 13dpc (Bossard *et al.*, 1997; Kuo *et al.*, 1990). HNF3α also cross-regulates HNF3β.

HNF4 expression is first detected when the foregut endoderm interacts with mesenchyme of the septum transversum, and it is expressed at higher levels during later differentiation phases. *HNF6* is also expressed at the onset of liver development; it binds to and activates the *HNF4* promoter and also stimulates *HNF3β* expression (see above). *HNF6* and *HNF3β* are downregulated transiently between E12.5 and E15 and reappear with further differentiation of hepatic cells, hence they may control later steps of differentiation into adult phenotype (Landry *et al.*, 1997).

These transcription factors therefore form a cross-regulatory network that controls expression of genes in the liver: HNF6 contributes to the control of *HNF4* and *HNF3β* genes, whose protein products stimulate transcription of the *HNF1* gene (Landry *et al.*, 1997). HNF1, along with HNFs 3, 4, and 6, has been shown to activate transcription of liver-specific differentiation products (Trouche and Yaniv, 1994).

ii. The Role of GATA Factors GATA factors are also critical for liver gene expression, but study of proteins bound to the serum albumin gene enhancer *in vivo* indicates that the HNF and GATA sites are occupied in the intestine, where the gene is not expressed, and other liver-specific proteins are essential for serum albumin expression in the liver (Bossard and Zaret, 1998; Gualdi *et al.*, 1996; Zaret, 1999). It is argued that HNF3β and GATA4-6 (principally GATA-4) create an open chromatin conformation, which then permits binding of the critical activating factors.

6. Pancreas Formation

The pancreas arises from foregut endoderm as dorsal and ventral buds that later fuse together to form a single organ. It has two functions: release of digestive enzymes into the intestine (exocrine) and secretion of hormones into the bloodstream to regulate glucose levels (endocrine). Mammals, birds, reptiles, and amphibians have a pancreas with similar histology and mode of development. In the early rodent embryo the endoderm of the future pancreas is a cup-shaped structure in which the notochord is embedded (reviews by Slack, 1995; St. Onge *et al.,* 1999), which may facilitate signaling from the notochord.

a. Transcription Factors Regulating Pancreas Development Pancreatic development depends on a variety of transcription factors, including homeodomain transcription factors such as Pdx1 (Ipf-1, Idx-1, *Xenopus* Xlhbox8), Hlxb9, and Isl1 (Edlund, 1998; Habener and Stoffers, 1998; St. Onge *et al.,* 1999). Gene inactivation shows that *Pdx-1* has a unique role, in that it is necessary for the development of both the endocrine and exocrine cells of dorsal and ventral pancreas and also for the ventral duodenum to form properly (Jonsson *et al.,* 1994; Offield *et al.,* 1996). However, the initial formation of the pancreatic buds does occur in the absence of *pdx-1,* and there is initial endocrine cell formation, with insulin and glucagon expression (Ahlgren *et al.,* 1996).

Dorsal pancreas requires expression of the Lim homeodomain gene *Isl1* in both endoderm and mesenchyme (Ahlgren *et al.,* 1997) and also *Hlxb9* (Harrison *et al.,* 1999; Li *et al.,* 1999). Pax4 is needed for insulin and somatostatin expression (Sosa-Pineda *et al.,* 1997), Pax6 for glucagon cells (Sander *et al.,* 1997; St. Onge *et al.,* 1997), and Nkx2.2 for insulin cells (Sussel *et al.,* 1998). NeuroD/Beta2 inactivation reduces endocrine cells (Naya *et al.,* 1997), while the only transcription factor so far shown to affect exclusively the exocrine cells is PTF1-p48, a bHLH protein (Krapp *et al.,* 1996). When p48 is inactivated exocrine cells are absent and, oddly, the endocrine cells populate the spleen (Krapp *et al.,* 1998), implying an influence of the exocrine cells on the organisation of the endocrine pancreas.

Roles for various HNF factors in pancreas formation and function have been revealed by identification of genes causing mature onset diabetes of the young (MODY), as already discussed (St. Onge *et al.,* 1999). *HNF6* expression is first detected once the pancreatic primordia have emerged from the endodermal lining of the gut, and is maintained through foetal development. HNF3β colocalizes with HNF6, and while HNF3β expression in the foregut endoderm precedes that of HNF6 (Landry *et al.,* 1997), HNF3β contains binding sites for HNF6 in its promoter and HNF6 probably plays a role in maintaining HNF3β expression during pancreas specification (Landry *et al.,* 1997; Rausa *et al.,* 1997; Samadani and Costa, 1996). HNF3β also regulates the transcription of the *pdx-1* gene in the insulin-producing β-cells (Wu *et al.,* 1997).

b. Signaling Pathways and Cell-Cell Interactions A scheme for *pdx-1* regulation has been presented by Rausa et al. (1997) and is summarized in Fig. 8. Two phases of cell interactions are necessary for the development of the pancreas. Early signals from the notochord to the endoderm allow dorsal pancreas morphogenesis by modulating Sonic Hedgehog (Shh) activity (Kim *et al.,* 1997). In early mouse and chicken embryos *Shh* and *Indian Hedgehog* (*Ihh*) are expressed in all the embryonic gut endoderm, except for the pancreatic endoderm. A signal from the notochord appears to suppress Shh signaling in this region. This signal cannot induce pancreatic gene expression in other endoderm, suggesting that the response is restricted to cells of the duodenal area. Hence, the notochord signal is permissive, rather than instructive.

The signal may include Activinβ_B and FGF2, which are expressed in the notochord during pancreas specification and experimentally mimic the effects of notochord on dorsal prepancreatic endoderm (Hebrok *et al.,* 1998; LeBras *et al.,* 1998). Their effect is through repressing Shh repression of pancreatic genes, indeed the same effect can be achieved by direct inhibition of Shh (Hebrok *et al.,* 1998; Kim and Melton, 1998). FGFs 1, 7, and 10 are expressed in the pancreas and appear to have an important role in pancreas development, operating through FGFR2IIIb (Miralles *et al.,* 1999).

It remains to be seen how general the notochord signaling is among vertebrates. It will be remembered that in *Xenopus* isolated vegetal poles regionally express the pancreas marker *Xlhbox8* in the absence of any mesoderm (although it is true that Xlhbox8 is also expressed in the duodenum).

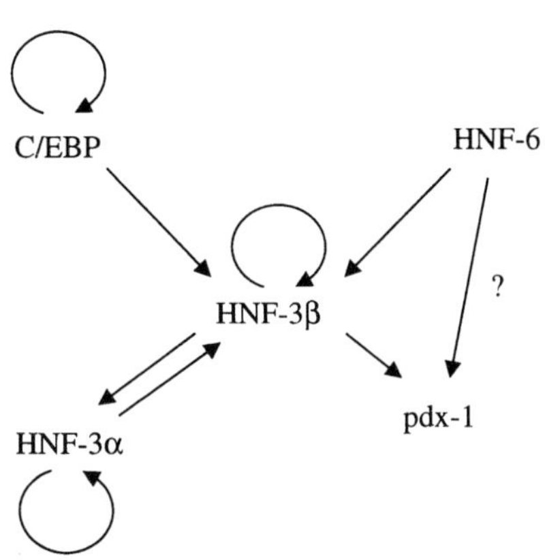

FIG. 8 Interactions between various transcription factors in murine pancreas development (Rausa *et al.,* 1997).

Ectopic expression of Shh in pancreatic endoderm converts pancreatic mesoderm into intestinal mesenchyme, forming smooth muscle and interstitial cells of Cajal, and inhibits spleen development (Apelqvist *et al.,* 1997). There is no effect on the fates of the pancreatic endoderm cells, which suggests that Shh acts as a signal from endoderm to mesoderm.
[Note: Shh expression seems to have wider roles in the foregut, as well as being involved in intestinal differentiation. This is particularly revealed by inactivation of the Shh effectors Gli2 and Gli3, which leads to failure of oesophagus, trachea and lung to form (Motoyama *et al.,* 1998).]

In a second phase of signaling the pancreatic mesoderm regulates the proportions of exocrine and endocrine tissues. If most of the mesenchyme is removed, endocrine cells fail to develop (Gittes *et al.,* 1996). Follistatin can replace the mesenchyme in organ culture (Miralles *et al.,* 1998), which suggests that the mesoderm inhibits activin and/or BMP7. Presumably these TGFβ class signals promote endocrine development. The balance between endocrine and exocrine cells is also apparently mediated by lateral inhibition through the Notch pathway, drawing parallels with patterning the nervous system (Apelqvist *et al.,* 1999)

V. Summary

In this review we have focused on a number of intensively studied model species, and because of space limitations we have largely omitted observations on a variety of other organisms. Of the three major bilaterian subdivisions, the Deuterostomia, Ecdysozoa, and Lophotrochozoa, the last (including molluscs, annelids, and platyhelminths) lacks a modern model organism (except to a limited extent the leech) and little is known of endoderm development in this major group. A particularly interesting feature of the annelids, at least, is that the endoderm is segmented through interaction with the mesoderm (Nardellihaefliger and Shankland, 1993; Wedeen and Shankland, 1997). It will therefore be interesting to see how endoderm formation in this situation compares with the two other groups, where it is not segmented. One feature of the presumptive endoderm that is apparently conserved in Lophotrochozoa is its action as a signaling center, revealed by classic polar lobe removal experiments in molluscs.

In the Deuterostomia there is considerable progress being made with basal groups. In some cases, like amphioxus, this primarily involves gene organization and expression patterns, since the experimental embryology is rudimentary. However, tunicates are rapidly being developed as experimental molecular embryological models, and recent advances in developing genetics should accelerate this work (Nakatani *et al.,* 1999). The echinoderms have long been an embryological model and their endodermal development has been reviewed by Wessel and Wikramanayake (1999). A particularly interesting aspect of their development is the extreme form of metamorphosis found in typical species.

The two most intensively studied invertebrate systems are *Drosophila* and *C. elegans,* both of which are highly derived examples of the major group. Although *C. elegans,* as a nematode, has long been thought to be a primitive bilaterian on morphological grounds, this is now thought to be incorrect and nematodes are placed with arthropods in the Ecdysozoa. Nematodes are likely to differ from arthropods primarily by character loss. Hence many of the features in common between the models, and there are less than one might expect, are not necessarily primitive bilaterian features. Therefore only those features that are also shared with deuterostomes are likely to be primitive. If one looks at the transcription factors that all groups share in regulating gene expression in the endoderm, there are clear parallels. Obvious examples are the Forkhead-related, winged helix domain proteins (Fkh, HNF3), and GATA and Hox proteins. Of course one problem in making these links is that there are a limited number of families of transcription factors, so the correlations could be chance rather than functional homology. The same could hold for the signaling molecules and signal transduction cassettes common to endodermal development, particularly the BMP/Dpp group of signaling molecules. Nevertheless, at the moment the homologies we have described seem more likely to be the result of common processes that originate in the Cambrian or Pre-Cambrian ancestors of modern groups. Only time will tell!

Acknowledgments

We thank Drs. Robert Old, Kevin Moffat, and David Stott for comments on the manuscript. Our work is supported by The Wellcome Trust.

References

Abel, T., Michelson, A. M., and Maniatis, T. (1993). A *Drosophila* GATA family member that binds to *Adh* regulatory sequences is expressed in the developing fat body. *Development* **119,** 623–633.

Acampora, D., Mazan, S., Lallemand, Y., Avantaggiato, V., Maury, M., Simeone, A., and Brulet, P. (1995). Forebrain and midbrain regions are deleted in *Otx2*(−/−) mutants due to a defective anterior neuroectoderm specification during gastrulation. *Development* **121,** 3279–3290.

Ahlgren, U., Jonsson, J., and Edlund, H. (1996). The morphogenesis of the pancreatic mesenchyme is uncoupled from that of the pancreatic epithelium in IPF1/PDX1-deficient mice. *Development* **122,** 1409–1416.

Ahlgren, U., Pfaff, S. L., Jessell, T. M., Edlund, T., and Edlund, H. (1997). Independent requirement for ISL1 in formation of pancreatic mesenchyme and islet cells. *Nature* **385,** 257–260.

Alexander, J., and Stainier, D. Y. R. (1999). A molecular pathway leading to endoderm formation in zebrafish. *Curr. Biol.* **9,** 1147–1157.

Altaba, A. R. (1996). Coexpression of HNF3 beta and Isl-1/2 and mixed distribution of ventral cell types in the early neural tube. *Int. J. Dev. Biol.* **40,** 1081–1088.

Altaba, A. R. I., Prezioso, V. R., Darnell, J. E., and Jessell, T. M. (1993). Sequential expression of HNF3-Beta and HNF3-Alpha by embryonic organizing centers—the dorsal lip/node, notochord and floor plate. *Mech. Dev.* **44,** 91–108.

Andree, B., Duprez, D., Vorbusch, B., Arnold, H. H., and Brand, T. (1998). BMP-2 induces ectopic expression of cardiac lineage markers and interferes with somite formation in chicken embryos. *Mech. Dev.* **70,** 119–131.

Ang, S. L., and Rossant, J. (1994). HNF3-Beta is essential for node and notochord formation in mouse development. *Cell* **78,** 561–574.

Ang, S. L., Wierda, A., Wong, D., Stevens, K. A., Cascio, S., Rossant, J., and Zaret, K. S. (1993). The formation and maintenance of the definitive endoderm lineage in the mouse—involvement of HNF3/Forkhead proteins. *Development* **119,** 1301–1315.

Apelqvist, A., Ahlgren, U., and Edlund, H. (1997). Sonic hedgehog directs specialised mesoderm differentiation in the intestine and pancreas. *Curr. Biol.* **7,** 801–804.

Apelqvist, A., Li, H., Sommer, L., Beatus, P., Anderson, D. J., Honjo, T., deAngelis, M. H., Lendahl, U., and Edlund, H. (1999). Notch signalling controls pancreatic cell differentiation. *Nature* **400,** 877–881.

Arceci, R. J., King, A. A. J., Simon, M. C., Orkin, S. H., and Wilson, D. B. (1993). Mouse GATA-4—a retinoic acid-inducible GATA-binding transcription factor expressed in endodermally derived tissues and heart. *Mol. Cell. Biol.* **13,** 2235–2246.

Arora, K., Dai, H., Kazuko, S. G., Jamal, J., O'Connor, M. B., Letsou, A., and Warrior, R. (1995). The Drosophila *schnurri* gene acts in the Dpp/TGF beta signaling pathway and encodes a transcription factor homologous to the human MBP family. *Cell* **81,** 781–790.

Asashima, M., Nakano, H., Uchiyama, H., Sugino, H., Nakamura, T., Eto, Y., Ejima, D., Nishimatsu, S. I., Ueno, N., and Kinoshita, K. (1991). Presence of activin (erythroid-differentiation factor) in unfertilized eggs and blastulas of *Xenopus laevis. Proc. Natl. Acad. Sci. USA* **88,** 6511–6514.

Aubin, J., Lemieux, M., Tremblay, M., Berard, J., and Jeannotte, L. (1997). Early postnatal lethality in *Hoxa-5* mutant mice is attributable to respiratory tract defects. *Dev. Biol.* **192,** 432–445.

Azzaria, M., Goszczynski, B., Chung, M. A., Kalb, J. M., and McGhee, J. D. (1996). A fork head/HNF3 homolog expressed in the pharynx and intestine of the *Caenorhabditis elegans* embryo. *Dev. Biol.* **178,** 289–303.

Balinsky, B. I., and Fabian, B. C. (1981). "An Introduction to Embryology." Saunders College Pub., Philadelphia.

Bartkowski, S., Zapp, D., Weber, H., Eberle, G., Zoidl, C., Senkel, S., Kleinhitpass, L., and Ryffel, G. U. (1993). Developmental Regulation and Tissue Distribution of the Liver Transcription Factor LFB1 (HNF1) in *Xenopus laevis. Mol. Cell. Biol.* **13,** 421–431.

Becker, S., Wang, Z. J., Massey, H., Arauz, A., Labosky, P., Hammerschmidt, M., St. Jacques, B., Bumcrot, D., McMahon, A., and Grabel, L. (1997). A role for Indian hedgehog in extraembryonic endoderm differentiation in F9 cells and the early mouse embryo. *Dev. Biol.* **187,** 298–310.

Beddington, R. S. P., and Robertson, E. J. (1998). Anterior patterning in mouse. *Trends Genet.* **14,** 277–284.

Belo, J. A., Bouwmeester, T., Leyns, L., Kertesz, N., Gallo, M., Follettie, M., and DeRobertis, E. M. (1997). Cerberus-like is a secreted factor with neuralizing activity expressed in the anterior primitive endoderm of the mouse gastrula. *Mech. Dev.* **68,** 45–57.

Bielinska, M., Narita, N., and Wilson, D. B. (1999). Distinct roles for visceral endoderm during embryonic mouse development. *Int. J. Dev. Biol.* **43,** 183–205.

Bielinska, M., and Wilson, D. B. (1997). Induction of yolk sac endoderm in GATA-4-deficient embryoid bodies by retinoic acid. *Mech. Dev.* **65,** 43–54.

Bienz, M. (1996). Induction of the endoderm in *Drosophila. Semin. Cell Dev. Biol.* **7,** 113–119.

Blackwell, T. K., Bowerman, B., Priess, J. R., and Weintraub, H. (1994). Formation of a monomeric DNA-binding domain by SKN-1 BZIP and homeodomain elements. *Science* **266,** 621–628.

Bolce, M. E., Hemmati-Brivanlou, A., and Harland, R. M. (1993). Xfkh2, a *Xenopus* Hnf-3-Alpha homolog, exhibits both Activin-inducible and autonomous phases of expression in early embryos. *Dev. Biol.* **160,** 413–423.

Bossard, P., McPherson, C. E., and Zaret, K. S. (1997). In vivo footprinting with limiting amounts of embryo tissues: A role for C/EBP beta in early hepatic development. *Methods-A Compan. Meth. Enzymol.* **11,** 180–188.

Bossard, P., and Zaret, K. S. (1998). GATA transcription factors as potentiators of gut endoderm differentiation. *Development* **125,** 4909–4917.

Bouwmeester, T., and Leyns, L. (1997). Vertebrate head induction by anterior primitive endoderm. *BioEssays* **19,** 855–863.

Bowerman, B., Draper, B. W., Mello, C. C., and Priess, J. R. (1993). The maternal gene *Skn-1* encodes a protein that is distributed unequally in early *C. elegans* embryos. *Cell* **74,** 443–452.

Bowerman, B., Eaton, B. A., and Priess, J. R. (1992). *Skn-1,* a maternally expressed gene required to specify the fate of ventral blastomeres in the early C. elegans embryo. *Cell* **68,** 1061–1075.

Brannon, M., and Kimelman, D. (1996). Activation of *Siamois* by the Wnt pathway. *Dev. Biol.* **180,** 344–347.

Briegel, K., Bartunek, P., Stengl, G., Lim, K. C., Beug, H., Engel, J. D., and Zenke, M. (1996). Regulation and function of transcription factor GATA-1 during red blood cell differentiation. *Development* **122,** 3839–3850.

Briegel, K., Lim, K. C., Plank, C., Beug, H., Engel, J. D., and Zenke, M. (1993). Ectopic expression of a conditional GATA-2 estrogen-receptor chimera arrests erythroid-differentiation in a hormone-dependent manner. *Genes Dev.* **7,** 1097–1109.

Britton, C., McKerrow, J. H., and Johnstone, L. L. (1998). Regulation of the *Caenorhabditis elegans* gut cysteine protease gene *cpr-1:* Requirement for GATA motifs. *J. Mol. Biol.* **283,** 15–27.

Bronner, G., Chu-LaGraff, Q., Doe, C. Q., Cohen, B., Weigel, D., Taubert, T., and Jackle, H. (1994). Spl/egr-like zinc-finger protein required for endoderm specification and germ-layer formation in *Drosophila. Nature* **369,** 664–668.

Bronner, G., and Jackle, H. (1991). Control and function of terminal gap gene activity in the posterior pole region of the *Drosophila* embryo. *Mech. Dev.* **35,** 205–211.

Brooke, N. M., Garcia Fernandez, J., and Holland, P. W. H. (1998). The Hox gene cluster and its extended family: Insights from amphioxus. *Dev. Biol.* **198,** 53.

Brunner, E., Peter, O., Schweizer, L., and Basler, K. (1997). *pangolin* encodes a Lef-1 homologue that acts downstream of *armadillo* to transduce the Wingless signal in *Drosophila. Nature* **385,** 829–833.

Campos-Ortega, J. A., and Hartenstein, V. (1997). "The Embryonic Development of *Drosophila melanogaster.*" Springer, Berlin/New York.

Casanova, J., Furriols, M., McCormick, C. A., and Struhl, G. (1995). Similarities between Trunk and Spatzle, putative extracellular ligands specifying body pattern in *Drosophila*. *Genes Dev.* **9,** 2539–2544.

Casey, E. S., Tada, M., Fairclough, L., Wylie, C. C., Heasman, J., and Smith, J. C. (1999). *Bix4* is activated directly by VegT and mediates endoderm formation in *Xenopus* development. *Development* **126,** 4193–4200.

Clements, D. C., Friday, R. V., and Woodland, H. R. (1999). Patterning of the *Xenopus* blastula by the maternal determinant VegT. *Development* **126,** 4903–4911.

Conlon, F. L., Lyons, K. M., Takaesu, N., Barth, K. S., Kispert, A., Herrmann, B., and Robertson, E. J. (1994). A primary requirement for nodal in the formation and maintenance of the primitive streak in the mouse. *Development* **120,** 1919–1928.

Coucouvanis, E., and Martin, G. R. (1999). BMP signaling plays a role in visceral endoderm differentiation and cavitation in the early mouse embryo. *Development* **126,** 535–546.

Dahl, E., Koseki, H., and Balling, R. (1997). *Pax* genes and organogenesis. *BioEssays* **19,** 755–765.

Dale, L., Matthews, G., and Colman, A. (1993). Secretion and mesoderm-inducing activity of the TGFβ-related domain of *Xenopus* Vg1. *EMBO J.* **12,** 4471–4480.

Dale, L., and Slack, J. M. W. (1987). Fate map for the 32-cell stage of *Xenopus laevis*. *Development* **99,** 527–551.

Darras, S., Marikawa, Y., Elinson, R. P., and Lemaire, P. (1997). Animal and vegetal pole cells of early *Xenopus* embryos respond differently to maternal dorsal determinants: Implications for the patterning of the organiser. *Development* **124,** 4275–4286.

Demartis, A., Maffei, M., Vignali, R., Barsacchi, G., and Desimone, V. (1994). Cloning and developmental expression of LFB3/HNF1-β transcription factor in *Xenopus laevis*. *Mech. Dev.* **47,** 19–28.

Deppe, U., Schierenberg, E., Cole, T., Krieg, C., Schmitt, D., Yoder, B., and von Ehrenstein, G. (1978). Cell lineages of the embryo of the nematode *Caenorhabditis elegans*. *Proc. Natl. Acad. Sci. USA* **75,** 376–380.

Ding, X. Y., Hausen, P., and Steinbeisser, H. (1998). Pre-MBT patterning of early gene regulation in *Xenopus:* The role of the cortical rotation and mesoderm induction. *Mech. Dev.* **70,** 15–24.

Dohrmann, C. E., Kessler, D. S., and Melton, D. A. (1996). Induction of axial mesoderm by Zdvr-1, the zebrafish orthologue of *Xenopus* Vg1. *Dev. Biol.* **175,** 108–117.

Drummond, F., Sowden, J., Morrison, K., and Edwards, Y. H. (1996). The caudal-type homeobox protein Cdx-2 binds to the colon promoter of the carbonic anhydrase 1 gene. *Eur. J. Biochem.* **236,** 670–681.

Dufort, D., Schwartz, L., Harpal, K., and Rossant, J. (1998). The transcription factor HNF3 beta is required in visceral endoderm for normal primitive streak morphogenesis. *Development* **125,** 3015–3025.

Duncan, S. A., Manova, K., Chen, W. S., Hoodless, P., Weinstein, D. C., Bachvarova, R. F., and Darnell, J. E. (1994). Expression of transcription factor HNF4 in the extraembryonic endoderm, gut, and nephrogenic tissue of the developing mouse embryo—HNF4 is a marker for primary endoderm in the implanting blastocyst. *Proc. Natl. Acad. Sci. USA* **91,** 7598–7602.

Duncan, S. A., Nagy, A., and Chan, W. (1997). Murine gastrulation requires HNF4 regulated gene expression in the visceral endoderm: Tetraploid rescue of *Hnf-4*(−/−) embryos. *Development* **124,** 279–287.

Duncan, S. A., Navas, M. A., Dufort, D., Rossant, J., and Stoffel, M. (1998). Regulation of a transcription factor network required for differentiation and metabolism. *Science* **281,** 692–695.

Ecochard, V., Cayrol, C., Rey, S., Foulquier, F., Caillol, D., Lemaire, P., and Duprat, A. M. (1998). A novel *Xenopus Mix*-like gene *milk* involved in the control of the endomesodermal fates. *Development* **125,** 2577–2585.

Edlund, H. (1998). Transcribing pancreas. *Diabetes* **47,** 1817–1823.

Egan, C. R., Chung, M. A., Allen, F. L., Heschl, M. F. P., Vanbuskirk, C. L., and McGhee, J. D. (1995). A gut-to-pharynx tail switch in embryonic expression of the *Caenorhabditis elegans Ges-1* gene centers on 2 GATA Sequences. *Dev. Biol.* **170,** 397–419.

Epstein, D. J., McMahon, A. P., and Joyner, A. L. (1999). Regionalization of Sonic hedgehog transcription along the anteroposterior axis of the mouse central nervous system is regulated by Hnf3-dependent and -independent mechanisms. *Development* **126,** 281–292.

Evans, S. M., Yan, W., Murillo, M. P., Ponce, J., and Papalopulu, N. (1995). *Tinman,* a *Drosophila* homeobox gene required for heart and visceral mesoderm specification, may be represented by a family of genes in vertebrates—XNkx-2.3, a second vertebrate homolog of *tinman. Development* **121,** 3889–3899.

Feldman, B., Gates, M. A., Egan, E. S., Dougan, S. T., Rennebeck, G., Sirotkin, H. I., Schier, A. F., and Talbot, W. S. (1998). Zebrafish organizer development and germ-layer formation require nodal-related signals. *Nature* **395,** 181–185.

Filosa, S., Rivera Perez, J. A., Gomez, A. P., Gansmuller, A., Sasaki, H., Behringer, R. R., and Ang, S. L. (1997). Goosecoid and HNF-3 beta genetically interact to regulate neural tube patterning during mouse embryogenesis. *Development* **124,** 2843–2854.

Finnerty, J. R., and Martindale, M. (1999). Early evolution of Hox and ParaHox genes: Evidence from the Cnidaria. *Dev. Biol.* **210,** 41.

Frumkin, A., Rangini, Z., Benyehuda, A., Gruenbaum, Y., and Fainsod, A. (1991). A chicken caudal homolog, Chox-Cad, is expressed in the epiblast with posterior localization and in the early endodermal lineage. *Development* **112,** 207–219.

Fukushige, T., Hawkins, M. G., and McGhee, J. D. (1998). The GATA-factor elt-2 is essential for formation of the *Caenorhabditis elegans* intestine. *Dev. Biol.* **198,** 286–302.

Furriols, M., Casali, A., and Casanova, J. (1998). Dissecting the mechanism of torso receptor activation. *Mech. Dev.* **70,** 111–118.

Fuss, B., and Hoch, M. (1998). *Drosophila* endoderm development requires a novel homebox gene which is a target of Wingless and Dpp signalling. *Mech. Dev.* **79,** 83–97.

Gamer, L. W., and Wright, C. V. E. (1995). Autonomous endodermal determination in *Xenopus*—Regulation of expression of the pancreatic gene *X1hbox-8. Dev. Biol.* **171,** 240–251.

Gao, X. P., Sedgwick, T., Shi, Y. B., and Evans, T. (1998). Distinct functions are implicated for the GATA-4, -5, and -6 transcription factors in the regulation of intestine epithelial cell differentiation. *Mol. Cell. Biol.* **18,** 2901–2911.

Gardner, R. L. (1983). Origin and differentiation of extraembryonic tissues in the mouse. *Int. Rev. Exp. Pathol.* **24,** 63–133.

Gerhart, J., and Kirschner, M. (1997). "Cells, Embryos, and Evolution: Toward a Cellular and Developmental Understanding of Phenotypic Variation and Evolutionary Adaptability." Blackwell Science, Malden, Mass.

Gittes, G. K., Galante, P. E., Hanahan, D., Rutter, W. J., and Debas, H. T. (1996). Lineage-specific morphogenesis in the developing pancreas: Role of mesenchymal factors. *Development* **122,** 439–447.

Goldstein, B. (1992). Induction of gut in *Caenorhabditis elegans* embryos. *Nature* **357,** 255–257.

Goldstein, B. (1993). Establishment of gut fate in the E-lineage of *C. elegans*—the roles of lineage-dependent mechanisms and cell interactions. *Development* **118,** 1267–1277.

Goldstein, B. (1995). An analysis of the response to gut induction in the *C. elegans* embryo. *Development* **121,** 1227–1236.

Grau, Y., Carteret, C., and Simpson, P. (1984). Mutations and chromosomal rearrangements affecting the expression of *snail,* a gene involved in embryonic patterning in *Drosophila melanogaster. Genetics* **108,** 347–360.

Grieder, N. C., Nellen, D., Burke, R., Basler, K., and Affolter, M. (1995). Schnurri is required for *Drosophila* Dpp signaling and encodes a zinc finger protein similar to the mammalian transcription factor PRDII-BF1. *Cell* **81,** 791–800.

Gritsman, K., Zhang, J. J., Cheng, S., Heckscher, E., Talbot, W. S., and Schier, A. F. (1999). The EGF-CFC protein one-eyed pinhead is essential for nodal signaling. *Cell* **97,** 121–132.

Gu, Z. Y., Nomura, M., Simpson, B. B., Lei, H., Feijen, A., van den Eijnden van Raaij, J., Donahoe, P. K., and Li, E. (1998). The type I activin receptor ActRIB is required for egg cylinder organization and gastrulation in the mouse. *Genes Dev.* **12,** 844–857.

Gu, Z. Y., Reynolds, E. M., Song, J. H., Lei, H., Feijen, A., Yu, L. A., He, W. W., MacLaughlin, D. T., van den Eijnden van Raaij, J., Donahoe, P. K., and Li, E. (1999). The type I serine threonine kinase receptor ActRIA (ALK2) is required for gastrulation of the mouse embryo. *Development* **126,** 2551–2561.

Gualdi, R., Bossard, P., Zheng, M. H., Hamada, Y., Coleman, J. R., and Zaret, K. S. (1996). Hepatic specification of the gut endoderm in vitro: Cell signaling and transcriptional control. *Genes Dev.* **10,** 1670–1682.

Habener, J. F., and Stoffers, D. A. (1998). A newly discovered role of transcription factors involved in pancreas development and the pathogenesis of diabetes mellitus. *Proc. Assoc. Am. Physicians* **110,** 12–21.

Harrison, K. A., Thaler, J., Pfaff, S. L., Gu, H., and Kehrl, J. H. (1999). Pancreas dorsal lobe agenesis and abnormal islets of Langerhans in *Hlxb9*-deficient mice. *Nature Genet.* **23,** 71–75.

Hawkins, M. G., and McGhee, J. D. (1995). Elt-2, a second GATA factor from the nematode *Caenorhabditis elegans. J. Biol. Chem.* **270,** 14666–14671.

Heasman, J., Wylie, C. C., Hausen, P., and Smith, J. C. (1984). Fates and states of determination of single vegetal pole blastomeres of *X. laevis. Cell* **37,** 185–194.

Hebrok, M., Kim, S. K., and Melton, D. A. (1998). Notochord repression of endodermal Sonic hedgehog permits pancreas development. *Genes Dev.* **12,** 1705–1713.

Hemmati-Brivanlou, A., and Melton, D. A. (1994). Inhibition of activin receptor signaling promotes neuralization in *Xenopus. Cell* **77,** 273–281.

Henry, G. L., Brivanlou, I. H., Kessler, D. S., Hemmati-Brivanlou, A., and Melton, D. A. (1996). TGF-Beta signals and a prepattern in *Xenopus laevis* endodermal development. *Development* **122,** 1007–1015.

Henry, G. L., and Melton, D. A. (1998). Mixer, a homeobox gene required for endoderm development. *Science* **281,** 91–96.

Hentsch, B., Lyons, I., Li, R. L., Hartley, L., Lints, T. J., Adams, J. M., and Harvey, R. P. (1996). *Hlx* homeo box gene is essential for an inductive tissue interaction that drives expansion of embryonic liver and gut. *Genes Dev.* **10,** 70–79.

Hiemisch, H., Schutz, G., and Kaestner, K. H. (1997). Transcriptional regulation in endoderm development: Characterization of an enhancer controlling HNF3 gamma expression by transgenesis and targeted mutagenesis. *EMBO J.* **16,** 3995–4006.

Hilberg, F., Aguzzi, A., Howells, N., and Wagner, E. F. (1993). C-Jun is essential for normal mouse development and hepatogenesis. *Nature* **365,** 179–181.

Hinman, V. F., and Degnan, B. M. (1998). Retinoic acid disrupts anterior ectodermal and endodermal development in ascidian larvae and postlarvae. *Dev. Genes Evol.* **208,** 336–345.

Hogan, B., Barlow, D., and Tilly, R. (1983). F9 teratocarcinoma cells as a model for the differentiation of parietal and visceral endoderm in the mouse embryo. *Cancer Surv.* **2,** 115–140.

Hoppler, S., and Bienz, M. (1994). Specification of a single-cell type by a *Drosophila* homeotic gene. *Cell* **76,** 689–702.

Hoppler, S., and Bienz, M. (1995). Two different thresholds of Wingless signaling with distinct developmental consequences in the *Drosophila* midgut. *EMBO J.* **14,** 5016–5026.

Horb, M. E., and Thomsen, G. H. (1997). A vegetally localized T-box transcription factor in *Xenopus* eggs specifies mesoderm and endoderm and is essential for embryonic mesoderm formation. *Development* **124,** 1689–1698.

Horikawa, Y., Iwasaki, N., Hara, M., Furuta, H., Hinokio, Y., Cockburn, B. N., Lindner, T., Yamagata, K., Ogata, M., Tomonga, O., Kuroki, H., T., K., Iwamoto, Y., and Bell, G. I.

(1997). Mutation in hepatocyte nuclear factor-1 beta gene (TCF2) associated with MODY. *Nat. Genet.* **17,** 384–385.

Huang, D., Chen, S. W., Langston, A. W., and Gudas, L. J. (1998). A conserved retinoic acid responsive element in the murine *Hoxb-1* gene is required for expression in the developing gut. *Development* **125,** 3235–3246.

Hudson, C., Clements, D., Friday, R. V., Stott, D., and Woodland, H. R. (1997). Xsox17alpha and -beta mediate endoderm formation in *Xenopus. Cell* **91,** 397–405.

Ip, Y. T., Park, R. E., Kosman, D., Yazdanbakhsh, K., and Levine, M. (1992). dorsal-twist interactions establish snail expression in the presumptive mesoderm of the *Drosophila* embryo. *Genes Dev.* **6,** 1518–1530.

Ishii, Y., Fukuda, K., Saiga, H., Matsushita, S., and Yasugi, S. (1997). Early specification of intestinal epithelium in the chicken embryo: A study on the localization and regulation of CdxA expression. *Dev. Growth Diff.* **39,** 643–653.

Ishii, Y., Rex, M., Scotting, P. J., and Yasugi, S. (1998). Region-specific expression of chicken Sox2 in the developing gut and lung epithelium: Regulation by epithelial-mesenchymal interactions. *Dev. Dyn.* **213,** 464–475.

Ishizuyaoka, A., Shimozawa, A., Takeda, H., and Shi, Y. B. (1994). Cell-specific and spatiotemporal expression of intestinal fatty-acid-binding protein gene during amphibian metamorphosis. *Rouxs Arch. Dev. Biol.* **204,** 150–155.

Jones, C. M., Broadbent, J., Thomas, P. Q., Smith, J. C., and Beddington, R. S. P. (1999). An anterior signalling centre in *Xenopus* revealed by the homeobox gene XHex. *Current Biol.* **9,** 946–954.

Jones, E. A., Abel, M. H., and Woodland, H. R. (1993). The possible role of mesodermal growth-factors in the formation of endoderm in *Xenopus laevis. Rouxs' Arch. Dev. Biol.* **202,** 233–239.

Jones, E. A., and Woodland, H. R. (1987). The development of animal cap cells in *Xenopus*—a measure of the start of animal cap competence to form mesoderm. *Development* **101,** 557–563.

Jonsson, J., Carlsson, L., Edlund, T., and Edlund, H. (1994). Insulin-Promoter-Factor-1 is required for pancreas development in mice. *Nature* **371,** 606–609.

Joseph, E. M., and Melton, D. A. (1998). Mutant Vg1 ligands disrupt endoderm and mesoderm formation in *Xenopus* embryos. *Development* **125,** 2677–2685.

Joubin, K., and Stern, C. D. (1999). Molecular interactions continuously define the organizer during the cell movements of gastrulation. *Cell* **98,** 559–571.

Jung, J. N., Zheng, M. H., Goldfarb, M., and Zaret, K. S. (1999). Initiation of mammalian liver development from endoderm by fibroblast growth factors. *Science* **284,** 1998–2003.

Jurgens, G., and Weigel, D. (1988). Terminal versus segmental development in the *drosophila* embryo—the role of the homeotic gene Fork Head. *Rouxs Arch. Dev. Biol.* **197,** 345–354.

Jurgens, G., Wieshaus, E., Nusslein-Volhard, C., and Kluding, H. (1984). Mutations affecting the pattern of the larval cuticle in *Drosophila melanogaster.* II. Zygotic loci on the third chromosome. *Roux's Arch. Dev. Biol.* **193,** 283–295.

Kadowaki, T., Wilder, E., Klingensmith, J., Zachary, K., and Perrimon, N. (1996). The segment polarity gene *porcupine* encodes a putative multitransmembrane protein involved in Wingless processing. *Genes Dev.* **10,** 3116–3128.

Kaestner, K. H., Hiemisch, H., and Schutz, G. (1998). Targeted disruption of the gene encoding hepatocyte nuclear factor 3 gamma results in reduced transcription of hepatocyte-specific genes. *Mol. Cell. Biol.* **18,** 4245–4251.

Kaestner, K. H., Katz, J., Liu, Y. F., Drucker, D. J., and Schutz, G. (1999). Inactivation of the winged helix transcription factor HNF3 alpha affects glucose homeostasis and islet glucagon gene expression in vivo. *Genes Dev.* **13,** 495–504.

Kalb, J. M., Lau, K. K., Goszczynski, B., Fukushige, T., Moons, D., Okkema, P. G., and McGhee, J. D. (1998). pha-4 is Ce-fkh-1, a fork head/HNF-3 alpha, beta, gamma homolog that functions in organogenesis of the *C. elegans* pharynx. *Development* **125,** 2171–2180.

Kanai, Y., Kanaiazuma, M., Noce, T., Saido, T. C., Shiroishi, T., Hayashi, Y., and Yazaki, K. (1996). Identification of two Sox17 messenger-RNA isoforms, with and without the High-Mobility Group box region, and their differential expression in mouse spermatogenesis. *J. Cell Biol.* **133,** 667–681.

Karperien, M., Lanser, P., DeLaat, S. W., Boonstra, J., and Defize, L. H. K. (1996). Parathyroid hormone related peptide mRNA expression during murine postimplantation development: Evidence for involvement in multiple differentiation processes. *Int. J. Dev. Biol.* **40,** 599–608.

Kasahara, H., Bartunkova, S., Schinke, M., Tanaka, M., and Izumo, S. (1998). Cardiac and extracardiac expression of Csx/Nkx2.5 homeodomain protein. *Circ. Res.* **82,** 936–946.

Keller, R. (1991). Early embryonic development of *Xenopus laevis. Meth. Cell Biol.* **36,** 61–113.

Keng, V. W., Fujimori, K. E., Myint, Z., Tamamaki, N., Nojyo, Y., and Noguchi, T. (1998). Expression of Hex mRNA in early murine postimplantation embryo development. *FEBS Lett.* **426,** 183–186.

Kim, S. K., Hebrok, M., and Melton, D. A. (1997). Notochord to endoderm signaling is required for pancreas development. *Development* **124,** 4243–4252.

Kim, S. K., and Melton, D. A. (1998). Pancreas development is promoted by cyclopamine, a Hedgehog signaling inhibitor. *Proc. Natl. Acad. Sci. USA* **95,** 13036–13041.

Kimmel, C. B., Warga, R. M., and Schilling, T. F. (1990). Origin and organization of the zebrafish fate map. *Development* **108,** 581–594.

Knoetgen, H., Viebahn, C., and Kessel, M. (1999). Head induction in the chick by primitive endoderm of mammalian, but not avian origin. *Development* **126,** 815–825.

Kondo, T., Dolle, P., Zakany, J., and Duboule, D. (1996). Function of posterior *Hoxd* genes in the morphogenesis of the anal sphincter. *Development* **122,** 2651–2659.

Koutsourakis, M., Langeveld, A., Patient, R., Beddington, R., and Grosveld, F. (1999). The transcription factor GATA6 is essential for early extraembryonic development. *Development* **126,** 723–732.

Krapp, A., Knofler, M., Frutiger, S., Hughes, G. J., Hagenbuchle, O., and Wellauer, P. K. (1996). The p48 DNA-binding subunit of transcription factor PTF1 is a new exocrine pancreas-specific basic helix-loop-helix protein. *EMBO J.* **15,** 4317–4329.

Krapp, A., Knofler, M., Ledermann, B., Burki, K., Berney, C., Zoerkler, N., Hagenbuchle, O., and Wellauer, P. K. (1998). The bHLH protein PTF1-p48 is essential for the formation of the exocrine and the correct spatial organization of the endocrine pancreas. *Genes Dev.* **12,** 3752–3763.

Kuo, C. F., Xanthopoulos, K. G., and Darnell, J. E. (1990). Fetal and adult localization of C/EBP—evidence for combinatorial action of transcription factors in cell-specific gene expression. *Development* **109,** 473–481.

Kuo, C. T., Morrisey, E. E., Anandappa, R., Sigrist, K., Lu, M. M., Parmacek, M. S., Soudais, C., and Leiden, J. M. (1997). GATA4 transcription factor is required for ventral morphogenesis and heart tube formation. *Genes Dev.* **11,** 1048–1060.

Ladd, A. N., Yatskievych, T. A., and Antin, P. B. (1998). Regulation of avian cardiac myogenesis by activin/TGF beta and bone morphogenetic proteins. *Dev. Biol.* **204,** 407–419.

Lambert, M., Colnot, S., Suh, E., Lhorset, F., Blin, C., Calliot, M. E., Raymondjean, M., Thomasset, M., Traber, P. G., and Perret, C. (1996). *cis*-acting elements and transcription factors involved in the intestinal specific expression of the rat calbindin-D9k gene—Binding of the intestine-specific transcription factor Cdx-2 to the TATA box. *Eur. J. Biochem.* **236,** 778–788.

Landry, C., Clotman, F., Hioki, T., Oda, H., Picard, J. J., Lemaigre, F. P., and Rousseau, G. G. (1997). HNF-6 is expressed in endoderm derivatives and nervous system of the mouse embryo and participates to the cross-regulatory network of liver-enriched transcription factors. *Dev. Biol.* **192,** 247–257.

Laser, B., Meda, P., Constant, I., and Philippe, J. (1996). The caudal-related homeodomain protein Cdx-2/3 regulates glucagon gene expression in islet cells. *J. Biol. Chem.* **271,** 28984–28994.

Laurent, M. N., Blitz, I. L., Hashimoto, C., Rothbacher, U., and Cho, K. W. Y. (1997). The *Xenopus* homeobox gene *Twin* mediates Wnt induction of *goosecoid* in establishment of Spemann's organizer. *Development* **124,** 4905–4916.

Laverriere, A. C., Macneill, C., Mueller, C., Poelmann, R. E., Burch, J. B. E., and Evans, T. (1994). GATA-4/5/6, a subfamily of 3 transcription factors transcribed in developing heart and gut. *J. Biol. Chem.* **269,** 23177–23184.

Lawson, K. A., Meneses, J. J., and Pedersen, R. A. (1991). Clonal analysis of epiblast fate during germ layer formation in the mouse embryo. *Development* **113,** 891–911.

LeBras, S., Miralles, F., Basmaciogullari, A., Czernichow, P., and Scharfmann, R. (1998). Fibroblast growth factor 2 promotes pancreatic epithelial cell proliferation via functional fibroblast growth factor receptors during embryonic life. *Diabetes* **47,** 1236–1242.

Lee, Y. J., Swencki, B., Shoichet, S., and Shivdasani, R. A. (1999). A possible role for the high mobility group box transcription factor Tcf-4 in vertebrate gut epithelial cell differentiation. *J. Biol. Chem.* **274,** 1566–1572.

Lemaire, P., Darras, S., Caillol, D., and Kodjabachian, L. (1998). A role for the vegetally expressed *Xenopus* gene *Mix. 1* in endoderm formation and in the restriction of mesoderm to the marginal zone. *Development* **125,** 2371–2380.

Lemaire, P., and Gurdon, J. B. (1994). A role for cytoplasmic determinants in mesoderm patterning—cell-autonomous activation of the *Goosecoid* and *Xwnt-8* genes along the dorsoventral axis of early *Xenopus* embryos. *Development* **120,** 1191–1199.

Leptin, M. (1991). twist and snail as positive and negative regulators during *Drosophila* mesoderm development. *Genes Dev.* **5,** 1568–1576.

Li, H., Arber, S., Jessell, T. M., and Edlund, H. (1999). Selective agenesis of the dorsal pancreas in mice lacking homeobox gene *Hlxb9. Nat. Genet.* **23,** 67–70.

Lin, K., Dorman, J. B., Rodan, A., and Kenyon, C. (1997). daf-16: An HNF-3/forkhead family member that can function to double the life-span of *Caenorhabditis elegans. Science* **278,** 1319–1322.

Lin, R. L., Hill, R. J., and Priess, J. R. (1998). POP-1 and anterior-posterior fate decisions in C. elegans embryos. *Cell* **92,** 229–239.

Lustig, K. D., Kroll, K. L., Sun, E. E., and Kirschner, M. W. (1996). Expression cloning of a *Xenopus* T-related gene (*Xombi*) involved in mesodermal patterning and blastopore lip formation. *Development* **122,** 4001–4012.

MacNeill, C., Ayres, B., Laverriere, A. C., and Burch, J. B. E. (1997). Transcripts for functionally distinct isoforms of chicken GATA-5 are differentially expressed from alternative first exons. *J. Biol. Chem.* **272,** 8396–8401.

Manley, N. R., and Capecchi, M. R. (1998). Hox group 3 paralogs regulate the development and migration of the thymus, thyroid, and parathyroid glands. *Dev. Biol.* **195,** 1–15.

Mead, P. E., Zhou, Y., Lustig, K. D., Huber, T. L., Kirschner, M. W., and Zon, L. I. (1998). Cloning of Mix-related homeodomain proteins using fast retrieval of gel shift activities, (FROGS), a technique for the isolation of DNA- binding proteins. *Proc. Natl. Acad. Sci. USA* **95,** 11251–11256.

Medina, A., Wendler, S. R., and Steinbeisser, H. (1997). Cortical rotation is required for the correct spatial expression of nr3, sia and gsc in *Xenopus* embryos. *Int. J. Dev. Biol.* **41,** 741–745.

Mello, C. C., Draper, B. W., Krause, M., Weintraub, H., and Priess, J. R. (1992). The *pie-1* and m*ex-1* genes and maternal control of blastomere identity in early *C. elegans* embryos. *Cell* **70,** 163–176.

Meneghini, M. D., Ishitani, T., Carter, J. C., Hisamoto, N., Ninomiya-Tsuji, J., Thorpe, C. J., Hamill, D. R., Matsumoto, K., and Bowerman, B. (1999). MAP kinase and Wnt pathways converge to downregulate an HMG-domain repressor in *Caenorhabditis elegans. Nature* **399,** 793–797.

Milewski, W. M., Duguay, S. J., Chan, S. J., and Steiner, D. F. (1998). Conservation of PDX-1 structure, function, and expression in zebrafish. *Endocrinology* **139,** 1440–1449.

Miralles, F., Czernichow, P., Ozaki, K., Itoh, N., and Scharfmann, R. (1999). Signaling through fibroblast growth factor receptor 2b plays a key role in the development of the exocrine pancreas. *Proc. Natl. Acad. Sci. USA* **96,** 6267–6272.

Miralles, F., Czernichow, P., and Scharfmann, R. (1998). Follistatin regulates the relative proportions of endocrine versus exocrine tissue during pancreatic development. *Development* **125,** 1017–1024.

Mishina, Y., Suzuki, A., Ueno, N., and Behringer, R. R. (1995). BMPR encodes a type I bone morphogenetic protein receptor that is essential for gastrulation during mouse embryogenesis. *Genes Dev.* **9,** 3027–3037.

Mizuno, T., Yamaha, E., Kuroiwa, A., and Takeda, H. (1999). Removal of vegetal yolk causes dorsal deficiencies and impairs dorsal-inducting ability of the yolk cell in zebrafish. *Mech. Dev.* **81,** 51–63.

Mizuno, T., Yamaha, E., Wakahara, M., Kuroiwa, A., and Takeda, H. (1996). Mesoderm induction in zebrafish. *Nature* **383,** 131–132.

Mizuno, T., Yamaha, E., and Yamazaki, F. (1997). Localized axis determinant in the early cleavage embryo of the goldfish, *Carassius auratus. Dev. Genes Evol.* **206,** 389–396.

Mohun, T., and Sparrow, D. (1997). Early steps in vertebrate cardiogenesis. *Curr. Opin. Genet. Dev.* **7,** 628–633.

Monaghan, A. P., Kaestner, K. H., Grau, E., and Schutz, G. (1993). Postimplantation expression patterns indicate a role for the mouse *forkhead/HNF-3 alpha, beta* and *gamma* genes in determination of the definitive endoderm, chordamesoderm and neuroectoderm. *Development* **119,** 567–578.

Moody, S. A. (1987). Fates of the blastomeres of the 32-cell stage *Xenopus* embryo. *Dev. Biol.* **122,** 300–319.

Moon, R. T., and Kimelman, D. (1998). From cortical rotation to organizer gene expression: Toward a molecular explanation of axis specification in *Xenopus. BioEssays* **20,** 536–545.

Morrisey, E. E., Tang, Z. H., Sigrist, K., Lu, M. M., Jiang, F., Ip, H. S., and Parmacek, M. S. (1998). GATA6 regulates HNF4 and is required for differentiation of visceral endoderm in the mouse embryo. *Genes Dev.* **12,** 3579–3590.

Motoyama, J., Liu, J., Mo, R., Ding, Q., Post, M., and Hui, C. C. (1998). Essential function of *Gli2* and *Gli3* in the formation of lung, trachea and oesophagus. *Nat. Genet.* **20,** 54–57.

Mulder, G. B., Manley, N., and Maggio-Price, L. (1998). Retinoic acid-induced thymic abnormalities in the mouse are associated with altered pharyngeal morphology, thymocyte maturation defects, and altered expression of *Hoxa3* and *Pax1. Teratology* **58,** 263–275.

Muller, F., Chang, B. E., Albert, S., Fischer, N., Tora, L., and Strahle, U. (1999). Intronic enhancers control expression of zebrafish sonic hedgehog in floor plate and notochord. *Development* **126,** 2103–2116.

Muller, T. S., Ebensperger, C., Neubuser, A., Koseki, H., Balling, R., Christ, B., and Wilting, J. (1996). Expression of avian *Pax1* and *Pax9* is intrinsically regulated in the pharyngeal endoderm, but depends on environmental influences in the paraxial mesoderm. *Dev. Biol.* **178,** 403–417.

Nakatani, Y., Moody, R., and Smith, W. C. (1999). Mutations affecting tail and notochord development in the ascidian *Ciona savignyi. Development* **126,** 3293–3301.

Nardellihaefliger, D., and Shankland, M. (1993). *Lox10,* a member of the *Nk-2* homeobox gene class, is expressed in a segmental pattern in the endoderm and in the cephalic nervous system of the leech *Helobdella. Development* **118,** 877–892.

Narita, N., Bielinska, M., and Wilson, D. B. (1997a). Cardiomyocyte differentiation by GATA-4-deficient embryonic stem cells. *Development* **124,** 3755–3764.

Narita, N., Bielinska, M., and Wilson, D. B. (1997b). Wild-type endoderm abrogates the ventral developmental defects associated with GATA-4 deficiency in the mouse. *Dev. Biol.* **189,** 270–274.

Narita, T., Ishii, Y., Nohno, T., Noji, S., and Yasugi, S. (1998). Sonic hedgehog expression in developing chicken digestive organs is regulated by epithelial-mesenchymal interactions. *Dev. Growth Diff.* **40,** 67–74.

Naya, F. J., Huang, H. P., Qiu, Y. H., Mutoh, H., DeMayo, F. J., Leiter, A. B., and Tsai, M. J. (1997). Diabetes, defective pancreatic morphogenesis, and abnormal enteroendocrine differentiation in BETA2/NeuroD-deficient mice. *Genes Dev.* **11,** 2323–2334.

Newman, C. S., Chia, F., and Krieg, P. A. (1997). The *XHex* homeobox gene is expressed during development of the vascular endothelium: Overexpression leads to an increase in vascular endothelial cell number. *Mech. Dev.* **66,** 83–93.

Newman-Smith, E., and Rothman, J. H. (1999). Evidence for the involvement of END-2, a member of the steroid hormone receptor family, in endoderm and mesoderm formation in *C. elegans. Dev. Biol.* **210,** 35.

Odenthal, J., and Nusslein-Volhard, C. (1998). fork head domain genes in zebrafish. *Dev. Genes Evol.* **208,** 245–258.

Offield, M. F., Jetton, T. L., Labosky, P. A., Ray, M., Stein, R. W., Magnuson, M. A., Hogan, B. L. M., and Wright, C. V. E. (1996). PDX-1 is required for pancreatic outgrowth and differentiation of the rostral duodenum. *Development* **122,** 983–995.

Ogasawara, M., Wada, H., Peters, H., and Satoh, N. (1999). Developmental expression of *Pax1/9* genes in urochordate and hemichordate gills: Insight into function and evolution of the pharyngeal epithelium. *Development* **126,** 2539–2550.

Ogg, S., Paradis, S., Gottlieb, S., Patterson, G. I., Lee, L., Tissenbaum, H. A., and Ruvkun, G. (1997). The Fork head transcription factor DAF-16 transduces insulin-like metabolic and longevity signals in *C. elegans. Nature* **389,** 994–999.

Okada, T. S. (1957). The pluripotency of the pharyngeal primordium in urodelan embryo. *J. Embryol. Exp. Morph.* **5,** 438–448.

Okada, T. S. (1960). Epithelio-mesenchymal relationships in the regional differentiation of the digestive tract in the amphibian embryo. *Roux' Arch. Entwick.* **152,** 1–21.

Osada, S. I., and Wright, C. V. E. (1999). *Xenopus* nodal-related signaling is essential for mesendodermal patterning during early embryogenesis. *Development* **126,** 3229–3240.

Peshavaria, M., Gamer, L., Henderson, E., Teitelman, G., Wright, C. V. E., and Stein, R. (1994). Xlhbox-8, an endoderm-specific *Xenopus* homeodomain protein, is closely-related to a mammalian insulin gene-transcription factor. *Mol. Endocrinol.* **8,** 806–816.

Pevny, L., Lin, C. S., Dagati, V., Simon, M. C., Orkin, S. H., and Costantini, F. (1995). Development of hematopoietic-cells lacking transcription factor GATA-1. *Development* **121,** 163–172.

Peyrieras, N., Strahle, U., and Rosa, F. (1998). Conversion of zebrafish blastomeres to an endodermal fate by TGF-beta-related signalling. *Curr. Biol.* **8,** 783–786.

Pontoglio, M., Barra, J., Hadchouel, M., Doyen, A., Kress, C., Bach, J. P., Babinet, C., and Yaniv, M. (1996). Hepatocyte nuclear factor 1 inactivation results in hepatic dysfunction, phenylketonuria, and renal Fanconi syndrome. *Cell* **84,** 575–585.

Poulson, D. F. (1950). Histogenesis, organogenesis and differentiation in the embryo of *Drosophila melanogaster* Meigen. *In* "Biology of *Drosophila*" (M. Demerec, Ed.), pp. 275–367. Hafner Publishing Company, New York.

Rausa, F., Samadani, U., Ye, H. G., Lim, L., Fletcher, C. F., Jenkins, N. A., Copeland, N. G., and Costa, R. H. (1997). The cut-homeodomain transcriptional activator HNF-6 is coexpressed with its target gene HNF-3 beta in the developing murine liver and pancreas. *Dev. Biol.* **192,** 228–246.

Reecy, J. M., Yamada, M., Cummings, K., Sosic, D., Chen, C. Y., Eichele, G., Olson, E. N., and Schwartz, R. J. (1997). Chicken *Nkx-2.8:* A novel homeobox gene expressed in early heart progenitor cells and pharyngeal pouch-2 and -3 endoderm. *Dev. Biol.* **188,** 295–311.

Rehorn, K. P., Thelen, H., Michelson, A. M., and Reuter, R. (1996). A molecular aspect of hematopoiesis and endoderm development common to vertebrates and *Drosophila. Development* **122,** 4023–4031.

Reiter, J. F., Alexander, J., Rodaway, A., Yelon, D., Patient, R., Holder, N., and Stainier, D. Y. R. (1999). Gata5 is required for the development of the heart and endoderm in zebrafish. *Genes Dev.* **13,** 2983–2995.

Reuter, R. (1994). The gene *serpent* has homeotic properties and specifies endoderm versus ectoderm within the *Drosophila* gut. *Development* **120,** 1123–1135.

Reuter, R., Grunewald, B., and Leptin, M. (1993). A role for the mesoderm in endodermal migration and morphogenesis in *Drosophila. Development* **119,** 1135–1145.

Reuter, R., and Leptin, M. (1994). Interacting functions of *snail, twist* and *huckebein* during the early development of germ layers in *Drosophila. Development* **120,** 1137–1150.

Riese, J., Tremml, G., and Bienz, M. (1997). D-Fos, a target gene of Decapentaplegic signalling with a critical role during *Drosophila* endoderm induction. *Development* **124,** 3353–3361.

Roberts, D. J., Johnson, R. L., Burke, A. C., Nelson, C. E., Morgan, B. A., and Tabin, C. (1995). Sonic hedgehog is an endodermal signal inducing *Bmp-4* and *Hox* genes during induction and regionalization of the chick hindgut. *Development* **121,** 3163–3174.

Roberts, D. J., Smith, D. M., Goff, D. J., and Tabin, C. J. (1998). Epithelial-mesenchymal signaling during the regionalization of the chick gut. *Development* **125,** 2791–2801.

Rocheleau, C. E., Downs, W. D., Lin, R. L., Wittmann, C., Bei, Y. X., Cha, Y. H., Ali, M., Priess, J. R., and Mello, C. C. (1997). Wnt signaling and an APC-related gene specify endoderm in early C. elegans embryos. *Cell* **90,** 707–716.

Rocheleau, C. E., Yasuda, J., Shin, T. H., Lin, R. L., Sawa, H., Okano, H., Priess, J. R., Davis, R. J., and Mello, C. C. (1999). WRM-1 activates the LIT-1 protein kinase to transduce anterior posterior polarity signals in C. elegans. *Cell* **97,** 717–726.

Rodaway, A., Takeda, H., Koshida, S., Broadbent, J., Price, B., Smith, J. C., Patient, R., and Holder, N. (1999). Induction of the mesendoderm in the zebrafish germ ring by yolk cell-derived TGF-beta family signals and discrimination of mesoderm and endoderm by FGF. *Development* **126,** 3067–3078.

Romanoff, A. L. (1960). "The Avian Embryo." The Macmillan Company, New York.

Rosa, F. M. (1989). *Mix.1,* a homeobox mRNA inducible by mesoderm inducers, is expressed mostly in the presumptive endodermal cells of *Xenopus* embryos. *Cell* **57,** 965–958.

Rose, L. S., and Kemphues, K. J. (1998). Early patterning of the *C. elegans* embryo. *Annu. Rev. Genet.* **32,** 521–545.

Samadani, U., and Costa, R. H. (1996). The transcriptional activator hepatocyte nuclear factor 6 regulates liver gene expression. *Mol. Cell. Biol.* **16,** 6273–6284.

Sander, M., Neubuser, A., Kalamaras, J., Ee, H. C., Martin, G. R., and German, M. S. (1997). Genetic analysis reveals that PAX6 is required for normal transcription of pancreatic hormone genes and islet development. *Genes Dev.* **11,** 1662–1673.

Sasai, Y., Lu, B., Piccolo, S., and De Robertis, E. M. (1996). Endoderm induction by the organizer-secreted factors chordin and noggin in *Xenopus* animal caps. *EMBO J.* **15,** 4547–4555.

Sasaki, H., and Hogan, B. L. M. (1993). Differential expression of multiple fork head-related genes during gastrulation and axial pattern-formation in the mouse embryo. *Development* **118,** 47–59.

Schier, A. F., Neuhauss, S. C. F., Helde, K. A., Talbot, W. S., and Driever, W. (1997). The one-eyed pinhead gene functions in mesoderm and endoderm formation in zebrafish and interacts with no tail. *Development* **124,** 327–342.

Schlesinger, A., Shelton, C. A., Maloof, J. N., Meneghini, M., and Bowerman, B. (1999). Wnt pathway components orient a mitotic spindle in the early *Caenorhabditis elegans* embryo without requiring gene transcription in the responding cell. *Genes Dev.* **13,** 2028–2038.

Schmidt, C., Bladt, F., Goedecke, S., Brinkmann, V., Zschiesche, W., Sharpe, M., Gherardi, E., and Birchmeier, C. (1995). Scatter factor/hepatocyte growth factor is essential for liver development. *Nature* **373,** 699–702.

Schroeder, D. F., and McGhee, J. D. (1998). Anterior-posterior patterning within the *Caenorhabditis elegans* endoderm. *Development* **125,** 4877–4887.
Schulte-Merker, S., Ho, R. K., Herrmann, B. G., and Nusslein-Volhard, C. (1992). The protein product of the zebrafish homolog of the mouse T-Gene is expressed in nuclei of the germ ring and the notochord of the early embryo. *Development* **116,** 1021.
Schultheiss, T. M., Burch, J. B. E., and Lassar, A. B. (1997). A role for bone morphogenetic proteins in the induction of cardiac myogenesis. *Genes Dev.* **11,** 451–462.
Sekimoto, T., Yoshinobu, K., Yoshida, M., Kuratani, S., Fujimoto, S., Araki, M., Tajima, N., Araki, K., and Yamamura, K. (1998). Region-specific expression of murine *Hox* genes implies the *Hox* code-mediated patterning of the digestive tract. *Genes Cells* **3,** 51–64.
Seleiro, E. A. P., Connolly, D. J., and Cooke, J. (1996). Early developmental expression and experimental axis determination by the chicken Vg1 gene. *Curr. Biol.* **6,** 1476–1486.
Selleck, M. A. J., and Stern, C. D. (1991). Fate mapping and cell lineage analysis of Hensen node in the chick embryo. *Development* **112,** 615 et seq.
Seydoux, G., and Dunn, M. A. (1997). Transcriptionally repressed germ cells lack a subpopulation of phosphorylated RNA polymerase II in early embryos of *Caenorhabditis elegans* and *Drosophila melanogaster. Development* **124,** 2191–2201.
Seydoux, G., Mello, C. C., Pettitt, J., Wood, W. B., Priess, J. R., and Fire, A. (1996). Repression of gene expression in the embryonic germ lineage of *C. elegans. Nature* **382,** 713–716.
Shah, S. B., Skromne, I., Hume, C. R., Kessler, D. S., Lee, K. J., Stern, C. D., and Dodd, J. (1997). Misexpression of chick Vg1 in the marginal zone induces primitive streak formation. *Development* **124,** 5127–5138.
Shawlot, W., and Behringer, R. R. (1995). Requirement for Lim1 in head organizer function. *Nature* **374,** 425–430.
Simpson, P. (1983). Maternal-zygotic interactions during formation of the dorso-ventral pattern in *Drosophila* embryos. *Genetics* **105,** 615–632.
Skaer, H. (1993). The alimentary canal. *In* "The Development of *Drosophila melanogaster*" (M. Bate and A. Martinez Arias, Eds.), pp. 941–1012. Cold Spring Harbor Laboratory Press, New York.
Slack, J. M. W. (1995). Developmental biology of the pancreas. *Development* **121,** 1569–1580.
Sosa-Pineda, B., Chowdhury, K., Torres, M., Oliver, G., and Gruss, P. (1997). The *Pax4* gene is essential for differentiation of insulin-producing beta cells in the mammalian pancreas. *Nature* **386,** 399–402.
Soudais, C., Bielinska, M., Heikinheimo, M., Macarthur, C. A., Narita, N., Saffitz, J. E., Simon, M. C., Leiden, J. M., and Wilson, D. B. (1995). Targeted mutagenesis of the transcription factor GATA-4 gene in mouse embryonic stem cells disrupts visceral endoderm differentiation in vitro. *Development* **121,** 3877–3888.
Spyropoulos, D. D., and Capecchi, M. R. (1994). Targeted disruption of the even-skipped gene, *Evx1,* causes early postimplantation lethality of the mouse conceptus. *Genes Dev.* **8,** 1949–1961.
St. Onge, L., SosaPineda, B., Chowdhury, K., Mansouri, A., and Gruss, P. (1997). Pax6 is required for differentiation of glucagon-producing alpha- cells in mouse pancreas. *Nature* **387,** 406–409.
St. Onge, L., Wehr, P., and Gruss, P. (1999). Pancreas development and diabetes. *Curr. Opin. Genet. Dev.* **9,** 295–300.
Stennard, F., Carnac, G., and Gurdon, J. B. (1996). The *Xenopus* T-box gene, *Antipodean,* encodes a vegetally localised maternal mRNA and can trigger mesoderm formation. *Development* **122,** 4179–4188.
Stern, C. D., and Canning, D. R. (1990). Origin of cells giving rise to mesoderm and endoderm in chick embryo. *Nature* **343,** 273–275.
Strahle, U., Blader, P., Henrique, D., and Ingham, P. W. (1993). *Axial,* a zebrafish gene expressed along the developing body axis, shows altered expression in *Cyclops* mutant embryos. *Genes Dev.* **7,** 1436–1446.

Sulston, J. E., and Horvitz, H. R. (1977). Post-embryonic cell lineages of the nematode *Caenorhabditis elegans. Dev. Biol.* **56,** 110–156.

Sulston, J. E., Schierenberg, E., White, J. G., and Thomson, J. N. (1983). The embryonic cell lineage of the nematode *Caenorhabditis elegans. Dev. Biol.* **100,** 64–119.

Sun, B., Hursh, D. A., Jackson, D., and Beachy, P. A. (1995). Ultrabithorax protein is necessary but not sufficient for full activation of Decapentaplegic expression in the visceral mesoderm. *EMBO J.* **14,** 520–535.

Sun, B. I., Bush, S. M., Collins-Racie, L. A., LaVallie, E. R., DiBlasio Smith, E. A., Wolfman, N. M., McCoy, J. M., and Sive, H. L. (1999). derriere: A TGF-beta family member required for posterior development in *Xenopus. Development* **126,** 1467–1482.

Sussel, L., Kalamaras, J., Hartigan-O'Connor, D. J., Meneses, J. J., Pedersen, R. A., Rubenstein, J. L. R., and German, M. S. (1998). Mice lacking the homeodomain transcription factor Nkx2.2 have diabetes due to arrested differentiation of pancreatic beta cells. *Development* **125,** 2213–2221.

Suzuki, T., Oohara, I., and Kurokawa, T. (1999). Retinoic acid given at late embryonic stage depresses sonic hedgehog and *Hoxd-4* expression in the pharyngeal area and induces skeletal malformation in flounder (*Paralichthys olivaceus*) embryos. *Dev. Growth Diff.* **41,** 143–152.

Tada, M., Casey, E. S., Fairclough, L., and Smith, J. C. (1998). *Bix1,* a direct target of *Xenopus* T-box genes, causes formation of ventral mesoderm and endoderm. *Development* **125,** 3997–4006.

Tamai, Y., Nakajima, R., Ishikawa, T., Takaku, K., Seldin, M. F., and Taketo, M. M. (1999). Colonic hamartoma development by anomalous duplication in *Cdx2* knockout mice. *Cancer Res.* **59,** 2965–2970.

Tannahill, D., and Melton, D. A. (1989). Localized synthesis of the Vg1 protein during early *Xenopus* development. *Development* **106,** 775–785.

Tenenhaus, C., Schubert, C., and Seydoux, G. (1998). Genetic requirements for PIE-1 localization and inhibition of gene expression in the embryonic germ lineage of *Caenorhabditis elegans. Dev. Biol.* **200,** 212–224.

Tepass, U., and Hartenstein, V. (1994). Epithelium formation in the *Drosophila* midgut depends on the interaction of endoderm and mesoderm. *Development* **120,** 579–590.

Tepass, U., and Hartenstein, V. (1995). Neurogenic and proneural genes control cell fate specification in the *Drosophila* endoderm. *Development* **121,** 393–405.

Thisse, C., Perrin-Schmitt, F., Stoetzel, C., and Thisse, B. (1991). Sequence-specific transactivation of the *Drosophila* twist gene by the dorsal gene product. *Cell* **65,** 1191–1201.

Thisse, C., and Thisse, B. (1992). Dorsoventral development of the *Drosophila* embryo is controlled by a cascade of transcriptional regulators. *Dev. Suppl.,* 173–181.

Thomas, P. Q., Brown, A., and Beddington, R. S. P. (1998). *Hex:* A homeobox gene revealing peri-implantation asymmetry in the mouse embryo and an early transient marker of endothelial cell precursors. *Development* **125,** 85–94.

Thomsen, G. H., and Melton, D. A. (1993). Processed Vg1 protein is an axial mesoderm inducer in *Xenopus. Cell* **74,** 433–441.

Thorpe, C. J., Schlesinger, A., Carter, J. C., and Bowerman, B. (1997). Wnt signaling polarizes an early *C. elegans* blastomere to distinguish endoderm from mesoderm. *Cell* **90,** 695–705.

Tissierseta, J. P., Mucchielli, M. L., Mark, M., Mattei, M. G., Goridis, C., and Brunet, J. F. (1995). *Barx1,* a new mouse homeodomain transcription factor expressed in craniofacial ectomesenchyme and the stomach. *Mech. Dev.* **51,** 3–15.

Tremml, G., and Bienz, M. (1989). Homeotic gene expression in the visceral mesoderm of *Drosophila* embryos. *EMBO J.* **8,** 2677–2685.

Tremml, G., and Bienz, M. (1992). Induction of *labial* expression in the *Drosophila* endoderm—Response elements for Dpp signaling and for autoregulation. *Development* **116,** 447–456.

Troelsen, J. T., Mitchelmore, C., Spodsberg, N., Jensen, A. M., Noren, O., and Sjostrom, H. (1997). Regulation of lactase-phlorizin hydrolase gene expression by the caudal-related homoeodomain protein Cdx-2. *Biochem. J.* **322,** 833–838.

Trouche, F., and Yaniv, M. (1994). "Liver Gene Expression." Landes, New York.

Varlet, I., Collignon, J., and Robertson, E. J. (1997). *nodal* expression in the primitive endoderm is required for specification of the anterior axis during mouse gastrulation. *Development* **124,** 1033–1044.

Verheijen, M. H. G., Karperien, M., Chung, U. I., van Wijuen, M., Heystek, H., Hendriks, J. A. A., Veltmaat, J. M., Lanske, B., Li, E., Lowik, C., de Laat, S. W., Kronenberg, H. M., and Defize, L. H. K. (1999). Parathyroid hormone-related peptide (PTHrP) induces parietal endoderm formation exclusively via the Type I PTH/PTHrP receptor. *Mech. Dev.* **81,** 151–161.

Vize, P. D. (1996). DNA sequences mediating the transcriptional response of the *Mix.2* homeobox gene to mesoderm induction. *Dev. Biol.* **177,** 226–231.

Vize, P. D., and Thomsen, G. H. (1994). Vg1 and regional specification in vertebrates—a new role for an old molecule. *Trends Genet.* **10,** 371–376.

Warga, R. M., and Kimmel, C. B. (1990). Cell movements during epiboly and gastrulation in zebrafish. *Development* **108,** 569–580.

Warga, R. M., and Nusslein-Volhard, C. (1999). Origin and development of the zebrafish endoderm. *Development* **126,** 827–838.

Warot, X., Fromental Ramain, C., Fraulob, V., Chambon, P., and Dolle, P. (1997). Gene dosage-dependent effects of the *Hoxa-13* and *Hoxd-13* mutations on morphogenesis of the terminal parts of the digestive and urogenital tracts. *Development* **124,** 4781–4791.

Weber, H., Strandmann, E. P. V., Holewa, B., Bartkowski, S., Zapp, D., Zoidl, C., and Ryffel, G. U. (1996). Regulation and function of the tissue-specific transcription factor HNF1 alpha (LFB1) during *Xenopus* development. *Int. J. Dev. Biol.* **40,** 297–304.

Wedeen, C. J., and Shankland, M. (1997). Mesoderm is required for the formation of a segmented endodermal cell layer in the leech *Helobdella. Dev. Biol.* **191,** 202–214.

Weeks, D. L., and Melton, D. A. (1987). A maternal messenger RNA localized to the vegetal hemisphere in *Xenopus* eggs codes for a growth factor related to TGF-β. *Cell* **51,** 861–867.

Weinstein, D. C., Altaba, A. R. I., Chen, W. S., Hoodless, P., Prezioso, V. R., Jessell, T. M., and Darnell, J. E. (1994). The winged-helix transcription factor HNF-3β is required for notochord development in the mouse embryo. *Cell* **78,** 575–588.

Wessel, G. M., and Wikramanayake, A. (1999). How to grow a gut: Ontogeny of the endoderm in the sea urchin embryo. *Bio Essays* **21,** 459–471.

Wharton, K. A., Cook, J. M., Torres-Schumann, S., de Castro, K., Borod, E., and Phillips, D. A. (1999). Genetic analysis of the bone morphogenetic protein-related gene, gbb, identifies multiple requirements during *Drosophila* development. *Genetics* **152,** 629–640.

Wood, H. B., and Episkopou, V. (1999). Comparative expression of the mouse *Sox1, Sox2* and *Sox3* genes from pre-gastrulation to early somite stages. *Mech. Dev.* **86,** 197–201.

Wood, W. B. (1988). "The nematode *Caenorhabditis elegans.*" Cold Spring Harbor Laboratory Press, New York.

Wright, C. V. E., Schnegelsberg, P., and De Robertis, E. M. (1989). Xlhbox 8—a novel *Xenopus* homeo protein restricted to a narrow-band of endoderm. *Development* **105,** 787.

Wu, K. L., Gannon, M., Peshavaria, M., Offield, M. F., Henderson, E., Ray, M., Marks, A., Gamer, L. W., Wright, C. V. E., and Stein, R. (1997). Hepatocyte nuclear factor 3 beta is involved in pancreatic beta-cell-specific transcription of the *pdx-1* gene. *Mol. Cell. Biol.* **17,** 6002–6013.

Yamagata, K., Furuta, H., Oda, N., Kaisaki, P. J., Menzel, S., Cox, N. J., Fajans, S. S., Signorini, S., Stoffel, M., and Bell, G. I. (1996). Mutations in the hepatocyte nuclear factor-4 alpha gene in maturity-onset diabetes of the young (MODY1). *Nature* **384,** 458–460.

Yasugi, S. (1993). Role of epithelial-mesenchymal interactions in differentiation of epithelium of vertebrate digestive organs. *Dev. Growth Diff.* **35,** 1–9.

Yasugi, S., Takeda, H., and Fukuda, K. (1991). Early determination of developmental fate in presumptive intestinal endoderm of the chicken embryo. *Dev. Growth Diff.* **33,** 235–241.

Yasuo, H., and Lemaire, P. (1999). A two-step model for the fate determination of presumptive endodermal blastomeres in *Xenopus* embryos. *Curr. Biol.* **9,** 869–879.

Yatskievych, T. A., Ladd, A. N., and Antin, P. B. (1997). Induction of cardiac myogenesis in avian pregastrula epiblast: The role of the hypoblast and activin. *Development* **124,** 2561–2570.

Yatskievych, T. A., Pascoe, S., and Antin, P. B. (1999). Expression of the homeobox gene *Hex* during early stages of chick embryo development. *Mech. Dev.* **80,** 107–109.

Yokouchi, Y., Sakiyama, J., and Kuroiwa, A. (1995). Coordinated expression of *Abd-B* subfamily genes of the *Hoxa* cluster in the developing disgestive-tract of chick embryo. *Dev. Biol.* **169,** 76–89.

Zakany, J., and Douboule, D. (1999). *Hox* genes and the making of sphincters. *Nature* **401,** 761–762.

Zaret, K. (1998). Early liver differentiation: Genetic potentiation and multilevel growth control. *Curr. Opin. Genet. Dev.* **8,** 526–531.

Zaret, K. (1999). Developmental competence of the gut endoderm: Genetic potentiation by GATA and HNF3 fork head proteins. *Dev. Biol.* **209,** 1–10.

Zarnegar, R., Defrances, M. C., and Michalopoloulos, G. K. (1994). Hepatocyte growth factor. *In* "The Liver: Biology and Pathobiology" (I. M. Arias, J. L. Boyer, N. Fausto, W. B. Jakoby, D. A. Schachter, and D. A. Shafritz, Eds.), pp. 1047–1057. Raven Press, New York.

Zeynali, B., and Dixon, K: E. (1998). Effects of retinoic acid on the endoderm in *Xenopus* embryos. *Dev. Genes Evol.* **208,** 318–326.

Zhang, J., Houston, D. W., King, M. L., Payne, C., Wylie, C., and Heasman, J. (1998a). The role of maternal VegT in establishing the primary germ layers in Xenopus embryos. *Cell* **94,** 515–524.

Zhang, J., and King, M. L. (1996). *Xenopus* VegT RNA is localized to the vegetal cortex during oogenesis and encodes a novel T-box transcription factor involved in mesodermal patterning. *Development* **122,** 4119–4129.

Zhang, J. J., Talbot, W. S., and Schier, A. F. (1998b). Positional cloning identifies zebrafish one-eyed pinhead as a permissive EGF-related ligand required during gastrulation. *Cell* **92,** 241–251.

Zhu, J. W., Fukushige, T., McGhee, J. D., and Rothman, J. H. (1998). Reprogramming of early embryonic blastomeres into endodermal progenitors by a *Caenorhabiditis elegans* GATA factor. *Genes Dev.* **12,** 3809–3814.

Zhu, J. W., Hill, R. J., Heid, P. J., Fukuyama, M., Sugimoto, A., Priess, J. R., and Rothman, J. H. (1997). end-1 encodes an apparent GATA factor that specifies the endoderm precursor in *Caenorhabiditis elegans* embryos. *Genes Dev.* **11,** 2883–2896.

Zorn, A. M., Butler, K., and Gurdon, J. B. (1999). Anterior endomesoderm specification in *Xenopus* by Wnt/beta-catenin and TGF-beta signalling pathways. *Dev. Biol.* **209,** 282–297.

Neural Patterning in the Vertebrate Embryo

Curtis R. Altmann and Ali H. Brivanlou
The Rockefeller University, New York, New York 10021

The embryonic central nervous system (CNS) is patterned along its antero-posterior, dorsal-ventral, and left-right axes. Along the dorsal-ventral axis, cell fate determination occurs during and following neural tube closure and involves the action of two opposing signaling pathways: SHH ventrally from the notochord and BMP/GDF dorsally from the boundary of neural and nonneural ectoderm and later from the roof plate. In addition, Wnt and retinoic acid signaling have been shown to act in dorsal-ventral patterning; however, their roles are understood in less detail. Along the antero-posterior axis, signals divide the neural tube into four major divisions: forebrain, midbrain, hindbrain, and spinal cord, and these differences can be detected soon after the formation of the neural plate. The FGF, Wnt, and retinoic acid signaling pathways have been implicated in the caudalization of neural tissue. Boundaries of Hox gene expression are observed along the antero-posterior axis and have been suggested to be involved in establishing different identities in the hindbrain and spinal cord. Complex gene expression patterns in the brain suggest the development of neuromeres dividing the brain into different regions that are elaborated further during development. Patterning along the left-right axis occurs concurrently with antero-posterior and dorsal-ventral patterning during gastrulation. A leading candidate for initiating asymmetry is activin, which acts through Nodal and Lefty before any morphological differences are observed. The big challenge will be understanding how these diverse signaling pathways interact both temporally and spatially to generate the complex adult nervous system.

KEY WORDS: Neural patterning, Embryonic axis, Cell fate, Cell signaling, Vertebrate embryos.

I. Introduction

The cellular and molecular basis underlying the formation of the nervous system in the vertebrate embryo is a topic that has generated passionate debate among developmental biologists. In recent years the tools of molecular biology have endowed experimental embryologists with the ability to provide molecular solutions to classical embryological problems. Among these problems were the molecular basis of embryonic induction, a subject of intense scrutiny for the last few decades of the twentieth century. In this chapter we discuss one aspect of embryonic induction: the processes that underlie the formation of the diverse regions and cell types of the nervous system, focusing specifically on the embryological, cellular, and molecular basis of early neural patterning in the vertebrate embryo. Neural patterning is a process that initiates during neural induction (or neuralization), the earliest phase of establishing a neuronal fate, and continues afterward throughout embryonic development. Most classical work in vertebrate embryology has been performed on the embryos of birds and amphibians, and these systems still play a leading role in allowing an understanding of the molecular aspects of embryonic development. Recent model systems such as the fish and the mouse have begun to contribute to our knowledge of embryonic development, and by taking advantage of genetic approaches, have joined the ranks of the classical studies in providing novel insights in the understanding of the molecular decisions of early vertebrate development.

In all vertebrates the presumptive neural tissue is specified during a very important process of early embryonic development called gastrulation. During gastrulation, the cells located in the dorsal side of the ectoderm receive neural-promoting signals from a dorsal organizing center. These signals inhibit ongoing signaling from the BMP/GDF subfamily of ligands [which belong to the transforming growth factor-β (TGF-β) super-family of signaling factors; Weinstein and Hemmati-Brivanlou, 1999]. In the ventral ectoderm, where the BMP/GDF signals escape these inhibitors, an epidermal fate is imposed (Wilson and Hemmati-Brivanlou, 1997). Therefore, while the induction of the epidermis clearly involves instructive signaling (in molecular terms, a signal that activates a transduction cascade), the genesis of the neurogenic ectoderm can be considered permissive (inhibition of signaling from outside of the cell which does not involve a direct activation of a signaling cascade). This molecular scenario for the formation of the embryonic nervous system has been named the *default model* of neurogenesis, and the word *neuralization* was suggested instead of neural induction (Hemmati-Brivanlou and Melton, 1997) to highlight the permissive nature of the signaling involved in the specification of the neuronal fate. It is important to emphasize, however, that while the default model has enjoyed the support of many lines of evidence and is generally accepted in the amphibian system, this model has been challenged in the amniotes and is still contested (Streit and Stern, 1999). Since this topic has been reviewed extensively elsewhere it will not be discussed in this chapter

(Eagleson *et al.*, 1995; Brivanlou and Melton, 1997; Gould and Grainger, 1997; Tiedemann *et al.*, 1998).

The first morphological manifestation of the future nervous system occurs shortly after gastrulation during neurulation, when a thickening of the ectoderm in the dorsal side demarcates the future neural plate. The neural plate undergoes morphogenetic movements that first create a neural groove by raising the lateral edges of the plate and ultimately the neural tube when the lateral edges of the plate meet and close dorsally. This tube will then further differentiate to create patterns of cell types in three axes, thus forming the brain in the anterior part and the spinal cord in the posterior end. At neural groove/neural tube stages, the neural crest, a complex population of cells derived from the dorsal side of the neural tube, detaches from the developing embryonic nervous system and begins migration toward specific targets (Mayor *et al.*, 1999; Bronner-Fraser, 1995a, b; Selleck *et al.*, 1993). Once the primary neurons have been specified, the cells divide once or twice before projecting axons to the appropriate targets (Hartenstein, 1993). The resultant vector of all these events ultimately gives rise to a mature nervous system that will allow the embryo to begin to interact with its environment.

In this chapter, whenever possible, we take a comparative approach in describing what is currently known about these important embryonic processes. As an introduction, we discuss briefly the origin of the nervous system in vertebrates as well as the spatial and temporal coordinates involved in the genesis of this tissue. We take a little historical tour by highlighting the crucial contribution of experimental embryology from its birth at the turn of the century. It will become obvious to the reader that the heritage of this early work had set the stage for the modern molecular characterization of neural patterning. We then explore the timing of formation as well as the spatial origin of the nervous system by discussing neurula fate maps. Following this is a description of the cellular and molecular basis of neural patterning along the antero-posterior (A-P) and dorsal-ventral (D-V) axes, and the right-left patterning of the embryonic brain and spinal cord. We then explore the lessons learned from genetic approaches in zebrafish and mouse regarding neural patterning.

II. Neural Patterning

A. Experimental Embryological View

The embryonic central nervous system (CNS) can be subdivided along its antero-posterior axis into four major divisions: forebrain, midbrain, hindbrain, and spinal cord. Classical experiments, which consisted of cutting and pasting different pieces of the neural tube in different orientations and locations, had afforded two discoveries: first, that A-P and D-V patterning are independent events, and second, that the establishment of the A-P axis precedes the establishment of fate in the D-V axis as demonstrated by experiments in chicks (Simon *et al.,* 1995; Graff *et al.,* 1989; Stumpo *et al.,* 1989).

During gastrulation, neural induction occurs in parallel with the earliest aspects of neural patterning. At the early neural plate stage, the axes of the future nervous system have been laid out. While not yet differentiated, the cells have received and continue to receive signals directing them to their fate. In *Xenopus,* this can be seen in the early neural plate by monitoring molecular markers such as the expression of OtxA (*otd*-type homeobox), En2 (*en* homeobox class), Krox20 (zinc finger containing transcription factor), and HoxB9 (*Antp*-type homeobox protein), which are localized to the forebrain, midbrain/hindbrain, hindbrain, and spinal cord, respectively (Lamb *et al.,* 1993; Brivanlou and Harland, 1989; Nieto *et al.,* 1992). Experiments to determine whether these tissues are specified (i.e. differentiate without further signals when explanted) are technically demanding due to the difficulty in separating the underlying mesoderm from the overlying neural tissue. By tracing the lineage of marked cells, a fate map of the embryo at neural plate stages has been prepared from *Xenopus* embryos (Eagleson and Harris, 1990) and is presented in Figure 1a (see color insert). These studies reveal that the more dorsal fates are derived from the lateral regions of the neural plate, while ventral cells arise from medial regions. This is a consequence of the folding of the neural plate to form the neural tube. The more rostral regions of the brain map to the anterior end of the neural plate, while caudal regions lie posteriorly. In addition, all subdivisions of the brain lay contiguously also at the neural plate stages, thus indicating that migration is not playing a major role in organizing the brain architecture (Eagleson *et al.,* 1995; Eagleson and Harris, 1990). However, the ventral forebrain arises from the antero-medial part of the neural ridge where descendents do not always lie contiguously. This indicates that in this region, cellular migration is important in determining the final spatial map. In the chick, a similar anterior to posterior arrangement of neural tissue has been identified (Figure 1b), though there appears to be more mixing of the cells, especially in the A-P axis (Bortier and Vakaet, 1992; Hatada and Stern, 1994; Garcia-Martinez *et al.,* 1993; Schoenwolf *et al.,* 1989; Schoenwolf and Sheard, 1990). These maps suggest a conservation of fate in both non-amniotes and amniotes, though differences are likely to be identified. To date, neurula fate maps in the mouse and zebrafish have not been determined at similar resolutions.

B. The D-V Patterning

1. Neural Tube

The initial patterning along the medio-lateral axis in the dorsal ectoderm occurs at the open neural plate stages. By the time the neural plate closes dorsally to form the neural tube, medial cells assume the ventral positions to develop into floor plate and ventral types of neurons, while the most lateral cells populate the dorsal structures and become neural crest, roof plate, and dorsal types of neurons. Signals from both the underlying mesoderm and the epidermis seem to influence dorsal-ventral patterning, and many factors have also been identified that play a role in this process (reviewed in Graham, 1997). Though most of our knowledge on regulation of D-V patterning comes from studies in chick or mouse at the spinal cord level, similar mechanisms seem to operate in the brain region and may be applied to other vertebrates.

2. Spinal Cord

The dorsal, lateral, and ventral differences in the embryonic spinal cord are manifested by the presence of different cell types. The most ventral cell type in the spinal cord is the floor plate. Immediately adjacent to these cells are the motor neurons. The lateral part of the neural tube is populated by different types of interneurons. The most dorsal cells in the neural tube belong to the roofplate, which itself is flanked on each side by sensory neurons (see Figure 2 on color insert and Table I). During neural tube closure (in amphibians) or immediately following closure (amniotes) a population of cells, called neural crest, migrate away from the most dorsal region of the neural tube (Mayor *et al.,* 1999; Bronner-Fraser, 1995a, b). Although these cells are originally multipotent, they differentiate in response to uncharacterized cues within the local environment into a wide variety of cell types including cartilage, pigmented cells, and neurons. In the case of the spinal cord these neural crest populations contribute to different trunk structures, while the more rostral neural crest cells contribute to a number of head structures, including the skeleton of the head as well as a diverse population of sensory neurons and glia. While these cells are pluripotent, they do become restricted at later times (Artinger and Bronner-Fraser, 1992b).

a. Cell Fate Determination in the Ventral Spinal Cord In recent years the molecular dissection of events underlying patterning across the D-V axis of the spinal cord has allowed the characterization of signaling factors involved in the specification of ventral cell fate. Elegant experiments performed first in chick embryos and subsequently on amphibians demon-

TABLE I

Spinal Nerves and Molecular Markers in Chick

Spinal nerve identity	Marker	Reference
MN_D	isl-1, isl-2, lim-3	(Ericson *et al.*, 1997; Tsuchida *et al.*, 1994)
MNv	isl-1	(Ericson *et al.*, 1997)
Dorsal limb	lim1, isl-2	(Jungbluth *et al.*, 1999)
Ventral limb/muscle	isl-1, isl-2	(Ruiz i Altaba, 1996)
Dorsal interneuron 1	mAth1	(Helms and Johnson, 1998)
Dorsal interneuron 2	Ngn-1	(Perez *et al.*, 1999)
Dorsal interneuron 3	D1b	(Lee *et al.*, 1998)
D1	Lim1/2-, LH2A/B	(Liem *et al.*, 1997)
V0	Evx1/2	(Ericson *et al.*, 1996; Burrill *et al.*, 1997)
V1	En1, Lim 1/2, Pax2	(Matise and Lance-Jones, 1996)
V2	Chx10, Lim 3	(Ericson *et al.*, 1997)
MN	Isl1/2	(Pfaff *et al.*, 1996)
V3	Nkx2.2	(Briscoe *et al.*, 1999)
FP	SHH	(Roelink *et al.*, 1994)
Ventral interneuron	lim-3, en-1	(Tsuchida *et al.*, 1994)

strated that the notochord (axial mesoderm) is the source of signals involved in the specification of the floor plate and secondarily to the formation of motor neurons and ventral interneurons. The notochord itself acts as an organizing center and hosts a number of secreted factors (Figure 2a). The leading candidates for the induction of the floor plate are currently the members of the hedgehog family of signaling factors. One such member, sonic hedgehog (SHH), is expressed in the notochord (Krauss *et al.*, 1993; Echelard *et al.*, 1993; Riddle *et al.*, 1993; Roelink *et al.*, 1994; Chang *et al.*, 1994). Loss-of-function as well as gain-of-function experiments have suggested that SHH is both necessary and sufficient to induce the floor plate. More interestingly, SHH has been shown to be able to act as a morphogen, eliciting different cell fates at different thresholds of concentration. While it is still unclear how many thresholds of SHH exist, there is evidence that at high concentration (10^{-8}M) SHH will strongly induce floor plate while at lower concentrations (10^{-9}M) it will induce motor neurons (Roelink *et al.*, 1995). Additionally, the induction of the floor plate seems to be contact-dependent whereas the induction of the motor neurons does not require contact (Placzek *et al.*, 1993, 1990; Tanabe *et al.*, 1995). Since SHH protein exists in both a membrane bound form as well as an N-terminally cleaved secreted protein (Lee *et al.*, 1994; Porter *et al.*, 1995; Bumcrot *et al.*, 1995), it is presumed that the membrane bound form of the protein is involved in floor plate induction while the soluble secreted form is in charge of motor neuron specification. The requirement for SHH is indicated by the inhibition of notochord-mediated induction of ventral types by neutralizing antibodies to SHH (Marti *et al.*, 1995; Ericson *et al.*, 1996). It is also important to note that while the induction of the floor plate by SHH has been suggested to be direct (Ruiz i Altaba *et al.*, 1995), the specification of motor neurons might require a signaling intermediate. In addition, there is evidence that SHH requires TGF-β signaling to elicit its effects (Bitgood and McMahon, 1995). Though still an open issue, recent experiments using inhibitory anti-SHH antibodies suggest that SHH directly induces and is required for motor neuron differentiation (Ericson *et al.*, 1996).

Through a signal transduction cascade (Figure 3a) elucidated by both genetic approaches as well as biochemical studies, SHH ultimately induces or represses the expression of specific set of transcription factors in the nucleus. Within the neural tube, SHH signaling acts to repress the expression of Pax3 and Pax7 (members of the paired box homeodomain family) as well as Msx1 and Msx2 (msh type of homeobox proteins, Goulding *et al.*, 1993a, b; Liem *et al.*, 1995). SHH signaling acts to upregulate nkx2.1, nkx2.2 (members of the NK-2 type homeobox class Barth and Wilson, 1995; Ericson *et al.*, 1995). By regulating specific downstream genes these transcription factors then proceed intrinsically to impose and maintain the fate dictated by SHH. Genes such as HNF3-β have been shown to be required for the formation of the notochord and the floor plate as HNF3-β null mice lack both of these tissues (Weinstein *et al.*, 1994; Ang and Rossant, 1994). Although not supported by direct evidence, the lack of the floor plate in these mutants is assumed to be due to the lack of notochord and therefore the absence of SHH signaling. Supporting this view are gain-of-function experiments performed in *Xenopus* embryos where ectopic expression of the *Xenopus* HNF3-β homologue (pintallavis) has been shown to induce ectopic floor plate (Ruiz i Altaba and Jessell, 1992; Ruiz i Altaba, 1992; Ruiz i Altaba *et al.*, 1993; Hynes *et al.*, 1995; Sasaki and Hogan, 1996). Another transcription factor, Isl-1, has been shown to be required for the formation of both motor neurons and a subpopulation of ventral interneurons (which are selectively expressing the marker En-1). In all vertebrates, Isl-1 is selectively expressed in the motor neurons (Karlsson *et al.*, 1990; Ericson *et al.*, 1992; Tsuchida *et al.*, 1994; Pfaff *et al.*, 1996). Both in the mouse and in the chick, elimination of the Isl-1 gene results in embryos lacking the entire population of motor neurons (Pfaff *et al.*, 1996). It is interesting that in addition to the lack of motor neurons, a population of ventral interneurons (which are also $En1^+$) is also missing in these homozygous $Isl\text{-}1^-$ mice. This suggests that the $En1^+$ interneurons are dependent on the presence of the motor neurons or dependent on a cell type generated by the motor neurons, and highlights a cascade of inductive interactions originally derived from the notochord. In these embryos the notochord and floor plate develop apparently

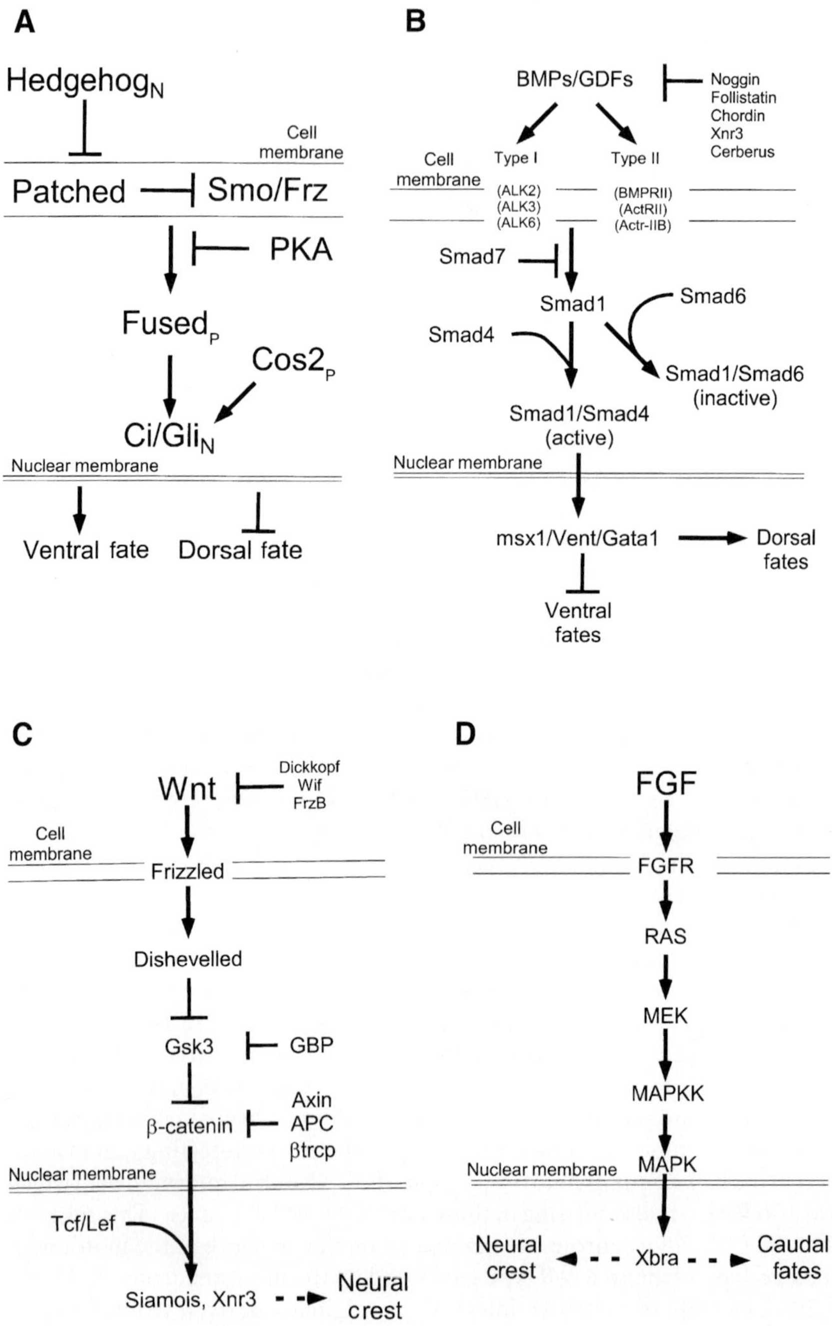
A
Hedgehog$_N$
Cell membrane
Patched
Smo/Frz
PKA
Fused$_P$
Cos2$_P$
Ci/Gli$_N$
Nuclear membrane
Ventral fate
Dorsal fate
B
BMPs/GDFs
Noggin
Follistatin
Chordin
Xnr3
Cerberus
Cell membrane
Type I
Type II
(ALK2)
(ALK3)
(ALK6)
(BMPRII)
(ActRII)
(Actr-IIB)
Smad7
Smad1
Smad6
Smad4
Smad1/Smad6
(inactive)
Smad1/Smad4
(active)
Nuclear membrane
msx1/Vent/Gata1
Dorsal fates
Ventral fates
C
Wnt
Dickkopf
Wif
FrzB
Cell membrane
Frizzled
Dishevelled
Gsk3
GBP
Axin
APC
βtrcp
β-catenin
Nuclear membrane
Tcf/Lef
Siamois, Xnr3
Neural crest
D
FGF
Cell membrane
FGFR
RAS
MEK
MAPKK
MAPK
Nuclear membrane
Neural crest
Xbra
Caudal fates

normally, thus suggesting that the lack of ventral cell types is not an indirect consequence of perturbing the differentiation of the midline. Secreted signals are not only necessary, but also sufficient for the specification of the $En1^+$ interneurons, as shown by experiments in which conditioned medium prepared from ventro-lateral neural tubes of wild-type mice restored the population of $En1^+$ interneurons in the neural tube of $Is1\text{-}1^{-/-}$ mice. It is also important to emphasize that not all the populations of ventral interneurons are dependent of SHH signaling. For example, the $Lim\text{-}2^+$ group of interneurons (V1) can be formed in embryos lacking notochord ($HNF3b^{-/-}$, Weinstein *et al.*, 1994) and this may be due to RA signaling. The inhibition of BMP/GDF ligands by Noggin is also required for the proper formation of ventral cell types in the presence of normal SHH signaling (McMahon *et al.*, 1998).

While SHH signaling accounts for some aspects of the patterning of the ventral spinal cord, a role for retinoic acid (RA) in D-V patterning has been established. SHH is sufficient to induce floor plate and motor neurons

---

FIG. 3 Signaling pathways in neural patterning. (A) Hedgehog signaling pathway: Hedgehog (Hh) binds to and inhibits the 12-membrane spanning Patched receptor (ptc1 or ptc2 in vertebrates) on the cell surface. Ligand binding acts to geneticaly de-repress the Smoothened/ Frizzled (Smo/Frz) receptor of the serpentine family, which has not been shown to interact directly with Hh but which is required for Hh function. Though the mechanism of action is not known, signaling results in the stable phosphorylation of the serine-threonine kinase Fused (Fu) and the kinesin-related protein Costal-2 (Cos-2) in the cytoplasm. Events in the cytoplasm affect the transcription factor Cubitus interruptus (Ci), which is related to the oncogene Gli (Gli1) in an uncharacterized manner. The Ci/Gli transcription factors can both activate and repress transcription leading to dorsal and ventral fate determination in the neural tube (Dierick and Bejsovec, 1999). (B) BMP/GDF signaling pathway: BMP/GDF ligands bind to heteromeric type I and type II receptors and induce the autophosphorylation of the receptors on serine and threonine residues. The active receptor complex phosphorylates the Smad1 protein, which associates with the Smad4 and translocates to the nucleus where it activates the transcription of Msx/Vent/Gata1 genes to promote dorsal fates and inhibit ventral fates. Smad6 acts as decoy and competes for heterodimerization of Smad1 with Smad4 and thus selectively inhibits Smad1. Smad7 is a general inhibitor of Smads and interferes with phosphorylation of Smads by the receptors (Weinstein and Hemmati-Brivanlou, 1999). (C) Wnt signaling pathway: Soluble Wnt ligands bind to the Frizzled receptor and activate disheveled (*dsh*). Dsh acts to inhibit the activity of GSK3, which in turn inhibits β-catenin. β-catenin translocates to the nucleus, where it associates with a member of the Tcf/Lef family of nuclear factors to activate the genes Siamois and Xnr3, thus leading to the formation of neural crest derivatives. Axin, APC, and Trcp act to inhibit β-catenin while GBP acts to inhibit GSK3 (Dierick and Bejsovec, 1999). (D) FGF signaling pathway: FGF ligands bind to the FGF receptor to induce dimerization and activate Ras signaling by tyrosine phosphorylation. Ras activates Raf and leads to the sequential phosphorylation of MEK, MAPKK, and MAPK. Activated MAPK translocates into the nucleus to activate brachyury, which later leads to the formation of caudal fates and the neural crest lineage (Weinstein and Hemmati-Brivanlou, 1999).

ventrally but is not required for all ventral cells types. While SHH knockout mice lack floorplate and motor neurons (Chiang *et al.,* 1996), four other classes of ventral neuron (as defined by patterns of homeobox expression) still develop. These markers are presented in Table 1. In the absence of SHH signaling, RA can induce the expression of the markers Dbx2 and Evx1/2 in tissue explants (Pierani *et al.,* 1999). This activity is inhibited by the expression of Pax7 (which is activated dorsally in response to BMP/GDF signaling). The source of the RA that might be responsible *in vivo* for this activity is not known. Possible sources include notochord precursors (Hogan *et al.,* 1992; Wagner *et al.,* 1992) and the paraxial mesoderm (Maden *et al.,* 1998). As has been noted in other sections, the strong conservation of signaling mechanisms along the neural tube suggests that other RA dependent D-V patterning pathways have yet to be characterized at other levels along the A-P axis.

b. Cell Fate Determination in the Dorsal Spinal Cord The anatomy of the dorsal part of the spinal cord resembles the ventral side in the sense that a specialized group of cells contributes to an anatomically distinct region called the roof plate. Flanking the roof plate are a population of dorsal sensory relay neurons. Moving more laterally are groups of dorsal interneurons. The dorsal side of the neural tube is also the birth place of the neural crest cells, which migrate following neural tube closure. There is evidence that the fate of these cell types is established independently of the signal derived from axial mesoderm. It is interesting that if the notochord is removed or if signaling from the notochord is blocked, the cells of the neural tube adopt a dorsal fate (Yamada *et al.,* 1991; Artinger and Bronner-Fraser, 1992a, b; Monsoro-Burq *et al.,* 1994). Evidence for this is provided by the behavior of two markers of dorsal cell types, Msx-1 and Pax-3. While in wild-type embryos these genes demarcate dorsal cell types within the neural tube, in embryos lacking signaling from the notochord these markers are expressed along the entire D-V axis of the neural tube, thus suggesting that the cells along this axis have adopted a dorsal fate (Goulding *et al.,* 1993b; Liem *et al.,* 1995). The action of BMP/GDF signaling dorsally, combined with the ventral signaling initiated by SHH, results in the differential expression of Pax7, Pax3, Msx1, and Mxs2 from the dorsal aspect and the expression of Pax6 and Gli1,2,3 genes ventrally.

What then is the origin of signals that specify the fate of the dorsal cells within the spinal cord? Molecular studies addressing these issues, mostly performed in chick embryos, have elegantly demonstrated that the specification of these cells occurs early during neural plate/neural groove stages of neural tube specification. Signals derived from the epidermis (immediately adjacent to the neural plate) can induce the differentiation of both roof plate as well as neural crest cells (Moury and Jacobson, 1990; Sechrist *et al.*, 1995; Dickinson *et al.*, 1995; Selleck and Bronner-Fraser, 1995; Liem *et al.*, 1997). Three types of signaling factors have been implicated in this biological activity. First, the BMP subgroup of the TGF-β superfamily of growth factors (specifically BMP 4/5/7 and dorsalin-1) has been shown to be able to induce both the roof plate as well as the neural crests in the chick embryo (Liem *et al.*, 1997; Liem *et al.*, 1995). The signal transduction cascade mediated by BMP/GDF signals has been recently identified (Massague, 1998) and characterized and is presented in Figure 3b. These BMP/GDF signaling factors are believed to be derived from the epidermis flanking the neural plate and are assumed to signal in a planar fashion, specifying fate at the border of neural and nonneural ectoderm. These recent molecular studies are also in agreement with experimental embryological approaches in which pieces of nonneural ectoderm (epidermis) were grafted in the neural plate and pieces of the neural plate were grafted within the epidermal environment (Moury and Jacobson, 1989; Moury and Jacobson, 1990; Selleck and Bronner-Fraser, 1995). In both cases neural crest cells were formed at the junction between the two tissues, suggesting that signaling for neural crest specification is derived from the interactions at the border between the two tissues. Later, at or after neural tube closure, the BMPs that are localized to the roof plate after the tube closes have also been shown to induce dorsal interneurons (Liem *et al.*, 1997). Since BMPs can induce mulitiple types of cells from a given source of signaling, it has been speculated that BMPs might act as a morphogen that would elicit different activities at different thresholds. In favor of this argument is the demonstration that BMP4 can act as a morphogen in the context of the embryonic ectoderm in the frog (Wilson *et al.*, 1997). There is evidence of at least three thresholds for BMP4: at low or zero BMP4 activity, cells will adopt an anterior neural fate (telencephalon); at intermediate activity, BMP4 will induce sensory placodes and cement gland (both structures arise at the anterior border of the neural plate in a region which overlaps both the neural as well as nonneural ectoderm). Finally, at the highest dose, BMP4 will induce an epidermal fate. It is important however to note that while these studies clearly indicate that BMP4 can act as a morphogen, there is to date no evidence that morphogens actually do exist in vertebrates. In parallel with this argument, there is also evidence in the chick against a morphogen type of effect of BMP4 for the specification of the dorsal cell types. The same dose of BMP4 has been shown to trigger all three cell types: roof plate, neural crest, and dorsal interneurons (Liem *et al.*, 1997). It could be speculated here that although the same amount of BMP4 protein elicited all three cell types the amount of activity that each cell has seen might be different, as secreted inhibitors of BMPs such as noggin and chordin have been shown to be present in the dorsal side of the chick neural tube. If BMP4 does not act as a morphogen in this context two other scenarios can be envisioned. It has been speculated that BMPs (as well as other members of the TGF-β family ligands) can heterodimerize and that these heterodimers display qualitatively and quantitatively different responses than their homodimer counterparts (Aono *et al.,* 1995). Therefore, different fates can be due to the formation of different heterodimers and the associated changes in activity. Finally, since a difference in the timing of the specification of the cell types exists (with roof plate being specified first and interneurons last) perhaps it is the difference in the competence of the responsive cells at different times that ultimately allows the generation of these three cell types. In favor of the timing arguments are experiments performed with young versus aged neural explants, which have shown that BMP induces different cell types as a function of time (Liem *et al.,* 1997).

The second type of signaling involved in neural crest specification has been attributed to specific members of the Wnt superfamily of signaling factors. The signaling pathway is presented in Figure 3c. Wnts may be involved in both cell fate determination and expansion of dorsal progenitor cells (through proliferation) within the embryonic neural tube. In *Xenopus,* expression of secreted BMP inhibitors such as noggin, chordin, or follistatin in embryonic ectodermal explants (animal caps) induces neural fate that is anterior in character. In the absence of these inhibitors the explants would solely differentiate as epidermis. Coexpression of Wnt1, Wnt3a, or Wnt7B with neural inducers in animal caps has been shown to induce expression of neural crest and dorsal neural markers (Saint-Jeannet *et al.,* 1997; Chang and Hemmati-Brivanlou, 1998a). In addition, *in vivo* overexpression of these Wnts enlarges the neural crest population flanking the neural plate. The induction of dorsal cell fates by Wnt1 and Wnt3a occurs in the absence of cell division, indicating that these Wnts (or other Wnt members) may specify dorsal cells directly instead of merely stimulating survival or multiplication of these cells (Saint-Jeannet *et al.,* 1997). Since experimental embryological evidence had demonstrated that juxtaposition of epidermis and neural tissue is sufficient to induce neural crest cells (see above and Selleck and Bronner-Fraser, 1995), it could have been possible that Wnts induce only epidermis, which would indirectly give rise to neural crest cells when in close proximity to the neural tissue formed in response to the neural inducers. Epidermal induction assays performed with the same Wnts have clearly demonstrated that Wnt activity alone is not sufficient to induce epidermis and that the induction of dorsal fate occurs without a prior commitment of epidermal induction by the Wnts in ectodermal explants (A. Suzuki and A. Hemmati-Brivanlou, unpublished observations). One unresolved issue in these experiments is that the same Wnts have strong caudalizing activity (see below), and it still can be argued that the induction of the dorsal fate and neural crest cells can be dependent, or secondary, to caudalization. In double knockout mice lacking both Wnt1 and Wnt3a, caudal structures seem to be normal but the embryos display a reduced (but not eliminated) neural crest cell population, suggesting that the caudalization and dorsalization activities of the Wnts can be uncoupled. Additional roles for Wnt1 and Wnt3a in expansion of dorsal neural cells as well as partial redundancy of activity with other Wnts has also been suggested (Ikeya *et al.,* 1997). The expression of these ligands in *Xenopus* and other vertebrate is also consistent with their described activities within the dorsal neural tube. In Xenopus, Wnt7B is expressed before the onset of neural crest induction and is later restricted to the dorsal midline of the neural tube. Wnt1 and Wnt3a are expressed in the dorsal midline of the spinal cord of many vertebrate embryos (Wolda *et al.,* 1993; Parr *et al.,* 1993; Hollyday *et al.,* 1995; Chang and Hemmati-Brivanlou, 1998a, b). It is presently unclear how BMPs and Wnts might work together to pattern the dorsal neural tube. It is possible that the two signals synergize with each other to fine tune the timing and/or spatial competence of the responding cells; alternatively, one pathway may work downstream of another to determine dorsal neural cell fates.

Finally, FGF has also been implicated in specification of neural crest cells. The FGF signaling pathway is presented in Figure 3d. A truncated FGF receptor blocks expression of the neural crest-specific gene, Slug, in embryos without affecting the pan-neural marker gene Sox-2, while co-expression of the neural inducer noggin and FGF stimulates Slug transcription (Mayor *et al.,* 1995, 1997). As mentioned in the case of Wnts, since FGF has been shown to have caudalizing activity (Cox and Hemmati-Brivanlou, 1995), it is presently not clear if the induction of neural crest by FGF is secondary to its caudalization activity (Kengaku and Okamoto, 1995; Lamb and Harland, 1995). Since the requirement for an active FGF pathway occurring during neural induction has been suggested, it is tempting to speculate that FGFs might have a general role in regulating the competence of the ectoderm to respond to these inducers in a similar scenario as the one suggested for FGF during mesoderm induction (Klein and Melton, 1994).

As is the case for cell fate determination in the ventral neural tube, a large number of transcription factors have been shown to have direct roles in the induction and maintenance of dorsal cell fates. Among these, some have been shown to be immediate early response genes to either BMPs or the Wnt signaling pathways. These include Msx1 and 2, which are expressed in a broad domain in the dorsal side of the neural tube in all vertebrates so far studies (Davidson, 1995). Msx genes expression has been demonstrated to be an immediate early response to BMP signaling (Suzuki *et al.,* 1997). Knock-out of msx1 and msx2 in mice leads to craniofacial defects which may be due to the elimination of cranial neural crest derivatives

(Foerst-Potts and Sadler, 1997). The paired homeobox genes Pax3 and Pax7 also demarcate the dorsal domain in the neural tube (Goulding *et al.*, 1993a, b; Jostes *et al.*, 1990). Consistent with a deterministic role of these transcription factors in the establishment of dorsal fate is the observation that elimination of these genes results in severe impairment of neural crest formation (Koblar *et al.*, 1999; Mansouri *et al.*, 1996). Among other transcription factors expressed in the dorsal neural tube are 1mx1 (Lim homeodomain gene) and slug (a vertebrate homologue of the *Drosophila* snail gene). The latter has been shown in the chick, using antisense approaches, to be necessary for the proper genesis of neural crest (Nieto *et al.*, 1994).

In summary, the establishment of fate along the medio-lateral (M-L) axis during neural plate stages and in the D-V axis during and following neural tube closure involves the concerted action of two opposing forces. One from the notochord is mediated by SHH, which suppresses dorsal fate and promotes ventral cells. The other, from the M-L boundary of neural and nonneural ectoderm at neural plate stages early and then from the roof plate itself, is mediated by BMPs and Wnts. These signals suppress ventral fates and impose a dorsal fate. It is imperative to remember, however, that although a relatively simple picture emerges for the patterning of the D-V axis of the neural tube, a large number of questions remain unresolved. Among the most burning issues are (a) What happens at the intermediate level of the neural tube at the boundary when the opposite signals intercept each other? and (b) What is the role of a large number of BMP and Wnt inhibitors present in the notochord? These include follistatin, chordin, cerberus, and frazzled. The future might have more surprises and additional levels of fine tuning for these important cell fate determination events.

3. D-V Patterning in the Brain

a. Forebrain The dorsal-ventral patterning of the forebrain seems to involve the same players seen in the spinal cord, with signals derived from axial mesoderm, as well as nonneural ectoderm. In all vertebrates, however, the anterior boundary of the notochord ends at approximately the telencephalic-diencephalic boundary. Immediately extending more anteriorly and adjacent to the notochord are mesodermal cells that were the leading edge of the involuting mesoderm during gastrulation, the prechordal plate (or head mesoderm). Signals from both of these mesodermal derivatives have been shown to be involved in the patterning of the ventral forebrain (Yamada *et al.*, 1993; Shimamura and Rubenstein, 1997). Perhaps not surprisingly, SHH, which is expressed both in the prechordal plate as well as in the notochord, has been shown to induce ventral cell fates in the forebrain. In addition, the fish mutant *Cyclops,* which lacks SHH expression, has severe deletions in the ventral forebrain. Dorsally, signals from nonneural ectoderm (epidermis) have been suggested to pattern the dorsal side of the forebrain. This suggestion is based on experiments in which the elimination of the anterior nonneural ectoderm tissue has led to the extension of forebrain markers such as BF1. This implies that signals from the ectoderm juxtaposed to the forebrain play a role in patterning of the forebrain. The candidate molecules to mediate this type of activity are BMPs and FGF8 (Muhr *et al.,* 1997). Fate mapping in amphibians (Figure 1a) has suggested that the anterior edge of the neural plate will give rise to the sensory placodes. These include, from the lateral edge from the posterior towards the anterior, otic vesicle, lens placode, olfactory placodes, and the pituitary gland. As described above, and in agreement with this hypothesis, studies in the frog embryo have demonstrated that at a given threshold of activity, BMP4 can induce both placodal fate as well as epidermal specific markers (Wilson *et al.,* 1997). The development of these sensory structures (which will not be discussed further here) will in turn involve further interactions between neural and nonneural ectoderm.

b. Midbrain In contrast to the spinal cord, little is known about the D-V patterning of the midbrain. As with other regions, it is expected that similar mechanisms operate there, including ventral SHH and RA signaling and dorsal BMP/GDF signaling. Wnt signaling, though not well defined outside of its activity on the neural crest, is also presumed to be effecting patterning in the midbrain.

c. Hindbrain Along the D-V axis, the hindbrain is patterned by the same cast of characters seen before. SHH, originally derived from the notochord, is later expressed in the floor plate underlying the entire hindbrain. Since the levels of SHH (at least at the RNA level) are homogenous along the length of the hindbrain, differential competence to respond to SHH in the ventral side of the hindbrain has been suggested as a possible mechanism for the generation of different rhombomeres (Jessell and Lumsden, 1997).

On the dorsal side, BMP4 effects mediated by its immediate early response genes msx1 and msx2 (Hollnagel *et al.,* 1999), as well as signaling from inter-rhombomeric segments, have been implicated in the specification (or absence) of specific populations of neural crest cells (Marazzi *et al.,* 1997). The hindbrain is also the site from which most of the cranial neural crest is derived; there is a direct relationship between rhombomere identity and the kind of cranial neural crest to which it contributes (Graham and Lumsden, 1996; Rubenstein *et al.,* 1998). In addition to specific populations of neural crest, the hindbrain contributes to a number of cranial nerves

(Kuratani and Eichele, 1993; Yamamoto and Schwarting, 1991). Table II shows the origin of cranial nerves and their future projections. While most rhombomeres contribute to cranial neural crest, this is not the case for rhombomeres 3 and 5. This is not because these rhombomeres cannot generate neural crest, but rather to the elimination of the neural crest they form by programmed cell death prior to migration (Graham *et al.,* 1993, 1996). The apoptosis of r3- and r5-derived neural crest cells is initiated by BMP4, which apparently is provided by the neighboring rhombomeres. In favor of this argument is the demonstration that when r3 and r5 are cultured alone, they can produce neural crest cells, but when cultured together with the neighbors these neural crest cells are eliminated. BMP4 cannot exert the same effect on r4. R4 can generate and maintain neural crest cells in the presence of BMP4. Since the Msx genes have been shown to be an immediate early response to BMP4, and since Msx2 expression correlates very well with the pattern of apoptosis in the chick, it has been suggested that selective BMP receptors expressed in these rhombomeres activate programmed cell death pathways via msx2 (Marazzi *et al.,* 1997). It is also important to emphasize that BMPs play additional patterning roles in the

TABLE II
Human Cranial Nerves

Name	Type	Origin	Distribution
I Olfactory	Sensory	Telencephalon/olfactory placode	Olfactory epithelium (telencephalon)
II Optic	Sensory	Telencephalon evagination	Eye
III Oculomotor	Motor	Midbrain	Eye muscles
IV Trochlear	Motor	Hindbrain r1	Eye muscles
V Trigeminal	Mixed	Hindbrain r2, r3	Face; sinuses; teeth; jaw muscles
VI Abducens	Motor	Hindbrain r5, r6	Eye muscles
VII Facial and intermediate	Mixed	Hindbrain r4, (branchial arch 2)	Face muscles; tongue; palate
VIII Vestibulocochlear	Sensory	Hindbrain r4, r5, r6 (otic placode)	Cochlea; vestibular organs
IX Glossopharyngeal	Mixed	Hindbrain r6, r7 (branchial arch 3)	Tongue; tonsil; pharnyx; pharnyx
X Vagus	Mixed	Hindbrain r7, r8 (branchial arch 4)	Heart; lung; digestive system; external ear
XI Spinal accessory	Motor	Hindbrain r7, r8 (branchial arch 4)	Neck/shoulder muscles
XII Hypoglossal	Motor	Hindbrain r8 (branchial arch 4)	Tongue

dorsal rhombencephalon and that their activity is not limited to programmed cell death. Experiments performed in the mouse embryo, for example, have demonstrated that BMPs can initiate the program for granule cell specification (Alder *et al.,* 1996). This is the most abundant class of CNS neurons and is generated at the dorsal border of the rhombic lip. Hence, in the hindbrain as well as in the spinal cord, SHH and BMPs regulate cell fate specification along the D-V axis.

C. The A-P Patterning

1. Neural Tube

Anterior-posterior (A-P) neural patterning occurs soon after neural induction at open neural plate stages. Neural cells with equivalent developmental potentials are induced to express position-specific genes along the A-P axis in broad domains, some of which are associated with morphological features such as bulges and restrictions. The prosencephalon, or forebrain, is the anterior bulge early in development and later forms two bulges called the telencephalon and diencephalon. The most anterior telencephalon gives rise to the paleocortex, corpus striatum, and neocortex while the diencephalon develops into the epithalamus, thalamus, hypothalamus, and infundibulum. The midbrain, or mesencephalon, lies posterior to the diencephalon and forms the tectum, tegmentum, and cerebral peduncles. The hindbrain, or rhombencephalon, gives rise to the metencephalon, which forms the cerebellum and pons, and the myelencephalon, from which the medulla develops. The spinal cord is the most posterior part of the CNS. Within the boundary of each domain historically defined by morphology, genes are being identified that correspond with morphological boundaries and further subdivide the CNS into distinct regions.

Classical embryological experiments performed in the amphibian embryo have suggested that the organizer (a group of cells localized in the dorsal marginal zone of an early gastrula) is a source of signals for A-P patterning of the neural tube (reviewed in Harland and Gerhart, 1997; Zoltewicz and Gerhart, 1997). When ectodermal explants are conjugated with the organizer *in vitro,* neural tissue is induced in the ectoderm, and it is patterned along its A-P axis. This is revealed not only by histological criteria but also by the expression of molecular markers demarcating different A-P levels. It is interesting that in these recombinants the organizer tissue from early gastrula embryos induces a fairly complete set of anterior and posterior markers, while organizers dissected from later gastrula induce progressively more posterior neural genes. Two models have been proposed to explain these observations (Chang and Hemmati-Brivanlou, 1998b). In

Mangold's (Mangold, 1933) model, different signals segregated in time (early vs. late organizer) have been suggested to independently induce the fate along the A-P axis. In this scenario the early organizer has all the signals and the late organizer has lost the anterior-specific signals. In Nieuwkoop's (Nieuwkoop, 1952a, b) model, two types of signals have been suggested that would both induce and pattern the embryonic CNS along its A-P axis. The first signal(s) induces the ectoderm to adopt a neural fate and at the same time imposes the most anterior (telencephalic) fate. This step is referred to as *activation.* The second signal(s) has been postulated not to have any neural-inducing activity, but instead to act on the first signal to add more caudal fates (midbrain, hindbrain, and spinal cord). This second step is referred to as *transformation.* Although most of the molecular evidence so far seems to favor the Nieuwkoop hypothesis, recently some evidence supporting the Mangold hypothesis has been suggested (Kengaku and Okamoto, 1995; Lamb and Harland, 1995; LeSueur and Graff, 1999). In addition, classical studies have suggested that signals transmitted both from the underlying mesoderm (vertical signals) and from the organizer region through the neuroectodermal layer (planar signal) are important to pattern the A-P neural axis (reviewed by Doniach, 1993; Ruiz i Altaba, 1992).

2. The Spinal Cord

Strict boundaries of gene expression as well as the distribution of different cell types highlights patterning along the A-P axis of the spinal cord. There are different rostral boundaries of expression of the genes belonging to the Hox complex, which subdivide the spinal cord into different segmental territories and presumably allow establishment of identity along the A-P axis of the spinal cord (Figure 4; Jungbluth *et al.,* 1999). In addition to these molecular differences, distinct subclasses of motor neurons distributed along the A-P axis innervate subsets of muscles (Tanaka *et al.,* 1997). The cell bodies of these subclasses occupy stereotypical positions in the spinal cord. Each subclass of motor neuron shares unique properties in terms of organization of projections of their axons and growth cones. In lower vertebrates (zebrafish) as well as higher vertebrates (chick), different primary motor neurons with distinct identities express a unique combination of LIM homeodomain proteins (Appel *et al.,* 1995). There are four LIM homeobox genes, and the unique combination of each specifies subtypes of motor neurons (Tosney *et al.,* 1995; Lumsden, 1995). Table I provides the code between LIM combination and motor neuron identity. There are suggestions that this combination of LIM transcription factors endows each subclass of motor neurons not only in terms of fate and identity but also by instructing distinct pathfinding routes for their axons and growth cones (Varela-Echavarria *et al.,* 1996;

Thor *et al.,* 1999). In favor of this hypothesis are experiments performed in zebrafish, where transplantation of single motor neurons in ectopic A-P positions within the spinal cord has led to both a reprogramming of the LIM combination (Appel *et al.,* 1995) and a reprogramming of the projections (Eisen and Pike, 1991) The unique demarcation of specific motor neuron subtype by specific combination of LIM proteins has allowed the use of these factors as molecular markers in studies that address the source and the nature of signals involved in the specification of these motor neuron subclasses in the spinal cord. Although the nature and origin of signals involved in the establishment of fate for these motor neurons are presently unknown, two hypothesis have been suggested. There is either a planar patterning signal diffusing from a given source imposing fate in the A-P axis, or a vertical signal from the axial and paraxial mesoderm specifying fate in the A-P axis of the spinal cord in synchrony with mesodermal patterning. This latter possibility is similar to the ability of anterior notochord to induce midbrain-hindbrain markers in the ectoderm. A role for somites (paraxial mesoderm) has been suggested for the mutant *spadetail (spt,* Ho and Kane, 1990). This mutation interferes with the proper development of the somites and also disturbs the acquisition of the proper identity of motor neurons, as judged by the changes in the combination of LIM marker expression. Hence, analysis of axonal projections in this mutant fish is complicated by the fact that the targets themselves are abnormal.

In addition to protein growth factors, retinoic acid has been shown to be a strong modifier of A-P pattern in the spinal cord (Clagett-Dame and Plum, 1997; Whiting, 1997; Lufkin, 1997; Eichele, 1997; Lufkin, 1996; Marshall *et al.,* 1996). In tissue culture, RA has been commonly used to examine the expression of Hox genes (Bürglin, 1994) while RA has been shown to affect A-P patterning both in the embryo and in tissue explants. RA acts through at least two receptors, RAR and RXR, which can interact with multiple putative coactivators and/or corepressors to yield a complex molecular pathway with a variety of pleiotropic effects (reviewed in Chambon, 1996; Ng *et al.,* 1995; Pemrick *et al.,* 1994). In general, the strongest effects of RA pathway activation or inhibition are observed in the developing brain (Old *et al.,* 1996; Lopez *et al.,* 1995).

3. The Brain

In all vertebrates the brain is anatomically subdivided into three main domains: the forebrain, midbrain, and hindbrain. Although the embryonic origins of the forebrain have been recently the topic of intense molecular and embryological scrutiny, the mechanisms underlying A-P patterning in the hindbrain are the best understood. The molecular dissection of the patterning of the midbrain is still in its infancy.

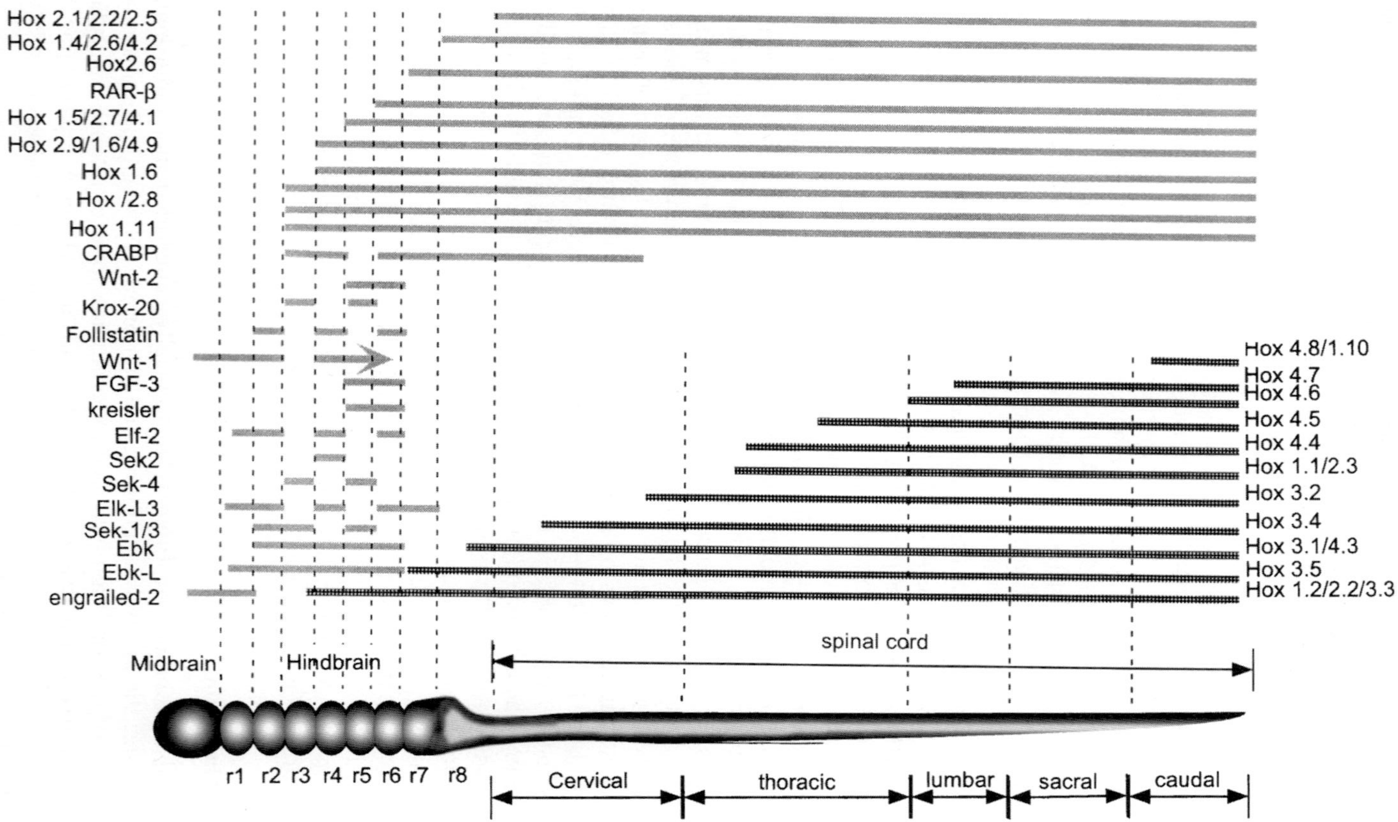
Hox 2.1/2.2/2.5
Hox 1.4/2.6/4.2
Hox2.6
RAR-β
Hox 1.5/2.7/4.1
Hox 2.9/1.6/4.9
Hox 1.6
Hox /2.8
Hox 1.11
CRABP
Wnt-2
Krox-20
Follistatin
Wnt-1
FGF-3
kreisler
Elf-2
Sek2
Sek-4
Elk-L3
Sek-1/3
Ebk
Ebk-L
engrailed-2
Hox 4.8/1.10
Hox 4.7
Hox 4.6
Hox 4.5
Hox 4.4
Hox 1.1/2.3
Hox 3.2
Hox 3.4
Hox 3.1/4.3
Hox 3.5
Hox 1.2/2.2/3.3
Midbrain
Hindbrain
spinal cord
r1 r2 r3 r4 r5 r6 r7 r8
Cervical
thoracic
lumbar
sacral
caudal

a. Forebrain In all vertebrates the forebrain is subdivided into two regions: an anterior region called the telencephalon and a posterior region called the diencephalon. As mentioned earlier, much of the available evidence seems to favor the activation-transformation model of Nieuwkoop with FGF, Wnt, and RA as the leading candidates for the transforming activity. But how does the A-P patterning within each part of the brain occur? For the forebrain, embryological, cellular and molecular studies performed mostly in amniotes have suggested the neuromeric (or segmentation) model, with each unit called a prosomere (Puelles and Rubenstein, 1993; Rubenstein *et al.,* 1994; Rubenstein and Puelles, 1994). This model postulates that A-P and D-V patterning mechanisms subdivide the embryonic forebrain into a "checker board" set of coordinates that will give rise to the different forebrain regions. This process allows a regional specification, which in turn establishes the further differentiation between each primordia. Six prosomeres, P1 to P6, have been suggested in the A-P axis (Rubenstein *et al.,* 1998). P1 and P2 are in the diencephalon while P3 to P6 are in the telencephalon (Figure 5, see color insert). Studies in the chick have suggested that the diencephalic prosomere P1 is itself subdivided into two domains (D3 and D4) (Rubenstein and Shimamura, 1997). The fact that both the forebrain and the midbrain undergo flexure ventrally has created some complications in referring to the axis in a traditional way, and sometimes these axes are referred to as longitudinal (for A-P) versus transverse (for D-V) axis (Figure 5, arrow; Rubenstein and Shimamura, 1997). Hence, after flexure, at the point where the most anterior level of the neural plate and the midline meet (P6), the A-P and D-V axis become the same. The neuromeric model of brain patterning is supported by the gene expression patterns of a variety of homeobox genes and components of signaling pathways. A number of these genes have been mapped onto the prosomeric diagram (Figure 5b) and have been extensively reviewed elsewhere (Rubenstein *et al.,* 1998; Shimamura *et al.,* 1995, 1997).

b. Midbrain The next major subdivision of the brain is the midbrain or mesencephalon. Among the various regions of the CNS, the midbrain is the

---

FIG. 4 Differential gene expression along the A-P axis. Gene expression patterns are indicated above a diagram of the caudal regions of the developing CNS. Hox genes are generally restricted to a boundary in the anterior, and expression extends to the most caudal regions. The expression patterns shown do not represent the expression at a single time but rather reflect the pattern of expression in general. Some genes, like the posterior genes, are not expressed until later while the more rostral genes are expressed earlier. The figure was compiled from a variety of sources (Puelles and Rubenstein, 1993; Duboule, 1994; Bürglin, 1994); the Hox nomenclature was changed to reflect current usage (Scott, 1993, 1992).

least characterized with respect to the formation of pattern. The midbrain contains three major regions, the tectum, the tegmentum, and the isthmus. The tectum receives projections from the visual system (in mammals, these target the forebrain), auditory, and somatosensory systems. The tegmentum is ventral to the tectum and is separated from the hindbrain by the isthmus, though some areas present in the hindbrain are continued into this region. The posterior aspect of the midbrain is called the isthmus and current evidence indicates that it plays an essential role in the patterning and formation of the midbrain and is a source of inductive signals. Experiments in chick and zebrafish reveal the presence of a signaling center at the midbrain-hindbrain boundary (Brand *et al.,* 1996a, b; Lun and Brand, 1998; Marin and Puelles, 1994). Two mutations in zebrafish, *no isthmus* (*noi*) and acerebellar (*ace*), result in the loss of midbrain structures (Brand *et al.,* 1996a, b). The *noi* mutation results in the inactivation of the paired box gene Pax2.1 (Pfeffer *et al.,* 1998) whereas the *ace* mutation inactivates FGF8 (Reifers *et al.,* 1998). Ectopic application of FGF8 results in ectopic midbrain formation (Crossley *et al.,* 1996). The examination of wnt1, FGF8, and her1 expression in noi mutant embryos reveals that in the absence of *noi* function, the expression of the genes is not maintained, though they are expressed early (Lun and Brand, 1998), thus suggesting that pax2 functions downstream of the polarizing signals. Hence, in the midbrain, as in other regions of the developing nervous system, FGF (FGF8 in zebrafish and FGF17 in mouse Hoshikawa *et al.,* 1998) signaling acts to pattern the A-P axis. A number of other genes involved in neural patterning have been shown to be expressed in the midbrain region, such as Otx, Mash1, Dbx, RAR 2, and Da H. Otx genes appear to play a role in the development of the diencephalic-mesencephalic boundary as otx1$^{-/+}$/otx2$^{-/+}$ heterozygotes result in a failure of this region to form properly while the more posterior metencephalon is expanded (Suda *et al.,* 1997). Since FGF8 expression is expanded dorsally, in these mutants, Otx1 and Otx2 may be acting to restrict the midbrain-hindbrain signaling center. Current midbrain markers are expressed in regions outside the midbrain leaving the question of further subdivision of the midbrain unanswered.

c. ***Hindbrain*** The development and patterning of the brain is best understood in the vertebrate embryonic hindbrain, specially along the A-P axis. In all vertebrates, after neural tube closure, the hindbrain gives rise to eight consecutive restrictions called rhombomeres (Figure 4, 5). The rhombomeres behave as restricted compartments. Single cell lineage tracing experiments, as well as experiments performed with chick-quail chimeras in which a single cell or a group of cells can be followed, have demonstrated that the progeny of a cell remains within a single rhombomere and does not cross boundaries. This is true for at least the first three days of embryonic development in the chick, and for as long as the ancestral cell is labeled after the rhombomeres are formed (Lumsden *et al.,* 1991, 1994; Sechrist *et al.,* 1994; Wingate and Lumsden, 1996). The boundaries between the rhombomeres become further specialized and generate new cell types at the junctions. The segmented pattern of the rhombomeres is transient and disappears later in development when groups of rhombomeric cells condense and migrate. These observations strongly suggest an analogy with compartments in the *Drosophila* embryo, which are also transient structures in which there is lineage restriction with no (or minimal) cell mixing (reviewed in Vincent, 1998; Blair; 1995). Based on boundary formation, rhombomeres appear to have alternate repeating units. When even-numbered rhombomeres are explanted together, no boundary forms whereas the juxtaposition of odd and even segments generates a new boundary (Guthrie and Lumsden, 1991). Lineage tracing experiments suggest that different classes of neurons, characterized with respect to axonal projection, are produced within the rhombomeres (Lumsden *et al.,* 1994). These results are consistent with the formation of diverse cranial nerves, many of which arise in the hindbrain (Table II). It is not clear, however, whether these early differences are sufficient for the formation and targeting of these important nerve fibers.

The molecular dissection of the events leading to the generation of rhombomeres has attracted considerable attention. The A-P axis within the hindbrain has been suggested to be established under the influence of two main driving forces: the Hox code and retinoic acid. First, there is evidence that the Hox genes encode A-P positional information in a way similar to that seen in the spinal cord (reviewed in Krumlauf *et al.,* 1993; Lumsden and Krumlauf, 1996). The anterior boundary of expression of a subset of these homeobox genes corresponds exactly to rhombomere boundaries (Figure 4). Hence, each rhombomere can be molecularly characterized by the expression of a specific combination of Hox genes. The unique combination of different Hox gene activity within each rhombomere has also been invoked as the molecular mechanism involved in the specification of individual rhombomeres. Loss-of-function approaches that selectively eliminate the activity of one or more Hox genes have resulted in aberrations consistent with a change of rhombomere identity. While this is clearly the case with the elimination of Hoxb-1 activity in which r4 is at least partially transformed to r2 and r6 (Studer *et al.,* 1996), elimination of other genes such as Hoxa-1 have more complicated effects that also include elimination of rhombomeres, not necessarily consistent with a simple role of Hox genes in assigning segmental identity within a segmented field (Carpenter *et al.,* 1996). Members of the Hox complex expressed in the rhombomeres are themselves induced by paraxial mesoderm (somites), highlighting again the influence of mesoderm in the patterning of the hindbrain (Gould *et al.,* 1998; Itasaki *et al.,* 1996). The target genes regulated by these homeobox-containing proteins are still relatively unknown, but have been shown to include the EPH receptor (Chen and Ruley, 1998) as well as other Hox genes (Maconochie *et al.,* 1997). Classical embryological studies, in which rhombomeres are rearranged, indicate that rhombomere identity exhibits some plasticity in some cases, while in others identity appears to be autonomous (Itasaki *et al.,* 1996; Grapin-Botton *et al.,* 1995; Guthrie *et al.,* 1992).

In addition to the potential role of somites in regulating Hox expression, there is also evidence of a role for the node (the amniote equivalent of the organizer) mediated by retinoic acid (RA). Both elimination of endogenous RA and addition of exogenous RA have been shown to influence the patterning of the hindbrain (Gould *et al.,* 1998; Marshall *et al.,* 1996; van der Wees *et al.,* 1998). Quail embryos that cannot produce RA because of a vitamin A deficiency lack posterior rhombomeres (Maden *et al.,* 1996, 1998a), and incubation of *Xenopus* embryos in RA leads to expanded hindbrain and midbrain and reduced forebrain (Kolm *et al.,* 1997; Ruiz i Altaba and Jessell, 1991). Additionally, both RA and its two major classes of receptors (RARs and RXRs) are expressed in the node (Berggren *et al.,* 1999; Hogan *et al.,* 1992; Maden *et al.,* 1998a, b). Addition or elimination of RA influences the expression of the Hox genes, which in turn are believed to modify the architecture of the hindbrain.

D. Right-Left Patterning

Like the patterning along the D-V and A-P axes, the determination of the L-R axis also occurs early in development during gastrulation, though the morphological effects are not observed until later (Izraeli *et al.,* 1999; Patel *et al.,* 1999; King and Brown, 1997). Most studies of L-R asymmetry have focused on mesodermal derivatives such as the heart and on the more obvious morphological asymmetries observed in the whole embryo and only recently has attention been focused to L-R asymmetry in the nervous system. The mechanisms involved in the formation of this asymmetric axis are likely to be well conserved among vertebrates. A number of genes have been shown to be asymmetrically expressed during gastrulation, including SHH, nodal, and the activin receptor ActRIIa (Levin *et al.,* 1997). In the chick, activin is first expressed on the right side of the primitive streak and developing node, and is the first sign of axis asymmetry prior to any morphological changes. Activation of this TGF-β family receptor leads to the asymmetric expression of both nodal and SHH. The result of this cascade is the expression of nodal in the lateral plate mesoderm on the left side, and it is this expression pattern that is conserved among vertebrates

(Adachi *et al.*, 1999, Bisgrove *et al.*, 1999). In the mouse, two TGF-β family members, lefty-1 (Lefty-A or EBAF in humans, Kosaki *et al.*, 1999; Stern and Foley, 1998; Hunt and Krumlauf, 1992; Kothapalli *et al.*, 1997) and lefty-2, act together with nodal to pattern the R-L axis in the early floor plate, where lefty-1 is highly expressed in four to five cells of the left floor plate (Meno *et al.*, 1996, 1997). The high level of lefty-1 expression in the floor plate persists only until the 12–14 somite stage, at which point it is down-regulated. The target of these genes appears to be Pitx2, a member of the bicoid class of homeodomain proteins, whose expression on the left side of the embryo is broader and persists longer (Piedra *et al.*, 1998; Ryan *et al.*, 1998; St. Amand *et al.*, 1998; Yoshioka *et al.*, 1998). Despite these early differences of left-right asymmetry in the neural tube, no link has been established between these changes and the complex functional asymmetries of the brain observed in the adult. These left-right differences include the well-known preference for one hand over the other and the less obvious cognitive differences in the brain.

III. Conclusions and Summary

The nervous system is perhaps the most complex organ in the vertebrate. It consists of hundreds of differentiated cell types highly organized spatially. The formation of this complex system begins early in development during gastrulation when the neuronal precursor cells in the ectoderm are deprived of BMP/GDF signaling. Signals from the mesoderm (follistatin, noggin, chordin, xnr-3, cerberus) inhibit the action of these soluble ligands and promote a neural fate (Weinstein and Hemmati-Brivanlou, 1999). At the same time, signals act to pattern the neural tissues along the A-P, D-V, and lateral axes. The main source of ventralizing signal in the neural tube is the notochord, and the prime candidate for this signaling is SHH. Anteriorly to the notochord, the prechordal mesoderm or head mesoderm acts to pattern the brain in a similar manner. In addition to SHH, retinoic acid also has ventralizing activity. During this same period, BMP/GDF signaling from the ectoderm acts to repress ventral signaling and promote dorsal fates. Members of the Wnt family also appear to play a role in the development of dorsal fates and are required for the formation of the neural crest (the most dorsal fate in the neural tube) in some regions. Along the A-P axis, neural induction leads to the generation of anterior cell types that are then caudalized by the action of FGFs and RA. In addition, various Wnt family ligands are implicated in the generation of posterior neuronal identities. Finally, the generation of L-R asymmetry in the vertebrate nervous system is just beginning to be elucidated with the first indications of L-R differences observed in the ventral neural tube. How these differences relate to the functional differences remain to be determined. Developmental biologists, of course, think that by understanding how an organ develops they will also ultimately learn how it works. If this turns out to be true, we still have a long way to go.

Acknowledgment

We thank Giorgio Lagna, Guojun Sheng, and Daniel Weinstein for critical reading of the manuscript and helpful suggestions. We especially thank Claudio Stern for his help preparing the chick neural plate fate maps.

References

Adachi, H., Saijoh, Y., Mochida, K., Ohishi, S., Hashiguchi, H., Hirao, A., and Hamada, H. (1999). Determination of left/right asymmetric expression of nodal by a left side-specific enhancer with sequence similarity to a lefty-2 enhancer. *Genes Dev.* **13,** 1589–1600.

Alder, J., Cho, N. K., and Hatten, M. E. (1996). Embryonic precursor cells from the rhombic lip are specified to a cerebellar granule neuron identity. *Neuron* **17,** 389–399.

Ang, S. L., and Rossant, J. (1994). HNF-3 beta is essential for node and notochord formation in mouse development. *Cell* **78,** 561–574.

Aono, A., Hazama, M., Notoya, K., Taketomi, S., Yamasaki, H., Tsukuda, R., Sasaki, S., and Fujisawa, Y. (1995). Potent ectopic bone-inducing activity of bone morphogenetic protein-4/7 heterodimer. *Biochem. Biophys. Res. Commun.* **210,** 670–677.

Appel, B., Korzh, V., Glasgow, E., Thor, S., Edlund, T., Dawid, I. B., and Eisen, J. S. (1995). Motoneuron fate specification revealed by patterned LIM homeobox gene expression in embryonic zebrafish. *Development* **121,** 4117–4125.

Artinger, K. B. and Bronner-Fraser, M. (1992a). Notochord grafts do not suppress formation of neural crest cells or commissural neurons. *Development* **116,** 877–886.

Artinger, K. B., and Bronner-Fraser, M. (1992b). Partial restriction in the developmental potential of late emigrating avian neural crest cells. *Dev. Biol.* **149,** 149–157.

Barth, K. A., and Wilson, S. W. (1995). Expression of zebrafish nk2.2 is influenced by sonic hedgehog/vertebrate hedgehog-1 and demarcates a zone of neuronal differentiation in the embryonic forebrain. *Development* **121,** 1755–1768.

Berggren, K., McCaffery, P., Drager, U., and Forehand, C. J. (1999). Differential distribution of retinoic acid synthesis in the chicken embryo as determined by immunolocalization of the retinoic acid synthetic enzyme, RALDH-2. *Dev. Biol.* **210,** 288–304.

Bisgrove, B. W., Essner, J. J., and Yost, H. J. (1999). Regulation of midline development by antagonism of lefty and nodal signaling. *Development* **126,** 3253–3262.

Bitgood, M. J., and McMahon, A. P. (1995). Hedgehog and Bmp genes are coexpressed at many diverse sites of cell-cell interaction in the mouse embryo. *Dev. Biol.* **172,** 126–138.

Blair, S. S. (1995). Compartments and appendage development in *Drosophila. BioEssays* **17,** 299–309.

Bortier, H., and Vakaet, L. C. (1992). Fate mapping the neural plate and the intraembryonic mesoblast in the upper layer of the chicken blastoderm with xenografting and time-lapse videography. *Dev. Suppl.* 93–97.

Brand M., Heisenberg, C. P., Jiang, Y. J., Beuchle, D., Lun, K., Furutani-Seiki, M., Granato, M., Haffter, P., Hammerschmidt, M., Kane, D. A., Kelsh, R. N., Mullins, M. C., Odenthal, J., van Eeden, F. J., and Nusslein-Volhard, C. (1996a). Mutations in zebrafish genes affecting the formation of the boundary between midbrain and hindbrain. *Development* **123,** 179–190.

Brand, M., Heisenberg, C. P., Warga, R. M., Pelegri, F., Karlstrom, R. O., Beuchle, D., Picker, A., Jiang, Y. J., Furutani-Seiki, M., van Eeden, F. J., Granato, M., Haffter, P., Hammerschmidt, M., Kane, D. A., Kelsh, R. N., Mullins, M. C., Odenthal, J., and Nusslein-Volhard, C. (1996b). Mutations affecting development of the midline and general body shape during zebrafish embryogenesis. *Development* **123,** 129–142.

Briscoe, J., Sussel, L., Serup, P., Hartigan-O' Connor, D., Jessell, T. M., Rubenstein, J. L., and Ericson, J. (1999). Homeobox gene Nkx2.2 and specification of neuronal identity by graded Sonic hedgehog signalling. *Nature* **398,** 622–627.

Brivanlou, A. H., and Harland, R. M. (1989). Expression of an engrailed-related protein is induced in the anterior neural ectoderm of early *Xenopus* embryos. *Development* **106,** 611–617.

Brivanlou, A. H., and Melton, D. (1997). Vertebrate neural induction. *Annu. Rev. Neurosci.* **20,** 43–60.

Bronner-Fraser, M. (1995a). Origin of the avian neural crest. *Stem. Cells (Dayt.)* **13,** 640–646.

Bronner-Fraser, M. (1995b). Origins and developmental potential of the neural crest. *Exp. Cell Res.* **218,** 405–417.

Bumcrot, D. A., Takada, R., and McMahon, A. P. (1995). Proteolytic processing yields two secreted forms of sonic hedgehog [published erratum appears in *Mol. Cell. Biol.* 1995 May, **15**(5), 2904]. *Mol. Cell Biol.* **15,** 2294–2303.

Burrill, J. D., Moran, L., Goulding, M. D., and Saueressing, H. (1997). PAX2 is expressed in multiple spinal cord interneurons, including a population of ENI+interneurons that require PAX6 for their development. *Development* **124,** 4493–4503.

Bürglin, T. R. (1994). Descriptions of homeobox genes. In "Guidebook to the Homeobox Genes" (D. Duboule, Ed.) pp. 73–272. Oxford University Press, Oxford.

Carpenter, E. M., Goddard, J. M., Chisaka, O., Manley, N. R., and Capecchi, M. R. (1993). Loss of Hox-A1 (Hox-1.6) function results in the reorganization of the murine hindbrain. *Development* **118,** 1063–1075.

Chambon, P. (1996). A decade of molecular biology of retinoic acid receptors. *FASEB J.* **10,** 940–954.

Chang, C., and Hemmati-Brivanlou, A. (1998a). Neural crest induction by Xwnt7B in Xenopus. *Dev. Biol.* **194,** 129–134.

Chang, C., and Hemmati-Brivanlou, A. (1998b). Cell fate determination in embryonic ectoderm. *J. Neurobiol.* **36,** 128–151.

Chang, D. T., Lopez, A., von Kessler, D. P., Chiang, C., Simandl, B. K., Zhao, R., Seldin, M. F., Fallon, J. F., and Beachy, P. A. (1994). Products, genetic linkage and limb patterning activity of a murine hedgehog gene. *Development* **120,** 3339–3353.

Chen, J., and Ruley, H. E. (1998). An enhancer element in the EphA2 (Eck) gene sufficient for rhombomere-specific expression is activated by HOXA1 and HOXB1 homeobox proteins. *J. Biol. Chem.* **273,** 24670–24675.

Chiang, C., Litingtung, Y., Lee, E., Young, K. E., Corden, J. L., Westphal, H., and Beachy, P. A. (1996). Cyclopia and defective axial patterning in mice lacking Sonic hedgehog gene function. *Nature* **383,** 407–413.

Clagett-Dame, M., and Plum, L. A. (1997). Retinoid-regulated gene expression in neural development. *Crit. Rev. Eukaryot. Gene Exp.* **7,** 299–342.

Cox, W. G., and Hemmati-Brivanlou, A. (1995). Caudalization of neural fate by tissue recombination and bFGF. *Development* **121,** 4349–4358.

Crossley, P. H., Martinez, S., and Martin, G. R. (1996). Midbrain development induced by FGF8 in the chick embryo. *Nature* **380,** 66–68.

Davidson, D. (1995). The function and evolution of Msx genes: Pointers and paradoxes. *Trends Genet.* **11,** 405–411.

Dickinson, M. E., Selleck, M. A., McMahon, A. P., and Bronner-Fraser, M. (1995). Dorsalization of the neural tube by the non-neural ectoderm. *Development* **121,** 2099–2106.

Dierick, H. and Bejsovec, A. (1999). Cellular mechanisms of wingless/Wnt signal transduction. *Curr. Top. Dev. Biol.* **43,** 153–190.

Doniach, T. (1993). Planar and vertical induction of anteroposterior pattern during the development of the amphibian central nervous system. *J. Neurobiol.* **24,** 1256–1275.

Duboule, D. (1994). "Guidebook to the homeobox genes." Sambrook & Tooze, Oxford.

Eagleson, G., Ferreiro, B., and Harris, W. A. (1995). Fate of the anterior neural ridge and the morphogenesis of the *Xenopus* forebrain. *J. Neurobiol.* **28,** 146–158.

Eagleson, G. W., and Harris, W. A. (1990). Mapping of the presumptive brain regions in the neural plate of *Xenopus* laevis. *J. Neurobiol.* **21,** 427–440.

Echelard, Y., Epstein, D. J., St-Jacques, B., Shen, L., Mohler, J., McMahon, J. A., and McMahon, A. P. (1993). Sonic hedgehog, a member of a family of putative signaling molecules, is implicated in the regulation of CNS polarity. *Cell* **75,** 1417–1430.

Eichele, G. (1997). Retinoids: From hindbrain patterning to Parkinson disease. *Trends Genet.* **13,** 343–345.

Eisen, J. S., and Pike, S. H. (1991). The spt-1 mutation alters segmental arrangement and axonal development of identified neurons in the spinal cord of the embryonic zebrafish. *Neuron* **6,** 767–776.

Ericson, J., Morton, S., Kawakami, A., Roelink, H., and Jessell, T. M. (1996). Two critical periods of Sonic Hedgehog signaling required for the specification of motor neuron identity. *Cell* **87,** 661–673.

Ericson, J., Muhr, J., Jessell, T. M., and Edlund, T. (1995). Sonic hedgehog: A common signal for ventral patterning along the rostrocaudal axis of the neural tube. *Int. J. Dev. Biol.* **39,** 809–816.

Ericson, J., Rashbass, P., Schedl, A., Brenner-Morton, S., Kawakami, A., van Heyningen, V., Jessell, T. M., and Briscoe, J. (1997). Pax6 controls progenitor cell identity and neuronal fate in response to graded Shh signaling. *Cell* **90,** 169–180.

Ericson, J., Thor, S., Edlund, T., Jessell, T. M., and Yamada, T. (1992). Early stages of motor neuron differentiation revealed by expression of homeobox gene Islet-1. *Science* **256,** 1555–1560.

Foerst-Potts, L. and Sadler, T. W. (1997). Disruption of Msx-1 and Msx-2 reveals roles for these genes in craniofacial, eye, and axial development. *Dev. Dyn.* **209,** 70–84.

Garcia-Martinez, V., Alvarez, I. S., and Schoenwolf, G. C. (1993). Locations of the ectodermal and nonectodermal subdivisions of the epiblast at stages 3 and 4 of avian gastrulation and neurulation. *J. Exp. Zool.* **267,** 431–446.

Gould, A., Itasaki, N., and Krumlauf, R. (1998). Initiation of rhombomeric Hoxb4 expression requires induction by somites and a retinoid pathway. *Neuron* **21,** 39–51.

Gould, S. E. and Grainger, R. M. (1997). Neural induction and antero-posterior patterning in the amphibian embryo: past, present and future. [Review] [136 refs]. *Cell. Mol. Life Sci.* **53,** 319–338.

Goulding, M., Sterrer, S., Fleming, J., Balling, R., Nadeau, J., Moore, K. J., Brown, S. D., Steel, K. P., and Gruss, P. (1993a). Analysis of the Pax-3 gene in the mouse mutant splotch. *Genomics* **17,** 355–363.

Goulding, M. D., Lumsden, A., and Gruss, P. (1993b). Signals from the notochord and floor plate regulate the region-specific expression of two Pax genes in the developing spinal cord. *Development* **117,** 1001–1016.

Graff, J. M., Stumpo, D. J., and Blackshear, P. J. (1989). Characterization of the phosphorylation sites in the chicken and bovine myristoylated alanine-rich C kinase substrate protein, a prominent cellular substrate for protein kinase C. *J. Biol. Chem.* **264,** 11912–11919.

Graham, A. (1997). The origin of dorsoventral patterning of the vertebrate nervous system. *Trends Genet.* **13,** 463–465.

Graham, A., Heyman, I., and Lumsden, A. (1993). Even-numbered rhombomeres control the apoptotic elimination of neural crest cells from odd-numbered rhombomeres in the chick hindbrain. *Development* **119,** 233–245.

Graham, A., Koentges, G., and Lumsden, A. (1996). Neural crest apoptosis and the establishment of craniofacial pattern: An honorable death. *Mol. Cell Neurosci.* **8,** 76–83.

Graham, A., and Lumsden, A. (1996). Patterning the cranial neural crest. *Biochem. Soc. Symp.* **62,** 77–83.

Grapin-Botton, A., Bonnin, M. A., McNaughton, L. A., Krumlauf, R., and Le Douarin, N. M. (1995). Plasticity of transposed rhombomeres: Hox gene induction is correlated with phenotypic modifications. *Development* **121,** 2707–2721.

Guthrie, S., and Lumsden, A. (1991). Formation and regeneration of rhombomere boundaries in the developing chick hindbrain. *Development* **112,** 221–229.

Guthrie, S., Muchamore, I., Kuroiwa, A., Marshall, H., Krumlauf, R., and Lumsden, A. (1992). Neuroectodermal autonomy of Hox-2.9 expression revealed by rhombomere transpositions. *Nature* **356,** 157–159.

Harland, R., and Gerhart, J. (1997). Formation and function of Spemann's organizer. *Annu. Rev. Cell Dev. Biol.* **13,** 611–667.

Hartenstein, V. (1993). Early pattern of neuronal differentiation in the *Xenopus* embryonic brainstem and spinal cord. *J. Comp. Neurol.* **328,** 213–231.

Hatada, Y., and Stern, C. D. (1994). A fate map of the epiblast of the early chick embryo. *Development* **120,** 2879–2889.

Helms, A. W., and Johnson, J. E. (1998). Progenitors of dorsal commissural interneurons are defined by MATH1 expression. *Development* **125,** 919–928.

Hemmati-Brivanlou, A., and Melton, D. (1997). Vertebrate neural induction. *Annu. Rev. Neurosci.* **20,** 43–60.

Ho, R. K., and Kane, D. A. (1990). Cell-autonomous action of zebrafish spt-1 mutation in specific mesodermal precursors. *Nature* **348,** 728–730.

Hogan, B. L., Thaller, C., and Eichele, G. (1992). Evidence that Hensen's node is a site of retinoic acid synthesis. *Nature* **359,** 237–241.

Hollnagel, A., Oehlmann, V., Heymer, J., Ruther, U., and Nordheim, A. (1999). Id genes are direct targets of bone morphogenetic protein induction in embryonic stem cells. *J. Biol. Chem.* **274,** 19838–19845.

Hollyday, M., McMahon, J. A., and McMahon, A. P. (1995). Wnt expression patterns in chick embryo nervous system. *Mech. Dev.* **52,** 9–25.

Hoshikawa, M., Ohbayashi, N., Yonamine, A., Konishi, M., Ozaki, K., Fukui, S., and Itoh, N. (1998). Structure and expression of a novel fibroblast growth factor, FGF-17, preferentially expressed in the embryonic brain. *Biochem. Biophys. Res. Commun.* **244,** 187–191.

Hunt, P., and Krumlauf, R. (1992). Hox codes and positional specification in vertebrate embryonic axes. *Annu. Rev. Cell Biol.* **8,** 227–256.

Hynes, M., Poulsen, K., Tessier-Lavigne, M., and Rosenthal, A. (1995). Control of neuronal diversity by the floor plate: Contact-mediated induction of midbrain dopaminergic neurons. *Cell* **80,** 95–101.

Ikeya, M., Lee, S. M., Johnson, J. E., McMahon, A. P., and Takada, S. (1997). Wnt signalling required for expansion of neural crest and CNS progenitors. *Nature* **389,** 966–970.

Itasaki, N., Sharpe, J., Morrison, A., and Krumlauf, R. (1996). Reprogramming Hox expression in the vertebrate hindbrain: Influence of paraxial mesoderm and rhombomere transposition. *Neuron* **16,** 487–500.

Izraeli, S., Lowe, L. A., Bertness, V. L., Good, D. J., Dorward, D. W., Kirsch, I. R., and Kuehn, M. R. (1999). The SIL gene is required for mouse embryonic axial development and left- right specification. *Nature* **399,** 691–694.

Jessell, T. M., and Lumsden, A. (1997). Inductive signals and the assignment of cell fate in the spinal cord and hindbrain. In "Molecular and Cellular Approaches to Neural Development" (W. M. Cowan, T. M. Jessell, and S. L. Zipursky, Eds.), pp. 290–333. Oxford University Press, New York.

Jostes, B., Walther, C., and Gruss, P. (1990). The murine paired box gene, Pax7, is expressed specifically during the development of the nervous and muscular system. *Mech. Dev.* **33,** 27–37.

Jungbluth, S., Bell, E., and Lumsden, A. (1999). Specification of distinct motor neuron identities by the singular activities of individual Hox genes. *Development* **126,** 2751–2758.

Karlsson, O., Thor, S., Norberg, T., Ohlsson, H., and Edlund, T. (1990). Insulin gene enhancer binding protein Is1-1 is a member of a novel class of proteins containing both a homeo- and a Cys-His domain. *Nature* **344,** 879–882.

Kengaku, M., and Okamoto, H. (1995). bFGF as a possible morphogen for the anteroposterior axis of the central nervous system in *Xenopus. Development* **121,** 3121–3130.

King, T., and Brown, N. A. (1997). Embryonic asymmetry: Left TGFbeta at the right time? *Curr. Biol.* **7,** R212–R215.

Klein, P. S. and Melton, D. A. (1994). Hormonal regulation of embryogenesis: The formation of mesoderm in *Xenopus laevis. Endocr. Rev.* **15,** 326–341.

Koblar, S. A., Murphy, M., Barrett, G. L., Underhill, A., Gros, P., and Bartlett, P. F. (1999). Pax-3 regulates neurogenesis in neural crest-derived precursor cells. *J. Neurosci. Res.* **56,** 518–530.

Kolm, P. J., Apekin, V., and Sive, H. (1997). *Xenopus* hindbrain patterning requires retinoid signaling. *Dev. Biol.* **192,** 1–16.

Kosaki, K., Bassi, M. T., Kosaki, R., Lewin, M., Belmont, J., Schauer, G., and Casey, B. (1999). Characterization and mutation analysis of human LEFTY A and LEFTY B, homologues of murine genes implicated in left-right axis development. *Am. J. Hum. Genet.* **64,** 712–721.

Kothapalli, R., Buyuksal, I., Wu, S. Q., Chegini, N., and Tabibzadeh, S. (1997). Detection of ebaf, a novel human gene of the transforming growth factor beta superfamily association of gene expression with endometrial bleeding. *J. Clin. Invest.* **99,** 2342–2350.

Krauss, S., Concordet, J. P., and Ingham, P. W. (1993). A functionally conserved homolog of the *Drosophila* segment polarity gene hh is expressed in tissues with polarizing activity in zebrafish embryos. *Cell* **75,** 1431–1444.

Krumlauf, R., Marshall, H., Studer, M., Nonchev, S., Sham, M. H., and Lumsden, A. (1993). Hox homeobox genes and regionalisation of the nervous system. *J. Neurobiol.* **24,** 1328–1340.

Kuratani, S. C., and Eichele, G. (1993). Rhombomere transplantation repatterns the segmental organization of cranial nerves and reveals cell-autonomous expression of a homeodomain protein. *Development* **117,** 105–117.

Lamb, T. M., and Harland, R. M. (1995). Fibroblast growth factor is a direct neural inducer, which combined with noggin generates anterior-posterior neural pattern. *Development* **121,** 3627–3636.

Lamb, T. M., Knecht, A. K., Smith, W. C., Stachel, S. E., Economides, A. N., Stahl, N., Yancopolous, G. D., and Harland, R. M. (1993). Neural induction by the secreted polypeptide noggin [see comments]. *Science* **262,** 713–718.

Lee, J. J., Ekker, S. C., von Kessler, D. P., Porter, J. A., Sun, B. I., and Beachy, P. A. (1994). Autoproteolysis in hedgehog protein biogenesis [see comments]. *Science* **266,** 1528–1537.

Lee, K. J., Mendelsohn, M., and Jessell, T. M. (1998). Neuronal patterning by BMPs: A requirement for GDF7 in the generation of a discrete class of commissural interneurons in the mouse spinal cord. *Genes Dev.* **12,** 3394–3407.

LeSueur, J. A., and Graff, J. M. (1999). Spemann organizer activity of Smad10. *Development* **126,** 137–146.

Levin, M., Pagan, S., Roberts, D. J., Cooke, J., Kuehn, M. R., and Tabin, C. J. (1997). Left/right patterning signals and the independent regulation of different aspects of situs in the chick embryo. *Dev. Biol.* **189,** 57–67.

Liem, K. F. J., Tremml, G., and Jessell, T. M. (1997). A role for the roof plate and its resident TGFbeta-related proteins in neuronal patterning in the dorsal spinal cord. *Cell* **91,** 127–138.

Liem, K. F. J., Tremml, G., Roelink, H., and Jessell, T. M. (1995). Dorsal differentiation of neural plate cells induced by BMP-mediated signals from epidermal ectoderm. *Cell* **82,** 969–979.

Lopez, S. L., Dono, R., Zeller, R., and Carrasco, A. E. (1995). Differential effects of retinoic acid and a retinoid antagonist on the spatial distribution of the homeoprotein Hoxb-7 in vertebrate embryos. *Dev. Dyn.* **204,** 457–471.

Lufkin, T. (1996). Transcriptional control of Hox genes in the vertebrate nervous system. *Curr. Opin. Genet. Dev.* **6,** 575–580.

Lufkin, T. (1997). Transcriptional regulation of vertebrate Hox genes during embryogenesis. *Crit. Rev. Eukaryot. Gene Expr.* **7,** 195–213.

Lumsden, A. (1995). Neural development. A 'LIM code' for motor neurons? *Curr. Biol.* **5,** 491–495.

Lumsden, A., Clarke, J. D., Keynes, R., and Fraser, S. (1994). Early phenotypic choices by neuronal precursors, revealed by clonal analysis of the chick embryo hindbrain. *Development* **120,** 1581–1589.

Lumsden, A., and Krumlauf, R. (1996). Patterning the vertebrate neuraxis. *Science* **274,** 1109–1115.

Lumsden, A., Sprawson, N., and Graham, A. (1991). Segmental origin and migration of neural crest cells in the hindbrain region of the chick embryo. *Development* **113,** 1281–1291.

Lun, K., and Brand, M. (1998). A series of no isthmus (noi) alleles of the zebrafish pax2.1 gene reveals multiple signaling events in development of the midbrain-hindbrain boundary. *Development* **125,** 3049–3062.

Maconochie, M. K., Nonchev, S., Studer, M., Chan, S. K., Popperl, H., Sham, M. H., Mann, R. S., and Krumlauf, R. (1997). Cross-regulation in the mouse HoxB complex: The expression of Hoxb2 in rhombomere 4 is regulated by Hoxb1. *Genes Dev.* **11,** 1885–1895.

Maden, M., Gale, E., Kostetskii, I., and Zile, M. (1996). Vitamin A-deficient quail embryos have half a hindbrain and other neural defects. *Curr. Biol.* **6,** 417–426.

Maden, M., Gale, E., and Zile, M. (1998a). The role of vitamin A in the development of the central nervous system. *J. Nutr.* **128,** 471S–475S.

Maden, M., Sonneveld, E., van der Saag, P. T., and Gale, E. (1998b). The distribution of endogenous retinoic acid in the chick embryo: Implications for developmental mechanisms. *Development* **125,** 4133–4144.

Mangold, O. (1933). Über die induktionsfähigkerider verschiedenen bezirke der neurula von urodelen. *Naturwissenschaften* **21,** 761–766.

Mansouri, A., Stoykova, A., Torres, M., and Gruss, P. (1996). Dysgenesis of cephalic neural crest derivatives in Pax7−/− mutant mice. *Development* **122,** 831–838.

Marazzi, G., Wang, Y., and Sassoon, D. (1997). Msx2 is a transcriptional regulator in the BMP4-mediated programmed cell death pathway. *Dev. Biol.* **186,** 127–138.

Marin, F., and Puelles, L. (1994). Patterning of the embryonic avian midbrain after experimental inversions: A polarizing activity from the isthmus. *Dev. Biol.* **163,** 19–37.

Marshall, H., Morrison, A., Studer, M., Popperl, H., and Krumlauf, R. (1996). Retinoids and Hox genes. *FASEB J.* **10,** 969–978.

Marti, E., Takada, R., Bumcrot, D. A., Sasaki, H., and McMahon, A. P. (1995). Distribution of Sonic hedgehog peptides in the developing chick and mouse embryo. *Development* **121,** 2537–2547.

Massague, J. (1998). TGF-beta signal transduction. *Annu. Rev. Biochem.* **67,** 753–791.

Matise, M. P., and Lance-Jones, C. (1996). A critical period for the specification of motor pools in the chick lumbosacral spinal cord. *Development* **122,** 659–669.

Mayor, R., Guerrero, N., and Martinez, C. (1997). Role of FGF and noggin in neural crest induction. *Dev. Biol.* **189,** 1–12.

Mayor, R., Morgan, R., and Sargent, M. G. (1995). Induction of the prospective neural crest of Xenopus. *Development* **121,** 767–777.

Mayor, R., Young, R., and Vargas, A. (1999). Development of neural crest in Xenopus. *Curr. Top. Dev. Biol.* **43,** 85–113.

McMahon, J. A., Takada, S., Zimmerman, L. B., Fan, C. M., Harland, R. M. and McMahon, A. P. (1998). Noggin-mediated antagonism of BMP signaling is required for growth and patterning of the neural tube and somite. *Genes Dev.* **12,** 1438–1452.

Meno, C., Ito, Y., Saijoh, Y., Matsuda, Y., Tashiro, K., Kuhara, S., and Hamada, H. (1997). Two closely-related left-right asymmetrically expressed genes, lefty-1 and lefty-2: Their distinct expression domains, chromosomal linkage and direct neuralizing activity in *Xenopus* embryos. *Genes Cells* **2,** 513–524.

Meno, C., Saijoh, Y., Fujii, H., Ikeda, M., Yokoyama, T., Yokoyama, M., Toyoda, Y., and Hamada, H. (1996). Left-right asymmetric expression of the TGF beta-family member lefty in mouse embryos. *Nature* **381,** 151–155.

Monsoro-Burq, A. H., Bontoux, M., Teillet, M. A., and Le Douarin, N. M. (1994). Heterogeneity in the development of the vertebra. *Proc. Natl. Acad. Sci. USA* **91,** 10435–10439.

Moury, J. D., and Jacobson, A. G. (1989). Neural fold formation at newly created boundaries between neural plate and epidermis in the axolotl. *Dev. Biol.* **133,** 44–57.

Moury, J. D., and Jacobson, A. G. (1990). The origins of neural crest cells in the axolotl. *Dev. Biol.* **141,** 243–253.

Muhr, J., Jessell, T. M., and Edlund, T. (1997). Assignment of early caudal identity to neural plate cells by a signal from caudal paraxial mesoderm. *Neuron* **19,** 487–502.

Ng, K. W., Zhou, H., Manji, S., and Martin, T. J. (1995). Regulation and regulatory role of the retinoids. *Crit. Rev. Eukaryot. Gene Expr.* **5,** 219–253.

Nieto, M. A., Bradley, L. C., Hunt, P., Das, G. R., Krumlauf, R., and Wilkinson, D. G. (1992). Molecular mechanisms of pattern formation in the vertebrate hindbrain. *Ciba Found. Symp.* **165,** 92–102; discussion 102–107.

Nieto, M. A., Sargent, M. G., Wilkinson, D. G., and Cooke, J. (1994). Control of cell behavior during vertebrate development by Slug, a zinc finger gene. *Science* **264,** 835–839.

Nieuwkoop, P. D. (1952a). Activation and organization of the central nervous system in amphibians. *J. Exp. Zool.* **120,** 1–31.

Nieuwkoop, P. D. (1952b). Activation and organization of the central nervous system in amphibians. Part II. Differentiation and organization. *J. Exp. Zool.* **120,** 33–81.

Old, R. W., Smith, D. P., Mason, C. S., Marklew, S., and Jones, E. A. (1996). Effects of retinoic acid on *Xenopus* embryos. *Biochem. Soc. Symp.* **62,** 157–174.

Parr, B. A., Shea, M. J., Vassileva, G., and McMahon, A. P. (1993). Mouse Wnt genes exhibit discrete domains of expression in the early embryonic CNS and limb buds. *Development* **119,** 247–261.

Patel, K., Isaac, A., and Cooke, J. (1999). Nodal signaling and the roles of the transcription factors SnR and Pitx2 in vertebrate left-right asymmetry. *Curr. Biol.* **9,** 609–612.

Pemrick, S. M., Lucas, D. A., and Grippo, J. F. (1994). The retinoid receptors. *Leukemia* **8,** 1797–1806.

Perez, S. E., Rebelo, S., and Anderson, D. J. (1999). Early specification of sensory neuron fate revealed by expression and function of neurogenins in the chick embryo. *Development* **126,** 1715–1728.

Pfaff, S. L., Mendelsohn, M., Stewart, C. L., Edlund, T., and Jessell, T. M. (1996). Requirement for LIM homeobox gene Isl1 in motor neuron generation reveals a motor neuron-dependent step in interneuron differentiation. *Cell* **84,** 309–320.

Pfeffer, P. L., Gerster, T., Lun, K., Brand, M., and Busslinger, M. (1998). Characterization of three novel members of the zebrafish Pax2/5/8 family: Dependency of Pax5 and Pax8 expression on the Pax2.1 (noi) function. *Development* **125,** 3063–3074.

Piedra, M. E., Icardo, J. M., Albajar, M., Rodriguez-Rey, J. C., and Ros, M. A. (1998). Pitx2 participates in the late phase of the pathway controlling left-right asymmetry. *Cell* **94,** 319–324.

Pierani, A., Brenner-Morton, S., Chiang, C., and Jessell, T.M. (1999). A sonic hedgehog-independent, retinoid-activated pathway of neurogenesis in the ventral spinal cord. *Cell* **97,** 903–915.

Placzek, M., Jessell, T. M., and Dodd, J. (1993). Induction of floor plate differentiation by contact-dependent, homeogenetic signals. *Development* **117,** 205–218.

Placzek, M., Tessier-Lavigne, M., Yamada, T., Jessell, T., and Dodd, J. (1990). Mesodermal control of neural cell identity: Floor plate induction by the notochord. *Science* **250,** 985–988.

Porter, J. A., von Kessler, D. P., Ekker, S. C., Young, K. E., Lee, J. J., Moses, K., and Beachy, P. A. (1995). The product of hedgehog autoproteolytic cleavage active in local and long-range signalling. *Nature* **374,** 363–366.

Puelles, L., and Rubenstein, J. L. (1993). Expression patterns of homeobox and other putative regulatory genes in the embryonic mouse forebrain suggest a neuromeric organization. *Trends Neurosci.* **16,** 472–479.

Reifers, F., Bohli, H., Walsh, E. C., Crossley, P. H., Stainier, D. Y., and Brand, M. (1998). Fgf8 is mutated in zebrafish acerebellar (ace) mutants and is required for maintenance of midbrain-hindbrain boundary development and somitogenesis. *Development* **125,** 2381–2395.

Riddle, R. D., Johnson, R. L., Laufer, E., and Tabin, C. (1993). Sonic hedgehog mediates the polarizing activity of the ZPA. *Cell* **75,** 1401–1416.

Roelink, H., Augsburger, A., Heemskerk, J., Korzh, V., Norlin, S., Ruiz i Altaba, A., Tanabe, Y., Placzek, M., Edlund, T., and Jessell, T.M. (1994). Floor plate and motor neuron induction by vhh-1, a vertebrate homolog of hedgehog expressed by the notochord. *Cell* **76,** 761–775.

Roelink, H., Porter, J. A., Chiang, C., Tanabe, Y., Chang, D. T., Beachy, P. A., and Jessell, T. M. (1995). Floor plate and motor neuron induction by different concentrations of the amino-terminal cleavage product of sonic hedgehog autoproteolysis. *Cell* **81,** 445–455.

Rosenquist, G. C. (1966). A radioautographic study of labeled grafts in the chick blastoderm. Development from primitive-streak stages to stage 12. *Contrib. Embryol. Carnegie Inst.* **38,** 71–110.

Rubenstein, J. L., Martinez, S., Shimamura, K., and Puelles, L. (1994). The embryonic vertebrate forebrain: The prosomeric model. *Science* **266,** 578–580.

Rubenstein, J. L., and Puelles, L. (1994). Homeobox gene expression during development of the vertebrate brain. *Curr. Top. Dev. Biol.* **29,** 1–63.

Rubenstein, J. L., and Shimamura, K. (1997). Regulation and patterning and differentiation in the embryonic vertebrate forebrain. In "Molecular and Cellular Approaches to Neural Development" (W. M. Cowan, T. M. Jessell, and S. L. Zipursky, Eds.) pp. 356–390. Oxford University Press, New York.

Rubenstein, J. L., Shimamura, K., Martinez, S., and Puelles, L. (1998). Regionalization of the prosencephalic neural plate. *Annu. Rev. Neurosci.* **21,** 445–477.

Rudnick, D. (1944). Early history and mechanics of the chick blastoderm. *Q. Rev. Biol.* **19,** 187–212.

Ruiz i Altaba, A. (1992). Planar and vertical signals in the induction and patterning of the Xenopus nervous system. *Development* **116,** 67–80.

Ruiz i Altaba, A. (1996). Coexpression of HNF-3 beta and Isl-1/2 and mixed distribution of ventral cell types in the early neural tube. *Int. J. Dev. Biol.* **40,** 1081–1088.

Ruiz i Altaba, A., and Jessell, T. (1991). Retinoic acid modifies mesodermal patterning in early *Xenopus* embryos. *Genes Dev.* **5,** 175–187.

Ruiz i Altaba, A., and Jessell, T. M. (1992). Pintallavis, a gene expressed in the organizer and midline cells of frog embryos: Involvement in the development of the neural axis. *Development* **116,** 81–93.

Ruiz i Altaba, A., Placzek, M., Baldassare, M., Dodd, J., and Jessell, T. M., (1995). Early stages of notochord and floor plate development in the chick embryo defined by normal and induced expression of HNF-3 beta. *Dev. Biol.* **170,** 299–313.

Ruiz i Altaba, A., Prezioso, V. R., Darnell, J. E., and Jessell, T. M. (1993). Sequential expression of HNF-3 beta and HNF-3 alpha by embryonic organizing centers: The dorsal lip/node, notochord and floor plate. *Mech. Dev.* **44,** 91–108.

Ryan, A. K., Blumberg, B., Rodriguez-Esteban, C., Yonei-Tamura, S., Tamura, K., Tsukui, T., de la Pena, J., Sabbagh, W., Greenwald, J., Choe, S., Norris, D. P., Robertson, E. J., Evans, R. M., Rosenfeld, M. G., and Izpisua, B. J. (1998). Pitx2 determines left-right asymmetry of internal organs in vertebrates. *Nature* **394,** 545–551.

Saint-Jeannet, J. P., He, X., Varmus, H. E., and Dawid, I. B. (1997). Regulation of dorsal fate in the neuraxis by Wnt-1 and Wnt-3a. *Proc. Natl. Acad. Sci. USA* **94,** 13713–13718.

Sasaki, H. and Hogan, B. L. (1996). Enhancer analysis of the mouse HNF-3 beta gene: Regulatory elements for node/notochord and floor plate are independent and consist of multiple sub-elements. *Genes Cells* **1,** 59–72.

Schoenwolf, G. C., Bortier, H., and Vakaet, L. (1989). Fate mapping the avian neural plate with quail/chick chimeras: Origin of prospective median wedge cells. *J. Exp. Zool.* **249,** 271–278.

Schoenwolf, G. C., and Sheard, P. (1990). Fate mapping the avian epiblast with focal injections of a fluorescent- histochemical marker: Ectodermal derivatives. *J. Exp. Zool.* **255,** 323–339.

Scott, M. P. (1992). Vertebrate homeobox gene nomenclature [letter]. *Cell* **71,** 551–553.

Scott, M. P. (1993). A rational nomenclature for vertebrate homeobox (HOX) genes. *Nucl. Acids Res.* **21,** 1687–1688.

Sechrist, J., Nieto, M. A., Zamanian, R. T., and Bronner-Fraser, M. (1995). Regulative response of the cranial neural tube after neural fold ablation: Spatiotemporal nature of neural crest regeneration and up-regulation of Slug. *Development* **121,** 4103–4115.

Sechrist, J., Scherson, T., and Bronner-Fraser, M. (1994). Rhombomere rotation reveals that multiple mechanisms contribute to the segmental pattern of hindbrain neural crest migration. *Development* **120,** 1777–1790.

Selleck, M. A., and Bronner-Fraser, M. (1995). Origins of the avian neural crest: The role of neural plate-epidermal interactions. *Development* **121,** 525–538.

Selleck, M. A., Scherson, T. Y., and Bronner-Fraser, M. (1993). Origins of neural crest cell diversity. *Dev. Biol.* **159,** 1–11.

Shimamura, K., Hartigan, D. J., Martinez, S., Puelles, L., and Rubenstein, J. L. (1995). Longitudinal organization of the anterior neural plate and neural tube. *Development* **121,** 3923–3933.

Shimamura, K., Martinez, S., Puelles, L., and Rubenstein, J. L. (1997). Patterns of gene expression in the neural plate and neural tube subdivide the embryonic forebrain into transverse and longitudinal domains. *Dev. Neurosci.* **19,** 88–96.

Shimamura, K., and Rubenstein, J. L. (1997). Inductive interactions direct early regionalization of the mouse forebrain. *Development* **124,** 2709–2718.

Simon, H., Hornbruch, A., and Lumsden, A. (1995). Independent assignment of antero-posterior and dorso-ventral positional values in the developing chick hindbrain. *Curr. Biol.* **5,** 205–214.

Spratt, N. T. (1952). Localization of the propective neural plate in the early chick blastoderm. *J. Exp. Zool.* **120,** 109–130.

St. Amand, T. R., Ra, J., Zhang, Y., Hu, Y., Baber, S. I., Qiu, M., and Chen, Y. (1998). Cloning and expression pattern of chicken Pitx2: A new component in the SHH signaling pathway controlling embryonic heart looping. *Biochem. Biophys. Res. Commun.* **247,** 100–105.

Stern, C. D., and Foley, A. C. (1998). Molecular dissection of Hox gene induction and maintenance in the hindbrain. *Cell* **94,** 143–145.

Streit, A., and Stern, C. D. (1999). Neural induction. A bird's eye view. *Trends Genet.* **15,** 20–24.

Studer, M., Lumsden, A., Ariza-McNaughton, L., Bradley, A., and Krumlauf, R. (1996). Altered segmental identity and abnormal migration of motor neurons in mice lacking Hoxb-1. *Nature* **384,** 630–634.

Stumpo, D. J., Graff, J. M., Albert, K. A., Greengard, P., and Blackshear, P. J. (1989). Molecular cloning, characterization, and expression of a cDNA encoding the "80- to 87-kDa" myristoylated alanine-rich C kinase substrate: A major cellular substrate for protein kinase C. *Proc. Natl. Acad. Sci. USA* **86,** 4012–4016.

Suda, Y., Matsuo, I., and Aizawa, S. (1997). Cooperation between Otx1 and Otx2 genes in developmental patterning of rostral brain. *Mech. Dev.* **69,** 125–141.

Suzuki, A., Ueno, N., and Hemmati-Brivanlou, A. (1997). Xenopus msx1 mediates epidermal induction and neural inhibition by BMP4. *Development* **124,** 3037–3044.

Tanabe, Y., Roelink, H., and Jessell, T. M. (1995). Induction of motor neurons by Sonic hedgehog is independent of floor plate differentiation. *Curr. Biol.* **5,** 651–658.

Tanaka, H., Fukushima, M., Ohta, K., Kimura, Y., Sudo, A., Shirabe, K., and Takeshita, M. (1997). Commitment of motoneuron progenitors of chick embryo along the anterior-posterior axis. *Dev. Neurosci.* **19,** 106–111.

Thor, S., Andersson, S. G., Tomlinson, A., and Thomas, J. B. (1999). A LIM-homeodomain combinatorial code for motor-neuron pathway selection. *Nature* **397,** 76–80.

Tiedemann, H., Asashima, M., Grunz, H., and Knochel, W. (1998). Review: Neural induction in embryos. *Dev. Growth Differ.* **40,** 363–376.

Tosney, K. W., Hotary, K. B., and Lance-Jones, C. (1995). Specifying the target identity of motoneurons. *BioEssays* **17,** 379–382.

Tsuchida, T., Ensini, M., Morton, S. B., Baldassare, M., Edlund, T., Jessell, T. M., and Pfaff, S. L. (1994). Topographic organization of embryonic motor neurons defined by expression of LIM homeobox genes [see comments]. *Cell* **79,** 957–970.

van der Wees, J., Schilthuis, J. G., Koster, C. H., Diesveld-Schipper, H., Folkers, G. E., van der Saag, P. T., Dawson, M. I., Shudo, K., van der Burg, B., and Durston, A. J. (1998). Inhibition of retinoic acid receptor-mediated signalling alters positional identity in the developing hindbrain. *Development* **125,** 545–556.

Varela-Echavarria, A., Pfaff, S. L., and Guthrie, S. (1996). Differential expression of LIM homeobox genes among motor neuron subpopulations in the developing chick brain stem. *Mol. Cell Neurosci.* **8,** 242–257.

Vincent, J. P. (1998). Compartment boundaries: Where, why and how? *Int. J. Dev. Biol.* **42,** 311–315.

Wagner, M., Han, B., and Jessell, T. M. (1992). Regional differences in retinoid release from embryonic neural tissue detected by an in vitro reporter assay. *Development* **116,** 55–56.

Weinstein, D. C., and Hemmati-Brivanlou, A. (1999). Neural induction. *Annu. Rev. Cell Dev. Biol.* **15,** 411–433.

Weinstein, D. C., Ruiz i Altaba, A., Chen, W. S., Hoodless, P., Prezioso, V. R., Jessell, T. M., and Darnell, J. E. J. (1994). The winged-helix transcription factor HNF-3 beta is required for notochord development in the mouse embryo. *Cell* **78,** 575–588.

Whiting, J. (1997). Craniofacial abnormalities induced by the ectopic expression of homeobox genes. *Mutat. Res.* **396,** 97–112.

Wilson, P. A., and Hemmati-Brivanlou, A. (1997). Vertebrate neural induction: Inducers, inhibitors, and a new synthesis *Neuron* **18,** 699–710.

Wilson, P. A., Lagna, G., Suzuki, A., and Hemmati-Brivanlou, A. (1997). Concentration-dependent patterning of the *Xenopus* ectoderm by BMP4 and its signal transducer Smad1. *Development* **124,** 3177–3184.

Wingate, R. J., and Lumsden, A. (1996). Persistence of rhombomeric organisation in the postsegmental hindbrain. *Development* **122,** 2143–2152.

Wolda, S. L., Moody, C. J., and Moon, R. T. (1993). Overlapping expression of Xwnt-3A and Xwnt-1 in neural tissue of *Xenopus laevis* embryos. *Dev. Biol.* **155,** 46–57.

Yamada, T., Pfaff, S. L., Edlund, T., and Jessell, T. M. (1993). Control of cell pattern in the neural tube: Motor neuron induction by diffusible factors from notochord and floor plate. *Cell* **73,** 673–686.

Yamada, T., Placzek, M., Tanaka, H., Dodd, J., and Jessell, T. M. (1991). Control of cell pattern in the developing nervous system: Polarizing activity of the floor plate and notochord. *Cell* **64,** 635–647.

Yamamoto, M., and Schwarting, G. (1991). The formation of axonal pathways in developing cranial nerves. *Neurosci. Res.* **11,** 229–260.

Yoshioka, H., Meno, C., Koshiba, K., Sugihara, M., Itoh, H., Ishimaru, Y., Inoue, T., Ohuchi, H., Semina, E. V., Murray, J. C., Hamada, H., and Noji, S. (1998). Pitx2, a bicoid-type homeobox gene, is involved in a lefty-signaling pathway in determination of left-right asymmetry. *Cell* **94,** 299–305.

Zoltewicz, J. S., and Gerhart, J. C. (1997). The Spemann organizer of *Xenopus* is patterned along its anteroposterior axis at the earliest gastrula stage. *Dev. Biol.* **192,** 482–491.

Cell Biology of Limb Patterning

Scott A. Schaller, Shaoguang Li,[1] Valerie Ngo-Muller,[2] Man-Jong Han, Minoru Omi, Rosalie Anderson, Ken Muneoka
Department of Cell and Molecular Biology, Tulane University, New Orleans, Lousiana 70118, [1]Center for Blood Research, Harvard Medical School, Boston, Massachusetts 02115, [2]INSERM U129, ICGM Cochin Port-Royal, 75014 Paris France

Of vertebrate organ systems, the developing limb has been especially well characterized. Morphological studies have yielded a wealth of information describing limb outgrowth and have allowed for the identification of a multitude of important factors. In terms of the latter, key signaling pathways are known to control numerous aspects of limb development, including establishment of the early limb field, determination of limb identity, elongation of the limb bud, specification of digit pattern, and sculpting of the digits. Modification of underlying signaling pathways can thus result in dramatic alterations of the limb phenotype, accounting for many of the diverse limb patterns observed in nature. Given this, it is clear that signaling pathways regulate the highly orchestrated and tightly controlled sequence of cellular events necessary for limb outgrowth; however, exactly how molecular signals interface with the cell biology of limb development remains largely a mystery. In this review we first provide an overview of a number of the morphogenetic signaling pathways that have been identified in the developing limb and then review how a subset of these signals are known to modify cell behaviors important for limb outgrowth.

KEY WORDS: Limb patterning, Embryo development, Signaling pathways, Morphogenesis.

I. Introduction

The past decade has witnessed an explosion in our understanding of the molecular mechanisms underlying how vertebrate organs develop. These advances are associated with major technical innovations that have increased our ability to detect spatial and temporal patterns of gene expression and have enabled the testing of gene function by targeted disruption and misexpression of specific genes. Although such technical advances have aided in our comprehension of all developing organ systems, this has been particularly true of the limb. Several reasons can be offered for why the vertebrate limb has been an exceptional target for these advances. First, the vertebrate limb is highly regulative, and its regulative behavior has been well-documented. This behavior allows for the combination of embryological approaches with functional studies to provide a powerful model for testing the biological role of specific gene products (Schwabe *et al.,* 1998). Second, the vertebrate limb is attractive as a genetic model, and a large collection of mutations in humans, mice, and birds are available for characterizing the molecular biology of limb patterning (Muragaki *et al.,* 1996). Third, progress in our understanding of vertebrate limb development is directly related to the rate that advances are made in understanding *Drosophilia* limb development. Many of the signaling pathways uncovered in insects also function in a similar manner in vertebrates (French *et al.,* 1976; Bryant, 1993). These various approaches have combined to make the developing limb perhaps the best understood vertebrate organ system, one that will continue to serve as a model for organ development.

This review is meant to provide an overview of what has become the enormous field of limb development. Many excellent reviews exist in the literature. Indeed, an entire journal volume was recently devoted to a large collection of individual reviews on various aspects of limb and muscle development (Zeller and Braun, 1999). Hence, it is unreasonable to think that a single review article might adequately represent the various component parts of this field. In this review we have taken an approach that reflects our own research philosophy, that the limb is the end product of a series of events in which cellular activities, i.e. proliferation, migration, adhesion, etc., are regulated in a spatial and temporal manner. How these cellular activities are controlled is key to our understanding of pattern formation. It is clear that these activities are regulated by intercellular signals, many of which have been identified. Therefore, the first part of this review is devoted to an overview of the developing limb in which we focus on its regulative capabilities and how studying pattern regulation has led to the identification of a number of morphogenetic signaling pathways. The second half is devoted to establishing an interface between these signaling pathways and the cell biology of limb development. In the latter, we focus on a number of cellular activities known to play key roles in limb outgrowth and how these activities are modulated by some of the known morphogenetic signals.

II. Developmental Anatomy, Pattern Regulation, and Morphogenetic Signals

Vertebrate limb morphogenesis encompasses a period during embryogenesis that begins prior to the initial emergence of the limb bud as a relatively homogeneous structure until achievement of its final, highly complex form. The limb initially appears as a bud of lateral plate mesoderm enclosed by a sheet of ectodermal cells. The limb bud elongates in the proximal-distal axis, eventually flattening in the dorsal-ventral axis and expanding in the anterior-posterior axis as the digit primordia emerge. The mesenchymal cells of the lateral plate that initially form the bud later differentiate to form skeletal components, tendons, and dermis. Once the limb bud has formed, somitic mesodermal cells invade and provide the source for future musculature (see Büscher and Izpisúa-Belmonté, 1999). Differentiation of tissues occurs in a clear proximal to distal sequence, a subtle posterior to anterior sequence, and no apparent dorsal-ventral sequence.

Limb patterning is most frequently related to final morphologies in terms of skeletal structures and can be described along three axes: proximal-distal, anterior-posterior, and dorsal-ventral. The early skeletal pattern is useful for morphological studies due to clear anatomical differences between the various skeletal components. Although final morphology differs extensively in tetrapod vertebrates, the general organization of tissue types is identical (Ettinger and Doljanski, 1992). Outgrowth occurs along the proximal-distal axis and is reflected in the sequence of skeletal structures that arise. The most proximal element (the stylopod) is termed either the humerus for the wing/forelimb/arm or the femur for the hindlimb/leg. Next, the zeugopod is composed of the radius and ulna in the wing/forelimb/arm or the tibia and fibula in the hindlimb/leg, with the ulna and the fibula being the longer of the two bones in their respective limbs. The distal-most skeletal elements are the bones of the autopod that include the digits and associated elements. Along the anterior-posterior limb axis, patterning is most often related to skeletal morphology in terms of the number and types of digits that form, hence discussion of anterior-posterior patterning is restricted to the autopod. In mice and humans, both fore- and hindlimbs contain five digits denoted as digits 1–5, with digit 1 representing the anterior-most digit (thumb or great toe in humans). In the chick, the forelimb or wing contains three digits (digits 2, 3, and 4), and the hindlimb or leg contains four digits (digits 1–4). Digit identity is characterized by the number and size of skeletal elements (metacarpals/metatarsals and phalangeal elements); however, in some instances multiple digits share a similar phenotype and cannot be distinguished (e.g. digits 2, 3, and 4 in the mouse).

In contrast to other axes, dorsal-ventral patterning of the limb is more subtle when viewed in terms of final skeletal morphology but can be described in terms of joint flexure, the positioning of tendons and muscle, and the differentiative characteristics of the skin. In humans, for example, the dorsal surface of the hand is the back of the hand, while the ventral surface is the palm.

Limb development is a continuous process that begins with the initial establishment of the limb field and ends with the differentiation of the various tissues that compose the mature limb. In understanding how this complex structure forms, it is instructive to fragment the process into temporally distinct stages that distinguish key events important for subsequent development. For the purposes of this review we describe limb development in three stages: the pre-bud stage when the limb field is established, the bud stage when proximal-distal elongation predominates, and the late bud stage when the final pattern of the autopod is elaborated. In Figure 1 we

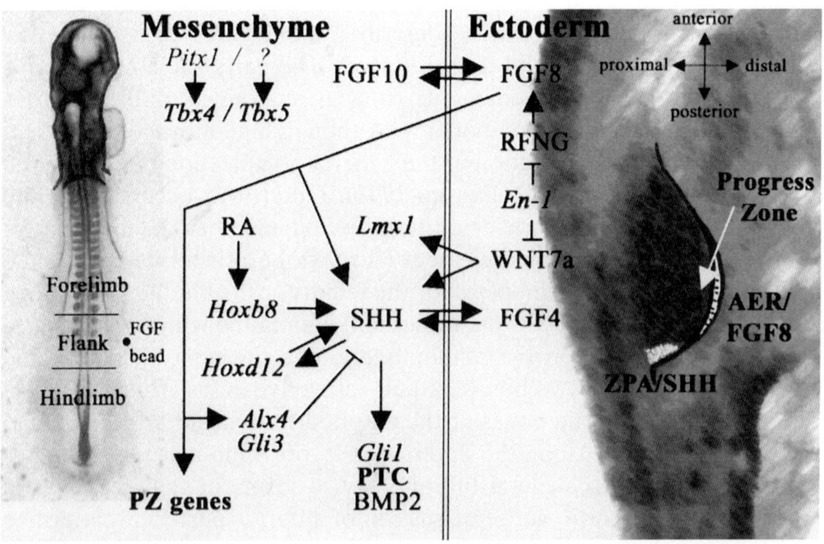

FIG. 1 Schematic illustration depicting some of the major signaling pathways that regulate limb patterning during pre-bud and bud stages. Mesenchymal and ectodermal signals are separated by the double vertical lines. A chick stage 13/14 embryo is shown on the left identifying the flank region where FGF bead implantation induces supernumerary limb formation. The prospective forelimb and hindlimb regions are also indicated. A chick stage 20/21 right wing bud in shown on the right identifying the apical ectodermal ridge (AER), the progress zone (PZ), and the zone of polarizing activity (ZPA). The limb is viewed from a dorsal prospective and the anterior-posterior and proximal-distal axes are indicated. Transcription factors are in italics. *Pitx1* and *Tbx4* are specific to the hindlimb, whereas *Tbx5* is forelimb-specific.

provide an outline of many of the signaling pathways described in this review highlighting the pre-bud and bud stages.

A. Pre-Bud Stage: The Initiation of Outgrowth

Prior to the appearance of a limb bud, a field of cells along the embryonic flank acquires the capacity to develop into a limb. Experimental studies of developmental events important for limb specification have been largely carried out in the chick embryo, characterized by stages established by Hamburger and Hamilton (1951). Whereas the limb bud is initially apparent by stage 17, explant studies indicate that the wing-forming region has the capacity to form limb structures by stage 12 (Stephens *et al.,* 1993). The stage 12 pre-bud region has been mapped to an area adjacent to somites 15 to 20 and is approximately 480 μm along the anterior-posterior axis, 200 μm along the dorsal-ventral axis, and 120 μm along the prospective proximal-distal axis. Fate mapping studies suggest that this pre-bud region expands in an organized manner. During stages 12 to 14, the anterior-posterior dimension more than doubles while the dorsal-ventral and proximal-distal dimensions remain constant (Searls and Janner, 1971). From stage 14, the anterior-posterior dimension remains relatively constant while the dorsal-ventral component increases (Chaube, 1959). The proximal-distal dimension expands as the limb bud forms between stages 16 and 17, and this expansion continues with bud elongation (Searls and Janner, 1971). Hence, changes in the size of the pre-bud region and the limb bud itself suggest highly coordinated patterns of growth.

1. *Tbx* Genes and Limb Identity

Recent evidence indicates that the characteristics distinguishing the forelimb from the hindlimb are controlled by the action of members of the *Tbx* family of transcriptional regulators (Simon, 1999). A group of 4 *Tbx* genes, *Tbx2, Tbx3, Tbx4,* and *Tbx5,* are expressed in the limb forming regions, and two of these genes, *Tbx4* and *Tbx5,* appear to be involved in identifying the forelimb from the hindlimb. *Tbx5* is expressed exclusively in the presumptive forelimb and *Tbx4* is expressed exclusively in the presumptive hindlimb. Expression of each is AER independent, and limb-type specific expression is maintained in chimeric limbs composed of forelimb and hindlimb tissues (Ohuchi *et al.,* 1998; Issac *et al.,* 1998; Gibson-Brown *et al.,* 1998; Logan *et al.,* 1998). The onset of *Tbx4* and *Tbx5* expression occurs after the initial specification of the limb field; however, there is evidence that misexpression of these genes can modulate the limb-specific phenotype (Logan and Tabin, 1999; Takeuchi *et al.,* 1999; Rodriguez-

Esteban *et al.,* 1999). In addition, misexpression studies suggest that the bicoid-type homeobox-containing gene *Pitx1* acts upstream of *Tbx4* in the hindlimb and is also involved in the regulation of the hindlimb phenotype (Logan and Tabin, 1999; Takeuchi *et al.,* 1999).

2. FGF Signaling Is Critical for the Pre-bud Field

The fibroblast growth factor (FGF) family of signaling proteins plays an important role in setting up the pre-bud field. FGFs are intercellular signaling molecules that display a strong binding affinity for the extracellular matrix and signal via the FGF receptor (FGFR), a member of the tyrosine kinase superfamily of cell surface receptors (Xu *et al.,* 1999). The *Fgf* gene family is very large and includes at least five members that are expressed during limb development, *Fgf2, Fgf4, Fgf8, Fgf9,* and *Fgf10* (see Martin, 1998; Ohuchi and Noji, 1999). The *FGFr* gene family includes four members, two of which are expressed during limb development, *FGFr1* and *FGFr2.* FGF signaling in the developing limb is complicated because some of the genes (e.g. *Fgf2, Fgf8, FGFr1,* and *FGFr2*) are differentially spliced to produce multiple gene products. In some cases there is direct evidence that the differentially spliced gene products have uniquely different functions, indicating that there is a high level of complexity associated with the FGF signaling pathways during limb development.

Fgf10 is expressed in the pre-bud region of the lateral plate mesoderm and is necessary for proper limb formation. In the chick, *Fgf10* is expressed in lateral plate mesoderm at stage 12 when the limb field becomes tissue autonomous (Ohuchi *et al.,* 1997). At this stage *Fgf10* is expressed beyond the mapped boundary of the limb; however, it is down-regulated in the surrounding tissue so that by stage 15 it is expressed only in the pre-bud mesoderm. Targeted disruption of the *Fgf10* gene in the mouse results in a limbless phenotype identifying it as an essential factor for limb formation (Min *et al.,* 1998; Sekine *et al.,* 1999). Similarly, interrupting the action of FGF10 either by over-expressing a soluble, dominant negative derivative of the *FGFr2B* gene or by the deletion of the FGF binding domain of the *FGFr2* gene results in a limbless or distally truncated phenotype (Celli *et al.,* 1998; Xu *et al.,* 1998). Since binding studies suggest that FGF10 is the endogenous ligand for FGFR2B, and *FGFr2b* expression in the limb bud is AER specific, the evidence indicates that a primary target for FGF10 action is the limb ectoderm (Xu *et al.,* 1998). One downstream target of FGF10 signaling is the AER-specific gene *Fgf8. Fgf8* expression in the pre-bud ectoderm in first observed at stage 16, some three hours after localization of *Fgf10* expression to the pre-limb mesenchymal tissue (Heikinheimo *et al.,* 1994; Ohuchi *et al.,* 1994; Crossley and Martin, 1995; Crossley *et al.,* 1996; Vogel *et al.,* 1996). The initial *Fgf8* expression domain encompasses a broad band of ectodermal cells that includes the future AER; however, once the limb bud forms, *Fgf8* expression becomes exclusively restricted to the AER. Additionally, expression of *Fgf8* in the limb ectoderm is FGF10-dependent (Min *et al.*, 1998; Sekine *et al.*, 1999) and can be induced by ectopic FGF10 application (Ohuchi *et al.*, 1997; Yonei-Tamura *et al.*, 1999). There is also evidence for a reciprocal regulatory loop involving control of *Fgf10* expression by FGF8 signaling through the FGFR2C receptor expressed in mesenchymal cells (see Xu *et al.*, 1999).

While the ability to induce ectopic limb structures from non-limb, embryonic flank tissue has been known for many years, the molecular basis for this inductive event has only recently been discovered (Ohuchi and Noji, 1999). In the chick, non-limb, embryonic flank tissue (stage 13–17) responds to an ectopic source of FGFs by initially forming an ectopic limb bud that later develops into identifiable limb structures (Cohn *et al.*, 1995; Ohuchi *et al.*, 1995). The ectopic limb is always of reverse handedness by comparison to the neighboring, endogenous fore- and hindlimbs, and the ectopic limb is generally a chimera of both tissues types (Ohuchi *et al.*, 1997). A number of FGFs have been tested using this assay, including FGF1, FGF2, FGF4, FGF7, FGF8, and FGF10. Of these, only FGF7 failed to induce the formation of ectopic limb structures (Cohn *et al.*, 1995; Ohuchi *et al.*, 1995, 1997; Crossley *et al.*, 1996; Vogel *et al.*, 1996). Ectopic limbs are generally induced by implants of microcarrier beads loaded with purified FGF protein, although implantation of cells expressing *Fgf4* can induce a similar response (Ohuchi *et al.*, 1995). What is interesting is that ectopic expression of *Fgf4* or *Fgf8* in flank cells through retroviral infection (Mima *et al.*, 1995; Vogel *et al.*, 1996) or ubiquitous expression of *Fgf2* or *Fgf4* in transgenic models (Abud *et al.*, 1996; Lightfoot *et al.*, 1997) does not result in ectopic limb formation, thus raising the possibility that a localized source of FGF is required for this response.

3. Retinoic Acid and the Pre-bud Field

Retinoic acid (RA) signaling is also critical for the initiation of limb bud outgrowth. RA activates a signaling pathway that is mediated by the RA receptors (RAR), a family of ligand-dependent transcription factors that are members of the superfamily of nuclear receptors (Chambon, 1996; Morriss-Kay and Ward, 1999). It is almost certain that RA acts as an intracellular signal, yet some evidence suggests that it is also secreted so as to influence neighboring cells (Maden *et al.*, 1998). RA production in the embryo involves the metabolic modification of all-trans-retinol to form a number of biologically active retinoids, including all-trans-RA, 4-oxo-RA, didehyrodro-RA, and 9-cis-RA (Means and Gudas, 1995). A fundamental enzyme in this metabolic transformation is retinaldehyde dehydrogenase

(RALDH), and its expression in the embryo provides an indirect measure of spatial and temporal patterns of RA signaling (Niederreither *et al.*, 1999). In addition, a number of genes expressed in the limb bud have been identified that are directly up-regulated by RA (e.g. *RARβ2, Bmp2, Hoxb8*), and their expression patterns provide evidence of RA signaling. Locally inhibiting RA synthesis during pre-bud stages results in a limbless phenotype and the absence of *Shh* expression, a gene important for proper outgrowth and patterning (see below; Helms *et al.*, 1996; Stratford *et al.*, 1996; Lu *et al.*, 1997). In addition, targeted disruption of the RALDH gene, which is involved in the production of RA, results in a similar phenotype (Niederreither *et al.*, 1999). During pre-bud stages in the chick embryo, RA is known to directly upregulate *Hoxb8* expression in the forelimb region (Stratford *et al.*, 1997; Lu *et al.*, 1997). In the mouse, misexpression of *Hoxb8* induces an ectopic anterior expression domain of *Shh* and the formation of duplicated forelimb structures (Charité *et al.*, 1994). Hence, RA induction of *Hoxb8* during pre-bud stages is linked to the initial induction of *Shh* and is necessary for limb outgrowth.

B. Bud Stage: Outgrowth

In the chick, the transition between pre-bud stages to limb bud stages is marked by the lateral bulging of the limb mesenchyme to form the limb bud, a homogeneous population of mesenchymal cells covered by ectoderm. From this initially homogeneous population, cells develop along several disparate fates with differentiation. Detailed fate maps of the chick wing bud have been established (Vargesson *et al.*, 1997), but fate maps of the mouse limb bud are only available for later stages (Muneoka *et al.*, 1989b). Fate maps are generated by marking a group of cells with a vital marker, such as carbon particles or DiI, to identify these cells at later stages. Fate mapping studies indicate that there is an anterior-posterior asymmetry to how the limb develops. This is particularly clear in the chick wing, where the posterior limb bud contributes to the majority of the limb proper, while the anterior limb bud contributes to the shoulder and parts of the stylopod. The significance of this anterior-posterior asymmetry to limb cell fate is presently unclear; however, there is evidence that growth along the anterior-posterior axis of the limb bud is spatially regulated (Aono and Ide, 1988; Shi and Muneoka, 1992).

1. The Apical Ectodermal Ridge

The apical ectodermal ridge (AER) is a developmentally transient ectodermal specialization at the distal tip of the limb bud in all amniote vertebrates.

It is typified by closely grouped, pseudostratified columnar epithelial cells separated from underlying mesenchymal cells by a basement membrane (Kelley and Fallon, 1976). Location of the AER is at the apex of the limb bud, where it runs along the distal boundary separating the dorsal and ventral ectodermal surfaces. In the chick, the AER is initially positioned symmetrically along the anterior-posterior axis and subsequently takes on a posterior bias. Although the AER forms soon after the bud is visible during chick development, in the mouse the AER does not form until limb bud outgrowth is well under way (Wanek *et al.,* 1989). The late appearance of the mouse AER as well as studies of the *limbless* mutation in the chick shows that initial formation of the limb bud is an AER-independent event (Carrington and Fallon, 1988).

While the AER is induced by FGF10 produced by the underlying mesenchyme, positioning of the AER is controlled by the expression of additional genes, including *Radical fringe* (*Rfng*) and *Engrailed-1* (*En-1*) (see Chen and Johnson, 1999). RFNG is a secreted signaling factor expressed in the dorsal ectoderm, and EN1 is a homeobox-containing transcription factor expressed in the ventral ectoderm. In the chick, *Rfng* is initially expressed in the presumptive dorsal ectoderm at stage 15. During the initiation of bud formation, its expression becomes concentrated along the dorsal/ventral boundary, and by stage 22–24 it is expressed exclusively in the AER. Ectopic expression of the *Rfng* gene in the limb bud results in the formation of ectopic AERs only in the ventral ectoderm, hence it is proposed that *Rfng* defines position of the AER at the interface between cells that are and are not expressing RFNG (Laufer *et al.,* 1997; Rodriguez *et al.,* 1997). Despite such evidence for RFNG involvement, targeted disruption of *Rfng* in the mouse results in the formation of normal limbs. This suggests that RFNG is not strictly required for limb development, possibly due to functional overlap among the *fringe* family (Moran *et al.,* 1999). With less ambiguity, disruption of the transcription factor *En-1* in the mouse results in a ventral expansion of the AER and in some cases the formation of an ectopic AER (Loomis *et al.,* 1996, 1998). In addition, misexpression of *En-1* in the dorsal ectoderm represses the expression of Wnt7a, a dorsal ectoderm marker, and *Rfng* (Laufer *et al.,* 1997; Logan *et al.,* 1997; Rodriguez *et al.,* 1997). The current model suggests that *Rfng* expression in the dorsal ectoderm and EN1 repression of *Rfng* in the ventral ectoderm creates an interface of *Rfng* expression at the apex of the limb bud (Chen and Johnson, 1999), with potential functional overlap among the *fringe* family (Moran *et al.,* 1999). This, in turn, allows for positioning of the AER along the dorsal-ventral ectodermal boundary (Chen and Johnson, 1999).

Dorsal-ventral pattern is imparted to the limb mesenchyme by signals derived from the limb ectoderm. During pre-bud stages the ectoderm first receives signals for dorsal-ventral patterning from somitic tissue (Michaud *et al.,* 1997). With limb outgrowth, the dorsal ectodermal signal WNT7a specifies underlying mesenchymal tissue to develop along a dorsal pathway (Parr and McMahon, 1995; Yang and Niswander, 1995) through induction of the transcription factor *Lmx1b* in dorsal mesenchymal cells (Riddle *et al.,* 1995; Vogel *et al.,* 1995b). Meanwhile, ventral patterning is regulated by *En-1* expression. EN1 apparently serves two independent roles in limb outgrowth; first, as described above, it acts to position the AER, and second, it is involved in ventral patterning by acting to repress dorsal specification. EN1 in the ventral ectoderm acts to repress *Wnt7a* expression, suggesting a model in which the ventral pattern is a default pathway (Loomis *et al.,* 1996, 1998; Logan *et al.,* 1997). An interesting aspect of dorsal-ventral patterning is that the WNT7a and EN1 signaling pathways are not active throughout the entire limb, but appear to function only in the distal region of the limb. The activity of *Lmx1b,* in contrast, is more widespread (Chen and Johnson, 1999). The dorsal-ventral patterning signals in proximal regions of the limb are presently unknown.

2. The Progress Zone

The AER achieves its function of providing for limb outgrowth by maintaining underlying mesenchymal cells in an undifferentiated, proliferative state known collectively as the progress zone. As the limb bud grows, cells leave the region of AER influence and differentiate as developmentally fated (Summerbell *et al.,* 1973). The progress zone is viewed as a region where active growth and pattern specification occur under the influence of the AER. Evidence linking AER function to limb outgrowth and pattern specification includes AER removal studies that result in stage-dependent distal limb truncations (Saunders, 1948, 1998) and the induction of supernumerary limb structures by an ectopic AER (Saunders and Errick, 1976; Goetinch, 1964). The progress zone model is supported by the localized expression of a number of developmentally important genes, many of which are regulated by the AER. Among these are the 5′ members of the *Hoxa* and *Hoxd* gene clusters, which play roles in the regional specification of the limb skeletal pattern (Zakany and Duboule, 1999). Another includes *Msx1,* which has been implicated in the control of cell proliferation (Song *et al.,* 1993; Wang and Sassoon, 1995).

As with limb initiation, the dependency of mesenchymal outgrowth on the AER is partially a function of FGF activity. Numerous studies have shown that outgrowth can proceed following AER removal in the presence of ectopically applied FGF. Though this function can be provided for by either FGF2, FGF4, or FGF8 (Fallon *et al.,* 1994; Niswander *et al.,* 1993; Mahmood *et al.,* 1995; Vogel *et al.,* 1996), FGF8 is assumed to be more relevant physiologically because it is expressed throughout the AER with no axial bias (Heikinheimo *et al.*, 1994; Ohuchi *et al.*, 1994). However, *Fgf9* is also expressed without axial bias in the AER (Colvin *et al.*, 1999). *Fgf2* is present in the dorsal ectoderm and peripheral mesoderm in addition to the AER (Savage and Fallon, 1995), and *Fgf4* transcripts are restricted to the posterior AER (Niswander and Martin, 1993) during early bud stages, later becoming localized distally. Ectopically applied FGF also induces distal outgrowth of amputated limb buds, thus implicating FGF signaling in the re-formation of the progress zone associated with a regeneration response (Taylor *et al.*, 1994; Kostakopoulou *et al.*, 1996).

3. The Zone of Polarizing Activity

The zone of polarizing activity (ZPA) describes a region in the posterior margin of the developing limb involved in anterior-posterior patterning, with the latter usually assessed in terms of digit pattern. The ZPA is induced early in bud formation by RA signaling from the flank (Wanek *et al.*, 1991; Noji *et al.*, 1991; Helms *et al.*, 1996; Stratford *et al.*, 1996; Lu *et al.*, 1997; Niederreither *et al.*, 1999). When grafted into the anterior margin of a host chick limb, ZPA tissue induces mirror symmetric digit duplications. With a fully symmetric duplication induced by ZPA grafts, the normal chick digit sequence 2, 3, 4, is respecified to form the digit sequence 4-3-2-2-3-4 (Saunders and Gasseling, 1968). It has been shown that the ability of the ZPA to respecify pattern (i.e. to signal) is a function of its cellular components (Saunders, 1972) and that the level of signaling correlates with the number of ZPA cells grafted (Tickle, 1981). The ZPA signal has been hypothesized to have a graded effect on digit specification with high levels of signaling associated with posterior digit specification and low levels of signaling associated with anterior digit specification (Tickle *et al.*, 1975). It has also been shown that respecification of digits induced by the ZPA is species-independent; grafts of the ZPA derived from avian, mammalian, or reptilian sources will induce formation of supernumerary digits in the chick embryo (see Fallon and Crosby, 1977). Hence, the ZPA signal is conserved among tetrapod vertebrates and elicits a graded response that is quantitatively dependent on the signaling level.

Embryological studies have demonstrated a relationship between signaling by the AER and signaling by the ZPA (Summerbell, 1974). This relationship has been shown even more directly in studies analyzing the maintenance of ZPA signaling in cultured limb bud cells (Anderson *et al.*, 1993, 1994; Vogel and Tickle, 1993; Vogel *et al.*, 1995a). Such studies demonstrate that ZPA signaling by posterior cells declines rapidly *in vitro* unless cocultured with the AER or with FGF2 or FGF4, in which cases signaling is maintained. Analogous studies *in vivo* demonstrate that ZPA signaling declines following AER removal, while FGF2 or FGF4 application can maintain ZPA signaling in the absence of the AER (Vogel and Tickle, 1993; Li *et al.,* 1996). In addition, there is evidence that FGF also plays a role in maintaining non-ZPA cells from transformation to a ZPA signaling state (Anderson *et al.,* 1994). The evidence suggests that limb bud cells have labile positional characteristics and that FGF acts to stabilize these position-specific characteristics.

The molecular cues for ZPA-induced respecification of anterior-posterior pattern have been the object of intensive scrutiny in recent years. *Sonic hedgehog* (*Shh*), the vertebrate homolog to the *Drosophila* segment polarity gene *hedeghog*, is mesenchymally expressed in a region of the posterior limb margin that colocalizes with the ZPA. Implantation of cells expressing *Shh* into the anterior margin beneath the AER induces supernumerary digit formation, thus suggesting that SHH is the functional ZPA signal (Riddle *et al.,* 1993). Support for this hypothesis comes from the finding that purified SHH protein can induce supernumerary digits in a dose-dependent manner (Yang *et al.,* 1997). In addition, targeted disruption of *Shh* results in the inhibition of distal limb structures, demonstrating a crucial role of SHH in limb outgrowth (Chiang *et al.,* 1996). Despite this, the concentration of SHH required to induce a typical ZPA response is very high, and exposure time required for respecification is very long (> 16 hours, Yang *et al.,* 1997). It is also significant that some cell lines expressing a functional SHH protein lack ZPA signaling ability, indicating that the endogenous ZPA signal is more complex than the simple production of SHH (Ogura *et al.,* 1996).

The mechanism of SHH signaling is largely conserved when compared to signal transduction of the Hedgehog (HH) signal in *Drosophila* (see Pearse and Tabin, 1998). SHH signaling is mediated through the interactions of two transmembrane protein genes, *Patched* (*Ptc*) and *Smoothened* (*Smo*), with PTC acting as the receptor for SHH (see Pearse and Tabin, 1998). Recently, an additional, non-PTC receptor for Hedgehogs, (*Hedgehog-interacting protein,* HIP) has been shown to negatively regulate HH activities (Chuang and McMahon, 1999). In *Drosophila, cubitus interruptus* (*Ci*) is thought to be the endpoint in HH transduction responsible for transcriptional regulation of HH target genes, and CI shares homology with members of the vertebrate GLI family of transcription factors, GLI1, GLI2, and GLI3 (Theil *et al.,* 1999). GLI1 can mediate SHH signaling through transcriptional activation of SHH target genes (Marigo *et al.,* 1996, Lee *et al.,* 1997), and GLI3 is a negative regulator of *Shh* expression (Masuya *et al.,* 1995, 1997; Marigo *et al.,* 1996; Büscher *et al.,* 1997).

Downstream gene targets of ZPA/SHH signaling include both growth factors and transcriptional regulators. These genes have been identified primarily by observing changes in gene expression associated with the induction of supernumerary digits in the ZPA signaling assay. In this manner, induced gene expression in the anterior can be correlated with normal interactions in the limb posterior. ZPA/SHH signaling induces anterior expression of *bmp2* (normally localized to the posterior mesenchyme; Francis *et al.,* 1994; Laufer *et al.,* 1994), suggesting that BMP2 may be involved in downstream SHH signaling. Ectopic *bmp2* expression in the anterior limb bud, however, does not recapitulate ZPA/SHH morphogenetic effects, although some putative downstream targets of SHH are induced (Duprez *et al.,* 1996). ZPA/SHH signaling modifies gene expression in the AER by inducing the expression of *Fgf4* in the anterior AER (Laufer *et al.,* 1994; Niswander *et al.,* 1994). This induction, along with FGF4 maintenance of endogenous *Shh* expression in the posterior bud, implicates SHH and FGF4 in a cross-regulatory signaling loop (Laufer *et al.,* 1994; Niswander *et al.,* 1994). Despite this, studies indicate that SHH is not strictly required for *Fgf4* expression (Duprez *et al.,* 1996; Pizette and Niswander, 1999). Whether FGF4 signaling is directly involved in the respecification of anterior cells or simply required for subsequent digit outgrowth is presently unknown.

ZPA grafts or ectopic SHH can also induce expression of members of the *Hox* gene clusters that play a regulatory role in the specification of limb patterns (see Zakany and Duboule, 1999). During limb development, members of the *Hoxa* and *Hoxd* clusters are expressed in highly complex patterns characterized by three independent phases (Nelson *et al.,* 1996). In terms of the *Hoxd* cluster, phase one is characterized by *Hoxd9* and *d10* expression throughout the limb and is associated with patterning of the prospective stylopod. Phase two is characterized by sequential induction of *Hoxd11-d13* in a nested pattern originating distally in the limb posterior in association with zeugopodal patterning. Phase three is characterized by an inversion in the *Hoxd11-d13* expression domains as compared to phase two. In phase three, *Hoxd13* occupies the widest domain with the most anterior expression, while *Hoxd11 and d12* are expressed more posteriorly in a shared domain that includes *d13* (Nelson *et al.,* 1996). This phase-dependent nesting of *Hoxd* expression can also be ectopically induced in the anterior by ectopic SHH (Riddle *et al.,* 1993; Nelson *et al.,* 1996) or grafts of ZPA tissue (Izpisúa-Belmonté *et al.,* 1991; Nohno *et al.,* 1991).

Though SHH is clearly involved in limb pattern formation, the exact nature of its involvement has not been precisely characterized. SHH has been proposed to be the ZPA signal hypothesized to have a graded distribution along the anterior-posterior limb bud axis (Tickle *et al.,* 1975; Riddle *et al.,* 1993). However, SHH is known to be tethered to membrane surfaces through cholesterol modification (Porter *et al.,* 1996), and it is clear that diffusion is not required for SHH signaling in the limb bud (López-Martínez *et al.,* 1995; Martí *et al.,* 1995; Yang *et al.,* 1997). A SHH-induced diffusible downstream signal that mediates ZPA signaling has been proposed (Yang *et al.,* 1997); however, at present there are no obvious candidates. Our understanding of the downstream SHH signaling pathway that leads to digit respecification is confused by the demonstration that a number of SHH-responsive genes have been shown to be inducible in the absence of SHH (Ros *et al.,* 1996, Ohsugi *et al.,* 1997; Pizette and Niswander, 1999). Moreover, supernumerary digit induction has been demonstrated without involvement of SHH, suggesting either an interface with the downstream SHH signaling pathway or the existence of SHH-independent mechanisms for pattern specification (Rodriguez *et al.,* 1996; Ohsugi *et al.,* 1997; Caruccio *et al.,* 1999). Finally, one major concern in approaching an understanding of ZPA or SHH signaling is the fact that respecification requires active signaling for an extended period ($>$ 16 hours, Smith, 1980; Yang *et al.,* 1997), indicating that pattern respecification is far from a simple event.

C. Late Bud—Making Digits

Later stages of limb development are characterized by the formation and separation of the digits. The autopodial plate is dorsal-ventrally flattened and fans out with chondrogenic rays that represent the forming digits. Fate maps of these digit rays indicate that they primarily form the metatarsal/metacarpal of the mature digit, with the phalangeal elements forming from only the distal-most region of each ray (Muneoka *et al.,* 1989b). Individual digit rudiments can develop at ectopic sites, indicating that at this stage each digit rudiment is an autonomously developing morphogenetic field (Reginelli *et al.,* 1995). The primary events of digit morphogenesis are distal elongation of the digit blastema that results from apical growth and programmed cell death within the interdigital regions that separate the digits.

1. The Digit Ray

When prechondrogenic cells of the autopod begin to aggregate to form the digit rays, the AER and the ZPA are no longer present, thus suggesting that digit specification is largely complete. However, many patterning genes that were expressed in the progress zone of the early limb bud (such as the *Hoxa* and *Hoxd* gene clusters and the *Msx* genes) remain expressed in the digit rudiments. This suggests that the patterning of individual digit rudiments is still incomplete. Furthermore, a number of experimental observations suggest that active patterning of digit rudiments continues into later limb bud stages. First, digit tip amputations *in utero* result in a rapid level-specific regeneration response that includes reexpression of the *Msx* genes (Reginelli *et al.,* 1995). Second, the digit rudiments of late stage mouse limb buds (Stage 7/8, E12.5) respond to implantation of a FGF4-releasing microcarrier bead by inducing specific bifurcation of digit IV, indicating that patterning of digit IV is incomplete (Ngo-Muller and Muneoka, 2000). Third, interdigital cells of the chick autopod are able to form digits in response to wounding or to implantation of transforming growth factor-β (TGFβ)-releasing microcarrier beads (Macias *et al.,* 1999). These studies support the view that patterning of individual digit rudiments continues during later stages of limb development, but the endogenous patterning signals that act during digit formation are presently unknown.

2. Interdigital Cell Death and Digit Separation

Apoptotic cell death in the interdigital region plays a key role in digit separation in animals with individualized digits (Saunders and Fallon, 1967). Interdigital cell death does not occur or is significantly reduced in animals with webbed autopods. Intercellular signals important for the induction of interdigital cell death involves BMP signaling. Misexpression of a dominant negative *BMP receptor* construct in interdigital cells of the chick leg results in the inhibition of cell death and a webbed phenotype (Zou and Niswander, 1996; Yokouchi *et al.,* 1996). During chick limb development, *Bmp2, Bmp4, Bmp7,* and *BMP receptor* genes are expressed in the interdigital region as well as in other limb regions where cell death occurs (i.e. anterior and posterior necrotic zones). This supports their involvement in the endogenous control of cell death (Francis *et al.,* 1994; Francis-West *et al.,* 1995; Luo *et al.,* 1995; Lyons *et al.,* 1995). *Bmp* gene expression in the interdigital region is regulated by the overlying ectoderm (Yokouchi *et al.,* 1996) and also by the neighboring digit ray (Lee *et al.,* 1994; Gañan *et al.,* 1994; Omi *et al.,* 2000), suggesting that signals emanating from tissues surrounding the interdigital region regulate BMP signaling. *Msx* genes are also expressed in the interdigital regions, and targeted disruption of both *Msx1* and *Msx2* results in a webbed phenotype, indicating an involvement of these genes in the apoptotic signaling pathway (Chen and Zhao, 1998). Interestingly, *Msx1*$^{-/-}$/Msx2$^{-/-}$ mutants display no *Bmp4* expression in the interdigital region, and interdigital cell death in these mutants is inducible by BMP4 treatment. These results suggest that apoptotic cell death in interdigital cells is induced by the BMP4 signaling pathway and that BMP4 production is regulated by the *Msx* genes. Furthermore, BMPs and/or *Msx* gene expression have been implicated in the control of cell death in other model systems (Marazzi *et al.,* 1997; Graham *et al.,* 1994; Glozak and Rogers, 1996).

RA is also implicated in the regulation of cell death during limb development. RA promotes cell death in both the interdigital mesenchyme of the mouse limb (Lussier *et al.,* 1993; Lee *et al.,* 1994) and the AER (Sulik and Dehart, 1988). RA administration can also rescue the limb phenotype in the *Hammertoe* mutant mouse, which displays polydactyly associated with the suppression of interdigital cell death (Zakeri *et al.,* 1994; Ahuja *et al.,* 1997). The RA-inducible gene *RAR-β* is expressed by interdigital cells, thus suggesting that RA and one of its receptors is expressed in the interdigital region (Dollé *et al.,* 1989). In addition, combined disruption of *RAR-β* and *RAR-γ* results in a webbed phenotype associated with the downregulation of *Bmp7* gene expression and the inhibition of cell death in the interdigital region (Ghyselinck *et al.,* 1997; Dupe *et al.,* 1999). These data support a model in which interdigital cell death is regulated by RA signaling that activates the BMP signaling pathway.

The inhibition of cell death interdigital regions can result in two distinct phenotypes, the formation of webbing or the formation of an ectopic digit. This observation indicates that additional morphogenetic signals are present in the autopod that can influence patterning late in limb development. The distinction between a webbed phenotype versus an ectopic digit appears to be controlled in part by the ectoderm. A webbed phenotype is induced by disruption of BMP signaling without damage to the ectoderm. BMP signaling has been disrupted by misexpression of a dominant negative *BMP receptor* construct (Zou and Niswander, 1996; Yokouchi *et al.,* 1996), as well as by misexpression of the BMP binding protein NOGGIN and ectopic application of FGF, which downregulates *BMPr-IB* expression (Merino *et al.,* 1998). Alternatively, removal of the ectoderm or wounding the interdigital region is known to induce the formation of ectopic digits (Macias *et al.,* 1999). Ectodermal wounding results in a localized downregulation of *Bmp4* expression and the inhibition of cell death (Yokouchi *et al.,* 1996). Implantation of ectopic TGFβ into the interdigital regions also induces the formation of an ectopic digit (Gañan *et al.,* 1996). It is interesting to note that embryonic wound healing is associated with the transient expression of TGFβs localized to the site of the wound (Martin, 1997), and such a transient rise in TGFβs may be responsible for ectopic digit formation after embryonic wounding.

III. Limb Morphogenesis: Cell Biology

Enormous advances have been made in recent years in the identification and characterization of a number of key morphogenetic signals, such as SHH, FGFs, BMPs, and RA, that regulate limb outgrowth. For many of these signals we know where, when, and how the signals are transmitted. Receptors have been identified and signal transduction pathways are beginning to be teased apart. Despite this, we know very little about how these signals modify some of the most basic of target cell behaviors, such as cell proliferation and cell migration, or how they alter the way neighboring cells interact with one another during limb outgrowth. There is no question that these basic cellular characteristics are integral to limb development, and characterizing the interface between morphogenetic signals and morphogenesis itself is one of the challenges for future generations of developmental biologists. In this section we focus on some of these basic behaviors in an effort to bring together the molecular genetics of limb development with the cell biology of limb development.

A. Cell Proliferation

The control of cell proliferation during limb development is a critical component of limb outgrowth, yet our understanding of its regulatory mechanics goes little beyond a description of growth patterns. It is often assumed that patterning mechanisms and the control of growth are largely independent; however, there is little evidence to accept or refute this assumption. Pattern cannot be elaborated without cell proliferation, so the two are related. How they are related is at present unclear.

The formation of the limb bud is associated with regions of the flank with a high proliferation rate by comparison to neighboring tissues. During the pre-bud stages, mesenchymal cell proliferation along the entire flank is uniform. With the formation of the limb bud, cell proliferation is maintained in limb forming regions but declines in adjacent non-limb regions (Searls and Janner, 1971). This proliferative decline in flank cells is the result of cells ceasing to cycle, rather than a slowing of the cell cycle. In terms of the ectoderm, there is surprisingly little known about growth throughout development, though the available evidence indicates that ectodermal cell proliferation is relatively constant (Hornbruch and Wolpert, 1970). Recent mutational studies demonstrate that the limb ectoderm requires the expression of the *p63* gene, a homolog of the *p53* tumor-suppressor gene, for proper maintenance of the AER (Yang *et al.,* 1999). Hence, the control of cell proliferation in the limb ectoderm is critical for normal morphogenesis.

Once formed, the AER exerts its influence on limb outgrowth by regulating bud elongation, and it is generally thought that the AER is mitogenic for progress zone cells (Globus and Vethamany-Globus, 1976; Reiter and Solursh, 1982). In spite of this, the endogenous patterns of cell proliferation in the limb bud indicate that the AER is not playing a unique role in regulating cell proliferation. Mesenchymal cell proliferation in the early limb bud is initially uniform, and only after significant elongation are gradients of proliferation evident (Summerbell and Wolpert, 1972). In the chick limb bud, a distal to proximal gradient of cell proliferation is discernible by stage 24, but this gradient is associated with a proximal decline in mitotic rate associated with overt differentiation of skeletal tissues. At these same stages, the dorsal-ventral axis also displays gradients of mesenchymal cell proliferation that is highest at the dorsal and ventral surfaces and lowest at the center of the limb bud where chondrogenesis is commencing. What is important is that growth differences are not apparent when comparing mesenchymal cells at the distal tip to cells at the ventral and dorsal surfaces at proximal levels. Hence, the growth profile of the limb bud is one in which there is a shell of cell proliferation surrounding a differentiating core. Given this, endogenous rates of cell proliferation suggest that the entire bud ectoderm, including the AER, is equally mitogenic for limb bud mesenchyme. However, cell proliferation measurements following AER removal indicate no change on the rate of ^{3}H-thymidine incorporation and only a transient depression in mitotic index, which returns to normal 24 hours later (Janner and Searls, 1971; Summerbell, 1977). Since AER removal inhibits limb outgrowth, the evidence indicates that the regulation of cell proliferation and the control of limb elongation by the AER can be dissociated from one another.

Members of the fibroblast growth factor (FGF) family of signaling molecules are produced by the AER, and it is generally viewed that FGFs produced by the AER are mitogenic for progress zone cells. *In vitro* studies indicate that FGFs stimulate limb bud cell proliferation (Aono and Ide, 1988; Niswander and Martin, 1993); however, *in vivo* studies provide compelling evidence that FGFs produced by the AER are not acting as mitogens for limb bud cells. First, the role of the AER and associated FGFs in controlling cell proliferation during bud formation is questioned based on studies of the *limbless* mutation that show bud formation in the absence of an AER or associated *Fgf4* and *Fgf8* expression (Carrington and Fallon, 1988; Ros *et al.,* 1996). Second, overexpression of *Fgf2* in a transgenic model results in a shortened limb phenotype due to an inhibitory influence on growth rates (Lightfoot *et al.,* 1997). Third, ectopic application of FGF2 or FGF4 into subapical regions of the limb bud inhibits bud outgrowth and results in a shortened limb phenotype (Li *et al.,* 1996; Li and Muneoka, 1999). A major conclusion that can be drawn is that the role of ectodermal FGFs in controlling limb outgrowth in not attributable to a mitogenic effect. We have proposed that any putative mitogenic activity connected with FGFs during limb development is an indirect effect resulting from the involvement of FGFs in the control of cell migration (Li and Muneoka, 1999; see below).

A second region of mitogenic activity in the limb bud is typically associated with the ZPA. While the ZPA is not associated with a posteriorly localized domain of enhanced proliferation, grafting of the ZPA into the limb anterior results in a rapid enhancement of cell proliferation in anterior cells. This enhancement is initially localized to the posterior graft/host interface and subsequently spreads to regions away from the graft (Cooke and Summerbell, 1980). It is interesting to note that all limb bud cells are proliferating at stages used for ZPA grafting, indicating that the enhancement of cell proliferation by the ZPA results from a shortening of the cell cycle (Cooke and Summerbell, 1980). The timing of the observed growth enhancement is very rapid (5 to 9 hours) by comparison to the induction of downstream ZPA/SHH genes, thus suggesting a direct effect on cell proliferation. This mitogenic activity is complicated by the fact that the effect is position-specific. All limb bud cells are proliferating, and there is no proliferative zone in the posterior where cell cycle time is shortened. This leaves us with a paradox in which cells neighboring the endogenous ZPA/SHH signal are unresponsive to its mitogenic effects, whereas cells far removed are responsive but do not respond during normal development. One potential solution takes into account position-dependent growth properties thought to play a regulatory role during limb regeneration in amphibians (see Bryant *et al.,* 1987; Muneoka *et al.,* 1989a). In this model, cell proliferation is regulated by position-specific interactions between neighboring cells rather than in response to a localized mitogenic source (see Bryant and Muneoka, 1986). At present the link between ZPA/SHH signaling and the regulation of endogenous growth rates in the limb bud is perplexing and is a topic in need of further study.

The regulation of cell proliferation during later stages of limb development has been linked to the expression of homeobox-containing genes. Mutational and misexpression studies suggest that *Hoxd* genes influence the formation of skeletal pattern by modulating their growth. Misexpression of *Hoxd11* and *Hoxd13* in combination with cell proliferation studies indicates that these genes regulate cell proliferation during skeletal morphogenesis (Goff and Tabin, 1997). A model linking *Hox* gene action to growth control has been proposed in which all *Hox* genes are growth promoters that differentially regulate the expression of the same target genes (Goff and Tabin, 1997). In this way growth is controlled by the combined action of all *Hox* genes expressed in any single cell. Interestingly, if this model is correct, then the demonstration that *Hoxd12* and *Shh* act in a positive feedback loop during limb outgrowth (Knezevic *et al.,* 1997) implicates SHH production by the ZPA in the control of cell proliferation. In support of this notion, SHH has been shown to control cell proliferation in other developing systems (see Fan *et al.,* 1997; Oro *et al.,* 1997).

Studies demonstrate that modification of the cell cycle length by transient inhibition of DNA polymerase A in limb mesenchymal cells has a profound effect on patterns of gene expression and limb morphogenesis (Ohsugi *et al.,* 1997). Surprisingly, slowing the cell cycle of anterior chick limb bud cells results in the induction of a supernumerary digit 2; whereas slowing the cell cycle in posterior cells has no effect on limb patterning. In addition, the expression of a number of posterior-specific genes is modified by lengthening the cell cycle, raising the possibility that the expression of patterning genes and, as such, limb morphology may involve regulation of cell cycle length (Ohsugi *et al.,* 1997). In this regard, it is interesting to note that the polydactylous mouse mutant *Strong's Luxoid* encodes for the paired-like homeodomain gene *Alx4* (Qu *et al.,* 1997, 1998; Takahashi *et al.,* 1998), a member of a class of transcriptional regulators that forms complexes with members of the retinoblastoma family of cell cycle regulators (Hudson *et al.,* 1998). Hence, *Alx4,* which is normally expressed in the anterior limb bud and is downregulated by SHH (Takahashi *et al.,* 1998), may influence limb patterning by direct modulation of the cell cycle.

B. Cell Migration

A growing number of studies indicate that the function of FGF signaling in the embryo is to modulate episodes of cell movements. In the limb bud, we have demonstrated that ectopic FGF2 and FGF4 modify cell movements and subsequent morphogenesis during limb development (Li *et al.,* 1996; Li and Muneoka, 1999). A migration role of FGF signaling is also indicated by genetic studies in *Drosophila* and *Caenorhabditis* where the FGF signaling pathway is directly linked to the control of specific episodes of cell migration (Sutherland *et al.,* 1996; Burdine *et al.,* 1997). In light of this, it seems likely that the *Fgf* gene family evolved from an ancestral gene that functioned in the control of cell migration. In vertebrate embryos, FGF signals are known to play a role in the regulation of cell migration during gastrulation, myogenesis, and lung development (Itoh *et al.,* 1996; Webb *et al.,* 1997; Min *et al.,* 1998; Park *et al.,* 1998; Sekine *et al.,* 1999; Sun *et al.,* 1999), suggesting that the family of vertebrate FGFs may have the general function of modulating multiple aspects of cell migration during development.

Migration by mesenchmal cells in normal limb development has been predicted based on computer modeling studies (Ede and Law, 1969); however, it has only recently been demonstrated. In cell marking studies we have shown that limb bud cells migrate to an ectopic FGF4 source in a rapid, dose-dependent and directionally unrestricted manner (Figure 2A,B,C; Li and Muneoka, 1999). We have also identified a subpopulation of limb cells that fail to respond to FGF4 but migrate to an AER-derived signal, thus suggesting additional chemotactic signals during limb bud outgrowth. Limb bud cells also modify their movements following FGF2 application; however, FGF2 is not chemoattractive (Li *et al.,* 1996). FGF2 and the limb bud ectoderm have been implicated in the control of cell movements through hyaluronic acid production by limb bud cells *in vitro* (Munaim *et al.,* 1991; Knudson *et al.,* 1995). Studies of supernumerary limb induction from flank tissue demonstrate that there is no chemoattractive activity associated with ectopic FGF8 (Cohn *et al.,* 1997); however, FGF8 is known to be required for cell migration during gastrulation (Sun *et al.,* 1999). The influence of FGF9 and FGF10 on cell migration during limb development has not been studied, although FGF10 is known to be chemotactic for developing lung epithelial cells (Min *et al.,* 1998; Park *et al.,* 1998; Sekine *et al.,* 1999).

Whether or not cell migration is critical for morphogenesis has been investigated in both experimental and theoretical analyses. Early computer modeling studies indicated that regionally localized cell proliferation in conjunction with the distal migration of cells is required to model shape changes associated with limb bud outgrowth (Ede and Law, 1969). In support of this, FGF2 and FGF4 bead implantation alter cell movements and cause dramatic changes in limb bud shape (Figure 2D, E; Li *et al.,* 1996; Li and Muneoka, 1999). Such FGF-treated limb buds develop skeletal defects that are associated with these shape changes, thus providing a link between early morphogenetic events and final patterning (Figure 2F). In terms of endogenous function, the chemotactic activity and the expression domain of *Fgf4* in the AER correlate with shape changes that occur during bud formation and elongation. *Fgf4* expression in the early limb bud is asymmetrically localized to the posterior AER. During early limb stages in the chick, the bud takes on an asymmetric shape with outgrowth visibly biased toward the posterior. This asymmetry correlates with the localized expression of *Fgf4* and not with changes in cell proliferation rates (Cooke and Summerbell, 1980; K. Ohsugi, personal communication). During later limb bud stages when *Fgf4* expression is apical and correlates with the direction of outgrowth, we have shown that subapical application of FGF4 can functionally compete with the AER to inhibit elongation (Li and Muneoka, 1999). These data suggest that that cell migration plays a key role in controlling shape changes in the developing limb bud.

C. Cell-Cell Interactions

Ultrastructural studies of the developing limb describe two major characteristics important for cell-cell interactions and limb patterning: (1) the existence of gap junctions between AER cells and between specific progress zone cells and (2) regions of direct cell-cell contact between the loosely arranged cells of the progress zone (Kelley and Fallon, 1978). The existence of gap junctional communication between cells of the limb bud has been demonstrated by electron microscopy (Kelley and Fallon, 1978), immunohistology (Green *et al.,* 1994), and dye coupling (Allen *et al.,* 1990). In the

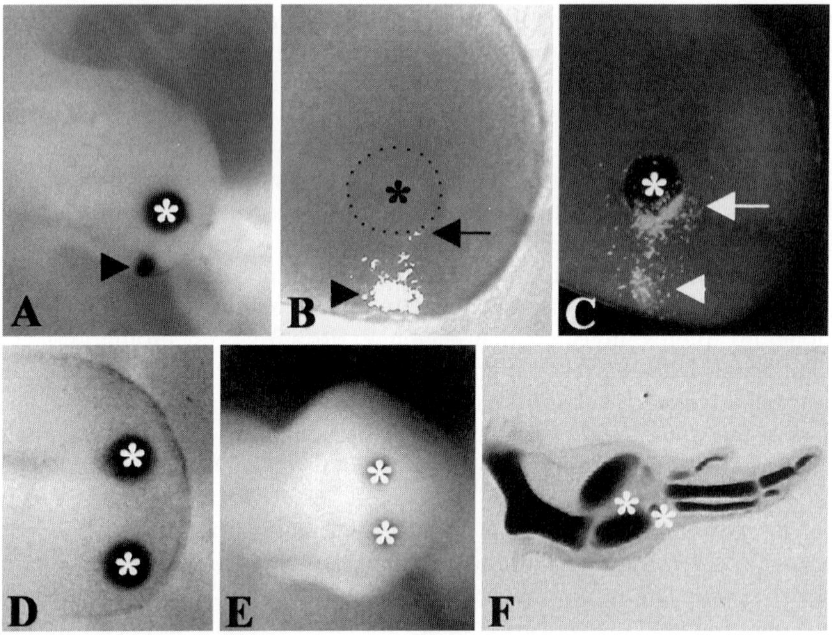

FIG. 2 FGF-4 is a chemoattractant for limb bud cells that can modify bud shape and skeletal morphogenesis. (A) The *in vivo* assay for migration consisted of DiI labeling of posterior-distal cells of a stage 24 limb bud (arrowhead) and implantation of a carrier bead (*) containing FGF-4 into the central region of the bud. (B) Six hours after implantation of a FGF-4 treated bead, DiI-labeled cells are observed migrating away from the injection site (arrowhead) toward the FGF-4 bead (*) and a few cells can be seen making contact with the bead (arrow). (C) Twelve hours following implantation of a FGF-4 treated bead (1000 μg/ml) two clusters of DiI labeled cells are apparent: one associated with the FGF-4 bead (*) located centrally in the limb bud (arrow) and a second at the posterior injection site (arrowhead). (D) An *in vivo* image showing two FGF-4 beads (*) implanted into the subapical region of a stage 24 limb bud. (E) Twenty-four hours after implantation, the limb bud shown in D was inhibited from elongating in the proximal-distal axis. Note that the inhibition of elongation was most severe in regions associated with the FGF4 beads (*) giving the distal limb bud an "arrowhead" appearance. (F) Skeletal preparation of the limb that resulted from the subapical implantation of two FGF-4 beads (*) shown in D and C. The overall skeletal pattern is complete, and the digit pattern is normal; however, proximal-distal elongation of many skeletal elements was inhibited. From Li and Muneoka (1999).

limb mesenchyme a gradient of gap junctional communication has been demonstrated with the highest level of coupling in the posterior and the lowest in the anterior (Coelho and Kosher, 1991). FGF4, which is known to maintain ZPA signaling *in vitro* (Anderson *et al.*, 1993; Vogel and Tickle, 1993), enhances the formation of gap junctions between posterior cells and is thus linked to ZPA signal maintenance (Makarenkova *et al.*, 1997).

Junctional communication between posterior and anterior cells appears to be required for ZPA signaling, since blocking antibodies to junctional proteins inhibits the formation of supernumerary digits resulting from ZPA grafting (Allen *et al.,* 1990). Recently, antisense oligonucleotide administration to inhibit *Connexin-43* expression in pre-limb bud stages was shown to result in severe limb malformations due to abnormalities in AER gene expression, thus suggesting a feedback loop between FGF and Connexin-43 mediated signaling during early limb development (Makarenkova and Patel, 1999).

Position-dependent differences in cell-cell adhesivity has been described in cell sorting experiments in which cells from different proximal-distal regions of the limb bud sort out in monolayer culture (Ide *et al.,* 1994) and in recombinate limb buds (Wada *et al.,* 1993). Retinoic acid treatment of distal limb bud cells induces these cells to acquire proximal adhesive characteristics (Tamura *et al.,* 1997) in a manner analogous to retinoic acid proximalization of blastema cells during limb regeneration (Bryant and Gardiner, 1992). Similarly, misexpression of the distalmost *Hoxa13* gene in proximal limb bud cells induces a proximal to distal transformation of skeletal elements associated with a similar change in cell sorting properties (Yokouchi *et al.,* 1995). These studies provide evidence that cell surface adhesion properties are associated with pattern specification. Significantly, a number of cell adhesion molecules are differentially expressed during limb development. Of interest are two cadherins family members, *N-Cadherin* and *Cadherin-11,* expressed in the distal region of the early limb bud. Both N-Cadherin and Cadherin-11 have been implicated in modulating the sorting of limb bud cells. Mesenchymal cells expressing *Cadherin-11* sort out from nonexpressing cells (Kimura *et al.,* 1995), while N-Cadherin antibodies disrupt position-dependent sorting of limb bud cells (Yajima *et al.,* 1999). In addition, the progress zone gene *Msx1* is implicated in the regulation of Cadherin-associated cell adhesion *in vitro* (Lincecum *et al.,* 1998). While it is unclear how cell-cell adhesivity is involved in pattern specification, there is a growing body of evidence indicating that position-specific differences in the adhesive properties of limb bud cells play a critical role in limb morphogenesis.

D. The Extracellular Environment

The mechanisms by which cells communicate with one another are an important aspect of the cell biology of patterning. Long-range signaling via a diffusible morphogen, paracrine signaling, and direct cell-cell contact have all been proposed to play a role in limb patterning. The extracellular environment is the medium through which intercellular signals must pass and, for direct cell-cell contact, cells must traverse. The composition of extracellular matrix (ECM) of the forming limb is complex, and a number of ECM components are regulated both spatially and temporally.

The developing limb ECM can elicit patterning effects through several means including regulation of growth factor activities, differentiation, and gene expression while also influencing cell and tissue shape (Gumbiner, 1996; Hata, 1996). The interaction of FGFs with ECM has been extensively studied. In particular, heparin/heparan sulfate has been shown to be necessary for FGF activity by mediating FGF presentation to and dimerization of FGF receptors (Spivak-Kroizman *et al.,* 1994). Immobilization of FGFs to sites in the ECM may also allow for prolongation of FGF bioavailability (Ruoslahti and Yamaguchi, 1991), providing a mechanism for establishing stable concentration gradients of FGF within the limb bud environment. This may explain the fact that FGF4 signaling by the AER appears to act at a distance of at least 75-100 μm (Li and Muneoka, 1999). Moreover, some species of heparan sulfate also appear to be inhibitory to FGF function (Guimond *et al.,* 1993), reflecting a complex degree of regulation that likely varies among tissues and developmental stages (Salmivirita and Jalkanen, 1995; San Antonio *et al.,* 1994). Given this, it is clear that ECM is a potent regulator of FGF activity.

Glycosaminoglycan (GAG) components of the ECM are also important to limb development. Heparan sulfate proteoglycans (HSPGs) are common in the early limb, with syndecans expressed in mesoderm and ectoderm (Solursh *et al.,* 1990; David *et al.,* 1993; Gould *et al.,* 1995). In addition, HSPGs such as glypican (Niu *et al.,* 1996) and CD44 splice variant (Yu *et al.,* 1996; Sherman *et al.,* 1998) are expressed in the AER. Given the numerous FGFs present in the limb during development, it is likely that HSPGs are important to the regulation of FGF activities. It is also known that GAGs are necessary to limb development in terms of differentiation. Individual syndecans in the limb bud have distinct and often nonoverlapping patterns of expression that coincide with differentiation, particularly chondrogenesis (Solursh *et al.,* 1990; David *et al.,* 1993; Gould *et al.,* 1995). In contrast, the GAG hyaluronic acid is thought to maintain the undifferentiated state in the early limb by separating cells within a hydrated matrix. This separation of cells is lost at later developmental stages when cell surface binding sites for hyaluronic acid become apparent. It is thought that these sites endocytose hyaluronic acid-rich matrices, bringing cells closer together and aiding in the condensation process characteristic of the first stages of chondrogenesis (see Maleski and Knudson, 1996).

IV. Conclusions

Classical embryological and recent molecular studies have combined to increase our understanding of the complex process of limb development.

It is known that the limb is specified much earlier than the initiation of overt limb outgrowth (Stephens *et al.,* 1993), owing in part to the activities of FGF10 (Ohuchi *et al.,* 1997) and the *Tbx* family of transcriptional regulators (Simon, 1999). With outgrowth, patterning regions in the nascent bud, including the AER, progress zone, and ZPA, have long been recognized as indispensable, and it can now be stated with some assurance that factors such as BMPs, FGFs, HOXs, MSXs, and SHH are particularly important representatives of these regions (see main text). However, while recent years have inarguably broadened our information base by identifying such morphogenetic signals, it is less clear how most of these signals influence development at the cellular level. Differences in cell adhesivity, cell proliferation, cell migration, gap junctional communication, and modification of activities by the extracellular environment are all integral to establishing correct pattern and are obvious targets of regulation by morphogenetic signals. Yet beyond initial studies, little in this regard is known. To properly model limb development with greater veracity, our understanding of patterning mechanisms must embrace more than descriptions of gene expression and regulation at the molecular level. Investigations of cellular mechanisms of outgrowth as governed by morphogenetic signals are required to yield a more integrated view of patterning, one whose contextural framework will embrace the full complexity of limb development.

References

Abud, H. E., Skinner, J. A., McDonald, F. J., Bedford, M. T., Lonai, P., and Heath, J. K. (1996). Ectopic expression of Fgf4 in chimeric mouse embryos induces the expression of early markers of limb development in the lateral ridge. *Dev. Genet.* **19,** 51–65.

Ahuja, H. S., James, W., and Zakeri, Z. (1997). Rescue of the limb deformity in hammertoe mutant mice by retinoic acid-induced cell death. *Dev. Dyn.* **208,** 466–481.

Allen, F., Tickle, C., and Warner, A. (1990). The role of gap junctions in patterning of the chick limb bud. *Development* **108,** 623–634.

Anderson, R., Landry, M., and Muneoka, K. (1993). Maintenance of ZPA signaling in cultured mouse limb bud cells. *Development* **117,** 1421–1433.

Anderson, R., Landry, M., Reginelli, A., Taylor, G., Achkar, C., Gudas, L., and Muneoka, K. (1994). Conversion of anterior limb bud cells to ZPA signaling cells *in vitro* and *in vivo*. *Dev. Biol.* **164,** 241–257.

Aono, H., and Ide, H. (1988). A gradient of responsiveness to the growth-promoting activity of ZPA (zone of polarizing activity) in the chick limb bud. *Dev. Biol.* **128,** 136–141.

Bryant, P. J. (1993). The polar coordinate model goes molecular. *Science* **259,** 471–472.

Bryant, S. V., and Gardiner, D. M. (1992). Retinoic acid, local cell-cell interactions, and pattern formation in vertebrate limbs. *Dev. Biol.* **152,** 1–25.

Bryant S. V., Gardiner, D. M., and Muneoka, K. (1987). Limb development and regeneration. *Am. Zool.* **27,** 675–696.

Bryant, S. V., and Muneoka, K. (1986). Views on limb development and regeneration. *Trends Genet.* **2,** 153–159.

Burdine, R. D., Chen, E. B., Kwok, S. F., and Stern, M. J. (1997). egl-17 encodes an invertebrate fibroblast growth factor family member required specifically for sex myoblast migration in Caenorhabditis elegans. *Proc. Natl. Acad. Sci. USA* **94,** 2433–2437.

Büscher, D., and Izpisúa-Belmonté, J. C. (1999). Muscle development during vertebrate limb outgrowth. *Cell Tissue Res.* **296,** 131–139.

Büscher, D., Bosse, B., Heymer, J., and Rüther, U. (1997). Evidence for genetic control of *Sonic hedgehog* by *Gli3* in mouse limb development. *Mech. Dev.* **62,** 175–182.

Carrington, J. L., and Fallon, J. F. (1988). Initial limb budding is independent of apical ectodermal ridge activity; evidence from a limbless mutant. *Development* **104,** 361–367.

Carrucio, N. C., Martinez-Lopez, A., Harris, M., Dvorak, L., Bitgood, J., Simandl, B. K., and Fallon, J. F. (1999). Constitutive activation of sonic hedgehog signaling in the chicken mutant talpid2: Shh-independent outgrowth and polarizing activity. *Dev. Biol.* **212,** 137–149.

Celli, G., LaRochelle, W. J., Mackem, S., Sharp, R., and Merlino, G. (1998). Soluble dominant-negative receptor uncovers essential roles for fibroblast growth factors in multi-organ induction and patterning. *EMBO J.* **17,** 1642–1655.

Chambon, P. (1996). A decade of molecular biology of retinoic acid receptors. *FASEB J.* **10,** 940–954.

Charité, J., De Graaff, W., Shen, S., and Deschamps, J. (1994). Ectopic expression of *Hoxb-8* causes duplication of the ZPA in the forelimb and homeotic transformation of axial structures. *Cell* **78,** 589–601.

Chaube, S. (1959). On axiation and symmetry in transplanted wing of the chick. *J. Exp. Zool.* **140,** 29–77.

Chen, H., and Johnson, R. L. (1999). Dorsoventral patterning of the vertebrate limb: A process governed by multiple events. *Cell Tissue Res.* **296,** 67–73.

Chen, Y., and Zhao, X. (1998). Shaping limbs by apoptosis. *J. Exp. Zool.* **282,** 691–702.

Chiang, C., Litingtung, Y., Lee, E., Young, K. E., Corden, J. L., Westphal, H., and Beachy, P. (1996). Cyclopia and defective axial patterning in mice lacking *Sonic hedgehog* gene function. *Nature* **383,** 407–413.

Chuang, P. T., and McMahon, A. P. (1999). Vertebrate Hedgehog signalling modulated by induction of a Hedgehog-binding protein. *Nature* **397,** 617–621.

Coelho, C. N., and Kosher, R. A. (1991). A gradient of gap junctional communication along the anterior-posterior axis of the developing chick limb bud. *Dev. Biol.* **148,** 529–535.

Cohn, M. J., Izpisúa-Belmonté, J.-C., Abud, H., Heath, J. K., and Tickle, C. (1995). Fibroblast growth factors induce additional limb development from the flank of chick embryos. *Cell* **80,** 739–746.

Cohn, M. J., Patel, K., Krumlauf, R., Wilkinson, D. G., Clarke, J. D., and Tickle, C. (1997). Hox9 genes and vertebrate limb specification. *Nature* **387,** 97–101.

Colvin, J. S., Feldmn, B., Nadeau, J. H., Goldfarb, M., and Ornitz, D. M. (1999). Genomic organization and embryonic expression of the mouse fibroblast growth factor 9 gene. *Dev. Dyn.* **216,** 72–88.

Cooke, J., and Summerbell, D. (1980). Cell cycle and experimental pattern duplication in the chick wing during embryonic development. *Nature* **287,** 697–701.

Crossley, P. H., Minowada, G., Macarthur, C. A., Martin, G. R. (1996). Roles for FGF8 in the induction, initiation, and maintenance of chick limb development. *Cell* **84,** 127–136.

Crossley, P. H., and Martin, G. R. (1995). The mouse *Fgf8* gene encodes a family of polypeptides and is expressed in regions that direct outgrowth and patterning in the developing embryo. *Development* **121,** 439–451.

David, G., Bai, X. M., Van der Schueren, B., Marynen, P., Cassiman, J. J., and Van den Berghe, H. (1993). Spatial and temporal changes in the expression of fibroglycan (syndecan-2) during mouse embryonic development. *Development* **119,** 841–854.

Dollé, P., Ruberte, E., Kastner, P., Petkovich, M., Stoner, C. M., Gudas, L. J., and Chambon, P. (1989). Differential expression of genes encoding α, β and γ retinoic acid receptors and CRABP in the developing limbs of the mouse. *Nature* **342,** 702–705.

Dupé, V., Ghyselinck, N. B., Thomazy, V., Nagy, L., Davies, P. J. A., Chambon, P., and Mark, M. (1999). Essential roles of retinoic acid signaling in interdigital apoptosis and control of BMP-7 expression in mouse autopods. *Dev. Biol.* **208,** 30–43.

Duprez, D. M., Kostakopoulou, K., Francis-West, P. H., Tickle, C., and Brickell, P. M. (1996). Activation of Fgf-4 and HoxD gene expression by BMP-2 expressing cells in the developing chick limb. *Development* **122,** 1821–1828.

Ede, D. A., and Law, J. T. (1969). Computer simulation of vertebrate limb morphogenesis. *Nature* **221,** 244–248.

Ettinger, L., and Doljanski, F. (1992). On the generation of form by the continuous interactions between cells and their extracellular matrix. *Biol. Rev. Cambridge Philos. Soc.* **67,** 459–489.

Fallon, J. F. and Crosby, G. M. (1977). Polarizing zone activity in limb buds of amnoites. *In* "Vertebrate Limb and Somite Morphogenesis" (D. A. Ede, J. R. Hinchliffe, and M. Balls, Eds.), pp. 55–69. Cambridge University Press, Cambridge.

Fallon, J. F., Lopez, A., Ros, M. A., Savage, M. P., Olwin, B. B., and Simandl, B. K. (1994). FGF-2: Apical ectodermal ridge growth signal for chick limb development. *Science* **264,** 104–107.

Fan, H., Oro, A. E., Scott, M. P., and Khavari, P. A. (1997). Induction of basal cell carcinoma features in transgenic human skin expressing Sonic Hedgehog. *Nat. Med.* **3,** 788–792.

Francis, P. H., Richardson, M. K., Brickell, P. M., and Tickle, C. (1994). Bone morphogenetic proteins and a signalling pathway that controls patterning in the developing chick limb. *Development* **120,** 209–218.

Francis-West, P. H., Robertson, K. E., Ede, D. A., Rodriguez, C., Izpisúa-Belmonté, J. C., Houston, B., Burt, D. W., Gribbin, C., Brickell, P. M., and Tickle, C. (1995). Expression of genes encoding bone morphogenetic proteins and sonic hedgehog in talpid (ta^3) limb buds: Their relationships in the signalling cascade involved in limb patterning. *Dev. Dyn.* **203,** 187–197.

French, V., Bryant, P. J., and Bryant, S. V. (1976). Pattern regulation in epimorphic fields. *Science* **193,** 969–981.

Gañan, Y., Macias, D., and Hurle, J. M. (1994). Pattern regulation in the chick autopodium at advanced stage of embryonic development. *Dev. Dyn.* **199,** 64–72.

Gañan, Y., Macias, D., Duterque-Coquillaud, M., Ros, M. A., and Hurle J. M. (1996). Role of TGFβs and BMPs as signals controlling the position of the digits and the areas of interdigital cell death in the developing chick limb autopod. *Development* **122,** 2349–2357.

Ghyselinck, N. B., Dupé, V., Dierich, A., Messaddeq, N., Garnier, J. M., Rochette-Egly, C., Chambon, P., and Mark, M. (1997). Role of the retinoic acid receptor beta (RARβ) during mouse development. *Int. J. Dev. Biol.* **41,** 425–447.

Gibson-Brown, J. J., Agulnik, S., Silver, L. M., and Papaioannou, V. E. (1998). Expression of T-box genes Tbx2-Tbx5 during chick organogenesis. *Mech. Dev.* **74,** 165–169.

Globus, M., and Vethamany-Globus, S. (1976). An *in vitro* analogue of early chick limb bud outgrowth. *Differentiation* **6,** 91–96.

Glozak, M. A., and Rogers, M. B. (1996). Specific induction of apoptosis in P19 embryonal carcinoma cells by retinoic acid and BMP2 or BMP4. *Dev. Biol.* **179,** 458–470.

Goetinck, P. F. (1964). Studies on limb morphogenesis. II. Experiments with the polydactylous mutant eudiplopodia. *Dev. Biol.* **10,** 71–91.

Goff, D. J., and Tabin, C. J. (1997). Analysis of Hoxd-13 and Hoxd-11 misexpression in chick limb buds reveals that Hox genes affect both bone condensation and growth. *Development* **124,** 627–636.

Gould, S. E., Upholt, W. B., and Kosher, R. A. (1995). Characterization of chicken syndecan-3 as a heparan sulfate proteoglycan and its expression during embryogenesis. *Dev. Biol.* **168,** 438–451.

Graham, A., Francis-West, P., Brickell, P., and Lumsden, A. (1994). The signalling molecule BMP4 mediates apoptosis in the rhombencephalic neural crest. *Nature* **372,** 684–686.

Green, C. R., Bowles, L., Crawley, A., and Tickle, C. (1994). Expression of the connexin43 gap junctional protein in tissues at the tip of the chick limb bud is related to the epithelial-mesenchymal interactions that mediate morphogenesis. *Dev. Biol.* **161,** 12–21.

Guimond, S., Maccarana, M., Olwin, B. B., Lindahl, U., and Rapraeger, A. C. (1993). Activating and inhibitory heparin sequences for FGF-2 (basic FGF). Distinct requirements for FGF-1, FGF-2, and FGF-4. *J. Biol. Chem.* **268,** 23906–23914.

Gumbiner, B. M. (1996). Cell adhesion: The molecular basis of tissue architecture and morphogenesis. *Cell* **84,** 345–357.

Hamburger, V., and Hamilton, H. L. (1951). A series of normal stages in the development of the chick embryo. *J. Morph.* **88,** 49–92.

Hata, R. I. (1996). Where am I? How a cell recognizes its positional information during morphogenesis. *Cell Biol. Int.* **20,** 59–65.

Heikinheimo, M., Lawshe, A., Shackleford, G. M., Wilson, D. B., and MacArthur, C. A. (1994). Fgf-8 expression in the post-gastrulation mouse suggests roles in the development of the face, limbs and central nervous system. *Mech. Dev.* **48,** 129–138.

Helms, J. A., Kim, C. H., Eichele, G., and Thaller, C. (1996). Retinoic acid signaling is required during early chick limb development. *Development* **122,** 1385–1394.

Hornbruch, A., and Wolpert, L. (1970). Cell division in the early growth and morphogenesis of the chick limb. *Nature* **226,** 764–766.

Hudson, R., Taniguchi-Sidle, A., Boras, K., Wiggan, O., and Hamel, P. A. (1998). Alx-4, a transcriptional activator whose expression is restricted to sites of epithelial-mesenchymal interactions. *Dev. Dyn.* **213,** 159–169.

Ide, H., Wada, N., and Uchiyama, K. (1994). Sorting out of cells from different parts and stages of the chick limb bud. *Dev. Biol.* **162,** 71–76.

Isaac, A., Rodriguez-Esteban, C., Ryan, A., Altabef, M., Tsukui, T., Patel, K., Tickle, C., Izpisúa-Belmonté, J. C. (1998). Tbx genes and limb identity in chick embryo development. *Development* **125,** 1867–1875.

Itoh, N., Mima, T., and Mikawa, T. (1996). Loss of fibroblast growth factor receptors is necessary for terminal differentiation of embryonic limb muscle. *Development* **122,** 291–300.

Izpisúa-Belmonté, J.-C., Tickle, C., Dolle, P., Wolpert, L., and Duboule, D. (1991). Expression of the homeobox *Hox-4* genes and the specification of position in chick wing development. *Nature* **350,** 585–589.

Janner, M. Y., and Searls, R. L. (1971). Effect of removal of the apical ectodermal ridge on the rate of cell division in the subridge mesenchyme of the embryonic chick wing. *Dev. Biol.* **24,** 465–476.

Kelley, R. O., and Fallon, J. F. (1976). Ultrastructural analysis of the apical ectodermal ridge during vertebrate limb morphogenesis. 1. The human forelimb with special reference to gap junctions. *Dev. Biol.* **51,** 241–256.

Kelley, R. O., and Fallon, J. F. (1978). Identification and distribution of gap junctions in the mesoderm of the developing chick limb bud. *J. Embryol. Exp. Morphol.* **46,** 99–110.

Kimura, Y., Matsunami, H., Inoue, T., Shimamura, K., Uchida, N., Ueno, T., Miyazaki, T., and Takeichi, M. (1995). Cadherin-11 expressed in association with mesenchymal morphogenesis in the head, somite, and limb bud of early mouse embryos. *Dev. Bio.* **169,** 347–358.

Knezevic, V., De Santo, R., Schughart, K., Huffstadt, U., Chiang, C., Mahon, K. A., and Mackem, S. (1997). Hoxd-12 differentially affects preaxial and postaxial chondrogenic branches in the limb and regulates Sonic hedgehog in a positive feedback loop. *Development* **124,** 4523–4536.

Knudson, C. B., Munaim, S. I., and Toole, B. P. (1995). Ectodermal stimulation of the production of hyaluronan-dependent pericellular matrix by embryonic limb mesodermal cells. *Dev. Dyn.* **204,** 186–191.

Kostakopoulou, K., Vogel, A., Birckell, P., and Tickle, C. (1996). "Regeneration" of wing bud stumps of chick embryos and reactivation of Msx-1 and *Shh* expression in response to FGF-4 and ridge signals. *Mech. Dev.* **55,** 119–131.

Laufer, E., Dahn, R., Orozco, O. E., Yeo, C. Y., Pisenti, J., Henrique, D., Abbott, U. K., Fallon, J. F., and Tabin, C. (1997). Expression of Radical fringe in limb-bud ectoderm regulates apical ectodermal ridge formation. *Nature* **386,** 366–373.

Laufer, E., Nelson, C. E., Johnson, R. L., Morgan, B. A., and Tabin, C. (1994). Sonic hedgehog and Fgf-4 act through a signaling cascade and feedback loop to integrate growth and patterning of the developing limb bud. *Cell* **79,** 993–1003.

Lee, J., Platt, K. A., Censullo, P., and Ruiz i Altaba, A. (1997). Gli1 is a target of Sonic hedgehog that induces ventral neural tube development. *Development* **124,** 2537–2552.

Lee, K. K. H., Li, F. C. H., Yung W. T., Kung, J. L. S., Ng, J. L., and Cheah, K. S. E. (1994). Influence of digits, ectoderm, and retinoic acid on chondrogenesis by mouse interdigital mesoderm in culture. *Dev. Dyn.* **201,** 297–309.

Li, S., and Muneoka K. (1999). Cell migration and chick limb development: Chemotactic action of FGF4 and the AER. *Dev. Biol.* **211,** 335–347.

Li, S., Anderson, R., Reginelli, A. D., and Muneoka, K. (1996). FGF-2 influences cell movements and gene expression during limb development. *J. Exp. Zool.* **274,** 234–247.

Lightfoot, P. S., Swisher, R., Coffin, J. D., Doetschman, T. C., and German, R. Z. (1997). Ontogenetic limb bone scaling in basic fibroblast growth factor (FGF-2) transgenic mice. *Growth Dev. Aging* **61,** 127–139.

Lincecum, J. M., Fannon, A., Song, K., Wang, Y., and Sassoon, D. A. (1998). Msh homeobox genes regulate cadherin-mediated cell adhesion and cell-cell sorting. *J. Cell Biochem.* **70,** 22–28.

Logan, C., Hornbruch, A., Campbell, I., and Lumsden, A. (1997). The role of Engrailed in establishing the dorsoventral axis of the chick limb. *Development* **124,** 2317–2324.

Logan, M., and Tabin, C. J. (1999). Role of pitx1 upstream of tbx4 in specification of hindlimb identity. *Science* **283,** 1736–1739.

Logan, M., Simon, H. G., and Tabin, C. (1998). Differential regulation of T-box and homeobox transcription factors suggests roles in controlling chick limb-type identity. *Development* **125,** 2825–2835.

Loomis, C. A., Harris, E., Michaud, J., Wurst, W., Hanks, M., and Joyner, A. L. (1996). The mouse Engrailed-1 gene and ventral limb patterning. *Nature* **382,** 360–363.

Loomis, C. A., Kimmel, R. A., Tong, C. X., Michaud, J., and Joyner, A. L. (1998). Analysis of the genetic pathway leading to formation of ectopic apical ectodermal ridges in mouse Engrailed-1 mutant limbs. *Development* **125,** 1137–1148.

López-Martínez, A., Chang, D. T., Chiang, C., Porter, J. A., Ros, M. A., Simandl, B. K., Beachy, P. A., and Fallon, J. F. (1995). Limb-patterning activity and restricted posterior localization of the amino-terminal product of *Sonic hedgehog* cleavage. *Curr. Biol.* **5,** 791–796.

Lu, H.-C., Revelli, J.-P., Goering, L., Thaller, C., and Eichele, G. (1997). Retinoid signaling is required for the establishment of a ZPA and for the expression of *Hoxb-8,* a mediator of ZPA formation. *Development* **124,** 1643–1651.

Luo, G., Hofmann, C., Bronckers, A. L. J. J., Sohocki, M., Bradley, A., and Karsenty, G. (1995). BMP-7 is an inducer of nephrogenesis, and is also required for eye development and skeletal patterning. *Genes Dev.* **9,** 2808–2820.

Lussier, M., Canoun, C., Ma, C., Sank, A., and Shuler, C. (1993). Interdigital soft tissue separation induced by retinoic acid in mouse limbs cultured in vitro. *Int. J. Dev. Biol.* **37,** 555–564.

Lyons, K. M., Hogan, B. L., and Robertson, E. J. (1995). Colocalization of BMP 7 and BMP 2 RNAs suggests that these factors cooperatively mediate tissue interactions during murine development. *Mech. Dev.* **50,** 71–83.

Macias, D., Gañan, Y., Rodiguez-Leon, J., Merino, R., and Hurle, J. M. (1999). Regulation by members of the transforming growth factor beta superfamily of the digital and interdigital fates of the autopodial limb mesoderm. *Cell Tissue Res.* **296,** 95–102.

Maden, M., Sonneveld, E., van der Saag, P. T., and Gale, E. (1998). The distribution of endogenous retinoic acid in the chick embryo: Implications for developmental mechanisms. *Development* **125,** 4133–4144.

Mahmood, R., Bresnick, J., Hornbruch, A., Mahony, C., Morton, N., Colquhoun, K., Martin, P., Lumsden, A., Dickson, C., and Mason, I. (1995). A role for FGF-8 in the initiation and maintenance of vertebrate limb bud outgrowth. *Curr. Biology* **5,** 797–806.

Makarenkova, H., and Patel, K. (1999). Gap junction signalling mediated through connexin-43 is required for chick limb development. *Dev. Biol.* **207,** 380–392.

Makarenkova, H., Becker, D. L., Tickle, C., and Warner, A. E. (1997). Fibroblast growth factor 4 directs gap junction expression in the mesenchyme of the vertebrate limb bud. *J. Cell. Biol.* **138,** 1125–1137.

Maleski, M. P., and Knudson, C. B. (1996). Hyaluronan-mediated aggregation of limb bud mesenchyme and mesenchymal condensation during chondrogenesis. *Exp. Cell Res.* **225,** 55–66.

Marazzi, G., Wang, Y., and Sassoon, D. (1997). Msx2 is a transcriptional regulator in the BMP4-mediated programmed cell death pathway. *Dev. Biol.* **186,** 127–138.

Marigo, V., Johnson, R. L., Vortkamp, A., and Tabin, C. (1996). Sonic hedgehog differentially regulates expression of *GLI* and *GLI3* during limb development. *Dev. Biol.* **180,** 273–283.

Martí, E., Takada, R., Bumcrot, D. A., Sasaki, H., and McMahon, A. P. (1995). Distribution of Sonic hedgehog peptides in the developing chick and mouse embryo. *Development* **121,** 2537–2547.

Martin, G. R. (1998). The roles of FGFs in the early development of vertebrate limbs. *Genes Dev.* **12,** 1571–1586.

Martin, P. (1997). Wound healing—aiming for perfect skin regeneration. *Science* **276,** 75–81.

Masuya, H., Sagai, T., Wakana, S., Moriwaki, K., and Shiroishi, T. (1995). A duplicated zone of polarizing activity in polydactylous mouse mutants. *Genes Dev.* **9,** 1645–1653.

Masuya, H., Sagai, T., Moriwaki, K., and Shiroishi, T. (1997). Multigenic control of the localization of the zone of polarizing activity in limb morphogenesis in the mouse. *Dev. Biol.* **182,** 42–51.

Means, A. L., and Gudas, L. J. (1995). The roles of retinoids in vertebrate development. *Annu. Rev. Biochem.* **64,** 201–233.

Merino, R., Gañan, Y., Macias, D., Economides, A. N., Sampath, K. T., and Hurle, J. M. (1998). Morphogenesis of digits in the avian limb is controlled by FGFs, TGFbetas, and noggin through BMP signaling. *Dev. Biol.* **200,** 34–45.

Michaud, J. L., Lapointe, F., and Le Douarin, N. M. (1997). The dorsoventral polarity of the presumptive limb is determined by signals produced by the somites and by the lateral somatopleure. *Development* **124,** 1453–1463.

Mima, T., Ohuchi, H., Noji, S., and Mikawa, T. (1995). FGF can induce outgrowth of somatic mesoderm both inside and outside of limb-forming regions. *Dev. Biol.* **167,** 617–620.

Min, H., Danilenko, D. M., Scully, S. A., Bolon, B., Ring, B. D., Tarpley, J. E., DeRose, M., and Simonet, W. S. (1998). Fgf-10 is required for both limb and lung development and exhibits striking functional similarity to *Drosophila branchless. Genes Dev.* **12,** 3156–3161.

Moran, J. L., Levorse, J. M., and Vogt, T. F. (1999). Limbs move beyond the radical fringe. *Nature* **399,** 742–743.

Morriss-Kay, G. M., and Ward, S. J. (1999). Retinoids and mammalian development. *Int. Rev. Cytol.* **188,** 73–131.

Munaim, S. I., Klagsbrun, M., and Toole, B. P. (1991). Hyaluronan-dependent pericellular coats of chick embryo limb mesodermal cells: Induction by basic fibroblast growth factor. *Dev. Biol.* **143,** 297–302.

Muneoka, K., Bryant, S. V., and Gardiner, D. M. (1989a). Growth control in limb regeneration. *In* "Developmental Biology of the Axolotl." (J. Armstrong and G. Malacinski, Eds.), pp. 143–156. Oxford University Press, Oxford, United Kingdom.

Muneoka, K., Wanek, N., and Bryant, S. V. (1989b). Mammalian limb bud development: *In situ* fate maps of early hindlimb buds. *J. Exp. Zool.* **249,** 50–54.

Muragaki Y., Mundlos S., Upton J., and Olsen B. R. (1996). Altered growth and branching patterns in synpolydactyly caused by mutations in HOXD13. *Science* **272**(5261), 548–551.

Nelson, C. E., Morgan, B. A., Burke, A. C., Laufer, E., DiMambro, E., Murtaugh, L. C., Gonzales, E., Tessarollo, L., Parada, L. F., and Tabin, C. (1996). Analysis of Hox gene expression in the chick limb bud. *Development* **122,** 1449–1466.

Ngo-Muller, V., and Muneoka, K. (1999). Influence of FGF–4 on digit morphogenesis during limb development in the mouse. *Dev. Biol.* **219,** 224–236.

Niederreither, K., Subbarayan, V., Dolle, P., and Chambon, P. (1999). Embryonic retinoic acid synthesis is essential for early mouse post-implantation development. *Nat. Genet.* **21,** 444–448.

Niswander, L., Jeffrey, S., Martin, G. R., and Tickle, C. (1994). A positive feedback loop coordinates growth and patterning in the vertebrate limb. *Nature* **371,** 609–612.

Niswander, L., and Martin, G. R. (1993). FGF-4 and BMP-2 have opposite effects on limb growth. *Nature* **361,** 68–71.

Niswander, L., Tickle, C., Vogel, A., Booth, I., and Martin, G. R., (1993). FGF-4 replaces the apical ectodermal ridge and directs outgrowth and patterning of the limb. *Cell* **75,** 579–587.

Niu, S., Antin, P. B., Akimoto, K., and Morkin, E. (1996). Expression of avian glypican is developmentally regulated. *Dev. Dyn.* **207,** 25–34.

Nohno, T., Noji, S., Koyama, E., Myokai, F., Kuroiwa, A., Saito, T., and Taniguchi, S. (1991). Involvement of the *Chox-4* chicken homeobox genes in determination of anteroposterior axial polarity during limb development. *Cell* **64,** 1197–1205.

Noji, S. Nohno, T., Koyama, E., Muto, K., Ohyama, K., Aoki, Y., Tamura, K., Ohsugi, K., Ide, H., Taniguchi, S., and Saito, T. (1991). Retinoic acid induces polarizing activity but is unlikely to be a morphogen in the chick limb bud. *Nature* **350,** 83–86.

Ogura, T., Alvarez, I. S., Vogel, A., Rodriguez, C., Evans, R. M., and Izpisúa-Belmonté, J. C. (1996). Evidence that Shh cooperates with a retinoic acid inducible co-factor to establish ZPA-like activity. *Development* **122,** 537–542.

Ohsugi, K., Gardiner, D. M., and Bryant, S. V. (1997). Cell cycle length affects gene expression and pattern formation in limbs. *Dev. Biol.* **189,** 13–21.

Ohuchi, H., Nakagawa, T., Yamamoto, A., Araga, A., Ohata, T., Ishimaru, Y., Yoshioka, H., Kuwana, T., Nohno, T., Yamasaki, M., Itoh, N., and Noji, S. (1997). The mesenchymal factor, FGF10, initiates and maintains the outgrowth of the chick limb bud through interaction with FGF8, and apical ectodermal factor. *Development* **124,** 2235–2244.

Ohuchi, H., Nakagawa, T., Yamauchi, M., Ohata, T., Yoshioka, H., Kuwana, T., Mima, T., Mikawa, T., Nohno, T., and Noji, S. (1995). An additional limb can be induced from the flank of the chick embryo by FGF4. *Biochem. Biophys. Res. Commun.* **209,** 809–816.

Ohuchi, H., Noji, S. (1999). Fibroblast-growth-factor-induced additional limbs in the study of initiation of limb formation, limb identity, myogenesis, and innervation. *Cell Tissue Res.* **296,** 45–56.

Ohuchi, H., Takeuchi, J., Yoshioka, H., Ishimaru, Y., Ogura, K., Takahashi, N., Ogura, T., and Noji, S. (1998). Correlation of wing-leg identity in ectopic FGF-induced chimeric limbs with the differential expression of chick Tbx5 and Tbx4. *Development* **125,** 51–60.

Ohuchi, H., Yoshioka, H., Tanaka, A., Kawakami, Y., Nohno, T., and Noji, S. (1994). Involvement of androgen-induced growth factor (FGF-8) gene in mouse embryogenesis and morphogenesis. *Biochem. Biophys. Res. Comm.* **204,** 882–888.

Omi, M., Sato-Maeda, M., and Ide, H. (2000). Role of chondrogenic tissue in programmed cell death and BMP expression in chick limb buds. *Int. J. Dev. Biol.* (in press).

Oro, A. E., Higgins, K. M., Hu, Z., Bonifas, J. M., Epstein, E. H., Jr., and Scott, M. P. (1997). Basal cell carcinomas in mice overexpressing sonic hedgehog. *Science* **276,** 817–821.

Park, W. Y., Miranda, B., Lebeche, D., Hashimoto, G., and Cardoso, W. V. (1998). FGF-10 is a chemotactic factor for distal epithelial buds during lung development. *Dev. Biol.* **201,** 125–134.

Parr, B. A., and McMahon, A. P. (1995). Dorsalizing signal *Wnt-7a* required for normal polarity of D-V and A-P axes of mouse limb. *Nature* **374,** 350–353.

Pearse, R. V., and Tabin, C. J. (1998). The molecular ZPA, *J. Exp. Zool.* **282,** 677–690.

Pizette, S., and Niswander, L. (1999). BMPs negatively regulate structure and function of the limb apical ectodermal ridge. *Development* **126,** 883–894.

Porter, J. A., Young, K. E., and Beachy, P. A. (1996). Cholesterol modification of hedgehog signaling proteins in animal development. *Science* **274,** 255–259.

Qu, S., Niswender, K. D., Qunsheng, J., van der Meer, R., Keeney, D., Magnuson, M. A., and Wisdom, R. (1997). Polydactyly and ectopic ZPA formation in *Alx-4* mutant mice. *Development* **124,** 3999–4008.

Qu, S., Tucker, S. C., Ehrlich, J. S., Levorse, J. M., Flaherty, L. A., Wisdom, R., and Vogt, T. F. (1998). Mutations in mouse *Aristaless-like4* cause *Strong's luxoid* polydactyly. *Development* **125,** 2711–2721.

Reginelli, A. D., Wang, Y., Sassoon, D., and Muneoka, K. (1995). Digit tip regeneration correlates with regions of msx1 (formerly Hox7.1) expression in fetal and newborn mice. *Development* **121,** 1065–1076.

Reiter, R. S., and Solursh, M. (1982). Mitogenic property of the apical ectodermal ridge. *Dev. Biol.* **93,** 28–35.

Riddle, R. D., Encini, M., Nelson, C., Tsuchida, T., Jessel, T. M., and Tabin, C. (1995). Induction of the LIM homeobox gene Lmx1 by WNT7a establishes dorsoventral pattern in the vertebrate limb. *Cell* **83,** 631–640.

Riddle, R. D., Johnson, R. L., Laufer, E., and Tabin, C. (1993). Sonic hedgehog mediates the polarizing activity of the ZPA. *Cell* **75,** 1401–1416.

Rodriguez, C., Kos, R., Macias, D., Abbott, U. K., and Izpisúa-Belmonté, J. C. (1996). Shh, HoxD, Bmp-2, and Fgf-4 gene expression during development of the polydactylous talpid2, diplopodia1, and diplopodia4 mutant chick limb buds. *Dev. Gen.* **19,** 26–32.

Rodriguez-Esteban, C., Schwabe, J. W., De La Pena, J., Foys, B., Eshelman, B., and Izpisúa Belmonté, J. C. (1997). Radical fringe positions the apical ectodermal ridge at the dorsoventral boundary of the vertebrate limb. *Nature* **386,** 360–366.

Rodriguez-Esteban, C., Tsukui, T., Yonei, S., Magallon, J., Tamura, K., and Izpisúa-Belmonté, J. C. (1999). The T-box genes Tbx4 and Tbx5 regulate limb outgrowth and identity. *Nature* **398,** 814–818.

Ros, M. A., López-Martínez, A., Simandl, B. K., Rodriguez, C., Izpisúa-Belmonté, J.-C., Dahn, R., and Fallon, J. F. (1996). The limb field mesoderm determines initial limb bud anteroposterior asymmetry and budding independent of *sonic hedgehog* or apical ectodermal gene expressions. *Development* **122,** 2319–2330.

Ruoslahti, E., and Yamaguchi, Y. (1991). Proteoglycans as modulators of growth factor activities. *Cell* **8:64(5),** 867–869.

Salmivirta, M., and Jalkanen, M. (1995). Syndecan family of cell surface proteoglycans: Developmentally regulated receptors for extracellular effector molecules. *Experientia* **51,** 863–872.

San Antonio, J. D., Lander, A. D., Karnovsky, M. J., and Slayter, H. S. (1994). Mapping the heparin-binding sites on type I collagen monomers and fibrils. *J. Cell Biol.* **125,** 1179–1188.

Saunders, J. W., Jr. (1948). The proximo-distal sequence of origin of the parts of the chick wing and the role of the ectoderm. *J. Exp. Zool.* **108,** 363–403.

Saunders, J. W., Jr. (1972). Development control of three-dimensional polarity in the avian limb. *Ann. NY Acad. Sci.* **193,** 29–42.

Saunders, J. W., Jr., (1998). Apical ectodermal ridge in retrospect. *J. Exp. Zool.* **282,** 669–676.

Saunders, J. W., Jr., and Errick, J. (1976). Inductive activity and enduring cellular constitution of a supernumerary apical ectodermal ridge grafted to the limb bud of the chick embryo. *Dev. Biol.,* **50,** 16–25.

Saunders, J. W., Jr., and Fallon, J. F. (1967). Cell death in morphogenesis. *In* "Major Problems in Developmental Biology" (M. Locke, Ed.), pp. 28–34. Academic Press, New York.

Saunders, J. W., Jr., and Gasseling, M. T. (1968). Ectodermal mesenchymal interations in the origin of limb symmetry. *In* "Epithelial-Mesenchymal Interactions" (R. Fleischmajer and R. E. Billingham, Eds.), pp. 78–97. Willams and Wilkins, Baltimore.

Savage, M. P., and Fallon, J. F. (1995). FGF-2 mRNA and its antisense message are expressed in a developmentally specific manner in the chick limb bud and mesonephros. *Dev. Dyn.* **202,** 343–353.

Schwabe, J. W., Rodriguez-Esteban, C., and Izpisúa-Belmonté, J. C. (1998). Limbs are moving: Where are they going? *Trends Genet.* **14**(6), 229–235.

Searls, R. L., and Janner, M. (1971). The initiation of limb bud outgrowth in the embryonic chick. *Dev. Biol.* **24,** 198–213.

Sekine, K., Ohuchi, H., Fujiwara, M., Yamasaki, M., Yoshizawa, T., Sato, T., Yagishita, N., Matsui, D., Koga, Y., Itoh, and N., Kato, S. (1999). Fgf10 is essential for limb and lung formation. *Nat. Genet.* **21,** 138–141.

Sherman, L., Wainwright, D., Ponta, H., and Herrlich, P. (1998). A splice variant of CD44 expressed in the apical ectodermal ridge presents fibroblast growth factors to limb mesenchyme and is required for limb outgrowth. *Genes Dev.* **12,** 1058–1071.

Shi, C., and Muneoka, K. (1992). Position specific growth of mouse limb bud cells *in vitro. Dev. Biol.* **151,** 9–17.

Simon, H. (1999). T-box genes and the formation of vertebrate forelimb and hindlimb specific pattern. *Cell Tissue Res.* **296,** 57–66.

Smith, J. C. (1980). The time required for positional signaling in the chick wing bud. *J. Embryol. Exp. Morph.* **60,** 321–328.

Solursh, M., Reiter, R. S., Jensen, K. L., Kato, M., and Bernfield, M. (1990). Transient expression of a cell surface heparan sulfate proteoglycan (syndecan) during limb development. *Dev. Biol.* **140,** 83–92.

Song, K., Wang, Y., and Sassoon, D. (1993). Expression of Hox-7.1 in myoblasts inhibits terminal differentiation and induces cell transformation. *Nature* **360,** 477–481.

Spivak-Kroizman, T., Lemmon, M. A., Dikic, I., Ladbury, J. E., Pinchasi, D., Huang, J., Jaye, M., Crumley, G., Schlessinger, J., and Lax, I. (1994). Heparin-induced oligomerization of FGF molecules is responsible for FGF receptor dimerization, activation, and cell proliferation. *Cell* **79,** 1015–1024.

Stephens, T. D., Baker, W. C., Cotterell, J. W., Edwards, D. R., Pugmire, D. S., Roberts, S. G., Shaker, M. R., Willis, H. J., and Winger, K. P., (1993). Evaluation of the chick wing territory as an equipotential self-differentiating system. *Dev. Dyn.* **197,** 157–168.

Stratford, T. H., Kostakopoulou, K., and Maden, M. (1997). *Hoxb-8* has a role in establishing early anterior-posterior polarity in chick forelimb but not hindlimb. *Development* **124,** 4225–4234.

Stratford, T. H., Horton, C., and Maden, M. (1996). Retinoic acid is required for the initiation of outgrowth in the chick limb bud. *Curr. Biol.* **6,** 1124–1133.

Sulik, K. K., and Dehart, D. B. (1988). Retinoic-acid-induced limb malformations resulting from apical ectodermal ridge cell death. *Teratology* **37,** 527–537.

Summerbell, D., Lewis, J. H., and Wolpert, L. (1973). Positional information in chick limb morphogenesis. *Nature* **244,** 492–496.

Summerbell, D. (1977). Reduction of the rate of outgrowth, cell density, and cell density, and cell division following removal of the apical ectodermal ridge of the chick limb-bud. *J. Embryol. Exp. Morph.* **40,** 1–21.

Summerbell, D. (1974). Interaction between the proximo-distal and antero-posterior coordinates of positional value during the specification of positional information in the early development of the chick limb bud. *J. Embryol. Exp. Morphol.* **32,** 227–237.

Summerbell, D., and Wolpert, L. (1972). Cell density and cell division in the early morphogenesis of the chick wing. *Nature* **239,** 24–26.

Sun, X., Meyers, E. N., Lewandoski, M., and Martin, G. R. (1999). Targeted disruption of Fgf8 causes failure of cell migration in the gastrulating mouse embryo. *Genes Dev.* **13,** 1834–1846.

Sutherland, D., Samakovlis, C., and Krasnow, M. A. (1996). branchless encodes a *Drosophila* FGF homolog that controls tracheal cell migration and the pattern of branching. *Cell* **87,** 1091–1101.

Takahashi, M., Tamura, K., Büscher, D., Masuya, H., Yonei-Tamura, S., Matsumoto, K., Naitoh-Matsuo, M., Takeuchi, J., Ogura, K., Shiroishi, T., Ogura, T., and Izpisúa-Belmonté, J. C. (1998). The role of *Alx-4* in the establishment of anteroposterior polarity during vertebrate limb development. *Development* **125,** 4417–4425.

Takeuchi, J. K., Koshiba-Takeuchi, K., Matsumoto, K., Vogel-Hopker, A., Naitoh-Matsuo, M., Ogura, K., Takahashi, N., Yasuda, K., and Ogura, T. (1999). Tbx5 and Tbx4 genes determine the wing/leg identity of limb buds. *Nature* **398,** 810–814.

Tamura, K., Yokouchi, Y., Kuroiwa, A., and Ide, H. (1997). Retinoic acid changes the proximodistal developmental competence and affinity of distal cells in the developing chick limb bud. *Dev. Biol.* **188,** 224–234.

Taylor, G., Anderson, R., Reginelli, A. D., and Muneoka, K. (1994). FGF-2 induces regeneration of the chick limb bud. *Dev. Biol.* **163,** 282–284.

Theil, T., Kaesler, S., Grotewold, L., Bose, J., and Ruther, U. (1999). Gli genes and limb development. *Cell Tissue Res.* **296,** 75–83.

Tickle, C. (1981). The number of polarizing region cells required to specify additional digits in the developing chick wing. *Nature* **289,** 295–298.

Tickle, C., Summerbell, D., and Wolpert, L. (1975). Positional signalling and specification of digits in chick limb morphogenesis. *Nature* **254,** 199–202.

Vargesson, N., Clarke, J. D., Vincent, K., Coles, C., Wolpert, L., and Tickle, C. (1997). Cell fate in the chick limb bud and relationship to gene expression. *Development* **124,** 1909–1918.

Vogel, A., Roberts-Clarke, D., and Niswander, L. (1995a). Effect of FGF on gene expression in chick limb bud cells *in vivo* and *in vitro. Dev. Biol.* **171,** 507–520.

Vogel, A., Rodriguez, C., and Izpisúa-Belmonté, J. C. (1996). Involvement of FGF-8 in initiation, outgrowth and patterning of the vertebrate limb. *Development* **122,** 1737–1750.

Vogel, A., Rodriguez, C., Warnken, W., and Izpisúa-Belmonté, J. C. (1995b). Dorsal cell fate specified by chick Lmx1 during vertebrate limb development. *Nature* **378,** 716–720.

Vogel, A., and Tickle, C. (1993). FGF-4 maintains polarizing activity of posterior limb bud cells *in vivo* and *in vitro. Development* **119,** 199–206.

Wada, N., Uchiyama, K., and Ide, H. (1993). Cell sorting and chondrogenic aggregate formation in limb bud recombinants and in culture. *Dev. Growth Differ.* **35,** 421–430.

Wanek, N., Gardiner, D. M., Muneoka, K., and Bryant, S. V. (1991). Conversion by retinoic acid of anterior cells into ZPA cells in the chick limb bud. *Nature* **350,** 81–83.

Wanek, N., Muneoka, K., Burton, R., Holler-Dinsmore, G., and Bryant, S. V. (1989). A staging system for mouse limb development. *J. Exp. Zool.* **249,** 41–49.

Wang, Y., and Sassoon, D. (1995). Ectoderm-mesenchyme and mesenchyme-mesenchyme interactions regulate *Msx-1* expression and cellular differentiation in the murine limb bud. *Dev. Biol.* **168,** 374–382.

Webb, S. E., Lee, K. K., Tang, M. K., and Ede, D. A. (1997). Fibroblast growth factors 2 and 4 stimulate migration of mouse embryonic limb myogenic cells. *Dev. Dyn.* **209,** 206–216.

Xu, X., Weinstein, M., Li, C., and Deng, C. (1999). Fibroblast growth factor receptors (FGFRs) and their roles in limb development. *Cell Tissue Res.* **296,** 33–43.

Xu, X., Weinstein, M., Li, C., Naski, M., Cohen, R. I., Ornitz, D. M., Leder, P., and Deng, C. (1998). Fibroblast growth factor receptor 2 (FGFR2)-mediated reciprocal regulation loop between FGF8 and FGF10 is essential for limb induction. *Development* **125,** 753–765.

Yajima, H., Yonei-Tamura, S., Watanabe, N., Tamura, K., and Ide, H. (1999). Role of N-Cadherin in the sorting-out of mesenchymal cells and in the positional identity along the proximodistal axis of the chick limb bud. *Dev. Dyn.* **216,** 274–284.

Yang, A., Schweitzer, R., Sun, D., Kaghad, M., Walker, N., Bronson, R. T., Tabin, C., Sharpe, A., Caput, D., Crum, C., and McKeon, F. (1999). p63 is essential for regenerative proliferation in limb, craniofacial and epithelial development. *Nature* **398,** 741–718.

Yang, Y., Drossopoulou, G., Chuang, P.-T., Duprez, D., Marti, E., Bumcrot, D., Vargesson, N., Clarke, J., Niswander, L., McMahon, A., and Tickle, C. (1997). Relationship between dose, distance and time in *Sonic Hedgehog*-mediated regulation of anteroposterior polarity in the chick limb. *Development* **124,** 4393–4404.

Yang, Y., and Niswander, L. (1995). Interaction between the signaling molecules WNT7a and SHH during vertebrate limb development: dorsal signals regulate anteroposterior patterning. *Cell* **80,** 939–947.

Yokouchi, Y., Nakazato, S., Yamamoto, M., Goto, Y., Kameda, T., Iba, H., and Kuroiwa, A. (1995). Misexpression of Hoxa-13 induces cartilage homeotic transformation and changes cell adhesiveness in chick limb buds. *Genes Dev.* **9,** 2509–2522.

Yokouchi, Y., Sakiyama, J., Kameda, T., Iba, H., Suzuki, A., Ueno, N., and Kuroiwa, A. (1996). BMP-2/-4 mediate programmed cell death in chicken limb buds. *Development* **122,** 3725–3734.

Yonei-Tamura, S., Endo, T., Yajima, H., Ohuchi, H., Ide, H., and Tamura, K. (1999). FGF7 and FGF10 directly induce the apical ectodermal ridge in chick in embryos. *Dev. Biol.* **211,** 133–143.

Yu, Q., Grammatikakis, N., and Toole, B. P. (1996). Expression of multiple CD44 isoforms in the apical ectodermal ridge of the embryonic mouse limb. *Dev. Dyn.* **207,** 204–214.

Zakany, J., and Duboule, D. (1999). Hox genes in digit development and evolution. *Cell Tissue Res.* **296,** 19–25.

Zakeri, Z., Quaglino, D., and Ahuja, H. S. (1994). Apoptotic cell death in the mouse limb and its suppression in the hammertoe mutant. *Dev. Biol.* **165,** 294–297.

Zeller, R., and Braun, T. (1999). Molecular bases of limb and muscle development. *Cell Tissue Res.* **296,** 1–220.

Zou, H., and Niswander, L. (1996). Requirement for BMP signaling in interdigital apoptosis and scale formation. *Science* **272,** 738–741.

Part III

Mechanisms to Establish Polarity and Initiate Cell Fate Determination

Defining Cis-Acting Elements and Trans-Acting Factors in RNA Localization

Karina Yaniv and Joel K. Yisraeli
Department of Anatomy and Cell Biology, Hebrew University—Hadassah Medical School, Jerusalem, Israel

Research over the last 10 to 15 years has revealed that intracellular RNA localization is a widespread phenomenon found in a large range of different cell types in an equally impressive number of different organisms (Bashirullah *et al.,* 1998; St. Johnston, 1995). Efforts have focused both on the molecular mechanisms involved in localizing RNAs to particular intracellular targets and on the functional importance (to the cell) of placing certain RNAs at particular cellular sites. In many cases, an understanding of the role of RNA localization seems to be predicated on a careful analysis of how a particular RNA achieves its characteristic distribution. A generalized model of RNA localization usually invokes cellular factors recognizing RNA target sequences. This review will focus on several systems in which cis-acting elements and trans-acting factors recognizing these elements are involved in RNA localization: how they have been defined, how they relate to each other, and how they interact and function to help achieve defined intracellular localization. Conservation of both RNA elements and protein factors across species suggests that RNA localization is probably a fundamental cellular process.
KEY WORDS: Intracellular RNA localization, RNA binding, Oocytes, Oligodendrocytes, Fibroblasts.

I. Introduction

In many somatic and germ cells, various processes occur in spatially restricted regions of the cell and are correlated with the heterogeneous distribution of particular mRNAs. Migratory fibroblasts polymerize actin microfilaments at the leading edge of the cell, and β-actin mRNA is localized to this site (Lawrence and Singer, 1986). Oligodendrocytes provide extensive myelin sheets for CNS neurons, and myelin basic protein (MBP) RNA is actively transported in these cells from the cell body out to the distal processes (Ainger *et al.,* 1993). In oocytes and embryos, a large number of RNAs have been found to be sorted to particular intracellular locations (Bashirullah *et al.,* 1998); early cleavages can then lead to enrichment or depletion of RNAs within specific cells or lineages.

In screens designed to identify localized RNAs, it was found that the large majority of RNA species in the cell are more or less homogeneously distributed (King and Barklis, 1985; Rebagliati *et al.,* 1985). These findings argue that those RNAs targeted for localization must be "marked" in some fashion, presumably by cis-acting sequences in the RNA molecule itself. In addition, it is theoretically possible for a nascent polypeptide chain to contain a localization element that could also lead to the localization of the encoding RNA. To the best of our knowledge, there are no examples in the literature, and in the cases in which this model has been tested, it has been ruled out (Sundell and Singer, 1990; Yisraeli and Melton, 1988). The sequences may specify unique intracellular locations, cytoskeletal associations, or sites for interacting factors (proteins or even other RNA molecules). Multiple sites, of the same or different types, may be present in a given RNA molecule. This marking must be read and/or interpreted by cellular factors that should ultimately lead to the asymmetric distribution of the RNA. Mechanisms resulting in RNA localization include active transport along cytoskeletal elements, anchoring of RNA at particular sites, and differential stability of the RNA. These mechanisms need not be mutually exclusive; for example, anchoring may be coupled with either active transport (Ainger *et al.,* 1997) or differential stability (Ding *et al.,* 1993).

Regardless of the mechanism utilized by a cell, interactions between specific RNA sequences and factors appear to be required. Understanding these interactions is critical to deciphering the precise roles these elements play in the cell. Numerous, excellent reviews have been published of late examining RNA localization in many different systems (Bashirullah *et al.,* 1998; Carson *et al.,* 1998; Hazelrigg, 1998; Jansen, 1999; Lasko, 1999; Mowry and Cote, 1999). In this chapter, we will review how cis-acting elements and trans-acting factors are being defined and examine systems in which both an RNA element and the protein factor recognizing it have been characterized. The unique nature of at least some of these protein factors, their high level of conservation across species, and the suggestion that some are involved in other processes in the cell, clearly argue the importance of these studies to our understanding of RNA biology in the cell. (See Cooperstock and Lipshitz, Chapter 14, and Kloc *et al.,* Chapter 2.)

II. Identifying Cis-Acting Elements

Cis-acting elements have been identified via a variety of methods, depending on the type of element being defined and the flexibility of the experimental system. In systems that are capable of localizing exogenous RNAs, localization elements have been mapped *in vivo* by assessing the ability of deleted or substituted RNA constructs to undergo localization. The RNA is tracked either by being fused to reporter sequences or otherwise labeled (via radioactive or nonradioactive substrates or fluorescent substrates or dyes). In several systems it has been possible to follow localization in real time (Bertrand *et al.,* 1998; Glotzer *et al.,* 1997; Knowles and Kosik, 1997). By attaching sequences to reporter constructs, it is theoretically possible to define whether elements are necessary and/or sufficient. Often, redundancy and overlapping of regions can complicate this analysis. In addition, the definition of localization can be somewhat fuzzy. There may be more than one step in the localization process, and intermediate intracellular targets may be specified by different signals (e. g., Kim-Ha *et al.,* 1993).

The method by which these RNA constructs are introduced into the cell is also an important consideration. Some RNAs appear to require nuclear factors for proper cytoplasmic localization; such factors would be normally acquired following transcription, before the RNA exits the nucleus. In these cases, genes encoding the RNAs must first be introduced into the cells. Other RNAs are capable of undergoing localization when injected directly into the cytoplasm. Even in these cases, however, it has been suggested that a rate-limiting step may be their association with nuclear factors that have found their way out to the cytoplasm.

An *in vitro* analysis of cis-acting elements can often complement *in vivo* work, both by helping to clarify the mechanism involved and by limiting the number of constructs that have to be tested *in vivo.* Such analyses consist of identifying protein-interacting domains by, for example, UV crosslinking or gel shift assays. Although there is a large number of general RNA binding proteins present in cells, the use of an *in vitro* system allows one to check for specificity by including different RNA transcripts as either specific or nonspecific competitors, as well as including multivalent inhibitors of nonspecific binding, such as heparin. As with any *in vitro* system, the way the extract is made, stored, and diluted can all have an effect. Nevertheless, this approach has been used successfully in a number of cases.

Once an element has been defined, it is possible to make use of the system to characterize the proteins binding the element (see below).

III. Defining Trans-Acting Factors

Several different approaches have been used to identify and characterize trans-acting factors potentially involved in RNA localization. When a cis-acting element is known or defined, an *in vitro* binding assay can help identify factors that interact specifically with the RNA element. Mutations in the RNA element that eliminate localization can aid in weeding out more general RNA binding proteins from specific interactors. Once an assay has been established, various biochemical and molecular biological methods may be employed to enrich and purify the protein. Generally, these approaches include an affinity chromatography step in which the RNA element is attached to a substrate and used to fish out the binding protein from a (usually) partially purified extract. Alternatively, a sib-selection approach has been used in conjunction with UV crosslinking to assay binding of induced proteins from increasingly smaller pools of expression library clones (Webster *et al.,* 1997). A genetic approach, involving the yeast three-hybrid system (Putz *et al.,* 1996; SenGupta *et al.,* 1996), is also possible, once the RNA element has been well characterized. Rather than starting from a known cis-acting element, genetic screens in *Drosophila* and yeast have proven to be effective in identifying factors involved in RNA localization (Jansen, 1999; Lasko, 1999). Of course, many of the mutations identified in these screens are not just in RNA binding factors but also in other important elements in the process.

IV. Analysis of Different Systems in which Cis-Acting Elements and Trans-Acting Factors for RNA Localization Have Been Identified

A. *Xenopus* Oocytes

Because of the relatively large size of *Xenopus* oocytes (up to 1.4 mm in diameter by the end of stage 6), their ease of injection and manipulation, and their clearly defined animal-vegetal polarity, this system represents onc of the most intensively studied systems for RNA localization. One of the first screens for localized RNAs was performed on *Xenopus* eggs and oocytes, and identified three animally and one vegetally localized RNAs

(Rebagliati *et al.,* 1985). Since then, several different techniques have identified a number of additional RNAs, whose functions are quite varied (see King *et al.,* 1999; Mowry and Cote, 1999).

Throughout oogenesis, the animal-vegetal axis represents the axis of symmetry around which the oocyte appears to be cylindrically symmetric. At least three distinct pathways of localization have been described in oocytes. Approximately half of all of the localized RNAs in oocytes are found in the animal hemisphere, distributed either uniformly throughout the hemisphere, along the animal cortex, or perinuclearly (Weeks *et al.,* 1995). Two different pathways are discernable for RNAs localized to the vegetal cortex. The earlier pathway, which occurs in stage I–III oocytes, involves a gradual condensation and coalescence of these RNAs concomitant, in time and location, with the formation of the mitochondrial cloud (MC) (Kloc *et al.,* 1996). The RNAs targeted to the MC take up characteristic positions within the MC, and then, as the MC disperses, these RNAs are apparently anchored at the vegetal cortex in a fairly narrow arc. Translocation of RNAs in this pathway is not susceptible to drugs that depolymerize microtubules (MTs) or microfilaments (MFs), although maintenance of the RNAs at the cortex appears to require intact MFs (Kloc and Etkin, 1995; Kloc *et al.,* 1996). The later pathway, which occurs in stage III–IV oocytes, involves a gradual disappearance of the localized RNAs from the animal hemisphere with a concomitant accumulation in the vegetal hemisphere and around the germinal vesicle (GV) in late stage III oocytes (Melton, 1987; Yisraeli and Melton, 1988). By the beginning of stage IV, these RNAs appear to be tightly anchored in a thin shell along a broad arc that extends essentially throughout the vegetal cortex. A transient wedge-like distribution of at least some of the localizing RNA, extending from the vegetal aspect of the GV to the cortex, has also been described (Kloc and Etkin, 1995). As opposed to the early pathway, translocation of RNAs along the late pathway is disrupted by MT depolymerization; maintenance of the RNA at the cortex requires intact MFs (Yisraeli *et al.,* 1990).

1. Cis-Acting Elements

The most extensively examined cis-acting element in RNAs localized in *Xenopus* oocytes is that of the late localizing Vg1 RNA. Both *in vivo* and *in vitro* analyses have been performed by several groups (Cote *et al.,* 1999; Deshler *et al.,* 1998, 1997; Gautreau *et al.,* 1997; Having *et al.,* 1998; Mowry, 1996; Mowry and Melton, 1992; Schwartz *et al.,* 1992). The results are summarized in Figure 1. A 340 nt element in the 3′ UTR that is both necessary and sufficient for Vg1 RNA localization in oocytes (termed the VLE) was identified by ligating Vg1 RNA sequences to a β-globin reporter gene and assaying localization *in vivo* (Mowry and Melton, 1992). A more

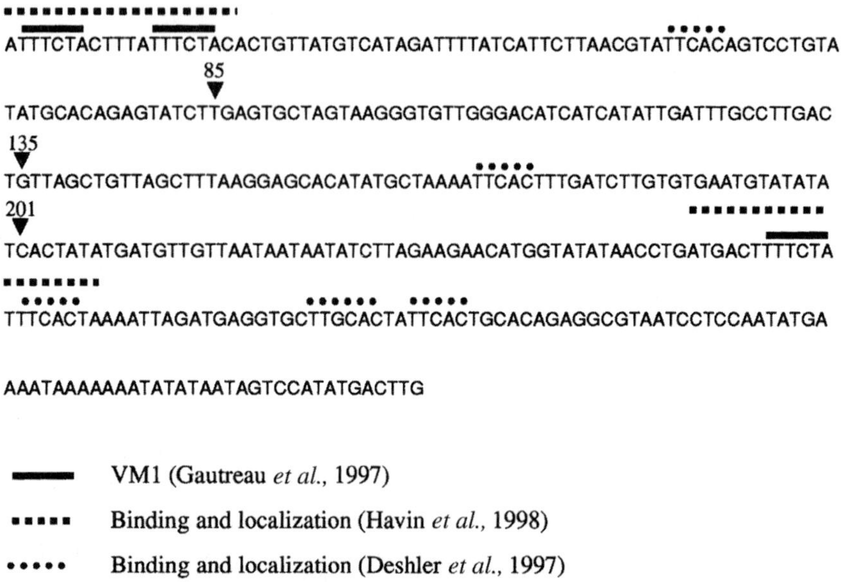

FIG. 1 The nucleotide sequence of the Vg1 VLE. The nucleotide sequence of the Vg1 VLE is shown, with the elements described in the text indicated above the sequence. The 5′ subelement, as mentioned in the text, consists of either sequences 1-85 (Cote *et al.*, 1999) or 1-135 (Gautreau *et al.*, 1997). The 3′ subelement, as mentioned in the text, consists of sequences from 201–340 (Gautreau *et al.*, 1997).

detailed mapping of the fine structure of this element has suggested that its composition is complex. Deletion analyses and linker scans spanning the entire VLE suggest that single, short deletions or replacements (of between 5 and 20 nt) tend to have, at best, marginal effects on localization (Deshler *et al.*, 1998; Gautreau *et al.*, 1997; Havin *et al.*, 1998). Larger, single deletions near either end of the VLE (of 40 to 65 nt) or a double replacement of two 20 nt fragments with vector sequences of the same length can successfully abolish localization. These results suggest that the localization signal is probably a compound element consisting of two subelements at or near the ends of the VLE. Furthermore, relatively large deletions between these subelements have no effect on localization. The precise nature of the cis-acting sequnces that are active in these subelements is still unclear. Deletion of three interspersed 5–6 nt repeat sequences (UUCAC or UUGCAC) in the downstream subelement nearly abolishes localization (Deshler *et al.*, 1998). Only the upstream, and not the downstream, subelement, however, is capable of directing localization when tandomly repeated. A 6 nt sequence (UUUCUA), termed VM1 (Gautreau *et al.*, 1997), is present in two copies in the first subelement and in a single copy in the second subelement. Mutating the four VM1 sequences in a tandomly repeated upstream subelement construct disrupts localization, suggesting that VM1 is a necessary component of the cis-acting signal. Tandomly repeated elements, however, do not necessarily fold or behave precisely like the larger sequence from which they are derived. It is important to assay these mutations in the context of the intact VLE in order to ensure that these sequences play a role *in vivo,* and so far this has not been shown.

Several other messages undergo localization along the late, Vg1-like pathway. VegT RNA is localized temporally and spatially in oocytes in a manner very similar to Vg1 RNA, although its release from the vegetal cortex appears to precede that of Vg1 RNA (Stennard *et al.,* 1996; Zhang and King, 1996). TGFβ-5 RNA is also distributed along the vegetal cortex in a manner similar to Vg1 RNA (Perry-O'Keefe *et al.,* 1990). Several RNAs (Xcat-2, Xlsirts, Xpat) localize normally along the early pathway but will also localize like Vg1 RNA when injected into stage III oocytes (Hudson and Woodland, 1998; Kloc *et al.,* 1993; Zhang and King, 1996). Rat tau RNA, which localizes to the proximal hillock of rat axons, also localizes to the vegetal cortex in a fashion that precisely mimics the localization of Vg1 RNA. As with almost all localized RNAs examined, the 3′ UTR of all of these RNAs contains their localization element. For only one of these RNAs, however (Xcat-2), is there more detailed localization data currently available (Zhou and King, 1996). Intriguingly, in this case, in the region required for late Xcat2 RNA localization, a sequence is present that is homologous to the VM1 repetitive element seen in the Vg1 VLE (see Kloc *et al.,* Chapter 2).

2. Trans-Acting Factors

Although, as indicated above, there is an impressive array of RNAs that localize to different regions of the oocyte at different stages of oogenesis, work in the *Xenopus* field on trans-acting factors has focused so far on those that interact with Vg1 RNA. Vg1 RBP was identified as a 69 kDa protein in S100 extracts that shows high affinity binding to the Vg1 VLE in UV crosslinking assays (Schwartz *et al.,* 1992). Vg1 RBP is enriched in oocyte MT preparations, pellets with semipurified, polymerized tubulin, and helps in the specific association of Vg1 RNA to reconstituted MTs *in vitro* (Litman *et al.,* 1996). A similarly sized protein (75 kDa) was identified in oocyte low-speed supernatants that also specifically interacted with the Vg1 VLE; this protein was termed Vera because of its ability to bind *V*g1 RNA and its apparent *e*ndoplasmic *r*eticulum *a*ssociation (Deshler *et al.,* 1997). Both Vg1 RBP and Vera have been purified and cloned with methods that included Vg1 VLE RNA affinity chromatography, and these proteins turn out to be 100% identical in sequence (Deshler *et al.,* 1998; Havin *et al.,* 1998). (It is currently unclear whether the difference in the observed size of the proteins is a result of posttranslational modifications or slight variations in electrophoresis conditions.) Vg1 RBP/Vera consists almost entirely of potential RNA binding domains: 2 RRM motifs, an RGG box, and 4 KH domains. The sequences mediating MT and/or endoplasmic reticulum (ER) associations have not yet been determined, although both RRM and KH domains can be involved in protein-protein interactions (Chen *et al.,* 1997; Samuels *et al.,* 1998). It is also unknown at the moment which domains are responsible for the specific interaction of Vg1 RBP/Vera with Vg1 VLE. It is intriguing that there are putative nuclear localization and export sequences present in the protein, suggesting that Vg1 RBP/Vera may spend at least some of its life in the nucleus. In fact, an allele, 97% identical at the protein level, was also isolated during the cloning of Vg1 RBP and is identical to a previously described factor B3 that binds to the promoter of TFIIIA and appears to participate in its transcriptional activation in early *Xenopus* oocytes (Pfaff and Taylor, 1992). Antibodies raised to Vg1 RBP/Vera and to B3 indicate that in early oocytes, these proteins are present in the nucleus as well as the cytoplasm (Zhang *et al.,* 1999b). Vg1 RBP/Vera and Vg1 RNA distributions are coincident throughout oogenesis, initially homogeneous and then localizing vegetally with the RNA as it translocates and gets anchored at the cortex.

UV crosslinking has revealed a number of additional proteins that interact with Vg1 VLE as well (Mowry, 1996). One of these, a 60 kDa protein termed VgRBP60, has recently been purified and cloned, based on its ability to bind specifically to the predominantly pyrimidine sequence, VM1 (Cote *et al.,* 1999). VgRBP60 contains 4 RRM domains and is highly related to the human hnRNP proteins, hnRNP I and PTB (polypyrimidine tract binding protein). In the case of the human hnRNP I protein, RRM3 and RRM4 are thought to be the important RNA binding domains, with the N-terminal half of the protein believed to be involved in protein-protein interactions (Oh *et al.,* 1998; Perez *et al.,* 1997). As does Vg1 RBP/Vera, Vg1RBP60 colocalizes with Vg1 RNA to the vegetal cortex (Cote *et al.,* 1999). A nuclear localization sequence (NLS), however, has also been detected in VgRBP60, a sequence conserved in hnRNP I. The ability of the hnRNP family proteins to shuttle between the nucleus and cytoplasm may be important for the role they play in RNA localization. It has recently been shown that *ftz* mRNA in *Drosophila* requires the association of the hnRNP *Squid* protein for proper apical/basal localization in syncitial blastoderm embryos (Lall *et al.,* 1999). *Squid* is also important for *grk* mRNA localization in *Drosophila* oocytes (Norvell *et al.,* 1999; Cooperstock and Lipshitz, Chapter 14). In addition, as discussed below, hnRNP A2 has been shown to associate with a cis-acting element in MBP mRNA that is necessary and sufficient for its transport into peripheral processes in oligodendrocytes (Hoek *et al.,* 1998).

How do these trans-acting factors interact specifically with the cis-acting VLE and what are their roles in localization? In the case of VgRBP60, mutagenesis evidence shows that the protein recognizes the VM1 sequences in the VLE, certainly the two copies in the 5′ subelement, and perhaps the third copy in the 3′ subelement (see Fig. 1). This correlates very nicely with the ability of constructs containing either wt 5′ and 3′ subelements or a tandomly duplicated 5′ subelement to be localized, but not a mutated, tandomly duplicated 5′ subelement (Gautreau *et al.,* 1997). In the case of Vg1 RBP/Vera, the story is less clear, with the answer somewhat dependent on the method used. A linker scan analysis, in which 20 nt-long sequences were progressively replaced by a 20 nt-long vector sequence throughout the length of the VLE, revealed that only two replacements affected UV crosslinking to Vg1 RBP/Vera (Havin *et al.,* 1998). (By preserving the total length of the VLE while making limited sequence changes, this approach may minimize the effects of the changes on the overall secondary and tertiary structure of the RNA.) Strikingly, one of these replacements removed both VM1 sequences from the 5′ subelement, and the other replacement removed the VM1 sequence from the 3′ subelement (see Fig. 1). Vg1 RBP/Vera was found to bind efficiently to the 5′ subelement only when it was a minimum length of 135 nt (Havin, 1998; Z. Elisha and J. K. Yisraeli, unpublished observations). There was a good correlation between the ability of these VLE constructs to bind the protein and their ability to localize upon injection into stage III oocytes. A good correlation between Vg1 RBP/Vera binding and localization was also found when the sequences UUCAC (or UUGCAC), which appear in 5 copies in the VLE, were either deleted entirely or in part (Deshler *et al.,* 1998, 1997). (Only one of these repeats is contained in the 20 nt sequences mentioned above.) Although a small RNA probe containing four copies of this sequence and consisting of only 23 nt can crosslink Vg1 RBP/Vera, it is hard to deduce from these results, with such an artificial substrate, the protein's specificity on the endogenous RNA. A definitive answer will have to await direct mapping of the binding sites by RNAse protection.

Neither the interaction between these proteins nor the contribution of each protein to Vg1 RNA localization is well understood. According to Cote *et al.* (1999), the wt tandemly repeated 5′ subelement gets localized and appears to bind well to VgRBP60 but poorly to Vg1 RBP/Vera. Furthermore, mutations in VM1 (in the context of the tandomly repeated 5′ element) prevent localization and severely reduce binding to VgRBP60, but have little effect on (the minimal) Vg1 RBP/Vera binding. On the basis of these results, Cote *et al.* argue that Vg1 RBP/Vera does not appear to interact with VM1. This argument seems premature, however. The size of the tandem repeat used in these constructs is 85 nt, which, as mentioned above (Z. Elisha and J. K. Yisraeli, unpublished observations) and as Cote *et al.* themselves show, does not support efficient Vg1 RBP/Vera binding. In fact, the longer, 135 nt tandem repeat localizes more efficiently (Gautreau *et al.*, 1997). It therefore appears that Vg1 RBP/Vera is unlikely to play a role in the (partial) localization of the tandomly repeated short 5′ subelement. With respect to Vg1 RBP/Vera interaction with VM1 sequences, however, the jury is still out. Given the potential for artifactual folding of RNA and the artificial nature of short, repeated constructs, only experiments performed in the context of the full-length VLE are likely to allow analysis of the full complement of the factors involved.

B. *Drosophila* Oocytes

Drosophila oocytes and embryos have been a particularly fruitful area for studying localized RNAs. A large number of RNAs involved in the determination of the oocyte, specification of the axes, and development of the germ cells have been identified as being localized in the oocyte, egg, and/or developing embryo (Bashirullah *et al.*, 1998; Lasko, 1999). Genetic analysis has revealed a number of genes that are likely to function at various points in the process, but the difficulty of performing biochemical assays in *Drosophila* has so far limited the number of trans-acting factors that have been clearly shown to interact directly with specific RNA targets.

1. Cis-Acting Elements

bicoid (bcd) mRNA is localized to the anterior pole of oocytes and early embryos in a process that appears to be composed of at least four distinct steps (Kim-Ha *et al.*, 1993; St. Johnston *et al.*, 1989). Using a combination of phylogenetic conservation considerations and the suboptimal RNA folding program of Zuker, a complex secondary structure model for the 3′ UTR of *bcd* mRNA has been proposed (Macdonald, 1990; Seeger and Kaufman, 1990; see Fig. 2). Although this model has not yet been proven biochemically, numerous predictions about mutations that would lead to compensatory and noncompensatory changes in stems suggest that the overall picture is probably correct (Ferrandon *et al.*, 1997; Macdonald and Kerr, 1998). Careful analyses of the cis-acting sequences involved in this process have been performed by using a transgenic approach to assay the ability of sequences to localize covalently linked reporters and also, in parallel, to rescue *bcd* null phenotypes when covalently linked to *bcd* coding sequences (which are localization-defective on their own; Berleth *et al.*, 1988; Macdonald and Kerr, 1997; Macdonald and Kerr, 1998; MacDonald *et al.*,

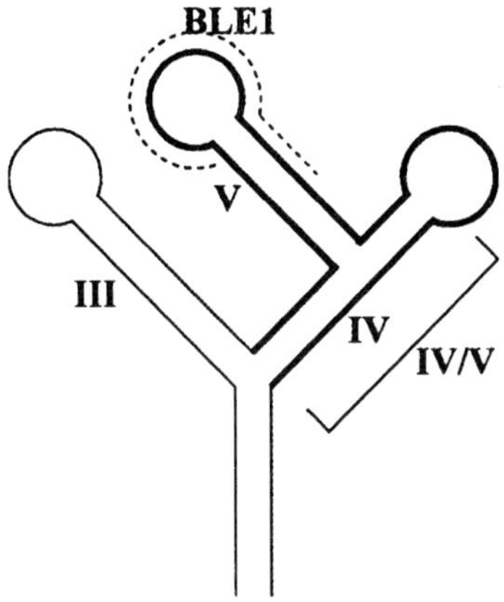

FIG. 2 A schematic representation of the putative secondary structure of the 3′UTR of *bcd* mRNA. The proposed secondary structure of the 3′ UTR of *bcd* mRNA is shown in a simplified form, with the stem/loop structures numbered according to Macdonald (1990). The location of the BLE1 region is shown as a dotted line.

1995; MacDonald and Struhl, 1988). Several cis-acting motifs have been mapped to parts of the *bcd* 3′ UTR that appear to mediate specific events in the localization process. Stem-loop IV/V contains all the information necessary for the earliest localization from nurse cells to oocytes (stage 4–5, termed event A). At least some sequences outside of stem-loop IV/V are required for a slightly later localization of bcd mRNA from nurse cells to the oocyte (stage 6–8, termed event B). Events A and B are redundant: it is possible to eliminate cis-acting elements that mediate one of the events without affecting the other. Sequences within stem-loop IV/V alone are insufficient for the later retention of *bcd* mRNA along the anterior border of stage 10–12 oocytes and in embryos; indirect evidence suggests that sequences in stem-loop III are likely to also be involved (see below) (Ferrandon *et al.,* 1997). Extensive deletion analysis has also revealed a 53 nt sequence in stem-loop V (termed BLE1) that is necessary for localization (Berleth *et al.,* 1988; Kim-Ha *et al.,* 1993). When present as a dimer, 2XBLE1 is sufficient to direct early localization in a manner indistinguishable from event A (although it cannot support localization beyond stage 10).

2. Trans-Acting Factors

Genetic screens for trans-acting factors have failed to identify any candidate genes involved in the earliest stages of *bcd* mRNA localization, perhaps in large measure because of the redundancy between events A and B. Three mutants shown to be defective in later steps of *bcd* mRNA localization have been further characterized. *Exuperentia* (*exu*) protein shows nonspecific RNA binding when assayed by UV-crosslinking (MacDonald *et al.*, 1995). *Swallow* (*sww*) protein, although it has a weak homology to an RNA binding motif, shows no evidence of RNA binding (Chao *et al.*, 1991). *Staufen* (*stau*) protein contains 5 putative double strand RNA binding domains (St. Johnston *et al.*, 1992). Studies assaying the ability of injected RNA to recruit endogenous *Stau* show that this process is very sequence-specific, and that sequences in stem-loop III of *bcd* mRNA, as well as those at the terminal portions of stem-loops IV and V, are involved (Ferrandon *et al.*, 1994, 1997). (Interestingly, these data also suggest that *intermolecular* interactions between different *bcd* mRNA molecules facilitates *Stau* binding.) UV-crosslinking experiments, using the 2XBLE1 probe that was shown to be capable of mediating localization until stage 10, indicate that there is a specific binding activity in oocyte protein extracts that recognizes this construct (MacDonald *et al.*, 1995). All point mutations in 2XBLE1 that affect the binding of this ~115 kDa protein (termed exl) *in vitro* are localized normally until stage 8 but cause precocious loss of localization in stage 9 oocytes (a phenotype similar to *exu* mutants). Hence, exl is a candidate for a trans-acting factor that is required for localization after events A and B, but before stage 9–10.

C. Oligodendrocytes

Oligodendrocytes can form up to 50 processes that wrap around axons and form myelin sheaths in the CNS. In addition to myelin basic protein (MBP), mRNAs encoding MBP, myelin-associated/oligodendrocytic basic protein (MOBP), and carbonic anhydrase II are also found in the distal myelin membranes in these cells (Colman *et al.*, 1982; Holmes *et al.*, 1988; Verity and Campagnoni, 1988). By injecting digoxigenin-labeled transcripts into oligodendrocytes, Carson and colleagues have been able to follow the transport of MBP mRNA from the perikaryon, into the processes, and out to the myelin membranes (Ainger *et al.*, 1993). This group has also mapped the sequences required for this transport and localization by assaying the ability of labeled transcripts made from different parts of the MBP mRNA to be localized into the processes and to the distal membranes (Ainger *et al.*, 1997). Using this approach, it was possible to identify a region in the 3′ UTR that is both necessary and sufficient for transport into the processes. Based on sequence comparisons with MBP mRNAs from other species, a 21 nucleotide sequence (termed the RTS) within this region was found to be fairly well conserved, including in another RNA that is transported into the processes (MOBP). Just the RTS alone, when linked to two different reporter sequences, is capable of directing translocation into the processes. An additional, more distal, region that is necessary for localization to the myelin sheath was also identified in these studies. It is important to note that how much of a role RNA stabilization plays in either of these processes (translocation or localization) is unclear from these studies. The potential importance to RNA intracellular localization of stabilization of RNA transcripts via specific cis-acting elements has been clearly shown, both in *Drosophila* oocytes, with hsp83 mRNA (Ding *et al.,* 1993), and in neurons, with tau RNA (Aronov *et al.,* 1999).

A biotinylated 35 base oligoribonucelotide containing the RTS formed the basis of the purification and cloning of a trans-acting factor that binds this sequence (Hoek *et al.,* 1998). hnRNP A2 was identified by Edmond degradation of tryptic peptides and confirmed by electron spray mass spectroscopy as the major protein in brain extracts that was retained on affinity columns loaded with the 35 nt oligo. This protein is present not only in the nucleus of oligodendrocytes, but also in the perikaryon and peripheral processes; significantly, no hnRNP A2 was detected in the myelin. As mentioned above, several recent reports have indicated a connection between sequence-specific hnRNP binding and intracellular transport/localization. The picture that emerges is that it may be necessary to properly package localized RNAs in the nucleus (or at least with protein components normally found in the nucleus) in order to allow them to become part of the localization machinery in the cytoplasm. This phenomenon is somewhat reminiscent of the requirement of a "nuclear history" for translational repression of RNAs injected into *Xenopus* oocytes (Bouvet and Wolffe, 1994). While necessary for localization, these factors, to borrow a concept from developmental biology, are most likely permissive, rather than instructive. The prediction would be that additional factors are required to help direct the RNA to the proper intracellular destination.

D. Fibroblasts

One of the earliest reports of RNA localization in somatic cells is the localization of β-actin RNA to the leading edge of lamellae in fibroblasts (Lawrence and Singer, 1986), the site of increased β-actin protein synthesis necessary for lamellipodia extension. This localization occurs in the presence of microtuble, but not microfilament inhibitors, implying that an intact actin cytoskeleton is required (Sundell and Singer, 1991); localization is independent of protein synthesis, indicating that cis-acting elements in the RNA are important. Kislauskis and colleagues (1994) assayed for the ability of different β-actin mRNA sequences to direct the expression of a chimeric lacZ/β-actin mRNA to the periphery of transfected chick embryo fibroblasts

(CEFs). A small region (54 nts) at the beginning of the β-actin 3′ UTR was identified as being necessary and sufficient for complete localization (although an additional downstream element may also be partially sufficient). This region has been termed a "zipcode" sequence because of its ability to direct localization to a particular intracellular site (Kislauskis *et al.,* 1994). Functional importance of this zipcode was suggested by the ability of antisense oligonucleotides, targeted to particular parts of the zipcode, to inhibit localization (Kislauskis *et al.,* 1994). The precise mechanism for this inhibition is unclear (both β-actin mRNA and protein levels are unaffected by the antisense treatments); nevertheless, the absence of any clear polarity in the antisense-treated, localization-impaired cells, as well as the altered site of actin protein accumulation (perinuclear as opposed to peripheral), suggests an important role for β-actin mRNA localization in the spatial organization of these cells.

A trans-acting factor recognizing the zipcode sequence was isolated by affinity chromatography (Ross *et al.,* 1997). Because specifically the proximal part of the zipcode (the first 27 nts), and not the distal part, was bound by proteins in both gel shift and UV crosslinking assays, this sequence was used to purify a 68 kDa protein from CEFs. This protein, called the zipcode binding protein-1 (ZBP-1), represented the first report of a highly conserved family of proteins, which now numbers at least 11 distinct members, all, so far, from vertebrate species. Other members include Vg1 RBP/Vera (see above), which has been implicated in Vg1 RNA localization in *Xenopus* oocytes (Deshler *et al.,* 1998; Havin *et al.,* 1998); CRD-BP, the c-myc coding region determinant binding protein involved in stabilizing c-myc transcripts in both mice and humans (Doyle *et al.,* 1998); and IMP 1, 2, and 3, RNA binding proteins involved in translational control of IGF-II mRNA (Nielsen *et al.,* 1999). In addition, there are two members of this family that were isolated on the basis of their overexpression in different types of human cancers: KOC, in pancreatic carcinoma lines (Mueller-Pillasch *et al.,* 1997), and p62, in hepatocarcinomas (Zhang *et al.,* 1999a). All of these proteins are greater than 70% identical in sequence, between 500 and 600 amino acids long, and organized in exactly the same way, with 2 RNA recognition motifs (RRMs) at the N-terminus and 4 hnRNP K homology (KH) domains at the C-terminus. The exact function of these proteins in the different cell types is still unclear, but the presence of the various RNA binding domains appears to have made these proteins well-suited for involvement in RNA translation, stabilization, and localization.

The ability of ZBP-1 to bind the proximal zipcode sequences correlates well with the ability of these sequences to direct peripheral localization in CEFs. In particular, a tandemly duplicated ACACCC sequence at the end of the proximal sequence is particularly important, and when both of these sequences are mutated, localization and binding are severely impaired.

Surprisingly, this sequence does not resemble any of the repeated sequences suggested to be involved in Vg1 RBP/Vera binding, even though these proteins are 78% identical/84% similar, particularly throughout the RNA binding motifs (Deshler *et al.,* 1998; Havin *et al.,* 1998).

V. Concluding Remarks

The study of RNA localization has revealed several highly conserved families of factors that appear to have roles in many different and disparate organisms and systems. The characteristics of these proteins may provide insights into their functions. The ZBP-1 family, so far found only in vertebrates, contains members that are involved in RNA localization in frogs and chicks. As mentioned above, ZBP-1 is a MF-associated protein believed to be involved in β-actin RNA localization (Ross *et al.,* 1997). A *Xenopus* homolog, Vg1 RBP/Vera, appears to associate with MTs, which, in stage III oocytes, are radially arrayed from the GV (germinal vesicle) to the cortex; Vg1 RBP/Vera also appears to associate with ER, which has been reported to be associated with and move along MTs (Allan and Vale, 1994; Dabora and Sheetz, 1988; Lane and Allan, 1999; Lee and Chen, 1988). Hence, the ZBP-1 family proteins are likely to play a role in mediating the association of RNA with a complex associated with cytoskeletal elements. The apparent ability of these proteins to interact with either MTs or MFs may allow flexibility for a given protein within a cell or for different family members in various cell types. Members of the hnRNP family appear to function in processing, packaging, and/or shuttling specific mRNAs out of the nucleus in a form that makes them recognized and/or accessible to the translocation machinery in the cytoplasm. Experiments aimed at knocking out or interfering with the individual proteins should show conclusively the role these proteins play in the localization process.

Acknowledgments

The work was partially supported by the Zalman Cohen-Mager Fund, administered by the Hebrew University Authority for Research and Development, and by a grant (to J. K. Y.) from the Israel Science Foundation, administered by Israel Academy of Arts and Sciences.

References

Ainger, K., Avossa, D., Diana, A. S., Barry, C., Barbarese, E. and Carson, J. H. (1997). Transport and localization elements in myelin basic protein mRNA. *J. Cell Biol.* **138,** 1077–1087.

Ainger, K., Avossa, D., Morgan, F., Hill, S. J., Barry, C., Barbarese, E. and Carson, J. H. (1993). Transport and localization of exogenous myelin basic protein mRNA microinjected into oligodendrocytes. *J. Cell Biol.* **123,** 431–441.

Allan, V. and Vale, R. (1994). Movement of membrane tubules along microtubules in vitro: Evidence for specialised sites of motor attachment. *J. Cell Sci.* **107,** 1885–1897.

Aronov, S., Marx, R., and Ginzburg, I. (1999). Identification of 3′UTR region implicated in tau mRNA stabilization in neuronal cells. *J. Mol. Neurosci.* **12,** 131–145.

Bashirullah, A., Cooperstock, R. L., and Lipshitz, H. D. (1998). RNA localization in development. *Annu. Rev. Biochem.* **67,** 335–394.

Berleth, T., Burri, M., Thoma, G., Bopp, D., Richstein, S., Frigerio, G., Noll, M., and Nusslein-Volhard, C. (1988). The role of localization of bicoid RNA in organizing the anterior pattern of the *Drosophila* embryo. *EMBO J.* **7,** 1749–1756.

Bertrand, E., Chartrand, P., Schaefer, M., Shenoy, S. M., Singer, R. H., and Long, R. M. (1998). Localization of ASH1 mRNA particles in living yeast. *Mol. Cell* **2,** 437–445.

Bouvet, P., and Wolffe, A. P. (1994). A role for transcription and FRGY2 in masking maternal mRNA within Xenopus oocytes. *Cell* **77,** 931–941.

Carson, J. H., Kwon, S., and Barbarese, E. (1998). RNA trafficking in myelinating cells. *Curr. Opin. Neurobiol.* **8,** 607–612.

Chao, Y. C., Donahue, K. M., Pokrywka, N. J., and Stephenson, E. C. (1991). Sequence of swallow, a gene required for the localization of bicoid message in *Drosophila* eggs. *Dev. Genet.* **12,** 333–341.

Chen, T., Damaj, B.B., Herrera, C., Lasko, P., and Richard, S. (1997). Self-association of the single-KH-domain family members Sam68, GRP33. GLD-1, and Qk1: Role of the KH domain. *Mol. Cell Biol.* **17,** 5707–5718.

Colman, D. R., Kreibich, G., Frey, A. B., and Sabatini, D. D. (1982). Synthesis and incorporation of myelin polypeptides into CNS myelin. *J. Cell Biol.* **95,** 598–608.

Cote, C. A., Gautreau, D., Denegre, J. M., Kress, T. L., Terry, N. A. and Mowry, K. L. (1999). A *Xenopus* protein related to hnRNP I has a role in cytoplasmic RNA localization. *Mol. Cell* **4,** 431–437.

Dabora, S. L., and Sheetz, M. P. (1988). The microtubule-dependent formation of a tubulovesicular network with characteristics of the ER from cultured cell extracts. *Cell* **54,** 27–35.

Deshler, J. O., Highett, M. I., Abramson, T., and Schnapp, B. J. (1998). A highly conserved RNA-binding protein for cytoplasmic mRNA localization in vertebrates. *Curr. Biol.* **8,** 489–496.

Deshler, J. O., Highett, M. I. and Schnapp, B. J. (1997). Localization of *Xenopus* Vg1 mRNA by Vera protein and the endoplasmic reticulum [see comments]. *Science* **276,** 1128–1131.

Ding, D., Parkhurst, S. M., Halsell, S. R. and Lipshitz, H. D. (1993). Dynamic Hsp83 RNA localization during Drosophila oogenesis and embryogenesis. *Mol. Cell Biol.* **13,** 3773–3781.

Doyle, G. A., Betz, N. A., Leeds, P. F., Fleisig, A. J., Prokipcak, R. D., and Ross, J. (1998). The c-myc coding region determinant-binding protein: A member of a family of KH domain RNA-binding proteins. *Nucleic Acids Res.* **26,** 5036–5044.

Ferrandon, D., Elphick, L., Nusslein-Volhard, C., and St. Johnston, D. (1994). Staufen protein associates with the 3′ UTR of bicoid mRNA to form particles that move in a microtubule-dependent manner. *Cell* **79,** 1221–1232.

Ferrandon, D., Koch, I., Westhof, E., and Nusslein-Volhard, C. (1997). RNA-RNA interaction is required for the formation of specific bicoid mRNA 5′ UTR-STAUFEN ribonucleoprotein particles. *EMBO J.* **16,** 1751–1758.

Gautreau, D., Cote, C. A., and Mowry, K. L. (1997). Two copies of a subelement from the Vg1 RNA localization sequence are sufficient to direct vegetal localization in *Xenopus* oocytes. *Development* **124,** 5013–5020.

Glotzer, J. B., Saffrich, R., Glotzer, M., and Ephrussi, A. (1997) Cytoplasmic flows localize injected oskar RNA in Drosophila oocytes. *Curr. Biol.* **7,** 326–337.

Havin, L., Git, A., Elisha, Z., Oberman, F., Yaniv, K., Schwartz, S. P., Standart, N., and Yisraeli, J. K. (1998). RNA-binding protein conserved in both microtubule- and microfilament-based RNA localization. *Genes Dev.* **12,** 1593–1598.

Hazelrigg, T. (1998). The destinies and destinations of RNAs. *Cell* **95,** 451–460.

Hoek, K. S., Kidd, G. J., Carson, J. H., and Smith, R. (1998). hnRNP A2 selectively binds the cytoplasmic transport sequence of myelin basic protein mRNA. *Biochemistry* **37,** 7021–7029.

Holmes, E., Hermanson, G., Cole, R., and de Vellis, J. (1988). Developmental expression of glial-specific mRNAs in primary cultures of rat brain visualized by in situ hybridization. *J. Neurosci. Res.* **19,** 389–396, 458–365.

Hudson, C., and Woodland, H. R. (1998). Xpat, a gene expressed specifically in germ plasm and primordial germ cells of *Xenopus laevis. Mech. Dev.* **73,** 159–168.

Jansen, R. P. (1999). RNA-cytoskeletal associations. *FASEB J.* **13,** 455–466.

Kim-Ha, J., Webster, P. J., Smith, J. L., and Macdonald, P. M. (1993). Multiple RNA regulatory elements mediate distinct steps in localization of oskar mRNA. *Development* **119,** 169–178.

King, M. L., and Barklis, E. (1985). Regional distribution of maternal messenger RNA in the amphibian oocyte. *Dev. Biol.* **112,** 203–212.

King, M. L., Zhou, Y., and Bubunenko, M. (1999). Polarizing genetic information in the egg: RNA localization in the frog oocyte. *Bioessays* **21,** 546–557.

Kislauskis, E. H., Zhu, X., and Singer, R. H. (1994). Sequences responsible for intracellular localization of beta-actin messenger RNA also affect cell phenotype. *J. Cell Biol.* **127,** 441–451.

Kloc, M., and Etkin, L. D. (1995). Two distinct pathways for the localization of RNAs at the vegetal cortex in Xenopus oocytes. *Development* **121,** 287–297.

Kloc, M., Larabell, C., and Etkin, L. D. (1996). Elaboration of the messenger transport organizer pathway for localization of RNA to the vegetal cortex of *Xenopus* oocytes. *Dev. Biol.* **180,** 119–130.

Kloc, M., Spohr, G., and Etkin, L. D. (1993). Translocation of repetitive RNA sequences with the germ plasm in Xenopus oocytes. *Science* **262,** 1712–1714.

Knowles, R. B., and Kosik, K. S. (1997). Neurotrophin-3 signals redistribute RNA in neurons. *Proc. Natl. Acad. Sci. USA* **94,** 14804–14808.

Lall, S., Francis-Lang, H., Flament, A., Norvell, A., Schupbach, T., and Ish-Horowicz, D. (1999). Squid hnRNP protein promotes apical cytoplasmic transport and localization of *Drosophila* pair-rule transcripts. *Cell* **98,** 171–180.

Lane, J. D., and Allan, V. J. (1999). Microtubule-based endoplasmic reticulum motility in *Xenopus* laevis: Activation of membrane-associated kinesin during development. *Mol. Biol. Cell* **10,** 1909–1922.

Lasko, P. (1999). RNA sorting in *Drosophila* oocytes and embryos. *FASEB J.* **13,** 421–433.

Lawrence, J. B., and Singer, R. H. (1986). Intracellular localization of messenger RNAs for cytoskeletal proteins. *Cell* **45,** 407–415.

Lee, C., and Chen, L. B. (1988). Dynamic behavior of endoplasmic reticulum in living cells. *Cell* **54,** 37–46.

Litman, P., Behar, L., Elisha, Z., Yisraeli, J. K., and Ginzburg, I. (1996). Exogenous tau RNA is localized in oocytes: Possible evidence for evolutionary conservation of localization mechanisms. *Devl. Biol.* **176,** 86–94.

Macdonald, P. M. (1990). bicoid mRNA localization signal: Phylogenetic conservation of function and RNA secondary structure. *Development* **110,** 161–171.

Macdonald, P. M., and Kerr, K. (1997). Redundant RNA recognition events in bicoid mRNA localization. *Rna* **3,** 1413–1420.

Macdonald, P. M., and Kerr, K. (1998). Mutational analysis of an RNA recognition element that mediates localization of bicoid mRNA. *Mol. Cell Biol.* **18,** 3788–3795.

MacDonald, P. M., Leask, A., and Kerr, K. (1995). exl protein specifically binds BLE1, a bicoid mRNA localization element, and is required for one phase of its activity. *Proc. Natl. Acad. Sci. USA* **92,** 10787–10791.

MacDonald, P. M., and Struhl, G. (1988). cis-acting sequences responsible for localizing bicoid mRNA at the anterior pole of *Drosophila* embryos. *Nature* **336,** 595–598.

Melton, D. A. (1987). Translocation of a localized maternal mRNA to the vegetal pole of *Xenopus* oocytes. *Nature* **328,** 80–82.

Mowry, K. L. (1996). Complex formation between stage-specific oocyte factors and a *Xenopus* mRNA localization element. *Proc. Natl. Acad. Sci. USA* **93,** 14608–14613.

Mowry, K. L., and Cote, C. A. (1999). RNA sorting in *Xenopus* oocytes and embryos. *FASEB J.* **13,** 435–445.

Mowry, K. L., and Melton, D. A. (1992). Vegetal messenger RNA localization directed by a 340-nt RNA sequence element in *Xenopus* oocytes. *Science* **255,** 991–994.

Mueller-Pillasch, F., Lacher, U., Wallrapp, C., Micha, A., Zimmerhackl, F., Hameister, H., Varga, G., Friess, H., Buchler, M., Beger, H. G., Vila, M. R., Adler, G., and Gress, T. M. (1997). Cloning of a gene highly overexpressed in cancer coding for a novel KH-domain containing protein. *Oncogene* **14,** 2729–2733.

Nielsen, J., Christiansen, J., Lykke-Andersen, J., Johnsen, A. H., Wewer, U. M., and Nielsen, F. C. (1999). A family of insulin-like growth factor II mRNA-binding proteins represses translation in late development. *Mol. Cell Biol.* **19,** 1262–1270.

Norvell, A., Kelley, R. L., Wehr, K., and Schupbach, T. (1999). Specific isoforms of squid, a *Drosophila* hnRNP, perform distinct roles in Gurken localization during oogenesis. *Genes Dev.* **13,** 864–876.

Oh, Y. L., Hahm, B., Kim, Y. K., Lee, H. K., Lee, J. W., Song, O., Tsukiyama-Kohara, K., Kohara, M., Nomoto, A. and Jang, S. K. (1998). Determination of functional domains in polypyrimidine-tract-binding protein. *Biochem. J.* **331,** 169–175.

Perez, I., McAfee, J. G., and Patton, J. G. (1997), Multiple RRMs contribute to RNA binding specificity and affinity for polypyrimidine tract binding protein. *Biochemistry* **36,** 11881–11890.

Perry-O'Keefe, H., Kintner, C., Yisraeli, J., and Melton, D. A. (1990) The use of *in situ* hybridization to study the localization of maternal mRNAs during *Xenopus* oogenesis. *In Situ Hybrid. Study Dev. Differen.* **43,** 115–130.

Pfaff, S. L. and Taylor, W. L. (1992). Characterization of a *Xenopus* oocyte factor that binds to a developmentally regulated cis-element in the TFIIIA gene. *Dev. Biol.* **151,** 306–316.

Putz, U., Skehel, P., and Kuhl, D. (1996). A tri-hybrid system for the analysis and detection of RNA-protein interactions. *Nucleic Acids Res.* **24,** 4838–4840.

Rebagliati, M. R., Weeks, D. L., Harvey, R. P., and Melton, D. A. (1985). Identification and cloning of localized maternal RNAs from *Xenopus* eggs. *Cell* **42,** 769–777.

Ross, A. F., Oleynikov, Y., Kislauskis, E. H., Taneja, K. L., and Singer, R. H. (1997). Characterization of a beta-actin mRNA zipcode-binding protein. *Mol. Cell Biol.* **17,** 2158–2165.

Samuels, M., Deshpande, G., and Schedl, P. (1998). Activities of the Sex-lethal protein in RNA binding and protein:protein interactions. *Nucleic Acids Res.* **26,** 2625–2637.

Schwartz, S. P., Aisenthal, L., Elisha, Z., Oberman, F., and Yisraeli, J. K. (1992). A 69 kDa RNA binding protein from *Xenopus* oocytes recognizes a common motif in two vegetally localized maternal mRNAs. *Proc. Natl. Acad. Sci. USA* **89,** 11895–11899.

Seeger, M. A., and Kaufman, T. C. (1990). Molecular analysis of the bicoid gene from *Drosophila* pseudoobscura: Of the bicoid mRNA. *EMBO J.* **9,** 2977–2987.

SenGupta, D. J., Zhang, B., Kraemer, B., Pochart, P., Fields, S., and Wickens, M. (1996). A three-hybrid system to detect RNA-protein interactions in vivo. *Proc. Natl. Acad. Sci. USA* **93,** 8496–8501.

St. Johnston, D., Brown, N. H., Gall, J. G., and Janstsch, M. (1992). A conserved double-stranded RNA-binding domain. *Proc. Natl. Acad. Sci. USA* **89,** 10979–10983.

St. Johnston, D. (1995). The intracellular localization of messenger RNAs. *Cell* **81,** 161–170.

St. Johnston, D., Driever, W., Berleth, T., Richstein, S., and Nusslein, V. C. (1989). Multiple steps in the localization of bicoid RNA to the anterior pole of the *Drosophila* oocyte. *Development* **107** (Supplement), 13–19.

Stennard, F., Carnac, G., and Gurdon, J. B. (1996). The *Xenopus* T-box gene, Antipodean, encodes a vegetally localized maternal mRNA and can trigger mesoderm formation. *Development* **122,** 4179–4188.

Sundell, C. L., and Singer, R. H. (1990). Actin mRNA localizes in the absence of protein synthesis. *J. Cell Biol.* **111,** 2397–2403.

Sundell, C. L., and Singer, R. H. (1991). Requirement of microfilaments in sorting of actin messenger RNA. *Science* **253,** 1275–1277.

Verity, A. N., and Campagnoni, A. T. (1988). Regional expression of myelin protein genes in the developing mouse brain: In situ hybridization studies. *J. Neurosci. Res.* **21,** 238–248.

Webster, P. J., Liang, L., Berg, C. A., Lasko, P., and Macdonald, P. M. (1997). Translational repressor bruno plays multiple roles in development and is widely conserved. *Genes Dev.* **11,** 2510–2521.

Weeks, D. L., Bailey, C., Bullock, E., Dagle, J., Gururajan, R., Linnen, J., and Longo, F. (1995). mRNAs Localized to the animal hemisphere of *Xenopus laevis* oocytes and early embryos and the proteins that they encode. *In* "Localized RNAs" (H. D. Lipshitz, Ed.,) pp. 173–183. R. G. Landes Company, Austin.

Yisraeli, J. K., and Melton, D. A. (1988). The maternal mRNA Vg1 is correctly localized following injection into *Xenopus* oocytes. *Nature* **336,** 592–595.

Yisraeli, J. K., Sokol, S., and Melton, D. A. (1990). A two-step model for the localization of a maternal mRNA in *Xenopus* oocytes: Involvement of microtubules and microfilaments in translocation and anchoring of Vg1 mRNA. *Development* **108,** 289–298.

Zhang, J., and King, M. L. (1996). *Xenopus* VegT RNA is localized to the vegetal cortex during oogenesis and encodes a novel T-box transcription factor involved in mesodermal patterning. *Development* **122,** 4119–4129.

Zhang, J. Y., Chan, E. K., Peng, X. X., and Tan, E. M. (1999a). A novel cytoplasmic protein with RNA-binding motifs is an autoantigen in human hepatocellular carcinoma. *J. Exp. Med.* **189,** 1101–1110.

Zhang, Q., Yaniv, K., Oberman, F., Wolke, U., Git, A., Fromer, M., Taylor, W. L., Meyer, D., Standart, N., Raz, E., and Yisraeli, J. K. (1999b). Vg1 RBP intracellular distribution and evolutionarily conserved expression at multiple stages during development. *Mech. Dev.* **88,** 101–106.

Zhou, Y., and King, M. L. (1996). RNA transport to the vegetal cortex of *Xenopus* oocytes. *Dev. Biol.* **179,** 173–183.

RNA Localization and Translational Regulation during Axis Specification in the *Drosophila* Oocyte

Ramona L. Cooperstock and Howard D. Lipshitz
Program in Developmental Biology, Research Institute, The Hospital for Sick Children and Department of Molecular & Medical Genetics, University of Toronto, Toronto, Ontario, M5G 1X8 Canada

The major axes of the oocyte—antero-posterior and dorso-ventral—are established over a one-day period during mid-oogenesis in *Drosophila.* The same molecule, GURKEN (GRK), functions to initiate signaling between the oocyte and the surrounding, somatically derived follicle cells. This results first in specification of the antero-posterior axis and, later, the dorso-ventral axis of the oocyte and surrounding follicle cells. Central to specification of both axes is a combination of cytoplasmic localization and translational regulation of the *grk* RNA. Here we discuss the mechanisms by which the *grk* RNA is localized within the oocyte and the role of translational regulation in spatially restricting the production of GRK protein. We then discuss the generality of these mechanisms during oogenesis by focusing on a second transcript, *oskar,* whose function is also regulated through a combination of transcript localization and translational control.

KEY WORDS: *Drosophila,* oocyte, RNA localization, Translational repression, Translational activation, *gurken, oskar.*

I. Introduction

Two posttranscriptional regulatory mechanisms—cytoplasmic RNA localization and translational regulation—have been studied in a variety of organisms (reviewed in Bashirullah *et al.,* 1998; Cooperstock and Lipshitz, 1997; Curtis *et al.,* 1995; Jacobson and Peltz, 1996; Macdonald and Smibert, 1996; Theodorakis and Cleveland, 1996; Wormington, 1994). Many classes of transcripts are subject to one or both of these mechanisms and several of the molecules that constitute the machinery that regulates these processes have been identified. In this review we will focus on RNA localization and translational control in *Drosophila,* since this is the only organism where it is understood how these mechanisms act *in vivo* to control axial specification of the oocyte.

The two bilaterally symmetric *Drosophila* ovaries each consist of 15 to 20 ovarioles. At the anterior tip of the ovariole, germ line stem cells divide asymmetrically to produce a stem cell and a committed cell, which is called a cystoblast (Spradling, 1993). Each cystoblast divides four times with incomplete cytokinesis to form 16 cystocyte cells interconnected by cytoplasmic bridges that run through specialized membrane cytoskeletal structures called ring canals. Only one of the 16 cystocytes is specified to be the oocyte while the remaining 15 become nurse cells. Each 16-cell germarial cyst becomes surrounded by somatically derived follicle cells to form a stage 1 egg chamber. The more posterior part of the ovariole comprises a connected series of progressively older egg chambers ordered such that the youngest is most anterior and the oldest (stage 14) most posterior relative to the body axis of the female. It takes three days for an egg chamber to produce a mature egg.

The focus here is on a one-day period that encompasses stages 6 to 10 of oogenesis. During this period, partly through accumulation of yolk and partly through directed transport of transcripts and proteins from the nurse cells into the oocyte, the oocyte increases in volume roughly a thousandfold. It is during these stages that signals between the oocyte and the surrounding follicle cells determine the antero-posterior and, later, the dorso-ventral axes of the egg chamber and future embryo. The antero-posterior and dorso-ventral axes of the oocyte and future embryo are established by two major signaling events. The molecule at the core of both activities is GURKEN (GRK), a transforming growth factor-α (TGF-α) homolog that acts as a ligand for the epidermal growth factor receptor (EGFR) which is present on the follicle cells that surround the oocyte. The proteins discussed in this review, homologies to known proteins/motifs, and the role of these proteins in axis specification are listed in Table I. *grk* RNA localization and translational regulation provide both spatial and temporal control of GRK protein synthesis, thus restricting activation of EGFR to only a subset of the follicle cells. Here we discuss the transcript localization and translational regulatory mechanisms that act to refine GRK expression, and show how this is crucial for correct axial specification. We then address the generality of these mechanisms by comparing the *grk* transcript localization and translation machinery with that which acts on another key transcript, *oskar,* which also functions to specify aspects of the antero-posterior axis of the oocyte.

II. Localization of *grk* RNA and Its Role in Establishing the Antero-posterior Axis of the Oocyte

A. Overview

In this section we discuss the role of GRK in establishment of the antero-posterior axis of the oocyte. During early oogenesis, *grk* mRNA is localized to the posterior pole of the oocyte (section II.B) where GRK protein is synthesized (section II.C) and signals to neighboring follicle cells to adopt posterior fates (section II.D.1). The product of the *cornichon* (*cni*) gene is required to transmit the signal, and EGFR is required to receive this signal in the follicle cells. An unknown molecule then signals from the posterior polar follicle cells back to the oocyte, specifying the region that receives this signal to be the oocyte's posterior pole (section II.D.2). The microtubule-based cytoskeleton within the oocyte then reorganizes such that the microtubule organizing center (MTOC) moves from the posterior to the anterior of the oocyte (section II.D.3). Subsequently, nuclear migration from the posterior of the oocyte to the future dorso-anterior pole is the first symmetry-breaking event along the dorso-ventral axis (section III.B). This leads to oocyte-to-follicle cell signaling—again mediated by GRK, CNI and EGFR—that specifies the dorso-ventral axis of the egg chamber.

B. *grk* RNA Localization

In the germarium, a cytoskeletal network develops with the microtubule organizing center localized in the oocyte. The MTOC represents the minus-ends of the microtubules. The microtubule network extends from the oocyte through the ring canals into the nurse cells (Cooley and Theurkauf, 1994). By stage 6 of oogenesis, the MTOC is localized at the posterior of the oocyte that resides at the posterior of the egg chamber. The mechanisms by which the oocyte moves to the posterior of the egg chamber will not be discussed here but have recently been elaborated (Godt and Tepass, 1998; González-Reyes and St. Johnston, 1998).

Specification of the antero-posterior (A-P) axis of the oocyte is initiated by localization of the *grk* transcript to the posterior pole of the oocyte during stages 1 to 7 of oogenesis (Fig. 1A) (Neuman-Silberberg and Schüpbach, 1993). There is some controversy as to the site of synthesis of the *grk* RNA that is localized to the posterior of the oocyte at these stages. In one analysis it has been suggested that *grk* is transcribed exclusively in the oocyte nucleus from where it is transported to the oocyte's posterior

TABLE I
Proteins that Function in Oocyte Axis Specification

Protein	Homology	Role	Reference
Apontic	RNA binding protein	translational repression	Lie and Macdonald, 1999
Aubergine	?	translational activation	Wilson *et al.,* 1996
Bicaudal-D	filamentous protein	oocyte determination microtubule formation nuclear migration	Mach and Lehmann, 1997; Swan *et al.,* 1999; Swan and Suter, 1996
Bruno (Arrest)	RNA binding protein	translational repression	Kim-Ha *et al.,* 1995; Webster *et al.,* 1997
Cornichon	–	cell-cell signaling	González-Reyes *et al.,* 1995; Roth *et al.,* 1995
Deadlock	?	oocyte nuclear DNA condensation	Ghabrial *et al.,* 1998
Delta	Notch ligand	cell-cell signaling	González-Reyes and St. Johnston, 1998; Ruohola *et al.,* 1991
Dicephalic	?	oocyte position	González-Reyes and St. Johnston, 1998
DLis-1	human Lissencephaly-1	oocyte identity nuclear migration	Swan *et al.,* 1999
Egalitarian	*H. influenzae* RNaseD	oocyte determination nuclear migration microtubule maintenance	Mach and Lehmann, 1997
EGFR	Epidermal growth factor receptor (Grk receptor)	cell-cell signaling	González-Reyes *et al.,* 1995; Neuman-Silberberg and Schüpbach, 1993; Roth *et al.,* 1995
Encore	?	germ line mitosis oocyte identity translational activation	Hawkins *et al.,* 1996; Hawkins *et al.,* 1997
Gurken	EGFR ligand (TGF-α)	cell-cell signaling	González-Reyes *et al.,* 1995; Neuman-Silberberg and Schüpbach, 1993; Roth *et al.,* 1995
K10	–	nuclear retention of hnRNPs (SQD). translational repression	Kelley, 1993; Norvell *et al.,* 1999

Maelstrom	?	RNA localization microtubule reorganization	Clegg *et al.*, 1997
Mago nashi	–	microtubule reorganization RNA localization	Micklem *et al.*, 1997; Newmark *et al.*, 1997
Notch	Delta receptor	cell-cell signaling	González-Reyes and St. Johnston, 1998; Ruohola *et al.*, 1991
Okra	ATP dependent DNA helicase (DmRad54)	meiotic checkpoint translational activation	Ghabrial *et al.*, 1998; Ghabrial and Schüpbach, 1999
Orb	RNA binding CPEB homolog	RNA localization translational activation	Chang *et al.*, 1999; Christerson and McKearin, 1994
Oskar	–	posterior pole plasm formation	Ephrussi *et al.*, 1991; Ephrussi and Lehmann, 1992; Smith *et al.*, 1992
Protein kinase A	kinase	cAMP signaling	Lane and Kalderon, 1994
Spn-A	?	translational activation	González-Reyes *et al.*, 1997
Spn-B	DMC/RAD51-like	meiotic checkpoint translational activation	Ghabrial *et al.*, 1998; Ghabrial and Schüpbach, 1999; González-Reyes *et al.*, 1997
Spn-C	Mus301	meiotic checkpoint translational activation	Ghabrial and Schüpbach, 1999; González-Reyes *et al.*, 1997
Spn-D	?	translational activation	Ghabrial *et al.*, 1998; González-Reyes *et al.*, 1997
Spn-E/Homeless	DE-H ATP-dependent RNA helicase	RNA localization translational activation	Gillespie and Berg, 1995; González-Reyes *et al.*, 1997
Squash	?	oocyte nuclear DNA condensation	Ghabrial *et al.*, 1998
Squid A	hnRNP	translational repression	Kelley, 1993; Norvell *et al.*, 1999
Squid S	hnRNP	nuclear export of RNA translational repression	Kelley, 1993; Norvell *et al.*, 1999
Staufen	double-stranded RNA binding protein	RNA localization	Ferrandon *et al.*, 1994
Vasa	DEAD-box RNA helicase	RNA localization translational activation	Styhler *et al.*, 1998; Tinker *et al.*, 1998; Tomancak *et al.*, 1998
Zucchini	?	oocyte nuclear DNA condensation	Ghabrial *et al.*, 1998

–, no striking molecular homology or protein motifs; ?, molecular identity unknown.

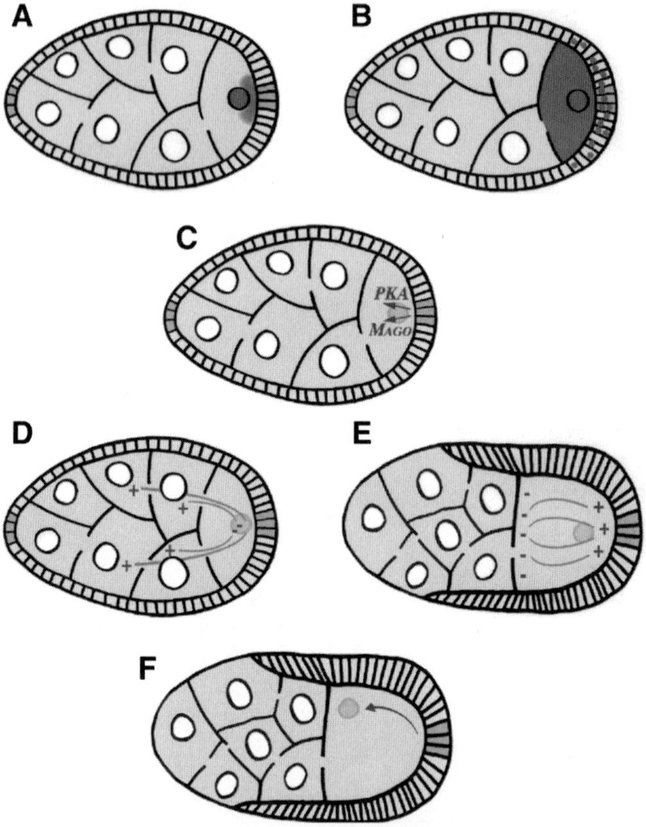

FIG. 1 Outline of events during A-P axis establishment. (A–C): Stage 6 egg chambers. (A) *grk* RNA (dark grey) is localized to the posterior pole of the oocyte. (B) GRK protein (dark grey shading) binds EGFR and the GRK-EGFR complex is internalized by adjacent follicle cells (dark grey dots). Only a subset of these follicle cells (the polar follicle cells) is competent to respond to the GRK signal. Main body follicle cells are unable to respond. (C) A posteriorizing signal is relayed from the polar follicle cells to the oocyte. PKA and MAGO are required in the oocyte to relay the signal. (D) Stage 6 egg chamber. The MTOC is at the posterior pole of the oocyte. Microtubules are oriented with their (+) ends in the nurse cells and their (−) ends at the MTOC. (E) Stage 8 egg chamber. The posteriorizing signal results in reorganization of the microtubule cytoskeleton. The posterior MTOC has disappeared, while a new MTOC is nucleated at the anterior of the oocyte. (F) Stage 8 egg chamber. Upon reorganization of the microtubule cytoskeleton, the oocyte nucleus migrates from the posterior pole to the presumptive dorso-anterior corner of the oocyte.

(Saunders and Cohen, 1999). These conclusions were based on the observation that *grk-βgal* transgenic transcripts driven by the *grk* promoter, and thought to be lacking nurse cell-to-oocyte transport signals, are detectable only in the oocyte. This suggests that the *grk* promoter directs transcription in the oocyte nucleus. Further, eggs from colchicine-fed flies show accumulation of *grk* RNA in the oocyte (colchicine inhibits microtubule polymerization and microtubules are required for nurse cell-to-oocyte transport).

A recent analysis suggests that at the early stages of oogenesis *grk* RNA is synthesized in both the nurse cell nuclei and the oocyte nucleus (Thio *et al.,* 2000). These conclusions derive from use of reporter transgenes carrying the endogenous *grk* promoter and all or part of the *grk* transcribed region in the absence of heterologous 5′ or 3′UTR sequences. [The first set of studies, described above (Saunders and Cohen, 1999), included heterologous sequences from the *K10* transcript that were incapable—when assayed alone—of directing transport from the nurse cells into the oocyte (Serano and Cohen, 1995).] Thio *et al.* showed that deletion of sequences in the 5′-region of the *grk* open reading frame results in low-level accumulation of transcripts in the oocyte while most RNA remains in the nurse cells at stages 1 to 7 of oogenesis. Hence at these stages *grk* RNA is indeed transcribed in nurse cells and this element is required for nurse cell-to-oocyte transport. The element is likely to be a functional RNA element, since translation is not required for transport (Saunders and Cohen, 1999; Thio *et al.,* 2000). It is unclear why *grk* transcripts do not accumulate in the nurse cells of colchicine-treated egg chambers (Saunders and Cohen, 1999). It is possible that endogenous *grk* transcripts are quite unstable and that accumulation in the nurse cells requires stabilization, perhaps by the *β-gal* tag used as a reporter (Thio *et al.,* 2000).

Whatever the source of the *grk* transcripts in stage 1–7 oocytes, it is clear that *grk* transcripts are enriched at the posterior pole (Fig. 1A) (Neuman-Silberberg and Schüpbach, 1993). *grk* transcript localization at the posterior of the oocyte is absolutely required for establishment of normal A-P polarity. For example, if *grk* RNA is displaced laterally or anteriorly, as in stage 3-6 *maelstrom* (*mael*) oocytes, the polar follicle cells fail to adopt posterior fates and this alters the overall polarity of the oocyte (Clegg *et al.,* 1997).

C. Grk Translation

Grk protein can be detected throughout the ooplasm at stages 1 to 6 of oogenesis (Neuman-Silberberg and Schüpbach, 1996; Peri *et al.,* 1999; Queenan *et al.,* 1999), and at stages 4 to 6 Grk is also found internalized into all of the follicle cells that abut the posterior portion of the oocyte (Fig. 1B) (Peri *et al.,* 1999; Queenan *et al.,* 1999). RNA localization may be required to target *grk* transcripts to the endoplasmic reticulum (ER) at the posterior of the oocyte, hence directing cell surface expression of Grk protein. Consistent with this, *grk* transcripts are found in close association with the ER at these stages (Saunders and Cohen, 1999). GRK protein is internalized into both the posterior polar follicle cells and the main body follicle cells (Peri *et al.*, 1999; Queenan *et al.*, 1999) (Fig. 1B). Prior differentiation of these two populations of follicle cells results in only the polar follicle cells responding to the GRK signal (González-Reyes and St. Johnston, 1998) (see II.D.1 below).

Several mutants result, among other defects, in production of double-anterior bipolar oocytes. These include *aubergine, deadlock, okra, spindle (A-E), squash, vasa* and *zucchini* (reviewed in van Eeden and St. Johnston, 1999). These axial abnormalities are likely to be caused, at least in part, by defects in translational regulation of *grk* mRNA resulting in a reduction in GRK protein production and thus in impairment of the oocyte-to-follicle cell signal. For example, as discussed below (section III.G) VASA is a DEAD-box RNA helicase that has been implicated in translational control of *grk* RNA (Styhler *et al.*, 1998; Tinker *et al.*, 1998; Tomancak *et al.*, 1998; Webster *et al.*, 1997). Similarly, SPINDLE-E/HLS resembles an RNA helicase and may also exert effects on translation (Gillespie and Berg, 1995).

Other mutants that result in GRK signaling defects and bipolar oocytes may affect translation more indirectly. For example, OKRA, SPINDLE-B (SPN-B), and SPN-C/MUS301 proteins are related to yeast proteins that regulate double-strand DNA break repair (Ghabrial *et al.*, 1998; Ghabrial and Schüpbach, 1999). In these mutants defective meiosis results from failure to complete recombination. It has been shown that *grk* translation is inhibited as an indirect consequence of activation of a meiotic checkpoint that controls translation of *grk* RNA via posttranslational regulation of the VASA protein (Ghabrial and Schüpbach, 1999).

D. Response to the GRK Signal

1. Posterior Follicle Cell Fate Determination

GRK signaling from the oocyte directs the posterior terminal follicle cells to adopt posterior rather than anterior fates. In the absence of GRK signaling, such as in mutants for GRK (the ligand), EGFR (the receptor), or CNI (whose molecular nature is not yet known), posterior follicle cell fate specification fails. This results in egg chambers in which both the anterior- and posterior-polar follicle cells adopt anterior fates, hence producing mirror-symmetric double-anterior oocytes (González-Reyes *et al.*, 1995; Roth *et al.*, 1995). Therefore, the default fate of the polar follicle cells in the absence of the GRK signal is anterior.

To determine which follicle cells are capable of responding to the GRK signal, an enhancer trap that labels posterior-fated follicle cells serves as a useful marker. Expression of the enhancer trap was examined in *dicephalic* (*dic*) mutant egg chambers in which the oocyte is not properly positioned within the egg chamber (González-Reyes and St. Johnston, 1998). When the oocyte is mispositioned at the anterior, the enhancer trap is expressed in the anterior follicle cells. This demonstrates that, prior to the GRK signal, anterior and posterior polar follicle cell populations are equipotent; each is capable of receiving and responding to the GRK signal (González-Reyes and St. Johnston, 1998). Furthermore, when the oocyte is mispositioned to the center of the egg chamber and thus contacts the main body rather than the polar follicle cells, the main body cells do not adopt polar fates. Therefore, only polar follicle cells are capable of responding to the GRK signal.

The details of the cell fate specification mechanisms that direct polar follicle cells to adopt posterior fates in response to the GRK signal have not yet been elucidated. However, it should be noted that the NOTCH-DELTA (N-DL) system has been shown to function in the polar follicle cells and to be necessary for their adoption of polar versus main body fates (González-Reyes and St. Johnston, 1998; Ruohola *et al.,* 1991). EGFR signaling and N-DL function together in cell fate specification in other equipotent populations of cells. For example, fates of vulval precursor cells in *C. elegans* are specified by a combination of TGF-α/EGFR and N-DL signaling (Sternberg, 1993).

2. Signaling Back from Follicle Cells to the Oocyte

Specification of the posterior polar follicle cells is followed by reciprocal signaling back to the oocyte from these cells (Fig. 1C). The nature of this signal remains unknown. However, protein kinase A (PKA) functions within the oocyte to transduce this signal (Lane and Kalderon, 1994). Since PKA functions in cyclic AMP-mediated signaling, it is possible that cyclic AMP levels are upregulated in the oocyte through an as yet unidentified receptor that receives the signal from the posterior polar follicle cells (Lane and Kalderon, 1994).

3. Readout of the Posterior Signal in the Oocyte: Microtubule Reorganization

Transduction of the follicle cell-to-oocyte signal directs microtubule reorganization within the oocyte (Figs. 1D, E). Consequently, at stages 7 to 10, the posterior localized MTOC gradually disappears while a new MTOC is nucleated at the anterior of the oocyte (Theurkauf *et al.,* 1992). Two *lacZ*-

fusion transgenes have been very useful in assaying the orientation of the microtubule-based cytoskeleton in the oocyte. One of these encodes a KINESIN-βGAL fusion protein that is translocated to the plus-ends of the microtubules (Clark *et al.,* 1994); the other encodes a NOD-βGAL fusion protein that is translocated to their minus-ends (Clark *et al.,* 1997). Upon reorganization of the cytoskeleton, KINESIN-βGAL is localized to the posterior pole while NOD-βGAL localizes to the anterior.

Possibly as a direct consequence of this reorganization, the oocyte nucleus begins to move from the posterior to the anterior (Fig. 1F and see III.B below). In addition, certain classes of transcripts begin to localize to either the anterior or posterior pole (Bashirullah *et al.,* 1998). Some of these RNAs encode key instructive proteins that function in axis specification either in the oocyte (e.g., *oskar*) or in the embryo (e.g., *bicoid*). For example, *oskar* transcripts localize to the posterior and play a key role during oogenesis in nucleating formation of the polar granules and in localization of other transcripts (see section IV). *bicoid* transcripts localize to the anterior pole and encode the anterior cell fate determinant in the early embryo. Transport of *bcd* and *osk* RNAs is microtubule dependent, as localization of these RNAs is disrupted by addition of microtubule depolymerizing drugs (Pokrywka and Stephenson, 1991).

Genes that function within the oocyte and are required for proper microtubule reorganization include *mago nashi* (*mago*) (Micklem *et al.,* 1997; Newmark *et al.,* 1997) and *PKA* (Lane and Kalderon, 1994) (Fig. 1C). In mutants that are defective in either of these genes, the posterior MTOC does not disappear but a new MTOC forms at the anterior. Consequently, the minus-ends of microtubules come to reside at both the anterior and posterior of the oocyte while the plus-ends cluster centrally. This ultimately produces mirror image double-anterior embryos because transcripts encoding the anterior determinant, BICOID are present at both poles while transcripts encoding the posterior determinant, OSKAR, are centrally positioned.

III. Localization of *grk* RNA and Its Role in Establishing the Dorso-Ventral Axis of the Oocyte

A. Overview

During stages 7 to 8 of oogenesis the oocyte nucleus moves from the posterior to an off-axis position in the anterior cortical rim of the oocyte (section III.B). There it synthesizes *grk* RNA, which in turn is localized to the presumptive dorso-anterior corner of the oocyte (sections III.C–E). Localized production and secretion of GRK is achieved through a combination of translational repression of *grk* transcripts that reside outside the dorso-anterior region and translational activation of those within this region (sections IIIF, G). GRK signals through EGFR to the overlying dorso-anterior follicle cells and instructs them to adopt dorsal fates.

B. Migration of the Oocyte Nucleus

The assembly of the anterior MTOC within the oocyte initiates at stage 7 of oogenesis (Fig. 1D, E) (Theurkauf *et al.*, 1992). At this stage the oocyte and surrounding follicle cells possess an A-P axis but are radially symmetric relative to this axis (i.e., no dorso-ventral axis exists). The first symmetry-breaking event in dorso-ventral (D-V) axis formation is migration of the oocyte nucleus from the posterior pole of the oocyte to an off-axis anterior cortical location (Fig. 1F). Several of the molecules involved in migration and localization of the nucleus have been identified, including BICAUDAL-D, EGALITARIAN, and DLIS-1 (Mach and Lehmann, 1997; Swan *et al.*, 1999; Swan and Suter, 1996). DLIS-1 is the *Drosophila* homolog of human LISSENCEPHALY-1, which may act as a cortical anchor for nuclear-localization factors (Swan *et al.*, 1999).

The particular position reached by the oocyte nucleus within the anterior perimeter is random (Ray and Schüpbach, 1996; Roth *et al.*, 1999). This has been elegantly demonstrated through analysis of binuclear oocytes (Roth *et al.*, 1999). In such oocytes, the relative position of the two nuclei within the anterior perimeter is random. Strikingly, *grk* RNA is localized to the cytoplasm adjacent to each nucleus and two D-V axes are specified (Roth *et al.*, 1999). Hence, there is no preexisting D-V axis within the oocyte or egg chamber. The specific off-axis position of the nucleus determines the D-V axis of the oocyte and, through GRK, signals the overlying follicle cells to adopt dorsal fates.

As nuclear migration is dependent on proper A-P specification, formation of the D-V axis is secondary to A-P axis formation. For example, in *mago* mutants, an MTOC is present at both the anterior and posterior poles of the oocyte. The oocyte nucleus may consequently be mislocalized to the posterior pole, resulting in coincidence of A-P and D-V axes (Micklem *et al.*, 1997; Newmark *et al.*, 1997).

C. *gurken* RNA Localization Dynamics

Although there is controversy regarding the source of *grk* RNA at stages 1 to 6 of oogenesis, there is agreement that subsequent synthesis of *grk* transcripts occurs in the oocyte nucleus (Fig. 2, see color insert) (Saunders and Cohen, 1999; Thio *et al.*, 2000). *grk* is the only example of a specific transcript that is synthesized in the oocyte nucleus; all other RNAs analyzed to date are synthesized in the nurse cell nuclei. Transcription in the oocyte nucleus at stage 9 was first detected using the low-sensitivity method of *in vivo* labeling with radioactive nucleotides (Mahowald and Tiefert, 1970); it is therefore likely that other classes of transcripts are also synthesized there at these stages.

As the oocyte nucleus migrates anteriorly, the concentration of *grk* RNA in the posterior decreases while levels increase at the anterior. The *grk* RNA that was localized to the posterior pole at earlier stages may accompany the nucleus to the anterior or may be degraded. Since transcription of *grk* has already initiated in the oocyte nucleus, at least a subset of the anterior *grk* transcripts must be newly synthesized (Saunders and Cohen, 1999; Thio *et al.*, 2000). The cortical arc of *grk* RNA that extends from the posterior of the oocyte to the leading edge of the nucleus may be formed by transcripts that were previously posterior localized and accompany the nucleus to the anterior (Saunders and Cohen, 1999). Alternatively, they may comprise newly synthesized transcripts that are exported from the nucleus as it migrates (Saunders and Cohen, 1999). By the end of stage 7, *grk* RNA forms a cortical ring around the anterior of the oocyte (Neuman-Silberberg and Schüpbach, 1993; Saunders and Cohen, 1999; Thio *et al.*, 2000). By stage 9, *grk* RNA is concentrated in a cap that overlies the nucleus at the dorso-anterior corner of the oocyte (Neuman-Silberberg and Schüpbach, 1993; Saunders and Cohen, 1999; Thio *et al.*, 2000). The transition from the anterior cortical ring to the dorso-anterior cap involves relocalization of *grk* transcripts and/or degradation of transcripts in the more lateral and ventral regions of the anterior cortex (Neuman-Silberberg and Schüpbach, 1993; Saunders and Cohen, 1999; Thio *et al.*, 2000).

A region within the 5′ portion of the *grk* open reading frame contains sequences that direct localization of *grk* transcripts in the anterior cortical ring (Thio *et al.*, 2000). Since translation is not required for *grk* RNA transport and localization (Saunders and Cohen, 1999; Thio *et al.*, 2000), this must be a functional RNA element. In addition, the *grk* 5′UTR may contain an element that is necessary for transcript stability (Saunders and Cohen, 1999). The *grk* 3′UTR is necessary for the redistribution of *grk* transcripts from the anterior cortical ring to the dorso-anterior of the oocyte (Saunders and Cohen, 1999; Thio *et al.*, 2000).

D. Export of *grk* RNA from the Oocyte Nucleus

To date no proteins have been identified that are required specifically for export of *grk* RNA from the oocyte nucleus. Two mutants with D-V axial defects—*squid* (*sqd*) and *K10*—encode proteins that accumulate in the oocyte nucleus, and Sqd has been shown to bind *grk* RNA directly (Norvell *et al.*, 1999). In both *K10* and *sqd* mutants, *grk* transcripts are exported to the cytoplasm and are found in the anterior cortex but do not localize to the dorso-anterior pole of the oocyte (Norvell *et al.*, 1999). Consistent with this, in *K10* or *sqd* single-mutants, as well as in the double-mutant, the egg chamber is dorsalized (Kelley, 1993). Hence, *grk* RNA is exported to the cytoplasm in these mutants but is not correctly localized and/or translationally regulated therein (see III.E and III.G).

E. Localization of *grk* Transcripts in the Cytoplasm

Drug inhibition experiments have been used to define which cytoskeletal structures function in localization of *grk* transcripts once they have been exported from the oocyte nucleus (Saunders and Cohen, 1999). Disruption of microfilaments (with cytochalasin D) has no effect on *grk* RNA localization. However, disruption of the microtubules (with colchicine) severely affects *grk* transcript localization. In colchicine-treated oocytes, *grk* transcripts remain in a few large clumps on the surface of the oocyte nucleus rather than translocating to the anterior cortical ring or the dorso-anterior of the oocyte.

Several mutants have been shown to affect *grk* transcript localization to the dorso-anterior corner of the oocyte (Fig. 2). In *K10, sqd, orb, okra, vas,* and *spn-A, B, C, D,* and *E/hls* mutants, while *grk* transcripts localize to the anterior at stages 7 to 8, the transcripts do not relocalize to the dorso-anterior region at stage 9 (Christerson and McKearin, 1994; González-Reyes *et al.,* 1997; Neuman-Silberberg and Schüpbach, 1993; Norvell *et al.,* 1999; Tinker *et al.,* 1998). The transcript localization phenotype observed in these mutants resembles that exhibited by transgenic *grk* transcripts that lack the 3′UTR, thus suggesting that the 3′UTR is essential for relocalization to the dorso-anterior (Saunders and Cohen, 1999; Thio *et al.,* 2000). Whether the proteins encoded by these loci act specifically in transcript localization within the cytoplasm or whether their role in localization is indirect (i.e., through effects on translational control or RNA stability) has not yet been determined definitively. It should be noted that these possibilities are not mutually exclusive; a protein may act both in RNA localization and in translational control. One such example may be the RNA-binding protein, Orb (Christerson and McKearin, 1994; Lantz *et al.,* 1992, 1994), which is required for tight dorso-anterior *grk* localization (Neuman-Silberberg and Schüpbach, 1996). Since Orb protein is present throughout the oocyte cortex (Christerson and McKearin, 1994; Lantz *et al.,* 1994), if Orb is an anchor for *grk* transcripts it must function along with other factors that confer spatial specificity to the anchoring. ORB also functions in translational activation of *grk* transcripts (see III.G.2).

F. GRK Protein Expression

Although *grk* transcripts are present in an anterior cortical ring at stages 7 to 8, GRK protein is found only in the dorso-anterior region of the oocyte (Neuman-Silberberg and Schüpbach, 1996; Peri *et al.,* 1999; Queenan *et al.,* 1999). At these stages, GRK protein is present in the dorso-anterior region of the oocyte, as well as in the overlying follicle cells in a punctate distribution that is presumed to represent protein internalized as a result of signaling through EGFR. Hence, prior to stage 9, the GRK protein distribution in the oocyte is more restricted than the *grk* RNA distribution, thus implicating translational control (in particular, translational repression of *grk* transcripts that reside outside of the dorso-anterior region of the oocyte). At stages 9 to 12 the *grk* RNA is tightly localized to the dorso-anterior corner of the oocyte (Fig. 2) (Neuman-Silberberg and Schüpbach, 1993), while GRK protein is present at stages 10 to 12 in an anterior-to-posterior stripe that extends over half the length of the dorsal midline of the oocyte (Neuman-Silberberg and Schüpbach, 1996). The distribution of GRK protein at these stages can be explained either by diffusion from the site of synthesis or by growth of the oocyte membrane at the anterior as GRK is being synthesized, thus displacing GRK posteriorly (Neuman-Silberberg and Schüpbach, 1996).

G. Translational Regulation of *grk* Transcripts

When *grk* transcripts enter the cytoplasm from the oocyte nucleus and are transported to the anterior, they are translationally regulated (Fig. 2). This is exemplified by the fact that although *grk* transcripts are present in an anterior cortical ring in wild-type oocytes at stages 7–8, GRK protein is only expressed at the dorso-anterior corner of the oocyte as discussed above. There is evidence that two mechanisms function in translational regulation of *grk:* repression and activation.

1. Translational Repression of *grk* Transcripts

Several gene products have been implicated in translational repression of *grk* transcripts, including K10, SQD, and BRUNO (Fig. 2). The *sqd* locus encodes a family of three protein isoforms—SQDA, B and S—related to vertebrate hnRNP proteins, which have been shown to shuttle RNAs from the nucleus to the cytoplasm (Kelley, 1993; Norvell *et al.*, 1999). As discussed above, in *sqd* mutants export of *grk* RNA from the oocyte nucleus occurs at stage 7, but *grk* RNA does not relocalize from the anterior cortical ring to the dorso-anterior corner of the oocyte. It has been shown that two of the SQD protein isoforms—SQDS and SQDA—function in posttranscriptional regulation of *grk* transcripts (Norvell *et al.*, 1999). One of these, SQDS, is localized to the oocyte nucleus (Fig. 2) (Norvell *et al.*, 1999). Nuclear accumulation of SQDS fails in *K10* mutants, thus implicating the K10 protein (which is also nuclear) in nuclear localization of SQDS. K10 and SQD have been shown to interact directly (Norvell *et al.*, 1999). Therefore, it is likely that SQDS is brought to the nucleus, or stabilized therein, in association with K10. Immunoprecipitation experiments have shown that SQD protein binds to the *grk* 3′UTR, but have not yet distinguished which specific SQD isoforms bind (Norvell *et al.*, 1999). It is likely, however, that both SQDS and SQDA bind to the *grk* 3′UTR. Since SQDS is not found in the cytoplasm, when *grk* transcripts are exported from the nucleus, SQDS must dissociate from the *grk* transcript (Fig. 2).

SQD proteins interact with BRUNO (BRU) (Norvell *et al.*, 1999), an RNA-binding protein that has been shown to translationally repress the *osk* RNA (Kim-Ha *et al.*, 1995; Webster *et al.*, 1997) (see IV below). Interestingly, the *grk* 3′UTR contains one or more BRU response elements (BREs), which can be bound by BRU protein (Kim-Ha *et al.*, 1995). A function for the BRE in the *grk* 3′UTR has not yet been reported. However, BREs have been shown to be essential for BRU binding to, and translational repression of, *osk* transcripts (Kim-Ha *et al.*, 1995; Webster *et al.*, 1997) (see section IV.C).

BRU protein is concentrated at the anterior of the oocyte at stages 7 to 8 (Webster *et al.*, 1997); hence, the spatial distribution of BRU is consistent with a role in translational repression of *grk* transcripts that are present in the anterior cortical ring. It has been proposed that a major role of SQDS is to recruit BRU to the *grk* 3′UTR as *grk* transcripts enter the cytoplasm, thus directing translational repression of *grk* in the anterior (Norvell *et al.*, 1999). It is likely that in the cytoplasm, SQDA replaces SQDS on the *grk* 3′UTR and translationally represses *grk* transcripts in association with BRU (Fig. 2) (Norvell *et al.*, 1999).

Genetic phenotypes and the GRK protein expression data are consistent with the postulated roles of SQD, K10, and BRU. *K10* and *sqd* mutant eggs are dorsalized rather than ventralized. This is a consequence of translation of GRK throughout the anterior cortex rather than just at the dorsal anterior (Neuman-Silberberg and Schüpbach, 1996; Serano *et al.*, 1995). In terms of the model (Fig. 2), in *K10* mutants, SQDS would not be retained in the oocyte nucleus and thus would not be recruited to the *grk* transcript (Norvell *et al.*, 1999). As a result, the translational repressor BRU would not be recruited by SqdS as *grk* transcripts enter the cytoplasm and so *grk* transcripts would be translated throughout the anterior cortex. This would lead to Grk signaling to ventrally located follicle cells, causing them to adopt dorsal fates. Similarly, in *sqd* mutants, Bru would not be recruited to the *grk* transcript, again abrogating translational repression in the anterior cortex. Consistent with this, the dorsalized egg phenotype of both *sqd* and *K10* is enhanced in *sqd; K10* double-mutants (Kelley, 1993).

As mentioned above, in *sqd* and *K10* mutants, *grk* transcripts do not relocalize from the anterior cortex to the dorsal anterior pole. One possible explanation for this is that in wild-type oocytes, translational repression of anterior *grk* transcripts leads to their destabilization (Saunders and Cohen, 1999). Precedent for this comes from analyses of *nanos* transcripts in the early embryo (Bashirullah *et al.,* 1999; Dahanukar and Wharton, 1996; Smibert *et al.,* 1996). Reciprocally, translational activation of *grk* transcripts in the dorso-anterior may lead to their stabilization (see III.G.2).

2. Translational Activation of *grk* Transcripts

Relief of translational repression of *grk* transcripts by Bru at the dorso-anterior pole of the oocyte is likely to be accomplished by several proteins, some of which are RNA binding proteins and a subset of these may also be translational activators. These include Orb (an RNA-binding protein; Christerson and McKearin, 1994; Lantz *et al.,* 1992, 1994), Vasa (a DEAD-box RNA helicase; Hay *et al.,* 1988; Lasko and Ashburner, 1988; Liang *et al.,* 1994), Spn-E/Hls (a DE-H-box putative RNA helicase; Gillespie and Berg, 1995), Encore (Hawkins *et al.,* 1997), Okra (Ghabrial *et al.,* 1998; Ghabrial and Schüpbach, 1999), and Spn-B, C, and D (Ghabrial *et al.,* 1998; Ghabrial and Schüpbach, 1999; González-Reyes *et al.,* 1997). Evidence in support of a role for these proteins in translational activation at the dorsal anterior again comes from a combination of genetic and protein expression analyses.

In *okra* and *spn-B, C,* and *D* mutants, Grk protein expression is reduced as expected for translational activators (Ghabrial *et al.,* 1998; Ghabrial and Schüpbach, 1999; González-Reyes *et al.,* 1997). Hence, dorsal fates cannot be specified and so the egg chambers are ventralized (Ghabrial *et al.,* 1998; González-Reyes *et al.,* 1997). If the role of the translational activators is to relieve Bru-mediated translational repression at the dorso-anterior, it would be expected that, in the absence of Bru, no translational activators would be required. This predicts that *K10* double-mutants with *spn-B, C,* or *D* would result in restoration of Grk expression since in *K10* mutants no Bru is recruited to the *grk* RNA and so no Spn-mediated activation would be required. Consistent with this, in *K10; spn* double-mutants, Grk protein expression is similar to that in *K10* alone and the resulting eggs are dorsalized (González-Reyes *et al.*, 1997). As discussed previously (II.C), OKRA, SPN-B, and SPN-C are likely to function indirectly in translational activation through posttranslational regulation of the VASA protein (Ghabrial *et al.*, 1998; Ghabrial and Schüpbach, 1999), whose role is described below. It should be noted that in *spn* mutants, since the *grk* transcripts present in the anterior cortical ring are translationally repressed but are stable, it is possible that transcript destabilization is unlinked from translational repression as has been suggested by studies in the early embryo (Bashirullah *et al.*, 1999).

vasa (*vas*) mutants produce eggs with both A-P (double anterior) and D-V (ventralization) defects that are consistent with a role for VASA in translation of *grk* during its earlier role in A-P specification as well as during D-V signaling (Styhler *et al.*, 1998; Tinker *et al.*, 1998; Tomancak *et al.*, 1998). In *vas* mutants GRK protein levels are severely reduced throughout oogenesis, consistent with a role for VAS in *grk* translational activation (Styhler *et al.*, 1998; Tinker *et al.*, 1998). It is possible that VAS is required to relieve BRU-mediated repression of *grk* translation. Consistent with this, VAS and BRU proteins have been shown to interact (Webster *et al.*, 1997). Furthermore, in the absence of BRU-mediated repression, VAS is no longer required: *K10; vas* double-mutants (in which the BRU translational repressor is not recruited to the *grk* transcript) are dorsalized, indicating that GRK is synthesized despite the reduction of VAS levels (Tinker *et al.*, 1998).

Additional proteins that may function in translational activation of *grk* are ORB (Christerson and McKearin, 1994; Lantz *et al.*, 1992, 1994), SPN-E/HLS (Gillespie and Berg, 1995), and the product of the *encore* (*enc*) locus (Hawkins *et al.*, 1996, 1997). In *orb, spn-E/hls,* and *enc* mutants, GRK protein levels are extremely reduced (González-Reyes *et al.*, 1997; Hawkins *et al.*, 1997; Neuman-Silberberg and Schüpbach, 1996), resulting in ventralized egg shells (Christerson and McKearin, 1994; Gillespie and Berg, 1995; Hawkins *et al.*, 1997; Roth and Schüpbach, 1994). There are no *grk* transcript localization defects in *enc* mutants, implicating it unambiguously in translational activation. However, ORB and SPN-E/HLS are also required for tight dorso-anterior *grk* localization (Gillespie and Berg, 1995; Neuman-Silberberg and Schüpbach, 1996). Therefore these proteins may function in both RNA localization and translational activation (Fig. 2).

A *cis*-element in the *grk* 5′UTR, GLE2 (*grk* localization element 2) has been implicated in translational activation (Saunders and Cohen, 1999). Transcripts that lack GLE2 are not translated and are also unstable at stage 8 (Saunders and Cohen, 1999). Proteins that bind GLE2 have not yet been identified. In addition, it is not yet known whether translational repression and transcript instability are mediated by the same element or whether

GLE2 contains two separable elements that independently mediate these processes.

H. Model for *grk* RNA Localization and Translational Control

The following hypothetical model (Fig. 2) summarizes all of the above data: The oocyte nucleus migrates to the dorso-anterior corner of the oocyte at stages 7–8. Within the nucleus, *grk* RNA is newly transcribed. K10 mediates SQDS retention in the oocyte nucleus, promoting SQDS's association with *grk* transcripts via their 3′UTR. As *grk* RNA bound to SQDS is exported from the nucleus, SQDS is replaced by SQDA and the cytoplasmic translational repressor BRU is recruited. The BRU-SQDA complex prevents translation of those *grk* transcripts that are outside of the dorso-anterior region of the oocyte. ORB and SPN-E/HLS are required for proper *grk* RNA localization to this region. Once *grk* transcripts are localized to the dorso-anterior of the oocyte, translational activators such as VAS and, possibly, ORB and SPN-E/HLS relieve BRU-mediated translational repression. OKRA, SPN-B, and SPN-C are required to posttranslationally modify VAS, thus enabling it to act as a translational activator. High levels of GRK protein are thus synthesized and subsequently translocated to the exterior of the oocyte at the dorso-anterior, where GRK binds and activates EGFR on the adjacent follicle cells. These cells internalize the GRK-EGFR complex, initiating an intracellular signaling cascade that confers dorsal fate on them. Since the focus of this review is on RNA localization and translational control during oogenesis, the details of subsequent cell-cell signaling within the follicle cell layer as well as the readout of this process as the D-V axis of the embryo will not be described here. Several recent reviews cover this subject (Nilson and Schüpbach, 1999; Ray and Schüpbach, 1996; Schweitzer and Shilo, 1997; van Eeden and St. Johnston, 1999).

IV. Generality of RNA Localization and Translational Control Mechanisms during Oogenesis: *oskar* RNA

A. Overview

The primary focus of this review has been on how *grk* RNA localization and translational control during *Drosophila* oogenesis is fundamental to specification of the A-P and D-V axes of the oocyte and egg chamber. How general are these mechanisms? A first step toward answering this question is to determine whether the mechanisms are shared by any other transcripts that are localized and translated during *Drosophila* oogenesis. Perhaps the best studied of these is *oskar* (*osk*). Here we address whether the mechanisms defined for *grk* transcript localization and translation may also apply at least in part to *osk* RNA.

osk RNA is concentrated at the posterior of the oocyte, where Osk protein is synthesized and initiates formation of the posterior pole plasm (Ephrussi *et al.,* 1991; Ephrussi and Lehmann, 1992; Smith *et al.,* 1992). The pole plasm functions in the early embryo to confer germ line fate on the cells that form at the posterior. The molecular nature of the germ plasm and its role in germ line specification are presented in Chapter 5 by A. Mahowald.

B. *osk* RNA Localization

As described above, *grk* is initially transcribed in both the nurse cell nuclei and the oocyte nucleus and, later, exclusively in the oocyte nucleus. In contrast, *osk* RNA is synthesized exclusively in the nurse cell nuclei (Ephrussi *et al.,* 1991; Ephrussi and Lehmann, 1992; Smith *et al.,* 1992). Hence, *osk* transcripts in the oocyte are first transported from the nurse cells into the oocyte and, subsequently, a subset of these transcripts is transported to, and anchored at, the posterior pole of the oocyte. Further, unlike *grk,* which is localized initially to the posterior of the oocyte (stages 1 to 7) and then to its anterior pole (stages 8 on), *osk* RNA is initially uniformly distributed in the oocyte (stages 1 to 6), then localized transiently to its anterior pole (stage 7) and subsequently to its posterior pole (from stage 8) (Kim-Ha *et al.,* 1991). Hence, especially during mid- to late oogenesis, *grk* and *osk* RNA localization patterns are reciprocal.

Despite these differences in source and localization, both *grk* and *osk* transcripts require microtubules for localization at the mid- to late-stages of oogenesis (Pokrywka and Stephenson, 1994, 1995, 1991; Saunders and Cohen, 1999). In addition, *osk* RNA requires microfilaments for localization to, and anchoring at, the posterior pole (Erdélyi *et al.,* 1995; Lantz *et al.,* 1999). In contrast, microfilaments do not function in *grk* transcript localization (Saunders and Cohen, 1999).

Several genes have been shown to function in *osk* localization, including *orb, vas, mago, stau,* and *TmII* (Bashirullah *et al.,* 1998). In these mutants, *osk* RNA is stable but is distributed throughout the oocyte. There is thus an overlap in genetic requirements for *grk* and osk transcript localization. For example, Orb has been shown to be involved directly in *grk* and *osk* transcript localization and thus may be a shared component of the transcript localization machinery (Christerson and McKearin, 1994; Lantz *et al.,* 1992;

Lantz *et al.,* 1994; Roth *et al.,* 1995). VAS may also function in localization and/or translational regulation of both *grk* and *osk* (Styhler *et al.,* 1998; Tinker *et al.,* 1998; Tomancak *et al.,* 1998) (see section IV.C). In contrast, other proteins such as STAU that are known to function in *osk* transcript localization (Ferrandon *et al.,* 1994) do not play a role in *grk* RNA localization.

C. OSK Translation

OSK protein is first synthesized at the posterior pole of the stage 8 oocyte, at the time of posterior *osk* RNA localization (Rongo *et al.,* 1995). OSK protein is found at the posterior through the remainder of oogenesis. There are two stages at which *osk* translation is repressed. First, *osk* transcripts are translationally repressed during transport from the anterior to the posterior pole of the oocyte (Kim-Ha *et al.,* 1995; Webster *et al.,* 1997). Second, it has been shown that although *osk* RNA is concentrated at the posterior, over 80% of *osk* RNA remains unlocalized (Bergsten and Gavis, 1999). Hence, the majority of the *osk* transcripts are translationally repressed throughout the oocyte during and after transport to the posterior. It is also likely that *osk* transcripts are translationally activated at the posterior of the oocyte, possibly through release from the translational repression mechanisms that function in the bulk cytoplasm. Therefore, both *osk* and *grk* transcripts are regulated by a combination of translational repression and activation.

Several translational repressors appear to be used for both *osk* and *grk*. Most strikingly, the BRU translational repressor may function in both cases. Both *osk* and *grk* transcripts contain BREs in their 3′UTRs and BRU has been shown to bind the *osk* BREs directly (Kim-Ha *et al.,* 1995; Webster *et al.,* 1997). Mutation of BREs in *osk* results in ectopic synthesis of OSK protein with dire phenotypic consequences (Kim-Ha *et al.,* 1995; Webster *et al.,* 1997). It has been shown that the RNA-binding protein, APONTIC, also binds to the *osk* 3′UTR, physically interacts with BRU, and functions in translational repression of *osk* (Lie and Macdonald, 1999). It is not yet known whether APONTIC functions in regulation of *grk* translation. Reciprocally, it is not yet known whether SQD, which recruits BRU to the *grk* 3′UTR (Norvell *et al.,* 1999), also functions in recruiting BRU to *osk* transcripts.

If unlocalized *osk* transcripts are translationally repressed by BRU, how do they become translationally active when localized to the posterior? VAS, which has been shown biochemically to interact with BRU (Webster *et al.,* 1997), is required for *osk* translational activation (Rongo *et al.,* 1995), possibly through release from BRU-mediated translational repression. Since

VAS-dependent activation is independent of the *osk* 3′UTR, to which BRU is bound, the mechanism by which VAS acts is uncertain (Rongo *et al.*, 1995). BRU may interfere with translational activation by preventing the *osk* 3′UTR from interacting productively with the 5′ region of the transcript. Consistent with this possibility, an element in the *osk* 5′UTR is required for translational derepression of posterior localized transcripts (Gunkel *et al.*, 1998). A protein that binds to this derepressor element also interacts with a BRE in the 3′UTR, providing a potential link between the 5′ and 3′UTRs and suggesting a mechanism for displacing the BRU protein (Gunkel *et al.*, 1998). It has recently been shown that ORB binds to the *osk* 3′UTR and that ORB may function to control translation through regulation of poly(A) tail length (Chang *et al.*, 1999). It is not yet clear whether ORB controls poly(A) tail length of *grk* transcripts (Chang *et al.*, 1999).

Hence, several of the proteins implicated in translational repression and derepression of *osk* may also carry out similar functions for the *grk* transcript: As described previously (section III.G), BRU is associated with the *grk* 3′UTR and functions to repress translation of transcripts that are not localized to the dorso-anterior corner of the oocyte. VAS and ORB both appear to be involved in translational activation of *grk*, VAS through release from BRU-mediated repression and ORB through unknown mechanisms. An additional gene that may function in translational activation of both *grk* and *osk* transcripts is *aubergine* (*aub*) (Wilson *et al.*, 1996).

To date the only translational activation element defined for *grk* is in the 5′UTR while translational repressor elements such as the BRE reside in the *osk* and *grk* 3′UTRs. Hence, for both *grk* and *osk* transcripts, the interacting proteins and the relative locations of *cis*-acting translational repressor and derepressor elements may be conserved.

V. Concluding Remarks

The roles of transcript localization and translational regulation in specification of the axes of the oocyte are rapidly being elucidated in *Drosophila* via a combination of the genetic and molecular biological tools available in this model system. We have focused here on two key axis-specifying transcripts, *grk* and *osk*, and have described how several of the molecular mechanisms involved in their localization and regulation are shared. However, at the same time, each class of transcripts is likely to be regulated by factors that are specific to the particular time and place at which it functions. Additional transcripts in *Drosophila* (see e.g., Lall *et al.*, 1999) as well as other organisms (see e.g., Havin *et al.*, 1998; Ross *et al.*, 1997) are likely to be regulated by a combination of common and specific factors that control their localization and/or translation.

Acknowledgments

We thank Amanda Pickup, Bruce Reed, and Ronit Wilk for comments on the manuscript. R.L.C. has been supported in part by graduate scholarships from the Medical Research Council of Canada (MRC) and, currently, the University of Toronto. Our research on transcript localization, stability, and translation is supported by an operating grant from the MRC to H.D.L.

References

Bashirullah, A., Cooperstock, R. L., and Lipshitz, H. D. (1998). RNA localization in development. *Annu. Rev. Biochem.* **67,** 335–394.

Bashirullah, A., Halsell, S. R., Cooperstock, R. L., Kloc, M., Karaiskakis, A., Fisher, W. W., Fu, W., Hamilton, J. K., Etkin, L. D., and Lipshitz, H. D. (1999). Joint action of two RNA degradation pathways controls the timing of maternal transcript elimination at the midblastula transition in *Drosophila melanogaster. EMBO J.* **18,** 2610–2620.

Bergsten, S. E., and Gavis, E. R. (1999). Role for mRNA localization in translational activation but not spatial restriction of *nanos* RNA. *Development* **126,** 659–669.

Chang, J. S., Tan, L., and Schedl, P. (1999). The *Drosophila* CPEB homolog, Orb, is required for Oskar protein expression in oocytes. *Dev. Biol.* **215,** 91–106.

Christerson, L. B., and McKearin, D. M. (1994). *orb* is required for anteroposterior and dorsoventral patterning during *Drosophila* oogenesis. *Genes Dev.* **8,** 614–628.

Clark, I., Giniger, E., Ruohola-Baker, H., Jan, L. Y., and Jan, Y. N. (1994). Transient posterior localization of a kinesin fusion protein reflects anteroposterior polarity of the *Drosophila* oocyte. *Curr. Biol.* **4,** 289–300.

Clark, I. E., Jan, L. Y., and Jan, Y. N. (1997). Reciprocal localization of Nod and kinesin fusion proteins indicates microtube polarity in the *Drosophila* oocyte, epithelium, neuron and muscle. *Development* **124,** 461–470.

Clegg, N. J., Frost, D. M., Keller Larkin, M., Subrahmanyan, L., Bryant, Z., and Ruohola-Baker, H. (1997). *maelstrom* is required for an early step in the establishment of *Drosophila* oocyte polarity: Posterior localization of *grk* mRNA. *Development* **124,** 4661–4671.

Cooley, L., and Theurkauf, W. (1994). Cytoskeletal functions during *Drosophila* oogenesis. *Science* **266,** 590–596.

Cooperstock, R. L., and Lipshitz, H. D. (1997). Control of mRNA stability and translation during *Drosophila* development. *Sem. Cell Dev. Biol.* **8,** 541–549.

Curtis, D., Lehmann, R., and Zamore, P. D. (1995). Translational regulation in development. *Cell* **81,** 171–178.

Dahanukar, A., and Wharton, R. P. (1996). The Nanos gradient in *Drosophila* embryos is generated by translational regulation. *Genes Dev.* **10,** 2610–2620.

Ephrussi, A., Dickinson, L. K., and Lehmann, R. (1991). *oskar* organizes the germ plasm and directs localization of the posterior determinant *nanos. Cell* **66,** 37–50.

Ephrussi, A., and Lehmann, R. (1992). Induction of germ cell formation by *oskar. Nature* **358,** 387–392.

Erdélyi, M., Michon, A.-M., Guichet, A., Glozter, J. B., and Ephrussi, A. (1995). Requirement for *Drosophila* cytoplasmic tropomyosin in *oskar* mRNA localization. *Nature* **377,** 524–527.

Ferrandon, D., Elphick, L., Nüsslein-Volhard, C., and St. Johnston, D. (1994). Staufen protein associates with the 3′UTR of *bicoid* mRNA to form particles that move in a microtubule-dependent manner. *Cell* **79,** 1221–1232.

Ghabrial, A., Ray, R. P., and Schüpbach, T. (1998). *okra* and *spindle-B* encode components of the RAD52 DNA repair pathway and affect meiosis and patterning in *Drosophila* oogenesis. *Genes Dev.* **12,** 2711–2723.

Ghabrial, A., and Schüpbach, T. (1999). Activation of a meiotic checkpoint regulates translation of Gurken during *Drosophila* oogenesis. *Nature Cell Biology* **1,** 354–357.

Gillespie, D., and Berg, C. (1995). *homeless* is required for RNA localization in *Drosophila* oogenesis and encodes a new member of the DE-H family of RNA-dependent ATPases. *Genes Dev.* **9,** 2495–2508.

Godt, D., and Tepass, U. (1998). *Drosophila* oocyte localization is mediated by differential cadherin- based adhesion. *Nature* **395,** 387–391.

González-Reyes, A., Elliot, H., and St Johnston, D. S. (1995). Polarization of both major body axes in *Drosophila* by *gurken-torpedo* signalling. *Nature* **375,** 654–658.

González-Reyes, A., Elliot, H., and St Johnston, D. (1997). Oocyte determination and the origin of polarity in *Drosophila:* The role of the *spindle* genes. *Development* **124,** 4927–4937.

González-Reyes, A., and St Johnston, D. (1998). Patterning of the follicle cell epithelium along the anterior-posterior axis during *Drosophila* oogenesis. *Development* **125,** 2837–2846.

Gunkel, N., Yano, T., Markussen, F.-H., Olsen, L. C., and Ephrussi, A. (1998). Localization-dependent translation requires a functional interaction between the 5′ and 3′ ends of *oskar* mRNA. *Genes Dev.* **12,** 1652–1664.

Havin, L., Git, A., Elisha, Z., Oberman, F., Yaniv, K., Pressman Schwartz, S., Standart, N., and Yisraeli, J. K. (1998). RNA-binding protein conserved in both microtubule- and microfilament-based RNA localization. *Genes Dev.* **12,** 1593–1598.

Hawkins, N. C., Thorpe, J., and Schüpbach, T. (1996). *encore,* a gene required for the regulation of germ line mitosis and oocyte differentiation during *Drosophila* oogenesis. *Development* **122,** 281–290.

Hawkins, N. C., Van Buskirk, C., Grossniklaus, U., and Schüpbach, T. (1997). Post-transcriptional regulation of *gurken* by *encore* is required for axis determination in *Drosophila. Development* **124,** 4801–4810.

Hay, B., Jan, L. Y., and Jan, Y. N. (1988). A protein component of *Drosophila* polar granules is encoded by *vasa* and has extensive sequence similarity to ATP-dependent helicases. *Cell* **55,** 577–587.

Jacobson, A., and Peltz, S. W. (1996). Interrelationships of the pathways of mRNA decay and translation in eukaryotic cells. *Annu. Rev. Biochem.* **65,** 693–739.

Kelley, R. L. (1993). Initial organization of the *Drosophila* dorsoventral axis depends on an RNA-binding protein encoded by the *squid* gene. *Genes Dev.* **7,** 948–960.

Kim-Ha, J., Kerr, K., and Macdonald, P. M. (1995). Translational regulation of *oskar* mRNA by Bruno, an ovarian RNA-binding protein, is essential. *Cell* **81,** 403–412.

Kim-Ha, J., Smith, J. L., and Macdonald, P. M. (1991), *oskar* mRNA is localized to the posterior pole of the *Drosophila* oocyte. *Cell* **66,** 23–35.

Lall, S., Francis-Lang, H., Flament, A., Norvell, A., Schüpbach, T., and Ish-Horowicz, D. (1999). Squid hnRNP protein promotes apical cytoplasmic transport and localization of *Drosophila* pair-rule transcripts. *Cell* **98,** 171–80.

Lane, M. E., and Kalderon, D. (1994). RNA localization along the anteroposterior axis of the *Drosophila* oocyte requires PKA-mediated signal transduction to direct normal microtubule organization. *Genes Dev.* **8,** 2986–2995.

Lantz, V., Ambrosio, L., and Schedl, P. (1992). The *Drosophila orb* gene is predicted to encode sex-specific germline RNA-binding proteins and has localized transcripts in ovaries and early embryos. *Development* **115,** 75–88.

Lantz, V., Chang, J. S., Horabin, J. I., Bopp, D., and Schedl, P. (1994). The *Drosophila orb* RNA-binding protein is required for the formation of the egg-chamber and establishment of polarity. *Genes Dev.* **8,** 598–613.

Lantz, V. A., Clemens, S. E., and Miller, K. G. (1999). The actin cytoskeleton is required for maintenance of posterior pole plasm components in the *Drosophila* embryo. *Mech. Dev.* **85,** 111–122.

Lasko, P. F., and Ashburner, M. (1988). The product of the *Drosophila* gene *vasa* is very similar to eukaryotic initiation factor 4A. *Nature* **335,** 611–616.

Liang, L., Diehl-Jones, W., and Lasko, P. (1994). Localization of *vasa* protein to the *Drosophila* pole plasm is independent of its RNA-binding and helicase activities. *Development* **120,** 1201–1211.

Lie, Y. S., and Macdonald, P. M. (1999). Apontic binds the translational repressor Bruno and is implicated in regulation of *oskar* mRNA translation. *Development* **126,** 1129–1138.

Macdonald, P. M., and Smibert, C. A. (1996). Translational regulation of maternal mRNAs. *Curr. Biol.* **6,** 403–407.

Mach, J. M., and Lehmann, R. (1997). An egalitarian-BicaudalD complex is essential for oocyte specification and axis determination in *Drosophila. Genes Dev.* **11,** 423–435.

Mahowald, A. P., and Tiefert, M. (1970). Fine structural changes in the *Drosophila* oocyte nucleus during a short period of RNA synthesis. *Wilhelm Roux's Arch. Dev. Biol.* **165,** 8–25.

Micklem, D. R., Dasgupta, R., Elliot, H., Gergely, F., Davidson, C., Brand, A., González-Reyes, A., and St. Johnston, D. (1997). The *mago nashi* gene is required for the polarisation of the oocyte and the formation of perpendicular axes in *Drosophila. Curr. Biol.* **7,** 468–478.

Neuman-Silberberg, F. S., and Schüpbach, T. (1993). The *Drosophila* dorsoventral patterning gene *gurken* produces a dorsally localized RNA and encodes a TGFα-like protein. *Cell* **75,** 165–174.

Neuman-Silberberg, F. S., and Schüpbach, T. (1996). The *Drosophila* TGF-α-like protein Gurken: Expression and cellular localization during *Drosophila* oogenesis. *Mech. Dev.* **59,** 105–113.

Newmark, P. A., Mohr, S. E., Gong, L., and Boswell, R. E. (1997). *mago nashi* mediates the posterior follicle cell-to-oocyte signal to organize axis formation in *Drosophila. Development* **124,** 3197–3207.

Nilson, L. A., and Schüpbach, T. (1999). EGF receptor signaling in *Drosophila* oogenesis. *Curr. Top. Dev. Biol.* **44,** 203–243.

Norvell, A., Kelley, R. L., Wehr, K., and Schüpbach, T. (1999). Specific isoforms of Squid, a *Drosophila* hnRNP, perform distinct roles in Gurken localization during oogenesis. *Genes Dev.* **13,** 864–876.

Peri, F., Bökel, C., and Roth, S. (1999). Local Gurken signaling and dynamic MAPK activation during *Drosophila* oogenesis. *Mech. Dev.* **81,** 75–88.

Pokrywka, N.-J., and Stephenson, E. C. (1991). Microtubules mediate the localization of *bicoid* RNA during *Drosophila* oogenesis. *Development* **113,** 55–66.

Pokrywka, N.-J., and Stephenson, E. C. (1994). Localized RNAs are enriched in cytoskeletal extracts of *Drosophila* oocytes. *Dev. Biol.* **166,** 210–219.

Pokrywka, N.-J., and Stephenson, E. C. (1995). Microtubules are a general component of mRNA localization systems in *Drosophila* oocytes. *Dev. Biol.* **167,** 363–370.

Queenan, A. M., Barcelo, G., Van Buskirk, C., and Schüpbach, T. (1999). The transmembrane region of Gurken is not required for biological activity, but is necessary for transport to the oocyte membrane in *Drosophila. Mech. Dev.* **89,** 35–42.

Ray, R. P., and Schüpbach, T. (1996). Intercellular signaling and the polarization of body axes during *Drosophila* oogenesis. *Genes Dev.* **10,** 1711–1723.

Rongo, C., Gavis, E., and Lehmann, R. (1995). Localization of *oskar* RNA regulates *oskar* translation and requires Oskar protein. *Development* **121,** 2737–2746.

Ross, A. F., Oleynikov, Y., Kislauskis, E. H., Taneja, K. L., and Singer, R. H. (1997). Characterization of a beta-actin mRNA zipcode-binding protein. *Mol. Cell. Biol.* **17,** 2158–2165.

Roth, S., Jordan, P., and Karess, R. (1999). Binuclear *Drosophila* oocytes: Consequences and implications for dorsal-ventral patterning in oogenesis and embryogenesis. *Development* **126,** 927–934.

Roth, S., Neuman-Silberberg, F. S., Barcelo, G., and Schüpbach, T. (1995). *cornichon* and the EGF receptor signaling process are necessary for both anterior-posterior and dorsal-ventral pattern formation in *Drosophila. Cell* **81,** 967–978.

Roth, S., and Schüpbach, F. S. (1994). The relationship between ovarian and embryonic dorsoventral patterning in *Drosophila. Development* **120,** 2245–2257.

Ruohola, H., Bremer, K. A., Baker, D., Swedlow, J. R., Jan, L. Y., and Jan, Y. N. (1991). Role of neurogenic genes in establishment of follicle cell fate and oocyte polarity during oogenesis in *Drosophila. Cell* **66,** 433–449.

Saunders, C., and Cohen, R. S. (1999). The role of oocyte transcription, the 5′UTR, and translation repression and derepression in *Drosophila gurken* mRNA and protein localization. *Mol. Cell* **3,** 43–54.

Schweitzer, R., and Shilo, B.-Z. (1997). A thousand and one roles for the *Drosophila* EGF receptor. *Trends Genet.* **13,** 191–196.

Serano, T. L., and Cohen, R. S. (1995). Gratuitous mRNA localization in the *Drosophila* oocyte. *Development* **121,** 3013–3021.

Serano, T. L., Karlin-McGinnes, M., and Cohen, R. S. (1995). The role of *fs(1)K10* in the localization of the mRNA of the TGFα homolog *gurken* within the *Drosophila* oocyte. *Mech. Dev.* **51,** 183–192.

Smibert, C. A., Wilson, J. E., Kerr, K., and Macdonald, P. M. (1996). Smaug protein represses translation of unlocalized *nanos* mRNA in the *Drosophila* embryo. *Genes Dev.* **10,** 2600–2609.

Smith, J. L., Wilson, J. E., and Macdonald, P. M. (1992). Overexpression of *oskar* directs ectopic activation of *nanos* and presumptive pole cell formation in *Drosophila* embryos. *Cell* **70,** 849–859.

Spradling, A. (1993). Developmental genetics of oogenesis. *In* "The Development of *Drosophila melanogaster*" (M. Bate and A. M. Arias, Eds.), Vol. 1, pp. 1–70. Cold Spring Harbor Laboratory Press, Cold Spring Harbor, NY.

Sternberg, P. W. (1993). Intercellular signaling and signal transduction in C. elegans. *Annu. Rev. Genet.* **27,** 497–521.

Styhler, S., Nakamura, A., Swan, A., Suter, B., and Lasko, P. (1998). *vasa* is required for GURKEN accumulation in the oocyte, and is involved in oocyte differentiation and germline cyst development. *Development* **125,** 1569–1578.

Swan, A., Nguyen, T., and Suter, B. (1999). *Drosophila* Lissencephaly-1 functions with Bic-D and dynein in oocyte determination and nuclear positioning. *Nature Cell Biol.* **1,** 444–449.

Swan, A., and Suter, B. (1996). Role of *Bicaudal-D* in patterning *Drosophila* egg chamber in mid-oogenesis. *Development* **122,** 3577–3586.

Theodorakis, N. G., and Cleveland, D. W. (1996). Translationally coupled degradation of mRNA in eukaryotes. *In* "Translational Control." Cold Spring Harbor Laborotory Press, Cold Spring Harbor.

Theurkauf, W. E., Smiley, S., Wong, M. L., and Alberts, B. M. (1992). Reorganization of the cytoskeleton during *Drosophila* oogenesis: Implications for axis specification and intercellular transport. *Development* **115,** 923–936.

Thio, G. L., Ray, R. P., Barcelo, G., and Schüpbach, T. (2000). Localization of *gurken* RNA in *Drosophila* oogenesis requires elements in the 5′- and 3′-regions of the transcript. *Dev. Biol.* **221,** 435–446.

Tinker, R., Silver, D., and Montell, D. J. (1998). Requirement for the Vasa RNA Helicase in *gurken* mRNA Localization. *Dev. Biol.* **199,** 1–10.

Tomancak, P., Guichet, A., Zavorszky, P., and Ephrussi, A. (1998). Oocyte polarity depends on regulation of *gurken* by Vasa. *Development* **125,** 1723–1732.

van Eeden, F., and St. Johnston, D. (1999). The polarisation of the anterior-posterior and dorsal-ventral axes during *Drosophila* oogenesis. *Curr. Opinion Genet. Dev.* **9,** 396–404.

Webster, P. J., Liang, L., Berg, C. A., Lasko, P., and Macdonald, P. M. (1997). Translational repressor Bruno plays multiple roles in development and is widely conserved. *Genes Dev.* **11,** 2510–2521.

Wilson, J. E., Connell, J. E., and Macdonald, P. M. (1996). *aubergine* enhances *oskar* translation in the *Drosophila* ovary. *Development* **122,** 1631–1639.

Wormington, M. (1994). Unmasking the role of the 3′UTR in the cytoplasmic polyadenylation and translational regulation of maternal mRNAs. *BioEssays* **16,** 533–535.

Translational Control in Vertebrate Development

Cornelia H. de Moor and Joel D. Richter
Department of Molecular Genetics and Microbiology, University of Massachusetts Medical School, Worcester, Massachusetts 01655

Translational control plays a large role in vertebrate oocyte maturation and contributes to the induction of the germ layers. Translational regulation is also observed in the regulation of cell proliferation and differentiation. The features of an mRNA that mediate translational control are found both in the 5′ and in the 3′ untranslated regions (UTRs). In the 5′ UTR, secondary structure, the binding of proteins, and the presence of upstream open reading frames can interfere with the association of initiation factors with the cap, or with scanning of the initiation complex. The 3′ UTR can mediate translational activation by directing cytoplasmic polyadenylation and can confer translational repression by interference with the assembly of initiation complexes. Besides mRNA-specific translational control elements, the nonspecific RNA-binding proteins contribute to the modulation of translation in development. This review discusses examples of translational control and their relevance for developmental regulation.

KEY WORDS: Translational control, Development, Cytoplasmic polyadenylation, Oocyte maturation, Differentiation, Cell cycle, Deadenylation.

I. Introduction

In the 30-odd years since the discovery of a population of untranslated mRNA in oocytes and embryos (Spirin, 1966), the field of translational control in development has moved from examining mixed pools of untranslated (masked) and translated mRNAs to the characterization of translational control elements in individual mRNAs. Advances in the last decade have included the characterization of some trans-acting factors and the implication of general translation initiation factors in mRNA specific translational control. Besides the classical translational control of masked

0074-7696/01 $35.00

mRNAs in oocytes and early embryos, translational regulation has also been implicated in later developmental events such as terminal differentiation. This review aims to illustrate the various types of translational regulation found in vertebrate development with some of the better characterized examples known to date.

Most translational control is exerted at the level of initiation. To facilitate understanding of the mechanisms involved in translational regulation, a description of the salient mechanisms is presented (Fig. 1). For more extensive discussions, the reader is referred to some recent reviews (Pain, 1996; Merrick and Hershey, 1996; Morley *et al.,* 1997; Kozak, 1999; Gingras *et al.,* 1999).

For most cellular mRNAs, efficient translation initiation is dependent on the presence of the cap, a guanosine methylated at the number 7 position, which is attached via three phosphate groups to the 5′ hydroxyl group of the first nucleotide of the mRNA (m^7GpppN). The initiation factor elF4E recognizes this structure and can recruit the central scaffold of the initiation complex, elF4G, to the mRNA (Figure 1). elF4G also binds elF4A, and this complex of elF4A, G, and E is called the cap-binding initiation complex, or elF4F. The association of elF4G with the mRNA is further stabilized by an interaction of elF4G with poly(A) binding protein (PABP). Hence, the 3′ poly(A) tails of mRNAs are thought to enhance translation by binding PABP and increasing the chances of an mRNA to recruit elF4G (reviewed in Sachs *et al.,* 1997; Gingras *et al.,* 1999). While this mechanism was first described in yeast, it has recently been shown to function in mammalian cells as well (Piron *et al.,* 1998, 1999; Imataka *et al.,* 1998). The interaction between factors binding the two ends of the mRNA leads to a circular configuration, the so-called "closed loop," first proposed by Jacobson and his coworkers (Jacobson and Favreau, 1983; Palatnik *et al.,* 1984). This closed loop is indeed observed when capped and polyadenylated mRNA is mixed with elF4E, elF4G, and PABP and examined via atomic force microscopy (Wells *et al.,* 1998). In addition to recruiting elF4G, PABP also binds a factor called PABP interacting factor (PAIP), which binds elF4A and may mediate the translation stimulation by poly(A) tail as well (Craig *et al.,* 1998).

There are some indications that the interaction of elF4G with elF4E and PABP is only necessary for the first initiation event on an mRNA. In mammalian tissue culture cells as well as in *Xenopus* oocytes, translation of newly available mRNAs is severely inhibited by a proteolytic cleavage that removes the elF4E- and PABP-binding domains from elF4G. In contrast, the ongoing translation of mRNA already in the polysomes is not impaired by this proteolysis (Keiper and Rhoads, 1997, 1999; Novoa and Carrasco, 1999). The data of Novoa and Carrasco (1999), using tissue culture cells, are particularly noteworthy because they show that both known iso-

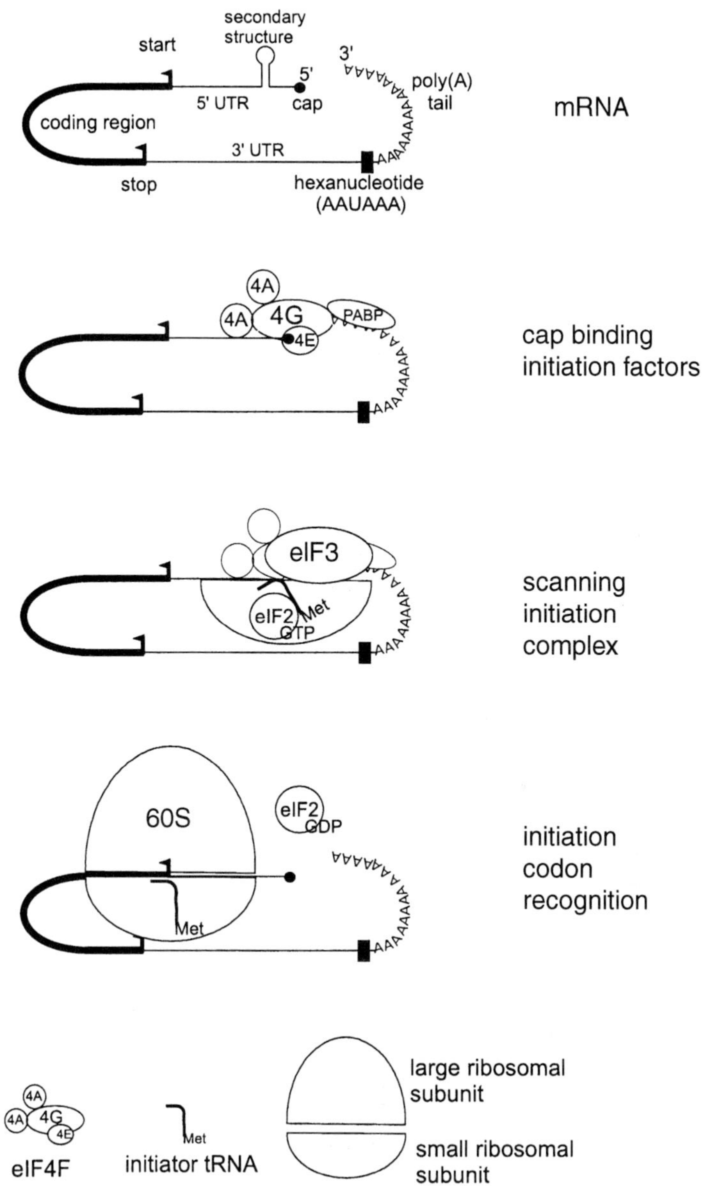

FIG. 1 A schematic representation of translation initiation. The cap of the mRNA is bound by the cap binding translation initiation factor eIF4E (4E), which, in conjunction with poly(A) binding protein (PABP), brings eIF4G (4G) to the mRNA. eIF4G recruits the RNA helicase eIF4A (4A), which can melt moderate secondary structure in the mRNA. The small ribosomal subunit is recruited to the mRNA by an interaction between eIF4G and eIF3. The initiatior tRNA is bound to the ribosomal subunit with eIF2-GTP. The small ribosomal subunit scans the 5′ UTR and the initiation codon is recognized with the help of eIF2, which converts the GTP to GDP. Subsequently, the large ribosomal subunit binds and translation ensues.

forms of eIF4G are cleaved in their system. Significantly, the translation of uncapped mRNAs, which is inefficient compared to capped mRNAs, was recently shown to still be dependent on a free 5′ end of the mRNA. This led to the "threading" model of translation initiation in which the small ribosomal subunit attaches to the mRNA like a bead being threaded on a string, regardless of the presence or absence of a cap (De Gregorio *et al.*, 1998). Possibly after the first initiation, which requires the formation of the closed loop mediated by the cap-eIF4E-eIF4G-PABP-poly(A) complex, the subsequent initiations can occur by threading, independently of this complex. The molecular basis for the difference between the first initiation and reinitiation is still far from clear and therefore is hard to incorporate into the known models for translation initiation, but it might involve structural changes in the mRNP, for instance in the mRNA secondary structure and/or protein association.

At least during the first initiation, eIF4G recruits the RNA helicase eIF4A, which is thought to unwind the 5′ UTR to facilitate assembly and passage of the initiation complex. eIF4A activity is enhanced by the binding of eIF4B. eIF4G also recruits the small subunit of the ribosome by binding eIF3. eIF2 brings the initiator Met-tRNA to the small ribosomal subunit and this complex starts searching the 5′ UTR for the first initiation codon in a process called scanning (Figure 1). eIF1 and eIF1A are necessary for this process (Pestova *et al.*, 1998). Once an initiation codon in a reasonably good sequence context is identified, the large ribosomal subunit will bind and elongation of the peptide ensues. The optimal context for an initiation codon is ccAcc<u>AUG</u>G (Kozak, 1997). The full optimal context is rarely observed in natural mRNAs, but functional initiation codons almost always have either the purine in position −3 (A of AUG is 1) or the G in +4. Because only very weak initiation codons (lacking both the −3 purine and the +4 G) are skipped under normal circumstances, the first AUG is usually the initiation codon. Both upstream open reading frames with an initiation codon in reasonable context and strong secondary structure in the 5′ UTR will inhibit cap-dependent translation by preventing the scanning complex from reaching the proper initiation codon.

The activity of the translation initiation factors is regulated by phosphorylation on many of the subunits under a variety of different circumstances. For a further treatment of this subject, the interested reader is referred to a number of reviews on this subject (Proud and Denton, 1997; Kleijn *et al.*, 1998; Gingras *et al.*, 1999). For the present discussion, only the regulation of eIF4E by the so-called 4E-BPs is relevant. The 4E-BPs are a family of small proteins that bind to eIF4E in the same site as eIF4G, thus preventing association of this essential initiation factor. This inhibition of eIF4G recruitment abrogates translation of most cellular mRNAs. Phosphorylation of the 4E-BPs, which is induced by mitogens, results in their release from eIF4E, thus allowing the binding of eIF4G to eIF4E (Raught and Gingras, 1999; Gingras *et al.,* 1999; Marcotrigiano *et al.,* 1999).

Not all translation initiation is dependent on the recruitment of eIF4G by eIF4E. Some viruses can inactivate cap-dependent translation of the host cell, for instance by expressing a protease that specifically cleaves the eIF4E binding domain from eIF4G. The viral mRNAs can bind the small ribosomal subunit without the benefit of a 5′ cap structure by utilizing special 5′ UTR elements called internal ribosome entry sites (IRES). Presumably the IRES is capable of recruiting initiation factors independently of eIF4E (Sachs *et al.,* 1997; Sizova *et al.,* 1998; Kolupaeva *et al.,* 1998; De Gregorio *et al.,* 1999).

II. Translational Control during Oocyte Maturation

During the growth of probably all animal oocytes, a pool of mRNA is synthesized and stored in an untranslated form. Many of these mRNAs are translationally activated in the late stages of oogenesis or in early embryogenesis when transcription is silent. This maternal mRNA is present in the oocyte in mRNP particles not bound to ribosomes (free mRNP). The free mRNP particles do not give translation products *in vitro,* although the protein-free mRNA is readily translatable in reticulocyte lysates. This form of translational repression is called "masking" and was the earliest evidence of translational regulation in development (Spirin, 1966). Later investigations showed that translation of different individual mRNAs occurs at different times in development (Stebbins-Boaz and Richter, 1997; Hake and Richter, 1997; Goodwin and Evans, 1997; Bashirullah *et al.,* 1998; Gray and Wickens, 1998; Richter, 1999). The translational regulation during oocyte maturation is among the most extensively studied control mechanisms in vertebrate development. To facilitate the understanding of translational control in oocyte maturation, a description of meiotic maturation in *Xenopus* oocytes follows (Gebauer and Richter, 1997; Sagata, 1997; Taieb *et al.,* 1997).

Vertebrate oocytes are arrested in prophase of meiosis I. During this period, the oocyte grows into a large cell and stores maternal mRNAs, as well as proteins, that are to be used in the embryo. A hormonal signal, progesterone in the case of *Xenopus,* induces the resumption of meiosis (maturation) in full-grown oocytes. An easily visible manifestation of maturation is the dissolution of the nuclear membrane, commonly referred to as germinal vesicle breakdown (GVBD). The oocyte progresses through meiosis I and arrests at the metaphase of meiosis II. This is called a mature oocyte or egg, which only completes meiosis II after fertilization.

In *Xenopus,* the binding of progesterone to a putative membrane receptor induces the synthesis of the protein kinase Mos. This kinase then sets off a kinase cascade that leads to activation of the M-phase promoting factor, consisting of cdc2 kinase/cyclin B, which leads to completion of meiosis I (Fig. 2). In addition, Mos kinase is necessary to prevent DNA synthesis between meiosis I and II and mediates the cell cycle arrest in the metaphase of meiosis II.

A. Cytoplasmic Polyadenylation

1. Cis-Acting Elements

Many of the masked maternal mRNAs have rather short poly(A) tails of 20 to 40 residues. Translational activation of mRNAs during meiotic maturation is often accompanied by an elongation of the poly(A) tail to 80 to 250 residues. This elongation takes place in the cytoplasm of the

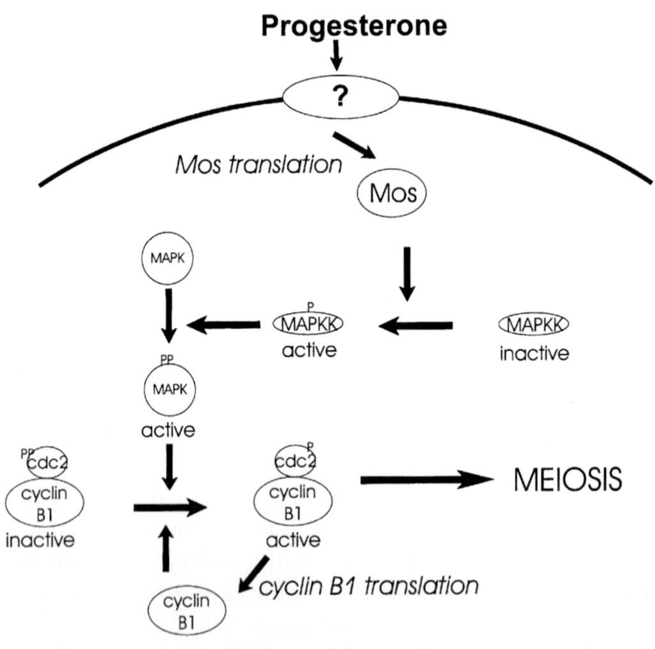

FIG. 2 Signal transduction during oocyte maturation. Progesterone binds to an unknown membrane receptor and induces Mos synthesis. Mos phosphorylates a MAP kinase kinase, which activates MAP kinase. MAP kinase activates cdc2 kinase/cyclin B indirectly and activation of cdc2 kinase leads to meiosis and the translation of cyclin B1 mRNA.

oocyte and is therefore referred to as cytoplasmic polyadenylation. Cytoplasmic polyadenylation requires the hexanucleotide polyadenylation signal (AAUAAA) and a U-rich cytoplasmic polyadenylation element, or CPE. Because it is required for nuclear polyadenylation, the hexanucleotide is present in almost all mRNAs. Consequently, the CPE determines which mRNAs are translationally regulated by cytoplasmic polyadenylation. In *Xenopus,* the CPE that regulates translation during meiotic maturation, the so-called maturation-type CPE, is typically UUUUA(A)U (McGrew *et al.,* 1989; Fox *et al.,* 1989). Maturation type CPEs can function from approximately 100 nt upstream and even downstream of the hexanucleotide (Fox *et al.,* 1989; McGrew and Richter, 1990; Paris and Richter, 1990). The CPE sequences in mouse oocyte mRNAs are less well defined but are also U-rich and contain one or more A residues (Vassalli *et al.,* 1989; Gebauer *et al.,* 1994; West *et al.,* 1996; Verrottl and Strickland, 1997).

One of the mRNAs that is translationally activated by cytoplasmic polyadenylation is encodes Mos, a serine-threonine kinase that is essential for the normal completion of meiosis in both mouse and *Xenopus* oocytes (Sheets *et al.,* 1994; Gebauer *et al.,* 1994; Sheets *et al.,* 1995). In *Xenopus,* the polyadenylation and translation of other oocyte mRNAs, including cyclin B1, is dependent upon the translation of Mos (Ballantyne *et al.,* 1997; De Moor and Richter, 1997). This hierarchy in cytoplasmic polyadenylation results in a difference in the timing of the translation of *c-mos* and cyclin B1 mRNAs.

2. Trans-Acting Factors

Cytoplasmic polyadenylation is mediated by RNA-bound proteins that recruit the poly(A) polymerase (PAP) to the mRNA in an active form. Only one RNA-binding cytoplasmic polyadenylation factor has been identified, CPE binding protein (CPEB), which has two RNA recognition motifs and a zinc finger (Hake and Richter, 1994; Gebauer and Richter, 1996; Hake *et al.,* 1998). Besides the CPEB genes in mouse and *Xenopus,* homologs of this gene have been identified in zebrafish, *Drosophila,* and clam; in all cases, the protein is expressed in the oocyte of these species and is likely to be implicated in translational regulation (Lantz *et al.,* 1994; Christerson and McKearin, 1994; Bally-Cuif *et al.,* 1998; Minshall *et al.,* 1999; Walker *et al.,* 1999; Katsu *et al.,* 1999). In *Xenopus,* CPEB binds to maturation-type CPEs and is essential for cytoplasmic polyadenylation (Hake and Richter, 1994; Stebbins-Boaz *et al.,* 1996). The activation of CPEB is controlled by Eg2, a serine/threonine kinase that is induced by progesterone early in the signal transduction cascade leading to oocyte maturation (Andrésson and Ruderman, 1998). Eg 2 phosphorylates CPEB on the serine at position 174 and this is required for polyadenylation and translation of *c-mos* mRNA (R. Mendez *et al.,* 2000). The mechanism by which this phosphorylation of CPEB mediates the recruitment or activation of poly(A) polymerase (PAP) is not known. However, phosphorylated CPEB may stabilize the binding of factors at the AAUAAA, which may ultimately recruit PAP (see below). Later in oocyte maturation, during germinal vesicle breakdown, CPEB becomes hyperphosphorylated and subsequently about 90% is degraded (Hake and Richter, 1994; De Moor and Richter, 1997). The functions of the late phosphorylation and degradation of CPEB are at present unclear, but since these events coincide with the Mos-dependent polyadenylation of cyclin B1 mRNA, they might play a role in regulating the timing of polyadenylation of different mRNAs.

The factor that recognizes the AAUAAA hexanucleotide is likely to be a distinct cytoplasmic form of the nuclear AAUAAA binding factor, cleavage, and polyadenylation specificity factor (CPSF), a complex consisting of 30, 70, 100, and 160 kDa subunits (Wahle and Rüegsegger, 1999). Immunodepletion of the 100 kDa subunit from cytoplasmic extracts abolishes polyadenylation, indicating that at least this subunit is present in the polyadenylation complex. Overexpression of influenza NS1 protein, which inhibits nuclear polyadenylation by interaction with the 30 kDa CPSF subunit, inhibits cytoplasmic polyadenylation as well. This indicates that the 30 kDa subunit or a related protein is involved in cytoplasmic polyadenylation. The 70 kDa subunit of CPSF is not essential for cytoplasmic polyadenylation, because this subunit is exclusively nuclear (Dickson *et al.,* 1999). The 160 kDa CPSF subunit has a central function in nuclear polyadenylation because it binds the AAUAAA and associates with PAP (Murthy and Manley, 1995). Unfortunately, no data are available concerning a possible role of the 160 kDa CPSF subunit in cytoplasmic polyadenylation.

Like many other cell types, *Xenopus* oocytes contain multiple forms of poly(A) polymerase (Ballantyne *et al.,* 1995; Gebauer and Richter, 1995; Zhao and Manley, 1996). One of these forms is predicted to be exclusively cytoplasmic (Gebauer and Richter, 1995), but only the "nuclear" forms have been detected in the oocyte cytoplasm (Ballantyne *et al.,* 1995). Thus it is unclear which PAP isoform mediates cytoplasmic polyadenylation. However, it has been demonstrated that phosphorylation of the nuclear forms of PAP by cdc2 kinase leads to inactivation of the enzyme during oocyte maturation (Colgan *et al.,* 1996, 1998). This is hard to reconcile with the ability of egg extract to conduct cytoplasmic polyadenylation. The proposed cytoplasmic form of PAP, however, lacks these phosphorylation sites and is therefore a good candidate for the enzyme mediating cytoplasmic polyadenylation. Figure 3 summarizes the factors known to participate in cytoplasmic polyadenylation.

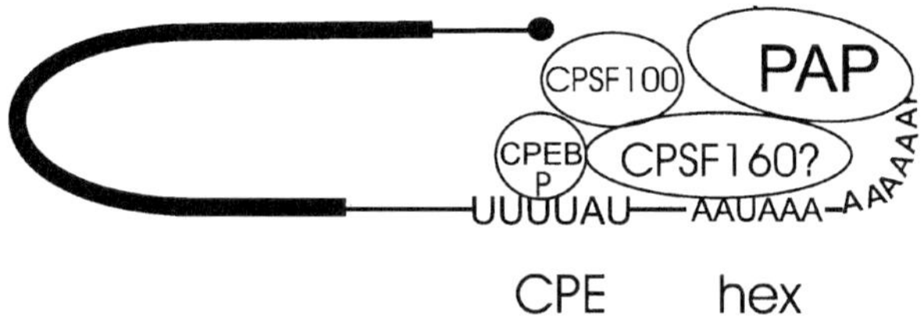

FIG. 3 Factors in cytoplasmic polyadenylation. Two mRNA elements are required for cytoplasmic polyadenylation, the cytoplasmic polyadenylation element or CPE (UUUUA(A)U) and the polyadenylation signal or hexanucleotide (hex, AAUAAA). The CPE is bound by the cytoplasmic polyadenylation element binding protein (CPEB), a factor that is required for cytoplasmic polyadenylation. Upon induction of meiotic maturation, CPEB becomes phosphorylated by Eg2 kinase, which is necessary for the induction of cytoplasmic polyadenylation. The hexanucleotide may be bound by the 160 kDa subunit of cleavage and polyadenylation specificity factor (CPSF), the nuclear hexanucleotide and poly(A) polymerase (PAP) binding factor. The 100 kDa subunit of CPSF is required for cytoplasmic polyadenylation, but its function is unkown. These factors, probably assisted by others, recruit PAP.

3. Translational Activation by Cytoplasmic Polyadenylation

How does cytoplasmic polyadenylation activate translation? The simplest explanation is that the longer poly(A) tail recruits more PABP, which stimulates translation by binding elF4G, as outlined in the Introduction. This seems plausible because the amount of available mRNA in the oocyte far exceeds the translational capacity of the oocyte, resulting in a fierce competition for the cap-binding initiation factor complexes (Laskey *et al.,* 1977; Audet *et al.,* 1987; Stambuck and Moon, 1992). In addition, this view is supported by the observation that two mRNAs, encoding histone B4 in *Xenopus* and tissue plasminogen activator in mouse, are translated after injection into immature oocytes if a they are appended with a long poly(A) tail prior to injection (Paris and Richter, 1990; Huarte *et al.,* 1992). However, two other *Xenopus* mRNAs, G10 and Cl2, do not get translated in immature oocytes if injected with long poly(A) tails (McGrew *et al.,* 1989; Simon and Richter, 1994). In these cases, mere PABP binding cannot be the primary activator. Instead, active cytoplasmic polyadenylation appears to be more important than the actual poly(A) tail for the translation of these mRNAs. Obviously, there must be a difference in the mRNP's of histone B4 and G10 to cause this difference in poly(A) dependence of translation in the *Xenopus* oocyte. Possibly, the mRNAs that need active polyadenylation contain specific translational repressor elements, while the ones that do not are only repressed by nonspecific RNA binding proteins (e.g. Y box proteins, see below). In this respect, the following experiment is significant: Barkoff *et al.* (1998) injected an antisense oligonucleotide to induce a sequence-specific cleavage in the *Xenopus c-mos* mRNA (this is mediated by endogenous RNase H). As expected, this mRNA was neither polyadenylated nor translated when oocytes were exposed to progesterone. To examine the influence of a poly(A) tail on the translation of this endogenous mRNA, a so-called prosthetic RNA was designed. This RNA contained a short sequence that can anneal to the remaining 3′UTR of the *c-mos* mRNA followed by a long poly(A) tail. Upon injection of this prosthetic RNA, Mos synthesis in response to progesterone was restored, presumably because the prosthetic poly(A) tail was sufficient for translational activation and active polyadenylation was not necessary. However, no Mos production was detected if the oocytes were not exposed to progesterone (Barkoff *et al.,* 1998). Although the lack of detectable Mos may be due to instability of the protein in the absence of progesterone (Nishizawa *et al.,* 1992), another possible explanation is that the mRNP structure is altered in response to progesterone, either by removal of a specific repressor, or by a modification of the nonspecific mRNP proteins.

An alternative mechanism for translational activation by cytoplasmic polyadenylation was proposed by Kuge and Richter (1995). They observed that cytoplasmic polyadenylation induced ribose methylation of the two nucleotides following the cap. Inhibitors of methylation specifically inhibited the translational induction of mRNAs by cytoplasmic polyadenylation without affecting either polyadenylation itself or general translation (Kuge and Richter, 1995). In addition, ribose methylation was sufficient to increase the translational efficiency of injected mRNAs considerably (Kuge *et al.,* 1998). However, it seems that not all mRNAs that are polyadenylated are modified by ribose methylation and this mechanism of translational induction may be specific for a subset of mRNAs (Gray and Wickens, 1998). Because cytoplasmic polyadenylation is required for ribose methylation to occur, this process may be involved in the translational activation of mRNAs that require cytoplasmic polyadenylation rather than a poly(A) tail *per se* for their translational activation.

B. Deadenylation

1. Deadenylation of Stored Maternal mRNAs

One important question is how the translationally repressed mRNAs in the oocyte come to have a short poly(A) tail. Is this due to the addition of a short poly(A) tail in the nucleus or to a deadenylation of a long poly(A) tail? In mouse, the second process is observed for the mRNA for tissue plasminogen activator (tPA). This mRNA is appended with a 300–400 nucleotide long poly(A) tail in the nucleus and gets deadenylated to 40–60 nucleotides in the cytoplasm. This deadenylation is dependent on a 3′ UTR sequence element of 42 nucleotides, the adenylation control element (ACE), which also contains the CPE (Huarte *et al.,* 1992). It is not known whether the two functions, i.e. poly(A) addition and removal, are mediated by the same elements within the ACE, because the minimal CPE and the deadenylation element of tPA mRNA have not been precisely delineated.

2. Default Deadenylation of Housekeeping mRNAs

During the maturation of the *Xenopus* oocyte, most mRNA species encoding "housekeeping" proteins, such as ribosomal protein mRNAs, are deadenylated and translationally repressed. This is caused by the release of a factor from the nucleus during germinal vesicle breakdown. No specific mRNA sequence directs this deadenylation, rather, it is a default mechanism that can be countered by cytoplasmic polyadenylation (Varnum and Wormington, 1990; Fox and Wickens, 1990; Varnum *et al.,* 1992). The default deadenylation is prevented by the overexpression of PABP (Wormington *et al.,* 1996). A deadenylating nuclease that is inhibited by PABP has been identified in *Xenopus* oocytes and is thought to be responsible for this deadenylation (Körner *et al.,* 1998).

C. Translational Repression by 3′ UTR Elements

Although a short poly(A) tail is expected to lead to inefficient translation due to decreased recruitment of elF4G, it is not likely to completely prevent translation. For instance, the mRNA for histone B4, which has a short poly(A) tail in the immature oocyte, is found partially in the polyribosome fraction (about 15% of the total) and the protein accumulates to significant levels during oogenesis (Dworkin *et al.,* 1985; Smith *et al.,* 1988; De Moor and Richter, 1997). In contrast, the translation of other maternal mRNAs with similarly short poly(A) tails, such as the cyclin B1 and *c-mos* mRNAs, is strongly repressed during translation. Consequently, special features of some mRNAs are probably involved where full masking is observed. Two such features are described below.

1. Translational Repression by CPEs

The translation of some mRNAs in the immature oocyte has to be strictly controlled to avoid inappropriate induction of maturation. Among these are the mRNAs encoding Mos and cyclin. Translational repression of cyclin mRNAs is observed in oocytes throughout the animal kingdom (Standart *et al.,* 1990; Kobayashi *et al.,* 1991; Dalby and Glover, 1992; Galas *et al.,*

1993; Sheets *et al.,* 1994; De Moor and Richter, 1997; Sakamoto *et al.,* 1998). Cyclin mRNAs in the *Xenopus* oocyte have short poly(A) tails (Sheets *et al.,* 1994). However, as indicated in the previously, a short poly(A) tail alone is not expected to give a strong repression of translation. Cyclin B1 mRNA, for instance, is absent from the polyribosomal fraction and the protein is virtually undetectable in immature *Xenopus* oocytes, thus indicating that its translation is not just inefficient, but strongly repressed. Injection of the cyclin B1 3′ UTR into oocytes induced translation of cyclin B1 in the absence of cytoplasmic polyadenylation, thus indicating that this RNA binds a translational repressor that is titrated by the injected RNA (De Moor and Richter, 1999). Mutational analysis of the injected RNA indicated that the CPE was sufficient to induce cyclin B1 synthesis in this assay. Moreover, if CPEs were placed behind a reporter coding region, they repressed translation in a dose-dependent manner. These data indicate that a translational repressor binds to the CPEs of cyclin B1 mRNA. Moreover, because the only known CPE-binding protein is CPEB, both in immature and mature oocytes, it seems likely that the switch from repression to activation is mediated by this protein.

It is not immediately clear how the CPEs in the 3′ UTR of an mRNA can interfere with translation initiation to cause repression. The translational repression is independent of the presence or absence of a short poly(A) tail, so inhibition of PABP function can be excluded as a repression mechanism. Consequently, the CPE associated repression complex has to interfere with the assembly of translation initiation complexes assembling at the 5′ end of the mRNA. To determine which step of translation initiation is inhibited by the CPE sequences, an internal ribosome entry site (IRES) was employed. An IRES has the capability of recruiting translation initiation factors to the 5′ UTR of an mRNA independently of the cap or elF4E. All other initiation factors, including elF4G, are still required for IRES mediated translation. Placing an IRES 5′ of a reporter coding region resulted in a marked reduction of CPE-mediated masking in injected oocytes, showing that the repression was cap-dependent and probably interfered with the binding of elF4E to the cap or the binding of elF4G to elF4E.

CPEB does not contain recognizable cap or elF4E binding sequences, suggesting that the translational repression was mediated by CPEB, associated proteins. A CPEB associated protein, named maskin, was isolated and cloned. It was found to contain a putative elF4E binding domain. The binding between maskin and elF4E was confirmed in oocyte extracts as well as in the yeast two-hybrid system. Maskin is proposed to act like an mRNA specific 4E-BP and to inhibit translation by excluding elF4G from the mRNP, as depicted in Fig. 4 (Stebbins-Boaz *et al.,* 1999).

Reporter mRNAs that cannot be polyadenylated because they lack the hexanucleotide retain their CPE mediated repression even after maturation

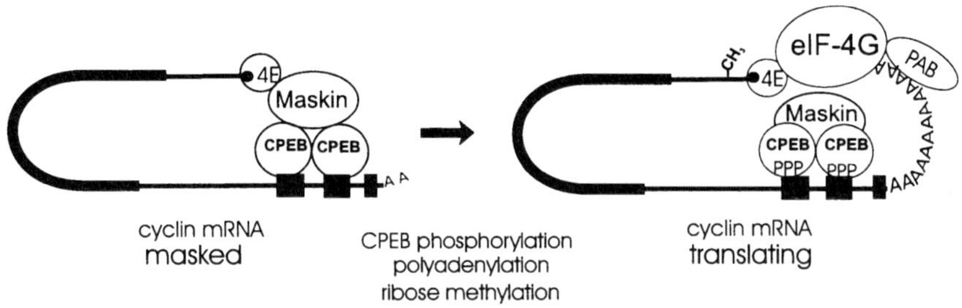

FIG. 4 A model for the masking and unmasking of cyclin B1 mRNA. In the immature oocyte, cyclin B1 mRNA is translationally repressed by the CPEB-binding factor maskin. Maskin interacts with the cap-binding initiation factor eIF4E and prevents the binding of eIF4G. During oocyte maturation, the process of cytoplasmic polyadenylation breaks the interaction of maskin with eIF4E and eIF4G can mediate translation initiation. For a legend of the features of the mRNAs see the top of Fig. 1.

is completed. This indicates that the relief from translational repression is dependent on cytoplasmic polyadenylation (De Moor and Richter, 1999). While the interaction between CPEB and maskin is retained during maturation, the eIF4E-maskin interaction is reduced (Stebbins-Boaz *et al.,* 1999). Possibly, the act of polyadenylation disrupts the eIF4E-maskin interaction, for instance by inducing ribose methylation. Alternatively, the presence of a long poly(A) tail may enhance the recruitment of eIF4G by PABP and bring eIF4G in a position to compete with maskin for the binding of eIF4E (Fig. 4).

2. Translational Regulation of the FGF Receptor-1

Another mRNA that is translationally repressed in immature *Xenopus* oocytes and activated upon maturation is the mRNA for the fibroblast growth factor (FGF) receptor-1 (FGFR-1). Because FGF signaling is required for mesoderm induction, this regulation is likely to be important for pattern formation (Amaya *et al.,* 1993). Although the FGFR-1 mRNA is polyadenylated during maturation, this event is uncoupled from translational activation in oocytes induced to mature by injection of cdc2/cyclin B1; polyadenylation occurs in these oocytes,but no FGFR-1 synthesis is detected (Culp and Musci, 1998). In addition, the translational regulation of FGFR-1 mRNA can be mimicked by a reporter mRNA containing a 3′ UTR element that lacks the CPE and the hexanucleotide. This translation inhibitory element (TIE) is 180 nt long and U-rich and it specifically binds a 43 kDa protein of unknown identity (Robbie *et al.,* 1995). Because polyadenylation seems not to be required for the translational activation of

TIE-containing mRNA, it is likely that a completely different set of proteins is involved in the translational control by the TIE and the CPE.

D. Translational Repression by the Y-Box RNA Binding Proteins

1. Y-Box Proteins as Nonspecific Translational Inhibitors

Among the major components of mRNP particles are the Y-box proteins. These proteins are sequence-specific DNA-binding transcription factors that bind to the Y-box in DNA. In addition, they are nonspecific RNA-binding proteins. All these proteins contain a nucleic acid binding domain called the cold shock domain, which forms a 5 stranded β barrel. The cold shock domain contains sequences reminiscent of the RNP1 and 2 sequences of the RNA recognition motif, which has a different domain structure ($\beta\alpha\beta\beta\alpha\beta$), but contacts the RNA with its antiparallel β-sheets. This indicates that these two nucleic acid binding motifs may be part of an ancient nucleic acid binding superfamily. In addition, the Y-box proteins contain a c-terminal domain with conserved basic/aromatic islands, which has some affinity for RNA and mediates the formation of homomultimers of these proteins (Matsumoto *et al.,* 1998; Evdokimova and Ovchinnikov, 1999; Sommerville, 1999). Germ cells contain specific Y-box proteins (FRGY2 in *Xenopus,* MSY1 in mouse), while somatic cells have their own Y-box proteins (p50 in reticulocytes).

The levels of Y-box proteins in untranslated mRNP are twice as high as in translated mRNP (Minich and Ovchinnikov, 1992) and moderately elevated levels of Y-box proteins lead to the inhibition of translation in reticulocyte lysates, oocytes, and somatic cells (Richter and Smith, 1984; Ranjan *et al.,* 1993; Minich *et al.,* 1993; Bouvet and Wolffe, 1994; Matsumoto *et al.,* 1996). In reticulocyte lysates, low levels of the Y-box protein p50 are necessary for translation initiation and faithful cap-dependent translation (Svitkin *et al.,* 1996; Evdokimova *et al.,* 1998), thus indicating a dual role for Y-box proteins; they function as translation factors at low levels and as inhibitors of translation at higher levels. These characteristics may be explained by the RNA-binding properties of Y-box proteins. They bind preferentially to single stranded RNA and may act as an RNA chaperone, keeping the RNA in a single stranded form, which would be expected to enhance translation (Evdokimova *et al.,* 1995). At high concentrations, the Y-box proteins can form multimers, which might contribute to "wrapping up" the mRNP into an inaccessible form (Spirin, 1996).

2. Translational Regulation by Y-Box Proteins

How do mRNP particles become assembled into mRNPs with either low or high levels of Y-box proteins? One possibility is that the Y-box protein content of mRNP particles is a reflection of their initial translational activity; efficient translation could prevent the accumulation of high levels of Y-box proteins while poor translation might permit it. This would result in an enhancement of specific translational control mechanisms by the silencing of poorly translating mRNAs. The levels of Y-box proteins in the cell could determine what translational efficiency is the threshold value. This mechanism could explain the switch of the bulk of the poly(A) mRNA from the free mRNP in oocytes to the polyribosomes in the embryo, where most of the FGRY2 is degraded (Tafuri and Wolffe, 1993). Alternatively, sequence-specific translational repressors may induce efficient binding of Y-box proteins. However, synthetic reporter mRNAs can still be masked by Y-box proteins. It is not known if Y-box proteins are assembled on the mRNA in the cytoplasm or in the nucleus, but several investigations suggest that the nuclear history of an mRNA can influence its packaging (Bouvet and Wolffe, 1994; Braddock *et al.,* 1994; Gunkel *et al.,* 1995; Sommerville and Ladomery, 1996; Meric *et al.,* 1996; Matsumoto *et al.,* 1998; Matsumoto and Wolffe, 1998).

The switch between an untranslated mRNP particle with a high Y-box protein content to a translated mRNP particle with lower levels of these proteins obviously will require some major remodeling of the particle, especially under conditions where the Y-box protein content of the cell does not change. One of the processes that is capable of inducing this change is cytoplasmic polyadenylation (Meric *et al.,* 1996). A possible mechanism is that cytoplasmic polyadenylation directly induces structural changes in the mRNP, possibly involving the activation of kinases, helicases, and chaperones found in these particles (Cummings and Sommerville, 1988; Cummings *et al.,* 1989; Ladomery *et al.,* 1997; Yurkova and Murray, 1997). Alternatively, the increased affinity of the mRNA for translation initiation factors may induce rearrangement of the mRNP by the translational machinery. Possibly, the difference between a first initiation and subsequent initiations is the Y-box protein content of the mRNP (see Introduction). In this context, it may also be significant that the initiation factor most likely to be required for mRNP remodelling, the helicase elF4A, has been reported to be the limiting factor for the translational capacity of *Xenopus* oocytes (Audet *et al.,* 1987).

Although the ability of Y box proteins to influence translation is firmly established, there is still little concrete knowledge concerning the interplay between these proteins, the translation initiation machinery, and mRNA-specific translational regulators. However, the indications are that Y-box proteins will prove to be important players in both translational repression and activation.

III. Translational Control during Embryogenesis

Translational control in embryogenesis is not as well characterized in vertebrates as in invertebrates. In *Drosophila,* localized mRNAs, which often are under translational control, account for most of the early pattern formation in the embryo (see the reviews by Cooperstock and Lipshitz, Chapter 14; Deng and Lin, Chapter 3; and Mahowald, Chapter 5). In *Xenopus,* several localized mRNAs have been characterized, although less is known of their functions than in *Drosophila* (see the review by Kloc *et al.,* Chapter 2). The mRNA for the transcription factor VegT, for instance, is vegetally localized in the oocyte and controls the pattern of primary germ layer specification in *Xenopus* embryos (Zhang *et al.,* 1998). The VegT protein cannot be detected in oocytes but is present in eggs, thus indicating that this mRNA is translationally regulated during oocyte maturation (Stennard *et al.,* 1999). Very little is known of the translational regulation of such localized mRNAs in the *Xenopus* embryo. In contrast to other vertebrates, mammals do not seem to utilize localized factors in the egg to determine pattern formation (Zernicka-Goetz, 1998). In addition, transcription of the zygotic genome starts after the two-cell stage, suggesting that translational control may be less essential in the mammalian embryo. However, extensive changes in the pattern of protein synthesis have been detected after fertilization of the egg and before the onset of transcription, thus indicating that translational control is likely to occur during this time (Latham *et al.,* 1991).

As mentioned above, the complement of Y-box mRNP proteins changes drastically after the fertilization of *Xenopus* eggs and most of the poly(A) mRNA commensurately shifts from the untranslated mRNP into the polyribosomes (Tafuri and Wolffe, 1993). This wholesale shift in translational expression is likely to reflect a need for housekeeping proteins in the cells of the developing embryo, but also includes the translation of dormant mRNAs, such as histone H1. Histone H1 replaces the maternal histone B4 in the chromatin during the cleavage stages. This leads to the silencing of certain genes and determines the fate of cells in the embryo (Wolffe, 1997; Wolffe and Hayes, 1999).

A. Cytoplasmic Polyadenylation

Translational control by cytoplasmic polyadenylation occurs in the early *Xenopus* embryo. Two cDNAs were selected for their ability to undergo cytoplasmic polyadenylation during the cleavage stage, Cl1 and Cl2 (Paris *et al.,* 1988; Paris and Philippe, 1990). Both mRNAs are translationally activated at the same time as they are polyadenylated. Cl1 was found to be the eukaryotic translation termination factor eRF1 (Frolova *et al.,* 1994), while the function of Cl2 is unknown. Similar to the cytoplasmic polyadenylation in oocytes, the polyadenylation of these mRNAs requires the presence of a hexanucleotide and a specific cytoplasmic polyadenylation element, referred to here as the embryonic CPE, or eCPE. The eCPE consists of a stretch of 12 or more uridine residues and mediates cytoplasmic polyadenylation at different time points in development dependent on the distance between the eCPE and the hexanucleotide (Simon *et al.,* 1992; Simon and Richter, 1994). The eCPE can also mediate cytoplasmic polyadenylation during oocyte maturation, but is normally prevented from doing so in its natural context because of repressive sequences in the 3′UTRs in which it resides.

The maternal mRNA for the activin receptor contains an eCPE and is polyadenylated during embryogenesis (Simon *et al.,* 1996). Because activin signaling is implicated in the induction of mesoderm, this could be important for pattern formation. Indeed, overexpression of activin receptor, which is known to induce severe abnormalities in embryos, requires the eCPE. Conversely, injection of a large excess of eCPE reduced the appearance of mesodermal markers (Simon *et al.,* 1996). The last result indicates that a titratable factor binds to the eCPE and is required for the translation of the activin receptor mRNA.

ElrA, a member of the ELAV family of RNA binding proteins, binds specifically to the eCPE (Wu *et al.,* 1997). The injection of a deletion mutant of ElrA caused a dominant exogastrula phenotype.

B. Deadenylation

1. Deadenylation of the Eg mRNAs

Several mRNAs that are polyadenylated during maturation are specifically deadenylated after fertilization (Paris *et al.,* 1988; Paris and Philippe, 1990). Many of these mRNAs encode cell cycle regulators, such as Eg1 (cdk2), Eg2 (a kinase of the AIM/AIK/Aurora family), Eg5 (a kinesin-like molecule), and Eg7 (a factor involved in chromosome condensation) (Paris *et al.,* 1991; Houliston *et al.,* 1994; Blangy *et al.,* 1995; Roghi *et al.,* 1998; Andrésson and Ruderman, 1998; Cubizolles *et al.,* 1998). These mRNAs contain maturation type CPEs, which by themselves would mediate polyadenylation in the embryo as well as in the maturing oocyte. In addition, however, the 3′UTRs of these mRNAs contain sequences that specifically inhibit polyadenylation and translation in the embryo. If these sequences are deleted, the RNA is polyadenylated upon injection into the embryo (Bouvet *et al.,* 1994; Stebbins-Boaz and Richter, 1994). At least two different deadenylation elements exist, one present in cdk2 mRNA (Stebbins-Boaz and Richter, 1994), and another one called the embryonic deadenylation control element (EDEN). This sequence is rich in UG/A repeats and was found in the Eg2, Eg5, and *c-mos* mRNAs (Bouvet *et al.,* 1994; Paillard *et al.,* 1998).

The factor that binds the EDEN has been named EDEN-BP. It contains three RNA recognition motifs, similar to the ELAV-like RNA binding proteins (Paillard *et al.,* 1998). Immunodepletion of EDEN-BP from embryo lysates abolished their capability for EDEN-specific deadenylation, indicating that EDEN-BP is essential for this process. A fairly close relative of EDEN-BP is the *Drosophila* protein Bruno, a translational repressor of the posterior determinant *oskar,* indicating that the translational repressor function of EDEN-BP may have been conserved during evolution (Webster *et al.,* 1997).

2. Deadenylation Mediated by AU-Rich Regions

Another sequence that can promote deadenylation in the *Xenopus* embryo is the well-known AU-rich mRNA destabilization element (ARE). The ARE, which typically contains several copies of the AUUUA core element, is found in several oncogene and cytokine mRNAs (Chen and Shyu, 1995; Voeltz and Steitz, 1998). The AREs are known to promote deadenylation in tissue culture cells as a first step toward mRNA degradation. This is then followed by decapping and rapid degradation of the mRNA. In *Xenopus* oocytes, however, deadenylated messengers are not degraded until the mid-blastula stage (Bouvet *et al.,* 1991; Audic *et al.,* 1997; Voeltz and Steitz, 1998). This may be due to the absence of decapping, the second step of mRNA degradation, in oocytes and early embryos (Zhang *et al.,* 1999).

AU-rich sequences have been reported to convey translational repression to reporter mRNAs in both oocytes and embryos, but this inhibition occurs also in the absence of a poly(A) tail (Kruys *et al.,* 1989; Marinx *et al.,* 1994). No mRNAs endogenous to the embryo have been reported to be regulated by ARE-mediated translational control, but translational regulation by AREs has been observed in the immune system, for instance for β interferon, granulocyte-macrophage colony stimulating factor, interleukin 3, and tumor necrosis factor α (TNFα) (Kruys *et al.,* 1989, 1990; Grafi *et al.,* 1993; Marinx *et al.,* 1994; Nair *et al.,* 1999). No clear link between ARE-mediated deadenylation and translational repression has been established for these mRNAs either. In at least in the case of TNFα, however, translational activation seems to correlate with a longer poly(A) tail (Crawford *et al.*, 1997).

C. eIF4E and 5′ UTR Structure

Secondary structure in the 5′ UTR usually inhibits translation by preventing the assembly or the scanning of the initiation complex. Many mRNAs encoding regulatory proteins such as oncogenes and growth factors have unusually long and/or structured 5′ UTRs. It has been demonstrated that elevated levels of eIF4E can induce translation of such mRNAs, presumably by recruiting more of the eIF4A RNA helicase to the mRNA (Koromilas *et al.*, 1992; Rosenwald *et al.*, 1993; Shantz and Pegg, 1994; Hoover *et al.*, 1997). Indeed, overexpression of eIF4E can lead to malignant transformation, presumably by inducing the translation of mRNAs encoding oncogenic proteins (De Benedetti and Harris, 1999; Willis, 1999; Clemens and Bommer, 1999). In the normal cell, eIF4E activity can be regulated by 4E-BP binding and phosphorylation, for instance in response to mitogens (Raught and Gingras, 1999). Therefore, it is conceivable that the translation of mRNAs with structured 5′ UTRs could be regulated by different levels or activity of eIF4E in the different germ layers of the embryo. To address this issue, Klein and Melton overexpressed eIF4E in *Xenopus* embryos, which led to the induction of mesoderm in tissues that would normally form ectoderm. The synthesis of the mesodermal inducer activin was found to be specifically enhanced in these cells (Klein and Melton, 1994). These data indicate that the regulation of translation by the modulation of eIF4E activity could play a role in mesoderm induction.

IV. Translational Control in Growth and Differentiation

In this section we discuss examples of translational control in the later stages of development, concentrating on mRNAs involved in growth and differentiation for which the regulatory mechanisms have been at least partially characterized.

A. Translational Control by the 5′ UTR and Coding Region

Because the initiation complex must bind in this region, the 5′ UTR is the most obvious location for translational regulation to occur. The average 5′ UTR is approximately 100 to 200 bases long and devoid of extensive secondary structure (Pesole *et al.,* 1994, 1997). In the majority of cases translation starts at the first AUG from the 5′ end, especially if it is in a favorable context (Kozak, 1991). Another prerequisite for accurate initiation codon recognition is a sufficient distance from the cap. A 5′ UTR of 12 nucleotides, for instance, leads to inefficient initiation codon recognition (Sedman *et al.,* 1990).

During normal cap-dependent initiation, the scanning initiation complex has to traverse the 5′ UTR. Any obstacle impeding the progress of this complex would reduce translation efficiency. A variety of such "road blocks" is employed as translational control switches in 5′ UTRs, including secondary structure, RNA-binding proteins, and small upstream open reading frames. The possibility of translational control mediated by secondary structure in the 5′ UTR and changes in eIF4E activity was discussed in the previous section; instances of translational control mediated by 5′ UTR-binding proteins and upstream initiation codons, as well as by internal initiation, will be discussed here.

1. Translational Regulation by the 5′ Terminal Oligopyrimidine Tract

Secondary structure is not necessarily the only feature of a 5′ UTR that can mediate a strong dependence of eIF4E activity. A situation in which a translational repressor protein competes with eIF4E for the cap could give a similar effect. One such case seems to be the translational control by the 5′ terminal oligopyrimidine tract (TOP), which comprises a cytidine at the transcription start site followed by an uninterrupted stretch of up to 13 pyrimidines. TOP mRNAs include all ribosomal protein mRNAs, the translation elongation factors EF1α and EF2 and PABP (Jefferies and Thomas, 1994; Terada *et al.,* 1994; Meyuhas *et al.,* 1996; Amaldi and Pierandrei-Amaldi, 1997; Avni *et al.,* 1997; Loreni and Amaldi, 1997; Hornstein *et al.,* 1999; Jefferies and Thomas, 1994). The translation of these mRNAs is repressed in quiescent cells, where eIF4E activity is low, and induced in cells stimulated with mitogens, which have high levels of eIF4E activity. In addition, translational repression in reticulocyte lysate can be relieved by the addition of eIF4F, the cap binding initiation complex of eIF4E/4G/4A. However, repression TOP mRNA translation is not always observed when eIF4E activity is low or prevented when eIF4E is kept high by overexpression (Shama *et al.,* 1995; Loreni and Amaldi, 1997). Recently, Biberman and Meyuhas reported that translation of TOP mRNAs was specifically enhanced in wheat germ extract and reticulocyte lysate by the addition of TOP polypyrimidine oligonucleotides (Biberman and Meyuhas, 1999). These data indicate that the translational regulation of TOP mRNAs is mediated by a translational repressor binding to the TOP sequence.

Although several different TOP-binding proteins have been reported, none have been demonstrated to actually function in the translational control of TOP mRNAs (Kaspar *et al.,* 1992; Severson *et al.,* 1995; Pellizzoni *et al.,* 1998).

Because the TOP sequence has to be adjacent to the cap and because an increase in eIF4F can relieve TOP mediated repression of translation in reticulocyte lysates, it seems likely that the unknown TOP-binding factor represses translation by preventing eIF4E binding to the cap (illustrated in Fig. 5). This is in contrast to the translational repression of ferritin mRNA

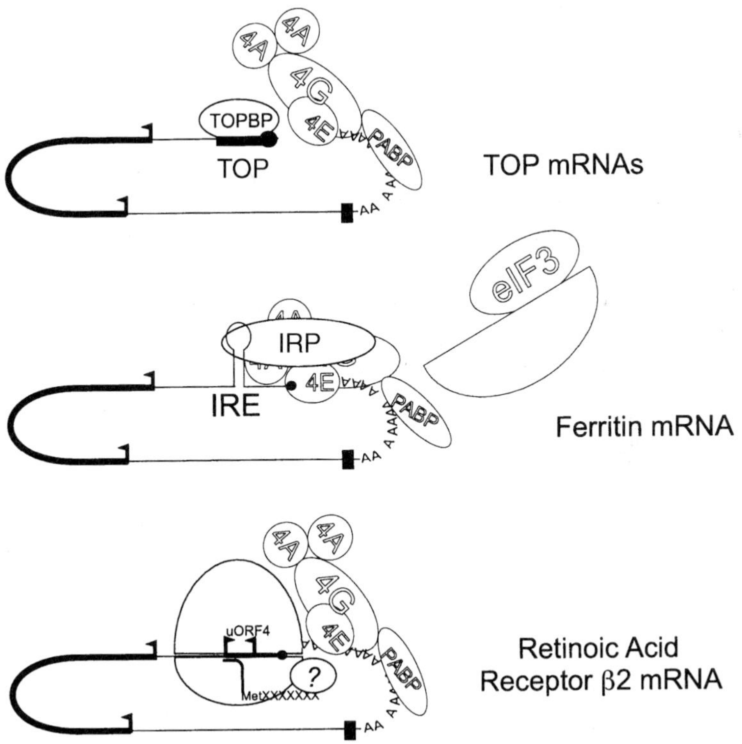

FIG. 5 Translational repression by 5′ UTR elements. A group of mRNAs containing 5′ terminal oligopyrimidine (TOP) tracts, which include the ribosomal protein and PABP mRNAs, is thought to be translationally repressed by a protein that binds to the TOP and prevents cap association of eIF4E (TOPBP). The iron response element (IRE) is a hairpin in the 5′ UTR of the ferritin mRNA, located close to the cap. Binding of iron response protein (IRP) to the IRE permits the association of the cap-binding initiation factors, but prevents association of eIF3 and recruitment of the small ribosomal subunit. A small upstream open reading frame in the retinoic acid receptor $\beta 2$ mRNA encodes a peptide that stalls the translating ribosome, possibly by binding a specific factor (?). Subsequent initiation complexes are prevented from forming or are blocked during scanning.

during iron starvation. This repression is mediated by the binding of the iron response protein 1 (IRP-1) to a hairpin close to the cap of ferritin mRNA, which inhibits the binding of elF3 and the small ribosomal subunit (Muckenthaler *et al.,* 1998), as illustrated in Fig. 5.

Two kinds of evidence suggest that p70 S6 kinase ($p70^{S6K}$) is likely to be involved in the translation of TOP mRNAs: (1) inhibition of this kinase by rapamycin leads to reduced translation of TOP mRNAs and (2) cells with disrupted $p70^{S6K}$ genes do not increase the translation of TOP mRNAs in response to mitogens (Jefferies *et al.,* 1994, 1997; Kawasome *et al.,* 1998; Schwab *et al.,* 1999). The phosphorylation of ribosomal protein S6 by p70 S6 kinase has been proposed to play a role in translational control by TOP sequences, but in the light of the recent findings outlined above, it seems more likely that this kinase is involved in the inactivation of the TOP binding inhibitor of translation.

Another mRNA that is translationally regulated by rapamycin is the most abundant of the human insulin-like growth factor 2 (IGF-2) mRNAs, the so-called leader 3 mRNA. This mRNA is translationally repressed in most tissues but gets actively translated in exponentially growing rhabdomyosarcoma cells (Nielsen *et al.,* 1990; De Moor *et al.,* 1994a, 1994b). Although the translation of leader 3 IGF2 mRNA is inhibited by rapamycin, it does not contain a TOP sequence. However, an internal polypyrimidine stretch has been implicated in its translational regulation (De Moor *et al.,* 1995). Recently, a family of leader 3 IGF2 specific binding proteins, called the IMPs, was identified. These proteins bind to the 5′ UTR and inhibit translation in a dose-dependent manner. They are highly homologous to the *Xenopus* Vera and chicken zip-code binding proteins, which are known to play a role in subcellular mRNA localization (Ross *et al.,* 1997; Deshler *et al.,* 1998; Havin *et al.,* 1998). The expression of the IMPs during mouse embryogenesis suggests a possible role in development (Nielsen *et al.,* 1999).

2. Translational Regulation by Upstream Open Reading Frames

While upstream open reading frames (uORFs) would be expected to repress the translation of a downstream coding region, several mRNAs employ mechanisms that modulate the repressive effect of such sequences to regulate their translation. One of the ways to avoid the repression mediated by a uORF is called leaky scanning. This occurs if the initiation codon of the uORF is poorly recognized by the initiation complex, for instance because of an unfavorable context or because it is too close to the cap. In this case, some of the scanning initiation complexes will pass the uORF and initiate at the correct initiation codon (Kozak, 1999). Another possibility is that the ribosome, after translating the uORF, will resume scanning for the next initiation codon. This is called reinitiation and has been documented only following the translation of short coding regions. The second initiation codon has to be some distance from the upstream stop codon, presumably to allow the binding of the initiator tRNA and elF2 (Hinnebusch, 1997). Most cases of translational regulation by uORFs seem to involve the regulation of the efficiency of initiation codon recognition (leaky scanning) and/or modulation of the reinitiation frequency. The most extensively studied example of upstream open reading frame-mediated translational regulation involves the yeast GCN4 mRNA translation during amino acid starvation. The reader is referred to an excellent review on this subject (Hinnebusch, 1997).

One of the better studied examples of translational control by uORFs during vertebrate development is the mRNA for the $\beta 2$ receptor for the morphogen retinoic acid (RAR $\beta 2$). The 5′ UTR of this mRNA contains five uORFs. The presence of these uORFs is essential for the repression of translation of RAR $\beta 2$ in the heart and brain of mouse embryos (Zimmer *et al.*, 1994). The initiation codon of the first uORF is in a very poor context and is probably skipped by most initiation complexes. The AUG of the second uORF (uORF2), which resides in a good context, is essential for optimal translation of the mRNA in the presence of the other three uORFs (Reynolds *et al.*, 1996). Presumably, RAR $\beta 2$ is translated by reinitiation after translation of uORF2 and the downstream uORFs are skipped because the ribosome is not ready to reinitiate. In brain and heart, however, uORF4 causes a strong repression of translation, which is dependent on the amino acids encoded by the open reading frame (Reynolds *et al.*, 1996). This is reminiscent of the translational control of S-adenosylmethionine decarboxylase (AdoMetDC), an enzyme involved in polyamine synthesis. The single uORF in the 5′ UTR of this mRNA is responsible for the translational repression of this mRNA in certain cell types and by high polyamine levels (Hill and Morris, 1992, 1993; Ruan *et al.*, 1994, 1996; Shantz *et al.*, 1994). The peptide encoded by the uORF of AdoMetDC (which is not related to the one encoded by uORF4) is thought to cause the translating ribosome to stall on the 5′ UTR and thus prevent the passage of subsequent initiation complexes (Geballe and Morris, 1994). Even a low number of ribosomes initiating on such a regulatory uORF would lead to a strong repression if the stalled ribosome was stable (Fig. 5). Another possibility is that the peptide encoded by uORF4 strongly inhibits reinitiation. In both models, the repression of RAR $\beta 2$ translation in embryonic heart and brain would be mediated either by an increase in the recognition of the initiation codon of the inhibitory uORF4 or by enhanced inhibitory activity of the encoded peptide.

3. Internal Initiation and Use of Alternative Initiation Codons

For certain viral mRNAs, the initiation complex does not assemble at the cap but instead binds in the 5′ UTR in the proximity of the initiation. Sequences that can mediate this internal initiation are called internal ribosome entry sites (IRES). A number of developmentally important mRNAs have been reported to be translated by internal initiation. The most common test for an IRES is to place the sequence between to coding regions in a bicistronic mRNA. The second coding region will only be translated to a significant extent if the inserted sequence is an IRES. Using this assay, IRES sequences have been reported in the *c-myc,* PDGF2, FGF2, and VEGF mRNAs (Vagner *et al.,* 1995; Bernstein *et al.,* 1997; Nanbru *et al.,* 1997; Stoneley *et al.,* 1998; Stein *et al.,* 1998; Huez *et al.,* 1998; Sella *et al.,* 1999). The cellular IRES have no clear similarity to the viral IRES and are generally less efficient in promoting internal initiation.

The *c-myc* and FGF2 mRNAs, besides encoding protein from the normal AUG initiation codon, generate N-terminally extended proteins by initiation on CUG codons in the 5′ UTR (Hann *et al.,* 1988; Nanbru *et al.,* 1997; Arnaud *et al.,* 1999). The N-terminally extended forms have different functional properties and are variably expressed in different cell types and under different conditions (Bugler *et al.,* 1991; Hann *et al.,* 1994; Vagner *et al.,* 1996; West *et al.,* 1998; Arnaud *et al.,* 1999; Galy *et al.,* 1999). Exactly how the choice of initiation codon is made in these cases is not clear yet, but it is likely to involve a balance among cap-dependent initiation, leaky scanning, and internal initiation. A number of other mRNAs employ alternative initiation codons to generate proteins with N-terminal extensions, among them the oncogenes Pim-1, WT-1, and int-2 (Saris *et al.,* 1991; Hoover *et al.,* 1997; Dickson and Acland, 1990; Bruening and Pelletier, 1996). It is not known if these mRNAs contain IRES.

The activity of several cellular IRES has been reported to be regulated during differentiation. For instance, PDGF2 mRNA is extremely poorly translated in most cells, but the internal initiation on the PDGF2 IRES is enhanced in megakaryocyte differentiation (Bernstein *et al.,* 1997). hnRNP C binds to a region of the PDGF2 IRES that is necessary for the induction, and this nuclear protein is released into the cytoplasm during differentiation, providing a possible mechanism for translational induction of PDGF2 translation (Sella *et al.,* 1999). Translational regulation by IRES during vertebrate embryogenesis has not yet been studied extensively. However, because many of the mRNAs that have been reported to contain an IRES and/or alternative initiation codons encode proteins that are important in early development, it seems likely that this type of regulation also occurs in the vertebrate embryo. In addition, in *Drosophila,* the mRNAs for the homeotic proteins Ultrabithorax and

Antennapedia are both regulated during fly development by IRES elements in their 5′ UTRs (Ye *et al.,* 1997).

B. Translational Control by 3′ UTR Elements during Differentiation

1. Translational Control of 15-Lipoxygenase during Erythrocyte Differentiation

Lipoxygenases are enzymes that are involved in the breakdown of membranes by lipid peroxidation. During erythrocyte differentiation, lipoxygenase mediates the breakdown of internal cell membranes, including the mitochondrial membranes. Obviously, such an irreversible process has to be tightly controlled (Kuhn *et al.,* 1999). 15-lipoxygenase (LOX) mRNA is abundant in bone marrow erythroid precursor cells, but it is translationally repressed until the final stages of reticulocyte maturation in the peripheral blood. This repression can be reconstituted in an *in vitro* translation system by the addition of LOX 3′ UTR binding proteins purified from reticulocytes. These proteins bind to a UC rich repeat sequence in the 3′ UTR called the differentiation control element (DICE) (Ostareck-Lederer *et al.,* 1994). Subsequently, the DICE binding proteins were identified as hnRNP E1 and hnRNP K (Ostareck *et al.,* 1997). hnRNP proteins are, of course, predominantly localized in the nucleus, which would seem to preclude a role for these proteins in the cytoplasm. However, a growing number of cytoplasmic functions have been ascribed to hnRNP proteins, such as mRNA stabilization, replication of RNA viruses, and internal initiation of translation (Ostareck-Lederer *et al.,* 1998; Krecic and Swanson, 1999). Although hnRNP K and E1 are present in all cell types, DICE-mediated translational repression was absent in tissue culture cells. However, if the hnRNP proteins were overexpressed, DICE-mediated translational repression was observed. This may mean that overexpression brings the cytoplasmic concentration of these proteins to a threshold level, or that a titratable factor inhibits their activity in these cells. In nonerythroid cells, the LOX mRNA has a 3′ extended 3′ UTR that does contain two DICE-like sequences and conveys hnRNP K and E1 mediated repression *in vitro.* The second DICE in this RNA (which is absent from the shorter 3′ UTR) binds a 93 kDa factor that was isolated from lung cytosol and can suppress the hnRNP E1 and K-mediated translational inhibition (Thiele *et al.,* 1999). These data suggest a mechanism whereby this protein interacts with hnRNP K and E1, countering their repressive action when it is bound to the same mRNA. It is not known how the repression of translation is relieved during erythrocyte differentiation.

In contrast to the CPE-mediated translational repression observed in oocytes (De Moor and Richter, 1999), the translational repression by DICE is not relieved in mRNAs translated by internal initiation (Ostareck *et al.,* 1997). This indicates that the DICE-mediated translational inhibition interacts with the translational machinery at a later, cap-independent step of translation initiation. In addition, the sedimentation of the repressed mRNA in sucrose gradients suggests it has bound a small, but not a large ribosomal subunit. This may point to a deficiency in scanning, initiation codon recognition, or recruitment of the large ribosomal subunit. Such a mechanism would require an interaction with the small ribosomal subunit or its bound factors, such as elF1, elF2, elF3, or even the initiator tRNA.

2. Translational Control of Protamine-1 mRNA during Spermiogenesis

During spermiogenesis, the postmeiotic stage of mouse spermatogenesis, the chromatin becomes tightly condensed. This compaction is mediated by the replacement of the main chromosomal proteins, the histones, with the transition proteins in the round spermatids. Subsequently, in the elongating spermatids, the transition proteins on the chromatin are exchanged for protamines. Transcription ceases several days after meiosis, making translational control an important regulatory mechanism during the late stages of spermiogenesis (Eddy, 1998; Braun, 1998; Steger, 1999).

The mRNAs for the transition proteins and the protamines are stored in a translationally repressed form and become translationally activated at different times during spermiogenesis. Translational activation is accompanied by a shortening of the long poly(A) tails (Kleene, 1989; Hake *et al.,* 1990; Domenjoud *et al.,* 1991). It is not known whether this deadenylation is the cause or the result of translational activation. In oocytes, such short poly(A) tails usually lead to inefficient translation. That this does not happen in spermatids is probably because their PABP levels are high and a short poly(A) tail is likely to sufficient for the recruitment of PABP-elF4G, as long as the translation of the mRNA is not otherwise repressed (Gu *et al.,* 1995).

Both the repression and the activation of protamine-1 mRNA is mediated its 3′ UTR, specifically by a 37 nt element immediately following the coding region and a 62 nt element that includes the hexanucleotide (Braun *et al.,* 1989). The upstream element alone confers repression but fails to activate translation of a reporter mRNA encoded by a transgene. This element specifically cross-links to a 48/50 kDa doublet of proteins in extracts from round spermatids, but is not observed in extracts from elongating spermatids (Fajardo *et al.,* 1997). The doublet probably contains a member of the Y-box family of RNA binding proteins (Braun, 1998). These data are suggestive, but paradoxical; the 37 nt element is repressive in both round and elongating spermatids, but the 48/50 kDa proteins bind to it only in round spermatids. Possibly, the binding of these proteins mediates a remodelling of the mRNP that persists after their binding is lost.

The second translational control element, located in the last 62 nt of the 3′ UTR, conveys both repression and activation to a reporter transgene (Braun, 1990). This RNA does not bind the 48/50 kDa doublet, but instead contains two binding sites for Prbp (protamine RNA-binding protein) (Lee *et al.,* 1996). Mouse Prbp is the ortholog of human TAR RNA binding protein (TRBP), a protein that binds to regulatory hairpin structures in the RNA of human immunodeficiency virus type 1 (Gatignol *et al.,* 1991; Park *et al.,* 1994). PRBP contains three double stranded RNA binding motifs, similar to the lysine and arginine rich motifs found in such double stranded RNA binding proteins as the double-stranded RNA activated kinase PKR (St. Johnston *et al.,* 1992). Other members of this family have been reported to exhibit RNA annealing activity and are proposed to be RNA chaperones (Hitti *et al.,* 1998). A knockout of the gene encoding PRBP (Tarbp2) leads to a high perinatal mortality rate and male infertility (Zhong *et al.,* 1999). The translational activation of a reporter transgene with the last 62 nt of the protamine-1 3′ UTR was partially impaired in the Tarbp2 knockout mice, which exhibited a mosaic expression pattern. The effect on protamine expression was similar. However, translational activation of a transition protein mRNA was not impaired. These data suggest that PRBP is necessary for the specific translational activation of protamine mRNAs, at least in a subset of the elongating spermatids (Zhong *et al.,* 1999).

3. Translational Control of Neurofilament mRNA and Neuronal Differentiation

In *Drosophila,* the gene *embryonic lethal abnormal vision* (*elav*) is essential for the development and maintenance of the nervous system (Campos *et al.,* 1985; Robinow and White, 1988). In vertebrates, four homologs of the elav protein have been described: HuA, B, C, and D. HuA (HuR, ElrA) is ubiquitously expressed, HuB (Hel-N1, ElrB) is found in ovary, testis, and neurons, while HuC and HuD (ElrC, ElrD) are exclusively neuronal (Good, 1995; Antic and Keene, 1997). All elav-like proteins contain 3 RNA recognition motifs and bind to U-rich sequences including the AREs of *c-myc, c-fos,* and GM-CSF (Levine *et al.,* 1993; Antic and Keene, 1997; Wu *et al.,* 1997). Elevated levels of these proteins can lead to stabilization of normally unstable ARE-containing mRNAs (Jain *et al.,* 1997; Fan and Steitz, 1998; Levy *et al.,* 1998; Peng *et al.,* 1998; Ford *et al.,* 1999).

Antic *et al.* (1999) characterized the role of Hel-N1 (HuB) in neuronal differentiation by studying hNT2 cells stably transfected with sense or antisense Hel-N1 constructs. Sense Hel-N1 transfected cells displayed neurite-like outgrowths, while vector transfected cells maintained their undifferentiated phenotype. Treatment of these cells with retinoic acid (RA) induces neuronal differentiation in untransfected cells. RA treatment of transfected cells resulted in enhanced neuronal differentiation in the cells transfected with a sense Hel-N1 construct. In cells transfected with the antisense Hel-N1 construct, RA-induced differentiation was inhibited. Hence, expression of Hel-N1 was necessary for differentiation and overexpression enhanced differentiation.

Neurofilament M (NF-M) protein expression was induced by the sense Hel-N1 transfection. The NF-M mRNA levels were the same before and after transfection, which excluded a stabilizing effect of Hel-N1 on the NF-M mRNA. Consequently, it seemed likely that Hel-N1 enhanced the translation of the NF-M mRNA. This was confirmed by the observation that NF-M mRNA was shifted into the heavy polyribosome fraction in sucrose gradients of extracts of Hel-N1 transfected cells. The 3′ UTR of NF-M contained several U-rich motifs that bound Hel-N1 *in vitro,* and cellular NF-M mRNA could be immunoprecipitated with the protein expressed from the Hel-N1 sense construct, thus making it likely that the translational stimulation of NF-M mRNA by Hel-N1 was a direct effect (Antic *et al.,* 1999). Both the lack of ribosome loading on NF-M mRNA and the absence of NF-M protein in untransfected hNT2 cells indicates that NF-M is translationally repressed in these cells. Consequently, it is possible that Hel-N1 induces translation of NF-M mRNA by displacing a translational repressor. Alternatively, Hel-N1 may be involved in recruiting translation initiation factors or changing the structure of the mRNP. In this respect, it would be informative to know what parts of the Hel-N1 protein are necessary for translational activation and which regions of the NF-M mRNA mediate the repression and activation of translation.

V. Concluding Remarks

This overview of translational control during development illustrates the diversity of mechanisms employed in this type of regulation. In the great majority of cases, translational control acts at the level of translation initiation, but the controlling elements can be either inhibitory or activating and are found both in the 5′ and the 3′ untranslated sequences. A large number of models have been developed to explain the different forms of translational repression. The mechanisms of translational activation are less well characterized, with the notable exception of translational activation by cytoplasmic polyadenylation.

The preponderance of U-rich elements in translational control (six out of the seven 3′ UTR elements discussed here) may reflect an evolutionary relatedness, a need for a loosely stacked single stranded recognition sequence, or a statistical accident that will disappear as more elements are discovered. In a similar vein, it is rather remarkable that translational control during the regulation of cell proliferation is almost exclusively mediated by 5′ UTR elements, while developmental systems seem to predominantly yield examples of 3′ UTR mediated translational control. Because a division between the regulation of proliferation and development seems rather artificial, this difference is likely to disappear as more instances of translational control are discovered in both experimental systems.

While the role of translational control in vertebrate development seems to be more subtle than in invertebrates, this review shows that there are many cases in which it does occur. Their are only a handful of cases in which a clear developmental function has been assigned to a vertebrate translational control system. This is probably due to the fact that the mechanisms for these systems are novel and still in the process of being clarified. Once the details of a translational control system have been worked out, experimental approaches that disrupt it should become easier to design. Such experiments are likely to define the importance of translational control in vertebrate development more clearly.

References

Amaldi, F., and Pierandrei-Amaldi, P. (1997). TOP genes: A translationally controlled class of genes including those coding for ribosomal protein genes. *Prog. Mol. Subcell. Biol.* **18,** 1–17.

Amaya, E., Stein, P., Musci, T., and Kirschner, M. (1993). FGF Signalling in the early specification of mesoderm in *Xenopus*. *Development* **118,** 477–487.

Andrésson, T., and Ruderman, J. V. (1998). The kinase Eg2 is a component of the *Xenopus* oocyte progesterone-activated signaling pathway. *EMBO J.* **17,** 5627–5637.

Antic, D., and Keene, J. D. (1997). Embryonic lethal abnormal visual RNA-binding proteins involved in growth, differentiation, and posttranscriptional gene expression. *Am. J. Hum. Genet.* **61,** 273–278.

Antic, D., Lu, N., and Keene, J. D. (1999). ELAV tumor antigen, Hel-N1, increases translation of neurofilament M mRNA and induces formation of neurites in human teratocarcinoma cells. *Genes Dev.* **13,** 449–461.

Arnaud, E., Touriol, C., Boutonnet, C., Gensac, M. C., Vagner, S., Prats, H., and Prats, A. C. (1999). A new 34-kilodalton isoform of human fibroblast growth factor 2 is cap dependently synthesized by using a non-AUG start codon and behaves as a survival factor. *Mol. Cell. Biol.* **19,** 505–514.

Audet, R. G., Goodchild, J., and Richter, J. D. (1987). Eukaryotic initiation factor 4A stimulates translation in microinjected *Xenopus* oocytes. *Dev. Biol.* **121,** 58–68.

Audic, Y., Omilli, F., and Osborne, H. B. (1997). Postfertilization deadenylation of mRNAs in *Xenopus laevis* embryos is sufficient to cause their degradation at the blastula stage. *Mol. Cell. Biol.* **17,** 209–218.

Avni, D., Biberman, Y., and Meyuhas, O. (1997). The 5′ terminal oligopyrimidine tract confers translational control on TOP mRNAs in a cell type- and sequence context-dependent manner. *Nucleic Acids Res.* **25,** 995–1001.

Ballantyne, S., Bilger, A., Astrom, J., Virtanen, A., and Wickens, M. (1995). Poly (A) polymerases in the nucleus and cytoplasm of frog oocytes: Dynamic changes during ocyte maturation and early development. *RNA* **1,** 64–78.

Ballantyne, S., Daniel, D. L., Jr., and Wickens, M. (1997). A dependent pathway of cytoplasmic polyadenylation reactions linked to cell cycle control by *c-mos* and CDK1 activation. *Mol. Biol. Cell* **8,** 1633–1648.

Bally-Cuif, L., Schatz, W. J., and Ho, R. K. (1998). Characterization of the zebrafish Orb/CPEB-related RNA binding protein and localization of maternal components in the zebrafish oocyte. *Mech. Dev.* **77,** 31–47.

Barkoff, A., Ballantyne, S., and Wickens, M. (1998). Meiotic maturation in *Xenopus* requires polyadenylation of multiple mRNAs. *EMBO J.* **17,** 3168–3175.

Bashirullah, A., Cooperstock, R. L., and Lipshitz, H. D. (1998). RNA localization in development. *Annu. Rev. Biochem.* **67,** 335–394.

Bernstein, J., Sella, O., Le, S. Y., and Elroy-Stein, O. (1997). PDGF2/c-*sis* mRNA leader contains a differentiation-linked internal ribosomal entry site (D-IRES). *J. Biol. Chem.* **272,** 9356–9362.

Biberman, Y., and Meyuhas, O. (1999). TOP mRNAs are translationally inhibited by a titratable repressor in both wheat germ extract and reticulocyte lysate. *FEBS Lett.* **456,** 357–360.

Blangy, A., Lane, H. A., D'Herin, P., Harper, M., Kress, M., and Nigg, E. A. (1995). Phosphorylation by p34cdc2 regulates spindle association of human Eg5, a kinesin-related motor essential for bipolar spindle formation *in vivo*. *Cell* **83,** 1159–1169.

Bouvet, P., Omilli, F., Arlot-Bonnemains, Y., Legagneux, V., Roghi, C., Bassez, T., and Osborne, H. B. (1994). The deadenylation conferred by the 3′ untranslated region of a developmentally controlled mRNA in *Xenopus* embryos is switched to polyadenylation by deletion of a short sequence element. *Mol. Cell. Biol.* **14,** 1893–1900.

Bouvet, P., Paris, J., Philippe, M., and Osborne, H. B. (1991). Degradation of a developmentally regulated mRNA in *Xenopus* embryos is controlled by the 3′ region and requires the translation of another maternal mRNA. *Mol. Cell. Biol.* **11,** 3115–3124.

Bouvet, P. and Wolffe, A. P. (1994). A role for transcription and FRGY2 in masking maternal mRNA within *Xenopus* oocytes. *Cell* **77,** 931–942.

Braddock, M., Muckenthaler, M., White, M. R. H., Thorburn, A. M., Sommerville, J., Kingsman, A. J., and Kingsman, S. M. (1994). Intron-less RNA injected into the nucleus of *Xenopus* oocytes accesses a regulated translation control pathway. *Nucleic Acids Res.* **22,** 5255–5264.

Braun, R. E. (1990). Temporal translational regulation of the protamine 1 gene during. mouse spermatogenesis. *Enzyme* **44,** 120–128.

Braun R. E. (1998). Repression and activation of protamine mRNA translation during murine spermatogenesis. *In* "Germ Cell Development, Division, Disruption and Death" (B. R. Zirkin, Eds.) pp. 105–117. Springer-Verlag, New York.

Braun, R. E., Peschon, J. J., Behringer, R. R., Brinster, R. L., and Palmiter, R. D. (1989). Protamine 3′-untranslated sequences regulate temporal translational control and subcellular localization of growth hormone in spermatids of transgenic mice. *Genes Dev.* **3,** 793–802.

Bruening, W. and Pelletier, J. (1996). A non-AUG translational initiation event generates novel WT1 isoforms. *J. Biol. Chem.* **271,** 8646–8654.

Bugler, B., Amalric, F., and Prats, H. (1991). Alternative initation of translation determines cytoplasmic or nuclear localization of basic fibroblast growth factor. *Mol. Cell. Biol.* **11,** 573–577.

Campos, A. R., Grossman, D., and White, K. (1985). Mutant alleles at the locus elav in *Drosophila melanogaster* lead to nervous system defects. A developmental-genetic analysis. *J. Neurogenet.* **2,** 197–218.

Chen, C. Y. A., and Shyu, A. B. (1995). AU-rich elements: Characterization and importance in mRNA degradation. *Trends Biochem. Sci.* **20,** 465–470.

Christerson, L. B. and McKearin, D. M. (1994). *orb* is required for anteroposterior and dorsoventral patterning during *Drosophila* oogenesis. *Genes Dev.* **8,** 614–628.

Clemens, M. J., and Bommer, U. A. (1999). Translational control: The cancer connection. *Int. J. Biochem. Cell Biol.* **31,** 1–23.

Colgan, D. F., Murthy, K. G., Zhao, W., Prives, C., and Manley, J. L. (1998). Inhibition of poly(A) polymerase requires p34cdc2/cyclin B phosphorylation of multiple consensus and non-consensus sites. *EMBO J.* **17,** 1053–1062.

Colgan, D. F., Murthy, K. G. K., Prives, C., and Manley, J. L. (1996). Cell-cycle related regulation poly(A) polymerase by phosphorylation. *Nature* **384,** 282–285.

Craig, A. W. B., Haghighat, A., Yu, A. T., and Sonenberg, N. (1998). Interaction of polyadenylate-binding protein with the eIF4G homologue PAIP enhances translation. *Nature* **392,** 520–523.

Crawford, E. K., Ensor, J. E., Kalvakolanu, I., and Hasday, J. D. (1997). The role of 3′ poly(A) tail metabolism in tumor necrosis factor-α regulation. *J. Biol. Chem.* **272,** 21120–21127.

Cubizolles, F., Legagneux, V., Le Guellec, R., Chartrain, I., Uzbekov, R., Ford, C., and Le Guellec, K. (1998). pEg7, a new *Xenopus* protein required for mitotic chromosome condensation in egg extracts. *J. Cell Biol.* **143,** 1437–1446.

Culp, P. A., and Musci, T. J. (1998). Translational activation and cytoplasmic polyadenylation of FGF receptor-1 are independently regulated during *Xenopus* oocyte maturation. *Dev. Biol.* **193,** 63–76.

Cummings, A., Barrett, P., and Sommerville, J. (1989). Multiple modifications in the phosphoproteins bound to stored messenger RNA in *Xenopus* oocytes. *Biochim. Biophys. Acta* **1014,** 319–326.

Cummings, A., and Sommerville, J. (1988). Protein kinase activity associated with stored messenger ribonucleoprotein particles of *Xenopus* oocytes. *J. Cell Biol.* **107,** 45–56.

Dalby, B., and Glover, D. M. (1992). 3′ Non-translated sequences in *Drosophila* cyclin B transcripts direct posterior pole accumulation late in oogenesis and peri-nuclear association in syncytial embryos. *Development* **115,** 989–997.

De Benedetti, A. and Harris, A. L. (1999). eIF4E expression in tumors: Its possible role in progression of malignancies. *Int. J. Biochem. Cell Biol.* **31,** 59–72.

De Gregorio, E., Preiss, T., and Hentze, M. W. (1998). Translational activation of uncapped mRNAs by the central part of human eIF4G is 5′ end-dependent. *RNA* **4,** 828–836.

De Gregorio, E., Preiss, T., and Hentze, M. W. (1999). Translation driven by an eIF4G core domain *in vivo. EMBO J.* **18,** 4865–4874.

De Moor, C. H., Jansen, M., Bonte, E. J., Thomas, A. A. M., Sussenbach, J. S., and Van den Brande, J. L. (1994b). Influence of the four leader sequences of the human insulin-like-growth-factor-2 mRNAs on the expression of reporter genes. *Eur. J. Biochem.* **226,** 1039–1047.

De Moor, C. H., Jansen, M., Bonte, E. J., Thomas, A. A. M., Sussenbach, J. S., and Van den Brande, J. L. (1995). Proteins binding to the leader of the 6.0 kb mRNA of human insulin-like growth factor 2 influence translation. *Biochem. J.* **307,** 225–231.

De Moor, C. H., Jansen, M., Sussenbach, J. S., and Van den Brande, J. L. (1994a). Differential polysomal localization of human insulin-like growth-factor-2 mRNAs in cell lines and foetal liver. *Eur. J. Biochem.* **222,** 1017–1024.

De Moor, C. H. and Richter, J. D. (1997). The Mos pathway regulates cytoplasmic polyadenylation in *Xenopus* oocytes. *Mol. Cell. Biol.* **17,** 6419–6426.

De Moor, C. H., and Richter, J. D. (1999). Cytoplasmic polyadenylation elements mediate masking and unmasking of cyclin B1 mRNA. *EMBO J.* **18,** 2294–2303.

Deshler, J. O., Highett, M. I., Abramson, T., and Schnapp, B. J. (1998). A highly conserved RNA-binding protein for cytoplasmic mRNA localization in vertebrates. *Curr. Biol.* **8,** 489–496.

Dickson, C., and Acland, P. (1990). Int-2: A member of the fibroblast growth factor family has different subcellular fates depending on the choice of initiation codon. *Enzyme* **44,** 225–234.

Dickson, K. S., Bilger, A., Ballantyne, S., and Wickens, M. P. (1999). The cleavage and polyadenylation specificity factor in *Xenopus laevis* oocytes is a cytoplasmic factor involved in regulated polyadenylation. *Mol. Cell Biol.* **19,** 5707–5717.

Domenjoud, L., Kremling, H., Burfeind, P., Maier, W. M., and Engel, W. (1991). On the expression of protamine genes in the testis of man and other mammals. *Andrologia* **23,** 333–337.

Dworkin, M. B., Shrutkovski, A., and Dworkin-Rastl, E. (1985). Mobilization of specific maternal RNA species into polysomes after fertilization in *Xenopus laevis. Proc. Natl. Acad. Sci. USA* **82,** 7636–7640.

Eddy, E. M. (1998). Regulation of gene expression during spermatogenesis. *Semin. Cell Dev. Biol.* **9,** 451–457.

Evdokimova, V. M., Kovrigina, E. A., Nashchekin, D. V., Davydova, E. K., Hershey, J. W., and Ovchinnikov, L. P. (1998). The major core protein of messenger ribonucleoprotein particles (p50) promotes initiation of protein biosynthesis *in vitro. J. Biol. Chem.* **273,** 3574–3581.

Evdokimova, V. M., and Ovchinnikov, L. P. (1999). Translational regulation by Y-box transcription factor: Involvement of the major mRNA-associated protein, p50. *Int. J. Biochem. Cell Biol.* **31,** 139–149.

Evdokimova, V. M., Wei, C.-L., Sitikov, A. S., Simonenko, P. N., Lazarev, O. A., Vasilenko, K. S., Ustinov, V. A., Hershey, J. W. B., and Ovchinnikov, L. P. (1995). The major protein of messenger ribonucleoprotein particles in somatic cells is a member of the Ybox binding transcription factor family. *J. Biol. Chem.* **270,** 3186–3192.

Fajardo, M. A., Haugen, H. S., Clegg, C. H., and Braun, R. E. (1997). Separate elements in the 3′ untranslated region of the mouse protamine 1 mRNA regulate translational repression and activation during murine spermatogenesis. *Dev. Biol.* **191,** 42–52.

Fan, X. C., and Steitz, J. A. (1998). Overexpression of HuR, a nuclear-cytoplasmic shutting protein, increases the in vivo stability of ARE-containing mRNAs. *EMBO J.* **17,** 3448–3460.

Ford, L. P., Watson, J., Keene, J. D., and Wilusz, J. (1999). ELAV proteins stabilize deadenylated intermediates in a novel *in vitro* mRNA deadenylation/degradation system. *Genes Dev.* **13,** 188–201.

Fox, C. A., Sheets, M. D., and Wickens, M. P. (1989). Poly(A) addition during maturation of frog oocytes: distinct nuclear and cytoplasmic activities and regulation by the sequence UUUUUAU. *Genes Dev.* **3,** 2151–2162.

Fox, C. A., and Wickens, M. (1990). Poly(A) removal during oocyte maturation: A default reaction selectively prevented by specific sequences in the 3′ UTR of certain maternal mRNAs. *Genes Dev.* **4,** 2287–2298.

Frolova, L., Le Goff, X., Rasmussen, H. H., Cheperegin, S., Drugeon, G., Kress, M., Arman, I., Haenni, A. L., Celis, J. E., Philippe, M., Justesen, J., and Kisselev. L. (1994). A highly conserved eukaryotic protein family possessing properties of polypeptide chain release factor. *Nature* **372,** 701–703.

Galas, S., Barakat, H., Dorée, M., and Picard, A. (1993). A nuclear factor required for specific translation of cyclin B may control the timing of first meiotic cleavage in starfish oocytes. *Mol. Biol. Cell* **4,** 1295–1306.

Galy, B., Maret, A., Prats, A. C., and Prats, H. (1999). Cell transformation results in the loss of the density-dependent translational regulation of the expression of fibroblast growth factor 2 isoforms. *Cancer Res.* **59,** 165–171.

Gatignol, A., Buckler-White, A., Berkhout, B., and Jeang, K.-T. (1991). Characterization of a human TAR RNA-binding protein that activates the HIV-1 LTR. *Science* **251,** 1597–1600.

Geballe, A. P., and Morris, D. R. (1994). Initiation codons within 5′-leaders of mRNAs as regulators of translation. *Trends Biochem. Sci.* **19,** 159–164.

Gebauer, F., and Richter, J. D. (1995). Cloning and characterization of a *Xenopus* poly(A) polymerase. *Mol. Cell. Biol.* **15,** 1422–1430.

Gebauer, F., and Richter, J. D. (1996). Mouse cytoplasmic polyadenylylation element binding protein: An evolutionarily conserved protein that interacts with the cytoplasmic polyadenylation elements of c-*mos* mRNA. *Proc. Natl. Acad. Sci. USA* **93,** 14602–14607.

Gebauer, F., and Richter, J. D. (1997). Synthesis and function of Mos: The control switch of vertebrate oocyte meiosis. *BioEssays* **19,** 23–28.

Gebauer, F., Xu, W., Cooper, G. M., and Richter, J. D. (1994). Translational control by cytoplasmic polyadenylation of *c-mos* mRNA is necessary for oocyte maturation in the mouse. *EMBO J.* **13,** 5712–5720.

Gingras, A. C., Gygi, S. P., Raught, B., Polakiewicz, R. D., Abraham, R. T., Hoekstra, M. F., Aebersold, R., and Sonenberg, N. (1999). Regulation of 4E-BP1 phosphorylation: A novel two-step mechanism. *Genes Dev.* **13,** 1422–1437.

Gingras, A. C., Raught, B., and Sonenberg, N. (1999). elF4 Initiation factors: Effectors of mRNA recruitment to ribosomes and regulators of translation. *Annu. Rev. Biochem.* **68,** 913–963.

Good, P. J. (1995). A conserved family of elav-like genes in vertebrates. *Proc. Natl. Acad. Sci. USA* **92,** 4557–4561.

Goodwin, E. B., and Evans, T. C. (1997). Translational control of development in *C. elegans*. *Semin. Cell Dev. Biol.* **8,** 551–559.

Grafi, G., Sela, I., and Galili, G. (1993). Translational regulation of human beta interferon mRNA: Association of the 3′ AU-rich sequence with the poly(A) tail reduces translation efficiency in vitro. *Mol. Cell. Biol.* **13,** 3487–3493.

Gray, N. K., and Wickens, M. (1998). Control of translation initiation in animals. *Annu. Rev. Cell Dev. Biol.* **14,** 399–458.

Gu, W., Kwon, Y., Oko, R., Hermo, L., and Hecht, N. B. (1995). Poly (A) binding protein is bound to both stored and polysomal mRNAs in the mammalian testis. *Mol. Reprod. Dev.* **40,** 273–285.

Gunkel, N., Braddock, M., Thorburn, A. M., Muckenthaler, M., Kingsman, A. J., and Kingsman, S. M. (1995). Promoter control of translation in *Xenopus* oocytes. *Nucleic Acids Res.* **23,** 405–412.

Hake, L. E., Alcivar, A. A., and Hecht, N. B. (1990). Changes in mRNA length accompany translational regulation of the somatic and testis-specific cytochrome c genes during spermatogenesis in the mouse. *Development* **110,** 249–257.

Hake, L. E., Mendez, R., and Richter, J. D. (1998). Specificity of RNA binding by CPEB: requirement for RNA recognition motifs and a novel zinc finger. *Mol. Cell. Biol.* **18,** 685–693.

Hake, L. E., and Richter, J. D. (1994). CPEB is a specificity factor that mediates cytoplasmic polyadenylation during *Xenopus* oocyte maturation. *Cell* **79,** 617–627.

Hake, L. E., and Richter, J. D. (1997). Translational regulation of maternal mRNA. *Biochim. Biophys. Acta Rev. Cancer* **1332,** M31–M38.

Hann, S. R., Dixit, M., Sears, R. C., and Sealy, L. (1994). The alternatively initiated c-Myc proteins differentially regulate transcription through a noncanonical DNA-binding site. *Genes Dev.* **8,** 2441–2452.

Hann, S. R., King, M. W., Bentley, D. L., Anderson, C. W., and Eisenman, R. N. (1988). A non-AUG translational initiation in *c-myc* exon 1 generates an N-terminally distinct protein whose synthesis is disrupted in Burkitt's lymphomas. *Cell* **52,** 185–195.

Havin, L., Git, A., Elisha, Z., Oberman, F., Yaniv, K., Schwartz, S. P., Standart, N., and Yisraeli, J. K. (1998). RNA-binding protein conserved in both microtubule- and microfilament-based RNA localization. *Genes Dev.* **12,** 1593–1598.

Hill, J. R., and Morris, D. R. (1992). Cell-specific translation of *S*-adenosylmethionine decarboxylase mRNA. Regulation by the 5′ transcript leader. *J. Biol. Chem.* **267,** 21886–21893.

Hill, J. R., and Morris, D. R. (1993). Cell-specific translational regulation of *S*-adenosylmethionine decarboxylase mRNA. Dependence on translation and coding capacity of the *cis*-acting upstream open reading frame. *J. Biol. Chem.* **268,** 726–731.

Hinnebusch, A. G. (1997). Translational regulation of yeast *GCN4*-A window on factors that control initiator-tRNA binding to the ribosome. *J. Biol. Chem.* **272,** 21661–21664.

Hitti, E., Neunteufl, A., and Jantsch, M. F. (1998). The double-stranded RNA-binding protein Xlrbpa promotes RNA strand annealing. *Nucleic Acids Res.* **26,** 4382–4388.

Hoover, D. S., Wingett, D. G., Zhang, J., Reeves, R., and Magnuson, N. S. (1997). Pim-1 protein expression is regulated by its 5′-untranslated region and translation initiation factor elF-4E. *Cell Growth Differ.* **8,** 1371–1380.

Hornstein, E., Git, A., Braunstein, I., Avni, D., and Meyuhas, O. (1999). The expression of poly(A)-binding protein gene is translationally regulated in a growth-dependent fashion through a 5′-terminal oligopyrimidine tract motif. *J. Biol. Chem.* **274,** 1708–1714.

Houliston, E., Le Guellec, R., Kress, M., Philippe, M., and Le Guellec, K. (1994). The kinesin-related protein Eg5 associates with both interphase and spindle microtubules during *Xenopus* early development. *Dev. Biol.* **164,** 147–159.

Huarte, J., Stutz, A., O'Connell, M. L., Gubler, P., Belin, D., Darrow, A. L., Strickland, S., and Vassalli, J.-D. (1992). Transient translational silencing by reversible mRNA deadenylation. *Cell* **69,** 1021–1030.

Huez, I., Creancier, L., Audigier, S., Gensac, M. C., Prats, A. C., and Prats, H. (1998): Two independent internal ribosome entry sites are involved in translation initiation of vascular endothelial growth factor mRNA. *Mol. Cell. Biol.* **18,** 6178–6190.

Imataka, H., Gradi, A., and Sonenberg, N. (1998). A newly identified N-terminal amino acid sequence of human elF4G binds poly(A)-binding protein and functions in poly(A)-dependent translation. *EMBO J.* **17,** 7480–7489.

Jacobson, A., and Favreau, M. (1983). Possible involvement of poly(A) in protein syntheisis. *Nucleic Acids Res.* **11,** 6353–6368.

Jain, R. G., Andrews, L. G., McGowan, K. M., Pekala, P. H., and Keene, J. D. (1997). Ectopic expression of Hel-N1, an RNA-binding protein, increases glucose transporter (GLUT1) expression in 3T3-L1 adipocytes. *Mol. Cell. Biol.* **17,** 954–962.

Jefferies, H. B. J., Fumagalli, S., Dennis, P. B., Reinhard, C., Pearson, R. B., and Thomas, G. (1997). Rapamycin suppresses 5′TOP mRNA translation through inhibition of $p70^{S6k}$. *EMBO J.* **16,** 3693–3704.

Jefferies, H. B. J., Reinhard, C., Kozma, S. C., and Thomas, G. (1994). Rapamycin selectively represses translation of the "polypyrimidine tract" mRNA family. *Proc. Natl. Acad. Sci. USA* **91,** 4441–4445.

Jefferies, H. B. J. and Thomas, G. (1994). Elongation factor-1α mRNA is selectively translated following mitogenic stimulation. *J. Biol. Chem.* **269,** 4367–4372.

Kaspar, R. L., Kakegawa, T., Cranston, H., Morris, D. R., and White, M. W. (1992). A regulatory cis element and a specific binding factor involved in the mitogenic control of murine ribosomal protein L32 translation. *J. Biol. Chem.* **267,** 508–514.

Katsu, Y., Minshall, N., Nagahama, Y., and Standart, N. (1999). Ca^{2+} is required for phosphorylation of clam p82/CPEB in vitro: Implications for dual and independent roles of MAP and Cdc2 kinases. *Dev. Biol.* **209,** 186–199.

Kawasome, H., Papst, P., Webb, S., Keller, G. M., Johnson, G. L., Gelfand, E. W., and Terada, N. (1998). Targeted disruption of p70(s6k) defines its role in protein synthesis and rapamycin sensitivity. *Proc. Natl. Acad. Sci. USA* **95,** 5033–5038.

Keiper, B. D., and Rhoads, R. E. (1997). Cap-independent translation initiation in *Xenopus* oocytes. *Nucleic Acids Res.* **25,** 395–402.

Keiper, B. D., and Rhoads, R. E. (1999). Translational recruitment of *Xenopus* maternal mRNAs in response to poly(A) elongation requires initiation factor elF4G-1. *Dev. Biol.* **206,** 1–14.

Kleene, K. C. (1989). Poly (A) shortening accompanies the activation of translation of five mRNAs during spermiogenesis in the mouse. *Development* **106,** 367–373.

Kleijn, M., Scheper, G. C., Voorma, H. O., and Thomas, A. A. (1998). Regulation of translation initiation factors by signal transduction. *Eur. J. Biochem.* **253,** 531–544.

Klein, P. S., and Melton, D. A. (1994). Induction of mesoderm in *Xenopus laevis* embryos by translation initiation factor 4E. *Science* **265,** 803–806.

Kobayashi, H., Minshull, J., Ford, C., Golsteyn, R., Poon, R., and Hunt, T. (1991). On the synthesis and destruction of A- and B-type cyclins during oogenesis and meiotic maturation in *Xenopus laevis. J. Cell Biol.* **114,** 755–765.

Kolupaeva, V. G., Pestova, T. V., Hellen, C. U., and Shatsky, I. N. (1998). Translation eukaryotic initiation factor 4G recognizes a specific structural element within the internal ribosome entry site of encephalomyocarditis virus RNA. *J. Biol. Chem.* **273,** 18599–18604.

Körner, C. G., Wormington, M., Muckenthaler, M., Schneider, S., Dehlin, E., and Wahle, E. (1998). The deadenylating nuclease (DAN) is involved in poly(A) tail removal during the meiotic maturation of *Xenopus* oocytes. *EMBO J.* **17,** 5427–5437.

Koromilas, A. E., Lazaris-Karatzas, A., and Sonenberg, N. (1992). mRNAs containing extensive secondary structure in their 5′ non-coding region translate efficiently in cells overexpressing initiation factor elF-4E. *EMBO J.* **11,** 4153–4158.

Kozak, M. (1991). An analysis of vertebrate mRNA sequences: Intimations of translational control. *J. Cell Biol.* **115,** 887–903.

Kozak, M. (1997). Recognition of AUG and alternative initiator codons is augmented by G in position +4 but is not generally affected by the nucleotides in positions +5 and +6. *EMBO J.* **16,** 2482–2492.

Kozak, M. (1999). Initiation of translation in prokaryotes and eukaryotes. *Gene* **234,** 187–208.

Krecic, A. M., and Swanson, M. S. (1999). hnRNP complexes: Composition, structure, and function. *Curr. Opin. Cell Biol.* **11,** 363–371.

Kruys, V., Beutler, B., and Huez, G. (1990). Translational control mediated by UA-rich sequences. *Enzyme* **44,** 193–202.

Kruys, V., Marinx, O., Shaw, G., Deschamps, J., and Huez, G. (1989). Translational blockade imposed by cytokine-derived UA-rich sequences. *Science* **245,** 852–855.

Kuge, H., Brownlee, G. G., Gershon, P. D., and Richter, J. D. (1998). Cap ribose methylation of *c-mos* mRNA stimulates translation and oocyte maturation in *Xenopus laevis. Nucleic Acids Res.* **26,** 3208–3214.

Kuge, H., and Richter, J. D. (1995). Cytoplasmic 3′ poly(A) addition induces 5′ cap ribose methylation: Implications for translational control of maternal mRNA. *EMBO J.* **14,** 6301–6310.

Kuhn, H., Heydeck, D., Brinckman, R., and Trebus, F. (1999). Regulation of cellular 15-lipoxygenase activity on pretranslational, translational, and posttranslational levels. *Lipids* **34 Suppl:S273–9,** S273–S279.

Ladomery, M., Wade, E., and Sommerville, J. (1997). Xp54, the *Xenopus* homologue of human RNA helicase p54, is an integral component of stored mRNP particles in oocytes. *Nucleic Acids Res.* **25,** 965–973.

Lantz, V., Chang, J. S., Horabin, J. I., Bopp, D., and Schedl, P. (1994). The *Drosophila orb* RNA-binding protein is required for the formation of the egg chamber and establishment of polarity. *Genes Dev.* **8,** 598–613.

Laskey, R. A., Mills, A. D., Gurdon, J. B., and Partington, G. A. (1977). Protein synthesis in oocytes of *Xenopus laevis* is not regulated by the supply of messenger RNA. *Cell* **11,** 345–351.

Latham, K. E., Garrels, J. I., Chang, C., and Solter, D. (1991). Quantitative analysis of protein synthesis in mouse embryos. I. Extensive reprogramming at the one- and two-cell stages. *Development* **112,** 921–932.

Lee, K., Fajardo, M. A., and Braun, R. E. (1996). A testis cytoplasmic RNA-binding protein that has the properties of a translational repressor. *Mol. Cell. Biol.* **16,** 3023–3034.

Levine, T. D., Gao, F., King, P. H., Andrews, L. G., and Keene, J. D. (1993). Hel-N1: An autoimmune RNA-binding protein with specificity for 3′ uridylate-rich untranslated regions of growth factor mRNAs. *Mol. Cell. Biol.* **13,** 3494–3504.

Levy, N. S., Chung, S., Furneaux, H., and Levy, A. P. (1998). Hypoxic stabilization of vascular endothelial growth factor mRNA by the RNA-binding protein HuR. *J. Biol. Chem.* **273,** 6417–6423.

Loreni, F., and Amaldi, F. (1997). Translational control of terminal oligopyrimidine mRNAs requires a specific regulator. *FEBS Lett.* **416,** 239–242.

Marcotrigiano, J., Gingras, A. C., Sonenberg, N., and Burley, S. K. (1999). Cap-dependent translation initiation in eukaryotes is regulated by a molecular mimic of eIF4G. *Mol. Cell* **3,** 707–716.

Marinx, O., Bertrand, S., Karsenti, E., Huez, G., and Kruys, V. (1994). Fertilization of *Xenopus* eggs imposes a complete translational arrest of mRNAs containing 3′UUAUUUAU elements. *FEBS Lett.* **345,** 107–112.

Matsumoto, K., Meric, F., and Wolffe, A. P. (1996). Translational repression dependent on the interaction of the *Xenopus* Y-box protein FRGY2 with mRNA-Role of the cold shock domain, tail domain, and selective RNA sequence recognition. *J. Biol. Chem.* **271,** 22706–22712.

Matsumoto, K., Wassarman, K. M., and Wolffe, A. P. (1998). Nuclear history of a pre-mRNA determines the translational activity of cytoplasmic mRNA. *EMBO J.* **17,** 2107–2121.

Matsumoto, K., and Wolffe, A. P. (1998). Gene regulation by Y-box proteins: Coupling control of transcription and translation. *Trends. Cell Biol.* **8,** 318–323.

McGrew, L. L., Dworkin-Rastl, E., Dworkin, M. B., and Richter, J. D. (1989). Poly (A) elongation during *Xenopus* oocyte maturation is required for translational recruitment and is mediated by a short sequence element. *Genes Dev.* **3,** 803–815.

McGrew, L. L., and Richter, J. D. (1990). Translational control by cytoplasmic polyadenylation during *Xenopus* oocyte maturation: Characterization of *cis* and *trans* elements and regulation by Cyclin/MPF. *EMBO J.* **9,** 3743–3751.

Mendez, R., Hake, L. E., Andrésson, T., Littlepage, L. E., Ruderman, J. V., and Richter, J. D. (2000). Phosphorylation of CPEB by Eg2 regulates *c-mos* mRNA translation. *Nature* **404,** 302–307.

Meric, F., Searfoss, A. M., Wormington, M., and Wolffe, A. P. (1996). Masking and unmasking maternal mRNA—The role of polyadenylation, transcription, splicing, and nuclear history. *J. Biol. Chem.* **271,** 30804–30810.

Merrick, W. C., and Hershey, J. W. B. (1996). The pathway and mechanism of eukaryotic protein synthesis. *In* "Translational Control" (J. W. B. Hershey, M. B. Mathews, and N. Sonenberg, Eds.), pp. 31–69. Cold Spring Harbor Laboratory Press, Cold Spring Harbor, New York.

Meyuhas, O., Avni, D., and Shama, S. (1996). Translational control of ribosomal protein mRNAs in eukaryotes. *In* "Translational Control" (J. W. B. Hershey, M. B. Mathews and N. Sonenberg, Eds.), pp. 363–388. Cold Spring Harbor Laboratory Press, Cold Spring Harbor, New York.

Minich, W. B., Maidebura, I. P., and Ovchinnikov, L. P. (1993). Purification and characterization of the major 50-kDa repressor protein from cytoplasmic mRNP of rabbit reticulocytes. *Eur. J. Biochem.* **212,** 633–638.

Minich, W. B., and Ovchinnikov, L. P. (1992). Role of cytoplasmic mRNP proteins in translation. *Biochimie* **74,** 477–483.

Minshall, N., Walker, J., Dale, M., and Standart, N. (1999). Dual roles of p82, the clam CPEB homolog, in cytoplasmic polyadenylation and translational masking. *RNA* **5,** 27–38.

Morley, S. J., Curtis, P. S., and Pain, V. M. (1997). elF4G: Translation's mystery factor begins to yield its secrets, *RNA Publ. RNA Soc.* **3,** 1085–1104.

Muckenthaler, M., Gray, N. K., and Hentze, M. W. (1998). IRP-1 binding to ferritin mRNA prevents the recruitment of the small ribosomal subunit by the cap-binding complex elF4F. *Mol. Cell* **2,** 383–388.

Murthy, K. G. K., and Manley, J. L. (1995). The 160-kD subunit of human cleavage polyadenylation specificity factor coordinates pre-mRNA 3′-end formation. *Genes Dev.* **9,** 2672–2683.

Nair, A. P., Hirsch, H. H., Colombi, M., and Moroni, C. (1999). Cyclosporin A promotes translational silencing of autocrine interleukin- 3 via ribosome-associated deadenylation. *Mol. Cell Biol.* **19,** 889–898.

Nanbru, C., Lafon, I., Audigier, S., Gensac, M. C., Vagner, S., Huez, G., and Prats, A. C. (1997). Alternative translation of the proto-oncogene *c-myc* by an internal ribosome entry site. *J. Biol. Chem.* **272,** 32061–32066.

Nielsen, F. C., Gammeltoft, S., and Christiansen, J. (1990). Translational discrimination of mRNAs coding for human insulin- like growth factor II. *J. Biol. Chem.* **265,** 13431–13434.

Nielsen, J., Christiansen, J., Lykke-Andersen, J., Johnsen, A. H., Wewer, U. M., and Nielsen, F. C. (1999). A family of insulin-like growth factor II mRNA-binding proteins represses translation in late development. *Mol. Cell Biol.* **19,** 1262–1270.

Nishizawa, M., Okazaki, K., Furuno, N., Watanabe, N., and Sagata, N. (1992). The 'second-codon rule' and autophosphorylation govern the stability and activity of Mos during the meiotic cell cycle in *Xenopus* oocytes. *EMBO J.* **11,** 2433–2446.

Novoa, I., and Carrasco, L. (1999). Cleavage of eukaryotic translation initiation factor 4G by exogenously added hybrid proteins containing poliovirus 2Apro in HeLa cells: effects on gene expression. *Mol. Cell Biol.* **19,** 2445–2454.

Ostareck-Lederer, A., Ostareck, D. H., and Hentze, M. W. (1998). Cytoplasmic regulatory functions of the KH-domain proteins hnRNPs K and E1/E2. *Trends. Biochem. Sci.* **23,** 409–411.

Ostareck-Lederer, A., Ostareck, D. H., Standart, N., and Thiele, B. J. (1994). Translation of 15-lipoxygenase mRNA is inhibited by a protein that binds to a repeated sequence in the 3′ untranslated region. *EMBO J.* **13,** 1476–1481.

Ostareck, D. H., Ostareck-Lederer, A., Wilm, M., Thiele, B. J., Mann, M., and Hentze, M. W. (1997). mRNA silencing in erythroid differentiation: hnRNP K and hnRNP E1 regulate 15-lipoxygenase translation from the 3′ end. *Cell* **89,** 597–606.

Paillard, L., Omilli, F., Legagneux, V., Bassez, T., Manjey, D., and Osborne, H. B. (1998). EDEN and EDEN-BP, a cis element and an associated factor that mediate sequence-specific mRNA deadenylation in Xenopus embryos. *EMBO J.* **17,** 278–287.

Pain, V. M. (1996). Initiation of protein synthesis in eukaryotic cells. *Eur. J. Biochem.* **236,** 747–771.

Palatnik, C. M., Wilkins, C., and Jacobson, A. (1984). Translational control during early *Dictyostelium* development: possible involvement of poly(A) sequences. *Cell* **36,** 1017–1025.

Paris, J., Le Guellec, R., Couturier, A., Le Guellec, K., Omilli, F., Camonis, J., MacNeill, S., and Philippe, M. (1991). Cloning by differential screening of a *Xenopus* cDNA coding for a protein highly homologous to cdc2. *Proc. Natl. Acad. Sci. USA* **88,** 1039–1043.

Paris, J., Osborne, H. B., Couturier, A., Le Guellec, R., and Philippe, M. (1988). Changes in the polyadenylation of specific stable RNA during the early development of *Xenopus laevis. Gene* **72,** 169–176.

Paris, J., and Philippe, M. (1990). Poly(A) metabolism and polysomal recruitment of maternal mRNAs during early *Xenopus* development. *Dev. Biol.* **140,** 221–224.

Paris, J., and Richter, J. D. (1990). Maturation-specific polyadenylation and translational control: Diversity of cytoplasmic polyadenylation elements, influence of poly(A) tail size, and formation of stable polyadenylation complexes. *Mol. Cell. Biol.* **10,** 5634–5645.

Park, H., Davies, M. V., Langland, J. O., Chang, H.-W., Nam, Y. S., Tartaglia, J., Paoletti, E., Jacobs, B. L., Kaufman, R. J., and Venkatesan, S. (1994). TAR RNA-binding protein is an inhibitor of the interferon-induced protein kinase PKR. *Proc. Natl. Acad. Sci. USA* **91,** 4713–4717.

Pellizzoni, L., Lotti, F., Rutjes, S. A., and Pierandrei-Amaldi, P. (1998). Involvement of the *Xenopus laevis* Ro60 autoantigen in the alternative interaction of La and CNBP proteins with the 5′UTR of L4 ribosomal protein mRNA. *J. Mol. Biol.* **281,** 593–608.

Peng, S. S. Y., Chen, C. Y., Xu, N., and Shyu, A. B. (1998). RNA stabilization by the AU-rich element binding protein, HuR, an ELAV protein. *EMBO J.* **17,** 3461–3470.

Pesole, G., Fiormarino, G., and Saccone, C. (1994). Sequence analysis and compositional properties of untranslated regions of human mRNAs. *Gene* **140,** 219–225.

Pesole, G., Liuni, S., Grillo, G., and Saccone, C. (1997). Structural and compositional features of untranslated regions of eukaryotic mRNAs. *Gene* **205,** 95–102.

Pestova, T. V., Borukhov, S. I., and Hellen, C. U. T. (1998). Eukaryotic ribosomes require initiation factors 1 and 1A to locate initiation codons. *Nature* **394,** 854–859.

Piron, M., Delaunay, T., Grosclaude, J., and Poncet, D. (1999). Identification of the RNA-binding, dimerization, and eIF4GI-binding domains of rotavirus nonstructural protein NSP3, *J. Virol.* **73,** 5411–5421.

Piron, M., Vende, P., Cohen, J., and Poncet, D. (1998). Rotavirus RNA-binding protein NSP3 interacts with eIF4GI and evicts the poly(A) binding protein from eIF4F. *EMBO J.* **17,** 5811–5821.

Proud, C. G., and Denton, R. M. (1997). Molecular mechanisms for the control of translation by insulin. *Biochem. J.* **328,** 329–341.

Ranjan, M., Tafuri, S. R., and Wolffe, A. P. (1993). Masking mRNA from translation in somatic cells. *Genes Dev.* **7,** 1725–1736.

Raught, B., and Gingras, A. C. (1999). eIF4E activity is regulated at multiple levels. *Int. J. Biochem. Cell Biol.* **31,** 43–57.

Reynolds, K., Zimmer, A. M., and Zimmer, A. (1996). Regulation of RARβ2 mRNA expression: Evidence for an inhibitory peptide encoded in the 5′-untranslated region. *J. Cell Biol.* **134,** 827–835.

Richter, J. D. (1999). Cytoplasmic polyadenylation in development and beyond. *Microbiol. Mol. Biol. Rev.* **63,** 446–456.

Richter, J. D., and Smith, L. D. (1984). The reversible inhibition of translation by *Xenopus* oocyte-specific proteins. *Nature* **309,** 378–380.

Robbie, E. P., Peterson, M., Amaya, E., and Musci, T. J. (1995). Temporal regulation of the *Xenopus* FGF receptor in development: A translation inhibitory element in the 3′ untranslated region. *Development* **121,** 1775–1785.

Robinow, S., and White, K. (1988). The locus elav of *Drosophila melanogaster* is expressed in neurons at all developmental stages. *Dev. Biol.* **126,** 294–303.

Roghi, C., Giet, R., Uzbekov, R., Morin, N., Chartrain, I., Le Guellec, R., Couturier, A., Doree, M., Philippe, M., and Prigent, C. (1998). The *Xenopus* protein kinase pEg2 associates with the centrosome in a cell cycle-dependent manner, binds to the spindle microtubules and is involved in bipolar mitotic spindle assembly. *J. Cell Sci.* **111,** 557–572.

Rosenwald, I. B., Lazaris-Karatzas, A., Sonenberg, N., and Schmidt, E. V. (1993). Elevated levels of cyclin D1 protein in response to increased expression of eukaryotic initiation factor 4E. *Mol. Cell. Biol.* **13,** 7358–7363.

Ross, A. F., Oleynikov, Y., Kislauskis, E. H., Taneja, K. L., and Singer, R. H. (1997). Characterization of a β-actin mRNA zipcode-binding protein. *Mol. Cell. Biol.* **17,** 2158–2165.

Ruan, H., Hill, J. R., Fatemie-Nainic, S., and Morris, D. R. (1994). Cell-specific translational regulation of S-adenosylmethionine decarboxylase mRNA. Influence of the structure of the 5′ transcript leader on regulation by the upstream open reading frame. *J. Biol. Chem.* **269,** 17905–17910.

Ruan, H., Shantz, L. M., Pegg, A. E., and Morris, D. R. (1996). The upstream open reading frame of the mRNA encoding S-adenosylmethionine decarboxylase synthesis through the 5′ untranslated regions of its mRNA. *J. Biol. Chem.* **271,** 29576–29582.

Sachs, A. B., Sarnow, P., and Hentze, M. W. (1997). Starting at the beginning, middle, and end: Translation initiation in eukaryotes. *Cell* **89,** 831–838.

Sagata, N. (1997). What does Mos do in oocytes and somatic cells. *BioEssays* **19,** 13–21.

Sakamoto, I., Takahara, K., Yamashita, M., and lwao, Y. (1998). Changes in cyclin B during oocyte maturation and early embryonic cell cycle in the newt, *Cynops pyrrhogaster:* Requirement of germinal vesicle for MPF activation. *Dev. Biol.* **195,** 60–69.

Saris, C. J., Domen, J., and Berns, A. (1991). The *pim-1* oncogene encodes two related serine/threonine kinases by alternative initiation at AUG and CUG. *EMBO J.* **10,** 655–664.

Schwab, M. S., Kim, S. H., Terada, N., Edfjall, C., Kozma, S. C., Thomas, G., and Maller, J. L. (1999). p70(S6K) controls selective mRNA translation during oocyte maturation and early embryogenesis in *Xenopus laevis. Mol. Cell Biol.* **19,** 2485–2494.

Sedman, S. A., Gelembiuk, G. W., and Mertz, J. E. (1990). Translation initiation at a downstream AUG occurs with increased efficiency when the upstream AUG is located very close to the 5′ cap. *J. Virol.* **64,** 453–457.

Sella, O., Gerlitz, G., Le, S. Y., and Elroy-Stein, O. (1999). Differentiation-induced internal translation of *c-sis* mRNA: analysis of the *cis* elements and their differentiation-linked binding to the hnRNP C protein. *Mol. Cell Biol.* **19,** 5429–5440.

Severson, W. E., Mascolo, P. L., and White, M. W. (1995). Lymphocyte p56L32 is a RNA/DNA-binding protein which interacts with conserved elements of the murine L32 ribosomal protein mRNA. *Eur. J. Biochem.* **229,** 426–432.

Shama, S., Avni, D., Frederickson, R. M., Sonenberg, N., and Meyuhas, O. (1995). Overexpression of initiation factor elF-4E does not relieve the translational repression of ribosomal protein mRNAs in quiescent cells. *Gene Expr.* **4,** 241–252.

Shantz, L. M., and Pegg, A. E. (1994). Overproduction of ornithine decarboxylase caused by relief of translational repression is associated with neoplastic transformation. *Cancer Res.* **54,** 2313–2316.

Shantz, L. M., Viswanath, R., and Pegg, A. E. (1994). Role of the 5′-untranslated region of mRNA in the synthesis of S-adenosylmethionine decarboxylase and its regulation by spermine. *Biochem. J.* **302,** 765–772.

Sheets, M. D., Fox, C. A., Hunt, T., Vande Woude, G., and Wickens, M. (1994). The 3′-untranslated regions of c-*mos* and cyclin mRNAs stimulate translation by regulating cytoplasmic polyadenylation. *Genes Dev.* **8,** 926–938.

Sheets, M. D., Wu, M., and Wickens, M. (1995). Polyadenylation of c-*mos* mRNA as a control point in *Xenopus* meiotic maturation. *Nature* **374,** 511–516.

Simon, R., and Richter, J. D. (1994). Further analysis of cytoplasmic polyadenylation in *Xenopus* embryos and identification of embryonic cytoplasmic polyadenylation element-binding proteins. *Mol. Cell. Biol.* **14,** 7867–7875.

Simon, R., Tassan, J.-P., and Richter, J. D. (1992). Translational control by poly(A) elongation during *Xenopus* development: Differential repression and enhancement by a novel cytoplasmic polyadenylation element. *Genes Dev.* **6,** 2580–2591.

Simon, R., Wu, L., and Richter, J. D. (1996). Cytoplasmic polyadenylation of activin receptor mRNA and the control of pattern formation in *Xenopus* development. *Dev. Biol.* **179,** 239–250.

Sizova, D. V., Kolupaeva, V. G., Pestova, T. V., Shatsky, I. N., and Hellen, C. U. (1998). Specific interaction of eukaryotic translation initiation factor 3 with the 5′ nontranslated regions of hepatitis C virus and classical swine fever virus RNAs. *J. Virol.* **72,** 4775–4782.

Smith, R. C., Dworkin-Rastl, E., and Dworkin, M. B. (1988). Expression of a histone H1-like protein is restricted to early *Xenopus* development. *Genes Dev.* **2,** 1284–1295.

Sommerville, J. (1999). Activities of cold-shock domain proteins in translation control. *BioEssays* **21,** 319–325.

Sommerville, J., and Ladomery, M. (1996). Transcription and masking of mRNA in germ cells: Involvement of Y-box proteins. *Chromosoma* **104,** 469–478.

Spirin, A. S. (1966). On "masked" forms of messenger RNA in early embryogenesis and in other differentiating systems. *Curr. Top. Dev. Biol.* **1,** 1–38.

Spirin, A. S. (1996). Masked and translatable messenger ribonucleoproteins in higher eukaryotes. *In* "Translational Control" (J. W. B. Hershey, M. B. Mathews, and N. Sonenberg, Eds.), pp. 319–334. Cold Spring Harbor Laboratory Press, New York.

St. Johnston, D., Brown, N. H., Gall, J. G., and Jantsch, M. (1992). A conserved double-stranded RNA-binding domain. *Proc. Natl. Acad. Sci. USA* **89,** 10979–10983.

Stambuck, R. A. and Moon, R. T. (1992). Purification and characterization of recombinant *Xenopus* poly(A)$^+$-binding protein expressed in a baculovirus system. *Biochem. J.* **287,** 761–766.

Standart, N., Dale, M., Stewart, E., and Hunt, T. (1990). Maternal mRNA from clam oocytes can be specifically unmasked in vitro by antisense RNA complementary to the 3′-untranslated region. *Genes Dev.* **4,** 2157–2168.

Stebbins-Boaz, B., Cao, Q., De Moor, C. H., Mendez, R., and Richter, J. D. (1999). Maskin is a CPEB associated factor that transiently interacts with elF-4E. *Mol. Cell.* **4,** 1017–1027.

Stebbins-Boaz, B., Hake, L. E., and Richter, J. D. (1996). CPEB controls the cytoplasmic polyadenylation of cyclin, Cdk2 and c-*mos* mRNAs and is necessary for oocyte maturation in *Xenopus. EMBO J.* **15,** 2582–2592.

Stebbins-Boaz, B., and Richter, J. D. (1994). Multiple sequence elements and a maternal mRNA product control cdk2 RNA polyadenylation and translation during early *Xenopus* development. *Mol. Cell. Biol.* **14,** 5870–5880.

Stebbins-Boaz, B., and Richter, J. D. (1997). Translational control during early development. *Crit. Rev. Eukaryotic Gene Express.* **7,** 73–94.

Steger, K. (1999). Transcriptional and translational regulation of gene expression in haploid spermatids. *Anat. Embryol. (Berlin)* **199,** 471–487.

Stein, I., Itin, A., Einat, P., Skaliter, R., Grossman, Z., and Keshet, E. (1998). Translation of vascular endothelial growth factor mRNA by internal ribosome entry: implications for translation under hypoxia. *Mol. Cell. Biol.* **18,** 3112–3119.

Stennard, F., Zorn, A. M., Ryan, K., Garrett, N., and Gurdon, J. B. (1999). Differential expression of VegT and antipodean protein isoforms in *Xenopus. Mech. Dev.* **86,** 87–89.

Stoneley, M., Paulin, F. E., Le Quesne, J. P., Chappell, S. A., and Willis, A. E. (1998). C-Myc 5′ untranslated region contains an internal ribosome entry segment. *Oncogene* **16,** 423–428.

Svitkin, Y. V., Ovchinnikov, L. P., Dreyfuss, G., and Sonenberg, N. (1996). General RNA binding proteins render translation cap dependent. *EMBO J.* **15,** 7147–7155.

Tafuri, S. R., and Wolffe, A. P. (1993). Selective recruitment of masked maternal mRNA from messenger ribonucleoprotein particles containing FRGY2 (mRNP4). *J. Biol. Chem.* **268,** 24255–24261.

Taieb, F., Thibier, C., and Jessus, C. (1997). On cyclins, oocytes, and eggs. *Mol. Reprod. Dev.* **48,** 397–411.

Terada, N., Patel, H. R., Takase, K., Kohno, K., Nairn, A. C., and Gelfand, E. W. (1994). Rapamycin selectively inhibits translation of mRNAs encoding elongation factors and ribosomal proteins. *Proc. Natl. Acad. Sci. USA* **91,** 11477–11481.

Thiele, B. J., Berger, M., Huth, A., Reimann, I., Schwarz, K., and Thiele, H. (1999). Tissue-specific translational regulation of alternative rabbit 15-lipoxygenase mRNAs differing in their 3′-untranslated regions. *Nucleic Acids Res.* **27,** 1828–1836.

Vagner, S., Gensac, M. C., Maret, A., Bayard, F., Amalric, F., Prats, H., and Prats, A. C. (1995). Alternative translation of human fibroblast growth factor 2 mRNA occurs by internal entry of ribosomes. *Mol. Cell. Biol.* **15,** 35–44.

Vagner, S., Touriol, C., Galy, B., Audigier, S., Gensac, M. C., Amalric, F., Bayard, F., Prats, H., and Prats, A. C. (1996). Translation of CUG- but not AUG-initiated forms of human fibroblast growth factor 2 is activated in transformed and stressed cells. *J. Cell Biol.* **135,** 1391–1402.

Varnum, S. M., Hurney, C. A., and Wormington, W. M. (1992). Maturation-specific deadenylation in *Xenopus* oocytes requires nuclear and cytoplasmic factors. *Dev. Biol.* **153,** 283–290.

Varnum, S. M., and Wormington, M. (1990). Deadenylation of maternal mRNAs during *Xenopus* oocyte maturation does not require specific *cis*-sequences: A default mechanism for translational control. *Genes Dev.* **4,** 2278–2286.

Vassalli, J.-D., Huarte, J., Belin, D., Gubler, P., Vassalli, A., O'Connell, M. L., Parton, L. A., Rickles, R. J., and Strickland, S. (1989). Regulated polyadenylation controls mRNA translation during meiotic maturation of mouse oocytes. *Genes Dev.* **3,** 2163–2171.

Verrotti, A. C., and Strickland, S. (1997). Oocyte selection of mutations affecting cytoplasmic polyadenylation of maternal mRNAs. *Mol. Reprod. Dev.* **46,** 482–488.

Voeltz, G. K., and Steitz, J. A. (1998). AUUUA sequences direct mRNA deadenylation uncoupled from decay during *Xenopus* early development. *Mol. Cell Biol.* **18,** 7537–7545.

Wahle, E. and Rüegsegger, U. (1999). 3′ end processing of pre-mRNA in eukaryotes. *FEMS Microbiol. Rev.* **23,** 277–295.

Walker, J., Minshall, N., Hake, L., Richter, J., and Standart, N. (1999). The clam 3′ UTR masking element-binding protein p82 is a member of the CPEB family. *RNA* **5,** 14–26.

Webster, P. J., Liang, L., Berg, C. A., Lasko, P., and Macdonald, P. M. (1997). Translational repressor *bruno* plays multiple roles in development and is widely conserved. *Genes Dev.* **11,** 2510–2521.

Wells, S. E., Hillner, P. E., Vale, R. D., and Sachs, A. B. (1998). Circularization of mRNA by eukaryotic translation initiation factors. *Mol. Cell* **2,** 135–140.

West, M. F., Verrotti, A. C., Sallés, F. J., Tsirka, S. E., and Strickland, S. (1996). Isolation and characterization of two novel, cytoplasmically polyadenylated, oocyte-specific, mouse maternal RNAs. *Dev. Biol.* **175,** 132–141.

West, M. J., Stoneley, M., and Willis, A. E. (1998). Translational induction of the *c-myc* oncogene via activation of the FRAP/TOR signalling pathway. *Oncogene* **17,** 769–780.

Willis, A. E. (1999). Translational control of growth factor and proto-oncogene expression. *Int. J. Biochem. Cell Biol.* **31,** 73–86.

Wolffe, A. P. (1997). Histone H1. *Int. J. Biochem. Cell Biol.* **29,** 1463–1466.

Wolffe, A. P., and Hayes, J. J. (1999). Chromatin disruption and modification. *Nucleic Acids Res.* **27,** 711–720.

Wormington, M., Searfoss, A. M., and Hurney, C. A. (1996). Overexpression of poly(A) binding protein prevents maturation-specific deadenylation and translational inactivation in *Xenopus* oocytes. *EMBO J.* **15,** 900–909.

Wu, L., Good, P. J., and Richter, J. D. (1997). The 36-kilodalton embryonic-type cytoplasmic polyadenylation element- binding protein in *Xenopus laevis* is ElrA, a member of the ELAV family of RNA-binding proteins. *Mol. Cell. Biol.* **17,** 6402–6409.

Ye, X., Fong, P., lizuka, N., Choate, D., and Cavener, D. R. (1997). *Ultrabithorax* and *Antennapedia* 5′ untranslated regions promote developmentally regulated internal translation initiation. *Mol. Cell. Biol.* **17,** 1714–1721.

Yurkova, M. S., and Murray, M. T. (1997). A translation regulatory particle containing the *Xenopus* oocyte Y box protein mRNP3+4. *J. Biol. Chem.* **272,** 10870–10876.

Zernicka-Goetz, M. (1998). Fertile offspring derived from mammalian eggs lacking either animal or vegetal poles. *Development* **125,** 4803–4808.

Zhang, J., Houston, D. W., King, M. L., Payne, C., Wylie, C., and Heasman, J. (1998). The role of maternal VegT in establishing the primary germ layers in *Xenopus* embryos. *Cell* **94,** 515–524.

Zhang, S., Williams, C. J., Wormington, M., Stevens, A., and Peltz, S. W. (1999). Monitoring mRNA decapping activity. *Methods* **17,** 46–51.

Zhao, W. Q., and Manley, J. L. (1996). Complex alternative RNA processing generates an unexpected diversity of poly(A) polymerase isoforms. *Mol. Cell. Biol.* **16,** 2378–2386.

Zhong, J., Peters, A. H., Lee, K., and Braun, R. E. (1999). A double-stranded RNA binding protein required for activation of repressed messages in mammalian germ cells. *Nat. Genet.* **22,** 171–174.

Zimmer, A., Zimmer, A. M., and Reynolds, K. (1994). Tissue specific expression of the retinoic acid receptor-beta 2: Regulation by short open reading frames in the 5′-noncoding region. *J. Cell Biol.* **127,** 1111–1119.

SUBJECT INDEX

A

D

E

F

H

N

O

P

R

W

X

Y

Z